FUNDAMENTAL EQUATIONS OF FLUID MECHANICS

KLAUS A. HOFFMANN

Professor of Aerospace Engineering

The Wichita State University

USA

STEVE T. L. CHIANG

Professor of Mechanical Engineering

National Ping-Tung Polytechnic Institute

Ping-Tung, Taiwan

M. SHAMOUN SIDDIQUI

Advanced Engineering Research Organization

Haripur, Pakistan

MICHAEL PAPADAKIS

Professor of Aerospace Engineering

The Wichita State University

USA

A Publication of Engineering Education System™, Wichita, Kansas 67208-1078, USA

The data and information published in this book are for information purposes only. The authors and publisher have used their best effort in preparing this book. The authors and publisher are not liable for any injury or damage due to use, reliance, or performance of materials appearing in this book.

ISBN 0-9623731-9-2
First Print: May 1996

This book is typeset by Jeanie Duvall dba SciTech Computer Typesetting of Austin, Texas.

To purchase additional copies, please write to:

Engineering Education System™
P.O. Box 20078
Wichita, Kansas 67208-1078
USA

CONTENTS

Chapter One:
Equations of Fluid Motion – Navier-Stokes Equations

Chapter Two:
Equations of Fluid Motion for Inviscid Flows

Chapter Three:
Boundary Layer Equations

Chapter Four:
Hypersonics/Chemistry Equations

Chapter Five:
Transformation of the Equations of Fluid Motion from Physical Space to Computational Space

Chapter Six:
Computational Fluid Dynamics

Chapter Seven:
Auxiliary Relations and Fluid Properties

Chapter Eight:
Conversion Factors

Chapter Nine:
Mathematical Relations

PREFACE

Fluids play an important role in virtually every aspect of life. From blood flow in the arteries to exhaust flow of jet engines, oil flow in pipelines, and chemical processes in fluids, all involve fluid mechanics. Students in such fields as aerospace, biomedical, civil, chemical, mechanical, and nuclear engineering must develop an understanding of the basic principles, mathematical representation of fluid mechanics and applications. Typically, a student may be required to take one course in basic fluid mechanics or several courses covering different aspects of fluid mechanics, depending on the area of specialization. Students are introduced to different equations and relations ranging from simple algebraic expressions to the complex system of partial differential equations. In many instances analytical solutions are not available, and experimental or numerical simulations must be performed.

This text has been prepared with the primary goal of assembling the broad range of fundamental equations of fluid mechanics and auxiliary relations in fluids. The text has been written as a reference text to be used by practicing engineers, scientists, students, and instructors. It is presumed that the reader has had reasonable exposure to fluid mechanics.

Formulations are presented with the necessary explanations, and, when appropriate, they are expressed in different coordinate systems. Furthermore, different forms of the equations, based on the imposed assumptions, are provided. The governing equations of fluid mechanics are presented in four chapters, whereas computational fluid mechanics is presented in Chapters Five and Six, and auxiliary relations and fluid properties are presented in Chapter Seven. An extensive set of conversion factors is presented in Chapter Eight, and some selected mathematical relations are presented in Chapter Nine.

While every attempt has been made to produce an error-free text, it is inevitable that some errors still exist. The authors would greatly appreciate the reader's input on any corrections, so that they may be incorporated into future printings. Furthermore, we would appreciate any comments and/or suggestions from the readers on the improvement of the text. Finally, due to the broad range of applications in fluid mechanics, certain segments of importance may have been inadvertently excluded from the text. Thus, the reader's suggestion to include additional materials would be much appreciated. Please forward your comments to: Klaus Hoffmann, P.O. Box 20078, Wichita, KS, 67208-1078, USA.

The authors greatly appreciate the support and help of many friends and colleagues—in particular, Dr. John J. Bertin of the U.S. Air Force Academy, Dr. Douglas Cline of the University of Texas at Austin, Mr. Stan Bouslog of Lockheed-Martin, Mr. John Buratti of IBM, Dr. Kamran Rokhsaz, Dr. Scott Miller, Dr. Walter Horn, Mr. Yildirim Suzen, Mr. Eric Hung, and Mr. Tom Wayman of the Wichita State University.

Finally, we greatly appreciate the efforts of Mrs. Ellen Horn for her editorial comments and Mrs. Jeanie Duvall for her skillful typing of the manuscript.

Chapter 1
Equations of Fluid Motion: Navier-Stokes Equations

1.1 Introductory Remarks

The equations of motion for a homogeneous fluid in the absence of a finite rate chemical reaction or mass diffusion are based on three physical conservation laws. If a fluid is composed of various chemical species with mass diffusion and/or chemical reaction, additional conservation laws must be included. Since, for most engineering applications, the average measurable values of the flow properties are desired, the assumption of continuous distribution of matter is imposed. This assumption is known as *continuum* and is valid as long as the characteristic length in a physical domain is much larger than the mean free path of molecules. For problems in which this assumption is violated the kinetic theory [1.1–1.5] approach must be utilized.

The basic equations of fluid motion are derived in either integral form or differential form from the

1. Conservation of mass (continuity),

2. Conservation of linear momentum (Newton's second law),

3. Conservation of energy (first law of thermodynamics).

The conservation of linear momentum is a vector equation and, therefore, provides three scalar equations for a three dimensional problem.

The system of equations contains nine unknowns which include ρ (density), u, v, w (components of the velocity vector), e_t (total energy [or h (enthalpy)]), p (thermodynamic pressure), T (temperature), μ (dynamic viscosity), and k (thermal conductivity). The system of equations is closed by introducing thermodynamic relations and auxiliary relations for the coefficient of viscosity and thermal conductivity. Functional relations for thermodynamic properties may be expressed as

$$\rho = \rho(p, T)$$

and

$$h = h(p, T)$$

which may be in the form of equations or tables (or charts). The transport properties μ and k are expressed as

$$\mu = \mu(p, T)$$

and

$$k = k(p, T)$$

Hence, a system of nine equations is available which must be solved simultaneously for the nine unknowns. Other variables may also appear in the governing equations, depending on the choice of dependent variables. For example, the energy equation may include the specific heats c_p and c_v. But in either case, there are expressions by which c_p and c_v are related to thermodynamic properties. In general, specific heats are functions of temperature and are slightly pressure dependent. The complexity of the system is further increased by the introduction of turbulence and chemistry.

The solution of such a system, even with the advancement of computer technology and numerical techniques, is extremely difficult and time consuming. Therefore, from a practical point of view, various assumptions are imposed on the system of equations in order to simplify the solution procedure.

The equations of fluid motion in most instances are expressed in either Cartesian, cylindrical, spherical, or general orthogonal curvilinear coordinate systems. A brief description of these coordinate systems is provided in the following section.

This chapter will review and summarize the equations of fluid motion for a continuum, homogenous flow.

1.2 Coordinate Systems

The governing equations may be expressed in various coordinate systems. The selection of a particular system depends on the configuration and domain of solution, as well as on how the physics of problems can be best expressed on that coordinate system. Appropriate selection of a coordinate system would clearly simplify the solution procedure. The most commonly used coordinate systems are Cartesian coordinates, cylindrical coordinates, and spherical coordinates. These coordinate systems are a subset of generalized orthogonal curvilinear coordinates.

1.2.1 Generalized Orthogonal Curvilinear Coordinate System

Let x_1, x_2, and x_3 denote the orthogonal coordinates, and let $\vec{e}_1$, $\vec{e}_2$, and $\vec{e}_3$ represent the corresponding unit vectors. Any length element ds may be expressed as

$$(ds)^2 = (h_1 dx_1)^2 + (h_2 dx_2)^2 + (h_3 dx_3)^2$$

and a volumetric element can be expressed as

$$d\,\text{Vol} = h_1 h_2 h_3 \, dx_1 \, dx_2 \, dx_3$$

where the coefficients h_1, h_2, and h_3 are known as *the metric coefficients*. The values of the metric coefficients for the commonly used coordinate systems are summarized in Table 1.1. The notation used for the velocity components is also shown. The generalized orthogonal curvilinear coordinate system is illustrated schematically in Figure 1.1.

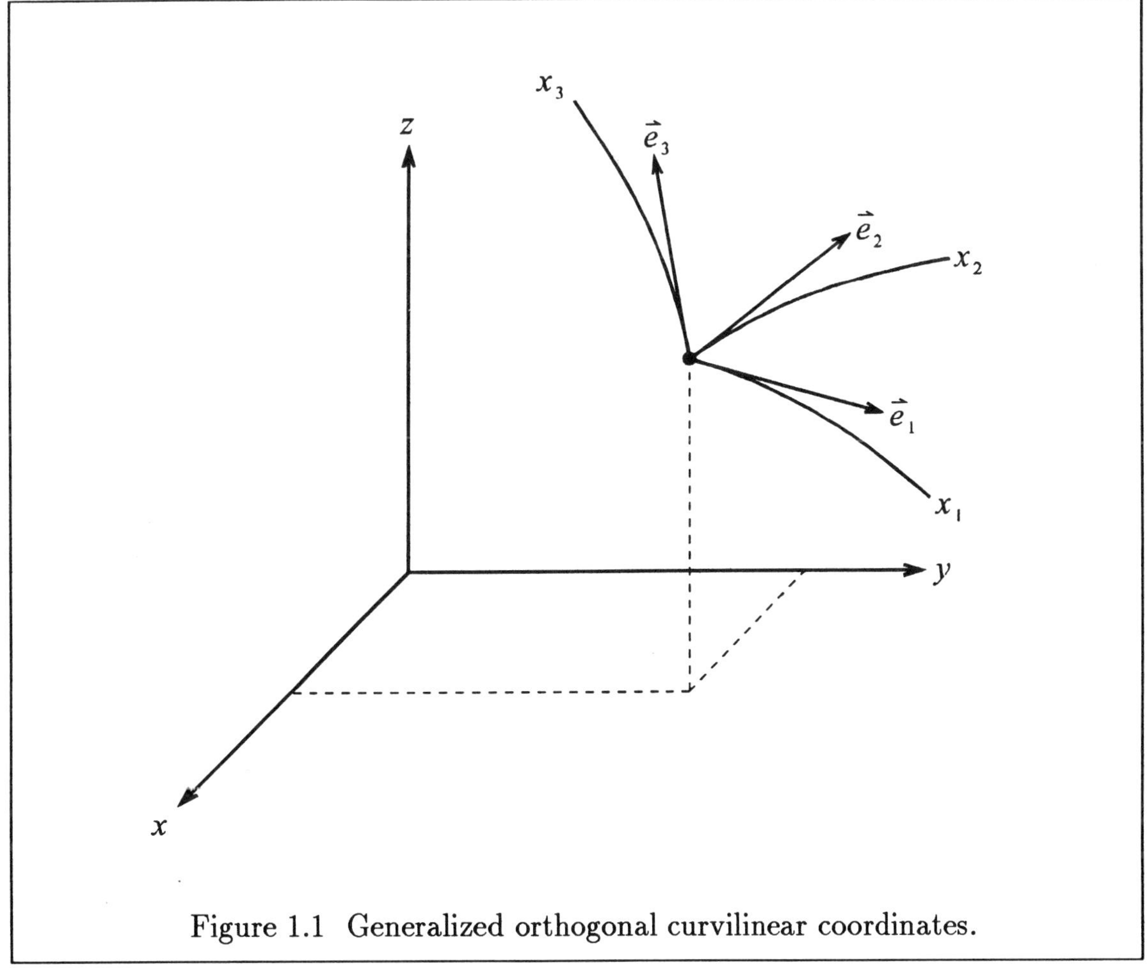

Figure 1.1 Generalized orthogonal curvilinear coordinates.

Table 1.1: Nomenclature for commonly used coordinate systems

		Cartesian (x, y, z)	Cylindrical (r, θ, z)	Cylindrical (x, r, ϕ)	Spherical (r, θ, ϕ)	Intrinsic coordinates for 2-D planar $(\alpha=0)$ or axisymmetric $(\alpha=1)$ (s, η, ϕ)
Coordinates	x_1	x	r	x	r	s
	x_2	y	θ	r	θ	η
	x_3	z	z	ϕ	ϕ	ϕ
Metric coefficients	h_1	1	1	1	1	$1 + \eta/R$
	h_2	1	r	1	r	1
	h_3	1	1	r	$r\sin\theta$	$(r + \eta\cos\beta)^\alpha$
Unit vectors	$\vec{e}_1$	$\vec{\imath}$	$\vec{e}_r$	$\vec{\imath}$	$\vec{e}_r$	$\vec{e}_s$
	$\vec{e}_2$	$\vec{\jmath}$	$\vec{e}_\theta$	$\vec{e}_r$	$\vec{e}_\theta$	$\vec{e}_\eta$
	$\vec{e}_3$	$\vec{k}$	$\vec{k}$	$\vec{e}_\phi$	$\vec{e}_\phi$	$\vec{e}_\phi$
Velocity components	u_1	u	u_r	u_x	u_r	u
	u_2	v	u_θ	u_r	u_θ	v
	u_3	w	u_z	u_ϕ	u_ϕ	$w = 0$

1.2.2 Commonly Used Coordinate Systems

The nomenclature used for the coordinates, velocity components, and unit vectors for commonly used coordinate systems were shown in Table 1.1. These coordinate systems are shown schematically in Figures 1.2a through 1.2e.

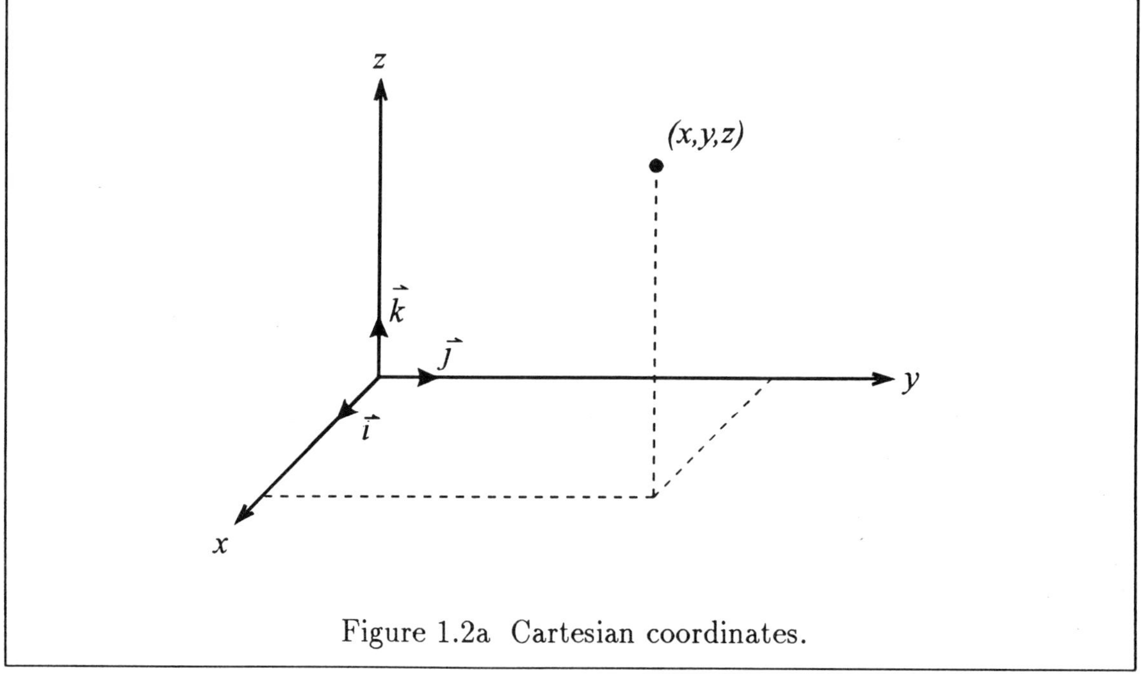

Figure 1.2a Cartesian coordinates.

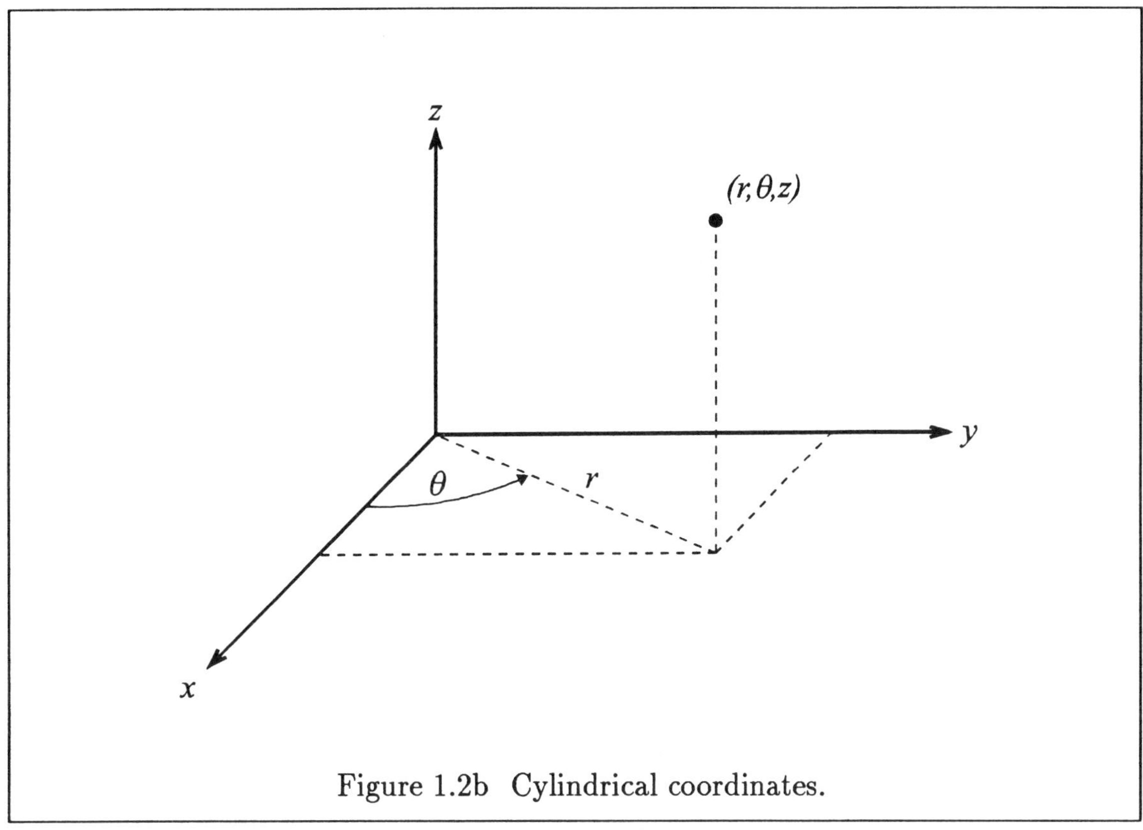

Figure 1.2b   Cylindrical coordinates.

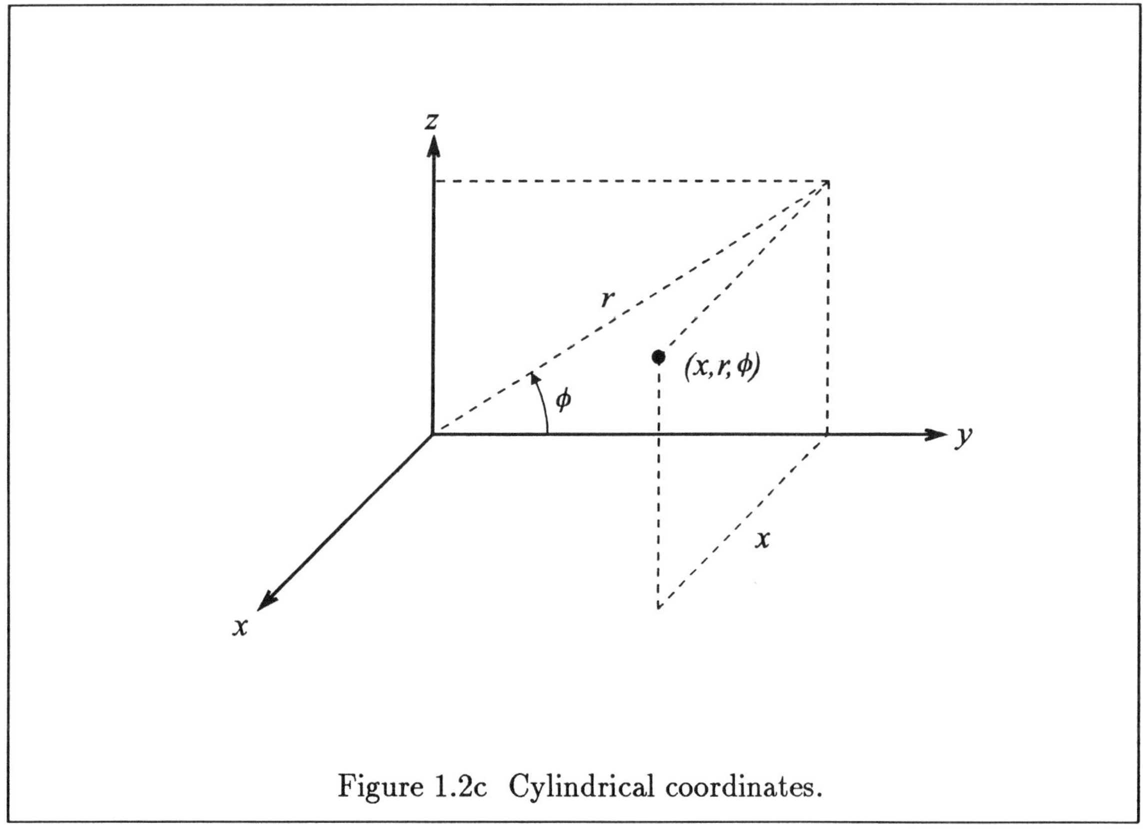

Figure 1.2c   Cylindrical coordinates.

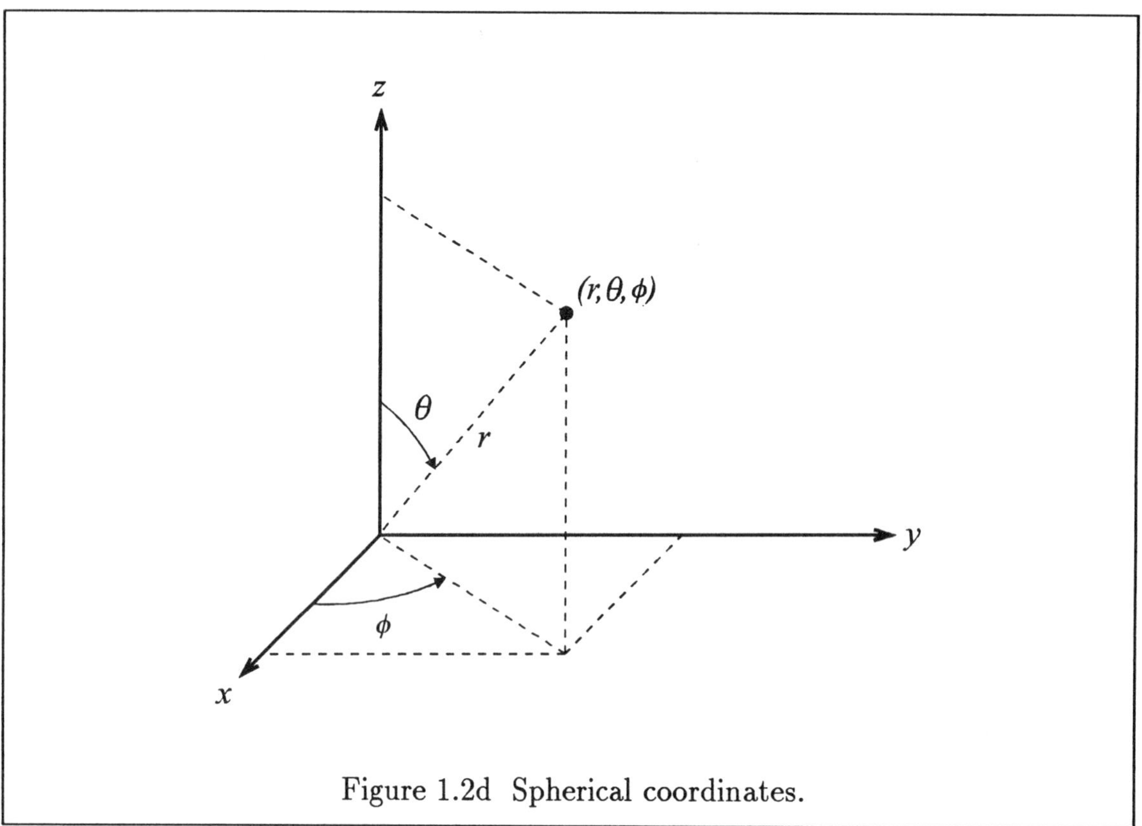

Figure 1.2d Spherical coordinates.

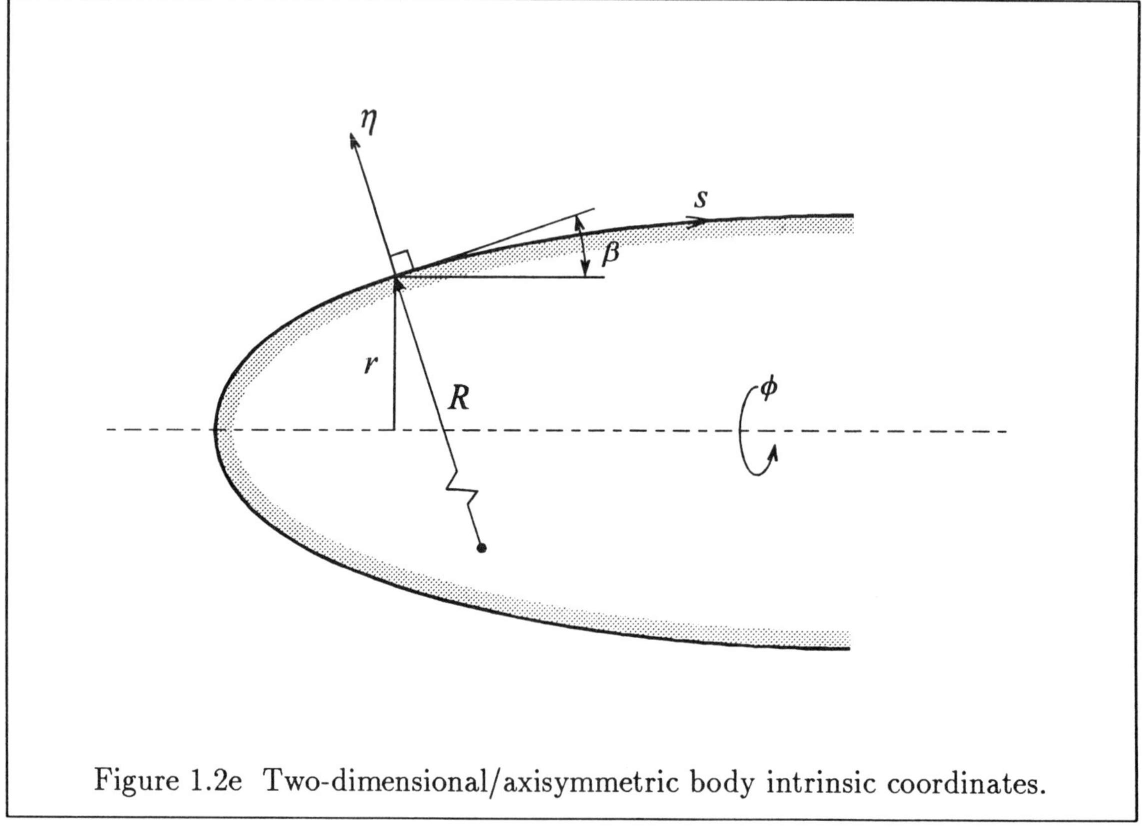

Figure 1.2e Two-dimensional/axisymmetric body intrinsic coordinates.

The relations between Cartesian, cylindrical, and spherical coordinates can be established as follows.

Cylindrical (b) $\quad x = r \cos \theta$
$\qquad\qquad\qquad y = r \sin \theta$
$\qquad\qquad\qquad z = z$

Cylindrical (c) $\quad x = x$
$\qquad\qquad\qquad y = r \cos \phi$
$\qquad\qquad\qquad z = r \sin \phi$

Spherical (d) $\quad x = r \sin \theta \cos \phi$
$\qquad\qquad\qquad y = r \sin \theta \sin \phi$
$\qquad\qquad\qquad z = r \cos \theta$

The cylindrical coordinate system illustrated in Figure 1.2b is used more commonly, and therefore it will be adapted in this text. Thus, the governing equations in cylindrical coordinates are expressed in that system.

The governing equations in the physical space expressed in various coordinate systems described above are typically transformed to a computational space to perform numerical computations. The computational domain and the transformation of the governing equations are addressed in Chapter 5.

1.3   Integral Formulations

Integral forms of the equations are used if an average value of the fluid properties at a cross-section is desired. This approach does not provide a detailed analysis of the flow field, however, the application is simple and is used extensively. In general, the integral form of the equation is derived for an extensive property and then the conservation laws are applied.

If N represents an extensive property, then a relation exists between the rate of change of extensive property for a system and the time rate of change of the property within the control volume plus net efflux of the property across the control surfaces. Defining η as the extensive property per unit mass, then

$$\frac{dN}{dt} = \frac{\partial}{\partial t} \int_{C.V.} \eta \rho \, d(\text{Vol}) + \int_{C.S.} \eta (\rho \vec{V} \cdot \vec{n}) dS \qquad (1.1)$$

where t represents time, ρ is the density of the fluid, $\vec{V}$ is the velocity vector and $\vec{n}$ is the unit vector normal to the control surface in the outward direction as shown in Figure 1.3. The derivation of Equation (1.1) may be found in any standard fluid mechanics text such as (1.6–1.8). $\frac{d}{dt}$ is used to represent the total or substantial derivative and is composed of a local derivative $\frac{\partial}{\partial t}$ and a convective derivative $\vec{V} \cdot \nabla$. Thus,

$$\frac{d}{dt} = \frac{\partial}{\partial t} + \vec{V} \cdot \nabla \tag{1.2}$$

Table 1.2 summarizes the total derivative in various coordinate systems.

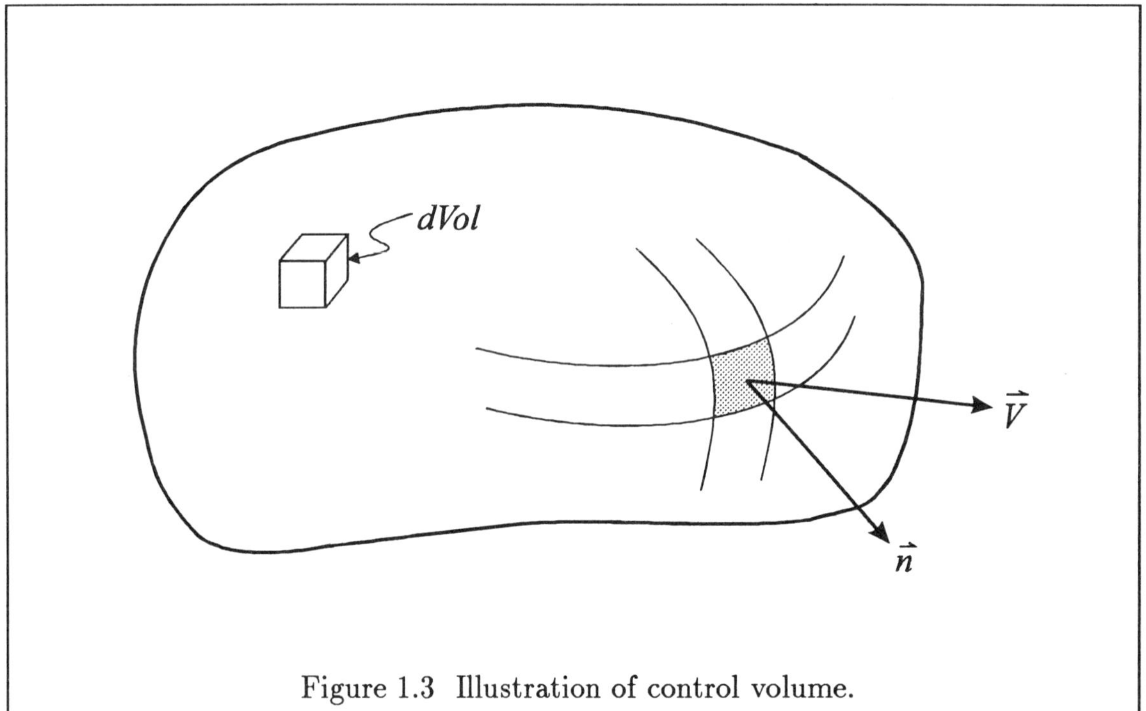

Figure 1.3 Illustration of control volume.

Table 1.2: Total derivative

Cartesian	$\dfrac{d}{dt} = \dfrac{\partial}{\partial t} + u\dfrac{\partial}{\partial x} + v\dfrac{\partial}{\partial y} + w\dfrac{\partial}{\partial z}$	(1.3)
Cylindrical	$\dfrac{d}{dt} = \dfrac{\partial}{\partial t} + u_r\dfrac{\partial}{\partial r} + \dfrac{u_\theta}{r}\dfrac{\partial}{\partial \theta} + u_z\dfrac{\partial}{\partial z}$	(1.4)
Spherical	$\dfrac{d}{dt} = \dfrac{\partial}{\partial t} + u_r\dfrac{\partial}{\partial r} + \dfrac{u_\theta}{r}\dfrac{\partial}{\partial \theta} + \dfrac{u_\phi}{r\sin\theta}\dfrac{\partial}{\partial \phi}$	(1.5)
General orthogonal curvilinear	$\dfrac{d}{dt} = \dfrac{\partial}{\partial t} + \dfrac{u_1}{h_1}\dfrac{\partial}{\partial x_1} + \dfrac{u_2}{h_2}\dfrac{\partial}{\partial x_2} + \dfrac{u_3}{h_3}\dfrac{\partial}{\partial x_3}$	(1.6)

To obtain the integral form of the equations of fluid motion, the conservation laws are applied to Equation (1.1).

1.3.1. *Conservation of Mass:* This conservation law requires that mass is neither created nor destroyed; mathematically this is expressed as $\frac{dM}{dt} = 0$. Using Equation (1.1), where $N = M$ and $\eta = 1$, the integral form of the conservation of mass is obtained as

$$\frac{\partial}{\partial t} \int_{C.V.} \rho \; d(\text{Vol}) + \int_{C.S.} \rho \vec{V} \cdot \vec{n} dS = 0 \qquad (1.7)$$

The physical interpretation of Equation (1.7) is as follows: The sum of the rate of change of mass within the control volume and net efflux of mass across the control surfaces is zero.

1.3.2. *Conservation of Linear Momentum:* Newton's second law applied to a nonaccelerating control volume which is referenced to a fixed coordinate system, will result in the integral form of the momentum equation once Equation (1.1) is utilized. In this case the linear momentum $\vec{G} = m\vec{V}$ is taken to be the extensive property and therefore $\eta = \vec{V}$. Newton's second law in an inertial reference is expressed as $\Sigma \vec{F} = \frac{d\vec{G}}{dt}$. Thus, use of Equation (1.1) yields

$$\Sigma \vec{F} = \frac{\partial}{\partial t} \int_{C.V.} \rho \vec{V} \; d(\text{Vol}) + \int_{C.S.} \vec{V}(\rho \vec{V} \cdot \vec{n}) dS \qquad (1.8)$$

This equation states that the sum of the forces acting on a control volume is equal to the sum of the rate of change of linear momentum within the control volume and net efflux of the linear momentum across the control surfaces. The forces acting on a control volume usually represent a combination of body force (such as gravity) and surface forces (such as viscous force). Note that Equation (1.8) is a vector equation from which scalar components may be obtained. For example, the x-component is expressed as

$$\Sigma F_x = \frac{\partial}{\partial t} \int_{C.V.} \rho u \; d(\text{Vol}) + \int_{C.S.} u(\rho \vec{V} \cdot \vec{n}) dS \qquad (1.9)$$

1.3.3. *Conservation of Energy:* This conservation law is based on the first law of thermodynamics, which may be expressed as

$$\frac{d(\rho e_t)}{dt} = \frac{\partial Q}{\partial t} + \frac{\partial W}{\partial t} \qquad (1.10)$$

In this relation, e_t represents the total energy of the system per unit mass, while $\frac{\partial Q}{\partial t}$ and $\frac{\partial W}{\partial t}$ represent the rate of heat transfer to the system and the rate of work done on the system, respectively. Usually heat added to the system and work done on the system are defined positive. In order to use Equation (1.1), ρe_t is selected to represent the extensive property N, and $\eta = e_t$, the total energy per unit mass. Hence,

$$\frac{\partial}{\partial t}\int_{C.V.}\rho e_t\, d(\text{Vol}) + \int_{C.S.}e_t(\rho\vec{V}\cdot\vec{n})dS = \dot{Q} + \dot{W} \qquad (1.11)$$

In general, the total energy is the sum of the internal energy, kinetic energy and potential energy, or in terms of energy per unit mass,

$$e_t = e + \tfrac{1}{2}V^2 + gz \qquad (1.12)$$

It is convenient to categorize the work as either flow work or shaft work. Flow work is due to stresses producing work on the control surfaces where there is fluid flow. If the control surfaces are selected to be perpendicular to the direction of the flow at the ports, the rate of flow work is shown to be

$$\dot{W}_f = \int_{C.S.}\tau_{nn}(\vec{V}\cdot\vec{n})dS \qquad (1.13)$$

The work transferred through the control surfaces is called the shaft work, $\dot{W}_S$. With such classification of work, we may express the energy equation as

$$\frac{\partial}{\partial t}\int_{C.V.}\rho e_t\, d(\text{Vol}) + \int_{C.S.}[e + \tfrac{1}{2}V^2 + gz - \tfrac{\tau_{nn}}{\rho}](\rho\vec{V}\cdot\vec{n})dS = \dot{Q} + \dot{W}_S \qquad (1.14)$$

For most engineering applications we may assume $\tau_{nn} \cong -p$, (note that for inviscid flow, τ_{nn} is $= p$) then (1.14) can be written in terms of specific enthalpy $h(h = e + \frac{p}{\rho})$ as:

$$\frac{\partial}{\partial t}\int_{C.V.}\rho e_t\, d(\text{Vol}) + \int_{C.S.}(h + \tfrac{1}{2}V^2 + gz)(\rho\vec{V}\cdot\vec{n})dS = \dot{Q} + \dot{W}_S \qquad (1.15)$$

The integral form of the equations of fluid motion are summarized in Table 1.3.

Table 1.3: Integral form of the equations

Conservation
of mass

$$\frac{\partial}{\partial t}\int_{C.V.}\rho\, d(\text{Vol}) + \int_{C.S.}\rho\vec{V}\cdot\vec{n}dS = 0 \qquad (1.16)$$

Conservation
of momentum

$$\Sigma\vec{F} = \frac{\partial}{\partial t}\int_{C.V.}\rho\vec{V}\, d(\text{Vol}) + \int_{C.S.}\vec{V}(\rho\vec{V}\cdot\vec{n})dS \qquad (1.17)$$

Conservation
of energy

$$\frac{\partial}{\partial t}\int_{C.V.}\rho e_t\, d(\text{Vol}) + \int_{C.S.}(h + \tfrac{1}{2}V^2 + gz)(\rho\vec{V}\cdot\vec{n})dS = \dot{Q} + \dot{W}_S \qquad (1.18)$$

Nomenclature for Table 1.3

<u>*Nomenclature*</u>   <u>*Units*</u>

e_t:	total energy per unit mass	m^2/sec^2	ft^2/sec^2
F:	force	N	lb_f
g:	gravitational acceleration	$9.81\ m/sec^2$	$32.17\ ft/sec^2$
h:	enthalpy per unit mass	m^2/sec^2	ft^2/sec^2
$\dot{Q}$:	heat transfer rate	$N \cdot m/sec$	$ft \cdot lb_f/sec$
$\vec{n}$:	unit vector normal to control surface		
S:	surface area	m^2	ft^2
t:	time	sec	sec
$\vec{V}$:	velocity vector	m/sec	ft/sec
Vol:	volume	m^3	ft^3
$\dot{W}_S$:	shaftwork	$N \cdot m/sec$	$ft \cdot lb_f/sec$
z:	elevation measured with respect to a reference plane	m	ft
ρ:	density	Kg/m^3	$slugs/ft^3$

Note: $1\ N \cdot m = 1\ J,\quad 1 Btu = 778.2\ ft \cdot lb_f$

$$1\ \tfrac{Btu}{lb_m} = 25038\ \tfrac{ft^2}{sec^2},\quad 1\ \tfrac{N \cdot m}{kg} = 1\ \tfrac{m^2}{sec^2}$$

1.4 Differential Formulations

The differential forms of the equations of motion are utilized for situations where a detailed solution of the flow field is required. These equations are obtained by the application of conservation laws to an infinitesimal fixed control volume. A typical differential element for a Cartesian coordinate system is shown in Figure 1.4. The flow properties at the control surfaces are expanded in Taylor series and integral forms of the equations are used. After a limiting process for which the differential element approaches zero, all the higher order terms are dropped and the differential form of the equations are produced. The differential equations which are derived based on fixed coordinate system and control volume are known as *Eulerian* approach. On the other hand, if we were to move with the fluid element, the approach is called *Lagrangian.* Clearly the Eulerian approach is much desirable in fluid mechanics, and in this text methods based on Eulerian formulation are reviewed.

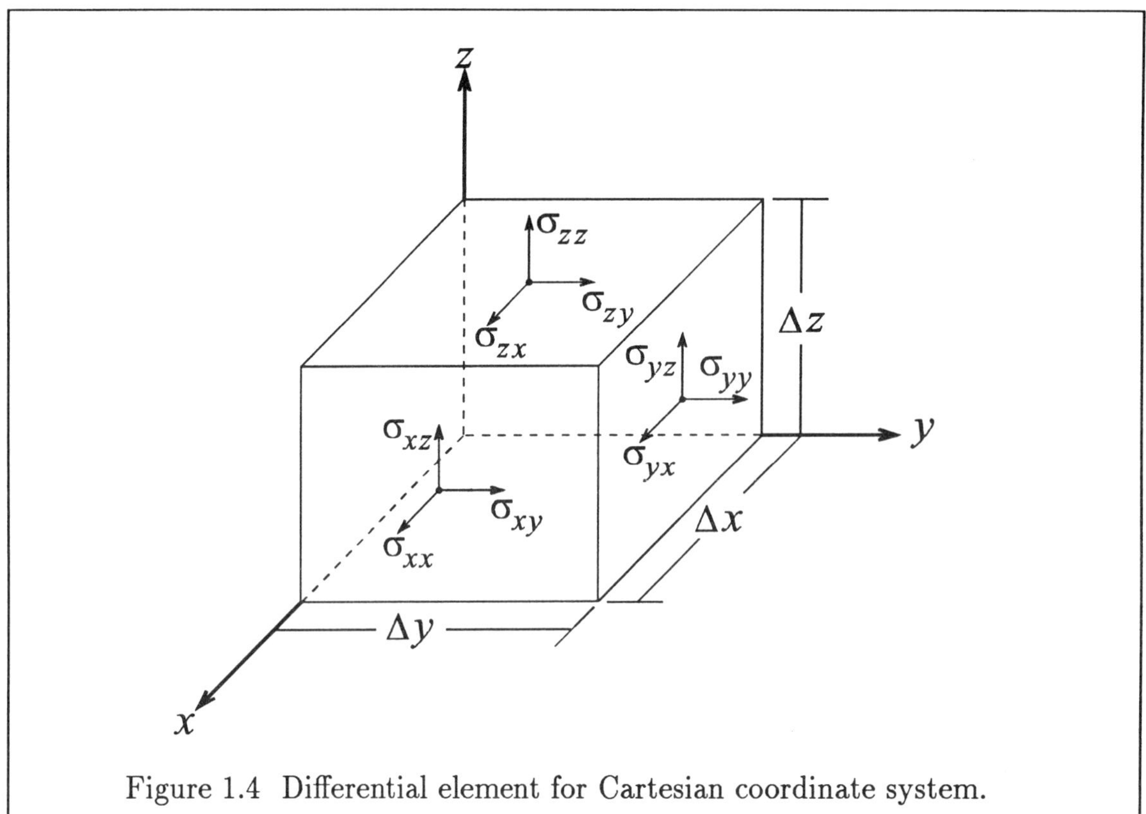

Figure 1.4 Differential element for Cartesian coordinate system.

1.4.1 Conservation of Mass

The differential form of the conservation of mass is known as *the continuity equation*, and in a vector form it is expressed as

$$\frac{\partial \rho}{\partial t} + \nabla \cdot (\rho \vec{V}) = 0 \tag{1.19}$$

The derivation of this equation may be found in any standard fluid mechanics text such as (1.6–1.9). This equation may be written in terms of the total derivative as

$$\frac{d\rho}{dt} + \rho \nabla \cdot \vec{V} = 0 \tag{1.20}$$

For a steady state problem $\frac{\partial}{\partial t} \equiv 0$, and, therefore,

$$\nabla \cdot (\rho \vec{V}) = 0$$

For an incompressible flow, where the density variations are considered negligible, continuity is reduced to

$$\nabla \cdot \vec{V} = 0 \tag{1.21}$$

Note that this equation is valid for unsteady problems as well. In Cartesian coordinate system, where the velocity vector is $\vec{V} = u\vec{i} + v\vec{j} + w\vec{k}$, Equation (1.19) becomes

$$\frac{\partial \rho}{\partial t} + \frac{\partial}{\partial x}(\rho u) + \frac{\partial}{\partial y}(\rho v) + \frac{\partial}{\partial z}(\rho w) = 0 \tag{1.22}$$

Various formulations of the continuity equation are summarized in Tables 1.4a and 1.4b.

Table 1.4a: Continuity equation

Vector form

$$\frac{\partial \rho}{\partial t} + \nabla \cdot (\rho \vec{V}) = 0 \tag{1.23}$$

$$\frac{d\rho}{dt} + \rho \nabla \cdot \vec{V} = 0 \tag{1.24}$$

Tensor form

$$\frac{\partial \rho}{\partial t} + \frac{\partial}{\partial x_i}(\rho u_i) = 0 \tag{1.25}$$

Cartesian coordinates

$$\frac{\partial \rho}{\partial t} + \frac{\partial}{\partial x}(\rho u) + \frac{\partial}{\partial y}(\rho v) + \frac{\partial}{\partial z}(\rho w) = 0 \tag{1.26}$$

Cylindrical coordinates

$$\frac{\partial \rho}{\partial t} + \frac{1}{r}\frac{\partial}{\partial r}(r\rho u_r) + \frac{1}{r}\frac{\partial}{\partial \theta}(\rho u_\theta) + \frac{\partial}{\partial z}(\rho u_z) = 0 \tag{1.27}$$

or

$$\frac{\partial \rho}{\partial t} + \frac{\partial}{\partial r}(\rho u_r) + \frac{1}{r}\frac{\partial}{\partial \theta}(\rho u_\theta) + \frac{\partial}{\partial z}(\rho u_z) + \frac{\rho u_r}{r} = 0 \tag{1.28}$$

Spherical coordinates

$$\frac{\partial \rho}{\partial t} + \frac{1}{r^2}\frac{\partial}{\partial r}(r^2 \rho u_r) + \frac{1}{r\sin\theta}\frac{\partial}{\partial \theta}(\rho u_\theta \sin\theta) + \frac{1}{r\sin\theta}\frac{\partial}{\partial \phi}(\rho u_\phi) = 0 \tag{1.29}$$

General orthogonal curvilinear coordinates

$$\frac{\partial \rho}{\partial t} + \frac{1}{h_1 h_2 h_3}\left[\frac{\partial}{\partial x_1}(h_2 h_3 \rho u_1) + \frac{\partial}{\partial x_2}(h_1 h_3 \rho u_2) + \frac{\partial}{\partial x_3}(h_1 h_2 \rho u_3)\right] = 0 \tag{1.30}$$

Table 1.4b: Continuity equation for an incompressible flow

Vector form

$$\nabla \cdot \vec{V} = 0 \tag{1.31}$$

Tensor form

$$\frac{\partial}{\partial x_i} u_i = 0 \tag{1.32}$$

Cartesian coordinates

$$\frac{\partial u}{\partial x} + \frac{\partial v}{\partial y} + \frac{\partial w}{\partial z} = 0 \tag{1.33}$$

Cylindrical coordinates

$$\frac{1}{r}\frac{\partial}{\partial r}(r u_r) + \frac{1}{r}\frac{\partial u_\theta}{\partial \theta} + \frac{\partial u_z}{\partial z} = 0 \tag{1.34}$$

or

$$\frac{\partial u_r}{\partial r} + \frac{1}{r}\frac{\partial u_\theta}{\partial \theta} + \frac{\partial u_z}{\partial z} + \frac{u_r}{r} = 0 \tag{1.35}$$

Spherical coordinates

$$\frac{1}{r}\frac{\partial}{\partial r}(r^2 u_r) + \frac{1}{\sin\theta}\frac{\partial}{\partial \theta}(u_\theta \sin\theta) + \frac{1}{\sin\theta}\frac{\partial u_\phi}{\partial \phi} = 0 \tag{1.36}$$

General orthogonal curvilinear coordinates

$$\frac{\partial}{\partial x_1}(h_2 h_3 u_1) + \frac{\partial}{\partial x_2}(h_1 h_3 u_2) + \frac{\partial}{\partial x_3}(h_1 h_2 u_3) = 0 \tag{1.37}$$

1.4.2   Conservation of Linear Momentum

The Navier-Stokes equations are obtained by the application of Newton's second law to a differential element in an inertial coordinate system. If σ is used to represent stresses acting on the differential element, the components of the Navier-Stokes equations in a Cartesian coordinate system takes the following form:

$$\rho\frac{du}{dt} = \rho f_x + \frac{\partial}{\partial x}(\sigma_{xx}) + \frac{\partial}{\partial y}(\sigma_{yx}) + \frac{\partial}{\partial z}(\sigma_{zx}) \tag{1.38}$$

$$\rho\frac{dv}{dt} = \rho f_y + \frac{\partial}{\partial x}(\sigma_{xy}) + \frac{\partial}{\partial y}(\sigma_{yy}) + \frac{\partial}{\partial z}(\sigma_{zy}) \tag{1.39}$$

$$\rho\frac{dw}{dt} = \rho f_z + \frac{\partial}{\partial x}(\sigma_{xz}) + \frac{\partial}{\partial y}(\sigma_{yz}) + \frac{\partial}{\partial z}(\sigma_{zz}) \tag{1.40}$$

Shear stress σ usually is written in terms of pressure p and viscous stress τ. In tensor notation this is expressed as

$$\sigma_{ij} = -p\delta_{ij} + \tau_{ij} \tag{1.41}$$

where δ_{ij} is the Kronecker delta,

$$\delta_{ij} = \begin{cases} 1 & \text{for } i = j \\ 0 & \text{for } i \neq j \end{cases}$$

The subscripts assigned to the components of stress are as follows. The first subscript denotes the direction of the normal to the surface on which stress acts, and the second denotes the direction in which the stress acts. This is illustrated in Figure 1.4.

Viscous stresses are related to the rates of strain by a physical law. For most fluids, this relation is linear and is known as *Newtonian fluid*. For a Newtonian fluid, viscous stresses in a Cartesian coordinate system are

$$\tau_{xx} = 2\mu\frac{\partial u}{\partial x} + \lambda \nabla \cdot \vec{V} \tag{1.42}$$

$$\tau_{yy} = 2\mu\frac{\partial v}{\partial y} + \lambda \nabla \cdot \vec{V} \tag{1.43}$$

$$\tau_{zz} = 2\mu\frac{\partial w}{\partial z} + \lambda \nabla \cdot \vec{V} \tag{1.44}$$

$$\tau_{xy} = \tau_{yx} = \mu\left(\frac{\partial u}{\partial y} + \frac{\partial v}{\partial x}\right) \tag{1.45}$$

$$\tau_{xz} = \tau_{zx} = \mu\left(\frac{\partial w}{\partial x} + \frac{\partial u}{\partial z}\right) \tag{1.46}$$

$$\tau_{yz} = \tau_{zy} = \mu\left(\frac{\partial v}{\partial z} + \frac{\partial w}{\partial y}\right) \tag{1.47}$$

where μ is known as *the coefficient of viscosity* or *dynamic viscosity* and λ is defined as *the second coefficient of viscosity*. The combination of μ and λ in the following form is known as *the bulk viscosity k*, i.e.,

$$k = \lambda + \tfrac{2}{3}\mu \tag{1.48}$$

If bulk viscosity of a fluid is assumed negligible, then

$$\lambda = -\tfrac{2}{3}\mu \tag{1.49}$$

This is known as *Stokes hypothesis*. Note that this assumption is not necessary for an incompressible flow, since $\nabla \cdot \vec{V}$ is equal to zero and thus λ does not appear in the stress term for an incompressible flow. Viscous stress terms in various coordinates are given in Table 1.5a for a Newtonian fluid and Table 1.5b for a Newtonian fluid with Stokes hypothesis invoked.

Table 1.5a: Viscous shear stresses for a Newtonian fluid

Tensor form

$$\tau_{ij} = \mu \left(\frac{\partial u_i}{\partial x_j} + \frac{\partial u_j}{\partial x_i} \right) + \delta_{ij}\lambda \frac{\partial u_k}{\partial x_k}$$

Cartesian coordinates (x,y,z)

$$\tau_{xx} = 2\mu\frac{\partial u}{\partial x} + \lambda \nabla \cdot \vec{V}$$

$$\tau_{yy} = 2\mu\frac{\partial v}{\partial y} + \lambda \nabla \cdot \vec{V}$$

$$\tau_{zz} = 2\mu\frac{\partial w}{\partial z} + \lambda \nabla \cdot \vec{V}$$

where $\nabla \cdot \vec{V} = \dfrac{\partial u}{\partial x} + \dfrac{\partial v}{\partial y} + \dfrac{\partial w}{\partial z}$

$$\tau_{xy} = \tau_{yx} = \mu \left(\frac{\partial u}{\partial y} + \frac{\partial v}{\partial x} \right)$$

$$\tau_{xz} = \tau_{zx} = \mu \left(\frac{\partial w}{\partial x} + \frac{\partial u}{\partial z} \right)$$

$$\tau_{yz} = \tau_{zy} = \mu \left(\frac{\partial v}{\partial z} + \frac{\partial w}{\partial y} \right)$$

Cylindrical coordinates (r,θ,z)

$$\tau_{rr} = 2\mu\frac{\partial u_r}{\partial r} + \lambda \nabla \cdot \vec{V}$$

$$\tau_{\theta\theta} = 2\mu \left(\frac{1}{r}\frac{\partial u_\theta}{\partial \theta} + \frac{u_r}{r} \right) + \lambda \nabla \cdot \vec{V}$$

$$\tau_{zz} = 2\mu\frac{\partial u_z}{\partial z} + \lambda \nabla \cdot \vec{V}$$

where $\nabla \cdot \vec{V} = \dfrac{1}{r}\dfrac{\partial}{\partial r}(ru_r) + \dfrac{1}{r}\dfrac{\partial u_\theta}{\partial \theta} + \dfrac{\partial u_z}{\partial z} = \dfrac{\partial u_r}{\partial r} + \dfrac{1}{r}\dfrac{\partial u_\theta}{\partial \theta} + \dfrac{\partial u_z}{\partial z} + \dfrac{u_r}{r}$

$$\tau_{r\theta} = \tau_{\theta r} = \mu \left[r\frac{\partial}{\partial r}\left(\frac{u_\theta}{r} \right) + \frac{1}{r}\frac{\partial u_r}{\partial \theta} \right]$$

$$\tau_{\theta z} = \tau_{z\theta} = \mu \left[\frac{\partial u_\theta}{\partial z} + \frac{1}{r}\frac{\partial u_z}{\partial \theta} \right]$$

$$\tau_{zr} = \tau_{rz} = \mu \left[\frac{\partial u_z}{\partial r} + \frac{\partial u_r}{\partial z} \right]$$

Spherical coordinates (r,θ,ϕ)

$$\tau_{rr} = 2\mu\frac{\partial u_r}{\partial r} + \lambda\nabla\cdot\vec{V}$$

$$\tau_{\theta\theta} = 2\mu\left(\frac{1}{r}\,\frac{\partial u_\theta}{\partial\theta} + \frac{u_r}{r}\right) + \lambda\nabla\cdot\vec{V}$$

$$\tau_{\phi\phi} = 2\mu\left(\frac{1}{r\sin\theta}\,\frac{\partial u_\phi}{\partial\phi} + \frac{u_r}{r} + \frac{u_\theta\cot\theta}{r}\right) + \lambda\nabla\cdot\vec{V}$$

where $\nabla\cdot\vec{V} = \dfrac{1}{r^2}\,\dfrac{\partial}{\partial r}(r^2 u_r) + \dfrac{1}{r\sin\theta}\,\dfrac{\partial}{\partial\theta}(u_\theta\sin\theta) + \dfrac{1}{r\sin\theta}\,\dfrac{\partial u_\phi}{\partial\phi}$

$$\tau_{r\theta} = \tau_{\theta r} = \mu\left[r\frac{\partial}{\partial r}\left(\frac{u_\theta}{r}\right) + \frac{1}{r}\frac{\partial u_r}{\partial\theta}\right]$$

$$\tau_{r\phi} = \tau_{\phi r} = \mu\left[\frac{1}{r\sin\theta}\,\frac{\partial u_r}{\partial\phi} + r\frac{\partial}{\partial r}\left(\frac{u_\phi}{r}\right)\right]$$

$$\tau_{\theta\phi} = \tau_{\phi\theta} = \mu\left[\frac{\sin\theta}{r}\,\frac{\partial}{\partial\theta}\left(\frac{u_\phi}{\sin\theta}\right) + \frac{1}{r\sin\theta}\,\frac{\partial u_\theta}{\partial\phi}\right]$$

General orthogonal curvilinear coordinates

$$\tau_{11} = 2\mu\left[\frac{1}{h_1}\,\frac{\partial u_1}{\partial x_1} + \frac{u_2}{h_1 h_2}\,\frac{\partial h_1}{\partial x_2} + \frac{u_3}{h_1 h_3}\,\frac{\partial h_1}{\partial x_3}\right] + \lambda\nabla\cdot\vec{V}$$

$$\tau_{22} = 2\mu\left[\frac{1}{h_2}\,\frac{\partial u_2}{\partial x_2} + \frac{u_3}{h_2 h_3}\,\frac{\partial h_2}{\partial x_3} + \frac{u_1}{h_1 h_2}\,\frac{\partial h_2}{\partial x_1}\right] + \lambda\nabla\cdot\vec{V}$$

$$\tau_{33} = 2\mu\left[\frac{1}{h_3}\,\frac{\partial u_3}{\partial x_3} + \frac{u_1}{h_1 h_3}\,\frac{\partial h_3}{\partial x_1} + \frac{u_2}{h_2 h_3}\,\frac{\partial h_3}{\partial x_2}\right] + \lambda\nabla\cdot\vec{V}$$

where $\nabla\cdot\vec{V} = \dfrac{1}{h_1 h_2 h_3}\left[\dfrac{\partial}{\partial x_1}(h_2 h_3 u_1) + \dfrac{\partial}{\partial x_2}(h_3 h_1 u_2) + \dfrac{\partial}{\partial x_3}(h_1 h_2 u_3)\right]$

$$\tau_{12} = \tau_{21} = \mu\left[\frac{h_2}{h_1}\,\frac{\partial}{\partial x_1}\left(\frac{u_2}{h_2}\right) + \frac{h_1}{h_2}\,\frac{\partial}{\partial x_2}\left(\frac{u_1}{h_1}\right)\right]$$

$$= \mu\left[\frac{1}{h_1}\,\frac{\partial u_2}{\partial x_1} + \frac{1}{h_2}\,\frac{\partial u_1}{\partial x_2} - \frac{1}{h_1 h_2}\left(u_1\frac{\partial h_1}{\partial x_2} + u_2\frac{\partial h_2}{\partial x_1}\right)\right]$$

$$\tau_{23} = \tau_{32} = \mu\left[\frac{h_3}{h_2}\,\frac{\partial}{\partial x_2}\left(\frac{u_3}{h_3}\right) + \frac{h_2}{h_3}\,\frac{\partial}{\partial x_3}\left(\frac{u_2}{h_2}\right)\right]$$

$$= \mu\left[\frac{1}{h_2}\,\frac{\partial u_3}{\partial x_2} + \frac{1}{h_3}\,\frac{\partial u_2}{\partial x_3} - \frac{1}{h_2 h_3}\left(u_2\frac{\partial h_2}{\partial x_3} + u_3\frac{\partial h_3}{\partial x_2}\right)\right]$$

$$\tau_{13} = \tau_{31} = \mu\left[\frac{h_1}{h_3}\,\frac{\partial}{\partial x_3}\left(\frac{u_1}{h_1}\right) + \frac{h_3}{h_1}\,\frac{\partial}{\partial x_1}\left(\frac{u_3}{h_3}\right)\right]$$

$$= \mu \left[\frac{1}{h_3} \frac{\partial u_1}{\partial x_3} + \frac{1}{h_1} \frac{\partial u_3}{\partial x_1} - \frac{1}{h_1 h_3} \left(u_3 \frac{\partial h_3}{\partial x_1} + u_1 \frac{\partial h_1}{\partial x_3} \right) \right]$$

Table 1.5b: Viscous shear stresses for a Newtonian fluid with Stokes hypothesis

Tensor form

$$\tau_{ij} = \mu \left(\frac{\partial u_i}{\partial x_j} + \frac{\partial u_j}{\partial x_i} \right) - \delta_{ij} \frac{2}{3} \mu \frac{\partial u_k}{\partial x_k}$$

Cartesian coordinates (x,y,z)

$$\tau_{xx} = \mu \left[\frac{4}{3} \frac{\partial u}{\partial x} - \frac{2}{3} \left(\frac{\partial v}{\partial y} + \frac{\partial w}{\partial z} \right) \right]$$

$$\tau_{yy} = \mu \left[\frac{4}{3} \frac{\partial v}{\partial y} - \frac{2}{3} \left(\frac{\partial u}{\partial x} + \frac{\partial w}{\partial z} \right) \right]$$

$$\tau_{zz} = \mu \left[\frac{4}{3} \frac{\partial w}{\partial z} - \frac{2}{3} \left(\frac{\partial u}{\partial x} + \frac{\partial v}{\partial y} \right) \right]$$

$$\tau_{xy} = \tau_{yx} = \mu \left(\frac{\partial u}{\partial y} + \frac{\partial v}{\partial x} \right)$$

$$\tau_{xz} = \tau_{zx} = \mu \left(\frac{\partial u}{\partial z} + \frac{\partial w}{\partial x} \right)$$

$$\tau_{yz} = \tau_{zy} = \mu \left(\frac{\partial v}{\partial z} + \frac{\partial w}{\partial y} \right)$$

Cylindrical coordinates (r,θ,z)

$$\tau_{rr} = \mu \left[\frac{4}{3} \frac{\partial u_r}{\partial r} - \frac{2}{3} \left(\frac{1}{r} \frac{\partial u_\theta}{\partial \theta} + \frac{\partial u_z}{\partial z} + \frac{u_r}{r} \right) \right]$$

$$\tau_{\theta\theta} = \mu \left[\frac{4}{3} \left(\frac{1}{r} \frac{\partial u_\theta}{\partial \theta} + \frac{u_r}{r} \right) - \frac{2}{3} \left(\frac{\partial u_r}{\partial r} + \frac{\partial u_z}{\partial z} \right) \right]$$

$$\tau_{zz} = \mu \left[\frac{4}{3} \frac{\partial u_z}{\partial z} - \frac{2}{3} \left(\frac{1}{r} \frac{\partial}{\partial r} (r u_r) + \frac{1}{r} \frac{\partial u_\theta}{\partial \theta} \right) \right]$$

$$\tau_{r\theta} = \tau_{\theta r} = \mu \left[r \frac{\partial}{\partial r} \left(\frac{u_\theta}{r} \right) + \frac{1}{r} \frac{\partial u_r}{\partial \theta} \right]$$

$$\tau_{\theta z} = \tau_{z\theta} = \mu \left[\frac{\partial u_\theta}{\partial z} + \frac{1}{r} \frac{\partial u_z}{\partial \theta} \right]$$

$$\tau_{zr} = \tau_{rz} = \mu \left[\frac{\partial u_z}{\partial r} + \frac{\partial u_r}{\partial z} \right]$$

After substitution of viscous stress terms for a Newtonian fluid given by Equations (1.42) through (1.47) into the Navier-Stokes Equations (1.38) through (1.40), we obtain

$$\rho\frac{du}{dt} \;=\; \rho f_x + \frac{\partial}{\partial x}\left[-p + 2\mu\frac{\partial u}{\partial x} + \lambda\nabla\cdot\vec{V}\right] + \frac{\partial}{\partial y}\left[\mu\left(\frac{\partial u}{\partial y} + \frac{\partial v}{\partial x}\right)\right]$$
$$+\frac{\partial}{\partial z}\left[\mu\left(\frac{\partial w}{\partial x} + \frac{\partial u}{\partial z}\right)\right] \tag{1.50}$$

$$\rho\frac{dv}{dt} \;=\; \rho f_y + \frac{\partial}{\partial x}\left[\mu\left(\frac{\partial u}{\partial y} + \frac{\partial v}{\partial x}\right)\right] + \frac{\partial}{\partial y}\left[-p + 2\mu\frac{\partial v}{\partial y} + \lambda\nabla\cdot\vec{V}\right]$$
$$+\frac{\partial}{\partial z}\left[\mu\left(\frac{\partial v}{\partial z} + \frac{\partial w}{\partial y}\right)\right] \tag{1.51}$$

$$\rho\frac{dw}{dt} \;=\; \rho f_z + \frac{\partial}{\partial x}\left[\mu\left(\frac{\partial w}{\partial x} + \frac{\partial u}{\partial z}\right)\right] + \frac{\partial}{\partial y}\left[\mu\left(\frac{\partial v}{\partial z} + \frac{\partial w}{\partial y}\right)\right]$$
$$+\frac{\partial}{\partial z}\left[-p + 2\mu\frac{\partial w}{\partial z} + \lambda\nabla\cdot\vec{V}\right] \tag{1.52}$$

In applications where the temperature variations are small, viscosity may be assumed constant and in addition to negligible changes in density, the flow is assumed incompressible. For such a flow, the Navier-Stokes equations are reduced to

$$\rho\frac{du}{dt} \;=\; \rho f_x - \frac{\partial p}{\partial x} + \mu\left(\frac{\partial^2 u}{\partial x^2} + \frac{\partial^2 u}{\partial y^2} + \frac{\partial^2 u}{\partial z^2}\right) \tag{1.53}$$

$$\rho\frac{dv}{dt} \;=\; \rho f_y - \frac{\partial p}{\partial y} + \mu\left(\frac{\partial^2 v}{\partial x^2} + \frac{\partial^2 v}{\partial y^2} + \frac{\partial^2 v}{\partial z^2}\right) \tag{1.54}$$

$$\rho\frac{dw}{dt} \;=\; \rho f_z - \frac{\partial p}{\partial z} + \mu\left(\frac{\partial^2 w}{\partial x^2} + \frac{\partial^2 w}{\partial y^2} + \frac{\partial^2 w}{\partial z^2}\right) \tag{1.55}$$

These scalar equations may be written in a vector form as

$$\rho\frac{d\vec{V}}{dt} = \rho\vec{f} - \nabla p + \mu\nabla^2\vec{V} \tag{1.56}$$

1.4.3   Conservative Form of the Navier-Stokes Equations

The Navier-Stokes equations may be written in a conservative form by addition of the continuity equation to the left hand side of the equation. For the x-component of the Navier-Stokes equation given by Equation (1.38) we may add the continuity equation multiplied by the u-component of the velocity to obtain

$$\rho\frac{\partial u}{\partial x} + \rho u\frac{\partial u}{\partial x} + \rho v\frac{\partial u}{\partial y} + \rho w\frac{\partial u}{\partial z} + u\left[\frac{\partial \rho}{\partial t} + \frac{\partial}{\partial x}(\rho u) + \right.$$

$$\left.\frac{\partial}{\partial y}(\rho v) + \frac{\partial}{\partial z}(\rho w)\right] = \rho f_x + \frac{\partial}{\partial x}(\sigma_{xx}) + \frac{\partial}{\partial y}(\sigma_{yx}) + \frac{\partial}{\partial z}(\sigma_{zx}) \qquad (1.57)$$

The terms on the left hand of Equation (1.57) may be combined to yield

$$\frac{\partial}{\partial t}(\rho u) + \frac{\partial}{\partial x}(\rho u^2) + \frac{\partial}{\partial y}(\rho uv) + \frac{\partial}{\partial z}(\rho uw) = \rho f_x + \frac{\partial}{\partial x}(\sigma_{xx}) + \frac{\partial}{\partial y}(\sigma_{yx}) + \frac{\partial}{\partial z}(\sigma_{zx}) \quad (1.58)$$

In terms of the viscous stress terms, we may write the scalar components of the Navier-Stokes equation in conservation law form as

x-component:

$$\frac{\partial}{\partial t}(\rho u) + \frac{\partial}{\partial x}(\rho u^2 + p) + \frac{\partial}{\partial y}(\rho uv) + \frac{\partial}{\partial z}(\rho uw) = \rho f_x + \frac{\partial}{\partial x}(\tau_{xx}) + \frac{\partial}{\partial y}(\tau_{xy}) + \frac{\partial}{\partial z}(\tau_{xz}) \quad (1.59)$$

y-component:

$$\frac{\partial}{\partial t}(\rho v) + \frac{\partial}{\partial x}(\rho uv) + \frac{\partial}{\partial y}(\rho v^2 + p) + \frac{\partial}{\partial z}(\rho vw) = \rho f_y + \frac{\partial}{\partial x}(\tau_{xy}) + \frac{\partial}{\partial y}(\tau_{yy}) + \frac{\partial}{\partial z}(\tau_{yz}) \quad (1.60)$$

z-component:

$$\frac{\partial}{\partial t}(\rho w) + \frac{\partial}{\partial x}(\rho uw) + \frac{\partial}{\partial y}(\rho vw) + \frac{\partial}{\partial z}(\rho w^2 + p) = \rho f_z + \frac{\partial}{\partial x}(\tau_{xz}) + \frac{\partial}{\partial y}(\tau_{yz}) + \frac{\partial}{\partial z}(\tau_{zz})$$
$$(1.61)$$

Various forms of the Navier-Stokes equation are summarized in Tables 1.6a–1.6f.

Table 1.6a: Navier-Stokes equations in vector form

$$\rho\frac{d\vec{V}}{dt} = \rho\vec{f} - \nabla p + \nabla\cdot\tau \qquad (1.62)$$

or

$$\rho\left(\frac{\partial\vec{V}}{\partial t} + \vec{V}\cdot\nabla\vec{V}\right) = \rho\vec{f} - \nabla p + \nabla\cdot\tau \qquad (1.63)$$

or

$$\frac{\partial}{\partial t}(\rho\vec{V}) + \nabla\cdot(\rho\vec{V}\vec{V}) = \rho\vec{f} - \nabla p + \nabla\cdot\tau \qquad (1.64)$$

For a Newtonian fluid

$$\rho\frac{d\vec{V}}{dt} = \rho\vec{f} - \nabla p - \nabla \times [\mu(\nabla \times \vec{V})] + \nabla[(\lambda + 2\mu)\nabla \cdot \vec{V}]$$

(1.65)

For a Newtonian fluid with Stokes hypothesis

$$\rho\frac{d\vec{V}}{dt} = \rho\bar{f} - \nabla p - \nabla \times [\mu(\nabla \times \vec{V})] + \nabla[(\tfrac{4}{3}\mu)\nabla \cdot \vec{V}]$$

(1.66)

Incompressible flow with constant properties

$$\rho\frac{d\vec{V}}{dt} = \rho\vec{f} - \nabla p + \mu\nabla^2\vec{V}$$

(1.67)

Steady incompressible flow with constant properties

$$\rho\vec{V} \cdot \nabla\vec{V} = \rho\vec{f} - \nabla p + \mu\nabla^2\vec{V}$$

(1.68)

Table 1.6b: Navier-Stokes equations in tensor notation

$$\rho\left(\frac{\partial u_i}{\partial t} + u_j\frac{\partial u_i}{\partial x_j}\right) = \rho f_i - \frac{\partial p}{\partial x_i} + \frac{\partial}{\partial x_j}\tau_{ij}$$

(1.69)

Conservative form

$$\frac{\partial}{\partial t}(\rho u_i) + \frac{\partial}{\partial x_j}(\rho u_i u_j) = \rho f_i - \frac{\partial p}{\partial x_i} + \frac{\partial}{\partial x_j}\tau_{ij}$$

(1.70)

For a Newtonian fluid

$$\rho\left(\frac{\partial u_i}{\partial t} + u_j\frac{\partial u_i}{\partial x_j}\right) = \rho f_i - \frac{\partial p}{\partial x_i} + \frac{\partial}{\partial x_j}\left[\mu\left(\frac{\partial u_i}{\partial x_j} + \frac{\partial u_j}{\partial x_i}\right)\right] + \frac{\partial}{\partial x_i}\left[\lambda\left(\frac{\partial u_k}{\partial x_k}\right)\right]$$

(1.71)

For a Newtonian fluid with Stokes hypothesis

$$\rho\left(\frac{\partial u_i}{\partial t} + u_j\frac{\partial u_i}{\partial x_j}\right) = \rho f_i - \frac{\partial p}{\partial x_i} + \frac{\partial}{\partial x_j}\left[\mu\left(\frac{\partial u_i}{\partial x_j} + \frac{\partial u_j}{\partial x_i}\right)\right] - \frac{\partial}{\partial x_i}\left[\frac{2}{3}\mu\left(\frac{\partial u_k}{\partial x_k}\right)\right]$$

(1.72)

Incompressible flow with constant properties

$$\rho\left(\frac{\partial u_i}{\partial t} + u_j\frac{\partial u_i}{\partial x_j}\right) = \rho f_i - \frac{\partial p}{\partial x_i} + \mu\frac{\partial^2 u_i}{\partial x_j^2}$$

(1.73)

Steady incompressible flow with constant properties

$$\rho u_j\frac{\partial u_i}{\partial x_j} = \rho f_i - \frac{\partial p}{\partial x_i} + \mu\frac{\partial^2 u_i}{\partial x_j^2}$$

(1.74)

Table 1.6c: Navier-Stokes equations in a Cartesian coordinate system

x-component

$$\rho\left(\frac{\partial u}{\partial t} + u\frac{\partial u}{\partial x} + v\frac{\partial u}{\partial y} + w\frac{\partial u}{\partial z}\right) = \rho f_x - \frac{\partial p}{\partial x} + \frac{\partial}{\partial x}\tau_{xx} + \frac{\partial}{\partial y}\tau_{yx} + \frac{\partial}{\partial z}\tau_{zx}$$

(1.75)

y-component

$$\rho\left(\frac{\partial v}{\partial t} + u\frac{\partial v}{\partial x} + v\frac{\partial v}{\partial y} + w\frac{\partial v}{\partial z}\right) = \rho f_y - \frac{\partial p}{\partial y} + \frac{\partial}{\partial x}\tau_{xy} + \frac{\partial}{\partial y}\tau_{yy} + \frac{\partial}{\partial z}\tau_{zy} \tag{1.76}$$

z-component

$$\rho\left(\frac{\partial w}{\partial t} + u\frac{\partial w}{\partial x} + v\frac{\partial w}{\partial y} + w\frac{\partial w}{\partial z}\right) = \rho f_z - \frac{\partial p}{\partial z} + \frac{\partial}{\partial x}\tau_{xz} + \frac{\partial}{\partial y}\tau_{yz} + \frac{\partial}{\partial z}\tau_{zz} \tag{1.77}$$

Conservative form

x-component

$$\frac{\partial}{\partial t}(\rho u) + \frac{\partial}{\partial x}(\rho u^2 + p) + \frac{\partial}{\partial y}(\rho uv) + \frac{\partial}{\partial z}(\rho uw) = \rho f_x + \frac{\partial}{\partial x}\tau_{xx} + \frac{\partial}{\partial y}\tau_{yx} + \frac{\partial}{\partial z}\tau_{zx} \tag{1.78}$$

y-component

$$\frac{\partial}{\partial t}(\rho v) + \frac{\partial}{\partial x}(\rho vu) + \frac{\partial}{\partial y}(\rho v^2 + p) + \frac{\partial}{\partial z}(\rho vw) = \rho f_y + \frac{\partial}{\partial x}\tau_{xy} + \frac{\partial}{\partial y}\tau_{yy} + \frac{\partial}{\partial z}\tau_{zy} \tag{1.79}$$

z-component

$$\frac{\partial}{\partial t}(\rho w) + \frac{\partial}{\partial x}(\rho wu) + \frac{\partial}{\partial y}(\rho wv) + \frac{\partial}{\partial z}(\rho w^2 + p) = \rho f_z + \frac{\partial}{\partial x}\tau_{xz} + \frac{\partial}{\partial y}\tau_{yz} + \frac{\partial}{\partial z}\tau_{zz} \tag{1.80}$$

For a Newtonian fluid with Stokes hypothesis

x-component

$$\rho\left(\frac{\partial u}{\partial t} + u\frac{\partial u}{\partial x} + v\frac{\partial u}{\partial y} + w\frac{\partial u}{\partial z}\right) = \rho f_x - \frac{\partial p}{\partial x} + \frac{\partial}{\partial x}\left[\frac{2}{3}\mu\left(2\frac{\partial u}{\partial x} - \frac{\partial v}{\partial y} - \frac{\partial w}{\partial z}\right)\right]$$

$$+ \frac{\partial}{\partial y}\left[\mu\left(\frac{\partial u}{\partial y} + \frac{\partial v}{\partial x}\right)\right] + \frac{\partial}{\partial z}\left[\mu\left(\frac{\partial w}{\partial x} + \frac{\partial u}{\partial z}\right)\right] \tag{1.81}$$

y-component

$$\rho\left(\frac{\partial v}{\partial t} + u\frac{\partial v}{\partial x} + v\frac{\partial v}{\partial y} + w\frac{\partial v}{\partial z}\right) = \rho f_y - \frac{\partial p}{\partial y} + \frac{\partial}{\partial x}\left[\mu\left(\frac{\partial u}{\partial y} + \frac{\partial v}{\partial x}\right)\right] +$$

$$+ \frac{\partial}{\partial y}\left[\frac{2}{3}\mu\left(2\frac{\partial v}{\partial y} - \frac{\partial u}{\partial x} - \frac{\partial w}{\partial z}\right)\right] + \frac{\partial}{\partial z}\left[\mu\left(\frac{\partial v}{\partial z} + \frac{\partial w}{\partial y}\right)\right] \tag{1.82}$$

z-component

$$\rho\left(\frac{\partial w}{\partial t} + u\frac{\partial w}{\partial x} + v\frac{\partial w}{\partial y} + w\frac{\partial w}{\partial z}\right) = \rho f_z - \frac{\partial p}{\partial z} + \frac{\partial}{\partial x}\left[\mu\left(\frac{\partial w}{\partial x} + \frac{\partial u}{\partial z}\right)\right] +$$

$$+\frac{\partial}{\partial y}\left[\mu\left(\frac{\partial v}{\partial z}+\frac{\partial w}{\partial y}\right)\right]+\frac{\partial}{\partial z}\left[\frac{2}{3}\mu\left(2\frac{\partial w}{\partial z}-\frac{\partial u}{\partial x}-\frac{\partial v}{\partial y}\right)\right] \tag{1.83}$$

For an incompressible flow with constant properties

x-component

$$\rho\left(\frac{\partial u}{\partial t}+u\frac{\partial u}{\partial x}+v\frac{\partial u}{\partial y}+w\frac{\partial u}{\partial z}\right)=\rho f_x-\frac{\partial p}{\partial x}+\mu\left(\frac{\partial^2 u}{\partial x^2}+\frac{\partial^2 u}{\partial y^2}+\frac{\partial^2 u}{\partial z^2}\right) \tag{1.84}$$

y-component

$$\rho\left(\frac{\partial v}{\partial t}+u\frac{\partial v}{\partial x}+v\frac{\partial v}{\partial y}+w\frac{\partial v}{\partial z}\right)=\rho f_y-\frac{\partial p}{\partial y}+\mu\left(\frac{\partial^2 v}{\partial x^2}+\frac{\partial^2 v}{\partial y^2}+\frac{\partial^2 v}{\partial z^2}\right) \tag{1.85}$$

z-component

$$\rho\left(\frac{\partial w}{\partial t}+u\frac{\partial w}{\partial x}+v\frac{\partial w}{\partial y}+w\frac{\partial w}{\partial z}\right)=\rho f_z-\frac{\partial p}{\partial z}+\mu\left(\frac{\partial^2 w}{\partial x^2}+\frac{\partial^2 w}{\partial y^2}+\frac{\partial^2 w}{\partial z^2}\right) \tag{1.86}$$

Table 1.6d: Navier-Stokes equations in a cylindrical coordinate system

r-component

$$\rho\left(\frac{\partial u_r}{\partial t}+u_r\frac{\partial u_r}{\partial r}+\frac{u_\theta}{r}\frac{\partial u_r}{\partial \theta}+u_z\frac{\partial u_r}{\partial z}-\frac{u_\theta^2}{r}\right)=\rho f_r-\frac{\partial p}{\partial r}+\frac{1}{r}\frac{\partial}{\partial r}(r\tau_{rr})+$$

$$\frac{1}{r}\frac{\partial}{\partial \theta}(\tau_{\theta r})+\frac{\partial}{\partial z}(\tau_{zr})-\frac{\tau_{\theta\theta}}{r} \tag{1.87}$$

θ-component

$$\rho\left(\frac{\partial u_\theta}{\partial t}+u_r\frac{\partial u_\theta}{\partial r}+\frac{u_\theta}{r}\frac{\partial u_\theta}{\partial \theta}+u_z\frac{\partial u_\theta}{\partial z}+\frac{u_r u_\theta}{r}\right)=\rho f_\theta-\frac{1}{r}\frac{\partial p}{\partial \theta}+\frac{1}{r}\frac{\partial}{\partial r}(r\tau_{r\theta})+$$

$$\frac{1}{r}\frac{\partial}{\partial \theta}(\tau_{\theta\theta})+\frac{\partial}{\partial z}(\tau_{z\theta})+\frac{\tau_{r\theta}}{r} \tag{1.88}$$

z-component

$$\rho\left(\frac{\partial u_z}{\partial t}+u_r\frac{\partial u_z}{\partial r}+\frac{u_\theta}{r}\frac{\partial u_z}{\partial \theta}+u_z\frac{\partial u_z}{\partial z}\right)=\rho f_z-\frac{\partial p}{\partial z}+\frac{1}{r}\frac{\partial}{\partial r}(r\tau_{rz})+$$

$$\frac{1}{r}\frac{\partial}{\partial \theta}(\tau_{\theta z})+\frac{\partial}{\partial z}(\tau_{zz}) \tag{1.89}$$

Conservative form

r-component

$$\frac{\partial}{\partial t}(\rho u_r) + \frac{\partial}{\partial r}(\rho u_r^2 + p) + \frac{1}{r\partial\theta}(\rho u_r u_\theta) + \frac{\partial}{\partial z}(\rho u_r u_z) = \rho f_r$$

$$+\frac{\partial}{\partial r}(\tau_{rr}) + \frac{1}{r\partial\theta}(\tau_{\theta r}) + \frac{\partial}{\partial z}(\tau_{zr}) - \frac{1}{r}\rho(u_r^2 - u_\theta^2) + \frac{1}{r}(\tau_{rr} - \tau_{\theta\theta}) \tag{1.90}$$

θ-component

$$\frac{\partial}{\partial t}(\rho u_\theta) + \frac{\partial}{\partial r}(\rho u_\theta u_r) + \frac{\partial}{\partial r\partial\theta}(\rho u_\theta^2 + p) + \frac{\partial}{\partial z}(\rho u_\theta u_z) = \rho f_\theta +$$

$$+\frac{\partial}{\partial r}(\tau_{r\theta}) + \frac{\partial}{\partial r\partial\theta}(\tau_{\theta\theta}) + \frac{\partial}{\partial z}(\tau_{z\theta}) - 2\frac{1}{r}\rho u_\theta u_r + 2\frac{1}{r}\tau_{r\theta} \tag{1.91}$$

z-component

$$\frac{\partial}{\partial t}(\rho u_z) + \frac{\partial}{\partial r}(\rho u_z u_r) + \frac{\partial}{\partial r\partial\theta}(\rho u_z u_\theta) + \frac{\partial}{\partial z}(\rho u_z^2 + p) = \rho f_z +$$

$$+\frac{\partial}{\partial r}(\tau_{rz}) + \frac{\partial}{\partial r\partial\theta}(\tau_{\theta z}) + \frac{\partial}{\partial z}(\tau_{zz}) - \frac{1}{r}\rho u_z u_r + \frac{1}{r}\tau_{rz} \tag{1.92}$$

For a Newtonian fluid with Stokes hypothesis

r-component

$$\rho\left(\frac{\partial u_r}{\partial t} + u_r\frac{\partial u_r}{\partial r} + \frac{u_\theta}{r}\frac{\partial u_r}{\partial\theta} + u_z\frac{\partial u_r}{\partial z} - \frac{u_\theta^2}{r}\right) = \rho f_r - \frac{\partial p}{\partial r}$$

$$+\frac{1}{r}\frac{\partial}{\partial r}\left\{r\mu\left[\frac{4}{3}\frac{\partial u_r}{\partial r} - \frac{2}{3}\left(\frac{1}{r}\frac{\partial u_\theta}{\partial\theta} + \frac{\partial u_z}{\partial z} + \frac{u_r}{r}\right)\right]\right\} + \frac{1}{r\partial\theta}\left\{\mu\left[r\frac{\partial}{\partial r}\left(\frac{u_\theta}{r}\right) + \right.\right.$$

$$\left.\left.\frac{1}{r}\frac{\partial u_r}{\partial\theta}\right]\right\} + \frac{\partial}{\partial z}\left[\mu\left(\frac{\partial u_z}{\partial r} + \frac{\partial u_r}{\partial z}\right)\right] - \frac{1}{r}\mu\left\{\frac{4}{3}\left(\frac{\partial u_\theta}{r\partial\theta} + \frac{u_r}{r}\right) - \frac{2}{3}\left(\frac{\partial u_r}{\partial r} + \frac{\partial u_z}{\partial z}\right)\right\} \tag{1.93}$$

θ-component

$$\rho\left(\frac{\partial u_\theta}{\partial t} + u_r\frac{\partial u_\theta}{\partial r} + \frac{u_\theta}{r}\frac{\partial u_\theta}{\partial\theta} + u_z\frac{\partial u_\theta}{\partial z} + \frac{u_r u_\theta}{r}\right) = \rho f_\theta - \frac{\partial p}{r\partial\theta}$$

$$+\frac{1}{r}\frac{\partial}{\partial r}\left\{r\mu\left[r\frac{\partial}{\partial r}\left(\frac{u_\theta}{r}\right) + \frac{1}{r}\frac{\partial u_r}{\partial\theta}\right]\right\} + \frac{\partial}{r\partial\theta}\left\{\mu\left[\frac{4}{3}\left(\frac{\partial u_\theta}{r\partial\theta} + \frac{u_r}{r}\right) - \frac{2}{3}\left(\frac{\partial u_r}{\partial r} + \right.\right.\right.$$

$$\left.\left.\left. + \frac{\partial u_z}{\partial z}\right)\right]\right\} + \frac{\partial}{\partial z}\left\{\mu\left(\frac{\partial u_\theta}{\partial z} + \frac{1}{r}\frac{\partial u_z}{\partial\theta}\right)\right\} + \frac{1}{r}\left\{\mu\left[r\frac{\partial}{\partial r}\left(\frac{u_\theta}{r}\right) + \frac{1}{r}\frac{\partial u_r}{\partial\theta}\right]\right\} \tag{1.94}$$

z-component

$$\rho \left(\frac{\partial u_z}{\partial t} + u_r \frac{\partial u_z}{\partial r} + u_\theta \frac{\partial u_z}{r \partial \theta} + u_z \frac{\partial u_z}{\partial z} \right) = \rho f_z - \frac{\partial p}{\partial z}$$

$$+ \frac{1}{r} \frac{\partial}{\partial r} \left\{ r\mu \left[\frac{\partial u_z}{\partial r} + \frac{\partial u_r}{\partial z} \right] \right\} + \frac{1}{r} \frac{\partial}{\partial \theta} \left\{ \mu \left[\frac{\partial u_\theta}{\partial z} + \frac{1}{r} \frac{\partial u_z}{\partial \theta} \right] \right\}$$

$$+ \frac{\partial}{\partial z} \left\{ \mu \left[\frac{4}{3} \frac{\partial u_z}{\partial z} - \frac{2}{3} \left(\frac{1}{r} \frac{\partial}{\partial r}(r u_r) + \frac{1}{r} \frac{\partial u_\theta}{\partial \theta} \right) \right] \right\} \tag{1.95}$$

For an incompressible flow with constant properties

r-component

$$\rho \left(\frac{\partial u_r}{\partial t} + u_r \frac{\partial u_r}{\partial r} + \frac{u_\theta}{r} \frac{\partial u_r}{\partial \theta} + u_z \frac{\partial u_r}{\partial z} - \frac{u_\theta^2}{r} \right) = \rho f_r - \frac{\partial p}{\partial r}$$

$$+ \mu \left[\frac{\partial^2 u_r}{\partial r^2} + \frac{1}{r} \frac{\partial u_r}{\partial r} - \frac{u_r}{r^2} + \frac{1}{r^2} \frac{\partial^2 u_r}{\partial \theta^2} + \frac{\partial^2 u_r}{\partial z^2} - \frac{2}{r^2} \frac{\partial u_\theta}{\partial \theta} \right] \tag{1.96}$$

θ-component

$$\rho \left(\frac{\partial u_\theta}{\partial t} + u_r \frac{\partial u_\theta}{\partial r} + u_\theta \frac{\partial u_\theta}{r \partial \theta} + u_z \frac{\partial u_\theta}{\partial z} + \frac{u_r u_\theta}{r} \right) = \rho f_\theta - \frac{1}{r} \frac{\partial p}{\partial \theta}$$

$$+ \mu \left[\frac{\partial^2 u_\theta}{\partial r^2} + \frac{1}{r} \frac{\partial u_\theta}{\partial r} - \frac{u_\theta}{r^2} + \frac{1}{r^2} \frac{\partial^2 u_\theta}{\partial \theta^2} + \frac{\partial^2 u_\theta}{\partial z^2} + \frac{2}{r^2} \frac{\partial u_r}{\partial \theta} \right] \tag{1.97}$$

z-component

$$\rho \left(\frac{\partial u_z}{\partial t} + u_r \frac{\partial u_z}{\partial r} + u_\theta \frac{\partial u_z}{r \partial \theta} + u_z \frac{\partial u_z}{\partial z} \right) = \rho f_z - \frac{\partial p}{\partial z}$$

$$+ \mu \left[\frac{\partial^2 u_z}{\partial r^2} + \frac{1}{r} \frac{\partial u_z}{\partial r} + \frac{1}{r^2} \frac{\partial^2 u_z}{\partial \theta^2} + \frac{\partial^2 u_z}{\partial z^2} \right] \tag{1.98}$$

Table 1.6e: Navier-Stokes equations in a spherical coordinate system

r-component

$$\rho \left(\frac{\partial u_r}{\partial t} + u_r \frac{\partial u_r}{\partial r} + \frac{u_\theta}{r} \frac{\partial u_r}{\partial \theta} + \frac{u_\phi}{r \sin \theta} \frac{\partial u_r}{\partial \phi} - \frac{u_\theta^2 + u_\phi^2}{r} \right)$$

$$= -\frac{\partial p}{\partial r} + \rho f_r + \frac{1}{r^2} \frac{\partial}{\partial r}(r^2 \tau_{rr}) + \frac{1}{r \sin \theta} \frac{\partial}{\partial \theta}(\tau_{\theta r} \sin \theta) + \frac{1}{r \sin \theta} \frac{\partial \tau_{\phi r}}{\partial \phi} - \frac{\tau_{\theta\theta} + \tau_{\phi\phi}}{r} \tag{1.99}$$

θ-component

$$\rho\left(\frac{\partial u_\theta}{\partial t} + u_r\frac{\partial u_\theta}{\partial r} + \frac{u_\theta}{r}\frac{\partial u_\theta}{\partial \theta} + \frac{u_\phi}{r\sin\theta}\frac{\partial u_\theta}{\partial \phi} + \frac{u_r u_\theta}{r} - \frac{u_\phi^2\cot\theta}{r}\right)$$

$$= -\frac{1}{r}\frac{\partial p}{\partial \theta} + \rho f_\theta + \frac{1}{r^2}\frac{\partial}{\partial r}(r^2\tau_{r\theta}) + \frac{1}{r\sin\theta}\frac{\partial}{\partial \theta}(\tau_{\theta\theta}\sin\theta)$$

$$+ \frac{1}{r\sin\theta}\frac{\partial \tau_{\phi\theta}}{\partial \phi} + \frac{\tau_{r\theta}}{r} - \frac{\tau_{\phi\phi}\cot\theta}{r} \tag{1.100}$$

ϕ-component

$$\rho\left(\frac{\partial u_\phi}{\partial t} + u_r\frac{\partial u_\phi}{\partial r} + \frac{u_\theta}{r}\frac{\partial u_\phi}{\partial \theta} + \frac{u_\phi}{r\sin\theta}\frac{\partial u_\phi}{\partial \phi} + \frac{u_r u_\phi}{r} + \frac{u_\theta u_\phi}{r}\cot\theta\right)$$

$$= \rho f_\phi - \frac{1}{r\sin\theta}\frac{\partial p}{\partial \phi} + \frac{1}{r^2}\frac{\partial}{\partial r}(r^2\tau_{r\phi}) + \frac{1}{r}\frac{\partial \tau_{\theta\phi}}{\partial \theta} + \frac{1}{r\sin\theta}\frac{\partial \tau_{\phi\phi}}{\partial \phi} + \frac{\tau_{r\phi}}{r} + 2\frac{\tau_{\theta\phi}}{r}\cot\theta \tag{1.101}$$

For a Newtonian fluid with Stokes hypothesis

r-component

$$\rho\left(\frac{\partial u_r}{\partial t} + u_r\frac{\partial u_r}{\partial r} + \frac{u_\theta}{r}\frac{\partial u_r}{\partial \theta} + \frac{u_\phi}{r\sin\theta}\frac{\partial u_r}{\partial \phi} - \frac{u_\theta^2 + u_\phi^2}{r}\right)$$

$$= \rho f_r - \frac{\partial p}{\partial r} + \frac{\partial}{\partial r}\left(2\mu\frac{\partial u_r}{\partial r} - \frac{2}{3}\mu\nabla\cdot\vec{V}\right) + \frac{1}{r}\frac{\partial}{\partial \theta}\left\{\mu\left[r\frac{\partial}{\partial r}\left(\frac{u_\theta}{r}\right) + \frac{1}{r}\frac{\partial u_r}{\partial \theta}\right]\right\}$$

$$+ \frac{1}{r\sin\theta}\frac{\partial}{\partial \phi}\left\{\mu\left[\frac{1}{r\sin\theta}\frac{\partial u_r}{\partial \phi} + r\frac{\partial}{\partial r}\left(\frac{u_\phi}{r}\right)\right]\right\}$$

$$+ \frac{\mu}{r}\left[4\frac{\partial u_r}{\partial r} - \frac{2}{r}\frac{\partial u_\theta}{\partial \theta} - \frac{2}{r\sin\theta}\frac{\partial u_\phi}{\partial \phi} + r\cot\theta\frac{\partial}{\partial r}\left(\frac{u_\theta}{r}\right)\right.$$

$$\left. + \frac{\cot\theta}{r}\frac{\partial u_r}{\partial \theta} - 4\frac{u_r}{r} - \frac{2u_\theta\cot\theta}{r}\right] \tag{1.102}$$

θ-component

$$\rho\left(\frac{\partial u_\theta}{\partial t} + u_r\frac{\partial u_\theta}{\partial r} + \frac{u_\theta}{r}\frac{\partial u_\theta}{\partial \theta} + \frac{u_\phi}{r\sin\theta}\frac{\partial u_\theta}{\partial \phi} + \frac{u_r u_\theta}{r} - \frac{u_\phi^2\cot\theta}{r}\right)$$

$$= \rho f_\theta - \frac{1}{r}\frac{\partial p}{\partial \theta} + \frac{1}{r}\frac{\partial}{\partial \theta}\left[2\frac{\mu}{r}\left(\frac{\partial u_\theta}{\partial \theta} + u_r\right) - \frac{2}{3}\mu\nabla\cdot\vec{V}\right]$$

$$+ \frac{1}{r\sin\theta}\frac{\partial}{\partial \phi}\left\{\mu\left[\frac{\sin\theta}{r}\frac{\partial}{\partial \theta}\left(\frac{u_\phi}{\sin\theta}\right) + \frac{1}{r\sin\theta}\frac{\partial u_\theta}{\partial \phi}\right]\right\} + \frac{\partial}{\partial r}\left\{\mu\left[r\frac{\partial}{\partial r}\left(\frac{u_\theta}{r}\right) + \frac{1}{r}\frac{\partial u_r}{\partial \theta}\right]\right\}$$

$$+ \frac{\mu}{r} \left\{ 2 \left(\frac{1}{r} \frac{\partial u_\theta}{\partial \theta} - \frac{1}{r \sin \theta} \frac{\partial u_\phi}{\partial \phi} - \frac{u_\theta \cot \theta}{r} \right) \cot \theta + 3 \left[r \frac{\partial}{\partial r} \left(\frac{u_\theta}{r} \right) + \frac{1}{r} \frac{\partial u_r}{\partial \theta} \right] \right\} \quad (1.103)$$

ϕ-component

$$\rho \left(\frac{\partial u_\phi}{\partial t} + u_r \frac{\partial u_\phi}{\partial r} + \frac{u_\theta}{r} \frac{\partial u_\phi}{\partial \theta} + \frac{u_\phi}{r \sin \theta} \frac{\partial u_\phi}{\partial \phi} + \frac{u_r u_\phi}{r} + \frac{u_\theta u_\phi}{r} \cot \theta \right)$$

$$= \rho f_\phi - \frac{1}{r \sin \theta} \frac{\partial p}{\partial \phi} + \frac{1}{r \sin \theta} \frac{\partial}{\partial \phi} \left[2 \frac{\mu}{r} \left(\frac{1}{\sin \theta} \frac{\partial u_\phi}{\partial \phi} + u_r + u_\theta \cot \theta \right) - \frac{2}{3} \mu \nabla \cdot \vec{V} \right]$$

$$+ \frac{\partial}{\partial r} \left\{ \mu \left[\frac{1}{r \sin \theta} \frac{\partial u_r}{\partial \phi} + r \frac{\partial}{\partial r} \left(\frac{u_\phi}{r} \right) \right] \right\} + \frac{1}{r} \frac{\partial}{\partial \theta} \left\{ \mu \left[\frac{\sin \theta}{r} \frac{\partial}{\partial \theta} \left(\frac{u_\phi}{\sin \theta} \right) + \frac{1}{r \sin \theta} \frac{\partial u_\theta}{\partial \phi} \right] \right\}$$

$$+ \frac{\mu}{r} \left\{ 3 \left[\frac{1}{r \sin \theta} \frac{\partial u_r}{\partial \phi} + r \frac{\partial}{\partial r} \left(\frac{u_\phi}{r} \right) \right] + 2 \cot \theta \left[\frac{\sin \theta}{r} \frac{\partial}{\partial \theta} \left(\frac{u_\phi}{\sin \theta} \right) + \frac{1}{r \sin \theta} \frac{\partial u_\theta}{\partial \phi} \right] \right\} \quad (1.104)$$

For an incompressible flow with constant properties

r-component

$$\rho \left(\frac{\partial u_r}{\partial t} + u_r \frac{\partial u_r}{\partial r} + \frac{u_\theta}{r} \frac{\partial u_r}{\partial \theta} + \frac{u_\phi}{r \sin \theta} \frac{\partial u_r}{\partial \phi} - \frac{u_\theta^2 + u_\phi^2}{r} \right)$$

$$= \rho f_r - \frac{\partial p}{\partial r} + \mu \left[\frac{1}{r^2} \frac{\partial}{\partial r} \left(r^2 \frac{\partial u_r}{\partial r} \right) + \frac{1}{r^2 \sin \theta} \frac{\partial}{\partial \theta} \left(\sin \theta \frac{\partial u_r}{\partial \theta} \right) \right.$$

$$\left. + \frac{1}{r^2 \sin^2 \theta} \frac{\partial^2 u_r}{\partial \phi^2} - 2 \frac{u_r}{r^2} - \frac{2}{r^2} \frac{\partial u_\theta}{\partial \theta} - \frac{2}{r^2} u_\theta \cot \theta - \frac{2}{r^2 \sin \theta} \frac{\partial u_\phi}{\partial \phi} \right] \quad (1.105)$$

θ-component

$$\rho \left(\frac{\partial u_\theta}{\partial t} + u_r \frac{\partial u_\theta}{\partial r} + \frac{u_\theta}{r} \frac{\partial u_\theta}{\partial \theta} + \frac{u_\phi}{r \sin \theta} \frac{\partial u_\theta}{\partial \phi} + \frac{u_r u_\theta}{r} - \frac{u_\phi^2 \cot \theta}{r} \right)$$

$$= \rho f_\theta - \frac{1}{r} \frac{\partial p}{\partial \theta} + \mu \left[\frac{1}{r^2} \frac{\partial}{\partial r} \left(r^2 \frac{\partial u_\theta}{\partial r} \right) + \frac{1}{r^2 \sin \theta} \frac{\partial}{\partial \theta} \left(\sin \theta \frac{\partial u_\theta}{\partial \theta} \right) \right.$$

$$\left. + \frac{1}{r^2 \sin^2 \theta} \frac{\partial^2 u_\theta}{\partial \phi^2} + \frac{2}{r^2} \frac{\partial u_r}{\partial \theta} - \frac{u_\theta}{r^2 \sin^2 \theta} - 2 \frac{\cos \theta}{r^2 \sin^2 \theta} \frac{\partial u_\phi}{\partial \phi} \right] \quad (1.106)$$

ϕ-component

$$\rho \left(\frac{\partial u_\phi}{\partial t} + u_r \frac{\partial u_\phi}{\partial r} + \frac{u_\theta}{r} \frac{\partial u_\phi}{\partial \theta} + \frac{u_\phi}{r \sin \theta} \frac{\partial u_\phi}{\partial \phi} + \frac{u_r u_\phi}{r} + \frac{u_\theta u_\phi}{r} \cot \theta \right)$$

$$= \rho f_\phi - \frac{1}{r \sin \theta} \frac{\partial p}{\partial \phi} + \mu \left[\frac{1}{r^2} \frac{\partial}{\partial r} \left(r^2 \frac{\partial u_\phi}{\partial r} \right) + \frac{1}{r^2 \sin \theta} \frac{\partial}{\partial \theta} \left(\sin \theta \frac{\partial u_\phi}{\partial \theta} \right) \right.$$

$$\left. + \frac{1}{r^2 \sin^2 \theta} \frac{\partial^2 u_\phi}{\partial \phi^2} - \frac{u_\phi}{r^2 \sin^2 \theta} + \frac{2}{r^2 \sin \theta} \frac{\partial u_r}{\partial \phi} + 2 \frac{\cos \theta}{r^2 \sin^2 \theta} \frac{\partial u_\theta}{\partial \phi} \right. \quad (1.107)$$

Table 1.6f: Navier-Stokes equations in a general orthogonal curvilinear coordinate system

x_1-component

$$\rho\left\{\frac{\partial u_1}{\partial t} + \frac{u_1}{h_1}\frac{\partial u_1}{\partial x_1} + \frac{u_2}{h_2}\frac{\partial u_1}{\partial x_2} + \frac{u_3}{h_3}\frac{\partial u_1}{\partial x_3} - u_2\left(\frac{u_2}{h_1 h_2}\frac{\partial h_2}{\partial x_1} - \frac{u_1}{h_1 h_2}\frac{\partial h_1}{\partial x_2}\right)\right.$$

$$\left.+u_3\left(\frac{u_1}{h_1 h_3}\frac{\partial h_1}{\partial x_3} - \frac{u_3}{h_3 h_1}\frac{\partial h_3}{\partial x_1}\right)\right\} = f_1 - \frac{1}{h_1}\frac{\partial p}{\partial x_1}$$

$$+\frac{1}{h_1 h_2 h_3}\left[\frac{\partial}{\partial x_1}(h_2 h_3 \tau_{11}) + \frac{\partial}{\partial x_2}(h_3 h_1 \tau_{21}) + \frac{\partial}{\partial x_3}(h_1 h_2 \tau_{31})\right]$$

$$+\frac{\tau_{12}}{h_1 h_2}\frac{\partial h_1}{\partial x_2} + \frac{\tau_{31}}{h_1 h_3}\frac{\partial h_1}{\partial x_3} - \frac{\tau_{22}}{h_1 h_2}\frac{\partial h_2}{\partial x_1} - \frac{\tau_{33}}{h_1 h_3}\frac{\partial h_3}{\partial x_1} \tag{1.108}$$

x_2-component

$$\rho\left\{\frac{\partial u_2}{\partial t} + \frac{u_1}{h_1}\frac{\partial u_2}{\partial x_1} + \frac{u_2}{h_2}\frac{\partial u_2}{\partial x_2} + \frac{u_3}{h_3}\frac{\partial u_2}{\partial x_3} - u_3\left(\frac{u_3}{h_3 h_2}\frac{\partial h_3}{\partial x_2} - \frac{u_2}{h_2 h_3}\frac{\partial h_2}{\partial x_3}\right)\right.$$

$$\left.+u_1\left(\frac{u_2}{h_2 h_1}\frac{\partial h_2}{\partial x_1} - \frac{u_1}{h_1 h_2}\frac{\partial h_1}{\partial x_2}\right)\right\} = f_2 - \frac{1}{h_2}\frac{\partial p}{\partial x_2}$$

$$+\frac{1}{h_1 h_2 h_3}\left[\frac{\partial}{\partial x_1}(h_2 h_3 \tau_{12}) + \frac{\partial}{\partial x_2}(h_3 h_1 \tau_{22}) + \frac{\partial}{\partial x_3}(h_1 h_2 \tau_{32})\right]$$

$$+\frac{\tau_{23}}{h_2 h_3}\frac{\partial h_2}{\partial x_3} + \frac{\tau_{12}}{h_1 h_2}\frac{\partial h_2}{\partial x_1} - \frac{\tau_{33}}{h_2 h_3}\frac{\partial h_3}{\partial x_2} - \frac{\tau_{11}}{h_1 h_2}\frac{\partial h_1}{\partial x_2} \tag{1.109}$$

x_3-component

$$\rho\left\{\frac{\partial u_3}{\partial t} + \frac{u_1}{h_1}\frac{\partial u_3}{\partial x_1} + \frac{u_2}{h_2}\frac{\partial u_3}{\partial x_2} + \frac{u_3}{h_3}\frac{\partial u_3}{\partial x_3} - u_1\left(\frac{u_1}{h_1 h_3}\frac{\partial h_1}{\partial x_3} - \frac{u_3}{h_1 h_3}\frac{\partial h_3}{\partial x_1}\right)\right.$$

$$\left.+u_2\left(\frac{u_3}{h_3 h_2}\frac{\partial h_3}{\partial x_2} - \frac{u_2}{h_2 h_3}\frac{\partial h_2}{\partial x_3}\right)\right\} = f_3 - \frac{1}{h_3}\frac{\partial p}{\partial x_3}$$

$$+\frac{1}{h_1 h_2 h_3}\left[\frac{\partial}{\partial x_1}(h_2 h_3 \tau_{13}) + \frac{\partial}{\partial x_2}(h_1 h_3 \tau_{23}) + \frac{\partial}{\partial x_3}(h_1 h_2 \tau_{33})\right]$$

$$+\frac{\tau_{31}}{h_1 h_3}\frac{\partial h_3}{\partial x_1} + \frac{\tau_{23}}{h_2 h_3}\frac{\partial h_3}{\partial x_2} - \frac{\tau_{11}}{h_3 h_1}\frac{\partial h_1}{\partial x_3} - \frac{\tau_{22}}{h_3 h_2}\frac{\partial h_2}{\partial x_3} \tag{1.110}$$

1.4.4 Energy Equation

The energy equation derived from the first law of thermodynamics may be written in various forms. One such formulation known as *the total energy equation* is as follows:

$$\frac{\partial}{\partial t}[\rho\,(\overbrace{e+\frac{1}{2}V^2)}^{e_t}]+\nabla\cdot[\rho(e+\frac{1}{2}V^2)\vec{V}] = \frac{\partial Q}{\partial t}-\nabla\cdot\vec{q}+\rho(\vec{V}\cdot\vec{f})-\nabla\cdot(p\vec{V})+\nabla\cdot(\tau\cdot\vec{V}) \quad (1.111)$$

where $\vec{q}$ represents heat flux vector and $\dfrac{\partial Q}{\partial t}$ represents any internal heat generation. Note that in this form of the energy equation, potential energy does not appear explicitly. This is so because the potential energy is included in the work term. If e_t is used to represent the sum of internal energy and the kinetic energy per unit mass, i.e., $e_t = e + \frac{1}{2}V^2$, then Equation (1.111) can be rearranged with the use of continuity and expressed as:

$$\rho\frac{d}{dt}\overbrace{(e+\frac{1}{2}V^2)}^{e_t} = \frac{\partial Q}{\partial t} - \nabla\cdot\vec{q} + \rho(\vec{V}\cdot\vec{f}) - \nabla\cdot(p\vec{V}) + \nabla\cdot(\tau\cdot\vec{V}) \quad (1.112)$$

The energy equation may be written in terms of the internal energy in which case it is known as the equation of *the thermal energy* and is expressed as:

$$\rho\frac{de}{dt} = \frac{\partial Q}{\partial t} - \nabla\cdot\vec{q} - p(\nabla\cdot\vec{V}) + \tau\cdot\nabla\vec{V} \quad (1.113)$$

If the energy equation is written in terms of the kinetic energy, it is known as *the equation of mechanical energy*, and the formulation is as follows:

$$\rho\frac{d}{dt}\left(\frac{1}{2}V^2\right) = \frac{\partial Q}{\partial t} - \vec{V}\cdot\nabla p + \rho\vec{V}\cdot\vec{f} + \vec{V}\cdot[\nabla\cdot\tau] \quad (1.114)$$

Various forms of the energy equation just reviewed are usually expressed in terms of temperature or enthalpy. At this point we will review these formulations. In the subsequent equations let us assume no internal heat generation and therefore $\dfrac{\partial Q}{\partial t}$ will be dropped from the equations. One form of the energy equation in terms of the temperature T is:

$$\rho c_v\frac{dT}{dt} = -\nabla\cdot\vec{q} - T\left(\frac{\partial p}{\partial T}\right)_\rho (\nabla\cdot\vec{V}) + \tau:\nabla\vec{V} \quad (1.115)$$

where c_v is the specific heat at constant volume.

In terms of enthalpy we may write:

$$\rho\frac{dh}{dt} = -\nabla\cdot\vec{q} + \frac{dp}{dt} + \tau:\nabla\vec{V} \quad (1.116)$$

The heat conduction term expressed by $\nabla \cdot \vec{q}$ is related to the temperature gradient by the Fourier heat conduction law as:

$$\vec{q} = -k\nabla T \tag{1.117}$$

Various forms of the energy equation may be reduced for specific applications by imposing pertinent assumptions. For example for an incompressible constant property flow, the term involving $\nabla \cdot \vec{V}$ will drop out and in addition the coefficients of viscosity and thermal conductivity will be treated as constants.

Various formulations of the energy equation and associated assumptions are summarized in Tables 1.7a through 1.7e. At this point we will consider some of the formulations of the energy equation and expand them in a Cartesian coordinate system. First consider Equation (1.112), which will take the following form

$$\rho\frac{d}{dt}\overbrace{\left(e + \frac{1}{2}V^2\right)}^{e_t} = \rho(uf_x + vf_y + wf_z) + \frac{\partial}{\partial x}[-pu + u\tau_{xx} + v\tau_{xy} + w\tau_{xz} - q_x]$$

$$+ \frac{\partial}{\partial y}[-pv + u\tau_{yx} + v\tau_{yy} + w\tau_{yz} - q_y] + \frac{\partial}{\partial z}[-pw + u\tau_{zx} + v\tau_{zy} + w\tau_{zz} - q_z] \tag{1.118}$$

Equation (1.118) may be expressed in a conservative form by the addition of the continuity equation as

$$\frac{\partial}{\partial t}(\rho e_t) + \frac{\partial}{\partial x}(\rho u e_t + pu) + \frac{\partial}{\partial y}(\rho v e_t + pv) + \frac{\partial}{\partial z}(\rho w e_t + pw) =$$

$$\frac{\partial}{\partial x}\left[u\tau_{xx} + v\tau_{xy} + w\tau_{xz} - q_x\right] + \frac{\partial}{\partial y}\left[u\tau_{yx} + v\tau_{yy} + w\tau_{yz} - q_y\right]$$

$$+ \frac{\partial}{\partial z}\left[u\tau_{zx} + v\tau_{zy} + w\tau_{zz} - q_z\right] \tag{1.119}$$

where the contribution due to the body force has been dropped. For a second case, consider Equation (1.116). Expanding the terms in Cartesian coordinates and substituting for the stresses relations defined by Newtonian fluid and after rearranging terms one has:

$$\rho\frac{dh}{dt} = \frac{\partial}{\partial x}\left(k\frac{\partial T}{\partial x}\right) + \frac{\partial}{\partial y}\left(k\frac{\partial T}{\partial y}\right) + \frac{\partial}{\partial z}\left(k\frac{\partial T}{\partial z}\right) + \frac{dp}{dt} + \Phi \tag{1.120}$$

where

$$\Phi = \mu\left\{2\left[\left(\frac{\partial u}{\partial x}\right)^2 + \left(\frac{\partial v}{\partial y}\right)^2 + \left(\frac{\partial w}{\partial z}\right)^2\right] + \left(\frac{\partial v}{\partial x} + \frac{\partial u}{\partial y}\right)^2\right.$$

$$\left. + \left(\frac{\partial w}{\partial y} + \frac{\partial v}{\partial z}\right)^2 + \left(\frac{\partial u}{\partial z} + \frac{\partial w}{\partial x}\right)^2 - \frac{2}{3}\left(\frac{\partial u}{\partial x} + \frac{\partial v}{\partial y} + \frac{\partial w}{\partial z}\right)^2\right\} \tag{1.121}$$

Φ is known as *the viscous dissipation term*. Other forms of the energy equation are given in Tables 1.7a through 1.7e.

Table 1.7a: Energy equation – Vector notation

Assumptions: No heat generation $\dfrac{\partial Q}{\partial t}$

$$\rho\frac{d}{dt}(e + \frac{1}{2}V^2) = \rho\vec{V}\cdot\vec{f} - \nabla\cdot\vec{q} - \nabla\cdot(p\vec{V}) + \nabla\cdot(\tau\cdot\vec{V}) \qquad (1.122)$$

$$\rho\frac{dh_t}{dt} = \rho\vec{V}\cdot f + \frac{\partial p}{\partial t} - \nabla\cdot\vec{q} + \nabla\cdot(\tau\cdot\vec{V}) \qquad (1.123)$$

where h_t is the total enthalpy $h_t = h + \dfrac{1}{2}V^2$

$$\rho\frac{de}{dt} = -\nabla\cdot\vec{q} - p\nabla\cdot\vec{V} + \tau : (\nabla\vec{V}) \qquad (1.124)$$

$$\rho\frac{d}{dt}(\frac{1}{2}V^2) = \rho\vec{V}\cdot\vec{f} - \vec{V}\cdot\nabla p + \vec{V}\cdot(\nabla\cdot\tau) \qquad (1.125)$$

$$\rho\frac{dh}{dt} = \frac{dp}{dt} - \nabla\cdot\vec{q} + \tau : (\nabla\vec{V}) \qquad (1.126)$$

$$\rho c_v\frac{dT}{dt} = -\nabla\cdot\vec{q} - T\left(\frac{\partial p}{\partial T}\right)_v (\nabla\cdot\vec{V}) + \tau : (\nabla\vec{V}) \qquad (1.127)$$

Ideal gas

$$\rho c_p\frac{dT}{dt} = \frac{dp}{dt} - \nabla\cdot\vec{q} + \tau : (\nabla\vec{V}) \qquad (1.128)$$

$$\rho c_v\frac{dT}{dt} = -\nabla\cdot\vec{q} - p\nabla\cdot\vec{V} + \tau : (\nabla\vec{V}) \qquad (1.129)$$

Table 1.7b: Energy equation – Tensor notation

Assumptions: No heat generation $\dfrac{\partial Q}{\partial t}$

$$\rho\frac{\partial e_t}{\partial t} + \rho u_i\frac{\partial e_t}{\partial x_i} = \rho u_i f_i - \frac{\partial q_i}{\partial x_i} - \frac{\partial}{\partial x_i}(pu_i) + \frac{\partial}{\partial x_i}(u_j\tau_{ij}) \tag{1.130}$$

where $\quad e_t = e + \dfrac{1}{2}V^2$

$$\rho\left(\frac{\partial e}{\partial t} + u_i\frac{\partial e}{\partial x_i}\right) = -\frac{\partial q_i}{\partial x_i} - p\frac{\partial u_i}{\partial x_i} + \tau_{ij}\frac{\partial u_i}{\partial x_j} \tag{1.131}$$

$$\rho\left[\frac{\partial}{\partial t}\left(\frac{1}{2}V^2\right) + u_i\frac{\partial}{\partial x_i}\left(\frac{1}{2}V^2\right)\right] = \rho u_i f_i - u_i\frac{\partial p}{\partial x_i} + u_j\frac{\partial}{\partial x_i}(\tau_{ij}) \tag{1.132}$$

$$\rho\left[\frac{\partial h}{\partial t} + u_i\frac{\partial h}{\partial x_i}\right] = \left[\frac{\partial p}{\partial t} + u_i\frac{\partial p}{\partial x_i}\right] - \frac{\partial q_i}{\partial x_i} + \tau_{ij}\frac{\partial u_i}{\partial x_j} \tag{1.133}$$

$$\rho c_v\left[\frac{\partial T}{\partial t} + u_i\frac{\partial T}{\partial x_i}\right] = -\frac{\partial q_i}{\partial x_i} - T\left(\frac{\partial p}{\partial T}\right)_v\frac{\partial u_i}{\partial x_i} + \tau_{ij}\frac{\partial u_i}{\partial x_j} \tag{1.134}$$

Conservative form

$$\frac{\partial}{\partial t}(\rho e_t) + \frac{\partial}{\partial x_i}(\rho u_i e_t) = \rho u_i f_i - \frac{\partial q_i}{\partial x_i} - \frac{\partial}{\partial x_i}(pu_i) + \frac{\partial}{\partial x_i}(u_j\tau_{ij}) \tag{1.135}$$

Ideal gas

$$\rho c_p\left(\frac{\partial T}{\partial t} + u_i\frac{\partial T}{\partial x_i}\right) = \left(\frac{\partial p}{\partial t} + u_i\frac{\partial p}{\partial x_i}\right) - \frac{\partial q_i}{\partial x_i} + \tau_{ij}\frac{\partial u_i}{\partial x_j} \tag{1.136}$$

$$\rho c_v\left(\frac{\partial T}{\partial t} + u_i\frac{\partial T}{\partial x_i}\right) = -\frac{\partial q_i}{\partial x_i} - p\frac{\partial u_i}{\partial x_i} + \tau_{ij}\frac{\partial u_i}{\partial x_j} \tag{1.137}$$

Table 1.7c: Energy equation – Cartesian coordinates

Assumption: No heat generation, Fourier heat conduction law

$$\rho\left(\frac{\partial e}{\partial t} + u\frac{\partial e}{\partial x} + v\frac{\partial e}{\partial y} + w\frac{\partial e}{\partial z}\right) = \frac{\partial}{\partial x}\left(k\frac{\partial T}{\partial x}\right) + \frac{\partial}{\partial y}\left(k\frac{\partial T}{\partial y}\right) + \frac{\partial}{\partial z}\left(k\frac{\partial T}{\partial z}\right)$$

$$-p\left(\frac{\partial u}{\partial x} + \frac{\partial v}{\partial y} + \frac{\partial w}{\partial z}\right) + \tau_{xx}\frac{\partial u}{\partial x} + \tau_{yy}\frac{\partial v}{\partial y} + \tau_{zz}\frac{\partial w}{\partial z} + \tau_{xy}\left(\frac{\partial u}{\partial y} + \frac{\partial v}{\partial x}\right)$$

$$+\tau_{xz}\left(\frac{\partial u}{\partial z} + \frac{\partial w}{\partial x}\right) + \tau_{yz}\left(\frac{\partial v}{\partial z} + \frac{\partial w}{\partial y}\right) \tag{1.138}$$

$$\rho\left(\frac{\partial h}{\partial t} + u\frac{\partial h}{\partial x} + v\frac{\partial h}{\partial y} + w\frac{\partial h}{\partial z}\right) = \frac{\partial}{\partial x}\left(k\frac{\partial T}{\partial x}\right) + \frac{\partial}{\partial y}\left(k\frac{\partial T}{\partial y}\right) + \frac{\partial}{\partial z}\left(k\frac{\partial T}{\partial z}\right)$$

$$+\left(\frac{\partial p}{\partial t} + u\frac{\partial p}{\partial x} + v\frac{\partial p}{\partial y} + w\frac{\partial p}{\partial z}\right) + \tau_{xx}\frac{\partial u}{\partial x} + \tau_{yy}\frac{\partial v}{\partial y} + \tau_{zz}\frac{\partial w}{\partial z}$$

$$+\tau_{xy}\left(\frac{\partial u}{\partial y} + \frac{\partial v}{\partial x}\right) + \tau_{xz}\left(\frac{\partial u}{\partial z} + \frac{\partial w}{\partial x}\right) + \tau_{yz}\left(\frac{\partial v}{\partial z} + \frac{\partial w}{\partial y}\right) \tag{1.139}$$

$$\rho c_v\left(\frac{\partial T}{\partial t} + u\frac{\partial T}{\partial x} + v\frac{\partial T}{\partial y} + w\frac{\partial T}{\partial z}\right) = \frac{\partial}{\partial x}\left(k\frac{\partial T}{\partial x}\right) + \frac{\partial}{\partial y}\left(k\frac{\partial T}{\partial y}\right) + \frac{\partial}{\partial z}\left(k\frac{\partial T}{\partial z}\right)$$

$$-T\left(\frac{\partial p}{\partial T}\right)_v\left(\frac{\partial u}{\partial x} + \frac{\partial v}{\partial y} + \frac{\partial w}{\partial z}\right) + \tau_{xx}\frac{\partial u}{\partial x} + \tau_{yy}\frac{\partial v}{\partial y} + \tau_{zz}\frac{\partial w}{\partial z}$$

$$+\tau_{xy}\left(\frac{\partial u}{\partial y} + \frac{\partial v}{\partial x}\right) + \tau_{xz}\left(\frac{\partial u}{\partial z} + \frac{\partial w}{\partial x}\right) + \tau_{yz}\left(\frac{\partial v}{\partial z} + \frac{\partial w}{\partial y}\right) \tag{1.140}$$

Assumption: No heat generation, Fourier heat conduction law, ideal gas

$$\rho c_p\left(\frac{\partial T}{\partial t} + u\frac{\partial T}{\partial x} + v\frac{\partial T}{\partial y} + w\frac{\partial T}{\partial z}\right) = \frac{\partial}{\partial x}\left(k\frac{\partial T}{\partial x}\right) + \frac{\partial}{\partial y}\left(k\frac{\partial T}{\partial y}\right) + \frac{\partial}{\partial z}\left(k\frac{\partial T}{\partial z}\right)$$

$$+\frac{\partial p}{\partial t} + u\frac{\partial p}{\partial x} + v\frac{\partial p}{\partial y} + w\frac{\partial p}{\partial z} + \tau_{xx}\frac{\partial u}{\partial x} + \tau_{yy}\frac{\partial v}{\partial y} + \tau_{zz}\frac{\partial w}{\partial z}$$

$$+\tau_{xy}\left(\frac{\partial u}{\partial y} + \frac{\partial v}{\partial x}\right) + \tau_{xz}\left(\frac{\partial u}{\partial z} + \frac{\partial w}{\partial x}\right) + \tau_{yz}\left(\frac{\partial v}{\partial z} + \frac{\partial w}{\partial y}\right) \tag{1.141}$$

Assumption: No heat generation, Fourier heat conduction, Newtonian fluid, Stokes hypothesis

$$\rho\left(\frac{\partial e}{\partial t} + u\frac{\partial e}{\partial x} + v\frac{\partial e}{\partial y} + w\frac{\partial e}{\partial z}\right) = \frac{\partial}{\partial x}\left(k\frac{\partial T}{\partial x}\right) + \frac{\partial}{\partial y}\left(k\frac{\partial T}{\partial y}\right) + \frac{\partial}{\partial z}\left(k\frac{\partial T}{\partial z}\right)$$

$$-p\left(\frac{\partial u}{\partial x} + \frac{\partial v}{\partial y} + \frac{\partial w}{\partial z}\right) + \Phi \qquad (1.142)$$

where

$$\Phi = \mu\left\{2\left[\left(\frac{\partial u}{\partial x}\right)^2 + \left(\frac{\partial v}{\partial y}\right)^2 + \left(\frac{\partial w}{\partial z}\right)^2\right] + \left(\frac{\partial u}{\partial y} + \frac{\partial v}{\partial x}\right)^2 + \left(\frac{\partial w}{\partial y} + \frac{\partial v}{\partial z}\right)^2\right.$$

$$\left. + \left(\frac{\partial u}{\partial z} + \frac{\partial w}{\partial x}\right)^2 - \frac{2}{3}\left(\frac{\partial u}{\partial x} + \frac{\partial v}{\partial y} + \frac{\partial w}{\partial z}\right)^2\right\}$$

$$\rho\left(\frac{\partial h}{\partial t} + u\frac{\partial h}{\partial x} + v\frac{\partial h}{\partial y} + w\frac{\partial h}{\partial z}\right) = \frac{\partial}{\partial x}\left(k\frac{\partial T}{\partial x}\right) + \frac{\partial}{\partial y}\left(k\frac{\partial T}{\partial y}\right) + \frac{\partial}{\partial z}\left(k\frac{\partial T}{\partial z}\right)$$

$$+ \left(\frac{\partial p}{\partial t} + u\frac{\partial p}{\partial x} + v\frac{\partial p}{\partial y} + w\frac{\partial p}{\partial z}\right) + \Phi \quad (1.143)$$

Assumption: No heat generation, Fourier heat conduction law, Newtonian fluid, Stokes hypothesis, ideal gas, constant μ, constant k, incompressible

$$\rho c_p\left(\frac{\partial T}{\partial t} + u\frac{\partial T}{\partial x} + v\frac{\partial T}{\partial y} + w\frac{\partial T}{\partial z}\right) = k\left(\frac{\partial^2 T}{\partial x^2} + \frac{\partial^2 T}{\partial y^2} + \frac{\partial^2 T}{\partial z^2}\right)$$

$$+\mu\left\{2\left[\left(\frac{\partial u}{\partial x}\right)^2 + \left(\frac{\partial v}{\partial y}\right)^2 + \left(\frac{\partial w}{\partial z}\right)^2\right] + \left(\frac{\partial u}{\partial y} + \frac{\partial v}{\partial x}\right)^2\right.$$

$$\left. + \left(\frac{\partial w}{\partial y} + \frac{\partial v}{\partial z}\right)^2 + \left(\frac{\partial u}{\partial z} + \frac{\partial w}{\partial x}\right)^2\right\} \qquad (1.144)$$

Table 1.7d: Energy equation – Cylindrical coordinates

Assumption: No heat generation, Fourier heat conduction law

$$\rho\left(\frac{\partial e}{\partial t} + u_r\frac{\partial e}{\partial r} + \frac{u_\theta}{r}\frac{\partial e}{\partial \theta} + u_z\frac{\partial e}{\partial z}\right) = \frac{1}{r}\frac{\partial}{\partial r}\left(rk\frac{\partial T}{\partial r}\right) + \frac{1}{r^2}\frac{\partial}{\partial \theta}\left(k\frac{\partial T}{\partial \theta}\right) + \frac{\partial}{\partial z}\left(k\frac{\partial T}{\partial z}\right)$$

$$- p\left[\frac{1}{r}\frac{\partial}{\partial r}(ru_r) + \frac{1}{r}\frac{\partial u_\theta}{\partial \theta} + \frac{\partial u_z}{\partial z}\right] + \tau_{rr}\left(\frac{\partial u_r}{\partial r}\right) + \tau_{\theta\theta}\left(\frac{1}{r}\frac{\partial u_\theta}{\partial \theta} + \frac{u_r}{r}\right) + \tau_{zz}\left(\frac{\partial u_z}{\partial z}\right)$$

$$+ \tau_{r\theta}\left[r\frac{\partial}{\partial r}\left(\frac{u_\theta}{r}\right) + \frac{1}{r}\frac{\partial u_r}{\partial \theta}\right] + \tau_{\theta z}\left(\frac{1}{r}\frac{\partial u_z}{\partial \theta} + \frac{\partial u_\theta}{\partial z}\right) + \tau_{rz}\left(\frac{\partial u_z}{\partial r} + \frac{\partial u_r}{\partial z}\right) \qquad (1.145)$$

Assumption: No heat generation, Fourier heat conduction law

$$\rho\left(\frac{\partial h}{\partial t} + u_r\frac{\partial h}{\partial r} + \frac{u_\theta}{\theta}\frac{\partial h}{\partial \theta} + u_z\frac{\partial h}{\partial z}\right) = \frac{1}{r}\frac{\partial}{\partial r}\left(rk\frac{\partial T}{\partial r}\right) + \frac{1}{r^2}\frac{\partial}{\partial \theta}\left(k\frac{\partial T}{\partial \theta}\right) + \frac{\partial}{\partial z}\left(k\frac{\partial T}{\partial z}\right)$$

$$+ \frac{\partial p}{\partial t} + u_r\frac{\partial p}{\partial r} + \frac{u_\theta}{r}\frac{\partial p}{\partial \theta} + u_z\frac{\partial p}{\partial z} + \tau_{rr}\left(\frac{\partial u_r}{\partial r}\right) + \tau_{\theta\theta}\left(\frac{1}{r}\frac{\partial u_\theta}{\partial \theta} + \frac{u_r}{r}\right) + \tau_{zz}\left(\frac{\partial u_z}{\partial z}\right)$$

$$+ \tau_{r\theta}\left[r\frac{\partial}{\partial r}\left(\frac{u_\theta}{r}\right) + \frac{1}{r}\frac{\partial u_r}{\partial \theta}\right] + \tau_{\theta z}\left(\frac{1}{r}\frac{\partial u_z}{\partial \theta} + \frac{\partial u_\theta}{\partial z}\right) + \tau_{rz}\left(\frac{\partial u_z}{\partial r} + \frac{\partial u_r}{\partial z}\right) \qquad (1.146)$$

Assumption: No heat generation, Fourier heat conduction law

$$\rho c_v\left(\frac{\partial T}{\partial t} + u_r\frac{\partial T}{\partial r} + \frac{u_\theta}{r}\frac{\partial T}{\partial \theta} + u_z\frac{\partial T}{\partial z}\right) = \frac{1}{r}\frac{\partial}{\partial r}\left(rk\frac{\partial T}{\partial r}\right) + \frac{1}{r^2}\frac{\partial}{\partial \theta}\left(k\frac{\partial T}{\partial \theta}\right)$$

$$+ \frac{\partial}{\partial z}\left(k\frac{\partial T}{\partial z}\right) - T\left(\frac{\partial p}{\partial T}\right)_V\left[\frac{1}{r}\frac{\partial}{\partial r}(ru_r) + \frac{1}{r}\frac{\partial u_\theta}{\partial \theta} + \frac{\partial u_z}{\partial z}\right] + \tau_{rr}\left(\frac{\partial u_r}{\partial r}\right)$$

$$+ \tau_{\theta\theta}\left(\frac{1}{r}\frac{\partial u_\theta}{\partial \theta} + \frac{u_r}{r}\right) + \tau_{zz}\left(\frac{\partial u_z}{\partial z}\right) + \tau_{r\theta}\left[r\frac{\partial}{\partial r}\left(\frac{u_\theta}{r}\right) + \frac{1}{r}\frac{\partial u_r}{\partial \theta}\right]$$

$$+ \tau_{\theta z}\left(\frac{1}{r}\frac{\partial u_z}{\partial \theta} + \frac{\partial u_\theta}{\partial z}\right) + \tau_{rz}\left(\frac{\partial u_z}{\partial r} + \frac{\partial u_r}{\partial z}\right) \qquad (1.147)$$

Assumption: No heat generation, Fourier heat conduction law, ideal gas

$$\rho c_p \left(\frac{\partial T}{\partial t} + u_r \frac{\partial T}{\partial r} + \frac{u_\theta}{r} \frac{\partial T}{\partial \theta} + u_z \frac{\partial T}{\partial z} \right) = \frac{1}{r} \frac{\partial}{\partial r} \left(r k \frac{\partial T}{\partial r} \right) + \frac{1}{r^2} \frac{\partial}{\partial \theta} \left(k \frac{\partial T}{\partial \theta} \right) + \frac{\partial}{\partial z} \left(k \frac{\partial T}{\partial z} \right)$$

$$+ \frac{\partial p}{\partial t} + u_r \frac{\partial p}{\partial r} + \frac{u_\theta}{r} \frac{\partial p}{\partial \theta} + u_z \frac{\partial p}{\partial z} + \tau_{rr} \left(\frac{\partial u_r}{\partial r} \right) + \tau_{\theta\theta} \left(\frac{1}{r} \frac{\partial u_\theta}{\partial \theta} + \frac{u_r}{r} \right) + \tau_{zz} \left(\frac{\partial u_z}{\partial z} \right)$$

$$+ \tau_{r\theta} \left[r \frac{\partial}{\partial r} \left(\frac{u_\theta}{r} \right) + \frac{1}{r} \frac{\partial u_r}{\partial \theta} \right] + \tau_{\theta z} \left(\frac{1}{r} \frac{\partial u_z}{\partial \theta} + \frac{\partial u_\theta}{\partial z} \right) + \tau_{rz} \left(\frac{\partial u_z}{\partial r} + \frac{\partial u_r}{\partial z} \right) \qquad (1.148)$$

Assumption: No heat generation, Fourier heat conduction law, Newtonian fluid, Stokes hypothesis

$$\rho \left(\frac{\partial e}{\partial t} + u_r \frac{\partial e}{\partial r} + \frac{u_\theta}{r} \frac{\partial e}{\partial \theta} + u_z \frac{\partial e}{\partial z} \right) = \frac{1}{r} \frac{\partial}{\partial r} \left(r k \frac{\partial T}{\partial r} \right) + \frac{1}{r^2} \frac{\partial}{\partial \theta} \left(k \frac{\partial T}{\partial \theta} \right) + \frac{\partial}{\partial z} \left(k \frac{\partial T}{\partial z} \right)$$

$$- p \left[\frac{1}{r} \frac{\partial}{\partial r} (r u_r) + \frac{1}{r} \frac{\partial u_\theta}{\partial \theta} + \frac{\partial u_z}{\partial z} \right] + \Phi \qquad (1.149)$$

where

$$\Phi = \mu \left\{ 2 \left[\left(\frac{\partial u_r}{\partial r} \right)^2 + \left(\frac{1}{r} \frac{\partial u_\theta}{\partial \theta} + \frac{u_r}{r} \right)^2 + \left(\frac{\partial u_z}{\partial z} \right)^2 \right] + \left[r \frac{\partial}{\partial r} \left(\frac{u_\theta}{r} \right) + \frac{1}{r} \frac{\partial u_r}{\partial \theta} \right]^2 \right.$$

$$+ \left[\frac{1}{r} \frac{\partial u_z}{\partial \theta} + \frac{\partial u_\theta}{\partial z} \right]^2 + \left[\frac{\partial u_r}{\partial z} + \frac{\partial u_z}{\partial r} \right]^2 - \frac{2}{3} \left[\frac{1}{r} \frac{\partial}{\partial r} (r u_r) + \frac{1}{r} \frac{\partial u_\theta}{\partial \theta} + \frac{\partial u_z}{\partial z} \right]^2$$

$$\rho \left(\frac{\partial h}{\partial t} + u_r \frac{\partial h}{\partial r} + \frac{u_\theta}{r} \frac{\partial h}{\partial \theta} + u_z \frac{\partial h}{\partial z} \right) = \frac{1}{r} \frac{\partial}{\partial r} \left(r k \frac{\partial T}{\partial r} \right) + \frac{1}{r^2} \frac{\partial}{\partial \theta} \left(k \frac{\partial T}{\partial \theta} \right)$$

$$+ \frac{\partial}{\partial z} \left(k \frac{\partial T}{\partial z} \right) + \frac{\partial p}{\partial t} + u_r \frac{\partial p}{\partial r} + \frac{u_\theta}{r} \frac{\partial p}{\partial \theta} + u_z \frac{\partial p}{\partial z} + \Phi \quad (1.150)$$

Assumption: No heat generation, Fourier heat conduction law, Newtonian fluid, Stokes hypothesis, ideal gas, constant μ, constant k, incompressible

$$\rho c_p \left(\frac{\partial T}{\partial t} + u_r \frac{\partial T}{\partial r} + \frac{u_\theta}{r} \frac{\partial T}{\partial \theta} + u_z \frac{\partial T}{\partial z} \right) = k \left(\frac{\partial^2 T}{\partial r^2} + \frac{1}{r} \frac{\partial T}{\partial r} + \frac{1}{r^2} \frac{\partial^2 T}{\partial \theta^2} + \frac{\partial^2 T}{\partial z^2} \right)$$

$$+ \mu \left\{ 2 \left[\left(\frac{\partial u_r}{\partial r} \right)^2 + \left(\frac{1}{r} \frac{\partial u_\theta}{\partial \theta} + \frac{u_r}{r} \right)^2 + \left(\frac{\partial u_z}{\partial z} \right)^2 \right] + \left[r \frac{\partial}{\partial r} \left(\frac{u_\theta}{r} \right) + \frac{1}{r} \frac{\partial u_r}{\partial \theta} \right]^2 \right.$$

$$\left. + \left[\frac{1}{r} \frac{\partial u_z}{\partial \theta} + \frac{\partial u_\theta}{\partial z} \right]^2 + \left[\frac{\partial u_r}{\partial z} + \frac{\partial u_z}{\partial r} \right]^2 \right\} \tag{1.151}$$

Table 1.7e: General orthogonal curvilinear coordinates

$$\frac{\partial}{\partial t}(\rho e_t) + \frac{1}{h_1 h_2 h_3} \left\{ \frac{\partial}{\partial x_1}(h_2 h_3 \rho e_t u_1) + \frac{\partial}{\partial x_2}(h_1 h_3 \rho e_t u_2) + \frac{\partial}{\partial x_3}(h_1 h_2 \rho e_t u_3) \right.$$

$$\left. + \frac{\partial}{\partial x_1}(h_2 h_3 p u_1) + \frac{\partial}{\partial x_2}(h_1 h_3 p u_2) + \frac{\partial}{\partial x_3}(h_1 h_2 p u_3) \right\}$$

$$= \frac{1}{h_1 h_2 h_3} \left\{ \frac{\partial}{\partial x_1} \left[h_2 h_3 (u_1 \tau_{11} + u_2 \tau_{12} + u_3 \tau_{13} - q_1) \right] \right.$$

$$+ \frac{\partial}{\partial x_2} \left[h_1 h_3 (u_1 \tau_{21} + u_2 \tau_{22} + u_3 \tau_{23} - q_2) \right]$$

$$\left. + \frac{\partial}{\partial x_3} \left[h_1 h_2 (u_1 \tau_{31} + u_2 \tau_{32} + u_3 \tau_{33} - q_3) \right] \right\} \tag{1.152}$$

$$q_1 = -k \frac{1}{h_1} \frac{\partial T}{\partial x_1} \quad , \quad q_2 = -k \frac{1}{h_2} \frac{\partial T}{\partial x_2} \quad , \quad q_3 = -k \frac{1}{h_3} \frac{\partial T}{\partial x_3}$$

1.5 Flux Vector Formulations

The conservative form of the equations of motion in a Cartesian coordinate system assuming negligible body forces is:

Continuity, Equation (1.22)

$$\frac{\partial \rho}{\partial t} + \frac{\partial}{\partial x}(\rho u) + \frac{\partial}{\partial y}(\rho v) + \frac{\partial}{\partial z}(\rho w) = 0$$

x-component of the momentum, Equation (1.59)

$$\frac{\partial}{\partial t}(\rho u) + \frac{\partial}{\partial x}(\rho u^2 + p) + \frac{\partial}{\partial y}(\rho uv) + \frac{\partial}{\partial z}(\rho uw) = \frac{\partial}{\partial x}(\tau_{xx}) + \frac{\partial}{\partial y}(\tau_{xy}) + \frac{\partial}{\partial z}(\tau_{xz})$$

y-component of the momentum, Equation (1.60)

$$\frac{\partial}{\partial t}(\rho v) + \frac{\partial}{\partial x}(\rho uv) + \frac{\partial}{\partial y}(\rho v^2 + p) + \frac{\partial}{\partial z}(\rho vw) = \frac{\partial}{\partial x}(\tau_{xy}) + \frac{\partial}{\partial y}(\tau_{yy}) + \frac{\partial}{\partial z}(\tau_{yz})$$

z-component of the momentum, Equation (1.61)

$$\frac{\partial}{\partial t}(\rho w) + \frac{\partial}{\partial x}(\rho uw) + \frac{\partial}{\partial y}(\rho vw) + \frac{\partial}{\partial z}(\rho w^2 + p) = \frac{\partial}{\partial x}(\tau_{xz}) + \frac{\partial}{\partial y}(\tau_{yz}) + \frac{\partial}{\partial z}(\tau_{zz})$$

Energy, Equation (1.119)

$$\frac{\partial}{\partial t}(\rho e_t) + \frac{\partial}{\partial x}(\rho u e_t + pu) + \frac{\partial}{\partial y}(\rho v e_t + pv) + \frac{\partial}{\partial z}(\rho w e_t + pw) =$$

$$+ \frac{\partial}{\partial x}\left[u\tau_{xx} + v\tau_{xy} + w\tau_{xz} - q_x \right] + \frac{\partial}{\partial y}\left[u\tau_{yx} + v\tau_{yy} + w\tau_{yz} - q_y \right]$$

$$+ \frac{\partial}{\partial z}\left[u\tau_{zx} + v\tau_{zy} + w\tau_{zz} - q_z \right]$$

It is convenient to write the Cartesian form of the equations of motion in a flux vector form. This formulation may be expressed in a Cartesian coordinate system as

$$\frac{\partial Q}{\partial t} + \frac{\partial E}{\partial x} + \frac{\partial F}{\partial y} + \frac{\partial G}{\partial z} = \frac{\partial E_v}{\partial x} + \frac{\partial F_v}{\partial y} + \frac{\partial G_v}{\partial z} \tag{1.153}$$

where

$$Q = \begin{bmatrix} \rho \\ \rho u \\ \rho v \\ \rho w \\ \rho e_t \end{bmatrix}$$

$$
E = \begin{bmatrix} \rho u \\ \rho u^2 + p \\ \rho u v \\ \rho u w \\ (\rho e_t + p)u \end{bmatrix}
\qquad
E_v = \begin{bmatrix} 0 \\ \tau_{xx} \\ \tau_{xy} \\ \tau_{xz} \\ u\tau_{xx} + v\tau_{xy} + w\tau_{xz} - q_x \end{bmatrix}
$$

$$
F = \begin{bmatrix} \rho v \\ \rho v u \\ \rho v^2 + p \\ \rho v w \\ (\rho e_t + p)v \end{bmatrix}
\qquad
F_v = \begin{bmatrix} 0 \\ \tau_{yx} \\ \tau_{yy} \\ \tau_{yz} \\ u\tau_{yx} + v\tau_{yy} + w\tau_{yz} - q_y \end{bmatrix}
$$

$$
G = \begin{bmatrix} \rho w \\ \rho w u \\ \rho w v \\ \rho w^2 + p \\ (\rho e_t + p)w \end{bmatrix}
\qquad
G_v = \begin{bmatrix} 0 \\ \tau_{zx} \\ \tau_{zy} \\ \tau_{zz} \\ u\tau_{zx} + v\tau_{zy} + w\tau_{zz} - q_z \end{bmatrix}
$$

In a cylindrical coordinate system the flux vector formulation is expressed as:

$$
\frac{\partial Q}{\partial t} + \frac{\partial E}{\partial z} + \frac{\partial F}{\partial r} + \frac{\partial G}{\partial \theta} + H = \frac{\partial E_v}{\partial z} + \frac{\partial F_v}{\partial r} + \frac{\partial G_v}{\partial \theta} + H_v \tag{1.154}
$$

where

$$
Q = \begin{bmatrix} \rho \\ \rho u_z \\ \rho u_r \\ \rho u_\theta \\ \rho e_t \end{bmatrix}
$$

$$
E = \begin{bmatrix} \rho u_z \\ \rho u_z^2 + p \\ \rho u_z u_r \\ \rho u_z u_\theta \\ (\rho e_t + p)u_z \end{bmatrix}
\qquad
E_v = \begin{bmatrix} 0 \\ \tau_{zz} \\ \tau_{zr} \\ \tau_{z\theta} \\ u_z\tau_{zz} + u_r\tau_{zr} + u_\theta\tau_{z\theta} - q_z \end{bmatrix}
$$

$$
F = \begin{bmatrix} \rho u_r \\ \rho u_r u_z \\ \rho u_r^2 + p \\ \rho u_r u_\theta \\ (\rho e_t + p)u_r \end{bmatrix}
\qquad
F_v = \begin{bmatrix} 0 \\ \tau_{rz} \\ \tau_{rr} \\ \tau_{r\theta} \\ u_z\tau_{rz} + u_r\tau_{rr} + u_\theta\tau_{r\theta} - q_r \end{bmatrix}
$$

$$G = \frac{1}{r}\begin{bmatrix} \rho u_\theta \\ \rho u_\theta u_z \\ \rho u_\theta u_r \\ \rho u_\theta^2 + p \\ (\rho e_t + p)u_\theta \end{bmatrix} \qquad G_v = \frac{1}{r}\begin{bmatrix} 0 \\ \tau_{\theta z} \\ \tau_{\theta r} \\ \tau_{\theta\theta} \\ u_z\tau_{\theta z} + u_r\tau_{\theta r} + u_\theta\tau_{\theta\theta} - q_\theta \end{bmatrix}$$

$$H = \frac{1}{r}\begin{bmatrix} \rho u_r \\ \rho u_z u_r \\ \rho(u_r^2 - u_\theta^2) \\ 2\rho u_\theta u_r \\ (\rho e_t + p)u_r \end{bmatrix} \qquad H_v = \frac{1}{r}\begin{bmatrix} 0 \\ \tau_{zr} \\ \tau_{rr} - \tau_{\theta\theta} \\ 2\tau_{\theta r} \\ u_z\tau_{rz} + u_r\tau_{rr} + u_\theta\tau_{r\theta} - q_r \end{bmatrix}$$

1.6 Two-Dimensional Planar and Axisymmetric Formulation

The equations of fluid motion may be expressed in a combined form for a two-dimensional planar flow and an axisymmetric flow. This form of the equation is particularly useful when numerical schemes are utilized for solution. Therefore one computer code may be developed for the solution of two different types of flows.

The equations of motion in a combined form are expressed as

$$\frac{\partial Q}{\partial t} + \frac{\partial E}{\partial x} + \frac{\partial F}{\partial y} + \alpha H = \frac{\partial E_v}{\partial x} + \frac{\partial F_v}{\partial y} + \alpha H_v \qquad (1.155)$$

where
$$\alpha = \begin{cases} 0 \text{ for 2-D planar flow} \\ 1 \text{ for 2-D axisymmetric flow} \end{cases}$$

Note that for a 2-D planar flow, the formulation is based on a Cartesian coordinate system whereas for a 2-D axisymmetric flow, the formulation is based on a cylindrical coordinate system. The flux vectors in Equation (1.154) are defined as

$$Q = \begin{bmatrix} \rho \\ \rho u \\ \rho v \\ \rho e_t \end{bmatrix} \qquad E = \begin{bmatrix} \rho u \\ \rho u^2 + p \\ \rho uv \\ (\rho e_t + p)u \end{bmatrix} \qquad F = \begin{bmatrix} \rho v \\ \rho vu \\ \rho v^2 + p \\ (\rho e_t + p)v \end{bmatrix}$$

$$H = \frac{1}{y}\begin{bmatrix} \rho v \\ \rho uv \\ \rho v^2 \\ (\rho e_t + p)v \end{bmatrix} \quad E_v = \begin{bmatrix} 0 \\ \tau_{xxp} \\ \tau_{xy} \\ u\tau_{xxp} + v\tau_{xy} - q_x \end{bmatrix} \quad F_v = \begin{bmatrix} 0 \\ \tau_{xy} \\ \tau_{yyp} \\ u\tau_{xy} + v\tau_{yyp} - q_y \end{bmatrix}$$

$$H_v = \frac{1}{y}\begin{bmatrix} 0 \\ \tau_{xy} - \frac{2}{3}y\frac{\partial}{\partial x}\left(\mu\frac{v}{y}\right) \\ \tau_{yyp} - \tau_{\theta\theta} - \frac{2}{3}\mu\frac{v}{y} - y\frac{2}{3}\frac{\partial}{\partial y}\left(\mu\frac{v}{y}\right) \\ u\tau_{xy} + v\tau_{yyp} - q_y - \frac{2}{3}\mu\frac{v^2}{y} - y\frac{\partial}{\partial y}\left(\frac{2}{3}\mu\frac{v^2}{y}\right) - y\frac{\partial}{\partial y}\left(\frac{2}{3}\mu\frac{uv}{y}\right) \end{bmatrix}$$

where for a Newtonian fluid with Stokes hypothesis,

$$\tau_{xxp} = \frac{4}{3}\mu\frac{\partial u}{\partial x} - \frac{2}{3}\mu\frac{\partial v}{\partial y}$$

$$\tau_{xx} = \tau_{xxp} - \frac{2}{3}\mu\frac{v}{y}$$

$$\tau_{yyp} = \frac{4}{3}\mu\frac{\partial v}{\partial y} - \frac{2}{3}\mu\frac{\partial u}{\partial x}$$

$$\tau_{yy} = \tau_{yyp} - \frac{2}{3}\mu\frac{v}{y}$$

$$\tau_{xy} = \mu\left(\frac{\partial u}{\partial y} + \frac{\partial v}{\partial x}\right)$$

$$\tau_{\theta\theta} = -\frac{2}{3}\mu\left(\frac{\partial u}{\partial x} + \frac{\partial v}{\partial y}\right) + \frac{4}{3}\mu\frac{v}{y}$$

$$q_x = -k\frac{\partial T}{\partial x} \qquad , \qquad q_y = -k\frac{\partial T}{\partial y}$$

Note: Subscript p represents 2-D planar flow.

1.7 Nondimensionalization of the Equations of Fluid Motion

Equations of fluid motion may be nondimensionalized to achieve certain objectives. For one, it would provide conditions upon which dynamic and energetic similarity may be obtained for geometrically similar situations. Second, the solution of such equations would usually provide values within limits between zero and one. Generally a characteristic dimension, such as the chord of an airfoil or the length of a vehicle, is selected to nondimensionalize the independent spatial variables. Free stream conditions are used to nondimensionalize the dependent variables. Among many choices available, the following will be used in this and in the subsequent chapters, unless otherwise specified:

$$t^* = \frac{tu_\infty}{L} \quad , \quad x^* = \frac{x}{L} \quad , \quad y^* = \frac{y}{L} \quad , \quad z^* = \frac{z}{L}$$

$$\mu^* = \frac{\mu}{\mu_\infty} \quad , \quad u^* = \frac{u}{u_\infty} \quad , \quad v^* = \frac{v}{u_\infty} \quad , \quad w^* = \frac{w}{u_\infty}$$

$$\rho^* = \frac{\rho}{\rho_\infty} \quad , \quad T^* = \frac{T}{T_\infty} \quad , \quad p^* = \frac{p}{\rho_\infty u_\infty^2} \quad , \quad e_t^* = \frac{e_t}{u_\infty^2}$$

The nondimensional parameters are defined as:

$$\text{Reynolds number:} \qquad Re_\infty = \frac{\rho_\infty u_\infty L}{\mu_\infty}$$

$$\text{Prandtl number:} \qquad Pr = \frac{\mu c_p}{k}$$

Using the nondimensional variables defined above, the equations of fluid motion in the Cartesian coordinate system are expressed as:

1. Continuity:

$$\frac{\partial \rho^*}{\partial t^*} + \frac{\partial}{\partial x^*}(\rho^* u^*) + \frac{\partial}{\partial y^*}(\rho^* v^*) + \frac{\partial}{\partial z^*}(\rho^* w^*) = 0 \qquad (1.156)$$

2. X-component of the Navier-Stokes equation:

$$\frac{\partial}{\partial t^*}(\rho^* u^*) + \frac{\partial}{\partial x^*}(\rho^* u^{*^2} + p^*) + \frac{\partial}{\partial y^*}(\rho^* u^* v^*) + \frac{\partial}{\partial z^*}(\rho^* u^* w^*) =$$

$$\frac{\partial}{\partial x^*}(\tau^*_{xx}) + \frac{\partial}{\partial y^*}(\tau^*_{xy}) + \frac{\partial}{\partial z^*}(\tau^*_{xz}) \qquad (1.157)$$

For a Newtonian fluid,

$$\tau^*_{xx} = \frac{1}{Re_\infty}\left[2\mu^* \frac{\partial u^*}{\partial x^*} + \lambda^* \nabla^* \cdot \vec{V}^*\right] \qquad (1.158)$$

$$\tau^*_{xy} = \tau^*_{yx} = \frac{1}{Re_\infty}\left[\mu^*(\frac{\partial u^*}{\partial y^*} + \frac{\partial v^*}{\partial x^*})\right] \qquad (1.159)$$

$$\tau^*_{xz} = \tau^*_{zx} = \frac{1}{Re_\infty}\left[\mu^*(\frac{\partial w^*}{\partial x^*} + \frac{\partial u^*}{\partial z^*})\right] \qquad (1.160)$$

3. Y-component of the Navier-Stokes equation:

$$\frac{\partial}{\partial t^*}(\rho^* v^*) + \frac{\partial}{\partial x^*}(\rho^* u^* v^*) + \frac{\partial}{\partial y^*}(\rho^* v^{*^2} + p^*) + \frac{\partial}{\partial z^*}(\rho^* v^* w^*) =$$

$$\frac{\partial}{\partial x^*}(\tau^*_{xy}) + \frac{\partial}{\partial y^*}(\tau^*_{yy}) + \frac{\partial}{\partial z^*}(\tau^*_{yz}) \qquad (1.161)$$

where

$$\tau^*_{yy} = \frac{1}{Re_\infty}\left[2\mu^* \frac{\partial v^*}{\partial y^*} + \lambda^* \nabla^* \cdot \vec{V}^*\right] \qquad (1.162)$$

$$\tau^*_{yz} = \tau^*_{zy} = \frac{1}{Re_\infty}\left[\mu^*(\frac{\partial v^*}{\partial z^*} + \frac{\partial w^*}{\partial y^*})\right] \qquad (1.163)$$

and τ_{xy}^* is given by Equation (1.159).

4. Z-component of the Navier-Stokes equation:

$$\frac{\partial}{\partial t^*}\left(\rho^* w^*\right) + \frac{\partial}{\partial x^*}\left(\rho^* u^* w^*\right) + \frac{\partial}{\partial y^*}\left(\rho^* v^* w^*\right) + \frac{\partial}{\partial z^*}\left(\rho^* w^{*2} + p^*\right) =$$

$$\frac{\partial}{\partial x^*}\left(\tau_{xz}^*\right) + \frac{\partial}{\partial y^*}\left(\tau_{yz}^*\right) + \frac{\partial}{\partial z^*}\left(\tau_{zz}^*\right) \tag{1.164}$$

where

$$\tau_{zz}^* = \frac{1}{Re_\infty}\left[2\mu^*\frac{\partial w^*}{\partial z^*} + \lambda^*\, \nabla^* \cdot \vec{V}^*\right] \tag{1.165}$$

and τ_{xz}^* and τ_{yz}^* are given by Equations (1.160) and (1.163), respectively.

5. Energy:

$$\frac{\partial}{\partial t^*}\left(\rho^* e_t^*\right) + \frac{\partial}{\partial x^*}\left(\rho^* u^* e_t^* + p^* u^*\right) + \frac{\partial}{\partial y^*}\left(\rho^* v^* e_t^* + p^* v^*\right) +$$

$$\frac{\partial}{\partial z^*}\left(\rho^* w^* e_t^* + p^* w^*\right) = \frac{\partial}{\partial x^*}[u^*\tau_{xx}^* + v^*\tau_{xy}^* + w^*\tau_{xz}^* - q_x^*] + \tag{1.166}$$

$$\frac{\partial}{\partial y^*}[u^*\tau_{yx}^* + v^*\tau_{yy}^* + w^*\tau_{yz}^* - q_y^*] + \frac{\partial}{\partial z^*}[u^*\tau_{zx}^* + v^*\tau_{zy}^* + w^*\tau_{zz}^* - q_z^*]$$

where

$$q_x^* = -\frac{\mu^*}{Re_\infty Pr\,(\gamma - 1)\, M_\infty^2}\,\frac{\partial T^*}{\partial x^*} \tag{1.167}$$

$$q_y^* = -\frac{\mu^*}{Re_\infty Pr\,(\gamma - 1)\, M_\infty^2}\,\frac{\partial T^*}{\partial y^*} \tag{1.168}$$

$$q_z^* = -\frac{\mu^*}{Re_\infty Pr\,(\gamma - 1)\, M_\infty^2}\,\frac{\partial T^*}{\partial z^*} \tag{1.169}$$

The nondimensional equations of fluid motion may be expressed in a flux vector form as:

$$\frac{\partial Q^*}{\partial t^*} + \frac{\partial E^*}{\partial x^*} + \frac{\partial F^*}{\partial y^*} + \frac{\partial G^*}{\partial z^*} = \frac{\partial E_v^*}{\partial x^*} + \frac{\partial F_v^*}{\partial y^*} + \frac{\partial G_v^*}{\partial z^*} \tag{1.170}$$

where

$$Q^* = \begin{bmatrix} \rho^* \\ \rho^* u^* \\ \rho^* v^* \\ \rho^* w^* \\ \rho^* e_t^* \end{bmatrix} \tag{1.171}$$

$$E^* = \begin{bmatrix} \rho^* u^* \\ \rho^* u^{*2} + p^* \\ \rho^* u^* v^* \\ \rho^* u^* w^* \\ (\rho^* e_t^* + p^*) u^* \end{bmatrix} \quad (1.172) \qquad E_v^* = \begin{bmatrix} 0 \\ \tau_{xx}^* \\ \tau_{xy}^* \\ \tau_{xz}^* \\ u^* \tau_{xx}^* + v^* \tau_{xy}^* + w^* \tau_{xz}^* - q_x^* \end{bmatrix} \quad (1.173)$$

$$F^* = \begin{bmatrix} \rho^* v^* \\ \rho^* v^* u^* \\ \rho^* v^{*2} + p^* \\ \rho^* v^* w^* \\ (\rho^* e_t^* + p^*) v^* \end{bmatrix} \quad (1.174) \qquad F_v^* = \begin{bmatrix} 0 \\ \tau_{yx}^* \\ \tau_{yy}^* \\ \tau_{yz}^* \\ u^* \tau_{yx}^* + v^* \tau_{yy}^* + w^* \tau_{yz}^* - q_y^* \end{bmatrix} \quad (1.175)$$

$$G^* = \begin{bmatrix} \rho^* w^* \\ \rho^* w^* u^* \\ \rho^* w^* v^* \\ \rho^* w^{*2} + p^* \\ (\rho^* e_t^* + p^*) w^* \end{bmatrix} \quad (1.176) \qquad G_v^* = \begin{bmatrix} 0 \\ \tau_{zx}^* \\ \tau_{zy}^* \\ \tau_{zz}^* \\ u^* \tau_{zx}^* + v^* \tau_{zy}^* + w^* \tau_{zz}^* - q_z^* \end{bmatrix} \quad (1.177)$$

1.8   Incompressible Navier-Stokes Equations

For applications for which the density remains uniform throughout the domain, the assumption of incompressible flow is invoked. Therefore, for an incompressible flow, the density is constant and no longer an unknown. Furthermore, variations in the coefficient of viscosity are essentially negligible and it is assumed a constant as well. The incompressible Navier-Stokes equations can be obtained from the Navier-Stokes equations for the limiting case of $M \to 0$ $(a \to \infty)$. It should be observed that the energy equation is decoupled from the system of equations composed of the continuity and momentum equations. Therefore, the velocity and pressure fields may be determined initially, and subsequently the energy equation may be solved for the temperature distribution if required.

Generally speaking, the governing equations for an incompressible flow may be expressed in two different formulations based on the dependent variables used. First is the primitive variable formulation expressed in terms of the pressure and velocity. The second form of the equations is the so-called vorticity–stream function formulation which is derived from the Navier-Stokes equations by incorporating the definitions for the vorticity and the stream function. This formulation is primarily used for two-dimensional applications. Obviously the implication is due to the definition of stream function, which exists for a two-dimensional (planar or axisymmetric) flow only. It should be noted that for a three-dimensional flow, it is possible to extend the approach of stream function by the use of the so-called vector potential [1.10]. However, additional complications arise. Therefore, extension of the vorticity–stream function formulation for three dimensions is not explored further. Interested readers are encouraged to consult Refs. [1.11] through [1.13]. For three-dimensional applications, the extension of the equations in primitive variable formulation is preferred.

Either one of the formulations described above can be expressed in dimensional or nondimensional form. Furthermore, they may be either in conservative or nonconservative form. For completeness, various forms of the incompressible Navier-Stokes equations without body forces are provided.

1.8.1 Primitive Variable Formulations

(I) Dimensional, conservative form

    1. Vector form

$$\nabla \cdot \vec{V} = 0 \tag{1.178}$$

$$\frac{\partial \vec{V}}{\partial t} + \nabla \cdot (\vec{V}\vec{V}) + \frac{\nabla p}{\rho} = \nu \nabla^2 \vec{V} \tag{1.179}$$

    2. Two-dimensional Cartesian coordinate

$$\frac{\partial u}{\partial x} + \frac{\partial v}{\partial y} = 0 \tag{1.180}$$

$$\frac{\partial u}{\partial t} + \frac{\partial}{\partial x}\left(u^2 + \frac{p}{\rho}\right) + \frac{\partial}{\partial y}(uv) = \nu \left(\frac{\partial^2 u}{\partial x^2} + \frac{\partial^2 u}{\partial y^2}\right) \tag{1.181}$$

$$\frac{\partial v}{\partial t} + \frac{\partial}{\partial x}(uv) + \frac{\partial}{\partial y}\left(v^2 + \frac{p}{\rho}\right) = \nu \left(\frac{\partial^2 v}{\partial x^2} + \frac{\partial^2 v}{\partial y^2}\right) \tag{1.182}$$

    3. Two-dimensional cylindrical coordinates

$$\frac{\partial u_r}{\partial r} + \frac{u_r}{r} + \frac{\partial u_z}{\partial z} = 0$$

$$\frac{\partial u_r}{\partial t} + \frac{\partial}{\partial r}\left(u_r^2 + \frac{p}{\rho}\right) + \frac{\partial}{\partial z}(u_r u_z) = \nu \left(\frac{\partial^2 u_r}{\partial r^2} + \frac{1}{r}\frac{\partial u_r}{\partial r} - \frac{u_r}{r} + \frac{\partial^2 u_r}{\partial z^2}\right)$$

$$\frac{\partial u_z}{\partial t} + \frac{\partial}{\partial r}(u_z u_r) + \frac{\partial}{\partial z}\left(u_z^2 + \frac{p}{\rho}\right) = \nu\left(\frac{\partial^2 u_z}{\partial r^2} + \frac{1}{r}\frac{\partial u_z}{\partial r} + \frac{\partial^2 u_z}{\partial z^2}\right)$$

(II) Dimensional, nonconservative form

    1. Vector form

$$\nabla \cdot \vec{V} = 0 \tag{1.183}$$

$$\frac{\partial \vec{V}}{\partial t} + (\vec{V} \cdot \nabla)\vec{V} + \frac{\nabla p}{\rho} = \nu\nabla^2\vec{V} \tag{1.184}$$

    2. Two-dimensional Cartesian coordinate

$$\frac{\partial u}{\partial x} + \frac{\partial v}{\partial y} = 0 \tag{1.185}$$

$$\frac{\partial u}{\partial t} + u\frac{\partial u}{\partial x} + v\frac{\partial u}{\partial y} + \frac{1}{\rho}\frac{\partial p}{\partial x} = \nu\left(\frac{\partial^2 u}{\partial x^2} + \frac{\partial^2 u}{\partial y^2}\right) \tag{1.186}$$

$$\frac{\partial v}{\partial t} + u\frac{\partial v}{\partial x} + v\frac{\partial v}{\partial y} + \frac{1}{\rho}\frac{\partial p}{\partial y} = \nu\left(\frac{\partial^2 v}{\partial x^2} + \frac{\partial^2 v}{\partial y^2}\right) \tag{1.187}$$

    3. Two-dimensional cylindrical coordinates

$$\frac{\partial u_r}{\partial r} + \frac{u_r}{r} + \frac{\partial u_z}{\partial z} = 0$$

$$\frac{\partial u_r}{\partial t} + u_r\frac{\partial u_r}{\partial r} + u_z\frac{\partial u_r}{\partial z} + \frac{1}{\rho}\frac{\partial p}{\partial r} = \nu\left(\frac{\partial^2 u_r}{\partial r^2} + \frac{1}{r}\frac{\partial u_r}{\partial r} - \frac{u_r}{r} + \frac{\partial^2 u_r}{\partial z^2}\right)$$

$$\frac{\partial u_z}{\partial t} + u_r\frac{\partial u_z}{\partial r} + u_z\frac{\partial u_z}{\partial z} + \frac{1}{\rho}\frac{\partial p}{\partial z} = \nu\left(\frac{\partial^2 u_z}{\partial r^2} + \frac{1}{r}\frac{\partial u_z}{\partial r} + \frac{\partial^2 u_z}{\partial z^2}\right)$$

(III) Nondimensional, conservative form

    1. Vector form

$$\nabla^* \cdot \vec{V}^* = 0 \tag{1.188}$$

$$\frac{\partial \vec{V}^*}{\partial t^*} + \nabla^* \cdot (\vec{V}^*\vec{V}^*) + \nabla^* p^* = \frac{1}{Re}\nabla^{*2}\vec{V}^* \tag{1.189}$$

    2. Two-dimensional Cartesian coordinate

$$\frac{\partial u^*}{\partial x^*} + \frac{\partial v^*}{\partial y^*} = 0 \tag{1.190}$$

$$\frac{\partial u^*}{\partial t^*} + \frac{\partial}{\partial x^*}(u^{*2} + p^*) + \frac{\partial}{\partial y^*}(u^* v^*) = \frac{1}{Re}\left(\frac{\partial^2 u^*}{\partial x^{*2}} + \frac{\partial^2 u^*}{\partial y^{*2}}\right) \tag{1.191}$$

$$\frac{\partial v^*}{\partial t^*} + \frac{\partial}{\partial x^*}(u^* v^*) + \frac{\partial}{\partial y^*}(v^{*2} + p^*) = \frac{1}{Re}\left(\frac{\partial^2 v^*}{\partial x^{*2}} + \frac{\partial^2 v^*}{\partial y^{*2}}\right) \tag{1.192}$$

3. Two-dimensional cylindrical coordinates

$$\frac{\partial u_r^*}{\partial r^*} + \frac{u_r^*}{r^*} + \frac{\partial u_z^*}{\partial z^*} = 0$$

$$\frac{\partial u_r^*}{\partial t^*} + \frac{\partial}{\partial r^*}\left(u_r^{*2} + p^*\right) + \frac{\partial}{\partial z^*}\left(u_r^* u_z^*\right) = \frac{1}{Re}\left(\frac{\partial^2 u_r^*}{\partial r^{*2}} + \frac{1}{r^*}\frac{\partial u_r^*}{\partial r^*} - \frac{u_r^*}{r^*} + \frac{\partial^2 u_r^*}{\partial z^{*2}}\right)$$

$$\frac{\partial u_z^*}{\partial t^*} + \frac{\partial}{\partial r^*}\left(u_z^* u_r^*\right) + \frac{\partial}{\partial z^*}\left(u_z^{*2} + p^*\right) = \frac{1}{Re}\left(\frac{\partial^2 u_z^*}{\partial r^{*2}} + \frac{1}{r^*}\frac{\partial u_z^*}{\partial r^*} + \frac{\partial^2 u_z^*}{\partial z^{*2}}\right)$$

(IV) Nondimensional, nonconservative form

1. Vector form

$$\nabla^* \cdot \vec{V}^* = 0 \tag{1.193}$$

$$\frac{\partial \vec{V}^*}{\partial t^*} + (\vec{V}^* \cdot \nabla^*)\vec{V}^* + \nabla^* p^* = \frac{1}{Re}\nabla^{*2}\vec{V}^* \tag{1.194}$$

2. Two-dimensional Cartesian coordinate

$$\frac{\partial u^*}{\partial x^*} + \frac{\partial v^*}{\partial y^*} = 0 \tag{1.195}$$

$$\frac{\partial u^*}{\partial t^*} + u^*\frac{\partial u^*}{\partial x^*} + v^*\frac{\partial u^*}{\partial y^*} + \frac{\partial p^*}{\partial x^*} = \frac{1}{Re}\left(\frac{\partial^2 u^*}{\partial x^{*2}} + \frac{\partial^2 u^*}{\partial y^{*2}}\right) \tag{1.196}$$

$$\frac{\partial v^*}{\partial t^*} + u^*\frac{\partial v^*}{\partial x^*} + v^*\frac{\partial v^*}{\partial y^*} + \frac{\partial p^*}{\partial y^*} = \frac{1}{Re}\left(\frac{\partial^2 v^*}{\partial x^{*2}} + \frac{\partial^2 v^*}{\partial y^{*2}}\right) \tag{1.197}$$

3. Two-dimensional cylindrical coordinates

$$\frac{\partial u_r^*}{\partial r^*} + \frac{u_r^*}{r^*} + \frac{\partial u_z^*}{\partial z^*} = 0$$

$$\frac{\partial u_r^*}{\partial t^*} + u_r^*\frac{\partial u_r^*}{\partial r^*} + u_z^*\frac{\partial u_r^*}{\partial z^*} + \frac{\partial p^*}{\partial r^*} = \frac{1}{Re}\left(\frac{\partial^2 u_r^*}{\partial r^{*2}} + \frac{1}{r^*}\frac{\partial u_r^*}{\partial r^*} - \frac{u_r^*}{r^*} + \frac{\partial^2 u_r^*}{\partial z^{*2}}\right)$$

$$\frac{\partial u_z^*}{\partial t^*} + u_r^*\frac{\partial u_z^*}{\partial r^*} + u_z^*\frac{\partial u_z^*}{\partial z^*} + \frac{\partial p^*}{\partial z^*} = \frac{1}{Re}\left(\frac{\partial^2 u_z^*}{\partial r^{*2}} + \frac{1}{r^*}\frac{\partial u_z^*}{\partial r^*} + \frac{\partial^2 u_z^*}{\partial z^{*2}}\right)$$

The variables in the equations above are nondimensionalized as follows,

$$t^* = \frac{tu_\infty}{L} \qquad x^* = \frac{x}{L} \qquad y^* = \frac{y}{L}$$

$$u^* = \frac{u}{u_\infty} \qquad v^* = \frac{v}{v_\infty} \qquad p^* = \frac{p}{\rho_\infty u_\infty^2}$$

$$r^* = \frac{r}{L} \qquad z^* = \frac{z}{L} \qquad u_r^* = \frac{u_r}{u_\infty}$$

$$u_z^* = \frac{u_z}{u_\infty}$$

where L is a characteristic length, and ρ_∞ and u_∞ are the reference (e.g., free stream) density and velocity, respectively. The nondimensional parameter Reynolds number is defined as

$$Re = \frac{\rho_\infty u_\infty L}{\mu_\infty} \tag{1.198}$$

It should be emphasized that other reference variables can be used to nondimensionalize the equations. For example, one may nondimensionalize the pressure with respect to the free stream pressure, p_∞, or nondimensionalize the velocity with respect to the free stream speed of sound (for high speed flows). Therefore, in reviewing various publications, one should pay close attention to the procedure by which the equations have been nondimensionalized. At this point, a comment with respect to nondimensionalization of time is in order. For applications where the Reynolds number is high (i.e., $Re \gg 1$), time is nondimensionalized with respect to $\frac{L}{u_\infty}$ as defined previously. That is due to the fact that in these types of problems, the convective term dominates the viscous term and $\frac{L}{u_\infty}$ is a natural representation of the time interval for the problem, i.e., the time by which a particle with the free stream velocity of u_∞ is convected a characteristic length of L. On the other hand, for a low Reynolds number flow (i.e., $Re \ll 1$), diffusion dominates over convection and $\frac{L^2}{\nu}$ is a more appropriate factor to nondimensionalize time. Thus, $t^* = t\frac{\nu}{L^2}$.

As an example, the nondimensional momentum equation in conservative form using the nondimensionalized time defined for diffusion-dominated problems becomes

$$\frac{\partial \vec{V}^*}{\partial t^*} + Re\left[\nabla^* \cdot (\vec{V}^*\vec{V}^*) + \nabla^* p^*\right] = \nabla^{*2}\vec{V}^* \tag{1.199}$$

Table 1.8: Components of the vorticity

a. Cartesian coordinates (x, y, z)

$$\Omega_x = \frac{\partial w}{\partial y} - \frac{\partial v}{\partial z} \qquad \Omega_y = \frac{\partial u}{\partial z} - \frac{\partial w}{\partial x} \qquad \Omega_z = \frac{\partial v}{\partial x} - \frac{\partial u}{\partial y}$$

b. Cylindrical coordinates (r, θ, z)

$$\Omega_r = \frac{1}{r}\frac{\partial u_z}{\partial \theta} - \frac{\partial u_\theta}{\partial z} \qquad \Omega_\theta = \frac{\partial u_r}{\partial z} - \frac{\partial u_z}{\partial r} \qquad \Omega_z = \frac{1}{r}\frac{\partial}{\partial r}(ru_\theta) - \frac{1}{r}\frac{\partial u_r}{\partial \theta}$$

c. Spherical coordinates (r,θ,ϕ)

$$\Omega_r = \frac{1}{r^2 \sin\theta}\left[\frac{\partial}{\partial r}(ru_\phi \sin\theta) - \frac{\partial}{\partial \phi}(ru_\theta)\right]$$

$$\Omega_\theta = \frac{1}{r \sin\theta}\left[\frac{\partial u_r}{\partial \phi} - \frac{\partial}{\partial r}(ru_\phi \sin\theta)\right]$$

$$\Omega_\phi = \frac{1}{r}\left[\frac{\partial}{\partial r}(ru_\theta) - \frac{\partial u_r}{\partial \theta}\right]$$

d. General orthogonal curvilinear coordinates (x_1, x_2, x_3)

$$\Omega_1 = \frac{1}{h_2 h_3}\left[\frac{\partial}{\partial x_2}(u_3 h_3) - \frac{\partial}{\partial x_3}(u_2 h_2)\right]$$

$$\Omega_2 = \frac{1}{h_3 h_1}\left[\frac{\partial}{\partial x_3}(u_1 h_1) - \frac{\partial}{\partial x_1}(u_3 h_3)\right]$$

$$\Omega_3 = \frac{1}{h_1 h_2}\left[\frac{\partial}{\partial x_1}(u_2 h_2) - \frac{\partial}{\partial x_2}(u_1 h_1)\right]$$

1.8.2 Vorticity–Stream Function Formulations

The vorticity at a fluid point is defined as twice the angular velocity and is

$$\vec{\Omega} = 2\vec{\omega} = \nabla \times \vec{V}$$

which, for a two-dimensional flow, is reduced to

$$\Omega_z = \frac{\partial v}{\partial x} - \frac{\partial u}{\partial y} \tag{1.200}$$

The components of vorticity in different coordinate systems are summarized in Table 1.8.

Now, for a two-dimensional, incompressible flow, a function may be defined which satisfies the continuity equation. Such a function is known as the *stream function* and, in a Cartesian coordinate system, is given by

$$u = \frac{\partial \psi}{\partial y} \tag{1.201}$$

$$v = -\frac{\partial \psi}{\partial x} \tag{1.202}$$

From a physical point of view, the lines of constant ψ represent streamlines, and the difference in the values of ψ between two streamlines gives the volumetric flow rate between the two.

The relations between the stream function and the velocity components in various coordinate systems for two-dimensional planar and axisymmetric flows are given in Table 1.9.

Table 1.9: Velocity components in terms of stream function

a. Planar flow in Cartesian coordinates (x, y)

$$u = \frac{\partial \psi}{\partial y} \quad , \quad v = -\frac{\partial \psi}{\partial x}$$

b. Planar flow in cylindrical coordinates (r, θ)

$$u_r = \frac{1}{r}\frac{\partial \psi}{\partial \theta} \quad , \quad u_\theta = -\frac{\partial \psi}{\partial r}$$

c. Axisymmetric flow in cylindrical coordinates (r, z)

$$u_r = -\frac{1}{r}\frac{\partial \psi}{\partial z} \quad , \quad u_z = \frac{1}{r}\frac{\partial \psi}{\partial r}$$

d. Axisymmetric flow in spherical coordinates (r, θ)

$$u_r = \frac{1}{r^2 \sin\theta}\frac{\partial \psi}{\partial \theta} \quad , \quad u_\theta = -\frac{1}{r \sin\theta}\frac{\partial \psi}{\partial r}$$

In order to derive the vorticity transport equation, the pressure is eliminated from the momentum equations by cross-differentiation, and, upon substitution of the vorticity defined by (1.200), one obtains

$$\frac{\partial \Omega}{\partial t} + u\frac{\partial \Omega}{\partial x} + v\frac{\partial \Omega}{\partial y} = \nu\left(\frac{\partial^2 \Omega}{\partial x^2} + \frac{\partial^2 \Omega}{\partial y^2}\right) \tag{1.203}$$

where the subscript z is dropped from Ω_z.

Substitution of relations (1.201) and (1.202) into the expression for vorticity yields

$$\frac{\partial^2 \psi}{\partial x^2} + \frac{\partial^2 \psi}{\partial y^2} = -\Omega \tag{1.204}$$

This equation is known as the *stream function equation* and is classified as an elliptic PDE.

The vorticity equation may be expressed in a nondimensional form by using the nondimensional quantities defined previously and a nondimensional vorticity defined as

$$\Omega^* = \frac{\Omega L}{u_\infty}$$

The nondimensional form of the vorticity equation can be expressed as

$$\frac{\partial \Omega^*}{\partial t^*} + u^* \frac{\partial \Omega^*}{\partial x^*} + v^* \frac{\partial \Omega^*}{\partial y^*} = \frac{1}{Re_\infty} \left(\frac{\partial^2 \Omega^*}{\partial x^{*2}} + \frac{\partial^2 \Omega^*}{\partial y^{*2}} \right) \tag{1.205}$$

Similarly, the nondimensional form of the stream function equation given by Equation (1.204) is

$$\frac{\partial^2 \psi^*}{\partial x^{*2}} + \frac{\partial^2 \psi^*}{\partial y^{*2}} = -\Omega^* \tag{1.206}$$

where

$$\psi^* = \frac{\psi}{u_\infty L}$$

A summary of the vorticity–stream function formulations is provided below.

(I) Dimensional, conservative form in Cartesian coordinates

$$\frac{\partial \Omega}{\partial t} + \frac{\partial}{\partial x}(u\Omega) + \frac{\partial}{\partial y}(v\Omega) = \nu \left(\frac{\partial^2 \Omega}{\partial x^2} + \frac{\partial^2 \Omega}{\partial y^2} \right) \tag{1.207}$$

$$\frac{\partial^2 \psi}{\partial x^2} + \frac{\partial^2 \psi}{\partial y^2} = -\Omega \tag{1.208}$$

(II) Dimensional, nonconservative form in Cartesian coordinates

$$\frac{\partial \Omega}{\partial t} + u \frac{\partial \Omega}{\partial x} + v \frac{\partial \Omega}{\partial y} = \nu \left(\frac{\partial^2 \Omega}{\partial x^2} + \frac{\partial^2 \Omega}{\partial y^2} \right) \tag{1.209}$$

$$\frac{\partial^2 \psi}{\partial x^2} + \frac{\partial^2 \psi}{\partial y^2} = -\Omega \tag{1.210}$$

(III) Nondimensional, conservative form in Cartesian coordinates

$$\frac{\partial \Omega^*}{\partial t^*} + \frac{\partial}{\partial x^*}(u^*\Omega^*) + \frac{\partial}{\partial y^*}(v^*\Omega^*) = \frac{1}{Re}\left(\frac{\partial^2 \Omega^*}{\partial x^{*2}} + \frac{\partial^2 \Omega^*}{\partial y^{*2}}\right) \qquad (1.211)$$

$$\frac{\partial^2 \psi^*}{\partial x^{*2}} + \frac{\partial^2 \psi^*}{\partial y^{*2}} = -\Omega^* \qquad (1.212)$$

(IV) Nondimensional, nonconservative form in Cartesian coordinates

$$\frac{\partial \Omega^*}{\partial t^*} + u^*\frac{\partial \Omega^*}{\partial x^*} + v^*\frac{\partial \Omega^*}{\partial y^*} = \frac{1}{Re}\left(\frac{\partial^2 \Omega^*}{\partial x^{*2}} + \frac{\partial^2 \Omega^*}{\partial y^{*2}}\right) \qquad (1.213)$$

$$\frac{\partial^2 \psi^*}{\partial x^{*2}} + \frac{\partial^2 \psi^*}{\partial y^{*2}} = -\Omega^* \qquad (1.214)$$

The vorticity–stream function equations in various coordinate systems are given in Table 1.10.

1.9 Poisson Equation for Pressure: Primitive Variables

In this section an equation is developed which may be used for the computation of the pressure field. The reason for incorporating the Poisson equation for pressure, which is usually used in lieu of the continuity equation, is the lack of a direct link for pressure between continuity and momentum equations. The steps required to obtain the Poisson equation for pressure are as follows. The conservative forms of the x- and y-components of the momentum equation obtained previously are

$$\frac{\partial u}{\partial t} + \frac{\partial}{\partial x}(u^2) + \frac{\partial p}{\partial x} + \frac{\partial}{\partial y}(uv) = \frac{1}{Re}\nabla^2 u \qquad (1.215)$$

$$\frac{\partial v}{\partial t} + \frac{\partial}{\partial x}(uv) + \frac{\partial}{\partial y}(v^2) + \frac{\partial p}{\partial y} = \frac{1}{Re}\nabla^2 v \qquad (1.216)$$

From cross-differentiation of the x and y components of momentum equation, one may obtain

$$\frac{\partial^2 p}{\partial x^2} + \frac{\partial^2 p}{\partial y^2} = -\frac{\partial D}{\partial t} - \frac{\partial^2}{\partial x^2}(u^2) - 2\frac{\partial^2}{\partial x \partial y}(uv) - \frac{\partial^2}{\partial y^2}(v^2) + \frac{1}{Re}\left[\frac{\partial^2}{\partial x^2}(D) + \frac{\partial^2}{\partial y^2}(D)\right] \qquad (1.217)$$

where

$$D = \frac{\partial u}{\partial x} + \frac{\partial v}{\partial y}$$

is known as *dilatation*. Equation (1.217) is known as *the Poisson equation for pressure*.

Table 1.10: Vorticity–stream function equations

a. Planar flow in Cartesian coordinates (x, y)

$$\frac{\partial \Omega_z}{\partial t} + u\frac{\partial \Omega_z}{\partial x} + v\frac{\partial \Omega_z}{\partial y} = \nu\left(\frac{\partial^2 \Omega_z}{\partial x^2} + \frac{\partial^2 \Omega_z}{\partial y^2}\right)$$

$$\frac{\partial^2 \psi}{\partial x^2} + \frac{\partial^2 \psi}{\partial y} = -\Omega_z$$

b. Planar flow in cylindrical coordinates (r, θ)

$$\frac{\partial \Omega_z}{\partial t} + u_r\frac{\partial \Omega_z}{\partial r} + u_\theta\frac{1}{r}\frac{\partial \Omega_z}{\partial \theta} = \nu\frac{1}{r}\frac{\partial}{\partial r}\left(r\frac{\partial^2 \Omega_z}{\partial r} + \nu\frac{1}{r^2}\frac{\partial^2 \Omega_z}{\partial \theta^2}\right)$$

$$\frac{\partial^2 \psi}{\partial r^2} + \frac{1}{r}\frac{\partial \psi}{\partial r} + \frac{1}{r^2}\frac{\partial^2 \psi}{\partial \theta^2} = -\Omega_z$$

c. Axisymmetric flow in cylindrical coordinates (r, z)

$$\frac{\partial \Omega_\theta}{\partial t} + u_r\frac{\partial \Omega_\theta}{\partial r} + u_z\frac{\partial \Omega_\theta}{\partial z} = \frac{1}{r}u_r\Omega_\theta + \nu\frac{\partial}{\partial r}\left[\frac{1}{r}\frac{\partial}{\partial r}(r\Omega_\theta)\right] + \nu\frac{\partial^2 \Omega_\theta}{\partial z^2}$$

$$\frac{\partial}{\partial r}\left(\frac{1}{r}\frac{\partial \psi}{\partial r}\right) + \frac{\partial^2}{\partial z^2}\left(\frac{\psi}{r}\right) = -\Omega_\theta$$

d. Axisymmetric flow in spherical coordinates (r, θ)

$$\frac{\partial \Omega_\phi}{\partial t} + u_r\frac{\partial \Omega_\phi}{\partial r} + u_\theta\frac{1}{r}\frac{\partial \Omega_\phi}{\partial \theta} = \Omega_\phi\frac{1}{r}(u_r + u_\theta\cot\theta)$$

$$+ \nu\frac{1}{r^2}\frac{\partial}{\partial r}\left(r^2\frac{\partial \Omega_\phi}{\partial r}\right) + \frac{\nu}{r^2}\frac{\partial}{\partial \theta}\left[\frac{1}{\sin\theta}\frac{\partial}{\partial \theta}(\Omega_\phi\sin\theta)\right]$$

$$\frac{1}{r^2}\frac{\partial}{\partial r}\left[r^2\frac{\partial}{\partial r}\left(\frac{\psi}{r\sin\theta}\right)\right] + \frac{1}{r^2}\frac{\partial}{\partial \theta}\left[\frac{1}{\sin\theta}\frac{\partial}{\partial \theta}\left(\frac{\psi}{r}\right)\right] = -\Omega_\phi$$

The Poisson equation for pressure in terms of the stream function is expressed as

$$\frac{\partial^2 p}{\partial x^2} + \frac{\partial^2 p}{\partial y^2} = 2\left[\left(\frac{\partial^2 \psi}{\partial x^2}\right)\left(\frac{\partial^2 \psi}{\partial y^2}\right) - \left(\frac{\partial^2 \psi}{\partial x \partial y}\right)^2\right] \tag{1.218}$$

Observe that Equation (1.218) is in nondimensional form. However, it may be expressed in a dimensional form as

$$\frac{\partial^2 p}{\partial x^2} + \frac{\partial^2 p}{\partial y^2} = 2\rho\left[\left(\frac{\partial^2 \psi}{\partial x^2}\right)\left(\frac{\partial^2 \psi}{\partial y^2}\right) - \left(\frac{\partial^2 \psi}{\partial x \partial y}\right)^2\right] \tag{1.219}$$

1.10   Irrotational, Incompressible Flow

For a two-dimensional flow, a function may be defined which satisfies the continuity equation. This function is known as the *stream function* and, for an incompressible flow in a Cartesian coordinate system, is given by

$$u = \frac{\partial \psi}{\partial y} \tag{1.220}$$

$$v = -\frac{\partial \psi}{\partial x} \tag{1.221}$$

If the flow is also irrotational, then

$$\frac{\partial v}{\partial x} - \frac{\partial u}{\partial y} = 0 \tag{1.222}$$

Substitution of relations (1.220) and (1.221) into (1.222) yields

$$\frac{\partial^2 \psi}{\partial x^2} + \frac{\partial^2 \psi}{\partial y^2} = 0 \tag{1.223}$$

which is known as the *stream function equation*. Since Equation (1.223) is linear, various solutions may be added to provide other solutions. Therefore, expressions for the stream function of simple flows are developed, from which more complex flowfields may be determined. The stream functions for elementary flows are provided in Table 1.11.

<table>
<tr><td colspan="3" align="center">Table 1.11: Stream functions for elementary flows</td></tr>
<tr><td align="center">Flow</td><td align="center">Cartesian (x,y)</td><td align="center">Polar (r,θ)</td></tr>
<tr><td>Uniform flow from left parallel to x-axis</td><td align="center">$\psi = u_\infty y$</td><td align="center">$\psi = u_\infty r \sin\theta$</td></tr>
<tr><td>Source of quantity Q located at the origin</td><td align="center">$\psi = \dfrac{Q}{2\pi} \tan^{-1}\dfrac{y}{x}$</td><td align="center">$\psi = \dfrac{Q}{2\pi}\theta$</td></tr>
<tr><td>Doublet of strength μ located at the origin</td><td align="center">$\psi = -\dfrac{\mu}{2\pi}\dfrac{y}{(x^2+y^2)}$</td><td align="center">$\psi = -\dfrac{\mu}{2\pi}\dfrac{\sin\theta}{r}$</td></tr>
<tr><td>Vortex of strength Γ located at the origin with clockwise circulation</td><td align="center">$\psi = \dfrac{\Gamma}{4\pi}\ell n(x^2+y^2)$</td><td align="center">$\psi = \dfrac{\Gamma}{2\pi}\ell n(r)$</td></tr>
</table>

1.11 Selected Solutions of the Navier-Stokes Equations for Internal Flows

The Navier-Stokes equations comprise a set of second-order, nonlinear partial differential equations which must be solved simultaneously for the unknowns. Unfortunately, analytical solution of the Navier-Stokes equations for a majority of complex problems does not exist. However, advancement in computer technology and in numerical algorithms allows us to obtain numerical solutions. These schemes vary both in accuracy and efficiency, and, obviously, cost and time must be considered in the selection of a particular scheme. The technology associated with the numerical solution of fluid problems has evolved into what is known as *computational fluid dynamics* (CFD). A summary of various numerical schemes used in CFD is presented in Chapter 6.

The Navier-Stokes equations, however, can be reduced for relatively simple problems where exact solutions can be achieved. In this section, selected exact solutions of the Navier-Stokes equations and some results from numerical solutions are presented.

Governing equations and results for turbulent flows are also included for completeness. However, it is noted that expressions for turbulent flows are obtained by semi-theoretical analysis and are primarily based on correlation of experimental data. The accuracy of these results could vary substantially because, first, the experimental data used in the development of these expressions are limited and, second, there could be large inaccuracies in the measurement of turbulence quantitites. Thus, these expressions should be used with care, recognizing the problems identified above.

An important class of Navier-Stokes solutions for internal flows associated with ducts is presented in this section. It is assumed that the flow is incompressible, steady, and fully developed within a duct of constant cross section. Furthermore, the effects of the hydrodynamic entry length are ignored.

1.11.1 Fully Developed Laminar Flow

Assuming that the streamwise direction is aligned along the x-axis, the only component of the velocity is then the streamwise component denoted by u, which can vary with y and z, but not with x. The nomenclature is illustrated in Figure 1.5. The Navier-Stokes equations are reduced to

Continuity: $\qquad\qquad\qquad\qquad \dfrac{\partial u}{\partial x} = 0$

x-component of the momentum equation: $\qquad -\dfrac{\partial p}{\partial x} + \mu\left(\dfrac{\partial^2 u}{\partial y^2} + \dfrac{\partial^2 u}{\partial z^2}\right) = 0$

y-component of the momentum equation: $\qquad -\dfrac{\partial p}{\partial y} = 0$

z-component of the momentum equation: $\qquad -\dfrac{\partial p}{\partial z} = 0$

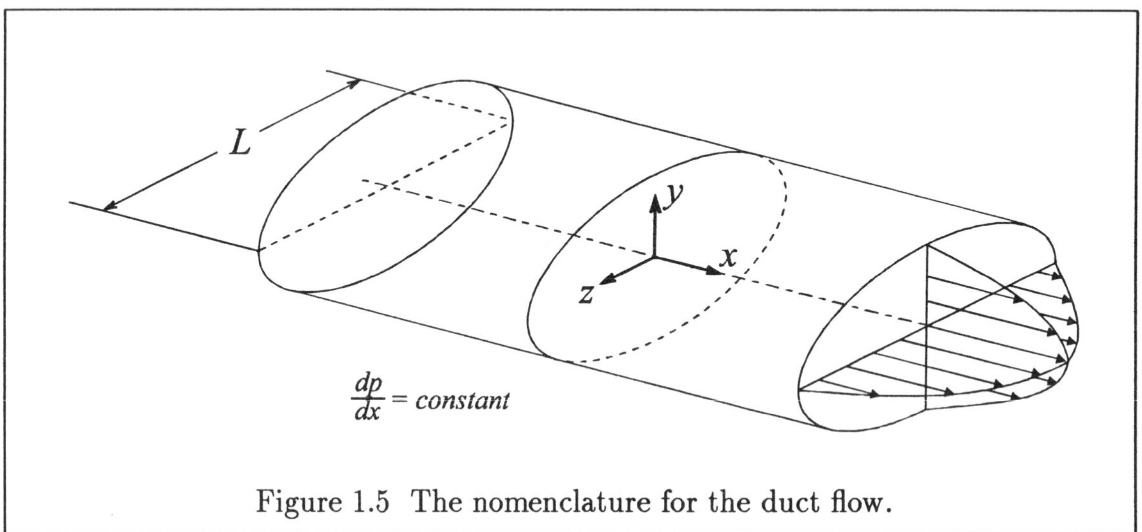

Figure 1.5 The nomenclature for the duct flow.

Introduction of the following nondimensional variables

$$y^* = \frac{y}{L}, \qquad z^* = \frac{z}{L}, \qquad u^* = \frac{\mu\, u}{L^2\left(-\frac{dp}{dx}\right)}$$

where L is a characteristic length, reduces the x-momentum equation to

$$\nabla^{*2} u^* = -1 \tag{1.224}$$

subject to the no-slip boundary condition of $u^* = 0$ at the surface. Equation (1.224) can be solved for a wide variety of duct cross sections and the results for laminar flows are summarized in Table 1.12.

Table 1.12: Steady, laminar, fully-developed duct flows

Cross section	Geometry/ Nomenclature	Velocity profile	Volumetric flow rate
Circular	$y^2 + z^2 = R^2$	$u = -\dfrac{1}{4\mu}\dfrac{dp}{dx}\left(R^2 - r^2\right)$	$Q = -\dfrac{dp}{dx}\left(\dfrac{\pi R^2}{8\mu}\right)$
Elliptic	$\dfrac{y^2}{a^2} + \dfrac{z^2}{b^2} \leq 1$	$u = -\dfrac{1}{2\mu}\dfrac{dp}{dx}\dfrac{a^2 b^2}{a^2 + b^2}\left(1 - \dfrac{y^2}{a^2} - \dfrac{z^2}{b^2}\right)$	$Q = -\dfrac{dp}{dx}\left(\dfrac{\pi}{4\mu}\dfrac{a^3 b^3}{a^2 + b^2}\right)$
Concentric circular annulus	$b \leq r \leq a$	$u = -\dfrac{1}{4\mu}\dfrac{dp}{dx}\left[a^2 - r^2 + (a^2 - b^2)\dfrac{\ln(a/r)}{\ln(b/a)}\right]$	$Q = -\dfrac{dp}{dx}\left(\dfrac{\pi}{8\mu}\right)\left[a^4 - b^4 - \dfrac{(a^2 - b^2)^2}{\ln(a/b)}\right]$
Equilateral triangle of side a		$u = -\dfrac{1}{2\sqrt{3}\mu a}\dfrac{dp}{dx}\left(z - \dfrac{\sqrt{3}}{2}a\right)\left(3y^2 - z^2\right)$	$Q = -\dfrac{dp}{dx}\left(\dfrac{\sqrt{3}a^4}{320\mu}\right)$

(Continued on next page)

Table 1.12: Continued

Cross section	Geometry/ Nomenclature	Velocity profile	Volumetric flow rate
Rectangular	$-a \leq y \leq a$ $-b \leq z \leq b$	$u = -\dfrac{dp}{dx}\dfrac{16a^2}{\mu\pi^3}\displaystyle\sum_{m=1,3,5,\ldots}^{\infty}(-1)^{(m-1)/2}\times$ $\left[1 - \dfrac{\cosh(m\pi z/2a)}{\cosh(m\pi b/2a)}\right]\dfrac{\cosh(m\pi y/2a)}{m^3}$	$Q = -\dfrac{dp}{dx}\dfrac{ba^3}{6\mu}\left[1 - \dfrac{192a}{\pi^5 b}\times\right.$ $\left.\displaystyle\sum_{m=1,3,5,\ldots}^{\infty}\dfrac{\tanh(m\pi b/2a)}{m^5}\right]$
Circular sector	$0 \leq r \leq a$ $-\dfrac{1}{2}\alpha \leq \theta \leq \dfrac{1}{2}\alpha$	$u(r,\theta) = -\dfrac{dp}{dx}\dfrac{1}{4\mu}\left[r^2\left(1 - \dfrac{\cos 2\theta}{\cos\alpha}\right) -\right.$ $\dfrac{16a^2\alpha^2}{\pi^3}\displaystyle\sum_{m=1,3,5,\ldots}^{\infty}(-1)^{(m+1)/2}\left(\dfrac{r}{a}\right)^m\times$ $\left.\dfrac{\cos(m\pi\theta/\alpha)}{m(m+2\alpha/\pi)(m-2\alpha/\pi)}\right]$	$Q = \dfrac{a^4}{4\mu}\left(-\dfrac{dp}{dx}\right)\left[\dfrac{\tan\alpha-\alpha}{4} - \dfrac{32\alpha^4}{\pi^5}\times\right.$ $\left.\displaystyle\sum_{m=1,3,5,\ldots}^{\infty}\dfrac{1}{m^2(m+2\alpha/\pi)^2(m-2\alpha/\pi)}\right]$

1.11.2 Fully Developed Flow in a Duct of Circular Cross Section

1.11.2.1 Skin Friction–Laminar Flow

Average velocity:
$$\bar{u} = \frac{1}{A}\int_A u\,dA = -\frac{dp}{dx}\frac{R^2}{8\mu} = \frac{1}{2}U_{\text{max}}$$

Shear stress:
$$\tau_w = \mu\left(-\frac{du}{dr}\right)_w = \frac{1}{2}R\left(-\frac{dp}{dx}\right) = 4\frac{\mu\bar{u}}{R}$$

Friction factor:
$$\lambda = \frac{8\tau_w}{\rho\bar{u}^2} = \frac{32\mu}{\rho R\bar{u}} = \frac{64}{Re_D}\ , Re_D = \frac{\rho\bar{u}D}{\mu}$$

Skin friction coefficient:
$$C_f = \frac{\tau_w}{\frac{1}{2}\rho\bar{u}^2} = \frac{1}{4}\lambda = \frac{16}{Re_D}$$

1.11.2.2 Skin Friction–Turbulent Flow

In the development of expressions for turbulent flows, the effect of roughness and roughness size/distribution must be considered. Some of these expressions are presented in this section as well as in Chapter 3 along with more detailed descriptions of turbulent boundary layers.

For a turbulent fully developed flow in a smooth circular pipe, the following expressions for skin friction coefficient have been proposed.

Karman-Nikuradse equation:

$$\frac{1}{\sqrt{4C_f}} = 0.87\,\ln(Re_D\sqrt{4C_f}) - 0.8 \tag{1.225}$$

This equation can be approximated for a Reynolds number range between 30,000 and 1,000,000 by a simpler expression given as

$$C_f = 0.046\,Re_D^{-1/5}$$

For a Reynolds number range between 5000 and 30,000, the Blasius equation given by the following can be used

$$C_f = 0.0791\,Re_D^{-0.25}$$

A discussion on rough pipes is provided in Section 3.11.3.

1.11.3 Hydraulic Diameter

The concept of hydraulic diameter has been introduced to describe flows through ducts with noncircular cross sections. It is generally defined as

$$D_h = \frac{4 \text{ cross-sectional area}}{\text{wetted perimeter}} = 4\frac{A}{p} \tag{1.226}$$

The friction factor for a noncircular duct is then

$$\lambda = \frac{2D_h}{\rho \bar{u}^2}\left(-\frac{dp}{dx}\right) \tag{1.227}$$

For each of the duct flows in Table 1.12, the expression for the hydraulic diameter is given by

$$D_h = 2\left[\frac{8\mu\bar{u}}{\left(-\frac{dp}{dx}\right)}\right]^{1/2} \tag{1.228}$$

1.11.4 Heat Transfer for Fully Developed Flows in Ducts of Circular Cross Section

The expressions for the heat transfer in terms of average Nusselt number for fully developed flows in ducts of circular cross section are provided in Table 1.13. The Nusselt number (based on the diameter as the characteristic length) is defined as

$$Nu_D = \frac{hD}{k}$$

Additional expressions for the Nusselt number for such flows subject to constant heat flux boundary as well as constant surface temperature boundary are given in the following sections.

Table 1.13: Heat transfer coefficients for forced convection of incompressible flow inside tubes and ducts of circular cross section

Equations	Restrictions/Remarks	Refs.
$Nu_D = 3.657$	Fully developed laminar flow, constant surface temperature, $Pr > 0.6$, (a)	
$Nu_D = 4.364$	Fully developed laminar flow, uniform heat flux, (a)	
$Nu_D = 1.86 \left[\left(\dfrac{D}{L} \right) Re_D Pr \right]^{1/3} \times \left(\dfrac{\mu}{\mu_s} \right)^{0.14}$	Fully developed laminar flow, constant wall temperature, $0.48 < Pr < 16700$, $0.004 < \dfrac{\mu}{\mu_s} < 10$, $\left(\dfrac{D}{L} Re_D Pr \right)^{1/3} \left(\dfrac{\mu}{\mu_s} \right)^{0.14} > 2$, (a)	[1.14]
$Nu_D = 3.66 + \dfrac{0.0668 \left(\dfrac{D}{L} \right) Re_D Pr}{1 + 0.4 \left[\left(\dfrac{D}{L} \right) Re_D Pr \right]^{2/3}}$	Thermally fully developed laminar flow, constant surface temperature, (a)	[1.15]
$Nu_D = 0.023\, Re_D^{0.8} Pr^n$	Fully developed turbulent flow, $0.7 < Pr < 160$, $10^4 < Re_D < 10^6$ Heating, $T_s > T_b$: $n = 0.4$ cooling, $T_s < T_b$: $n = 0.3$ $|T_s - T_b| < 6\,°\mathrm{C}$ for liquids $|T_s - T_b| > 60\,°\mathrm{C}$ for gases, (a)	[1.16]
$Nu_D = 0.027\, Re_D^{0.8} Pr^{1/3} \left(\dfrac{\mu}{\mu_s} \right)^{0.14}$	Fully developed turbulent flow, $0.7 < Pr < 160$, $10^4 < Re_D < 10^6$, $L/D > 60$	[1.14]
$Nu_D = 0.036\, Re_D^{0.8} Pr^{1/3} \left(\dfrac{D}{L} \right)^{1/18}$	Turbulent flow, relatively short tubes, $10 < \dfrac{L}{D} < 400$, (a)	[1.17]

(a) Fluid properties evaluated at the average of inlet and outlet bulk temperature. s: value at the surface, T_b: bulk temperature

1.11.4.1 Fully Developed Laminar Velocity and Temperature Profiles in Ducts of Circular Cross Section: Constant Heat Flux Boundary

The problem of constant heat rate per unit length of tube occurs in several practical applications such as heat exchangers, electric heating, and nuclear heating. Mathematically, it may be expressed as

$$\frac{q_w}{k} = \frac{\partial T}{\partial r} = \text{constant}$$

The temperature difference $(T_w - T)$ is independent of x, and, thus

$$\frac{\partial}{\partial x}(T_w - T) = 0$$

or

$$\frac{\partial T}{\partial x} = \frac{\partial T_w}{\partial x} = \text{constant}$$

The energy equation is then written as

$$\rho c_p u \frac{\partial T_w}{\partial x} = \frac{k}{r} \frac{\partial}{\partial r}\left(r \frac{\partial T}{\partial r}\right) \tag{1.229}$$

where viscous dissipation has been neglected. The solution of Equation (1.229) provides the temperature distribution as

$$T = T_w - \frac{\rho c_p \bar{u} R^2}{8k} \frac{\partial T_w}{\partial x}\left(3 - 4\frac{r^2}{R^2} + \frac{r^4}{R^4}\right) \tag{1.230}$$

The Nusselt number for this type flow is defined based on an average temperature referred to as the *cup-mixing temperature*, defined as

$$T_m = \frac{\int T \, d\overset{\circ}{m}}{\int d\overset{\circ}{m}} = \frac{\int_o^R T \rho u \, dA}{\int_o^R \rho u \, dA} \tag{1.231}$$

where $dA = 2\pi r dr$ for circular cross section. Thus, T_m is computed by averaging T over the mass flow rate of the fluid inside the duct. Obviously, for an incompressible flow, the density in expression (1.231) will cancel out. Using the temperature distribution given by Equation (1.230) and the appropriate velocity profile from Table 1.12, one obtains

$$T_w - T_m = \frac{11}{18}\left(\frac{3}{8} \frac{\rho c_p}{k} \bar{u} R^2 \frac{\partial T_w}{\partial x}\right)$$

It is upon the difference $(T_w - T_m)$ that most engineering approximations to the Nusselt number are based. Finally, the Nusselt number is given by

$$Nu = \frac{q_w(2R)}{k(T_w - T_m)} = 4.364$$

1.11.4.2 Fully Developed Turbulent Velocity and Temperature Profiles in Ducts of Circular Cross Section: Constant Heat Flux Boundary

For a fully developed turbulent flow, the Nusselt number varies widely depending on the Reynolds number and Prandtl number. A sample of these empirical expressions and their range of applicability is provided below.

$$Nu = 0.0167 Re^{0.85} Pr^{0.93} + 6.3$$

Liquid metals, $Pr < 0.03$ (1.232)
Ref. [1.18]

$$Nu = 0.0165 Re^{0.85} Pr^{0.86} + 5.6$$

Liquid metals, $Pr < 0.03$ (1.233)
Ref. [1.19]

$$Nu = 0.015 Re^a Pr^b + 5.0$$

Other liquids,   (1.234)

$$a = 0.88 - \frac{0.24}{Pr + 4}$$

$0.1 < Pr < 10^4$
$10^4 < Re < 10^6$

$$b = 0.333 + 0.5 \exp(-0.6 Pr)$$

Ref. [1.18]

$$Nu = \frac{RePr\dfrac{C_f}{2}}{12.7\sqrt{\dfrac{C_f}{2}}(Pr^{2/3} - 1) + 1.07}$$

Other liquids,   (1.235)
$0.5 < Pr < 2000$
$10^4 < Re < 5 \times 10^6$

where Equation (1.225) can be used for C_f.   Ref. [1.20]

$$Nu = \frac{Pr(Re - 1000)\dfrac{C_f}{2}}{12.7\sqrt{\dfrac{C_f}{2}}(Pr^{2/3} - 1) + 1.0}$$

Modified from (1.235)   (1.236)
Other liquids
$0.5 < Pr < 2000$
$2300 < Re < 5 \times 10^6$
Ref. [1.21]

$$Nu = 0.023 Re^{0.8} Pr^{0.4}$$

Other liquids   (1.237)
$0.7 < Pr < 120$
$10^4 < Re < 1.2 \times 10^5$
$\dfrac{L}{D} > 60$,   Ref. [1.22]

$$Nu = 0.023 Re^{0.8} Pr^{1/3}$$

Same as above   (1.238)
Ref. [1.23]

$$Nu = 0.022 Pr^{0.5} Re^{0.8}$$

Gases,   (1.239)
$0.6 < Pr < 0.8$, $Re < 10^5$
Ref. [1.24]

1.11.4.3   Fully Developed Laminar Velocity and Temperature Profiles in Ducts of Circular Cross Section: Constant Surface Temperature Boundary

In this case, the governing equation is given by

$$\frac{k}{r}\frac{\partial}{\partial r}\left(r\frac{\partial T}{\partial r}\right) = \rho c_p u \frac{T_w - T}{T_w - T_m}\frac{dT_m}{dx}$$

and the solution is given by infinite series [1.25] as

$$\frac{T_w - T}{T_w - T_m} = \sum_{n=0}^{\infty} C_{2n}\left(\frac{r}{R}\right)^{2n}$$

where the coefficients are

$$C_0 = 1$$

$$C_2 = -\frac{1}{4}\lambda_0^2 = -1.828397$$

$$C_{2n} = \frac{\lambda_0^2}{(2n)^2}\left(C_{2n-4} - C_{2n-2}\right)$$

$$\lambda_0 = 2.704364$$

The Nusselt number becomes

$$Nu = 3.657$$

1.11.4.4   Fully Developed Turbulent Velocity and Temperature Profiles in Ducts of Circular Cross Section: Constant Surface Temperature Boundary

Some of the expressions developed for constant heat flux are applicable to constant temperature conditions as well. It turns out that these empirical expressions are essentially independent of wall boundary conditions. Thus, Equations (1.234), (1.235), and (1.236) can be used for constant temperature wall condition. The modifications for remaining equations are as follows:

$$Nu = 0.0156 Re^{0.85} Pr^{0.93} + 4.8 \qquad \text{Liquid metals, } Pr < 0.03 \qquad (1.240)$$
$$\text{Ref. [1.18]}$$

$$Nu = 0.0156 Re^{0.85} Pr^{0.86} + 4.5 \qquad \text{Liquid metals, } Pr < 0.03 \qquad (1.241)$$
$$\text{Ref. [1.19]}$$

$$Nu = 0.021 Pr^{0.5} Re^{0.8} \qquad \text{Gases, } 0.6 < Pr < 0.8 \qquad (1.242)$$
$$Re < 10^5$$
$$\text{Ref. [1.24]}$$

1.11.5 Fully Developed Flow for Noncircular Cross Sections

1.11.5.1 Skin Friction–Laminar Flow

An approximate expression for the skin friction coefficient of fully developed laminar flow in a duct of rectangular cross section has been proposed by Shah and London [1.26]. The Reynolds number is again based on the hydraulic diameter $D_h = 4\frac{A}{P}$, and the aspect ratio of the duct is defined as short side/long side $= \beta$. The proposed approximation is

$$C_f Re = 24(1 - 1.3553\beta + 1.9467\beta^2 - 1.7012\beta^3 + 0.9564\beta^4 - 0.2537\beta^5) \quad (1.243)$$

1.11.5.2 Skin Friction–Turbulent Flow

The skin friction coefficient of turbulent flow in a noncircular cross section differs only slightly from those of circular ducts. Therefore, the results presented previously for circular ducts can be used for most engineering applications.

1.11.5.3 Heat Transfer–Laminar Flow

Solutions for heat transfer in ducts of noncircular cross sections can be obtained by numerical schemes. In fact, the results of these solutions, as presented by [1.25], are given in Table 1.14. These solutions are either for constant heat flux or constant wall temperature, as provided in Table 1.14. Additional cross-sectional shapes and the corresponding solutions may be found in Refs. [1.25] and [1.26].

1.11.6 Fully Developed Flow in Circular Tube Annulus

1.11.6.1 Skin Friction–Laminar Flow

The solutions by Lundgren et al. [1.27] provide the skin friction coefficient for fully developed laminar flow. The Reynolds number is based on hydraulic diameter or equivalent diameter defined as

$$D_h = 4r_h = 2(r_o - r_i) = D_o - D_i$$

where D_o and D_i represent the outer and inner diameters, respectively. A summary of solutions is provided in Table 1.15.

1.11.6.2 Skin Friction–Turbulent Flow

Experimental data on fully turbulent flow suggest that skin friction dependency on the radius ratio is small and, in some cases, is in fact independent of the radius ratio. A simple approximation for the skin friction in the Reynolds number range between 6,000 to 300,000 is

$$C_f = 0.08 Re_{D_h}^{-0.25}$$

Again, as before, $D_h = D_o - D_i$.

Table 1.14: Nusselt number for fully developed laminar flows

Cross-sectional shape	b/a	Nu^*	Nu^{**}	Nu^{***}
Circular		4.364	3.66	4.364
Rectangular	1.0	3.61	2.98	3.091
	1.43	3.73	3.08	
	2.0	4.12	3.39	3.017
	3.0	4.79	3.96	
	4.0	5.33	4.44	2.930
	8.0	6.49	5.60	2.904
Parallel plate	∞	8.235	7.54	8.235
Parallel plate, one side insulated		5.385	4.86	
Triangular		3.11	2.49	1.892

* Constant heat flux in the flow direction and constant wall temperature at any cross section

** Constant wall temperature

*** Constant heat flux both axially and circumferentially

Table 1.15: Skin friction coefficient for fully developed laminar flow in circular tube annulus

r_i/r_o	$C_f Re$
0.001	18.67
0.01	20.03
0.05	21.56
0.10	22.34
0.20	23.09
0.30	23.46
0.40	23.68
0.60	23.89
0.80	23.98
0.90	23.99
1.00	24.00

1.11.6.3 Heat Transfer–Laminar Flow

For a fully developed temperature profile, the energy equation is reduced to

$$\frac{1}{r}\frac{\partial}{\partial r}\left(r\frac{\partial T}{\partial x}\right) = \frac{u}{\alpha}\left(\frac{\partial T_m}{\partial x}\right)$$

Furthermore, the fully developed velocity profile for concentric circular tube is known as given in Table 1.12. Two special cases on the imposed boundary conditions are used to determine fundamental solutions. Subsequently, the principle of superposition is used to obtain other solutions by adding the fundamental solutions. This is, of course, due to linearity of the governing equation. The two fundamental cases considered are (a) outer wall is subject to a specified heat flux while the inner wall is insulated, and (b) the heat flux is imposed on the inner wall where the outer wall is insulated. The following definitions are used in the solution, and the nomenclature is illustrated in Figure 1.6.

Heat flux for inner tube: $\quad \dot{q}_i = h_i(T_i - T_m)$

Heat flux for outer tube: $\quad \dot{q}_o = h_o(T_o - T_m)$

where h_i and h_o are the heat transfer coefficients. Note that heat flux is specified as positive when heat is into the fluid.

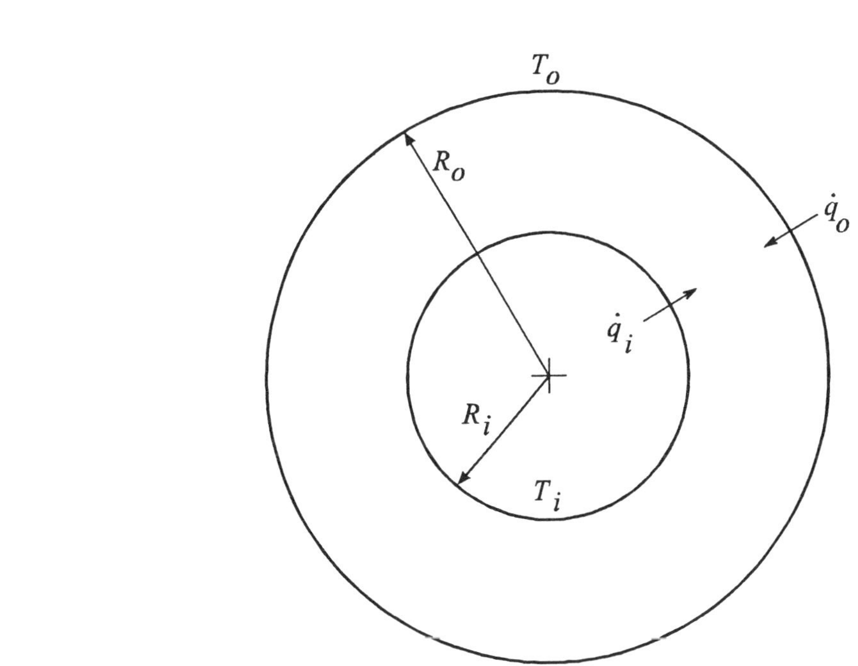

Figure 1.6 Nomenclature used in the definitions of parameters and solutions of fully developed flow in a circular tube annulus.

The Nusselt numbers associated with cases (a) and (b) are denoted by Nu_{ii} and Nu_{oo}. Furthermore, the influence coefficients which are obtained from fundamental solutions are denoted by θ_i^* and θ_o^*. Now a general solution for a specified heat flux ratio $\dot{q}_o/\dot{q}_i$ is given by

$$Nu_i = \frac{Nu_{ii}}{1 - \dfrac{\dot{q}_o}{\dot{q}_i}\theta_i^*} \tag{1.244}$$

and

$$Nu_o = \frac{Nu_{oo}}{1 - \dfrac{\dot{q}_i}{\dot{q}_o}\theta_o^*}$$ (1.245)

where the Nusselt numbers are defined as

$$Nu_i = \frac{h_i D_h}{k}$$

$$Nu_o = \frac{h_o D_h}{k}$$

and

$$D_h = 4R_h = 2(R_o - R_i)$$

The fundamental solutions and influence coefficients, as provided in Ref. [1.24] are given in Table 1.16.

Table 1.16. Influence coefficients and fundamental solutions for fully developed flow in circular tube annulus with constant heat flux

r_i/r_o	θ_i^*	θ_o^*	Nu_{ii}	Nu_{oo}
0.00	∞	0.00	∞	4.364
0.05	2.18	0.0294	17.81	4.792
0.10	1.383	0.0562	11.91	4.834
0.20	0.905	0.1041	8.499	4.883
0.40	0.603	0.1823	6.583	4.979
0.60	0.473	0.2455	5.912	5.099
0.80	0.401	0.299	5.58	5.24
1.00	0.346	0.346	5.385	5.385

The solution for constant surface temperature on one surface and the other surface insulated is computed by Lundberg et al. [1.28] and is provided in Table 1.17.

1.11.6.4 Heat Transfer–Turbulent Flow

Several experimental efforts have been performed to provide the necessary data used in expressions which determine heat transfer in an annulus system. The investigation of Kays and Leung [1.29] for constant heat transfer rate is such an effort. Their results provide the values of Nu_{ii}, Nu_{oo}, and the influence coefficients for different radius ratio, Reynolds number, and Prandtl number. Selected values of these parameters are presented in Table 1.18. Equations (1.244) and (1.245) can now be used to compute the heat transfer rates.

Table 1.17. The Nusselt number for the constant surface temperature on one surface and other surface insulated, fully developed laminar velocity and temperature profiles

r_i/r_o	Nu_o	Nu_i
0.0	3.66	∞
0.05	4.06	17.46
0.10	4.11	11.56
0.25	4.23	7.37
0.50	4.43	5.74
1.0	4.86	4.86

Table 1.18. Influence coefficients and Nusselt numbers for fully developed turbulent flow in a circular-tube annulus with constant heat rate

$r_i/r_o = 0.2$						
	$Re = 10^4$		$Re = 10^5$		$Re = 10^6$	
Pr	Nu_{ii}	θ_i^*	Nu_{ii}	θ_i^*	Nu_{ii}	θ_i^*
0.	8.40	1.009	8.30	1.020	8.30	1.020
0.001	8.40	1.009	8.30	1.020	8.90	0.976
0.003	8.40	1.009	8.50	1.025	12.5	0.834
0.01	8.50	1.000	9.70	0.944	33.6	0.748
0.03	9.00	1.012	15.8	0.771	81.0	0.374
0.5	31.2	0.520	157.	0.333	980.	0.262
0.7	38.6	0.412	196.	0.286	1,270.	0.235
1.0	46.8	0.339	247.	0.248	1,640.	0.209
3.	77.4	0.172	465.	0.143	3,250.	0.135
10.	120.	0.120	800.	0.072	6,000.	0.077
30.	172.	0.036	1,210.	0.035	9,300.	0.038
100.	243.	0.014	1,760.	0.015	13,800.	0.016
1,000.	448.	0.004	3,280.	0.002	26,000.	0.003
$r_i/r_o = 0.5$						
0.	6.28	0.620	6.30	0.651	6.30	0.654
0.001	6.28	0.620	6.30	0.651	6.75	0.644
0.003	6.28	0.620	6.40	0.656	9.40	0.585
0.01	6.37	0.622	7.30	0.623	23.2	0.427
0.03	6.75	0.627	12.0	0.533	65.5	0.333
0.5	24.6	0.343	130.	0.253	835.	0.208
0.7	30.9	0.300	166.	0.225	1,080.	0.185
1.0	38.2	0.247	212.	0.208	1,420.	0.170
3.	66.8	0.129	402.	0.115	2,870.	0.111
10.	106.	0.059	715.	0.059	5,400.	0.061
30.	153.	0.028	1,080.	0.028	8,400.	0.032
100.	220.	0.006	1,600.	0.006	12,600.	0.007
1,000.	408.	0.002	3,000.0	0.002	24,000.	0.002

(Continued on next page)

Table 1.18. Continued

$r_i/r_o = 0.8$

Pr	$Re = 10^4$		$Re = 10^5$		$Re = 10^6$	
	Nu_{ii}	θ_i^*	Nu_{ii}	θ_i^*	Nu_{ii}	θ_i^*
0.	5.87	0.489	5.92	0.515	5.97	0.528
0.001	5.87	0.489	5.92	0.515	6.33	0.516
0.003	5.87	0.489	6.03	0.485	8.80	0.468
0.01	5.95	0.485	6.80	0.493	21.7	0.382
0.03	6.20	0.478	11.4	0.445	61.0	0.276
0.5	22.9	0.268	123.	0.214	800.	0.174
0.7	28.5	0.244	157.	0.186	1,050.	0.160
1.0	35.5	0.200	202.	0.166	1,350.	0.140
3.	63.0	0.108	386.	0.097	2,750.	0.093
10.	102.	0.051	693.	0.052	5,150.	0.051
30.	147.	0.027	1,050.	0.028	8,100.	0.030
100.	215.	0.010	1,540.	0.010	12,100.	0.012
1,000.	393.	0.002	2,890.	0.002	23,000.	0.002

$r_i/r_o = 0.2$

Pr	Nu_{oo}	θ_o^*	Nu_{oo}	θ_o^*	Nu_{oo}	θ_o^*
0.	5.83	0.140	6.10	0.151	6.35	0.157
0.001	5.83	0.140	6.10	0.151	6.92	0.153
0.003	5.83	0.140	6.22	0.150	10.2	0.136
0.01	5.95	0.140	7.40	0.144	24.6	0.102
0.03	6.22	0.140	12.7	0.125	80.0	0.074
0.5	22.5	0.071	130.	0.055	823.	0.044
0.7	29.4	0.063	165.	0.049	1,070.	0.040
1.0	35.5	0.051	206.	0.042	1,390.	0.035
3.	60.0	0.026	390.	0.024	2,760.	0.023
10.	98.0	0.013	680.	0.012	4,980.	0.012
30.	142.	0.004	1,030.	0.006	7,850.	0.006
100.	205.	0.003	1,500.	0.003	12,000.	0.003
1,000.	380.	0.001	2,830.	0.001	22,500.	0.001

(Continued on next page)

<table>
<tr><td colspan="7" align="center">Table 1.18. Continued</td></tr>
<tr><td colspan="7" align="center">$r_i/r_o = 0.5$</td></tr>
<tr><td></td><td colspan="2" align="center">$Re = 10^4$</td><td colspan="2" align="center">$Re = 10^5$</td><td colspan="2" align="center">$Re = 10^6$</td></tr>
<tr><td>Pr</td><td>Nu_{oo}</td><td>θ_o^*</td><td>Nu_{oo}</td><td>θ_o^*</td><td>Nu_{oo}</td><td>θ_o^*</td></tr>
<tr><td>0.</td><td>5.66</td><td>0.281</td><td>5.80</td><td>0.296</td><td>5.95</td><td>0.310</td></tr>
<tr><td>0.001</td><td>5.66</td><td>0.281</td><td>5.80</td><td>0.296</td><td>6.40</td><td>0.304</td></tr>
<tr><td>0.003</td><td>5.66</td><td>0.281</td><td>5.85</td><td>0.294</td><td>9.00</td><td>0.278</td></tr>
<tr><td>0.01</td><td>5.73</td><td>0.281</td><td>6.80</td><td>0.289</td><td>22.6</td><td>0.217</td></tr>
<tr><td>0.03</td><td>6.03</td><td>0.279</td><td>11.6</td><td>0.258</td><td>64.0</td><td>0.163</td></tr>
<tr><td>0.5</td><td>22.6</td><td>0.162</td><td>125.</td><td>0.123</td><td>795.</td><td>0.098</td></tr>
<tr><td>0.7</td><td>28.3</td><td>0.137</td><td>158.</td><td>0.107</td><td>1,040.</td><td>0.090</td></tr>
<tr><td>1.0</td><td>34.8</td><td>0.111</td><td>200.</td><td>0.092</td><td>1,340.</td><td>0.078</td></tr>
<tr><td>3.</td><td>60.5</td><td>0.059</td><td>384.</td><td>0.055</td><td>2,730.</td><td>0.052</td></tr>
<tr><td>10.</td><td>100.</td><td>0.028</td><td>680.</td><td>0.028</td><td>5,030.</td><td>0.028</td></tr>
<tr><td>30.</td><td>143.</td><td>0.013</td><td>1,030.</td><td>0.014</td><td>8,000.</td><td>0.015</td></tr>
<tr><td>100.</td><td>207.</td><td>0.006</td><td>1,500.</td><td>0.006</td><td>12,000.</td><td>0.006</td></tr>
<tr><td>1,000.</td><td>387.</td><td>0.001</td><td>2,830.</td><td>0.001</td><td>23,000.</td><td>0.001</td></tr>
<tr><td colspan="7" align="center">$r_i/r_o = 0.8$</td></tr>
<tr><td>0.</td><td>5.65</td><td>0.379</td><td>5.75</td><td>0.398</td><td>5.85</td><td>0.409</td></tr>
<tr><td>0.001</td><td>5.65</td><td>0.379</td><td>5.75</td><td>0.398</td><td>6.25</td><td>0.407</td></tr>
<tr><td>0.003</td><td>5.65</td><td>0.379</td><td>5.84</td><td>0.397</td><td>8.80</td><td>0.374</td></tr>
<tr><td>0.01</td><td>5.75</td><td>0.381</td><td>6.72</td><td>0.390</td><td>21.0</td><td>0.286</td></tr>
<tr><td>0.03</td><td>6.10</td><td>0.388</td><td>11.1</td><td>0.339</td><td>62.0</td><td>0.216</td></tr>
<tr><td>0.5</td><td>22.4</td><td>0.225</td><td>121.</td><td>0.169</td><td>790.</td><td>0.136</td></tr>
<tr><td>0.7</td><td>28.0</td><td>0.192</td><td>156.</td><td>0.150</td><td>1,020.</td><td>0.122</td></tr>
<tr><td>1.0</td><td>34.8</td><td>0.159</td><td>197.</td><td>0.129</td><td>1,330.</td><td>0.111</td></tr>
<tr><td>3.</td><td>61.3</td><td>0.083</td><td>382.</td><td>0.078</td><td>2,730.</td><td>0.073</td></tr>
<tr><td>10.</td><td>100.</td><td>0.039</td><td>670.</td><td>0.039</td><td>5,050.</td><td>0.040</td></tr>
<tr><td>30.</td><td>146.</td><td>0.019</td><td>1,040.</td><td>0.020</td><td>8,000.</td><td>0.022</td></tr>
<tr><td>100.</td><td>209.</td><td>0.008</td><td>1,500.</td><td>0.009</td><td>12,000.</td><td>0.010</td></tr>
<tr><td>1,000.</td><td>385.</td><td>0.002</td><td>2,870.</td><td>0.002</td><td>23,000.</td><td>0.002</td></tr>
</table>

1.11.7 Fully Developed Flow Between Parallel Plates

Relations (1.244) and (1.245) can be used for the flow between parallel plates. Note that this case is the limiting situation of circular tube annulus. In this special case, both equations become identical and are reduced to

$$Nu_1 = \frac{5.385}{1 - 0.346(\dot{q}_2/\dot{q}_1)} \tag{1.246}$$

where the subscripts 1 and 2 are used to denote the two surfaces.

1.12 One-Dimensional Flow with Friction (Fanno Line)

The conservation laws applied to a cylindrical control volume for an adiabatic, constant area flow and the subsequent integration yield

$$\int_{x_1}^{x_2} \frac{4f\,dx}{D} = \left[-\frac{1}{\gamma M^2} - \frac{\gamma+1}{2\gamma} \ln\left(\frac{M^2}{1 + \frac{\gamma-1}{2}M^2} \right) \right]\Bigg|_{M_1}^{M_2}$$

where D is the diameter, and f is the friction coefficient. For a noncircular cross-section, D is replaced by the hydraulic diameter defined as

$$D_h = \frac{4A}{p}$$

where A is the cross-sectional area and p is the perimeter.

The ratio of flow properties in terms of the upstream and downstream Mach numbers may be written as

$$\frac{p_2}{p_1} = \frac{M_1}{M_2} \left[\frac{2 + (\gamma-1)M_1^2}{2 + (\gamma-1)M_2^2} \right]^{\frac{1}{2}}$$

$$\frac{T_2}{T_1} = \frac{2 + (\gamma-1)M_1^2}{2 + (\gamma-1)M_2^2}$$

$$\frac{\rho_2}{\rho_1} = \frac{M_1}{M_2} \left[\frac{2 + (\gamma-1)M_2^2}{2 + (\gamma-1)M_1^2} \right]^{\frac{1}{2}}$$

$$\frac{p_{t2}}{p_{t1}} = \frac{M_1}{M_2} \left[\frac{2 + (\gamma-1)M_2^2}{2 + (\gamma-1)M_1^2} \right]^{\frac{\gamma+1}{2(\gamma-1)}}$$

$$\frac{T_{t2}}{T_{t1}} = 1$$

The equations above may be expressed in terms of the reference properties at Mach number of 1 (designated by an asterisk) as follows

$$\frac{p}{p^*} = \frac{1}{M} \left[\frac{\gamma + 1}{2 + (\gamma - 1)M^2} \right]^{\frac{1}{2}} \tag{1.247}$$

$$\frac{T}{T^*} = \frac{\gamma + 1}{2 + (\gamma - 1)M^2} \tag{1.248}$$

$$\frac{\rho}{\rho^*} = \frac{1}{M} \left[\frac{2 + (\gamma - 1)M^2}{(\gamma + 1)} \right]^{\frac{1}{2}} \tag{1.249}$$

$$\frac{p_t}{p_t^*} = \frac{1}{M} \left[\frac{2 + (\gamma - 1)M^2}{\gamma + 1} \right]^{\frac{\gamma + 1}{2(\gamma - 1)}} \tag{1.250}$$

$$\frac{s - s^*}{c_p} = \ell n \left\{ M^2 \left[\frac{\gamma + 1}{M^2 [2 + (\gamma - 1)M^2]} \right]^{\frac{\gamma + 1}{2\gamma}} \right\} \tag{1.251}$$

If L^* represents the location where $M = 1$, then

$$\frac{4 \bar{f} L^*}{D} = \frac{1 - M^2}{\gamma M^2} + \frac{\gamma + 1}{2\gamma} \ell n \left[\frac{(\gamma + 1)M^2}{2 + (\gamma - 1)M^2} \right] \tag{1.252}$$

where $\bar{f}$ is an average friction coefficient defined as

$$\bar{f} = \frac{1}{L^*} \int_o^{L^*} f \, dx$$

Expressions (1.247) through (1.250) and (1.252) are tabulated for $\gamma = 1.4$ in Table 1.19.

Table 1.19: One-dimensional flow with friction

M	P/P^*	T/T^*	ρ/ρ^*	P_t/P_t^*	$4fL^*/D$
0.200E−01	0.5477E+02	0.1200E+01	0.4565E+02	0.2894E+02	0.1778E+04
0.400E−01	0.2738E+02	0.1200E+01	0.2283E+02	0.1448E+02	0.4404E+03
0.600E−01	0.1825E+02	0.1199E+01	0.1522E+02	0.9666E+01	0.1930E+03
0.800E−01	0.1368E+02	0.1198E+01	0.1142E+02	0.7262E+01	0.1067E+03
0.100E+00	0.1094E+02	0.1198E+01	0.9138E+01	0.5822E+01	0.6692E+02
0.120E+00	0.9116E+01	0.1197E+01	0.7618E+01	0.4864E+01	0.4541E+02
0.140E+00	0.7809E+01	0.1195E+01	0.6533E+01	0.4182E+01	0.3251E+02
0.160E+00	0.6829E+01	0.1194E+01	0.5720E+01	0.3673E+01	0.2420E+02
0.180E+00	0.6066E+01	0.1192E+01	0.5088E+01	0.3278E+01	0.1854E+02
0.200E+00	0.5455E+01	0.1190E+01	0.4583E+01	0.2964E+01	0.1453E+02
0.220E+00	0.4955E+01	0.1188E+01	0.4169E+01	0.2708E+01	0.1160E+02
0.240E+00	0.4538E+01	0.1186E+01	0.3825E+01	0.2496E+01	0.9386E+01
0.260E+00	0.4185E+01	0.1184E+01	0.3535E+01	0.2317E+01	0.7688E+01
0.280E+00	0.3882E+01	0.1181E+01	0.3286E+01	0.2166E+01	0.6357E+01
0.300E+00	0.3619E+01	0.1179E+01	0.3070E+01	0.2035E+01	0.5299E+01
0.320E+00	0.3389E+01	0.1176E+01	0.2882E+01	0.1922E+01	0.4447E+01
0.340E+00	0.3185E+01	0.1173E+01	0.2716E+01	0.1823E+01	0.3752E+01
0.360E+00	0.3004E+01	0.1170E+01	0.2568E+01	0.1736E+01	0.3180E+01
0.380E+00	0.2842E+01	0.1166E+01	0.2437E+01	0.1659E+01	0.2705E+01
0.400E+00	0.2696E+01	0.1163E+01	0.2318E+01	0.1590E+01	0.2308E+01
0.420E+00	0.2563E+01	0.1159E+01	0.2212E+01	0.1529E+01	0.1974E+01
0.440E+00	0.2443E+01	0.1155E+01	0.2114E+01	0.1474E+01	0.1692E+01
0.460E+00	0.2333E+01	0.1151E+01	0.2026E+01	0.1425E+01	0.1451E+01
0.480E+00	0.2231E+01	0.1147E+01	0.1945E+01	0.1380E+01	0.1245E+01
0.500E+00	0.2138E+01	0.1143E+01	0.1871E+01	0.1340E+01	0.1069E+01
0.520E+00	0.2052E+01	0.1138E+01	0.1802E+01	0.1303E+01	0.9174E+00
0.540E+00	0.1972E+01	0.1134E+01	0.1739E+01	0.1270E+01	0.7866E+00
0.560E+00	0.1898E+01	0.1129E+01	0.1680E+01	0.1240E+01	0.6736E+00
0.580E+00	0.1828E+01	0.1124E+01	0.1626E+01	0.1213E+01	0.5757E+00
0.600E+00	0.1763E+01	0.1119E+01	0.1575E+01	0.1188E+01	0.4908E+00
0.620E+00	0.1703E+01	0.1114E+01	0.1528E+01	0.1166E+01	0.4172E+00
0.640E+00	0.1646E+01	0.1109E+01	0.1484E+01	0.1145E+01	0.3533E+00
0.660E+00	0.1592E+01	0.1104E+01	0.1442E+01	0.1127E+01	0.2979E+00
0.680E+00	0.1541E+01	0.1098E+01	0.1403E+01	0.1110E+01	0.2498E+00
0.700E+00	0.1493E+01	0.1093E+01	0.1367E+01	0.1094E+01	0.2081E+00
0.720E+00	0.1448E+01	0.1087E+01	0.1332E+01	0.1081E+01	0.1721E+00
0.740E+00	0.1405E+01	0.1082E+01	0.1299E+01	0.1068E+01	0.1411E+00
0.760E+00	0.1365E+01	0.1076E+01	0.1269E+01	0.1057E+01	0.1145E+00
0.780E+00	0.1326E+01	0.1070E+01	0.1240E+01	0.1047E+01	0.9167E−01
0.800E+00	0.1289E+01	0.1064E+01	0.1212E+01	0.1038E+01	0.7229E−01
0.820E+00	0.1254E+01	0.1058E+01	0.1186E+01	0.1030E+01	0.5593E−01
0.840E+00	0.1221E+01	0.1052E+01	0.1161E+01	0.1024E+01	0.4226E−01
0.860E+00	0.1189E+01	0.1045E+01	0.1137E+01	0.1018E+01	0.3097E−01
0.880E+00	0.1158E+01	0.1039E+01	0.1115E+01	0.1013E+01	0.2179E−01
0.900E+00	0.1129E+01	0.1033E+01	0.1093E+01	0.1009E+01	0.1451E−01
0.920E+00	0.1101E+01	0.1026E+01	0.1073E+01	0.1006E+01	0.8913E−02
0.940E+00	0.1074E+01	0.1020E+01	0.1053E+01	0.1003E+01	0.4815E−02
0.960E+00	0.1049E+01	0.1013E+01	0.1035E+01	0.1001E+01	0.2057E−02
0.980E+00	0.1024E+01	0.1007E+01	0.1017E+01	0.1000E+01	0.4947E−03
0.100E+01	0.1000E+01	0.1000E+01	0.1000E+01	0.1000E+01	0.8515E−08

Table 1.19: One-dimensional flow with friction (continued)

M	P/P^*	T/T^*	ρ/ρ^*	P_t/P_t^*	$4fL^*/D$
0.102E+01	0.9771E+00	0.9933E+00	0.9837E+00	0.1000E+01	0.4587E−03
0.104E+01	0.9551E+00	0.9866E+00	0.9681E+00	0.1001E+01	0.1769E−02
0.106E+01	0.9338E+00	0.9798E+00	0.9531E+00	0.1003E+01	0.3838E−02
0.108E+01	0.9133E+00	0.9730E+00	0.9387E+00	0.1005E+01	0.6585E−02
0.110E+01	0.8936E+00	0.9662E+00	0.9249E+00	0.1008E+01	0.9935E−02
0.112E+01	0.8745E+00	0.9593E+00	0.9116E+00	0.1011E+01	0.1382E−01
0.114E+01	0.8561E+00	0.9524E+00	0.8988E+00	0.1015E+01	0.1819E−01
0.116E+01	0.8383E+00	0.9455E+00	0.8865E+00	0.1020E+01	0.2298E−01
0.118E+01	0.8210E+00	0.9386E+00	0.8747E+00	0.1025E+01	0.2814E−01
0.120E+01	0.8044E+00	0.9317E+00	0.8633E+00	0.1030E+01	0.3364E−01
0.122E+01	0.7882E+00	0.9247E+00	0.8524E+00	0.1037E+01	0.3943E−01
0.124E+01	0.7726E+00	0.9178E+00	0.8418E+00	0.1043E+01	0.4547E−01
0.126E+01	0.7574E+00	0.9108E+00	0.8316E+00	0.1050E+01	0.5174E−01
0.128E+01	0.7427E+00	0.9038E+00	0.8218E+00	0.1058E+01	0.5820E−01
0.130E+01	0.7285E+00	0.8969E+00	0.8123E+00	0.1066E+01	0.6483E−01
0.132E+01	0.7147E+00	0.8899E+00	0.8031E+00	0.1075E+01	0.7161E−01
0.134E+01	0.7012E+00	0.8829E+00	0.7942E+00	0.1084E+01	0.7850E−01
0.136E+01	0.6882E+00	0.8760E+00	0.7856E+00	0.1094E+01	0.8550E−01
0.138E+01	0.6755E+00	0.8690E+00	0.7773E+00	0.1104E+01	0.9259E−01
0.140E+01	0.6632E+00	0.8621E+00	0.7693E+00	0.1115E+01	0.9974E−01
0.142E+01	0.6512E+00	0.8551E+00	0.7615E+00	0.1126E+01	0.1069E+00
0.144E+01	0.6396E+00	0.8482E+00	0.7540E+00	0.1138E+01	0.1142E+00
0.146E+01	0.6282E+00	0.8413E+00	0.7467E+00	0.1150E+01	0.1215E+00
0.148E+01	0.6172E+00	0.8344E+00	0.7397E+00	0.1163E+01	0.1288E+00
0.150E+01	0.6065E+00	0.8276E+00	0.7328E+00	0.1176E+01	0.1360E+00
0.152E+01	0.5960E+00	0.8207E+00	0.7262E+00	0.1190E+01	0.1433E+00
0.154E+01	0.5858E+00	0.8139E+00	0.7198E+00	0.1204E+01	0.1506E+00
0.156E+01	0.5759E+00	0.8071E+00	0.7135E+00	0.1219E+01	0.1579E+00
0.158E+01	0.5662E+00	0.8004E+00	0.7074E+00	0.1234E+01	0.1651E+00
0.160E+01	0.5568E+00	0.7937E+00	0.7016E+00	0.1250E+01	0.1724E+00
0.162E+01	0.5476E+00	0.7869E+00	0.6958E+00	0.1267E+01	0.1795E+00
0.164E+01	0.5386E+00	0.7803E+00	0.6903E+00	0.1284E+01	0.1867E+00
0.166E+01	0.5299E+00	0.7736E+00	0.6849E+00	0.1301E+01	0.1938E+00
0.168E+01	0.5213E+00	0.7670E+00	0.6796E+00	0.1319E+01	0.2008E+00
0.170E+01	0.5130E+00	0.7605E+00	0.6745E+00	0.1338E+01	0.2078E+00
0.172E+01	0.5048E+00	0.7539E+00	0.6696E+00	0.1357E+01	0.2147E+00
0.174E+01	0.4969E+00	0.7474E+00	0.6648E+00	0.1376E+01	0.2216E+00
0.176E+01	0.4891E+00	0.7410E+00	0.6601E+00	0.1397E+01	0.2284E+00
0.178E+01	0.4815E+00	0.7345E+00	0.6555E+00	0.1418E+01	0.2352E+00
0.180E+01	0.4741E+00	0.7282E+00	0.6511E+00	0.1439E+01	0.2419E+00
0.182E+01	0.4668E+00	0.7218E+00	0.6467E+00	0.1461E+01	0.2485E+00
0.184E+01	0.4597E+00	0.7155E+00	0.6425E+00	0.1484E+01	0.2551E+00
0.186E+01	0.4528E+00	0.7093E+00	0.6384E+00	0.1507E+01	0.2616E+00
0.188E+01	0.4460E+00	0.7030E+00	0.6344E+00	0.1531E+01	0.2680E+00
0.190E+01	0.4394E+00	0.6969E+00	0.6305E+00	0.1555E+01	0.2743E+00
0.192E+01	0.4329E+00	0.6907E+00	0.6267E+00	0.1580E+01	0.2806E+00
0.194E+01	0.4265E+00	0.6847E+00	0.6230E+00	0.1606E+01	0.2868E+00
0.196E+01	0.4203E+00	0.6786E+00	0.6193E+00	0.1633E+01	0.2929E+00
0.198E+01	0.4142E+00	0.6726E+00	0.6158E+00	0.1660E+01	0.2990E+00
0.200E+01	0.4082E+00	0.6667E+00	0.6124E+00	0.1687E+01	0.3050E+00

Table 1.19: One-dimensional flow with friction (continued)

M	P/P^*	T/T^*	ρ/ρ^*	P_t/P_t^*	$4fL^*/D$
0.205E+01	0.3939E+00	0.6520E+00	0.6041E+00	0.1760E+01	0.3196E+00
0.210E+01	0.3802E+00	0.6376E+00	0.5963E+00	0.1837E+01	0.3339E+00
0.215E+01	0.3673E+00	0.6235E+00	0.5890E+00	0.1919E+01	0.3476E+00
0.220E+01	0.3549E+00	0.6098E+00	0.5821E+00	0.2005E+01	0.3609E+00
0.225E+01	0.3432E+00	0.5963E+00	0.5756E+00	0.2096E+01	0.3738E+00
0.230E+01	0.3320E+00	0.5831E+00	0.5694E+00	0.2193E+01	0.3862E+00
0.235E+01	0.3213E+00	0.5702E+00	0.5635E+00	0.2295E+01	0.3983E+00
0.240E+01	0.3111E+00	0.5576E+00	0.5580E+00	0.2403E+01	0.4099E+00
0.245E+01	0.3014E+00	0.5453E+00	0.5527E+00	0.2517E+01	0.4211E+00
0.250E+01	0.2921E+00	0.5333E+00	0.5477E+00	0.2637E+01	0.4320E+00
0.255E+01	0.2832E+00	0.5216E+00	0.5430E+00	0.2763E+01	0.4425E+00
0.260E+01	0.2747E+00	0.5102E+00	0.5385E+00	0.2896E+01	0.4526E+00
0.265E+01	0.2666E+00	0.4991E+00	0.5342E+00	0.3036E+01	0.4624E+00
0.270E+01	0.2588E+00	0.4882E+00	0.5301E+00	0.3183E+01	0.4718E+00
0.275E+01	0.2513E+00	0.4776E+00	0.5262E+00	0.3338E+01	0.4809E+00
0.280E+01	0.2441E+00	0.4673E+00	0.5225E+00	0.3500E+01	0.4898E+00
0.285E+01	0.2373E+00	0.4572E+00	0.5189E+00	0.3671E+01	0.4983E+00
0.290E+01	0.2307E+00	0.4474E+00	0.5155E+00	0.3850E+01	0.5065E+00
0.295E+01	0.2243E+00	0.4379E+00	0.5123E+00	0.4038E+01	0.5145E+00
0.300E+01	0.2182E+00	0.4286E+00	0.5092E+00	0.4235E+01	0.5222E+00
0.305E+01	0.2124E+00	0.4195E+00	0.5062E+00	0.4441E+01	0.5296E+00
0.310E+01	0.2067E+00	0.4107E+00	0.5034E+00	0.4657E+01	0.5368E+00
0.315E+01	0.2013E+00	0.4021E+00	0.5007E+00	0.4884E+01	0.5437E+00
0.320E+01	0.1961E+00	0.3937E+00	0.4980E+00	0.5121E+01	0.5504E+00
0.325E+01	0.1911E+00	0.3855E+00	0.4955E+00	0.5369E+01	0.5569E+00
0.330E+01	0.1862E+00	0.3776E+00	0.4931E+00	0.5629E+01	0.5632E+00
0.335E+01	0.1815E+00	0.3699E+00	0.4908E+00	0.5900E+01	0.5693E+00
0.340E+01	0.1770E+00	0.3623E+00	0.4886E+00	0.6184E+01	0.5752E+00
0.345E+01	0.1727E+00	0.3550E+00	0.4865E+00	0.6480E+01	0.5809E+00
0.350E+01	0.1685E+00	0.3478E+00	0.4845E+00	0.6790E+01	0.5864E+00
0.355E+01	0.1645E+00	0.3409E+00	0.4825E+00	0.7113E+01	0.5918E+00
0.360E+01	0.1606E+00	0.3341E+00	0.4806E+00	0.7450E+01	0.5970E+00
0.365E+01	0.1568E+00	0.3275E+00	0.4788E+00	0.7802E+01	0.6020E+00
0.370E+01	0.1531E+00	0.3210E+00	0.4770E+00	0.8169E+01	0.6068E+00
0.375E+01	0.1496E+00	0.3148E+00	0.4753E+00	0.8552E+01	0.6115E+00
0.380E+01	0.1462E+00	0.3086E+00	0.4737E+00	0.8951E+01	0.6161E+00
0.385E+01	0.1429E+00	0.3027E+00	0.4721E+00	0.9366E+01	0.6206E+00
0.390E+01	0.1397E+00	0.2969E+00	0.4706E+00	0.9799E+01	0.6248E+00
0.395E+01	0.1366E+00	0.2912E+00	0.4691E+00	0.1025E+02	0.6290E+00
0.400E+01	0.1336E+00	0.2857E+00	0.4677E+00	0.1072E+02	0.6331E+00
0.405E+01	0.1307E+00	0.2803E+00	0.4663E+00	0.1121E+02	0.6370E+00
0.410E+01	0.1279E+00	0.2751E+00	0.4650E+00	0.1171E+02	0.6408E+00
0.415E+01	0.1252E+00	0.2700E+00	0.4637E+00	0.1224E+02	0.6445E+00
0.420E+01	0.1226E+00	0.2650E+00	0.4625E+00	0.1279E+02	0.6481E+00
0.425E+01	0.1200E+00	0.2602E+00	0.4613E+00	0.1336E+02	0.6516E+00
0.430E+01	0.1175E+00	0.2554E+00	0.4601E+00	0.1395E+02	0.6550E+00
0.435E+01	0.1151E+00	0.2508E+00	0.4590E+00	0.1457E+02	0.6583E+00
0.440E+01	0.1128E+00	0.2463E+00	0.4579E+00	0.1521E+02	0.6615E+00
0.445E+01	0.1105E+00	0.2419E+00	0.4569E+00	0.1587E+02	0.6646E+00
0.450E+01	0.1083E+00	0.2376E+00	0.4559E+00	0.1656E+02	0.6676E+00

Table 1.19: One-dimensional flow with friction (continued)

M	P/P^*	T/T^*	ρ/ρ^*	P_t/P_t^*	$4fL^*/D$
0.455E+01	0.1062E+00	0.2334E+00	0.4549E+00	0.1728E+02	0.6706E+00
0.460E+01	0.1041E+00	0.2294E+00	0.4539E+00	0.1802E+02	0.6734E+00
0.465E+01	0.1021E+00	0.2254E+00	0.4530E+00	0.1879E+02	0.6762E+00
0.470E+01	0.1001E+00	0.2215E+00	0.4521E+00	0.1958E+02	0.6790E+00
0.475E+01	0.9823E−01	0.2177E+00	0.4512E+00	0.2041E+02	0.6816E+00
0.480E+01	0.9637E−01	0.2140E+00	0.4504E+00	0.2126E+02	0.6842E+00
0.485E+01	0.9457E−01	0.2104E+00	0.4495E+00	0.2215E+02	0.6867E+00
0.490E+01	0.9281E−01	0.2068E+00	0.4487E+00	0.2307E+02	0.6891E+00
0.495E+01	0.9110E−01	0.2034E+00	0.4480E+00	0.2402E+02	0.6915E+00
0.500E+01	0.8944E−01	0.2000E+00	0.4472E+00	0.2500E+02	0.6938E+00
0.510E+01	0.8625E−01	0.1935E+00	0.4458E+00	0.2707E+02	0.6983E+00
0.520E+01	0.8322E−01	0.1873E+00	0.4444E+00	0.2928E+02	0.7025E+00
0.530E+01	0.8034E−01	0.1813E+00	0.4431E+00	0.3165E+02	0.7065E+00
0.540E+01	0.7761E−01	0.1756E+00	0.4419E+00	0.3417E+02	0.7104E+00
0.550E+01	0.7501E−01	0.1702E+00	0.4407E+00	0.3687E+02	0.7140E+00
0.560E+01	0.7254E−01	0.1650E+00	0.4396E+00	0.3974E+02	0.7175E+00
0.570E+01	0.7018E−01	0.1600E+00	0.4385E+00	0.4280E+02	0.7208E+00
0.580E+01	0.6794E−01	0.1553E+00	0.4375E+00	0.4605E+02	0.7240E+00
0.590E+01	0.6580E−01	0.1507E+00	0.4366E+00	0.4951E+02	0.7270E+00
0.600E+01	0.6376E−01	0.1463E+00	0.4357E+00	0.5318E+02	0.7299E+00
0.610E+01	0.6181E−01	0.1421E+00	0.4348E+00	0.5708E+02	0.7326E+00
0.620E+01	0.5994E−01	0.1381E+00	0.4340E+00	0.6121E+02	0.7353E+00
0.630E+01	0.5816E−01	0.1343E+00	0.4332E+00	0.6559E+02	0.7378E+00
0.640E+01	0.5646E−01	0.1305E+00	0.4324E+00	0.7023E+02	0.7402E+00
0.650E+01	0.5482E−01	0.1270E+00	0.4317E+00	0.7513E+02	0.7425E+00
0.660E+01	0.5326E−01	0.1236E+00	0.4310E+00	0.8032E+02	0.7448E+00
0.670E+01	0.5176E−01	0.1203E+00	0.4304E+00	0.8580E+02	0.7469E+00
0.680E+01	0.5032E−01	0.1171E+00	0.4298E+00	0.9159E+02	0.7489E+00
0.690E+01	0.4894E−01	0.1140E+00	0.4292E+00	0.9770E+02	0.7509E+00
0.700E+01	0.4762E−01	0.1111E+00	0.4286E+00	0.1041E+03	0.7528E+00
0.710E+01	0.4635E−01	0.1083E+00	0.4280E+00	0.1109E+03	0.7546E+00
0.720E+01	0.4512E−01	0.1056E+00	0.4275E+00	0.1181E+03	0.7564E+00
0.730E+01	0.4395E−01	0.1029E+00	0.4270E+00	0.1256E+03	0.7580E+00
0.740E+01	0.4282E−01	0.1004E+00	0.4265E+00	0.1335E+03	0.7597E+00
0.750E+01	0.4173E−01	0.9796E−01	0.4260E+00	0.1418E+03	0.7612E+00
0.760E+01	0.4068E−01	0.9560E−01	0.4256E+00	0.1506E+03	0.7627E+00
0.770E+01	0.3967E−01	0.9333E−01	0.4251E+00	0.1598E+03	0.7642E+00
0.780E+01	0.3870E−01	0.9113E−01	0.4247E+00	0.1694E+03	0.7656E+00
0.790E+01	0.3776E−01	0.8901E−01	0.4243E+00	0.1795E+03	0.7669E+00
0.800E+01	0.3686E−01	0.8696E−01	0.4239E+00	0.1901E+03	0.7682E+00
0.850E+01	0.3279E−01	0.7767E−01	0.4221E+00	0.2511E+03	0.7740E+00
0.900E+01	0.2935E−01	0.6977E−01	0.4207E+00	0.3272E+03	0.7790E+00
0.950E+01	0.2642E−01	0.6299E−01	0.4194E+00	0.4211E+03	0.7832E+00
0.100E+02	0.2390E−01	0.5714E−01	0.4183E+00	0.5359E+03	0.7868E+00
0.105E+02	0.2173E−01	0.5206E−01	0.4174E+00	0.6750E+03	0.7900E+00
0.110E+02	0.1984E−01	0.4762E−01	0.4166E+00	0.8419E+03	0.7927E+00
0.115E+02	0.1818E−01	0.4372E−01	0.4159E+00	0.1041E+04	0.7951E+00
0.120E+02	0.1672E−01	0.4027E−01	0.4153E+00	0.1276E+04	0.7972E+00
0.125E+02	0.1543E−01	0.3721E−01	0.4147E+00	0.1553E+04	0.7991E+00
0.130E+02	0.1428E−01	0.3448E−01	0.4142E+00	0.1876E+04	0.8007E+00

Table 1.19: One-dimensional flow with friction (continued)

M	P/P^*	T/T^*	ρ/ρ^*	P_t/P_t^*	$4fL^*/D$
0.135E+02	0.1326E−01	0.3204E−01	0.4138E+00	0.2252E+04	0.8022E+00
0.140E+02	0.1234E−01	0.2985E−01	0.4134E+00	0.2685E+04	0.8036E+00
0.145E+02	0.1151E−01	0.2787E−01	0.4131E+00	0.3184E+04	0.8048E+00
0.150E+02	0.1077E−01	0.2609E−01	0.4128E+00	0.3755E+04	0.8058E+00
0.155E+02	0.1009E−01	0.2446E−01	0.4125E+00	0.4406E+04	0.8068E+00
0.160E+02	0.9476E−02	0.2299E−01	0.4122E+00	0.5145E+04	0.8077E+00
0.165E+02	0.8916E−02	0.2164E−01	0.4120E+00	0.5980E+04	0.8085E+00
0.170E+02	0.8403E−02	0.2041E−01	0.4118E+00	0.6921E+04	0.8093E+00
0.175E+02	0.7934E−02	0.1928E−01	0.4116E+00	0.7977E+04	0.8100E+00
0.180E+02	0.7502E−02	0.1824E−01	0.4114E+00	0.9159E+04	0.8106E+00
0.185E+02	0.7105E−02	0.1728E−01	0.4112E+00	0.1048E+05	0.8112E+00
0.190E+02	0.6739E−02	0.1639E−01	0.4111E+00	0.1195E+05	0.8117E+00
0.195E+02	0.6400E−02	0.1557E−01	0.4109E+00	0.1357E+05	0.8122E+00
0.200E+02	0.6086E−02	0.1481E−01	0.4108E+00	0.1538E+05	0.8126E+00

1.13 References

[1.1] Vincenti, W. G., and Kruger, C. H. Jr., "Introduction to Physical Gas Dynamics," Robert E. Krieger Publishing Co., 1965.

[1.2] Anderson, J. D., Jr., "Hypersonic and High Temperature Gas Dynamics," McGraw-Hill, 1989.

[1.3] Eu, B. C., "Kinetic Theory and Irreversible Thermodynamics," John Wiley & Sons, 1992.

[1.4] Gatignol, R., and Soubbaramayer (Eds.), "Advances in Kinetic Theory and Continuum Mechanics," Springer-Verlag, 1990.

[1.5] Heer, C. V., "Statistical Mechanics, Kinetic Theory, and Stochastic Processes," Academic Press, 1972.

[1.6] Fox, R. W., and McDonald, A. T., "Introduction to Fluid Mechanics," John Wiley & Sons, 1985.

[1.7] Kreider, J. F, "Principles of Fluid Mechanics," Allyn and Bacon, 1985.

[1.8] Shames, I. H., "Mechanics of Fluids," McGraw-Hill, 1962.

[1.9] Bertin, J. J., "Engineering Fluid Mechanics," 2nd ed., Prentice-Hall, 1987.

[1.10] Aziz, K., and Hellums J. D., "Numerical Solution of the Three-Dimensional Equations of Motion for Laminar Natural Convection," *Phys. Fluids,* Vol. 10, pp. 314–324, 1967.

[1.11] Mallinson, G. D., and deVahl, Davis, "Three-Dimensional Natural Convection in a Box: A Numerical Study," *J. of Fluid Mechanics,* Vol. 83, pp. 1–31, 1977.

[1.12] Aregbesola, Y. A. S., and Burley, D. M., "The Vector and Scalar Potential Method for the Numerical Solution of Two- and Three-Dimensional Navier-Stokes Equations," *J. Comput. Physics,* Vol. 24, p. 398, 1977.

[1.13] Wong, A. K., and Reizer, J. A., "An Effective Vorticity – Vector Potential Formulation for Numerical Solution of Three-Dimensional Duct Flow Problems," *J. of Comput. Physics,* Vol. 55, pp. 98–114, 1984.

[1.14] Sieder, E. N., and Tate, E. G., "Heat Transfer and Pressure Drop of Liquids in Tubes," *Ind. Eng. Chem.,* Vol. 28, p. 1429, 1936.

[1.15] Hausen, H., "Darstellung des Wärmeüberganges in Rohren durch vergallgemeinerte Potenzbeziehungen," *Z.A.V.D.I. Beihefte Verfahrenstech.,* No. 4, p. 91, 1943.

[1.16] Dittus, F. W., and Boelter, L. M. K., *Univ. Calif., Berkeley, Publ. Eng.,* Vol. 2, p. 443, 1930.

[1.17] Nusselt, W., "Der Wärmeaustausch Zwischen Wand und Wasser im Roher," *Forsch. Geb. Ingenieurwes.,* Vol. 2, p. 309, 1931.

[1.18] Sleicher, C. A., and Rouse, M. W., "A Convenient Correlation for Heat Transfer to Constant and Variable Property Fluids in Turbulent Pipe Flow," *Int. J. Heat Mass Transfer,* Vol. 18, pp. 677–683, 1975.

[1.19] Chen, C. J., and Chiou, J. S., "Laminar and Turbulent Heat Transfer in the Pipe Entrance for Liquid Metals," *Int. J. Heat Mass Transfer,* Vol. 24, pp. 1179–1190, 1981.

[1.20] Petukhov, B. S., "Heat Transfer and Friction in Turbulent Pipe Flow with Variable Physical Properties," *Advances in Heat Transfer,* T. F. Irvine and J. P. Hartnett (eds.), Vol. 6, pp. 503–564, 1970.

[1.21] Gnielinski, V., "New Equations for Heat and Mass Transfer in Turbulent Pipe and Channel Flow," *Int. Chem. Eng.,* Vol. 16, pp. 359–368, 1976.

[1.22] Dittus, F. W., and Boelter, L. M. K., "Heat Transfer in Automobile Radiators of the Tubular Type," *Univ. Calif. Publ. Eng.,* Vol. 2, pp. 443–461, 1930.

[1.23] Colburn, A. P., "A Method of Correlating Forced Convection Heat Transfer Data and a Comparison with Fluid Friction," *Trans. AIChE,* Vol. 29, pp. 174–210, 1933.

[1.24] Kays, W. M., and Crawford, M. E., *Convective Heat and Mass Transfer,* 2nd ed., McGraw-Hill, New York, 1980.

[1.25] Kakac, S., Shah, R. K., and Aung, W., *Handbook of Single-Phase Convective Heat Transfer,* John Wiley, New York, 1987.

[1.26] Shah, R. K., and London, A. L., "Laminar Flow Forced Convention in Ducts," *Advances in Heat Transfer,* Academic Press, New York, 1978.

[1.27] Lundgren, T. S., Sparrow, E. M., and Starr, J. B., "Pressure Drop Due to the Entrance Region in Ducts of Arbitrary Cross Section," *J. Basic Eng.,* Vol. 86, pp. 620–626, 1964.

[1.28] Lundberg, R. E., McCuen, P. A., and Reynolds, W. C., "Heat Transfer in Annular Passages: Hydrodynamically Developed Laminar Flow with Arbitrarily Prescribed Wall Temperatures or Heat Fluxes," *Int. J. Heat Mass Transfer,* Vol. 6, pp. 495–529, 1963.

[1.29] Kays, W. M., and Leung, E. Y., "Heat Transfer in Annular Passages: Hydrodynamically Developed Turbulent Flow with Arbitrarily Prescribed Heat Flux," *Int. J. Heat Mass Transfer,* Vol. 6, pp. 537–557, 1963.

Chapter 2

Equations of Fluid Motion for Inviscid Flows

2.1 Introductory Remarks

The equations of fluid motion reviewed in Chapter 1 are valid over the entire domain, with limited restrictions. The governing equations are classified as nonlinear and, in most cases, second-order partial differential equations. For a majority of problems of interest, analytical solutions of the equations of fluid motion do not exist, and the numerical solutions of equations in general require large computation time and computer memory. From a practical point of view, these equations may be reduced by imposing additional assumptions, limiting the applicability of the equations to a certain class of problems, and thereby simplifying the solution procedure. One traditional approach is to divide the flowfield into two regions. A region near the surface where viscous forces are important and the corresponding terms must be retained in the governing equations. This region is known as the *boundary layer*. The boundary layer equations and associated assumptions will be reviewed in Chapter 3. A second region, where the viscous forces are small compared to inertia and pressure forces, is called the *inviscid region*. In this region, viscous forces, which are represented by viscous stresses in the equations of motion, are dropped. Note that the fluid has the same viscosity everywhere; however, the product of viscosity and velocity gradient which produces shear stresses is negligible in the inviscid region of the flowfield.

2.2 Euler's Equations

The continuity equation does not include any viscous term, and, therefore, it is not affected by the assumption of inviscid flow. Hence, it remains unchanged, i.e.,

$$\frac{\partial \rho}{\partial t} + \nabla \cdot (\rho \, \vec{V}) = 0 \tag{2.1}$$

However, the momentum equation is reduced to

$$\rho \frac{d\vec{V}}{dt} = \rho \vec{f} - \nabla p \tag{2.2}$$

This equation is known as the *Euler's equation*. Once the viscous terms are dropped, the energy equation may be expressed as

$$\frac{\partial}{\partial t}(\rho e_t) + \nabla \cdot (\rho e_t \vec{V}) = \rho(\vec{V} \cdot \vec{f}) - \nabla \cdot (p\vec{V}) \tag{2.3}$$

Note that, in addition to dropping the viscous terms, it is assumed that the internal heat generation term is zero.

The equations of motion for an inviscid flow may be expressed in a compact vector form in various coordinate systems. In the following formulations it is assumed that the body force is negligible. The compact vector form in Cartesian coordinates is

$$\frac{\partial Q}{\partial t} + \frac{\partial E}{\partial x} + \frac{\partial F}{\partial y} + \frac{\partial G}{\partial z} = 0 \tag{2.4}$$

where

$$Q = \begin{bmatrix} \rho \\ \rho u \\ \rho v \\ \rho w \\ \rho e_t \end{bmatrix} \tag{2.5a} \qquad\qquad E = \begin{bmatrix} \rho u \\ \rho u^2 + p \\ \rho u v \\ \rho u w \\ \rho u e_t + p u \end{bmatrix} \tag{2.5b}$$

$$F = \begin{bmatrix} \rho v \\ \rho u v \\ \rho v^2 + p \\ \rho v w \\ \rho v e_t + p v \end{bmatrix} \tag{2.5c} \qquad\qquad G = \begin{bmatrix} \rho w \\ \rho u w \\ \rho v w \\ \rho w^2 + p \\ \rho w e_t + p w \end{bmatrix} \tag{2.5d}$$

In cylindrical coordinates we have

$$\frac{\partial Q}{\partial t} + \frac{\partial E}{\partial z} + \frac{\partial F}{\partial r} + \frac{\partial G}{\partial \theta} + H = 0 \tag{2.6}$$

where

$$Q = \begin{bmatrix} \rho \\ \rho u_z \\ \rho u_r \\ \rho u_\theta \\ \rho e_t \end{bmatrix} \tag{2.7a} \qquad\qquad E = \begin{bmatrix} \rho u_z \\ \rho u_z^2 + p \\ \rho u_z u_r \\ \rho u_z u_\theta \\ \rho u_z e_t + p u_z \end{bmatrix} \tag{2.7b}$$

$$F = \begin{bmatrix} \rho u_r \\ \rho u_r u_z \\ \rho u_r^2 + p \\ \rho u_r u_\theta \\ \rho u_r e_t + p u_r \end{bmatrix} \qquad (2.7\text{c})$$

$$G = \frac{1}{r} \begin{bmatrix} \rho u_\theta \\ \rho u_\theta u_z \\ \rho u_\theta u_r \\ \rho u_\theta^2 + p \\ \rho u_\theta e_t + p u_\theta \end{bmatrix} \qquad (2.7\text{d})$$

$$H = \frac{1}{r} \begin{bmatrix} \rho u_r \\ \rho u_r u_z \\ \rho(u_r^2 - u_\theta^2) \\ 2\rho u_\theta u_r \\ \rho u_r e_t + p u_r \end{bmatrix} \qquad (2.7\text{e})$$

2.3 Bernoulli's Equation

The Euler's equation may be further simplified if additional assumptions are imposed. When the body force is assumed to be a conservative force, then $\vec{f}$ can be expressed as a gradient of a scalar as:

$$\vec{f} = -\nabla F$$

Recall that a force is defined as conservative if the work produced is independent of the path, an example of which is the gravitational force. If we further assume steady flow, Euler's equation can be integrated along a streamline to yield

$$\int d\left(\frac{V^2}{2}\right) + \int dF + \int \frac{dp}{\rho} = \text{constant} \qquad (2.8)$$

At this point, two additional conditions are imposed. First, assume gravity is the only body force and that it is acting along the z-direction; then $f = gz$. Second, the flow is baratropic, i.e., density is a function of pressure only. Now, one may integrate Equation (2.8) for two cases.

1. *Incompressible flow,* where ρ is constant, then (2.8) yields

$$\frac{V^2}{2} + gz + \frac{p}{\rho} = \text{constant} \qquad (2.9)$$

2. *Isentropic flow,* where $\dfrac{p}{\rho^\gamma} = \text{constant}$, for which (2.8) yields

$$\frac{V^2}{2} + gz + \frac{\gamma}{\gamma - 1}\frac{p}{\rho} = \text{constant} \qquad (2.10)$$

Equation (2.10) is valid for compressible flow, while Equation (2.9) is restricted to incompressible flows only. Recall that both Equations (2.9) and (2.10) are valid along a streamline only. This restriction can be removed if one assumes irrotational flow, in which case Bernoulli's equation is valid everywhere. With the assumption of irrotational flow, one may express the velocity vector as a gradient of a scalar known as the *velocity potential function*. If one chooses Φ to represent the potential function, then $\vec{V} = \nabla\Phi$, where Φ, in general, is a function of time and space. Now, one may write the unsteady Bernoulli's equation for an incompressible flow as

$$\frac{\partial \Phi}{\partial t} + \frac{1}{2}V^2 + gz + \frac{p}{\rho} = F(t)$$

A summary of various forms of the Bernoulli's equation is given in Table 2.1.

Table 2.1: Bernoulli's equation

1. Steady, constant gravitational force

   (a) Along a streamline

   $$\int \frac{dp}{\rho} + \frac{1}{2}V^2 + gz = C$$

   (b) Irrotational flow

   $$\int \frac{dp}{\rho} + \frac{1}{2}V^2 + gz = C$$

2. Steady, constant gravitational force, isentropic, along a streamline

   $$\frac{\gamma}{\gamma - 1}\frac{p}{\rho} + \frac{1}{2}V^2 + gz = C$$

3. Steady, constant gravitational force, incompressible

   $$\frac{p}{\rho} + \frac{1}{2}V^2 + gz = C$$

4. Constant gravitational force, incompressible, irrotational

   $$\frac{\partial \Phi}{\partial t} + \frac{p}{\rho} + \frac{1}{2}V^2 + gz = F(t)$$

where $\vec{V} = \nabla\Phi$

2.4 Velocity Potential Function

For an irrotational flow defined by $\nabla \times \vec{V} = 0$, one may introduce a velocity potential function Φ as follows

$$\vec{V} = \nabla \Phi$$

The components of the velocity vector in various coordinate systems are provided in Table 2.2.

The velocity potential functions for various elementary flows are given in Table 2.3, and the corresponding velocity components are provided in Table 2.4. The coordinate systems are illustrated in Figure 2.1.

Table 2.2: Velocity components in terms of velocity potential function

1. Cartesian coordinates (x,y,z)

$$u = \frac{\partial \Phi}{\partial x} = \Phi_x \qquad v = \frac{\partial \Phi}{\partial y} = \Phi_y \qquad w = \frac{\partial \Phi}{\partial z} = \Phi_z$$

2. Cylindrical coordinates (r,θ,z)

$$u_r = \frac{\partial \Phi}{\partial r} \qquad u_\theta = \frac{1}{r}\frac{\partial \Phi}{\partial \theta} \qquad u_z = \frac{\partial \Phi}{\partial z}$$

3. Spherical coordinates (r,θ,ϕ)

$$u_r = \frac{\partial \Phi}{\partial r} \qquad u_\theta = \frac{1}{r}\frac{\partial \Phi}{\partial \theta} \qquad u_\phi = \frac{1}{r \sin\theta}\frac{\partial \Phi}{\partial \phi}$$

Table 2.3: Velocity potential functions for elementary flows

Flow	Planar Two-Dimensional		Axisymmetric
	Cartesian (x, y)	Polar (r, θ)	Spherical (r, θ)
Uniform flow from left	$\Phi = u_\infty x$	$\Phi = u_\infty r \cos\theta$	$\Phi = u_\infty r \cos\theta$
Source of quantity Q located at the origin	$\Phi = \dfrac{Q}{2\pi} \ln(x^2 + y^2)^{1/2}$	$\Phi = \dfrac{Q}{2\pi} \ln r$	$\Phi = \dfrac{Q}{4\pi} \dfrac{1}{r}$
Doublet of strength μ located at the origin	$\Phi = \dfrac{\mu}{2\pi} \dfrac{x}{(x^2 + y^2)}$	$\Phi = \dfrac{\mu}{2\pi} \dfrac{\cos\theta}{r}$	$\Phi = \dfrac{\mu}{4\pi} \dfrac{\cos\theta}{r^2}$
Vortex of strength Γ located at the origin with clockwise circulation	$\Phi = -\dfrac{\Gamma}{2\pi} \tan^{-1}\dfrac{y}{x}$	$\Phi = -\dfrac{\Gamma}{2\pi}\theta$	

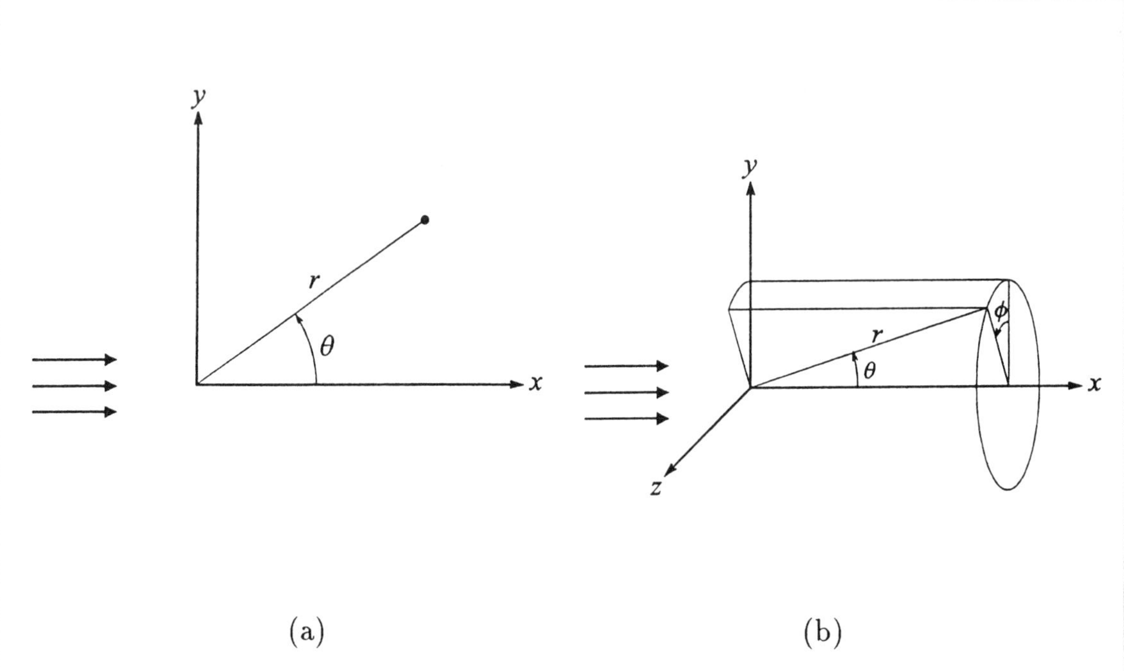

Figure 2.1   Schematic of (a) planar two-dimensional Cartesian/polar coordinate system, and (b) axisymmetric spherical coordinate system.

Table 2.4: Velocity components for elementary flows

Flow	Planar Two-Dimensional		Axisymmetric
	Cartesian (x, y)	Polar (r, θ)	Spherical (r, θ)
Uniform flow from left	$u = u_\infty$ $v = 0$	$u_r = u_\infty \cos\theta$ $u_\theta = -u_\infty \sin\theta$	$u_r = u_\infty \cos\theta$ $u_\theta = -u_\infty \sin\theta$
Source of quantity Q located at the origin	$u = \dfrac{Q}{2\pi}\dfrac{x}{(x^2+y^2)}$ $v = \dfrac{Q}{2\pi}\dfrac{y}{x^2+y^2}$	$u_r = \dfrac{Q}{2\pi r}$ $u_\theta = 0$	$u_r = -\dfrac{Q}{4\pi}\dfrac{1}{r^2}$ $u_\theta = 0$
Doublet of strength μ located at the origin	$u = -\dfrac{\mu}{2\pi}\dfrac{x^2-y^2}{(x^2+y^2)^2}$ $v = -\dfrac{\mu}{\pi}\dfrac{xy}{(x^2+y^2)^2}$	$u_r = -\dfrac{\mu}{2\pi}\dfrac{\cos\theta}{r^2}$ $u_\theta = -\dfrac{\mu}{2\pi}\dfrac{\sin\theta}{r^2}$	$u_r = -\dfrac{\mu}{2\pi}\dfrac{\cos\theta}{r^3}$ $u_\theta = -\dfrac{\mu}{4\pi}\dfrac{\sin\theta}{r^3}$
Vortex of strength Γ located at the origin with clockwise circulation	$u = \dfrac{\Gamma}{2\pi}\dfrac{y}{x^2+y^2}$ $v = -\dfrac{\Gamma}{2\pi}\dfrac{x}{x^2+y^2}$	$u_r = 0$ $u_\theta = -\dfrac{\Gamma}{2\pi r}$	

2.4.1 Velocity Potential Equation

For an irrotational flow, one may combine the continuity equation with the inviscid Euler's equation to form an equation in terms of the velocity potential. The assumptions made in the process of obtaining the velocity potential equation are: steady-state, irrotational, inviscid, and isentropic. In Cartesian coordinates, the velocity potential equation is expressed as

$$\left(1 - \frac{\Phi_x^2}{a^2}\right)\Phi_{xx} + \left(1 - \frac{\Phi_y^2}{a^2}\right)\Phi_{yy} + \left(1 - \frac{\Phi_z^2}{a^2}\right)\Phi_{zz} - 2\frac{\Phi_x \Phi_y}{a^2}\Phi_{xy}$$

$$- 2\frac{\Phi_y \Phi_z}{a^2}\Phi_{yz} - 2\frac{\Phi_x \Phi_z}{a^2}\Phi_{xz} = 0 \tag{2.11}$$

where $\Phi_x = u$, $\Phi_y = v$, and $\Phi_z = w$. The local speed of sound a in Equation (2.11) may be obtained from the energy equation for isentropic perfect gas and is

$$a^2 = a_t^2 - \frac{\gamma - 1}{2}\left(\Phi_x^2 + \Phi_y^2 + \Phi_z^2\right) \tag{2.12}$$

The velocity potential equation may be further reduced with the assumption of incompressible flow, for which Laplace's equation will result. Thus, for an incompressible flow,

$$\frac{\partial^2 \Phi}{\partial x^2} + \frac{\partial^2 \Phi}{\partial y^2} + \frac{\partial^2 \Phi}{\partial z^2} = 0 \qquad (2.13)$$

The velocity potential equation in different coordinate systems is provided in Tables 2.5, 2.6, and 2.7 for three-dimensional compressible flows, for two-dimensional planar and axisymmetric flows, and for incompressible flows, respectively.

Table 2.5: Velocity potential equation for a compressible flow

1. Cartesian coordinates (x,y,z)

$$\left(1 - \frac{\Phi_x^2}{a^2}\right)\Phi_{xx} + \left(1 - \frac{\Phi_y^2}{a^2}\right)\Phi_{yy} + \left(1 - \frac{\Phi_z^2}{a^2}\right)\Phi_{zz} - 2\frac{\Phi_x\Phi_y}{a^2}\Phi_{xy}$$

$$- 2\frac{\Phi_y\Phi_z}{a^2}\Phi_{yz} - 2\frac{\Phi_x\Phi_z}{a^2}\Phi_{xz} = 0$$

2. Cylindrical coordinates (r,θ,z)

$$\left(1 - \frac{\Phi_r^2}{a^2}\right)\Phi_{rr} + \left(1 - \frac{\Phi_\theta^2}{r^2a^2}\right)\frac{\Phi_{\theta\theta}}{r^2} + \left(1 - \frac{\Phi_z^2}{a^2}\right)\Phi_{zz}$$

$$- 2\frac{\Phi_r\Phi_\theta}{r^2a^2}\Phi_{r\theta} - 2\frac{\Phi_\theta\Phi_z}{r^2a^2}\Phi_{\theta z} - 2\frac{\Phi_z\Phi_r}{a^2}\Phi_{zr} + \left(1 + \frac{\Phi_\theta^2}{r^2a^2}\right)\frac{\Phi_r}{r} = 0$$

$$a^2 = a_t^2 - \frac{\gamma-1}{2}\left(\Phi_r^2 + \frac{\Phi_\theta^2}{r^2} + \Phi_z^2\right)$$

3. Spherical coordinates (r,θ,ϕ)

$$\left(1 - \frac{\Phi_r^2}{a^2}\right)\Phi_{rr} + \left(1 - \frac{\Phi_\theta^2}{r^2a^2}\right)\frac{\Phi_{\theta\theta}}{r^2} + \left(1 - \frac{\Phi_\phi^2}{a^2r^2\sin^2\theta}\right)\frac{\Phi_{\phi\phi}}{r^2\sin^2\theta}$$

$$- 2\frac{\Phi_r\Phi_\theta}{a^2r^2}\Phi_{r\theta} - 2\frac{\Phi_r\Phi_\phi}{a^2r^2\sin^2\theta}\Phi_{r\phi} - 2\frac{\Phi_\theta\Phi_\phi}{a^2r^4\sin^2\theta}\Phi_{\theta\phi}$$

$$+ \left(2 + \frac{\Phi_\theta^2}{a^2r^2} + \frac{\Phi_\phi^2}{a^2r^2\sin^2\theta}\right)\frac{\Phi_r}{r} + \left(1 + \frac{\Phi_\phi^2}{a^2r^2\sin^2\theta}\right)\frac{\Phi_\theta\cot\theta}{r^2} = 0$$

$$a^2 = a_t^2 - \frac{\gamma-1}{2}\left(\Phi_r^2 + \frac{\Phi_\theta^2}{r^2} + \frac{\Phi_\phi^2}{r^2\sin^2\theta}\right)$$

Table 2.6: Velocity potential equation for two-dimensional planar and axisymmetric flows

1. Planar flow, Cartesian coordinates (x,y)

$$\left(1 - \frac{\Phi_x^2}{a^2}\right)\Phi_{xx} + \left(1 - \frac{\Phi_y^2}{a^2}\right)\Phi_{yy} - 2\frac{\Phi_x\Phi_y}{a^2}\Phi_{xy} = 0$$

$$a^2 = a_t^2 - \frac{\gamma - 1}{2}\left(\Phi_x^2 + \Phi_y^2\right)$$

2. Planar flow, cylindrical coordinates (r,θ)

$$\left(1 - \frac{\Phi_r^2}{a^2}\right)\Phi_{rr} + \left(1 - \frac{\Phi_\theta^2}{r^2 a^2}\right)\frac{\Phi_{\theta\theta}}{r^2} - 2\frac{\Phi_r\Phi_\theta}{r^2 a^2}\Phi_{r\theta} + \frac{\Phi_r}{r}\left(1 + \frac{\Phi_\theta^2}{r^2 a^2}\right) = 0$$

$$a^2 = a_t^2 - \frac{\gamma - 1}{2}\left(\Phi_r^2 + \frac{\Phi_\theta^2}{r^2}\right)$$

3. Axisymmetric flow, cylindrical coordinates (r,z), flow along the z direction.

$$\left(1 - \frac{\Phi_r^2}{a^2}\right)\Phi_{rr} + \left(1 - \frac{\Phi_z^2}{a^2}\right)\Phi_{zz} - 2\frac{\Phi_r\Phi_z}{a^2}\Phi_{rz} + \frac{\Phi_r}{r} = 0$$

$$a^2 = a_t^2 - \frac{\gamma - 1}{2}\left(\Phi_r^2 + \Phi_z^2\right)$$

Table 2.7: Velocity potential equation for an incompressible flow

1. Cartesian coordinates (x,y,z)

$$\frac{\partial^2\Phi}{\partial x^2} + \frac{\partial^2\Phi}{\partial y^2} + \frac{\partial^2\Phi}{\partial z^2} = 0$$

2. Cylindrical coordinates (r,θ,z)

$$\frac{\partial^2\Phi}{\partial r^2} + \frac{\partial^2\Phi}{r^2\partial\theta^2} + \frac{\partial^2\Phi}{\partial z^2} + \frac{\partial\Phi}{r\partial r} = 0$$

3. Spherical coordinates (r,θ,ϕ)

$$\frac{\partial^2\Phi}{\partial r^2} + \frac{\partial^2\Phi}{r^2\partial\theta^2} + \frac{1}{r^2\sin^2\theta}\frac{\partial^2\Phi}{\partial\phi^2} + 2\frac{\partial\Phi}{r\partial r} + \frac{\cot\theta}{r^2}\frac{\partial\Phi}{\partial\theta} = 0$$

2.4.2 Velocity Potential Equation for Small Perturbations

The full velocity potential equation can be simplified for slender bodies. The assumption imposed for the reduction of the equation is that the flow is perturbed slightly due to the presence of the body. If we denote the disturbed velocity components by a prime and the corresponding perturbed velocity potential by φ, we may write

$$u = u_\infty + u' = u_\infty + \varphi_x \tag{2.14a}$$

$$v = v' = \varphi_y \tag{2.14b}$$

$$w = w' = \varphi_z \tag{2.14c}$$

The energy equation may be expressed in terms of Mach number as

$$\frac{a^2}{a_\infty^2} = 1 - \frac{\gamma - 1}{2} M_\infty^2 \left(2\frac{u'}{u_\infty} + \frac{u'^2 + v'^2 + w'^2}{u_\infty^2} \right) \tag{2.15}$$

Substitution of the velocity components given by relations (2.14a) through (2.14c) into Equation (2.15) and application of the binomial theorem yield

$$\frac{a_\infty^2}{a^2} = 1 + \frac{\gamma - 1}{2} M_\infty^2 \left(2\frac{u'}{u_\infty} + \frac{u'^2 + v'^2 + w'^2}{u_\infty^2} \right) + O\left(M_\infty^4 \frac{u'^2}{u_\infty^2} \right) + \cdots$$

With the substitution of this equation and the perturbation velocity components into the full velocity potential equation, we obtain, after some rearrangements

$$(1 - M_\infty^2) \frac{\partial u'}{\partial x} + \frac{\partial v'}{\partial y} + \frac{\partial w'}{\partial z} = M_\infty^2 \left[(\gamma + 1) \frac{u'}{u_\infty} + \frac{\gamma + 1}{2} \frac{u'^2}{u_\infty^2} \right.$$

$$\left. + \frac{\gamma - 1}{2} \frac{v'^2 + w'^2}{u_\infty^2} \right] \frac{\partial u'}{\partial x} +$$

$$M_\infty^2 \left[(\gamma - 1) \frac{u'}{u_\infty} + \frac{\gamma + 1}{2} \frac{v'^2}{u_\infty^2} + \frac{\gamma - 1}{2} \frac{w'^2 + u'^2}{u_\infty^2} \right] \frac{\partial v'}{\partial y} +$$

$$M_\infty^2 \left[(\gamma - 1) \frac{u'}{u_\infty} + \frac{\gamma + 1}{2} \frac{w'^2}{u_\infty^2} + \frac{\gamma - 1}{2} \frac{u'^2 + v'^2}{u_\infty^2} \right] \frac{\partial w'}{\partial z} +$$

$$M_\infty^2 \left[\frac{v'}{u_\infty} \left(1 + \frac{u'}{u_\infty} \right) \left(\frac{\partial u'}{\partial y} + \frac{\partial v'}{\partial x} \right) + \frac{w'}{u_\infty} \left(1 + \frac{u'}{u_\infty} \right) \left(\frac{\partial u'}{\partial z} + \frac{\partial w'}{\partial x} \right) + \right.$$

$$\left. \frac{v'w'}{u_\infty^2} \left(\frac{\partial w'}{\partial y} + \frac{\partial v'}{\partial x} \right) \right] + \cdots$$

If higher order terms are dropped, the equation above is reduced to

$$(1 - M_\infty^2)\frac{\partial u'}{\partial x} + \frac{\partial v'}{\partial y} + \frac{\partial w'}{\partial z} = M_\infty^2(\gamma + 1)\frac{u'}{u_\infty}\frac{\partial u'}{\partial x} \tag{2.16}$$

This equation is valid for subsonic, transonic, and supersonic flows. In terms of the perturbed velocity potential function, one may write

$$(1 - M_\infty^2)\varphi_{xx} + \varphi_{yy} + \varphi_{zz} = \frac{M_\infty^2}{u_\infty}(\gamma + 1)\varphi_x\varphi_{xx} \tag{2.17}$$

This equation is still nonlinear due to the term $\varphi_x\varphi_{xx}$. If an additional condition is imposed where one assumes

$$(1 - M_\infty^2)\varphi_{xx} \gg \frac{M_\infty^2}{u_\infty}(\gamma + 1)\varphi_x\varphi_{xx}$$

then Equation (2.17) is reduced to a linear form as

$$(1 - M_\infty^2)\varphi_{xx} + \varphi_{yy} + \varphi_{zz} = 0 \tag{2.18}$$

This is known as the *linearized velocity potential equation* and is valid only for subsonic flow ($M_\infty \leq 0.8$) and supersonic flow ($M_\infty \geq 1.2$).

2.5   Thin Airfoil Theory

Simple expressions for aerodynamic parameters of an airfoil can be developed by superimposing the potential functions for a uniform flow and a vortex sheet. In the following sections, some definitions and governing equations are presented.

2.5.1 Definitions

2.5.1.1 Circulation: Circulation is defined as the line integral of velocity around a closed curve, mathematically

$$\Gamma = -\oint \vec{V} \cdot \vec{dr}$$

2.5.1.2 Vorticity: Vorticity is defined as twice the angular velocity,

$$\vec{\Omega} = 2\,\vec{\omega} = \nabla \times \vec{V}$$

2.5.1.3 Irrotational flow: A flow for which vorticity is zero is called an irrotational flow. For an irrotational flow $\nabla \times \vec{V} = 0$

2.5.1.4 Stoke's theorem: A relation between line integral and surface integral given by the expression below is known as the Stoke's theorem.

$$\oint \vec{V} \cdot \vec{dr} = \int_S (\nabla \times \vec{V}) \cdot \vec{n} \, dS$$

2.5.1.5 Vortex sheet: A sheet composed of an infinite number of line vortices with each having an infinitesimally small strength. The local strength of vortex sheet (γ) is equal to the difference in tangential velocities across the sheet, mathematically,

$$\gamma = u_1 - u_2$$

2.5.1.6 Kutta-Joukowski theorem: The Kutta-Joukowski theorem establishes a relation between lift and circulation for an inviscid, incompressible flow as follows

$$L = \rho_\infty u_\infty \Gamma$$

where Γ is the circulation.

2.5.1.7 Kutta condition: In order to establish a steady flow around an airfoil, a specific value of circulation must be developed which results in flow leaving the trailing edge smoothly. This requirement is called the *Kutta condition* and is mathematically expressed as

$$\gamma(TE) = 0.0$$

2.5.2 Governing Equations

Thin airfoil theory is developed based on the following assumptions: (1) steady, incompressible flow; (2) two-dimensional; (3) inviscid; (4) airfoil operating at small angles of attack, ($|\alpha| < 10° \sim 12°$); (5) flowfield slightly influenced by the airfoil thickness, small flap, and slat deflections; and (6) small camber and thickness. To derive the appropriate governing equations, the airfoil camber line is represented by a vortex sheet. The strength of the vortex sheet is determined such that the camber line must be a streamline in a uniform flow of velocity u_∞ and at an angle of attack α. The governing equation is given by

$$\frac{1}{2\pi} \int_o^c \frac{\gamma(\xi)d\xi}{x - \xi} = u_\infty \left(\alpha - \frac{dz}{dx} \right) \tag{2.19}$$

where $z = z(x)$ describes the equation for the camber line. Equation (2.19) is solved subject to Kutta condition, $\gamma(c) = 0.0$. Thin airfoil theory provides the following results.

2.5.2.1 Symmetric airfoil:

$$C_l = 2\pi\alpha$$

$$x_{cp} = x_{ac} = \frac{c}{4}$$

$$Cm_{c/4} = Cm_{ac} = 0.0$$

2.5.2.2 Cambered airfoil:

$$C_l = \pi(2A_0 + A_1)$$

$$x_{ac} = \frac{c}{4}$$

$$x_{cp} = \frac{c}{4}\left[1 + \frac{\pi}{C_l}(A_1 - A_2)\right]$$

$$Cm_{l.e.} = -\frac{\pi}{2}\left(A_0 + A_1 - \frac{A_2}{2}\right) , \quad Cm_{l.e.} = Cm_{c/4} - \frac{C_l}{4}$$

$$Cm_{c/4} = \frac{\pi}{4}(A_2 - A_1)$$

$$\alpha_{OL} = -\frac{1}{\pi}\int_o^\pi \frac{dz}{dx}(\cos\theta_o - 1)d\theta_o , \quad \alpha_{OL} = \alpha - \frac{C_l}{2\pi}$$

$$A_0 = \alpha - \frac{1}{\pi}\int_o^\pi \frac{dz}{dx}d\theta_o$$

$$A_n = \frac{2}{\pi}\int_o^\pi \frac{dz}{dx}\cos n\theta_o d\theta_o$$

where

$$x = \frac{c}{2}(1 - \cos\theta_o)$$

2.5.2.3 Thin airfoil with flap/slat:

Define by θ_s and θ_f the locations of hinge point for slat and hinge point for flap, respectively, then

$$E_s c = \frac{c}{2}(1 - \cos\theta_s)$$

$$(1 - E_f)c = \frac{c}{2}(1 - \cos\theta_f)$$

$$A_0 = \delta_f - \frac{1}{\pi}(\delta_s\theta_s + \delta_f\theta_f)$$

$$A_1 = \frac{2}{\pi}(\delta_s \sin \theta_s + \delta_f \sin \theta_f)$$

$$A_2 = \frac{1}{\pi}(\delta_s \sin 2\theta_s + \delta_f \sin 2\theta_f)$$

All angles are in radians, and the nomenclature is illustrated in Figure 2.2.

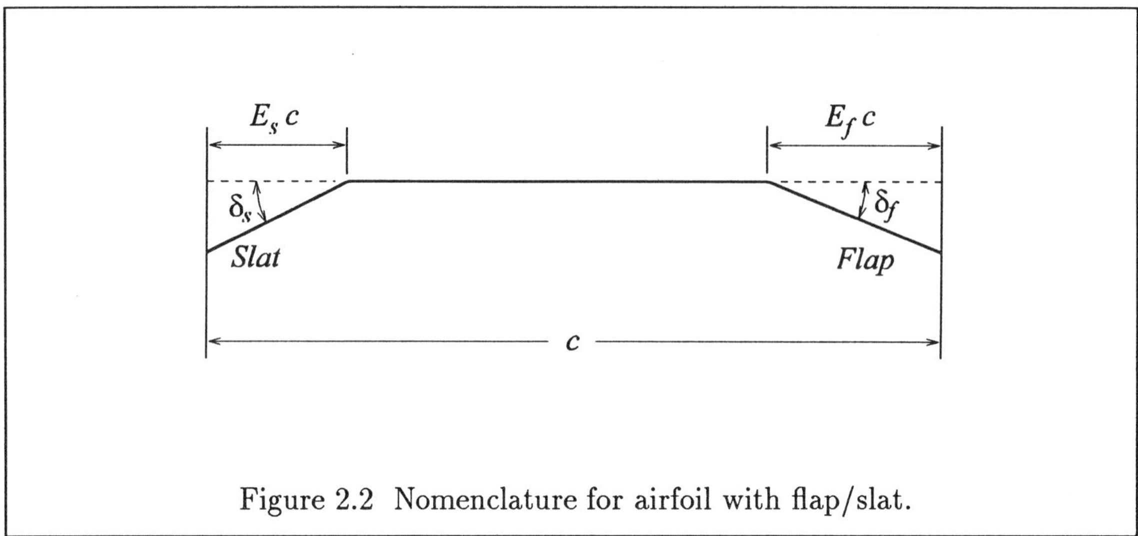

Figure 2.2 Nomenclature for airfoil with flap/slat.

2.6   Finite Wing Theory

The flowfield around a wing of finite span is three dimensional due to variations of flow properties in the chordwise and spanwise directions. Flow rolls up at the wing tips due to high pressure beneath the wing and form the tip vortices. The tip vortices create a velocity component in the downward direction along the span. In the following sections, some definitions and governing equations are reviewed.

2.6.1 Definitions

2.6.1.1 Downwash: The induced velocity in the downward direction due to the tip vortices is called *downwash* and is usually denoted by w.

2.6.1.2 Downwash angle: The angle between the free stream velocity and the local relative velocity is known as the *downwash angle* and is denoted by ε.

2.6.1.3 Effective angle of attack: The angle of attack seen by the local airfoil section, denoted by α_e.

2.6.1.4 Biot-Savart law: The velocity induced by a segment of vortex filament $\vec{dl}$ at a distance $\vec{r}$ is given by

$$\vec{dV} = \frac{\Gamma}{4\pi} \frac{\vec{dl} \times \vec{r}}{|\vec{r}|^3}$$

where Γ is the strength of vortex filament. Velocity induced at a distance h by an infinite vortex filament is given by

$$V = \frac{\Gamma}{2\pi h}$$

and the velocity induced by a semi-infinite vortex filament is

$$V = \frac{\Gamma}{4\pi h}$$

2.6.1.5 Helmholtz vortex theorem: (1) The strength of a vortex line is constant along its length. (2) A vortex line cannot end in a fluid; it must form a closed path or extend to the boundaries of the fluid (which can be at infinity). (3) A fluid which is originally irrotational will remain irrotational in the absence of rotational external forces.

2.6.2 Governing Equations

An infinite number of horseshoe vortices with their bound vortices aligned at quarter chord is used to model a finite span wing. The governing equation is known as the *fundamental equation of Prandtl's lifting line*, and is given by

$$\alpha(y_o) = \frac{\Gamma(y_o)}{\pi u_\infty c(y_o)} + \alpha_{OL}(y_o) + \frac{1}{4\pi u_\infty} \int_{-\frac{b}{2}}^{+\frac{b}{2}} \frac{\left(\dfrac{d\Gamma}{dy}\right) dy}{y_o - y} \tag{2.20}$$

The Γ distribution is computed from the equation above, given the free stream properties, operating conditions, and wing geometry. Once the Γ distribution is determined, the following parameters may be calculated.

Lift:
$$L = \rho_\infty u_\infty \int_{-\frac{b}{2}}^{+\frac{b}{2}} \Gamma(y) dy$$

Lift coefficient:
$$C_L = \frac{2}{u_\infty S} \int_{-\frac{b}{2}}^{+\frac{b}{2}} \Gamma(y) dy$$

Induced drag:
$$D_i \cong \rho_\infty u_\infty \int_{-\frac{b}{2}}^{+\frac{b}{2}} \Gamma(y)\varepsilon(y)dy$$

Induced drag coefficient:
$$C_{D_i} \cong \frac{2}{u_\infty S} \int_{-\frac{b}{2}}^{+\frac{b}{2}} \Gamma(y)\varepsilon(y)dy$$

Downwash:
$$w(y_o) = -\frac{1}{4\pi} \int_{-\frac{b}{2}}^{+\frac{b}{2}} \frac{\left(\dfrac{d\Gamma}{dy}\right)dy}{y_o - y}$$

where b is the wing span.

2.6.2.1 *Elliptic spanwise circulation distribution*

For an elliptic Γ distribution, the following results are obtained. During the mathematical manipulations, a coordinate transformation is introduced by the following relation, $y = \frac{b}{2}\cos\theta$

Circulation:
$$\Gamma(y) = \Gamma_o \left[1 - \left(\frac{2y}{b}\right)^2\right]^{1/2}$$

Downwash:
$$w(\theta_o) = -\frac{\Gamma_o}{2b}$$

Downwash angle:
$$\varepsilon = \frac{\Gamma_o}{2bu_\infty} = \frac{C_L}{\pi \mathbb{R}}$$

where $\mathbb{R} = b^2/S$ is the aspect ratio, b is the wing span, and S is the surface area of the wing including sections covered by fuselage.

Lift coefficient:
$$C_L = \frac{\pi b \Gamma_o}{2u_\infty S}$$

Induced drag coefficient:
$$C_{D_i} = \frac{C_L^2}{\pi \mathbb{R}}$$

Lift curve slopes:
$$a = \frac{a_o}{1 + \dfrac{a_o}{\pi \mathbb{R}}}$$

where a and a_o are the lift curve slopes for the finite wing and the equivalent infinite wing, respectively.

2.6.2.2 *General spanwise circulation distribution*

Circulation:
$$\Gamma(\theta) = 2bu_\infty \sum_{n=1}^{N} A_n \sin n\theta \ , \quad y = -\frac{b}{2}\cos\theta$$

Lifting line theory:
$$\alpha(\theta_o) = \frac{2b}{\pi c(\theta_o)} \sum_{n=1}^{N} A_n \sin n\theta_o + \alpha_{OL}(\theta_o) + \sum_{n=1}^{N} nA_n \frac{\sin n\theta_o}{\sin\theta_o}$$

Monoplane equation:
$$\mu(\alpha - \alpha_{OL})\sin\theta = \sum_{n=1}^{N} A_n \sin n\,\theta(\mu n + \sin\theta)$$

where $\mu = \dfrac{a_o c}{4b}$

Downwash:
$$w(\theta) = -u_\infty \frac{\sum_{n=1}^{N} n A_n \sin n\theta}{\sin\theta}$$

Downwash angle:
$$\varepsilon(\theta) = \sum_{n=1}^{N} n A_n \frac{\sin n\theta}{\sin\theta}$$

Lift coefficient: $C_L = A_1 \pi\!R$

Induced drag coefficient :
$$C_{D_i} = (1 + \delta)\frac{C_L^2}{\pi\!R} = \frac{C_L^2}{\pi e\!R}$$

where $\delta = \sum_{n=2}^{N} n\left(\dfrac{A_n}{A_1}\right)^2$ and $e = \dfrac{1}{1+\delta}$: The span efficiency factor

Lift curve slopes: $a = \dfrac{a_o}{1 + (\tau + 1)\left(\dfrac{a_o}{\pi\!R}\right)}$

τ: Glauert efficiency factor, typically in the range of 0.05 to 0.25.

The coefficients A_n are calculated from the monoplane equation (or equivalently from the lifting line theory).

2.7 Isentropic Relations

A flow which is both adiabatic and reversible is called an *isentropic flow*. For an isentropic flow, the change in entropy is zero. A set of relations between states 1 and 2 for an isentropic flow can be established as follows.

$$\frac{p_2}{p_1} = \left(\frac{T_2}{T_1}\right)^{\gamma/(\gamma-1)} \tag{2.21}$$

$$\frac{T_2}{T_1} = \left(\frac{\rho_2}{\rho_1}\right)^{\gamma-1} \tag{2.22}$$

$$\frac{p_2}{p_1} = \left(\frac{\rho_2}{\rho_1}\right)^{\gamma} \tag{2.23}$$

2.8 Stagnation Properties

The location where the flow is brought to rest isentropically is known as the *stagnation point*. Subscripts t or o are typically used to designate the flow properties at the stagnation point. The flow properties at the stagnation point can be determined from the following relations. Relations (2.24) through (2.26) hold for adiabatic flow, whereas isentropic relations have been used in Equations (2.27) and (2.28). Perfect gas assumption has been incorporated in relations (2.25) through (2.28).

Stagnation enthalpy

$$h_t = h + \frac{1}{2}V^2 \tag{2.24}$$

Stagnation temperature

$$T_t = T\left(1 + \frac{\gamma - 1}{2}M^2\right) \tag{2.25}$$

Stagnation acoustic speed

$$a_t^2 = a^2 + \frac{\gamma - 1}{2}V^2 \tag{2.26}$$

Stagnation pressure

$$p_t = p\left(1 + \frac{\gamma - 1}{2}M^2\right)^{\gamma/(\gamma-1)} \tag{2.27}$$

Stagnation density

$$\rho_t = \rho\left(1 + \frac{\gamma - 1}{2}M^2\right)^{1/(\gamma-1)} \tag{2.28}$$

The ratio of static to stagnation properties for air where $\gamma = 1.4$ is tabulated and given in Tables 2.13 and 2.14 at the end of this chapter.

2.9 Mach Number

Mach number is defined as the velocity over the speed of sound,

$$M = \frac{V}{a}$$

2.9.1 Characteristic Mach Number

It is convenient to use the flow properties at the stagnation point or at the sonic point as reference properties. It is common to denote the properties at sonic condition by an asterisk. For example, the temperature at sonic condition is denoted by T^*, and the speed of sound for a perfect gas is $a^* = \sqrt{\gamma R T^*}$.

Now, one may define a characteristic Mach number based on a^* as follows.

$$\overline{M} = \frac{V}{a^*}$$

The characteristic Mach number can be related to the Mach number M by the following relation.

$$\overline{M}^2 = \frac{(\gamma+1)M^2}{2+(\gamma-1)M^2}$$

Note that, as $M \to \infty$, $\overline{M} \to \sqrt{\dfrac{\gamma+1}{\gamma-1}}$

Expressions for M, p/p_t, T/T_t, and ρ/ρ_t in terms of $\overline{M}$ are given by the following.

$$M^2 = \frac{\left(\dfrac{2}{\gamma+1}\right)\overline{M}^2}{1-\left(\dfrac{\gamma-1}{\gamma+1}\right)\overline{M}^2} \tag{2.29}$$

$$\frac{T}{T_t} = \left(1 - \frac{\gamma-1}{\gamma+1}\overline{M}^2\right) \tag{2.30}$$

$$\frac{p}{p_t} = \left(1 - \frac{\gamma-1}{\gamma+1}\overline{M}^2\right)^{\gamma/(\gamma-1)} \tag{2.31}$$

$$\frac{\rho}{\rho_t} = \left(1 - \frac{\gamma-1}{\gamma+1}\overline{M}^2\right)^{1/(\gamma-1)} \tag{2.32}$$

The relations between the flow properties at sonic and stagnation conditions and the corresponding values for air where $\gamma = 1.4$ are

$$\frac{T^*}{T_t} = \frac{a^{*2}}{a_t^2} = \frac{2}{\gamma+1}\,, \quad \frac{T^*}{T_t} = 0.8333 \tag{2.33}$$

$$\frac{p^*}{p_t} = \left(\frac{2}{\gamma+1}\right)^{\gamma/(\gamma-1)}\,, \quad \frac{p^*}{p_t} = 0.5283 \tag{2.34}$$

$$\frac{\rho^*}{\rho_t} = \left(\frac{2}{\gamma+1}\right)^{1/(\gamma-1)}\,, \quad \frac{\rho^*}{\rho_t} = 0.6339 \tag{2.35}$$

2.10   Crocco's Theorem

A relation for the variation of entropy along a streamline can be established by the combination of Euler's equation, the energy equation, and the second law of thermodynamics. This equation in a vector form is given by

$$T\nabla s = \frac{\partial \vec{V}}{\partial t} + \nabla h_t - \vec{V} \times (\nabla \times \vec{V}) \tag{2.36}$$

and is called *Crocco's theorem*. The stagnation enthalpy is given by

$$h_t = h + \frac{1}{2}V^2 \tag{2.37}$$

For a steady flow, Equation (2.36) is reduced to

$$T\nabla s = \nabla h_t - \vec{V} \times (\nabla \times \vec{V}) \tag{2.38}$$

Crocco's theorem for a two-dimensional flow, where n designates the direction normal to the streamline, is given by

$$\Omega_z = \frac{1}{V}\left(T\frac{\partial s}{\partial n} - \frac{\partial h_t}{\partial n}\right)$$

2.11   Entropy

The change in entropy for a thermally perfect gas is given by

$$ds = c_p\frac{dT}{T} - R\frac{dp}{p} \tag{2.39}$$

Equation (2.39) can be integrated between states 1 and 2 to provide a change in entropy as follows.

$$\Delta s = \int_{T_1}^{T_2} c_p\frac{dT}{T} - R\ln\frac{p_2}{p_1} \tag{2.40}$$

where, for a calorically perfect gas ($c_p = $ constant), it is reduced to

$$\Delta s = c_p\ln\frac{T_2}{T_1} - R\ln\frac{p_2}{p_1} \tag{2.41}$$

For a steady adiabatic flow, the change in entropy is given by

$$\Delta s = -R\ln\frac{p_{t2}}{p_{t1}} \tag{2.42}$$

2.12　Pressure Coefficient

The pressure coefficient is defined as

$$C_p = \frac{p - p_\infty}{\frac{1}{2}\rho_\infty u_\infty^2} = \frac{p - p_\infty}{q_\infty} \tag{2.43}$$

where q_∞ is the dynamic pressure, $q_\infty = \frac{1}{2}\rho_\infty u_\infty^2$.

The pressure coefficient for a perfect gas may be expressed in terms of the free stream Mach number as follows.

$$C_p = \frac{p - p_\infty}{\frac{1}{2}\gamma p_\infty M_\infty^2} \tag{2.44}$$

For an isentropic flow, the pressure distribution can be determined from the following relations

$$C_p = \frac{2}{\gamma M_\infty^2}\left\{\left[\frac{2 + (\gamma - 1)M_\infty^2}{2 + (\gamma - 1)M^2}\right]^{\gamma/(\gamma - 1)} - 1\right\} \tag{2.45}$$

or

$$C_p = \frac{2}{\gamma M_\infty^2}\left\{\left[1 + \frac{\gamma - 1}{2}M_\infty^2\left(1 - \frac{V^2}{V_\infty^2}\right)\right]^{\gamma/(\gamma - 1)} - 1\right\} \tag{2.46}$$

Various expressions for the pressure coefficient are summarized in Table 2.8.

Table 2.8: Expressions for the pressure coefficient

Pressure Coefficient	Remarks
$C_p = \dfrac{p - p_\infty}{\frac{1}{2}\rho_\infty u_\infty^2}$	Definition
$C_p = \dfrac{p - p_\infty}{\frac{1}{2}\gamma p_\infty M_\infty^2} = \dfrac{2}{\gamma M_\infty^2}\left(\dfrac{p}{p_\infty} - 1\right)$	In terms of M_∞
$C_p = 1 - \dfrac{V^2}{V_\infty^2}$	Incompressible flow
$C_p = \dfrac{2}{\gamma M_\infty^2}\left\{\left[\dfrac{2 + (\gamma - 1)M_\infty^2}{2 + (\gamma - 1)M^2}\right]^{\gamma/(\gamma-1)} - 1\right\}$	Isentropic flow, perfect gas
$C_p = \dfrac{2}{\gamma M_\infty^2}\left\{\left[1 + \dfrac{\gamma - 1}{2}M_\infty^2\left(1 - \dfrac{V^2}{V_\infty^2}\right)\right]^{\gamma/(\gamma-1)} - 1\right\}$	Isentropic flow, perfect gas
$C_p = \dfrac{C_{p_i}}{\sqrt{1 - M_\infty^2}}$	Prandtl-Glauert rule, subsonic flow, accounts for compressibility effect; C_{p_i} is the incompressible pressure coefficient.
$C_p = (C_{p_i})/\left\{\sqrt{1 - M_\infty^2} + \left[\frac{1}{2}M_\infty^2\left(1 + \dfrac{\gamma - 1}{2}M_\infty^2\right)\left(1 - M_\infty^2\right)^{-1/2}\right]C_{p_i}\right\}$	Laiton's rule, subsonic flow, accounts for compressibility effect; C_{p_i} is the incompressible pressure coefficient.
$C_p = \dfrac{C_{p_i}}{\sqrt{1 - M_\infty^2} + \frac{1}{2}\left[\dfrac{M_\infty^2}{1 + \sqrt{1 - M_\infty^2}}\right]C_{p_i}}$	Karman-Tsien rule, subsonic flow, accounts for compressibility effect; C_{p_i} is the incompressible pressure coefficient.

(continued on next page)

Pressure Coefficient	Remarks
$$C_p = C_{p_i}\left[1 + \frac{\ln(1 - M_\infty^2)^{1/2}}{1 + \ln 2n}\right]$$	Ellipsoid of revolution at zero degree angle of attack, subsonic flow; n is the thickness ratio, i.e., maximum diameter to length; C_{p_i} is the incompressible pressure coefficient.
$$C_p = \frac{2\theta}{\sqrt{M_\infty^2 - 1}}$$	Linearized supersonic flow; two-dimensional flow, θ is the surface inclination with respect to the free stream in radians; C_p is positive $(+)$ for compression and negative $(-)$ for expansion.
$$C_p = \frac{2\theta}{\sqrt{M_\infty^2 - 1}} + \left[\frac{(\gamma + 1)M_\infty^2 - 4M_\infty^2 + 4}{2(M_\infty^2 - 1)^2}\right]\theta^2$$	Busemann's theory, same as above.
$$C_p = 2\sin^2\theta$$	Newtonian pressure, hypersonic flow.
$$C_p = C_{p_{\max}}\sin^2\theta$$	Modified Newtonian pressure, hypersonic flow.
where $$C_{p_{\max}} = \frac{2}{\gamma M_\infty^2}\left\{\left(\frac{1 - \gamma + 2\gamma M_\infty^2}{\gamma + 1}\right) \times \left[\frac{(\gamma + 1)^2 M_\infty^2}{4\gamma M_\infty^2 - 2(\gamma - 1)}\right]^{\gamma/(\gamma-1)} - 1\right\}$$	

(continued on next page)

Pressure Coefficient	Remarks
$C_{p_j} = 2\sin^2\theta_j + 2\sin\theta_j\left(\dfrac{d\theta}{dy}\right)_j \displaystyle\int_0^{y_j}\cos\theta\,dy$ 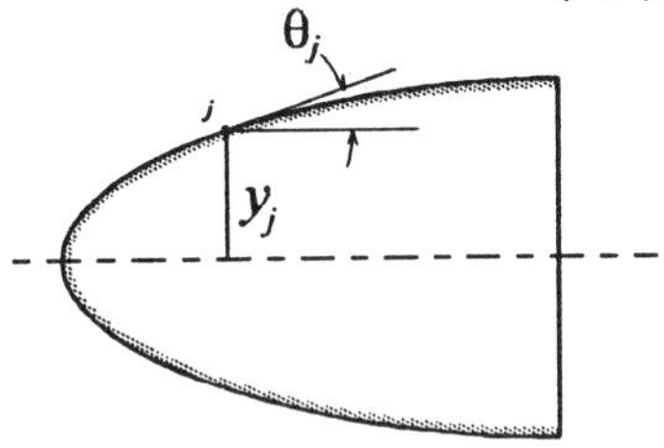	Newtonian Busemann, two-dimensional flow, hypersonic flow.
$C_{p_j} = 2\sin^2\theta_j + \dfrac{1}{y_j}2\sin\theta_j\left(\dfrac{d\theta}{dy}\right)_j \displaystyle\int_0^{y_j}y\cos\theta\,dy$ or $C_{p_j} = 2\sin^2\theta_j + 2\sin\theta_j\left(\dfrac{d\theta}{dA}\right)_j \displaystyle\int_{A_j}\cos\theta\,dA$	Newtonian Busemann, hypersonic flow, axisymmetric, y_j is the distance from the body axis to the surface point j. $A_j = \pi y_j^2$
$C_{p_j} = 2(\theta_j^2 + k_j y_j)$	Slender body, two-dimensional, hypersonic flow, k_j is the curvature of the surface at point j.
$C_{p_j} = 2\theta_j^2 + k_j y_j$	Slender body, hypersonic flow, axisymmetric, k_j is the curvature of the surface at point j.
$C_p = \theta^2\left[1 + \dfrac{(\gamma+1)M_\infty^2\theta^2 + 2}{(\gamma-1)M_\infty^2\theta^2 + 2}\times \right.$ $\left. \ln\left(\dfrac{\gamma+1}{2} + \dfrac{1}{M_\infty^2\theta^2}\right)\right]$	Slender cone at zero degree angle of attack, hypersonic flow, θ is the cone angle.

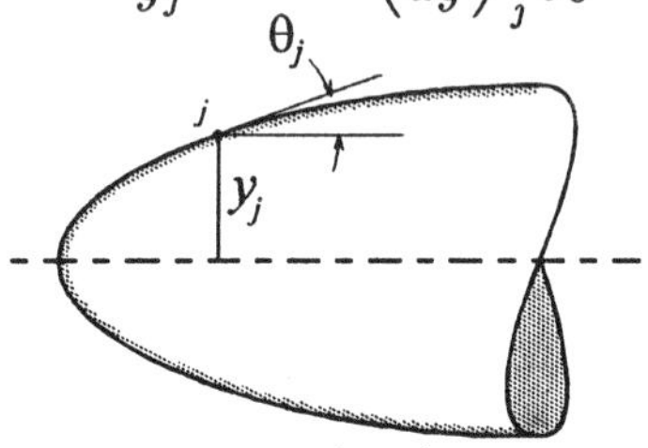

2.13 One-Dimensional Flow

A flow may be considered as one-dimensional if the flow properties are uniform over any cross-section and may vary only in the streamwise direction. Now consider a domain with a constant cross-section and the control volume shown in Figure 2.3.

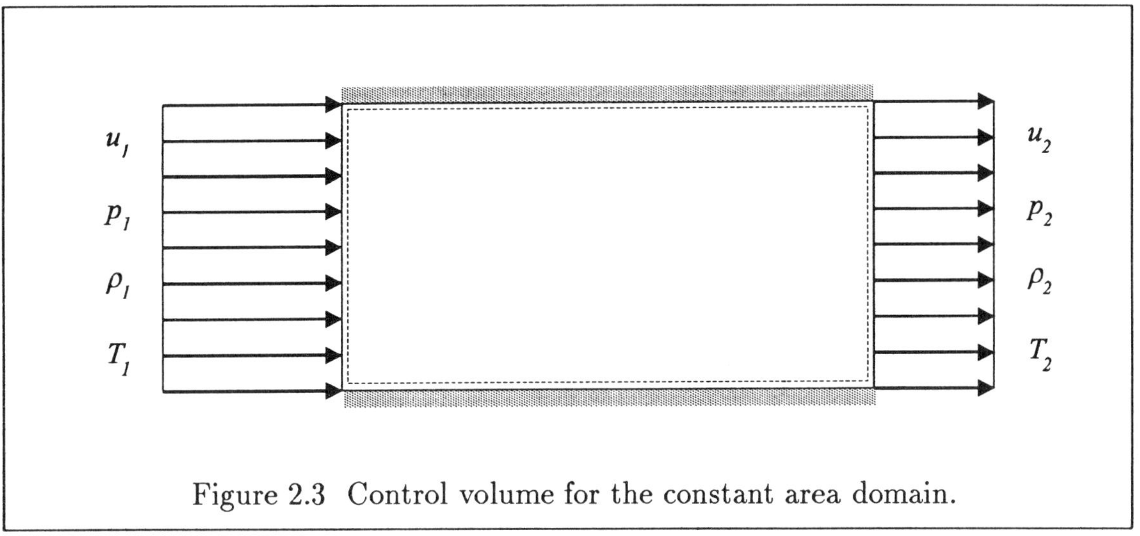

Figure 2.3 Control volume for the constant area domain.

Various factors could cause variation in the flow properties. It is possible to have a normal shock within the domain; as a consequence, the flow properties at section 2 would be different from those at the inflow. Normal shock relations are addressed in Section 2.14. Two other factors may cause a change in the flow properties. One is the addition of heat (heating) or the extraction of heat (cooling), which is considered in the next section. The other is the friction effect, which was addressed in Chapter 1. Note that the effect of each is considered separately.

2.13.1 One-Dimensional Flow with Heat Transfer

The conservation laws applied to the control volume shown in Figure 2.3 provide the following equations.

$$\rho_2 u_2 = \rho_1 u_1 \tag{2.47}$$

$$p_2 + \rho_2 u_2^2 = p_1 + \rho_1 u_1^2 \tag{2.48}$$

$$h_2 + \frac{1}{2}u_2^2 = h_1 + \frac{1}{2}u_1^2 + q \tag{2.49}$$

where q represents heat transfer per unit mass. Note that Equation (2.49) for a calorically perfect gas may be expressed as

$$q = c_p(T_{t2} - T_{t1})$$

A plot representing the entropy–enthalpy relation due to heat transfer is called the *Rayleigh* line.

The governing equations may be modified and expressed in terms of the Mach numbers as follows.

$$\frac{p_2}{p_1} = \frac{1 + \gamma M_1^2}{1 + \gamma M_2^2} \tag{2.50}$$

$$\frac{T_2}{T_1} = \left(\frac{1 + \gamma M_1^2}{1 + \gamma M_2^2}\right)^2 \left(\frac{M_2}{M_1}\right)^2 \tag{2.51}$$

$$\frac{\rho_2}{\rho_1} = \left(\frac{1 + \gamma M_2^2}{1 + \gamma M_1^2}\right)^2 \left(\frac{M_1}{M_2}\right)^2 \tag{2.52}$$

$$\frac{p_{t2}}{p_{t1}} = \left(\frac{1 + \gamma M_1^2}{1 + \gamma M_2^2}\right) \left[\frac{1 + \dfrac{\gamma - 1}{2} M_2^2}{1 + \dfrac{\gamma - 1}{2} M_1^2}\right]^{\frac{\gamma}{\gamma - 1}} \tag{2.53}$$

$$\frac{T_{t2}}{T_{t1}} = \left(\frac{1 + \gamma M_1^2}{1 + \gamma M_2^2}\right) \left(\frac{1 + \dfrac{\gamma - 1}{2} M_2^2}{1 + \dfrac{\gamma - 1}{2} M_1^2}\right) \left(\frac{M_2}{M_1}\right)^2 \tag{2.54}$$

$$\Delta s = s_2 - s_1 = \frac{\gamma R}{\gamma - 1} \ln \left[\left(\frac{M_2}{M_1}\right)^{\frac{2\gamma}{\gamma - 1}} \left(\frac{1 + \gamma M_1^2}{1 + \gamma M_2^2}\right)^{\frac{\gamma + 1}{\gamma - 1}}\right] \tag{2.55}$$

The solution procedure is simplified if one introduces reference properties at $M = 1$. Typically, an asterisk is used to denote the reference properties. Thus the equations above may be written as

$$\frac{p}{p^*} = \frac{\gamma + 1}{1 + \gamma M^2} \tag{2.56}$$

$$\frac{T}{T^*} = \frac{(\gamma+1)^2 M^2}{(1+\gamma M^2)^2} \tag{2.57}$$

$$\frac{\rho}{\rho^*} = \frac{1+\gamma M^2}{(\gamma+1)M^2} \tag{2.58}$$

$$\frac{p_t}{p_t^*} = \frac{(\gamma+1)}{1+\gamma M^2} \left[\frac{2\left(1+\frac{\gamma-1}{2}M^2\right)}{\gamma+1} \right]^{\frac{\gamma}{\gamma-1}} \tag{2.59}$$

$$\frac{T_t}{T_t^*} = \frac{2(\gamma+1)\left(1+\frac{\gamma-1}{2}M^2\right)M^2}{(1+\gamma M^2)^2} \tag{2.60}$$

$$\Delta s = s - s^* = \frac{\gamma R}{\gamma-1} \ln\left[M^2 \left(\frac{\gamma+1}{1+\gamma M^2}\right)^{\frac{\gamma+1}{\gamma}} \right] \tag{2.61}$$

Expressions (2.56) through (2.60) can be tabulated as a function of M and presented in tables. For air with a γ of 1.4, they are provided in Table 2.12 at the end of this chapter.

2.14 Shock Waves

Any disturbance in a flowfield emanates signals which travel with the speed of sound to influence the incoming flow. In a subsonic flow, the velocity of fluid particles is less than the speed of sound, and, therefore, the existence of an obstacle downstream is felt by the fluid. Subsequently, fluid particles adjust their path and maneuver themselves around the object. On the other hand, for a supersonic flow, the existence of an obstacle cannot be communicated upstream. Thus, a phenomena must take place to decelerate the supersonic flow as well as to provide a mechanism for the turning of a supersonic flow. As a consequence of this physical requirement, shock waves form in nature. Shock waves are extremely thin regions (mathematical discontinuity) in the flowfield across which large variations in the flow properties take place. The actual thickness of a shock wave is typically in the order of the mean free path of molecules (about 10^{-5} cm at atmospheric conditions). The velocity and temperature gradients are large within this region in the flowfield. In fact, the flow is highly dissipative, and heat transfer takes place within the shock wave. To simplify the mathematical approach, shock waves are treated as a discontinuity within the flowfield. Subsequently, conservation laws are used to provide the appropriate relations for the flow properties across the shock wave. The shape of a shock wave is determined by the shape of the object in the supersonic stream. Typically, shock waves are curved and three-dimensional. However, for some applications shock waves may be considered two-dimensional and straight, which is inclined at some angle to the incoming flow. This type of shock wave is known as *oblique shock*. If the shock wave is perpendicular to the incoming flow, then it is called a *normal shock*. One-dimensional analysis is used for normal shocks, whereas two-dimensional analysis is used for oblique shocks.

2.14.1 Normal Shock Equations

The conservation laws in the integral form are applied to a control volume which includes the normal shock to provide the governing equations. Using the index 1 to represent the flow properties ahead of the shock and index 2 to represent the flow properties after the normal shock, the following equations are obtained.

1. Conservation of mass

$$\rho_2 u_2 = \rho_1 u_1 \tag{2.62}$$

2. Conservation of x-momentum

$$p_2 + \rho_2 u_2^2 = p_1 + \rho_1 u_1^2 \tag{2.63}$$

3. Conservation of energy

$$h_2 + \frac{1}{2}u_2^2 = h_1 + \frac{1}{2}u_1^2 \tag{2.64}$$

To close the system of equations, one introduces an equation of state and a relation for enthalpy. Typically, these relations are in the form of tables and/or charts for real gas and in the form of equations for a perfect gas. For a real gas, an iterative scheme is used to solve the system of equations. For a perfect gas, the following equations may be incorporated.

4. Equation of state for a perfect gas

$$p_2 = \rho_2 R T_2$$

5. Relation for the enthalpy for a calorically perfect gas

$$h_2 = c_p T_2$$

where

$$c_p = \frac{\gamma R}{\gamma - 1}$$

Thus, a sufficient number of equations has been established to solve the system for all the flow properties after the normal shock wave. The solution process is simplified if the equations are modified such that all the unknowns are written directly in terms of the flow properties ahead of the shock. The result of such mathematical manipulation provides the following relations

$$M_2^2 = \frac{(\gamma - 1)M_1^2 + 2}{2\gamma M_1^2 - (\gamma - 1)} \tag{2.65}$$

$$\frac{p_2}{p_1} = \frac{2\gamma}{\gamma + 1}M_1^2 - \frac{\gamma - 1}{\gamma + 1} \tag{2.66}$$

$$\frac{T_2}{T_1} = \frac{[2\gamma M_1^2 - (\gamma - 1)]\,[(\gamma - 1)M_1^2 + 2]}{(\gamma + 1)^2 M_1^2} \tag{2.67}$$

$$\frac{\rho_2}{\rho_1} = \frac{u_1}{u_2} = \frac{(\gamma + 1)M_1^2}{(\gamma - 1)M_1^2 + 2} \tag{2.68}$$

$$\frac{p_{t2}}{p_{t1}} = \left[\frac{(\gamma + 1)M_1^2}{(\gamma - 1)M_1^2 + 2}\right]^{\frac{\gamma}{\gamma - 1}} \left[\frac{\gamma + 1}{2\gamma M_1^2 - (\gamma - 1)}\right]^{\frac{1}{\gamma - 1}} \tag{2.69}$$

$$\frac{p_{t2}}{p_1} = \left[\frac{(\gamma + 1)M_1^2}{2}\right]^{\frac{\gamma}{\gamma - 1}} \left[\frac{\gamma + 1}{2\gamma M_1^2 - (\gamma - 1)}\right]^{\frac{1}{\gamma - 1}} \tag{2.70}$$

$$\frac{p_2}{p_{t1}} = \left[\frac{2\gamma M_1^2 - (\gamma - 1)}{\gamma + 1}\right] \left[\frac{2}{(\gamma - 1)M_1^2 + 2}\right]^{\frac{\gamma}{\gamma - 1}} \tag{2.71}$$

$$\frac{p_2}{p_{t2}} = \left[\frac{4\gamma M_1^2 - 2(\gamma - 1)}{(\gamma + 1)^2 M_1^2}\right]^{\frac{\gamma}{\gamma - 1}} \tag{2.72}$$

$$\frac{\Delta s}{R} = -\ln\frac{p_{t2}}{p_{t1}} = \frac{\gamma}{\gamma - 1}\ln\left[\frac{2}{(\gamma + 1)M_1^2} + \frac{\gamma - 1}{\gamma + 1}\right]$$
$$+ \frac{1}{\gamma - 1}\ln\left[\frac{2\gamma}{\gamma + 1}M_1^2 - \frac{\gamma - 1}{\gamma + 1}\right] \tag{2.73}$$

Equations (2.65) through (2.70) for γ of 1.4 are evaluated and given in Table 2.14 at the end of this chapter.

It is noted that the governing equations are established for a steady-state problem, i.e., a stationary normal shock. However, the governing equations and subsequent relations given by Equations (2.66) through (2.68) can be used for moving shocks if the instantaneous velocities are adjusted such that they are measured with respect to the shock wave.

The limiting values of the flow properties may be established as follows. The corresponding values for γ of 1.4 are provided as well.

$$\lim_{M_1 \to \infty} M_2 = \sqrt{\frac{\gamma - 1}{2\gamma}} = 0.378$$

$$\lim_{M_1 \to \infty} \frac{\rho_2}{\rho_1} = \frac{\gamma + 1}{\gamma - 1} = 6.0$$

$$\lim_{M_1 \to \infty} \frac{p_2}{p_1} = \infty$$

$$\lim_{M_1 \to \infty} \frac{T_2}{T_1} = \infty$$

$$\lim_{M_1 \to 1} M_2 = 1$$

$$\lim_{M_1 \to 1} \frac{\rho_2}{\rho_1} = 1$$

$$\lim_{M_1 \to 1} \frac{p_2}{p_1} = 1$$

$$\lim_{M_1 \to 1} \frac{T_2}{T_1} = 1$$

It should be emphasized that, as the Mach number increases beyond that of $5 \sim 8$ where the flow temperature after the shock could exceed 1,000 K, chemistry effects must be considered. In fact, these effects would prevent the temperature from approaching extremely high values. Chemistry effects and hypersonic flow are addressed in Chapter 4.

2.14.1.1 Shock Thickness: In the previous discussions, shock waves were treated as discontinuities. In reality, the effects of viscosity and thermal conductivity tend to resist steepening of compression waves. An order of magnitude analysis along with the use of relations from kinetic theory of gases provide an approximate expression for the shock thickness δ [2.1] as

$$\delta \cong \frac{5}{8}\ell^* Re \tag{2.74}$$

where ℓ is the mean free path of molecules, and $*$ denotes the value at temperature of T^* (temperature at $M = 1$). As seen from Equation (2.74), the shock thickness increases at higher altitudes where the value of ℓ becomes larger.

A relation for the shock thickness of a perfect gas is given by Shapiro and Kline [2.2] as follows.

$$\delta = \left\{ \left(\frac{\mu_1}{\rho_1 a_1}\right) D \frac{1 + \overline{M}_1}{(\gamma + 1)\,(\overline{M}_1 - 1)\overline{M}_1} \left[\left(\frac{\gamma + 1}{2}\right) \left(1 - \frac{\gamma - 1}{\gamma + 1}\overline{M}_1^2\right)\right]^{(1/2)-n} \right\}$$

$$\left\{ 1 \pm \left[1 + \frac{8\gamma(\gamma + 1)\,(\overline{M}_1 - 1)^2}{(3Pr^*)\,(\overline{M}_1)D^2}\right]^{1/2} \right\} \tag{2.75}$$

where $(+)$ is used if $D > 0$, and $(-)$ is used if $D < 0$, and

$$D = \frac{4}{3} + \frac{2\gamma}{Pr^*} - \frac{(\gamma + 1)\,(\overline{M}_1^2 + 1)}{2Pr^*\overline{M}_1} \tag{2.76}$$

$\overline{M}$ is the characteristic Mach number, defined in Section 2.9, and n is the exponent in the viscosity-temperature relation, typically $n = \frac{1}{2}$ for perfect gas. However, experimental values provide $n = 0.768$ for air at standard temperatures.

2.14.2 Rankine-Hugoniot Relations

A relation between the pressure ratio and the density ratio is known as the *Rankine-Hugoniot* relation, which may be expressed as either

$$\frac{\rho_2}{\rho_1} = \frac{\left(\dfrac{\gamma + 1}{\gamma - 1}\right) \dfrac{p_2}{p_1} + 1}{\dfrac{\gamma + 1}{\gamma - 1} + \dfrac{p_2}{p_1}} \tag{2.77}$$

or

$$\frac{p_2}{p_1} = \frac{\left(\dfrac{\gamma + 1}{\gamma - 1}\right) \dfrac{\rho_2}{\rho_1} - 1}{\dfrac{\gamma + 1}{\gamma - 1} - \dfrac{\rho_2}{\rho_1}} \tag{2.78}$$

2.14.3 Oblique Shock Equations

The analysis for an oblique shock is simplified due to the fact that normal shock relations may be used. In fact, an oblique shock with a free stream Mach number of M_1 may be reduced to a normal shock if $M_1 \sin \theta$ is used as the corresponding free stream Mach number. The nomenclature used in the following equations is shown in Figure 2.4 and is as follows. The turning angle is denoted by δ, and the shock angle is denoted by θ. The subscripts t and n are used to designate the tangential and normal components of the velocity, respectively.

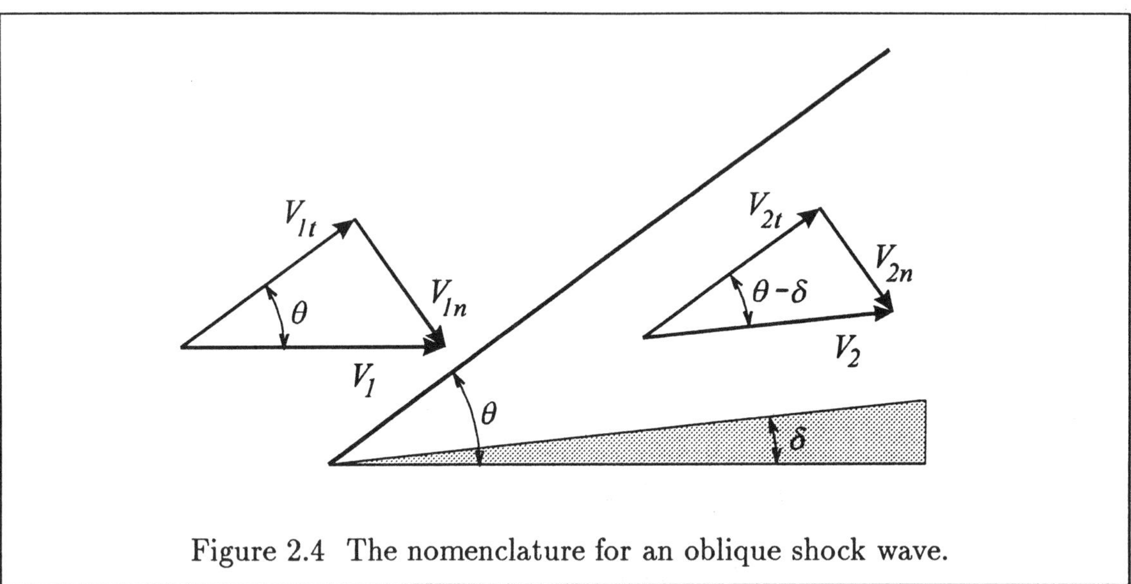

Figure 2.4 The nomenclature for an oblique shock wave.

The conservation laws provide the following equations.

1. Conservation of mass

$$\rho_2 V_{2n} = \rho_1 V_{1n} \tag{2.79}$$

2. Conservation of normal component of momentum equation

$$p_2 + \rho_2 V_{2n}^2 = p_1 + \rho_1 V_{1n}^2 \tag{2.80}$$

3. Conservation of tangential component of momentum equation

$$V_{2t} = V_{1t} \tag{2.81}$$

4. Conservation of energy

$$h_2 + \frac{1}{2} V_{2n}^2 = h_1 + \frac{1}{2} V_{1n}^2 \tag{2.82}$$

Furthermore, geometrical consideration yields

$$M_{1n} = M_1 \sin \theta$$

$$M_{1t} = M_1 \cos \theta$$

$$M_{2n} = M_2 \sin(\theta - \delta)$$

$$M_{2t} = M_2 \cos(\theta - \delta)$$

A relation between the turning angle δ, the shock angle θ, and the free stream Mach number can be established as follows

$$\tan \delta = \frac{2 \cot \theta (M_1^2 \sin^2 \theta - 1)}{2 + (\gamma + \cos 2\theta) M_1^2} \tag{2.83}$$

An iterative scheme must be used to solve Equation (2.83) for the shock angle θ when the turning angle δ and the Mach number M_1 are provided. Typically, Equation (2.83) is presented graphically, and, therefore, a graphical solution is sought. Once the shock angle is determined, M_{1n} is calculated and the flow properties are computed from the normal shock relations (or tables). The appropriate charts for oblique shock solutions for γ of 1.4, adopted from Ref. [2.3], are provided in Figures 2.5 through 2.7 at the end of this chapter.

Observe that, for a given M_1 and δ, there are either two solutions or no solution at all, in which case the shock is a detached curved shock. In the case of an attached shock, typically a weak solution is observed to occur for external flows, and, thus, given M_1 and δ, θ can be uniquely determined.

2.14.4 Conical Flow

A circular cone at zero-degree angle of attack in a supersonic stream will generate an attached shock wave in the form of a circular cross-section cone for appropriate values of M_1 and δ. The flow properties at the cone surface may be determined by charts shown in Figures 2.8 through 2.10, which are adopted from Ref. [2.3] and provided at the end of this chapter. As in the case of two-dimensional oblique shock, weak solutions are believed to exist for external flows.

2.15   Expansion Waves

2.15.1 Mach Wave and Mach Angle

When a supersonic flow is turned through an infinitesimal angle, a Mach wave is generated. The angle between the Mach wave and the direction of flow is defined as *the Mach angle*, typically denoted by μ. The Mach angle can be determined in terms of the Mach number from the following relation

$$\mu = \sin^{-1} \frac{1}{M}$$

The values of Mach angle are provided in Table 2.14 at the end of this chapter.

2.15.2 Prandtl-Meyer Expansion Waves

If the flow is turned away from itself through a finite angle, a series of waves (each wave being a Mach wave) is generated, which is referred to as a *Prandtl-Meyer expansion wave*. The Prandtl-Meyer expansion is characterized by the following: (1) the changes in the flowfield across the wave are continuous; (2) streamlines are curved and smooth; and (3) the expansion process is isentropic.

The change in flow direction is related to change in the flow Mach number by the following relation.

$$\Delta\theta = \int_{M_1}^{M_2} \frac{\sqrt{M^2 - 1}}{M} \frac{dM}{1 + \dfrac{\gamma - 1}{2} M^2} \tag{2.84}$$

Typically, a flow at Mach number of 1 is used as a reference condition, and, thus, Equation (2.84) is integrated to provide

$$\nu(M) = \sqrt{\frac{\gamma + 1}{\gamma - 1}} \tan^{-1} \sqrt{\frac{\gamma - 1}{\gamma + 1}(M^2 - 1)} - \tan^{-1} \sqrt{M^2 - 1} \tag{2.85}$$

where ν is known as the Prandtl-Meyer function. Observe that it represents the angle through which a stream initially at Mach number of 1 must be expanded to reach a supersonic Mach number of M. Given a supersonic flow at Mach number of M_1 and a turning angle of θ, the following relation holds

$$\theta = \nu(M_2) - \nu(M_1)$$

Once $\nu(M_2)$ is determined, the downstream Mach number M_2 can be computed. However, observe that Equation (2.85) cannot be solved directly for M. Thus, an iterative procedure must be used. For practical engineering problems involving air, where $\gamma = 1.4$, relation (2.85) is tabulated and provided in tables which may be used to facilitate the solution procedure. The Prandtl-Meyer function for γ of 1.4 is provided in Table 2.14 at the end of this chapter. Since the flow is considered to be isentropic, therefore,

$$T_{t2} = T_{t1}$$

and

$$p_{t2} = p_{t1}$$

and, subsequently,

$$\frac{T_2}{T_1} = \frac{1 + \dfrac{\gamma - 1}{2}M_1^2}{1 + \dfrac{\gamma - 1}{2}M_2^2} \tag{2.86}$$

and

$$\frac{p_2}{p_1} = \left(\frac{1 + \dfrac{\gamma - 1}{2}M_1^2}{1 + \dfrac{\gamma - 1}{2}M_2^2}\right)^{\frac{\gamma}{\gamma-1}} \tag{2.87}$$

2.16 Ouasi One-Dimensional Flow

Conservation laws applied to a control volume where cross-sectional surfaces at the inflow and outflow differ by an amount dA yield the following relations. Note that the flow is assumed to be one-dimensional and isentropic.

1. Conservation of mass

$$\frac{du}{u} + \frac{d\rho}{\rho} + \frac{dA}{A} = 0 \tag{2.88}$$

2. Conservation of x-momentum (inviscid flow)

$$dp + \rho u \, du = 0 \tag{2.89}$$

3. Conservation of energy (adiabatic)

$$dh + u \, du = 0 \tag{2.90}$$

Mathematical manipulation of the equations above and the introduction of Mach number provide

$$(1 - M^2)dp = \rho u^2 \frac{dA}{A} \tag{2.91}$$

which is known as the *area–pressure relation*. Similarly,

$$(M^2 - 1)\frac{du}{u} = \frac{dA}{A} \tag{2.92}$$

which is known as the *area–velocity relation*.

Mass flow rate at a cross-section may be determined from the following relation

$$\overset{\circ}{m} = \frac{p_t A M}{\sqrt{T_t}} \sqrt{\frac{\gamma}{R}} \left(1 + \frac{\gamma - 1}{2}M^2\right)^{-\frac{\gamma+1}{2(\gamma-1)}} \tag{2.93}$$

where M is the Mach number at the cross-section, and p_t and T_t are the stagnation pressure and temperature, respectively. An area ratio may be established where $M = 1$ is used as a reference and the corresponding area is designated by A^*. This relation is known as the area–Mach number relation and is given by

$$\frac{A}{A^*} = \frac{1}{M}\left[\left(\frac{2}{\gamma+1}\right)\left(1+\frac{\gamma-1}{2}M^2\right)\right]^{\frac{\gamma+1}{2(\gamma-1)}} \tag{2.94}$$

For a given A/A^*, there are two solutions: one subsonic and one supersonic. The unique solution is determined by the imposed pressure. The values of A/A^* for γ of 1.4 are provided in Table 2.14.

2.17   Lift Coefficient

The lift coefficient is defined as

$$C_L = \frac{L}{\frac{1}{2}\rho_\infty u_\infty^2 S} \tag{2.95}$$

where S is the reference area. For a two-dimensional problem, a section lift coefficient is defined as follows.

$$C_\ell = \frac{\ell}{\frac{1}{2}\rho_\infty u_\infty^2 c} \tag{2.96}$$

where c is a characteristic length, e.g., the chord for an airfoil or diameter for a cylinder. Simple expressions for the section lift coefficient and the lift coefficient are summarized in Tables 2.9 and 2.10, respectively.

Table 2.9: Expressions for the section lift coefficient

Configuration	Formulation	Remarks
Symmetric airfoil	$C_\ell = 2\pi\alpha$	Incompressible, inviscid flow, α is the angle of attack.
Cambered airfoil	$C_\ell = \pi(2A_0 + A_1)$	$A_0 = \alpha - \dfrac{1}{\pi}\displaystyle\int_0^\pi \dfrac{dz}{dx}d\theta_o$ $A_1 = \dfrac{2}{\pi}\displaystyle\int_0^\pi \dfrac{dz}{dx}\cos\theta_o d\theta_o$ $x = \dfrac{c}{2}(1 - \cos\theta_o)$
	$C_\ell = \dfrac{C_{\ell_i}}{\sqrt{1 - M_\infty^2}}$	Subsonic flow, accounts for compressibility effect; C_{ℓ_i} is the incompressible lift coefficient.
	$C_\ell = C_{\ell_i}\left\{\dfrac{1}{\sqrt{1 - M_\infty^2}} + \dfrac{\delta}{1-\delta}\left[\dfrac{1}{1-M_\infty^2} - \dfrac{1}{\sqrt{1-M_\infty^2}} + \dfrac{\gamma+1}{4}\left(\dfrac{M_\infty^2}{1-M_\infty^2}\right)^2\right]\right\}$	Subsonic flow, accounts for compressibility and thickness effects; C_{ℓ_i} is the incompressible lift coefficient; δ is the thickness ratio.
	$C_\ell = \dfrac{4\alpha}{\sqrt{M_\infty^2 - 1}}$	Supersonic airfoil, α is the angle of attack
	$C_\ell = \dfrac{4\alpha\cos\Lambda}{\sqrt{M_\infty^2\cos^2\Lambda - 1}}$	Supersonic airfoil with a swept angle of Λ, supersonic leading edge, α is the angle of attack.

Table 2.10: Expressions for the lift coefficient

Configuration	Formulation	Remarks
Finite wing, elliptic	$C_L = \dfrac{\pi b \Gamma_0}{2 u_\infty S}$	Incompressible, inviscid flow, elliptic spanwise circulation distribution, $\Gamma(y) = \Gamma_0 \left[1 - \left(\frac{2y}{b} \right)^2 \right]^{1/2}$, b is the span.
Finite wing	$C_L = A_1 \pi R\!\!\!/\,$	Incompressible, inviscid flow, general spanwise circulation distribution, $R\!\!\!/\,$ is the aspect ratio.
Flat plate, rectangular wing	$C_L = \dfrac{4\alpha}{\sqrt{M_\infty^2 - 1}} \left(1 - \dfrac{1}{2 R\!\!\!/\,\sqrt{M_\infty^2 - 1}} \right)$	Supersonic flow, $R\!\!\!/\, = \frac{b^2}{S} = \frac{b}{c}$, α in radians
Rectangular wing with thickness	$C_L = \dfrac{4\alpha}{\sqrt{M_\infty^2 - 1}} \left[1 - \dfrac{1}{2 R\!\!\!/\,\sqrt{M_\infty^2 - 1}} \times (1 - c_1 A') \right]$	Supersonic flow, A' is the airfoil cross-sectional area/chord squared, and $c_1 = \dfrac{\gamma M_\infty^4 + (M_\infty^2 - 2)^2}{2(M_\infty^2 - 1)^{3/2}}$

2.18 Drag Coefficient

The drag coefficient is defined as

$$C_D = \frac{D}{\frac{1}{2}\rho_\infty u_\infty^2 S} \tag{2.97}$$

Again, for a two-dimensional problem, $S = (c)(1)$, where c is a characteristic length, and therefore

$$C_d = \frac{d}{\frac{1}{2}\rho_\infty u_\infty^2 c} \qquad (2.98)$$

The total drag may be composed of pressure (or form) drag, skin friction drag, induced drag, and wave drag.

Pressure (or form) drag: The pressure drag is the component of drag due to the pressure distribution around the configuration.

Skin friction drag: The skin friction drag is the component of drag due to viscous forces. Expressions for skin friction drag coefficient are provided in Chapter 3.

Induced drag: The induced drag or drag due to lift is the component of drag generated by the presence of downwash.

Wave drag: The wave drag is the component of drag due to compressibility effects and is typically related to the formation of shock waves.

Profile drag: A combination of pressure drag and skin friction drag is called the profile drag.

The (total) drag coefficient may be expressed as the sum of profile drag, C_{D_o}, induced drag, $C_{D_i} = kC_L^2$, and wave drag, C_{D_w}. Thus,

$$C_D = C_{D_o} + C_{D_i} + C_{D_w}$$

or

$$C_D = C_{D_o} + kC_L^2 + C_{D_w}$$

Drag polar: A plot of C_d versus C_ℓ is referred to as a *drag polar*.

Some simple expressions for various drag coefficients are provided in Table 2.11, whereas expressions for the skin friction coefficients are given in Chapter 3.

Table 2.11: Expressions for drag coefficients

Configuration	Formulation	Remarks
Finite wing, elliptic	$C_{D_i} = \dfrac{\pi \Gamma_0^2}{4 u_\infty^2 S} = \dfrac{C_L^2}{\pi /\!\!R}$	Incompressible, inviscid flow, elliptic spanwise circulation distribution.
Finite wing	$C_{D_i} = \dfrac{C_L^2}{\pi /\!\!R}(1 + \delta)$ $\delta = \dfrac{3 A_3^2}{A_1^2} + \dfrac{5 A_5^2}{A_1^2} + \dfrac{7 A_5^2}{A_1^2} + \ldots$	Incompressible, inviscid flow, general spanwise circulation distribution.
Two-dimensional, supersonic airfoil	$C_D = C_{D_i} + C_{D_{Th}}$ $C_{D_i} = \dfrac{4\alpha^2}{\sqrt{M_\infty^2 - 1}}$ $C_{D_{Th}} = \dfrac{4}{\sqrt{M_\infty^2 - 1}}\left(\dfrac{\overline{\sigma_L^2} + \overline{\sigma_u^2}}{2}\right)$	Inviscid $\overline{\sigma_L^2} = \frac{1}{L}\displaystyle\int_0^L \sigma_L^2\, dx$ $\overline{\sigma_u^2} = \frac{1}{L}\displaystyle\int_0^L \sigma_u^2\, dx$ σ_L and σ_u are the lower and upper surface inclinations, respectively, Fig. T2.11
Flat plate, rectangular wing	$C_D = \dfrac{4\alpha^2}{\sqrt{M_\infty^2 - 1}}\left(1 - \dfrac{1}{2/\!\!R\sqrt{M_\infty^2 - 1}}\right)$	Inviscid

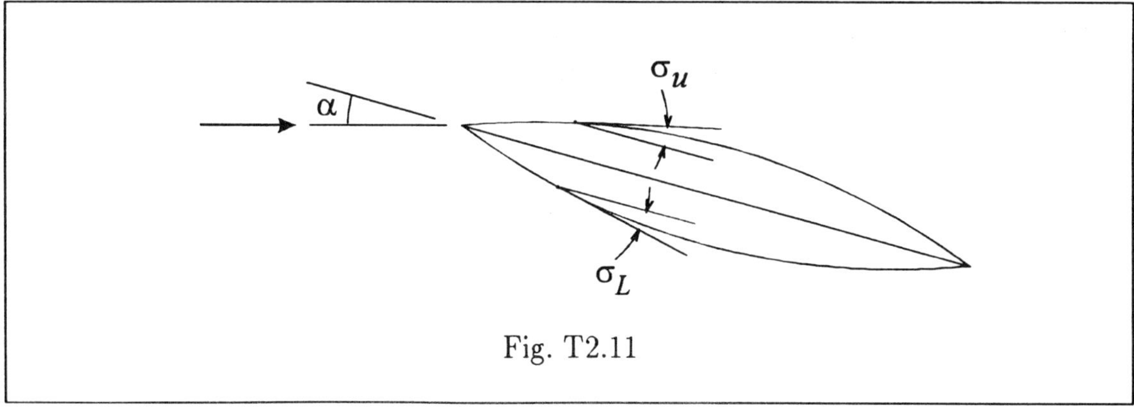

Fig. T2.11

Table 2.12: One-dimensional flow with heat transfer.

M	P/P^*	T/T^*	ρ/ρ^*	P_t/P_t^*	T_t/T_t^*
0.200E−01	0.2399E+01	0.2301E−02	0.1042E+04	0.1268E+01	0.1918E−02
0.400E−01	0.2395E+01	0.9175E−02	0.2610E+03	0.1266E+01	0.7648E−02
0.600E−01	0.2388E+01	0.2053E−01	0.1163E+03	0.1265E+01	0.1712E−01
0.800E−01	0.2379E+01	0.3621E−01	0.6569E+02	0.1262E+01	0.3022E−01
0.100E+00	0.2367E+01	0.5602E−01	0.4225E+02	0.1259E+01	0.4678E−01
0.120E+00	0.2353E+01	0.7970E−01	0.2952E+02	0.1255E+01	0.6661E−01
0.140E+00	0.2336E+01	0.1069E+00	0.2184E+02	0.1251E+01	0.8947E−01
0.160E+00	0.2317E+01	0.1374E+00	0.1686E+02	0.1246E+01	0.1151E+00
0.180E+00	0.2296E+01	0.1708E+00	0.1344E+02	0.1241E+01	0.1432E+00
0.200E+00	0.2273E+01	0.2066E+00	0.1100E+02	0.1235E+01	0.1736E+00
0.220E+00	0.2248E+01	0.2445E+00	0.9192E+01	0.1228E+01	0.2057E+00
0.240E+00	0.2221E+01	0.2841E+00	0.7817E+01	0.1221E+01	0.2395E+00
0.260E+00	0.2193E+01	0.3250E+00	0.6747E+01	0.1214E+01	0.2745E+00
0.280E+00	0.2163E+01	0.3667E+00	0.5898E+01	0.1206E+01	0.3104E+00
0.300E+00	0.2131E+01	0.4089E+00	0.5213E+01	0.1199E+01	0.3469E+00
0.320E+00	0.2099E+01	0.4512E+00	0.4652E+01	0.1190E+01	0.3837E+00
0.340E+00	0.2066E+01	0.4933E+00	0.4188E+01	0.1182E+01	0.4206E+00
0.360E+00	0.2031E+01	0.5348E+00	0.3798E+01	0.1174E+01	0.4572E+00
0.380E+00	0.1996E+01	0.5755E+00	0.3469E+01	0.1165E+01	0.4935E+00
0.400E+00	0.1961E+01	0.6151E+00	0.3187E+01	0.1157E+01	0.5290E+00
0.420E+00	0.1925E+01	0.6535E+00	0.2945E+01	0.1148E+01	0.5638E+00
0.440E+00	0.1888E+01	0.6903E+00	0.2736E+01	0.1139E+01	0.5975E+00
0.460E+00	0.1852E+01	0.7254E+00	0.2552E+01	0.1131E+01	0.6301E+00
0.480E+00	0.1815E+01	0.7587E+00	0.2392E+01	0.1122E+01	0.6614E+00
0.500E+00	0.1778E+01	0.7901E+00	0.2250E+01	0.1114E+01	0.6914E+00
0.520E+00	0.1741E+01	0.8196E+00	0.2124E+01	0.1106E+01	0.7199E+00
0.540E+00	0.1704E+01	0.8469E+00	0.2012E+01	0.1098E+01	0.7470E+00
0.560E+00	0.1668E+01	0.8723E+00	0.1912E+01	0.1090E+01	0.7725E+00
0.580E+00	0.1632E+01	0.8955E+00	0.1822E+01	0.1083E+01	0.7965E+00
0.600E+00	0.1596E+01	0.9167E+00	0.1741E+01	0.1075E+01	0.8189E+00
0.620E+00	0.1560E+01	0.9358E+00	0.1667E+01	0.1068E+01	0.8398E+00
0.640E+00	0.1525E+01	0.9530E+00	0.1601E+01	0.1061E+01	0.8592E+00
0.660E+00	0.1491E+01	0.9682E+00	0.1540E+01	0.1055E+01	0.8771E+00
0.680E+00	0.1457E+01	0.9814E+00	0.1484E+01	0.1049E+01	0.8935E+00
0.700E+00	0.1423E+01	0.9929E+00	0.1434E+01	0.1043E+01	0.9085E+00
0.720E+00	0.1391E+01	0.1003E+01	0.1387E+01	0.1038E+01	0.9221E+00
0.740E+00	0.1359E+01	0.1011E+01	0.1344E+01	0.1033E+01	0.9344E+00
0.760E+00	0.1327E+01	0.1017E+01	0.1305E+01	0.1028E+01	0.9455E+00
0.780E+00	0.1296E+01	0.1022E+01	0.1268E+01	0.1023E+01	0.9553E+00
0.800E+00	0.1266E+01	0.1025E+01	0.1234E+01	0.1019E+01	0.9639E+00
0.820E+00	0.1236E+01	0.1028E+01	0.1203E+01	0.1016E+01	0.9715E+00
0.840E+00	0.1207E+01	0.1029E+01	0.1174E+01	0.1012E+01	0.9781E+00
0.860E+00	0.1179E+01	0.1028E+01	0.1147E+01	0.1010E+01	0.9836E+00
0.880E+00	0.1152E+01	0.1027E+01	0.1121E+01	0.1007E+01	0.9883E+00
0.900E+00	0.1125E+01	0.1025E+01	0.1098E+01	0.1005E+01	0.9921E+00
0.920E+00	0.1098E+01	0.1021E+01	0.1076E+01	0.1003E+01	0.9951E+00
0.940E+00	0.1073E+01	0.1017E+01	0.1055E+01	0.1002E+01	0.9973E+00
0.960E+00	0.1048E+01	0.1012E+01	0.1035E+01	0.1001E+01	0.9988E+00
0.980E+00	0.1024E+01	0.1006E+01	0.1017E+01	0.1000E+01	0.9997E+00
0.100E+01	0.1000E+01	0.1000E+01	0.1000E+01	0.1000E+01	0.1000E+01

Table 2.12: One-dimensional flow with heat transfer (continued).

M	P/P^*	T/T^*	ρ/ρ^*	P_t/P_t^*	T_t/T_t^*
0.102E+01	0.9770E+00	0.9930E+00	0.9838E+00	0.1000E+01	0.9997E+00
0.104E+01	0.9546E+00	0.9855E+00	0.9686E+00	0.1001E+01	0.9989E+00
0.106E+01	0.9327E+00	0.9776E+00	0.9542E+00	0.1002E+01	0.9977E+00
0.108E+01	0.9115E+00	0.9691E+00	0.9406E+00	0.1003E+01	0.9960E+00
0.110E+01	0.8909E+00	0.9603E+00	0.9277E+00	0.1005E+01	0.9939E+00
0.112E+01	0.8708E+00	0.9512E+00	0.9155E+00	0.1007E+01	0.9915E+00
0.114E+01	0.8512E+00	0.9417E+00	0.9039E+00	0.1010E+01	0.9887E+00
0.116E+01	0.8322E+00	0.9320E+00	0.8930E+00	0.1012E+01	0.9856E+00
0.118E+01	0.8137E+00	0.9220E+00	0.8826E+00	0.1016E+01	0.9823E+00
0.120E+01	0.7958E+00	0.9118E+00	0.8727E+00	0.1019E+01	0.9787E+00
0.122E+01	0.7783E+00	0.9015E+00	0.8633E+00	0.1023E+01	0.9749E+00
0.124E+01	0.7613E+00	0.8911E+00	0.8543E+00	0.1028E+01	0.9709E+00
0.126E+01	0.7447E+00	0.8805E+00	0.8458E+00	0.1033E+01	0.9668E+00
0.128E+01	0.7287E+00	0.8699E+00	0.8376E+00	0.1038E+01	0.9624E+00
0.130E+01	0.7130E+00	0.8592E+00	0.8299E+00	0.1044E+01	0.9580E+00
0.132E+01	0.6978E+00	0.8484E+00	0.8225E+00	0.1050E+01	0.9534E+00
0.134E+01	0.6830E+00	0.8377E+00	0.8154E+00	0.1056E+01	0.9487E+00
0.136E+01	0.6686E+00	0.8269E+00	0.8086E+00	0.1063E+01	0.9440E+00
0.138E+01	0.6546E+00	0.8161E+00	0.8021E+00	0.1070E+01	0.9391E+00
0.140E+01	0.6410E+00	0.8054E+00	0.7959E+00	0.1078E+01	0.9343E+00
0.142E+01	0.6278E+00	0.7947E+00	0.7900E+00	0.1086E+01	0.9293E+00
0.144E+01	0.6149E+00	0.7840E+00	0.7843E+00	0.1094E+01	0.9243E+00
0.146E+01	0.6024E+00	0.7735E+00	0.7788E+00	0.1103E+01	0.9193E+00
0.148E+01	0.5902E+00	0.7629E+00	0.7736E+00	0.1112E+01	0.9143E+00
0.150E+01	0.5783E+00	0.7525E+00	0.7685E+00	0.1122E+01	0.9093E+00
0.152E+01	0.5668E+00	0.7422E+00	0.7637E+00	0.1132E+01	0.9042E+00
0.154E+01	0.5555E+00	0.7319E+00	0.7590E+00	0.1142E+01	0.8992E+00
0.156E+01	0.5446E+00	0.7217E+00	0.7545E+00	0.1153E+01	0.8942E+00
0.158E+01	0.5339E+00	0.7117E+00	0.7502E+00	0.1164E+01	0.8892E+00
0.160E+01	0.5236E+00	0.7017E+00	0.7461E+00	0.1176E+01	0.8842E+00
0.162E+01	0.5135E+00	0.6919E+00	0.7421E+00	0.1188E+01	0.8792E+00
0.164E+01	0.5036E+00	0.6822E+00	0.7383E+00	0.1200E+01	0.8743E+00
0.166E+01	0.4940E+00	0.6726E+00	0.7345E+00	0.1213E+01	0.8694E+00
0.168E+01	0.4847E+00	0.6631E+00	0.7310E+00	0.1226E+01	0.8645E+00
0.170E+01	0.4756E+00	0.6538E+00	0.7275E+00	0.1240E+01	0.8597E+00
0.172E+01	0.4668E+00	0.6445E+00	0.7242E+00	0.1254E+01	0.8549E+00
0.174E+01	0.4581E+00	0.6355E+00	0.7210E+00	0.1269E+01	0.8502E+00
0.176E+01	0.4497E+00	0.6265E+00	0.7178E+00	0.1284E+01	0.8455E+00
0.178E+01	0.4415E+00	0.6176E+00	0.7148E+00	0.1300E+01	0.8409E+00
0.180E+01	0.4335E+00	0.6089E+00	0.7119E+00	0.1316E+01	0.8363E+00
0.182E+01	0.4257E+00	0.6004E+00	0.7091E+00	0.1332E+01	0.8317E+00
0.184E+01	0.4181E+00	0.5919E+00	0.7064E+00	0.1349E+01	0.8273E+00
0.186E+01	0.4107E+00	0.5836E+00	0.7038E+00	0.1367E+01	0.8228E+00
0.188E+01	0.4035E+00	0.5754E+00	0.7012E+00	0.1385E+01	0.8185E+00
0.190E+01	0.3964E+00	0.5673E+00	0.6988E+00	0.1403E+01	0.8141E+00
0.192E+01	0.3896E+00	0.5594E+00	0.6964E+00	0.1422E+01	0.8099E+00
0.194E+01	0.3828E+00	0.5516E+00	0.6940E+00	0.1442E+01	0.8057E+00
0.196E+01	0.3763E+00	0.5439E+00	0.6918E+00	0.1462E+01	0.8015E+00
0.198E+01	0.3699E+00	0.5364E+00	0.6896E+00	0.1482E+01	0.7974E+00
0.200E+01	0.3636E+00	0.5289E+00	0.6875E+00	0.1503E+01	0.7934E+00

Table 2.12: One-dimensional flow with heat transfer (continued).

M	P/P^*	T/T^*	ρ/ρ^*	P_t/P_t^*	T_t/T_t^*
0.205E+01	0.3487E+00	0.5109E+00	0.6825E+00	0.1558E+01	0.7835E+00
0.210E+01	0.3345E+00	0.4936E+00	0.6778E+00	0.1616E+01	0.7741E+00
0.215E+01	0.3212E+00	0.4770E+00	0.6735E+00	0.1678E+01	0.7649E+00
0.220E+01	0.3086E+00	0.4611E+00	0.6694E+00	0.1743E+01	0.7561E+00
0.225E+01	0.2968E+00	0.4458E+00	0.6656E+00	0.1813E+01	0.7477E+00
0.230E+01	0.2855E+00	0.4312E+00	0.6621E+00	0.1886E+01	0.7395E+00
0.235E+01	0.2749E+00	0.4172E+00	0.6588E+00	0.1963E+01	0.7317E+00
0.240E+01	0.2648E+00	0.4038E+00	0.6557E+00	0.2045E+01	0.7242E+00
0.245E+01	0.2552E+00	0.3910E+00	0.6527E+00	0.2131E+01	0.7170E+00
0.250E+01	0.2462E+00	0.3787E+00	0.6500E+00	0.2222E+01	0.7101E+00
0.255E+01	0.2375E+00	0.3669E+00	0.6474E+00	0.2317E+01	0.7034E+00
0.260E+01	0.2294E+00	0.3556E+00	0.6450E+00	0.2418E+01	0.6970E+00
0.265E+01	0.2216E+00	0.3448E+00	0.6427E+00	0.2523E+01	0.6908E+00
0.270E+01	0.2142E+00	0.3344E+00	0.6405E+00	0.2634E+01	0.6849E+00
0.275E+01	0.2071E+00	0.3244E+00	0.6384E+00	0.2751E+01	0.6793E+00
0.280E+01	0.2004E+00	0.3149E+00	0.6365E+00	0.2873E+01	0.6738E+00
0.285E+01	0.1940E+00	0.3057E+00	0.6346E+00	0.3001E+01	0.6685E+00
0.290E+01	0.1879E+00	0.2969E+00	0.6329E+00	0.3136E+01	0.6635E+00
0.295E+01	0.1820E+00	0.2884E+00	0.6312E+00	0.3277E+01	0.6586E+00
0.300E+01	0.1765E+00	0.2803E+00	0.6296E+00	0.3424E+01	0.6540E+00
0.305E+01	0.1711E+00	0.2725E+00	0.6281E+00	0.3579E+01	0.6495E+00
0.310E+01	0.1660E+00	0.2650E+00	0.6267E+00	0.3741E+01	0.6452E+00
0.315E+01	0.1612E+00	0.2577E+00	0.6253E+00	0.3910E+01	0.6410E+00
0.320E+01	0.1565E+00	0.2508E+00	0.6240E+00	0.4087E+01	0.6370E+00
0.325E+01	0.1520E+00	0.2441E+00	0.6228E+00	0.4272E+01	0.6331E+00
0.330E+01	0.1477E+00	0.2377E+00	0.6216E+00	0.4465E+01	0.6294E+00
0.335E+01	0.1436E+00	0.2315E+00	0.6205E+00	0.4667E+01	0.6258E+00
0.340E+01	0.1397E+00	0.2255E+00	0.6194E+00	0.4878E+01	0.6224E+00
0.345E+01	0.1359E+00	0.2197E+00	0.6183E+00	0.5098E+01	0.6190E+00
0.350E+01	0.1322E+00	0.2142E+00	0.6173E+00	0.5328E+01	0.6158E+00
0.355E+01	0.1287E+00	0.2088E+00	0.6164E+00	0.5568E+01	0.6127E+00
0.360E+01	0.1254E+00	0.2037E+00	0.6155E+00	0.5817E+01	0.6097E+00
0.365E+01	0.1221E+00	0.1987E+00	0.6146E+00	0.6078E+01	0.6068E+00
0.370E+01	0.1190E+00	0.1939E+00	0.6138E+00	0.6349E+01	0.6040E+00
0.375E+01	0.1160E+00	0.1893E+00	0.6130E+00	0.6631E+01	0.6013E+00
0.380E+01	0.1131E+00	0.1848E+00	0.6122E+00	0.6926E+01	0.5987E+00
0.385E+01	0.1103E+00	0.1805E+00	0.6114E+00	0.7232E+01	0.5962E+00
0.390E+01	0.1077E+00	0.1763E+00	0.6107E+00	0.7550E+01	0.5937E+00
0.395E+01	0.1051E+00	0.1722E+00	0.6100E+00	0.7882E+01	0.5914E+00
0.400E+01	0.1026E+00	0.1683E+00	0.6094E+00	0.8227E+01	0.5891E+00
0.405E+01	0.1002E+00	0.1645E+00	0.6087E+00	0.8585E+01	0.5869E+00
0.410E+01	0.9782E−01	0.1609E+00	0.6081E+00	0.8958E+01	0.5847E+00
0.415E+01	0.9557E−01	0.1573E+00	0.6075E+00	0.9345E+01	0.5827E+00
0.420E+01	0.9340E−01	0.1539E+00	0.6070E+00	0.9747E+01	0.5807E+00
0.425E+01	0.9130E−01	0.1506E+00	0.6064E+00	0.1016E+02	0.5787E+00
0.430E+01	0.8927E−01	0.1473E+00	0.6059E+00	0.1060E+02	0.5768E+00
0.435E+01	0.8730E−01	0.1442E+00	0.6054E+00	0.1105E+02	0.5750E+00
0.440E+01	0.8540E−01	0.1412E+00	0.6049E+00	0.1152E+02	0.5732E+00
0.445E+01	0.8356E−01	0.1383E+00	0.6044E+00	0.1200E+02	0.5715E+00
0.450E+01	0.8177E−01	0.1354E+00	0.6039E+00	0.1250E+02	0.5698E+00

Table 2.12: One-dimensional flow with heat transfer (continued).

M	P/P^*	T/T^*	ρ/ρ^*	P_t/P_t^*	T_t/T_t^*
0.455E+01	0.8004E−01	0.1326E+00	0.6035E+00	0.1302E+02	0.5682E+00
0.460E+01	0.7837E−01	0.1300E+00	0.6030E+00	0.1356E+02	0.5666E+00
0.465E+01	0.7675E−01	0.1274E+00	0.6026E+00	0.1412E+02	0.5651E+00
0.470E+01	0.7517E−01	0.1248E+00	0.6022E+00	0.1470E+02	0.5636E+00
0.475E+01	0.7365E−01	0.1224E+00	0.6018E+00	0.1530E+02	0.5622E+00
0.480E+01	0.7217E−01	0.1200E+00	0.6014E+00	0.1592E+02	0.5608E+00
0.485E+01	0.7073E−01	0.1177E+00	0.6010E+00	0.1657E+02	0.5594E+00
0.490E+01	0.6934E−01	0.1154E+00	0.6007E+00	0.1723E+02	0.5581E+00
0.495E+01	0.6798E−01	0.1132E+00	0.6003E+00	0.1792E+02	0.5568E+00
0.500E+01	0.6667E−01	0.1111E+00	0.6000E+00	0.1863E+02	0.5556E+00
0.510E+01	0.6415E−01	0.1070E+00	0.5994E+00	0.2013E+02	0.5532E+00
0.520E+01	0.6177E−01	0.1032E+00	0.5987E+00	0.2173E+02	0.5509E+00
0.530E+01	0.5951E−01	0.9950E−01	0.5982E+00	0.2344E+02	0.5487E+00
0.540E+01	0.5738E−01	0.9602E−01	0.5976E+00	0.2527E+02	0.5467E+00
0.550E+01	0.5536E−01	0.9272E−01	0.5971E+00	0.2721E+02	0.5447E+00
0.560E+01	0.5345E−01	0.8958E−01	0.5966E+00	0.2928E+02	0.5429E+00
0.570E+01	0.5163E−01	0.8660E−01	0.5962E+00	0.3148E+02	0.5411E+00
0.580E+01	0.4990E−01	0.8376E−01	0.5957E+00	0.3382E+02	0.5394E+00
0.590E+01	0.4826E−01	0.8106E−01	0.5953E+00	0.3631E+02	0.5378E+00
0.600E+01	0.4669E−01	0.7849E−01	0.5949E+00	0.3895E+02	0.5363E+00
0.610E+01	0.4520E−01	0.7603E−01	0.5945E+00	0.4174E+02	0.5349E+00
0.620E+01	0.4378E−01	0.7369E−01	0.5942E+00	0.4471E+02	0.5335E+00
0.630E+01	0.4243E−01	0.7145E−01	0.5938E+00	0.4785E+02	0.5322E+00
0.640E+01	0.4114E−01	0.6931E−01	0.5935E+00	0.5117E+02	0.5309E+00
0.650E+01	0.3990E−01	0.6726E−01	0.5932E+00	0.5468E+02	0.5297E+00
0.660E+01	0.3872E−01	0.6531E−01	0.5929E+00	0.5840E+02	0.5285E+00
0.670E+01	0.3759E−01	0.6343E−01	0.5926E+00	0.6232E+02	0.5274E+00
0.680E+01	0.3651E−01	0.6164E−01	0.5923E+00	0.6645E+02	0.5264E+00
0.690E+01	0.3547E−01	0.5991E−01	0.5921E+00	0.7082E+02	0.5254E+00
0.700E+01	0.3448E−01	0.5826E−01	0.5918E+00	0.7541E+02	0.5244E+00
0.710E+01	0.3353E−01	0.5668E−01	0.5916E+00	0.8026E+02	0.5234E+00
0.720E+01	0.3262E−01	0.5516E−01	0.5914E+00	0.8536E+02	0.5225E+00
0.730E+01	0.3174E−01	0.5370E−01	0.5912E+00	0.9072E+02	0.5217E+00
0.740E+01	0.3090E−01	0.5229E−01	0.5909E+00	0.9636E+02	0.5208E+00
0.750E+01	0.3009E−01	0.5094E−01	0.5907E+00	0.1023E+03	0.5200E+00
0.760E+01	0.2932E−01	0.4964E−01	0.5905E+00	0.1085E+03	0.5193E+00
0.770E+01	0.2857E−01	0.4839E−01	0.5904E+00	0.1150E+03	0.5185E+00
0.780E+01	0.2785E−01	0.4719E−01	0.5902E+00	0.1219E+03	0.5178E+00
0.790E+01	0.2716E−01	0.4603E−01	0.5900E+00	0.1291E+03	0.5171E+00
0.800E+01	0.2649E−01	0.4491E−01	0.5898E+00	0.1366E+03	0.5165E+00
0.850E+01	0.2349E−01	0.3988E−01	0.5891E+00	0.1799E+03	0.5135E+00
0.900E+01	0.2098E−01	0.3565E−01	0.5885E+00	0.2339E+03	0.5110E+00
0.950E+01	0.1885E−01	0.3205E−01	0.5880E+00	0.3004E+03	0.5088E+00
0.100E+02	0.1702E−01	0.2897E−01	0.5875E+00	0.3816E+03	0.5070E+00
0.105E+02	0.1545E−01	0.2631E−01	0.5871E+00	0.4799E+03	0.5054E+00
0.110E+02	0.1408E−01	0.2400E−01	0.5868E+00	0.5977E+03	0.5041E+00
0.115E+02	0.1289E−01	0.2198E−01	0.5865E+00	0.7381E+03	0.5029E+00
0.120E+02	0.1185E−01	0.2021E−01	0.5862E+00	0.9041E+03	0.5018E+00
0.125E+02	0.1092E−01	0.1864E−01	0.5860E+00	0.1099E+04	0.5009E+00
0.130E+02	0.1010E−01	0.1724E−01	0.5858E+00	0.1327E+04	0.5001E+00

Table 2.12: One-dimensional flow with heat transfer (continued).

M	P/P^*	T/T^*	ρ/ρ^*	P_t/P_t^*	T_t/T_t^*
0.135E+02	0.9370E−02	0.1600E−01	0.5856E+00	0.1591E+04	0.4993E+00
0.140E+02	0.8715E−02	0.1489E−01	0.5855E+00	0.1896E+04	0.4986E+00
0.145E+02	0.8126E−02	0.1388E−01	0.5853E+00	0.2247E+04	0.4981E+00
0.150E+02	0.7595E−02	0.1298E−01	0.5852E+00	0.2649E+04	0.4975E+00
0.155E+02	0.7114E−02	0.1216E−01	0.5851E+00	0.3106E+04	0.4970E+00
0.160E+02	0.6678E−02	0.1142E−01	0.5850E+00	0.3625E+04	0.4966E+00
0.165E+02	0.6280E−02	0.1074E−01	0.5849E+00	0.4212E+04	0.4962E+00
0.170E+02	0.5917E−02	0.1012E−01	0.5848E+00	0.4873E+04	0.4958E+00
0.175E+02	0.5585E−02	0.9551E−02	0.5847E+00	0.5615E+04	0.4955E+00
0.180E+02	0.5279E−02	0.9030E−02	0.5846E+00	0.6445E+04	0.4952E+00
0.185E+02	0.4998E−02	0.8551E−02	0.5846E+00	0.7371E+04	0.4949E+00
0.190E+02	0.4739E−02	0.8109E−02	0.5845E+00	0.8402E+04	0.4946E+00
0.195E+02	0.4500E−02	0.7700E−02	0.5844E+00	0.9545E+04	0.4944E+00
0.200E+02	0.4278E−02	0.7321E−02	0.5844E+00	0.1081E+05	0.4942E+00

Table 2.13: Isentropic relations for air, $\gamma = 1.4$, subsonic flow.

M	$\frac{p}{p_t}$	$\frac{\rho}{\rho_t}$	$\frac{T}{T_t}$	$\frac{A}{A^*}$	M	$\frac{p}{p_t}$	$\frac{\rho}{\rho_t}$	$\frac{T}{T_t}$	$\frac{A}{A^*}$
0.00	1.0000	1.0000	1.0000	∞	0.50	0.8430	0.8852	0.9524	1.3398
0.01	0.9999	0.9999	1.0000	57.8738	0.51	0.8374	0.8809	0.9506	1.3212
0.02	0.9997	0.9998	0.9999	28.9421	0.52	0.8317	0.8766	0.9487	1.3034
0.03	0.9994	0.9996	0.9998	19.3005	0.53	0.8259	0.8723	0.9468	1.2865
0.04	0.9989	0.9992	0.9997	14.4815	0.54	0.8201	0.8679	0.9449	1.2703
0.05	0.9983	0.9988	0.9995	11.5914	0.55	0.8142	0.8634	0.9430	1.2549
0.06	0.9975	0.9982	0.9993	9.6659	0.56	0.8082	0.8589	0.9410	1.2403
0.07	0.9966	0.9976	0.9990	8.2915	0.57	0.8022	0.8544	0.9390	1.2263
0.08	0.9955	0.9968	0.9987	7.2616	0.58	0.7962	0.8498	0.9370	1.2130
0.09	0.9944	0.9960	0.9984	6.4613	0.59	0.7901	0.8451	0.9349	1.2003
0.10	0.9930	0.9950	0.9980	5.2188	0.60	0.7840	0.8405	0.9328	1.1882
0.11	0.9916	0.9940	0.9976	5.2992	0.61	0.7778	0.8357	0.9307	1.1767
0.12	0.9900	0.9928	0.9971	4.8643	0.62	0.7716	0.8310	0.9286	1.1656
0.13	0.9883	0.9916	0.9966	4.4969	0.63	0.7654	0.8262	0.9265	1.1552
0.14	0.9864	0.9903	0.9961	4.1824	0.64	0.7591	0.8213	0.9243	1.1451
0.15	0.9844	0.9888	0.9955	3.9103	0.65	0.7528	0.8164	0.9221	1.1356
0.16	0.9823	0.9873	0.9949	3.6727	0.66	0.7465	0.8115	0.9199	1.1265
0.17	0.9800	0.9857	0.9943	3.4635	0.67	0.7401	0.8066	0.9176	1.1179
0.18	0.9776	0.9840	0.9936	3.2779	0.68	0.7338	0.8016	0.9153	1.1097
0.19	0.9751	0.9822	0.9928	3.1123	0.69	0.7274	0.7966	0.9131	1.1018
0.20	0.9668	0.9762	0.9904	2.7076	0.70	0.7209	0.7916	0.9107	1.0944
0.21	0.9638	0.9740	0.9895	2.5968	0.71	0.7145	0.7865	0.9084	1.0873
0.22	0.9607	0.9718	0.9886	2.4956	0.72	0.7080	0.7814	0.9061	1.0806
0.23	0.9638	0.9740	0.9895	2.5968	0.73	0.7016	0.7763	0.9037	1.0742
0.24	0.9607	0.9718	0.9886	2.4956	0.74	0.6951	0.7712	0.9013	1.0681
0.25	0.9575	0.9694	0.9877	2.4027	0.75	0.6886	0.7660	0.8989	1.0624
0.26	0.9541	0.9670	0.9867	2.3173	0.76	0.6821	0.7609	0.8964	1.0570
0.27	0.9506	0.9645	0.9856	2.2385	0.77	0.6756	0.7557	0.8940	1.0519
0.28	0.9470	0.9619	0.9846	2.1656	0.78	0.6691	0.7505	0.8915	1.0471
0.29	0.9433	0.9592	0.9835	2.0979	0.79	0.6625	0.7452	0.8890	1.0425
0.30	0.9395	0.9564	0.9823	2.0351	0.80	0.6560	0.7400	0.8865	1.0382
0.31	0.9355	0.9535	0.9811	1.9765	0.81	0.6495	0.7347	0.8840	1.0342
0.32	0.9315	0.9506	0.9799	1.9219	0.82	0.6430	0.7295	0.8815	1.0305
0.33	0.9274	0.9476	0.9787	1.8707	0.83	0.6365	0.7242	0.8789	1.0270
0.34	0.9231	0.9445	0.9774	1.8229	0.84	0.6300	0.7189	0.8763	1.0237
0.35	0.9188	0.9413	0.9761	1.7780	0.85	0.6235	0.7136	0.8737	1.0207
0.36	0.9143	0.9380	0.9747	1.7358	0.86	0.6170	0.7083	0.8711	1.0179
0.37	0.9098	0.9347	0.9733	1.6961	0.87	0.6106	0.7030	0.8685	1.0153
0.38	0.9052	0.9313	0.9719	1.6587	0.88	0.6041	0.6977	0.8659	1.0129
0.39	0.9004	0.9278	0.9705	1.6234	0.89	0.5977	0.6924	0.8632	1.0108
0.40	0.8956	0.9243	0.9690	1.5901	0.90	0.5913	0.6870	0.8606	1.0089
0.41	0.8907	0.9207	0.9675	1.5587	0.91	0.5849	0.6817	0.8579	1.0071
0.42	0.8857	0.9170	0.9659	1.5289	0.92	0.5785	0.6764	0.8552	1.0056
0.43	0.8807	0.9132	0.9643	1.5007	0.93	0.5721	0.6711	0.8525	1.0043
0.44	0.8755	0.9094	0.9627	1.4740	0.94	0.5658	0.6658	0.8498	1.0031
0.45	0.8703	0.9055	0.9611	1.4487	0.95	0.5595	0.6604	0.8471	1.0021
0.46	0.8650	0.9016	0.9594	1.4246	0.96	0.5532	0.6551	0.8444	1.0014
0.47	0.8596	0.8976	0.9577	1.4018	0.97	0.5469	0.6498	0.8416	1.0008
0.48	0.8541	0.8935	0.9559	1.3801	0.98	0.5407	0.6445	0.8389	1.0003
0.49	0.8486	0.8894	0.9542	1.3595	0.99	0.5345	0.6392	0.8361	1.0001
					1.00	0.5283	0.6339	0.8333	1.0000

Table 2.14: Isentropic and normal shock properties for air, $\gamma = 1.4$.

M or M_1	$\frac{P}{P_t}$	$\frac{\rho}{\rho_t}$	$\frac{T}{T_t}$	$\frac{A}{A^*}$	ν	μ	M_2	$\frac{P_2}{P_1}$	$\frac{\rho_2}{\rho_1}$	$\frac{T_2}{T_1}$	$\frac{P_{t_2}}{P_{t_1}}$	$\frac{P_1}{P_{t_2}}$
1.00	0.5283	0.6339	0.8333	1.000	0.000	90.00	1.0000	1.000	1.000	1.000	1.0000	0.5283
1.01	0.5221	0.6287	0.8306	1.000	0.045	81.93	0.9901	1.023	1.017	1.007	1.0000	0.5221
1.02	0.5160	0.6234	0.8278	1.000	0.126	78.64	0.9805	1.047	1.033	1.013	1.0000	0.5160
1.03	0.5099	0.6181	0.8250	1.001	0.229	76.14	0.9712	1.071	1.050	1.020	1.0000	0.5100
1.04	0.5039	0.6129	0.8222	1.001	0.351	74.06	0.9620	1.095	1.067	1.026	0.9999	0.5039
1.05	0.4979	0.6077	0.8193	1.002	0.487	72.25	0.9531	1.120	1.084	1.033	0.9999	0.4979
1.06	0.4919	0.6024	0.8165	1.003	0.637	70.63	0.9444	1.144	1.101	1.039	0.9998	0.4920
1.07	0.4860	0.5972	0.8137	1.004	0.797	69.16	0.9360	1.169	1.118	1.046	0.9996	0.4861
1.08	0.4800	0.5920	0.8108	1.005	0.968	67.81	0.9277	1.194	1.135	1.052	0.9994	0.4803
1.09	0.4742	0.5869	0.8080	1.006	1.148	66.55	0.9196	1.219	1.152	1.059	0.9992	0.4746
1.10	0.4684	0.5817	0.8052	1.008	1.336	65.38	0.9118	1.245	1.169	1.065	0.9989	0.4689
1.11	0.4626	0.5766	0.8023	1.010	1.532	64.28	0.9041	1.271	1.186	1.071	0.9986	0.4632
1.12	0.4568	0.5714	0.7994	1.011	1.735	63.23	0.8966	1.297	1.203	1.078	0.9982	0.4576
1.13	0.4511	0.5663	0.7966	1.013	1.944	62.25	0.8892	1.323	1.221	1.084	0.9978	0.4521
1.14	0.4455	0.5612	0.7937	1.015	2.160	61.31	0.8820	1.350	1.238	1.090	0.9973	0.4467
1.15	0.4398	0.5562	0.7908	1.017	2.381	60.41	0.8750	1.376	1.255	1.097	0.9967	0.4413
1.16	0.4343	0.5511	0.7879	1.020	2.607	59.55	0.8682	1.403	1.272	1.103	0.9961	0.4360
1.17	0.4287	0.5461	0.7851	1.022	2.839	58.73	0.8615	1.430	1.290	1.109	0.9953	0.4307
1.18	0.4232	0.5411	0.7822	1.025	3.074	57.94	0.8549	1.458	1.307	1.115	0.9946	0.4255
1.19	0.4178	0.5361	0.7793	1.028	3.314	57.18	0.8485	1.485	1.324	1.122	0.9937	0.4204
1.20	0.4124	0.5311	0.7764	1.030	3.558	56.44	0.8422	1.513	1.342	1.128	0.9928	0.4154
1.21	0.4070	0.5262	0.7735	1.033	3.806	55.74	0.8360	1.541	1.359	1.134	0.9918	0.4104
1.22	0.4017	0.5213	0.7706	1.037	4.057	55.05	0.8300	1.570	1.376	1.141	0.9907	0.4055
1.23	0.3964	0.5164	0.7677	1.040	4.312	54.39	0.8241	1.598	1.394	1.147	0.9896	0.4006
1.24	0.3912	0.5115	0.7648	1.043	4.569	53.75	0.8183	1.627	1.411	1.153	0.9884	0.3958
1.25	0.3861	0.5067	0.7619	1.047	4.830	53.13	0.8126	1.656	1.429	1.159	0.9871	0.3911
1.26	0.3809	0.5019	0.7590	1.050	5.093	52.53	0.8071	1.686	1.446	1.166	0.9857	0.3865
1.27	0.3759	0.4971	0.7561	1.054	5.359	51.94	0.8016	1.715	1.463	1.172	0.9842	0.3819
1.28	0.3708	0.4923	0.7532	1.058	5.627	51.38	0.7963	1.745	1.481	1.178	0.9827	0.3774
1.29	0.3658	0.4876	0.7503	1.062	5.898	50.82	0.7911	1.775	1.498	1.185	0.9811	0.3729
1.30	0.3609	0.4829	0.7474	1.066	6.170	50.28	0.7860	1.805	1.516	1.191	0.9794	0.3685
1.31	0.3560	0.4782	0.7445	1.071	6.445	49.76	0.7809	1.835	1.533	1.197	0.9776	0.3642
1.32	0.3512	0.4736	0.7416	1.075	6.721	49.25	0.7760	1.866	1.551	1.204	0.9758	0.3599
1.33	0.3464	0.4690	0.7387	1.080	7.000	48.75	0.7712	1.897	1.568	1.210	0.9738	0.3557
1.34	0.3417	0.4644	0.7358	1.084	7.279	48.27	0.7664	1.928	1.585	1.216	0.9718	0.3516
1.35	0.3370	0.4598	0.7329	1.089	7.561	47.79	0.7618	1.960	1.603	1.223	0.9697	0.3475
1.36	0.3323	0.4553	0.7300	1.094	7.843	47.33	0.7572	1.991	1.620	1.229	0.9676	0.3435
1.37	0.3277	0.4508	0.7271	1.099	8.128	46.88	0.7527	2.023	1.638	1.235	0.9653	0.3395
1.38	0.3232	0.4463	0.7242	1.104	8.413	46.44	0.7483	2.055	1.655	1.242	0.9630	0.3356
1.39	0.3187	0.4418	0.7213	1.109	8.699	46.01	0.7440	2.087	1.672	1.248	0.9607	0.3317
1.40	0.3142	0.4374	0.7184	1.115	8.987	45.58	0.7397	2.120	1.690	1.255	0.9582	0.3280
1.41	0.3098	0.4330	0.7155	1.120	9.276	45.17	0.7355	2.153	1.707	1.261	0.9557	0.3242
1.42	0.3055	0.4287	0.7126	1.126	9.565	44.77	0.7314	2.186	1.724	1.268	0.9531	0.3205
1.43	0.3012	0.4244	0.7097	1.132	9.855	44.37	0.7274	2.219	1.742	1.274	0.9504	0.3169
1.44	0.2969	0.4201	0.7069	1.138	10.146	43.98	0.7235	2.253	1.759	1.281	0.9476	0.3133
1.45	0.2927	0.4158	0.7040	1.144	10.438	43.60	0.7196	2.286	1.776	1.287	0.9448	0.3098
1.46	0.2886	0.4116	0.7011	1.150	10.730	43.23	0.7157	2.320	1.793	1.294	0.9420	0.3063
1.47	0.2845	0.4074	0.6982	1.156	11.023	42.86	0.7120	2.354	1.811	1.300	0.9390	0.3029
1.48	0.2804	0.4032	0.6954	1.163	11.317	42.51	0.7083	2.389	1.828	1.307	0.9360	0.2996
1.49	0.2764	0.3991	0.6925	1.169	11.611	42.16	0.7047	2.423	1.845	1.314	0.9329	0.2962
1.50	0.2724	0.3950	0.6897	1.176	11.905	41.81	0.7011	2.458	1.862	1.320	0.9298	0.2930
1.51	0.2685	0.3909	0.6868	1.183	12.200	41.47	0.6976	2.493	1.879	1.327	0.9266	0.2898
1.52	0.2646	0.3869	0.6840	1.190	12.495	41.14	0.6941	2.529	1.896	1.334	0.9233	0.2866
1.53	0.2608	0.3829	0.6811	1.197	12.790	40.81	0.6907	2.564	1.913	1.340	0.9200	0.2835
1.54	0.2570	0.3789	0.6783	1.204	13.086	40.49	0.6874	2.600	1.930	1.347	0.9166	0.2804

Table 2.14: Isentropic and normal shock properties for air, $\gamma = 1.4$ (continued).

M or M_1	$\dfrac{P}{P_t}$	$\dfrac{\rho}{\rho_t}$	$\dfrac{T}{T_t}$	$\dfrac{A}{A^*}$	ν	μ	M_2	$\dfrac{P_2}{P_1}$	$\dfrac{\rho_2}{\rho_1}$	$\dfrac{T_2}{T_1}$	$\dfrac{P_{t_2}}{P_{t_1}}$	$\dfrac{P_1}{P_{t_2}}$
1.55	0.2533	0.3750	0.6754	1.212	13.381	40.18	0.6841	2.636	1.947	1.354	0.9132	0.2773
1.56	0.2496	0.3710	0.6726	1.219	13.677	39.87	0.6809	2.673	1.964	1.361	0.9097	0.2743
1.57	0.2459	0.3672	0.6698	1.227	13.973	39.56	0.6777	2.709	1.981	1.367	0.9062	0.2714
1.58	0.2423	0.3633	0.6670	1.234	14.269	39.27	0.6746	2.746	1.998	1.374	0.9026	0.2685
1.59	0.2388	0.3595	0.6642	1.242	14.564	38.97	0.6715	2.783	2.015	1.381	0.8989	0.2656
1.60	0.2353	0.3557	0.6614	1.250	14.860	38.68	0.6684	2.820	2.032	1.388	0.8952	0.2628
1.61	0.2318	0.3520	0.6586	1.258	15.156	38.40	0.6655	2.857	2.049	1.395	0.8915	0.2600
1.62	0.2284	0.3483	0.6558	1.267	15.452	38.12	0.6625	2.895	2.065	1.402	0.8877	0.2573
1.63	0.2250	0.3446	0.6530	1.275	15.747	37.84	0.6596	2.933	2.082	1.409	0.8838	0.2546
1.64	0.2217	0.3409	0.6502	1.284	16.043	37.57	0.6568	2.971	2.099	1.416	0.8799	0.2519
1.65	0.2184	0.3373	0.6475	1.292	16.338	37.31	0.6540	3.010	2.115	1.423	0.8760	0.2493
1.66	0.2152	0.3337	0.6447	1.301	16.633	37.04	0.6512	3.048	2.132	1.430	0.8720	0.2467
1.67	0.2119	0.3302	0.6419	1.310	16.928	36.78	0.6485	3.087	2.148	1.437	0.8680	0.2442
1.68	0.2088	0.3266	0.6392	1.319	17.222	36.53	0.6458	3.126	2.165	1.444	0.8639	0.2417
1.69	0.2057	0.3232	0.6364	1.328	17.516	36.28	0.6431	3.165	2.181	1.451	0.8599	0.2392
1.70	0.2026	0.3197	0.6337	1.338	17.810	36.03	0.6405	3.205	2.198	1.458	0.8557	0.2368
1.71	0.1996	0.3163	0.6310	1.347	18.103	35.79	0.6380	3.245	2.214	1.466	0.8516	0.2343
1.72	0.1966	0.3129	0.6283	1.357	18.396	35.55	0.6355	3.285	2.230	1.473	0.8474	0.2320
1.73	0.1936	0.3095	0.6256	1.367	18.689	35.31	0.6330	3.325	2.247	1.480	0.8431	0.2296
1.74	0.1907	0.3062	0.6229	1.376	18.981	35.08	0.6305	3.366	2.263	1.487	0.8389	0.2273
1.75	0.1878	0.3029	0.6202	1.386	19.273	34.85	0.6281	3.406	2.279	1.495	0.8346	0.2251
1.76	0.1850	0.2996	0.6175	1.397	19.565	34.62	0.6257	3.447	2.295	1.502	0.8302	0.2228
1.77	0.1822	0.2964	0.6148	1.407	19.855	34.40	0.6234	3.488	2.311	1.509	0.8259	0.2206
1.78	0.1794	0.2931	0.6121	1.418	20.146	34.18	0.6210	3.530	2.327	1.517	0.8215	0.2184
1.79	0.1767	0.2900	0.6095	1.428	20.436	33.96	0.6188	3.571	2.343	1.524	0.8171	0.2163
1.80	0.1740	0.2868	0.6068	1.439	20.725	33.75	0.6165	3.613	2.359	1.532	0.8127	0.2142
1.81	0.1714	0.2837	0.6041	1.450	21.014	33.54	0.6143	3.655	2.375	1.539	0.8082	0.2121
1.82	0.1688	0.2806	0.6015	1.461	21.302	33.33	0.6121	3.698	2.391	1.547	0.8038	0.2100
1.83	0.1662	0.2776	0.5989	1.472	21.590	33.12	0.6099	3.740	2.407	1.554	0.7993	0.2080
1.84	0.1637	0.2745	0.5963	1.484	21.877	32.92	0.6078	3.783	2.422	1.562	0.7948	0.2060
1.85	0.1612	0.2715	0.5936	1.495	22.163	32.72	0.6057	3.826	2.438	1.569	0.7902	0.2040
1.86	0.1587	0.2686	0.5910	1.507	22.449	32.52	0.6036	3.870	2.454	1.577	0.7857	0.2020
1.87	0.1563	0.2656	0.5885	1.519	22.734	32.33	0.6016	3.913	2.469	1.585	0.7811	0.2001
1.88	0.1539	0.2627	0.5859	1.531	23.019	32.13	0.5996	3.957	2.485	1.592	0.7765	0.1982
1.89	0.1516	0.2598	0.5833	1.543	23.303	31.94	0.5976	4.001	2.500	1.600	0.7720	0.1963
1.90	0.1492	0.2570	0.5807	1.555	23.586	31.76	0.5956	4.045	2.516	1.608	0.7674	0.1945
1.91	0.1470	0.2542	0.5782	1.568	23.869	31.57	0.5937	4.089	2.531	1.616	0.7627	0.1927
1.92	0.1447	0.2514	0.5756	1.580	24.151	31.39	0.5918	4.134	2.546	1.624	0.7581	0.1909
1.93	0.1425	0.2486	0.5731	1.593	24.432	31.21	0.5899	4.179	2.562	1.631	0.7535	0.1891
1.94	0.1403	0.2459	0.5705	1.606	24.712	31.03	0.5880	4.224	2.577	1.639	0.7488	0.1873
1.95	0.1381	0.2432	0.5680	1.619	24.992	30.85	0.5862	4.270	2.592	1.647	0.7442	0.1856
1.96	0.1360	0.2405	0.5655	1.633	25.271	30.68	0.5844	4.315	2.607	1.655	0.7395	0.1839
1.97	0.1339	0.2378	0.5630	1.646	25.549	30.51	0.5826	4.361	2.622	1.663	0.7349	0.1822
1.98	0.1318	0.2352	0.5605	1.660	25.827	30.33	0.5808	4.407	2.637	1.671	0.7302	0.1806
1.99	0.1298	0.2326	0.5580	1.674	26.104	30.17	0.5791	4.453	2.652	1.679	0.7255	0.1789
2.00	0.1278	0.2300	0.5556	1.687	26.380	30.00	0.5774	4.500	2.667	1.687	0.7209	0.1773
2.01	0.1258	0.2275	0.5531	1.702	26.655	29.84	0.5757	4.547	2.681	1.696	0.7162	0.1757
2.02	0.1239	0.2250	0.5506	1.716	26.930	29.67	0.5740	4.594	2.696	1.704	0.7115	0.1741
2.03	0.1220	0.2225	0.5482	1.730	27.203	29.51	0.5723	4.641	2.711	1.712	0.7069	0.1726
2.04	0.1201	0.2200	0.5458	1.745	27.476	29.35	0.5707	4.689	2.725	1.720	0.7022	0.1710
2.05	0.1182	0.2176	0.5433	1.760	27.748	29.20	0.5691	4.736	2.740	1.729	0.6975	0.1695
2.06	0.1164	0.2152	0.5409	1.775	28.020	29.04	0.5675	4.784	2.755	1.737	0.6928	0.1680
2.07	0.1146	0.2128	0.5385	1.790	28.290	28.89	0.5659	4.832	2.769	1.745	0.6882	0.1665
2.08	0.1128	0.2104	0.5361	1.806	28.560	28.74	0.5643	4.881	2.783	1.754	0.6835	0.1651
2.09	0.1111	0.2081	0.5337	1.821	28.829	28.59	0.5628	4.929	2.798	1.762	0.6789	0.1636

Table 2.14: Isentropic and normal shock properties for air, $\gamma = 1.4$ (continued).

M or M_1	$\frac{P}{P_t}$	$\frac{\rho}{\rho_t}$	$\frac{T}{T_t}$	$\frac{A}{A^*}$	ν	μ	M_2	$\frac{P_2}{P_1}$	$\frac{\rho_2}{\rho_1}$	$\frac{T_2}{T_1}$	$\frac{P_{t_2}}{P_{t_1}}$	$\frac{P_1}{P_{t_2}}$
2.10	0.1094	0.2058	0.5314	1.837	29.097	28.44	0.5613	4.978	2.812	1.770	0.6742	0.1622
2.11	0.1077	0.2035	0.5290	1.853	29.364	28.29	0.5598	5.027	2.826	1.779	0.6696	0.1608
2.12	0.1060	0.2013	0.5266	1.869	29.631	28.14	0.5583	5.077	2.840	1.787	0.6649	0.1594
2.13	0.1043	0.1990	0.5243	1.885	29.896	28.00	0.5568	5.126	2.854	1.796	0.6603	0.1580
2.14	0.1027	0.1968	0.5219	1.902	30.161	27.86	0.5554	5.176	2.868	1.805	0.6557	0.1567
2.15	0.1011	0.1946	0.5196	1.919	30.425	27.72	0.5540	5.226	2.882	1.813	0.6511	0.1553
2.16	0.0996	0.1925	0.5173	1.935	30.688	27.58	0.5525	5.277	2.896	1.822	0.6464	0.1540
2.17	0.9802 −01	0.1903	0.5150	1.953	30.951	27.44	0.5511	5.327	2.910	1.831	0.6419	0.1527
2.18	0.9650 −01	0.1882	0.5127	1.970	31.212	27.30	0.5498	5.378	2.924	1.839	0.6373	0.1514
2.19	0.9500 −01	0.1861	0.5104	1.987	31.473	27.17	0.5484	5.429	2.938	1.848	0.6327	0.1501
2.20	0.9352 −01	0.1841	0.5081	2.005	31.732	27.04	0.5471	5.480	2.951	1.857	0.6281	0.1489
2.21	0.9207 −01	0.1820	0.5059	2.023	31.991	26.90	0.5457	5.531	2.965	1.866	0.6236	0.1476
2.22	0.9064 −01	0.1800	0.5036	2.041	32.249	26.77	0.5444	5.583	2.978	1.875	0.6191	0.1464
2.23	0.8923 −01	0.1780	0.5014	2.059	32.507	26.64	0.5431	5.635	2.992	1.883	0.6145	0.1452
2.24	0.8785 −01	0.1760	0.4991	2.078	32.763	26.51	0.5418	5.687	3.005	1.892	0.6100	0.1440
2.25	0.8648 −01	0.1740	0.4969	2.096	33.018	26.39	0.5406	5.740	3.019	1.901	0.6055	0.1428
2.26	0.8514 −01	0.1721	0.4947	2.115	33.273	26.26	0.5393	5.792	3.032	1.910	0.6011	0.1416
2.27	0.8382 −01	0.1702	0.4925	2.134	33.527	26.14	0.5381	5.845	3.045	1.919	0.5966	0.1405
2.28	0.8252 −01	0.1683	0.4903	2.154	33.780	26.01	0.5368	5.898	3.058	1.929	0.5921	0.1394
2.29	0.8123 −01	0.1664	0.4881	2.173	34.032	25.89	0.5356	5.951	3.071	1.938	0.5877	0.1382
2.30	0.7997 −01	0.1646	0.4859	2.193	34.283	25.77	0.5344	6.005	3.085	1.947	0.5833	0.1371
2.31	0.7873 −01	0.1628	0.4837	2.213	34.533	25.65	0.5332	6.059	3.098	1.956	0.5789	0.1360
2.32	0.7751 −01	0.1609	0.4816	2.233	34.782	25.53	0.5321	6.113	3.110	1.965	0.5745	0.1349
2.33	0.7631 −01	0.1592	0.4794	2.254	35.031	25.42	0.5309	6.167	3.123	1.974	0.5702	0.1338
2.34	0.7512 −01	0.1574	0.4773	2.274	35.279	25.30	0.5297	6.222	3.136	1.984	0.5658	0.1328
2.35	0.7396 −01	0.1556	0.4752	2.295	35.525	25.18	0.5286	6.276	3.149	1.993	0.5615	0.1317
2.36	0.7281 −01	0.1539	0.4731	2.316	35.771	25.07	0.5275	6.331	3.162	2.002	0.5572	0.1307
2.37	0.7168 −01	0.1522	0.4709	2.338	36.016	24.96	0.5264	6.386	3.174	2.012	0.5529	0.1296
2.38	0.7057 −01	0.1505	0.4689	2.359	36.261	24.85	0.5253	6.442	3.187	2.021	0.5486	0.1286
2.39	0.6948 −01	0.1488	0.4668	2.381	36.504	24.73	0.5242	6.497	3.199	2.031	0.5444	0.1276
2.40	0.6840 −01	0.1472	0.4647	2.403	36.746	24.62	0.5231	6.553	3.212	2.040	0.5401	0.1266
2.41	0.6734 −01	0.1456	0.4626	2.425	36.988	24.52	0.5221	6.609	3.224	2.050	0.5359	0.1256
2.42	0.6630 −01	0.1439	0.4606	2.448	37.229	24.41	0.5210	6.666	3.237	2.059	0.5317	0.1247
2.43	0.6527 −01	0.1424	0.4585	2.471	37.469	24.30	0.5200	6.722	3.249	2.069	0.5276	0.1237
2.44	0.6426 −01	0.1408	0.4565	2.494	37.708	24.19	0.5189	6.779	3.261	2.079	0.5234	0.1228
2.45	0.6327 −01	0.1392	0.4544	2.517	37.946	24.09	0.5179	6.836	3.273	2.088	0.5193	0.1218
2.46	0.6229 −01	0.1377	0.4524	2.540	38.183	23.99	0.5169	6.894	3.285	2.098	0.5152	0.1209
2.47	0.6133 −01	0.1362	0.4504	2.564	38.419	23.88	0.5159	6.951	3.298	2.108	0.5111	0.1200
2.48	0.6038 −01	0.1346	0.4484	2.588	38.655	23.78	0.5149	7.009	3.310	2.118	0.5071	0.1191
2.49	0.5945 −01	0.1332	0.4464	2.612	38.890	23.68	0.5140	7.067	3.321	2.128	0.5030	0.1182
2.50	0.5853 −01	0.1317	0.4444	2.637	39.124	23.58	0.5130	7.125	3.333	2.137	0.4990	0.1173
2.51	0.5762 −01	0.1302	0.4425	2.661	39.356	23.48	0.5120	7.183	3.345	2.147	0.4950	0.1164
2.52	0.5674 −01	0.1288	0.4405	2.686	39.589	23.38	0.5111	7.242	3.357	2.157	0.4911	0.1155
2.53	0.5586 −01	0.1274	0.4386	2.712	39.820	23.28	0.5102	7.301	3.369	2.167	0.4871	0.1147
2.54	0.5500 −01	0.1260	0.4366	2.737	40.050	23.18	0.5092	7.360	3.380	2.177	0.4832	0.1138
2.55	0.5415 −01	0.1246	0.4347	2.763	40.280	23.09	0.5083	7.420	3.392	2.187	0.4793	0.1130
2.56	0.5332 −01	0.1232	0.4328	2.789	40.508	22.99	0.5074	7.479	3.403	2.198	0.4754	0.1122
2.57	0.5250 −01	0.1218	0.4309	2.815	40.736	22.90	0.5065	7.539	3.415	2.208	0.4715	0.1113
2.58	0.5169 −01	0.1205	0.4289	2.842	40.963	22.81	0.5056	7.599	3.426	2.218	0.4677	0.1105
2.59	0.5090 −01	0.1192	0.4271	2.869	41.189	22.71	0.5047	7.659	3.438	2.228	0.4639	0.1097
2.60	0.5012 −01	0.1179	0.4252	2.896	41.415	22.62	0.5039	7.720	3.449	2.238	0.4601	0.1089
2.61	0.4935 −01	0.1166	0.4233	2.923	41.639	22.53	0.5030	7.781	3.460	2.249	0.4564	0.1081
2.62	0.4859 −01	0.1153	0.4214	2.951	41.863	22.44	0.5022	7.842	3.471	2.259	0.4526	0.1073
2.63	0.4784 −01	0.1140	0.4196	2.979	42.085	22.35	0.5013	7.903	3.483	2.269	0.4489	0.1066
2.64	0.4711 −01	0.1128	0.4177	3.007	42.307	22.26	0.5005	7.965	3.494	2.280	0.4452	0.1058

Table 2.14: Isentropic and normal shock properties for air, $\gamma = 1.4$ (continued).

M or M_1	$\dfrac{p}{p_t}$	$\dfrac{\rho}{\rho_t}$	$\dfrac{T}{T_t}$	$\dfrac{A}{A^*}$	ν	μ	M_2	$\dfrac{P_2}{P_1}$	$\dfrac{\rho_2}{\rho_1}$	$\dfrac{T_2}{T_1}$	$\dfrac{P_{t_2}}{P_{t_1}}$	$\dfrac{P_1}{P_{t_2}}$
2.65	0.4639 −01	0.1115	0.4159	3.036	42.528	22.17	0.4996	8.026	3.505	2.290	0.4416	0.1051
2.66	0.4568 −01	0.1103	0.4141	3.065	42.749	22.08	0.4988	8.088	3.516	2.301	0.4379	0.1043
2.67	0.4498 −01	0.1091	0.4122	3.094	42.968	22.00	0.4980	8.150	3.527	2.311	0.4343	0.1036
2.68	0.4429 −01	0.1079	0.4104	3.123	43.187	21.91	0.4972	8.213	3.537	2.322	0.4307	0.1028
2.69	0.4362 −01	0.1067	0.4086	3.153	43.404	21.82	0.4964	8.275	3.548	2.332	0.4271	0.1021
2.70	0.4295 −01	0.1056	0.4068	3.183	43.621	21.74	0.4956	8.338	3.559	2.343	0.4236	0.1014
2.71	0.4229 −01	0.1044	0.4051	3.213	43.838	21.65	0.4949	8.401	3.570	2.354	0.4201	0.1007
2.72	0.4165 −01	0.1033	0.4033	3.244	44.053	21.57	0.4941	8.465	3.580	2.364	0.4166	0.1000
2.73	0.4102 −01	0.1022	0.4015	3.275	44.267	21.49	0.4933	8.528	3.591	2.375	0.4131	0.9929 −01
2.74	0.4039 −01	0.1010	0.3998	3.306	44.481	21.41	0.4926	8.592	3.601	2.386	0.4097	0.9860 −01
2.75	0.3978 −01	0.0999	0.3980	3.338	44.694	21.32	0.4918	8.656	3.612	2.397	0.4062	0.9792 −01
2.76	0.3917 −01	0.9885 −01	0.3963	3.370	44.906	21.24	0.4911	8.721	3.622	2.407	0.4028	0.9724 −01
2.77	0.3858 −01	0.9778 −01	0.3945	3.402	45.117	21.16	0.4903	8.785	3.633	2.418	0.3994	0.9658 −01
2.78	0.3799 −01	0.9671 −01	0.3928	3.434	45.327	21.08	0.4896	8.850	3.643	2.429	0.3961	0.9591 −01
2.79	0.3742 −01	0.9566 −01	0.3911	3.467	45.537	21.00	0.4889	8.915	3.653	2.440	0.3928	0.9526 −01
2.80	0.3685 −01	0.9463 −01	0.3894	3.500	45.746	20.92	0.4882	8.980	3.664	2.451	0.3895	0.9461 −01
2.81	0.3629 −01	0.9360 −01	0.3877	3.534	45.954	20.85	0.4875	9.045	3.674	2.462	0.3862	0.9397 −01
2.82	0.3574 −01	0.9259 −01	0.3860	3.567	46.161	20.77	0.4868	9.111	3.684	2.473	0.3829	0.9334 −01
2.83	0.3520 −01	0.9159 −01	0.3844	3.601	46.367	20.69	0.4861	9.177	3.694	2.484	0.3797	0.9271 −01
2.84	0.3467 −01	0.9059 −01	0.3827	3.636	46.573	20.62	0.4854	9.243	3.704	2.496	0.3765	0.9209 −01
2.85	0.3415 −01	0.8962 −01	0.3810	3.671	46.778	20.54	0.4847	9.310	3.714	2.507	0.3733	0.9147 −01
2.86	0.3363 −01	0.8865 −01	0.3794	3.706	46.982	20.47	0.4840	9.376	3.724	2.518	0.3701	0.9086 −01
2.87	0.3312 −01	0.8769 −01	0.3777	3.741	47.185	20.39	0.4833	9.443	3.734	2.529	0.3670	0.9026 −01
2.88	0.3263 −01	0.8675 −01	0.3761	3.777	47.388	20.32	0.4827	9.510	3.743	2.540	0.3639	0.8966 −01
2.89	0.3213 −01	0.8581 −01	0.3745	3.813	47.589	20.24	0.4820	9.577	3.753	2.552	0.3608	0.8906 −01
2.90	0.3165 −01	0.8489 −01	0.3729	3.850	47.790	20.17	0.4814	9.645	3.763	2.563	0.3577	0.8848 −01
2.91	0.3118 −01	0.8398 −01	0.3712	3.887	47.990	20.10	0.4807	9.713	3.773	2.575	0.3547	0.8790 −01
2.92	0.3071 −01	0.8308 −01	0.3696	3.924	48.190	20.03	0.4801	9.781	3.782	2.586	0.3517	0.8732 −01
2.93	0.3025 −01	0.8218 −01	0.3681	3.961	48.388	19.96	0.4795	9.849	3.792	2.598	0.3487	0.8675 −01
2.94	0.2980 −01	0.8130 −01	0.3665	3.999	48.586	19.89	0.4788	9.918	3.801	2.609	0.3457	0.8619 −01
2.95	0.2935 −01	0.8043 −01	0.3649	4.038	48.783	19.81	0.4782	9.986	3.811	2.621	0.3428	0.8563 −01
2.96	0.2891 −01	0.7957 −01	0.3633	4.076	48.980	19.75	0.4776	10.06	3.820	2.632	0.3398	0.8507 −01
2.97	0.2848 −01	0.7872 −01	0.3618	4.115	49.175	19.68	0.4770	10.12	3.829	2.644	0.3369	0.8453 −01
2.98	0.2805 −01	0.7788 −01	0.3602	4.155	49.370	19.61	0.4764	10.19	3.839	2.656	0.3340	0.8398 −01
2.99	0.2764 −01	0.7705 −01	0.3587	4.194	49.564	19.54	0.4758	10.26	3.848	2.667	0.3312	0.8345 −01
3.00	0.2722 −01	0.7623 −01	0.3571	4.235	49.757	19.47	0.4752	10.33	3.857	2.679	0.3283	0.8291 −01
3.01	0.2682 −01	0.7541 −01	0.3556	4.275	49.950	19.40	0.4746	10.40	3.866	2.691	0.3255	0.8238 −01
3.02	0.2642 −01	0.7461 −01	0.3541	4.316	50.142	19.34	0.4740	10.47	3.875	2.703	0.3227	0.8186 −01
3.03	0.2603 −01	0.7382 −01	0.3526	4.357	50.333	19.27	0.4734	10.54	3.884	2.714	0.3200	0.8134 −01
3.04	0.2564 −01	0.7303 −01	0.3511	4.399	50.523	19.20	0.4729	10.62	3.893	2.726	0.3172	0.8083 −01
3.05	0.2526 −01	0.7226 −01	0.3496	4.441	50.713	19.14	0.4723	10.69	3.902	2.738	0.3145	0.8032 −01
3.06	0.2489 −01	0.7149 −01	0.3481	4.483	50.902	19.07	0.4717	10.76	3.911	2.750	0.3118	0.7982 −01
3.07	0.2452 −01	0.7074 −01	0.3466	4.526	51.090	19.01	0.4712	10.83	3.920	2.762	0.3091	0.7932 −01
3.08	0.2416 −01	0.6999 −01	0.3452	4.570	51.277	18.95	0.4706	10.90	3.929	2.774	0.3065	0.7882 −01
3.09	0.2380 −01	0.6925 −01	0.3437	4.613	51.464	18.88	0.4701	10.97	3.938	2.786	0.3038	0.7833 −01
3.10	0.2345 −01	0.6852 −01	0.3422	4.657	51.650	18.82	0.4695	11.05	3.947	2.799	0.3012	0.7785 −01
3.11	0.2310 −01	0.6779 −01	0.3408	4.702	51.835	18.76	0.4690	11.12	3.955	2.811	0.2986	0.7737 −01
3.12	0.2276 −01	0.6708 −01	0.3393	4.747	52.019	18.69	0.4685	11.19	3.964	2.823	0.2960	0.7689 −01
3.13	0.2243 −01	0.6637 −01	0.3379	4.792	52.203	18.63	0.4679	11.26	3.973	2.835	0.2935	0.7642 −01
3.14	0.2210 −01	0.6568 −01	0.3365	4.838	52.386	18.57	0.4674	11.34	3.981	2.847	0.2910	0.7595 −01
3.15	0.2177 −01	0.6499 −01	0.3351	4.884	52.569	18.51	0.4669	11.41	3.990	2.860	0.2885	0.7549 −01
3.16	0.2146 −01	0.6430 −01	0.3337	4.930	52.750	18.45	0.4664	11.48	3.998	2.872	0.2860	0.7503 −01
3.17	0.2114 −01	0.6363 −01	0.3323	4.977	52.931	18.39	0.4659	11.56	4.006	2.885	0.2835	0.7457 −01
3.18	0.2083 −01	0.6296 −01	0.3309	5.025	53.112	18.33	0.4654	11.63	4.015	2.897	0.2811	0.7412 −01
3.19	0.2053 −01	0.6231 −01	0.3295	5.073	53.291	18.27	0.4648	11.71	4.023	2.909	0.2786	0.7367 −01

Table 2.14: Isentropic and normal shock properties for air, $\gamma = 1.4$ (continued).

M or M_1	$\dfrac{P}{P_t}$	$\dfrac{\rho}{\rho_t}$	$\dfrac{T}{T_t}$	$\dfrac{A}{A^*}$	ν	μ	M_2	$\dfrac{P_2}{P_1}$	$\dfrac{\rho_2}{\rho_1}$	$\dfrac{T_2}{T_1}$	$\dfrac{P_{t_2}}{P_{t_1}}$	$\dfrac{P_1}{P_{t_2}}$
3.20	0.2023 −01	0.6165 −01	0.3281	5.121	53.470	18.21	0.4643	11.78	4.031	2.922	0.2762	0.7323 −01
3.21	0.1993 −01	0.6101 −01	0.3267	5.170	53.649	18.15	0.4639	11.86	4.040	2.935	0.2738	0.7279 −01
3.22	0.1964 −01	0.6037 −01	0.3253	5.219	53.826	18.09	0.4634	11.93	4.048	2.947	0.2715	0.7235 −01
3.23	0.1936 −01	0.5975 −01	0.3240	5.268	54.003	18.03	0.4629	12.01	4.056	2.960	0.2691	0.7192 −01
3.24	0.1908 −01	0.5912 −01	0.3226	5.319	54.179	17.98	0.4624	12.08	4.064	2.972	0.2668	0.7149 −01
3.25	0.1880 −01	0.5851 −01	0.3213	5.369	54.355	17.92	0.4619	12.16	4.072	2.985	0.2645	0.7107 −01
3.26	0.1853 −01	0.5790 −01	0.3199	5.420	54.529	17.86	0.4614	12.23	4.080	2.998	0.2622	0.7065 −01
3.27	0.1826 −01	0.5730 −01	0.3186	5.472	54.704	17.81	0.4610	12.31	4.088	3.011	0.2600	0.7023 −01
3.28	0.1799 −01	0.5671 −01	0.3173	5.523	54.877	17.75	0.4605	12.39	4.096	3.023	0.2577	0.6982 −01
3.29	0.1773 −01	0.5612 −01	0.3160	5.576	55.050	17.70	0.4600	12.46	4.104	3.036	0.2555	0.6941 −01
3.30	0.1748 −01	0.5554 −01	0.3147	5.629	55.222	17.64	0.4596	12.54	4.112	3.049	0.2533	0.6900 −01
3.31	0.1722 −01	0.5497 −01	0.3134	5.682	55.393	17.58	0.4591	12.62	4.120	3.062	0.2511	0.6860 −01
3.32	0.1698 −01	0.5440 −01	0.3121	5.736	55.564	17.53	0.4587	12.69	4.128	3.075	0.2489	0.6820 −01
3.33	0.1673 −01	0.5384 −01	0.3108	5.790	55.734	17.48	0.4582	12.77	4.135	3.088	0.2468	0.6781 −01
3.34	0.1649 −01	0.5329 −01	0.3095	5.845	55.904	17.42	0.4578	12.85	4.143	3.101	0.2446	0.6741 −01
3.35	0.1625 −01	0.5274 −01	0.3082	5.900	56.073	17.37	0.4573	12.93	4.151	3.114	0.2425	0.6702 −01
3.36	0.1602 −01	0.5220 −01	0.3069	5.956	56.241	17.31	0.4569	13.01	4.158	3.127	0.2404	0.6664 −01
3.37	0.1579 −01	0.5166 −01	0.3057	6.012	56.409	17.26	0.4565	13.08	4.166	3.140	0.2383	0.6626 −01
3.38	0.1557 −01	0.5113 −01	0.3044	6.069	56.576	17.21	0.4560	13.16	4.173	3.154	0.2363	0.6588 −01
3.39	0.1534 −01	0.5061 −01	0.3032	6.126	56.742	17.16	0.4556	13.24	4.181	3.167	0.2342	0.6550 −01
3.40	0.1512 −01	0.5009 −01	0.3019	6.184	56.907	17.10	0.4552	13.32	4.188	3.180	0.2322	0.6513 −01
3.41	0.1491 −01	0.4958 −01	0.3007	6.242	57.072	17.05	0.4548	13.40	4.196	3.194	0.2302	0.6476 −01
3.42	0.1470 −01	0.4908 −01	0.2995	6.301	57.237	17.00	0.4544	13.48	4.203	3.207	0.2282	0.6439 −01
3.43	0.1449 −01	0.4858 −01	0.2982	6.360	57.401	16.95	0.4540	13.56	4.211	3.220	0.2263	0.6403 −01
3.44	0.1428 −01	0.4808 −01	0.2970	6.420	57.564	16.90	0.4535	13.64	4.218	3.234	0.2243	0.6367 −01
3.45	0.1408 −01	0.4759 −01	0.2958	6.480	57.726	16.85	0.4531	13.72	4.225	3.247	0.2224	0.6331 −01
3.46	0.1388 −01	0.4711 −01	0.2946	6.541	57.888	16.80	0.4527	13.80	4.232	3.261	0.2205	0.6296 −01
3.47	0.1368 −01	0.4663 −01	0.2934	6.602	58.050	16.75	0.4523	13.88	4.240	3.274	0.2186	0.6261 −01
3.48	0.1349 −01	0.4616 −01	0.2922	6.664	58.210	16.70	0.4519	13.96	4.247	3.288	0.2167	0.6226 −01
3.49	0.1330 −01	0.4569 −01	0.2910	6.727	58.370	16.65	0.4515	14.04	4.254	3.301	0.2148	0.6191 −01
3.50	0.1311 −01	0.4523 −01	0.2899	6.790	58.530	16.60	0.4512	14.13	4.261	3.315	0.2129	0.6157 −01
3.51	0.1293 −01	0.4478 −01	0.2887	6.853	58.689	16.55	0.4508	14.21	4.268	3.329	0.2111	0.6123 −01
3.52	0.1274 −01	0.4433 −01	0.2875	6.917	58.847	16.50	0.4504	14.29	4.275	3.342	0.2093	0.6089 −01
3.53	0.1257 −01	0.4388 −01	0.2864	6.982	59.004	16.46	0.4500	14.37	4.282	3.356	0.2075	0.6056 −01
3.54	0.1239 −01	0.4344 −01	0.2852	7.047	59.162	16.41	0.4496	14.45	4.289	3.370	0.2057	0.6023 −01
3.55	0.1221 −01	0.4300 −01	0.2841	7.113	59.318	16.36	0.4492	14.54	4.296	3.384	0.2039	0.5990 −01
3.56	0.1204 −01	0.4257 −01	0.2829	7.179	59.474	16.31	0.4489	14.62	4.303	3.398	0.2022	0.5957 −01
3.57	0.1188 −01	0.4214 −01	0.2818	7.246	59.629	16.27	0.4485	14.70	4.309	3.412	0.2004	0.5925 −01
3.58	0.1171 −01	0.4172 −01	0.2806	7.313	59.784	16.22	0.4481	14.79	4.316	3.426	0.1987	0.5892 −01
3.59	0.1155 −01	0.4131 −01	0.2795	7.382	59.938	16.17	0.4478	14.87	4.323	3.440	0.1970	0.5861 −01
3.60	0.1138 −01	0.4089 −01	0.2784	7.450	60.091	16.13	0.4474	14.95	4.330	3.454	0.1953	0.5829 −01
3.61	0.1123 −01	0.4049 −01	0.2773	7.519	60.244	16.08	0.4471	15.04	4.336	3.468	0.1936	0.5798 −01
3.62	0.1107 −01	0.4008 −01	0.2762	7.589	60.397	16.04	0.4467	15.12	4.343	3.482	0.1920	0.5767 −01
3.63	0.1092 −01	0.3969 −01	0.2751	7.659	60.549	15.99	0.4463	15.21	4.350	3.496	0.1903	0.5736 −01
3.64	0.1076 −01	0.3929 −01	0.2740	7.730	60.700	15.95	0.4460	15.29	4.356	3.510	0.1887	0.5705 −01
3.65	0.1062 −01	0.3890 −01	0.2729	7.802	60.850	15.90	0.4456	15.38	4.363	3.525	0.1871	0.5675 −01
3.66	0.1047 −01	0.3852 −01	0.2718	7.874	61.000	15.86	0.4453	15.46	4.369	3.539	0.1855	0.5645 −01
3.67	0.1032 −01	0.3813 −01	0.2707	7.947	61.150	15.81	0.4450	15.55	4.376	3.553	0.1839	0.5615 −01
3.68	0.1018 −01	0.3776 −01	0.2697	8.020	61.299	15.77	0.4446	15.63	4.382	3.567	0.1823	0.5585 −01
3.69	0.1004 −01	0.3739 −01	0.2686	8.094	61.447	15.72	0.4443	15.72	4.388	3.582	0.1807	0.5556 −01
3.70	0.9903 −02	0.3702 −01	0.2675	8.169	61.595	15.68	0.4439	15.81	4.395	3.596	0.1792	0.5526 −01
3.71	0.9767 −02	0.3665 −01	0.2665	8.244	61.743	15.64	0.4436	15.89	4.401	3.611	0.1777	0.5497 −01
3.72	0.9633 −02	0.3629 −01	0.2654	8.320	61.889	15.59	0.4433	15.98	4.408	3.625	0.1761	0.5469 −01
3.73	0.9500 −02	0.3594 −01	0.2644	8.397	62.035	15.55	0.4430	16.07	4.414	3.640	0.1746	0.5440 −01
3.74	0.9370 −02	0.3558 −01	0.2633	8.474	62.181	15.51	0.4426	16.15	4.420	3.654	0.1731	0.5412 −01

Table 2.14: Isentropic and normal shock properties for air, $\gamma = 1.4$ (continued).

$\dfrac{M}{\text{or}}\ M_1$	$\dfrac{p}{p_t}$	$\dfrac{\rho}{\rho_t}$	$\dfrac{T}{T_t}$	$\dfrac{A}{A^*}$	ν	μ	M_2	$\dfrac{p_2}{p_1}$	$\dfrac{\rho_2}{\rho_1}$	$\dfrac{T_2}{T_1}$	$\dfrac{p_{t_2}}{p_{t_1}}$	$\dfrac{p_1}{p_{t_2}}$
3.75	0.9242 −02	0.3524 −01	0.2623	8.552	62.326	15.47	0.4423	16.24	4.426	3.669	0.1717	0.5384 −01
3.76	0.9116 −02	0.3489 −01	0.2613	8.630	62.471	15.42	0.4420	16.33	4.432	3.684	0.1702	0.5356 −01
3.77	0.8991 −02	0.3455 −01	0.2602	8.709	62.615	15.38	0.4417	16.42	4.439	3.698	0.1687	0.5328 −01
3.78	0.8869 −02	0.3421 −01	0.2592	8.789	62.758	15.34	0.4414	16.50	4.445	3.713	0.1673	0.5301 −01
3.79	0.8748 −02	0.3388 −01	0.2582	8.870	62.901	15.30	0.4410	16.59	4.451	3.728	0.1659	0.5274 −01
3.80	0.8629 −02	0.3355 −01	0.2572	8.951	63.044	15.26	0.4407	16.68	4.457	3.743	0.1645	0.5247 −01
3.81	0.8512 −02	0.3322 −01	0.2562	9.032	63.186	15.22	0.4404	16.77	4.463	3.757	0.1631	0.5220 −01
3.82	0.8396 −02	0.3290 −01	0.2552	9.115	63.327	15.18	0.4401	16.86	4.469	3.772	0.1617	0.5193 −01
3.83	0.8283 −02	0.3258 −01	0.2542	9.198	63.468	15.14	0.4398	16.95	4.475	3.787	0.1603	0.5167 −01
3.84	0.8171 −02	0.3227 −01	0.2532	9.282	63.608	15.09	0.4395	17.04	4.481	3.802	0.1589	0.5140 −01
3.85	0.8060 −02	0.3195 −01	0.2522	9.366	63.748	15.05	0.4392	17.13	4.487	3.817	0.1576	0.5114 −01
3.86	0.7951 −02	0.3165 −01	0.2513	9.451	63.887	15.01	0.4389	17.22	4.492	3.832	0.1563	0.5089 −01
3.87	0.7844 −02	0.3134 −01	0.2503	9.537	64.026	14.98	0.4386	17.31	4.498	3.847	0.1549	0.5063 −01
3.88	0.7739 −02	0.3104 −01	0.2493	9.622	64.164	14.94	0.4383	17.40	4.504	3.862	0.1536	0.5038 −01
3.89	0.7635 −02	0.3074 −01	0.2484	9.711	64.302	14.90	0.4380	17.49	4.510	3.878	0.1523	0.5012 −01
3.90	0.7532 −02	0.3044 −01	0.2474	9.799	64.439	14.86	0.4377	17.58	4.516	3.893	0.1510	0.4987 −01
3.91	0.7431 −02	0.3015 −01	0.2465	9.888	64.576	14.82	0.4375	17.67	4.521	3.908	0.1497	0.4962 −01
3.92	0.7332 −02	0.2986 −01	0.2455	9.977	64.712	14.78	0.4372	17.76	4.527	3.923	0.1485	0.4938 −01
3.93	0.7234 −02	0.2958 −01	0.2446	10.07	64.848	14.74	0.4369	17.85	4.533	3.939	0.1472	0.4913 −01
3.94	0.7137 −02	0.2929 −01	0.2436	10.16	64.984	14.70	0.4366	17.94	4.538	3.954	0.1460	0.4889 −01
3.95	0.7042 −02	0.2902 −01	0.2427	10.25	65.118	14.66	0.4363	18.04	4.544	3.969	0.1448	0.4865 −01
3.96	0.6948 −02	0.2874 −01	0.2418	10.34	65.253	14.63	0.4360	18.13	4.549	3.985	0.1435	0.4841 −01
3.97	0.6855 −02	0.2846 −01	0.2408	10.44	65.386	14.59	0.4358	18.22	4.555	4.000	0.1423	0.4817 −01
3.98	0.6764 −02	0.2819 −01	0.2399	10.53	65.520	14.55	0.4355	18.31	4.560	4.016	0.1411	0.4793 −01
3.99	0.6675 −02	0.2793 −01	0.2390	10.62	65.652	14.51	0.4352	18.41	4.566	4.031	0.1399	0.4770 −01
4.00	0.6586 −02	0.2766 −01	0.2381	10.72	65.785	14.48	0.4350	18.50	4.571	4.047	0.1388	0.4747 −01
4.01	0.6499 −02	0.2740 −01	0.2372	10.81	65.917	14.44	0.4347	18.59	4.577	4.062	0.1376	0.4723 −01
4.02	0.6413 −02	0.2714 −01	0.2363	10.91	66.048	14.40	0.4344	18.69	4.582	4.078	0.1364	0.4700 −01
4.03	0.6328 −02	0.2688 −01	0.2354	11.01	66.179	14.37	0.4342	18.78	4.588	4.094	0.1353	0.4678 −01
4.04	0.6245 −02	0.2663 −01	0.2345	11.11	66.309	14.33	0.4339	18.88	4.593	4.110	0.1342	0.4655 −01
4.05	0.6163 −02	0.2638 −01	0.2336	11.21	66.439	14.29	0.4336	18.97	4.598	4.125	0.1330	0.4633 −01
4.06	0.6082 −02	0.2613 −01	0.2327	11.31	66.569	14.26	0.4334	19.06	4.604	4.141	0.1319	0.4610 −01
4.07	0.6002 −02	0.2589 −01	0.2319	11.41	66.698	14.22	0.4331	19.16	4.609	4.157	0.1308	0.4588 −01
4.08	0.5923 −02	0.2564 −01	0.2310	11.51	66.826	14.19	0.4329	19.25	4.614	4.173	0.1297	0.4566 −01
4.09	0.5845 −02	0.2540 −01	0.2301	11.61	66.954	14.15	0.4326	19.35	4.619	4.189	0.1286	0.4544 −01
4.10	0.5769 −02	0.2516 −01	0.2293	11.71	67.082	14.12	0.4324	19.44	4.624	4.205	0.1276	0.4523 −01
4.11	0.5694 −02	0.2493 −01	0.2284	11.82	67.209	14.08	0.4321	19.54	4.630	4.221	0.1265	0.4501 −01
4.12	0.5619 −02	0.2470 −01	0.2275	11.92	67.336	14.05	0.4319	19.64	4.635	4.237	0.1254	0.4480 −01
4.13	0.5546 −02	0.2447 −01	0.2267	12.03	67.462	14.01	0.4316	19.73	4.640	4.253	0.1244	0.4459 −01
4.14	0.5474 −02	0.2424 −01	0.2258	12.14	67.588	13.98	0.4314	19.83	4.645	4.269	0.1234	0.4438 −01
4.15	0.5403 −02	0.2401 −01	0.2250	12.24	67.713	13.94	0.4311	19.93	4.650	4.285	0.1223	0.4417 −01
4.16	0.5333 −02	0.2379 −01	0.2242	12.35	67.838	13.91	0.4309	20.02	4.655	4.301	0.1213	0.4396 −01
4.17	0.5264 −02	0.2357 −01	0.2233	12.46	67.963	13.88	0.4306	20.12	4.660	4.318	0.1203	0.4375 −01
4.18	0.5195 −02	0.2335 −01	0.2225	12.57	68.087	13.84	0.4304	20.22	4.665	4.334	0.1193	0.4355 −01
4.19	0.5128 −02	0.2313 −01	0.2217	12.68	68.210	13.81	0.4302	20.32	4.670	4.350	0.1183	0.4334 −01
4.20	0.5062 −02	0.2292 −01	0.2208	12.79	68.333	13.77	0.4299	20.41	4.675	4.367	0.1173	0.4314 −01
4.21	0.4997 −02	0.2271 −01	0.2200	12.90	68.456	13.74	0.4297	20.51	4.680	4.383	0.1164	0.4294 −01
4.22	0.4932 −02	0.2250 −01	0.2192	13.02	68.578	13.71	0.4295	20.61	4.685	4.399	0.1154	0.4274 −01
4.23	0.4869 −02	0.2229 −01	0.2184	13.13	68.700	13.67	0.4292	20.71	4.690	4.416	0.1144	0.4254 −01
4.24	0.4806 −02	0.2209 −01	0.2176	13.25	68.821	13.64	0.4290	20.81	4.694	4.432	0.1135	0.4235 −01
4.25	0.4745 −02	0.2189 −01	0.2168	13.36	68.942	13.61	0.4288	20.91	4.699	4.449	0.1126	0.4215 −01
4.26	0.4684 −02	0.2168 −01	0.2160	13.48	69.063	13.58	0.4286	21.01	4.704	4.465	0.1116	0.4196 −01
4.27	0.4624 −02	0.2149 −01	0.2152	13.60	69.183	13.54	0.4283	21.11	4.709	4.482	0.1107	0.4177 −01
4.28	0.4565 −02	0.2129 −01	0.2144	13.72	69.303	13.51	0.4281	21.20	4.713	4.499	0.1098	0.4158 −01
4.29	0.4507 −02	0.2110 −01	0.2136	13.83	69.422	13.48	0.4279	21.30	4.718	4.515	0.1089	0.4139 −01

Table 2.14: Isentropic and normal shock properties for air, $\gamma = 1.4$ (continued).

M or M_1	$\frac{P}{P_t}$	$\frac{\rho}{\rho_t}$	$\frac{T}{T_t}$	$\frac{A}{A^*}$	ν	μ	M_2	$\frac{P_2}{P_1}$	$\frac{\rho_2}{\rho_1}$	$\frac{T_2}{T_1}$	$\frac{P_{t_2}}{P_{t_1}}$	$\frac{P_1}{P_{t_2}}$
4.30	0.4449 −02	0.2090 −01	0.2129	13.95	69.541	13.45	0.4277	21.41	4.723	4.532	0.1080	0.4120 −01
4.31	0.4393 −02	0.2071 −01	0.2121	14.08	69.659	13.42	0.4275	21.51	4.728	4.549	0.1071	0.4101 −01
4.32	0.4337 −02	0.2052 −01	0.2113	14.20	69.777	13.38	0.4272	21.61	4.732	4.566	0.1062	0.4082 −01
4.33	0.4282 −02	0.2034 −01	0.2105	14.32	69.895	13.35	0.4270	21.71	4.737	4.583	0.1054	0.4064 −01
4.34	0.4228 −02	0.2015 −01	0.2098	14.45	70.012	13.32	0.4268	21.81	4.741	4.600	0.1045	0.4046 −01
4.35	0.4174 −02	0.1997 −01	0.2090	14.57	70.129	13.29	0.4266	21.91	4.746	4.616	0.1036	0.4027 −01
4.36	0.4121 −02	0.1979 −01	0.2082	14.70	70.245	13.26	0.4264	22.01	4.751	4.633	0.1028	0.4009 −01
4.37	0.4069 −02	0.1961 −01	0.2075	14.82	70.361	13.23	0.4262	22.11	4.755	4.650	0.1020	0.3991 −01
4.38	0.4018 −02	0.1943 −01	0.2067	14.95	70.476	13.20	0.4260	22.22	4.760	4.668	0.1011	0.3973 −01
4.39	0.3967 −02	0.1926 −01	0.2060	15.08	70.591	13.17	0.4258	22.32	4.764	4.685	0.1003	0.3956 −01
4.40	0.3918 −02	0.1909 −01	0.2053	15.21	70.706	13.14	0.4255	22.42	4.768	4.702	0.9948 −01	0.3938 −01
4.41	0.3868 −02	0.1892 −01	0.2045	15.34	70.821	13.11	0.4253	22.52	4.773	4.719	0.9867 −01	0.3921 −01
4.42	0.3820 −02	0.1875 −01	0.2038	15.47	70.934	13.08	0.4251	22.63	4.777	4.736	0.9787 −01	0.3903 −01
4.43	0.3772 −02	0.1858 −01	0.2030	15.61	71.048	13.05	0.4249	22.73	4.782	4.753	0.9707 −01	0.3886 −01
4.44	0.3725 −02	0.1841 −01	0.2023	15.74	71.161	13.02	0.4247	22.83	4.786	4.771	0.9628 −01	0.3869 −01
4.45	0.3678 −02	0.1825 −01	0.2016	15.87	71.274	12.99	0.4245	22.94	4.790	4.788	0.9550 −01	0.3852 −01
4.46	0.3633 −02	0.1808 −01	0.2009	16.01	71.386	12.96	0.4243	23.04	4.795	4.805	0.9473 −01	0.3835 −01
4.47	0.3587 −02	0.1792 −01	0.2002	16.15	71.498	12.93	0.4241	23.14	4.799	4.823	0.9396 −01	0.3818 −01
4.48	0.3543 −02	0.1776 −01	0.1994	16.28	71.610	12.90	0.4239	23.25	4.803	4.840	0.9320 −01	0.3801 −01
4.49	0.3499 −02	0.1761 −01	0.1987	16.42	71.721	12.87	0.4237	23.35	4.808	4.858	0.9244 −01	0.3785 −01
4.50	0.3455 −02	0.1745 −01	0.1980	16.56	71.832	12.84	0.4236	23.46	4.812	4.875	0.9170 −01	0.3768 −01
4.51	0.3412 −02	0.1729 −01	0.1973	16.70	71.942	12.81	0.4234	23.56	4.816	4.893	0.9096 −01	0.3752 −01
4.52	0.3370 −02	0.1714 −01	0.1966	16.84	72.052	12.78	0.4232	23.67	4.820	4.910	0.9022 −01	0.3735 −01
4.53	0.3329 −02	0.1699 −01	0.1959	16.99	72.162	12.75	0.4230	23.77	4.824	4.928	0.8950 −01	0.3719 −01
4.54	0.3287 −02	0.1684 −01	0.1952	17.13	72.271	12.72	0.4228	23.88	4.829	4.946	0.8878 −01	0.3703 −01
4.55	0.3247 −02	0.1669 −01	0.1945	17.28	72.380	12.70	0.4226	23.99	4.833	4.963	0.8806 −01	0.3687 −01
4.56	0.3207 −02	0.1654 −01	0.1938	17.42	72.489	12.67	0.4224	24.09	4.837	4.981	0.8735 −01	0.3671 −01
4.57	0.3168 −02	0.1640 −01	0.1932	17.57	72.597	12.64	0.4222	24.20	4.841	4.999	0.8665 −01	0.3655 −01
4.58	0.3129 −02	0.1625 −01	0.1925	17.72	72.705	12.61	0.4220	24.31	4.845	5.017	0.8596 −01	0.3640 −01
4.59	0.3090 −02	0.1611 −01	0.1918	17.87	72.812	12.58	0.4219	24.41	4.849	5.034	0.8527 −01	0.3624 −01
4.60	0.3053 −02	0.1597 −01	0.1911	18.02	72.919	12.56	0.4217	24.52	4.853	5.052	0.8459 −01	0.3609 −01
4.61	0.3015 −02	0.1583 −01	0.1905	18.17	73.026	12.53	0.4215	24.63	4.857	5.070	0.8391 −01	0.3593 −01
4.62	0.2978 −02	0.1569 −01	0.1898	18.32	73.132	12.50	0.4213	24.74	4.861	5.088	0.8324 −01	0.3578 −01
4.63	0.2942 −02	0.1556 −01	0.1891	18.48	73.238	12.47	0.4211	24.84	4.865	5.106	0.8257 −01	0.3563 −01
4.64	0.2906 −02	0.1542 −01	0.1885	18.63	73.344	12.45	0.4210	24.95	4.869	5.124	0.8192 −01	0.3548 −01
4.65	0.2871 −02	0.1529 −01	0.1878	18.79	73.449	12.42	0.4208	25.06	4.873	5.142	0.8126 −01	0.3533 −01
4.66	0.2836 −02	0.1515 −01	0.1872	18.94	73.554	12.39	0.4206	25.17	4.877	5.161	0.8062 −01	0.3518 −01
4.67	0.2802 −02	0.1502 −01	0.1865	19.10	73.659	12.36	0.4204	25.28	4.881	5.179	0.7997 −01	0.3503 −01
4.68	0.2768 −02	0.1489 −01	0.1859	19.26	73.763	12.34	0.4203	25.39	4.885	5.197	0.7934 −01	0.3488 −01
4.69	0.2734 −02	0.1476 −01	0.1852	19.42	73.867	12.31	0.4201	25.50	4.889	5.215	0.7871 −01	0.3474 −01
4.70	0.2701 −02	0.1464 −01	0.1846	19.58	73.970	12.28	0.4199	25.61	4.893	5.233	0.7808 −01	0.3459 −01
4.71	0.2669 −02	0.1451 −01	0.1839	19.75	74.073	12.26	0.4197	25.71	4.896	5.252	0.7747 −01	0.3445 −01
4.72	0.2636 −02	0.1438 −01	0.1833	19.91	74.176	12.23	0.4196	25.82	4.900	5.270	0.7685 −01	0.3431 −01
4.73	0.2605 −02	0.1426 −01	0.1827	20.07	74.279	12.21	0.4194	25.94	4.904	5.289	0.7625 −01	0.3416 −01
4.74	0.2573 −02	0.1414 −01	0.1820	20.24	74.381	12.18	0.4192	26.05	4.908	5.307	0.7564 −01	0.3402 −01
4.75	0.2543 −02	0.1402 −01	0.1814	20.41	74.483	12.15	0.4191	26.16	4.912	5.325	0.7505 −01	0.3388 −01
4.76	0.2512 −02	0.1390 −01	0.1808	20.58	74.584	12.13	0.4189	26.27	4.915	5.344	0.7445 −01	0.3374 −01
4.77	0.2482 −02	0.1378 −01	0.1802	20.75	74.685	12.10	0.4187	26.38	4.919	5.363	0.7387 −01	0.3360 −01
4.78	0.2452 −02	0.1366 −01	0.1795	20.92	74.786	12.08	0.4186	26.49	4.923	5.381	0.7329 −01	0.3346 −01
4.79	0.2423 −02	0.1354 −01	0.1789	21.09	74.886	12.05	0.4184	26.60	4.926	5.400	0.7271 −01	0.3333 −01
4.80	0.2394 −02	0.1343 −01	0.1783	21.26	74.986	12.02	0.4183	26.71	4.930	5.418	0.7214 −01	0.3319 −01
4.81	0.2366 −02	0.1331 −01	0.1777	21.44	75.086	12.00	0.4181	26.83	4.934	5.437	0.7157 −01	0.3305 −01
4.82	0.2338 −02	0.1320 −01	0.1771	21.61	75.186	11.97	0.4179	26.94	4.937	5.456	0.7101 −01	0.3292 −01
4.83	0.2310 −02	0.1309 −01	0.1765	21.79	75.285	11.95	0.4178	27.05	4.941	5.475	0.7046 −01	0.3278 −01
4.84	0.2282 −02	0.1298 −01	0.1759	21.97	75.383	11.92	0.4176	27.16	4.945	5.494	0.6990 −01	0.3265 −01

Table 2.14: Isentropic and normal shock properties for air, $\gamma = 1.4$ (continued).

M or M_1	$\dfrac{P}{P_t}$	$\dfrac{\rho}{\rho_t}$	$\dfrac{T}{T_t}$	$\dfrac{A}{A^*}$	ν	μ	M_2	$\dfrac{P_2}{P_1}$	$\dfrac{\rho_2}{\rho_1}$	$\dfrac{T_2}{T_1}$	$\dfrac{P_{t_2}}{P_{t_1}}$	$\dfrac{P_1}{P_{t_2}}$
4.85	0.2255 −02	0.1287 −01	0.1753	22.15	75.482	11.90	0.4175	27.28	4.948	5.512	0.6936 −01	0.3252 −01
4.86	0.2229 −02	0.1276 −01	0.1747	22.33	75.580	11.87	0.4173	27.39	4.952	5.531	0.6882 −01	0.3239 −01
4.87	0.2202 −02	0.1265 −01	0.1741	22.51	75.678	11.85	0.4172	27.50	4.955	5.550	0.6828 −01	0.3226 −01
4.88	0.2176 −02	0.1254 −01	0.1735	22.70	75.775	11.82	0.4170	27.62	4.959	5.569	0.6775 −01	0.3213 −01
4.89	0.2151 −02	0.1244 −01	0.1729	22.88	75.872	11.80	0.4169	27.73	4.962	5.588	0.6722 −01	0.3200 −01
4.90	0.2126 −02	0.1233 −01	0.1724	23.07	75.969	11.78	0.4167	27.85	4.966	5.607	0.6670 −01	0.3187 −01
4.91	0.2101 −02	0.1223 −01	0.1718	23.25	76.066	11.75	0.4165	27.96	4.969	5.626	0.6618 −01	0.3174 −01
4.92	0.2076 −02	0.1213 −01	0.1712	23.44	76.162	11.73	0.4164	28.07	4.973	5.646	0.6567 −01	0.3161 −01
4.93	0.2052 −02	0.1202 −01	0.1706	23.63	76.258	11.70	0.4162	28.19	4.976	5.665	0.6516 −01	0.3149 −01
4.94	0.2028 −02	0.1192 −01	0.1700	23.82	76.353	11.68	0.4161	28.30	4.980	5.684	0.6465 −01	0.3136 −01
4.95	0.2004 −02	0.1182 −01	0.1695	24.02	76.449	11.66	0.4160	28.42	4.983	5.703	0.6415 −01	0.3124 −01
4.96	0.1981 −02	0.1173 −01	0.1689	24.21	76.544	11.63	0.4158	28.54	4.987	5.722	0.6366 −01	0.3111 −01
4.97	0.1957 −02	0.1163 −01	0.1683	24.41	76.638	11.61	0.4157	28.65	4.990	5.742	0.6316 −01	0.3099 −01
4.98	0.1935 −02	0.1153 −01	0.1678	24.60	76.733	11.58	0.4155	28.77	4.993	5.761	0.6268 −01	0.3087 −01
4.99	0.1912 −02	0.1144 −01	0.1672	24.80	76.827	11.56	0.4154	28.88	4.997	5.781	0.6219 −01	0.3075 −01
5.00	0.1890 −02	0.1134 −01	0.1667	25.00	76.920	11.54	0.4152	29.00	5.000	5.800	0.6172 −01	0.3062 −01
5.01	0.1868 −02	0.1125 −01	0.1661	25.20	77.014	11.51	0.4151	29.12	5.003	5.820	0.6124 −01	0.3050 −01
5.02	0.1846 −02	0.1115 −01	0.1656	25.40	77.107	11.49	0.4149	29.23	5.007	5.839	0.6077 −01	0.3038 −01
5.03	0.1825 −02	0.1106 −01	0.1650	25.61	77.200	11.47	0.4148	29.35	5.010	5.859	0.6030 −01	0.3027 −01
5.04	0.1804 −02	0.1097 −01	0.1645	25.81	77.292	11.44	0.4147	29.47	5.013	5.878	0.5984 −01	0.3015 −01
5.05	0.1783 −02	0.1088 −01	0.1639	26.02	77.384	11.42	0.4145	29.59	5.016	5.898	0.5938 −01	0.3003 −01
5.06	0.1763 −02	0.1079 −01	0.1634	26.23	77.476	11.40	0.4144	29.70	5.020	5.918	0.5893 −01	0.2991 −01
5.07	0.1742 −02	0.1070 −01	0.1628	26.43	77.568	11.38	0.4142	29.82	5.023	5.937	0.5848 −01	0.2980 −01
5.08	0.1722 −02	0.1061 −01	0.1623	26.64	77.659	11.35	0.4141	29.94	5.026	5.957	0.5803 −01	0.2968 −01
5.09	0.1703 −02	0.1053 −01	0.1618	26.86	77.750	11.33	0.4140	30.06	5.029	5.977	0.5759 −01	0.2957 −01
5.10	0.1683 −02	0.1044 −01	0.1612	27.07	77.841	11.31	0.4138	30.18	5.033	5.997	0.5715 −01	0.2945 −01
5.11	0.1664 −02	0.1035 −01	0.1607	27.28	77.932	11.29	0.4137	30.30	5.036	6.017	0.5672 −01	0.2934 −01
5.12	0.1645 −02	0.1027 −01	0.1602	27.50	78.022	11.26	0.4136	30.42	5.039	6.036	0.5628 −01	0.2923 −01
5.13	0.1626 −02	0.1019 −01	0.1597	27.72	78.112	11.24	0.4134	30.54	5.042	6.056	0.5586 −01	0.2911 −01
5.14	0.1608 −02	0.1010 −01	0.1591	27.94	78.201	11.22	0.4133	30.66	5.045	6.076	0.5543 −01	0.2900 −01
5.15	0.1589 −02	0.1002 −01	0.1586	28.16	78.290	11.20	0.4132	30.78	5.048	6.096	0.5501 −01	0.2889 −01
5.16	0.1571 −02	0.9938 −02	0.1581	28.38	78.379	11.17	0.4130	30.90	5.051	6.116	0.5460 −01	0.2878 −01
5.17	0.1553 −02	0.9858 −02	0.1576	28.60	78.468	11.15	0.4129	31.02	5.054	6.137	0.5418 −01	0.2867 −01
5.18	0.1536 −02	0.9778 −02	0.1571	28.83	78.557	11.13	0.4128	31.14	5.058	6.157	0.5377 −01	0.2856 −01
5.19	0.1518 −02	0.9699 −02	0.1566	29.06	78.645	11.11	0.4126	31.26	5.061	6.177	0.5337 −01	0.2845 −01
5.20	0.1501 −02	0.9620 −02	0.1561	29.28	78.733	11.09	0.4125	31.38	5.064	6.197	0.5296 −01	0.2834 −01
5.21	0.1484 −02	0.9543 −02	0.1555	29.51	78.820	11.07	0.4124	31.50	5.067	6.217	0.5257 −01	0.2824 −01
5.22	0.1468 −02	0.9466 −02	0.1550	29.74	78.908	11.04	0.4123	31.62	5.070	6.238	0.5217 −01	0.2813 −01
5.23	0.1451 −02	0.9389 −02	0.1545	29.98	78.995	11.02	0.4121	31.75	5.073	6.258	0.5178 −01	0.2802 −01
5.24	0.1435 −02	0.9314 −02	0.1540	30.21	79.081	11.00	0.4120	31.87	5.076	6.278	0.5139 −01	0.2792 −01
5.25	0.1419 −02	0.9239 −02	0.1535	30.45	79.168	10.98	0.4119	31.99	5.079	6.299	0.5100 −01	0.2781 −01
5.26	0.1403 −02	0.9165 −02	0.1531	30.68	79.254	10.96	0.4118	32.11	5.082	6.319	0.5062 −01	0.2771 −01
5.27	0.1387 −02	0.9091 −02	0.1526	30.92	79.340	10.94	0.4116	32.24	5.085	6.340	0.5024 −01	0.2761 −01
5.28	0.1371 −02	0.9019 −02	0.1521	31.16	79.426	10.92	0.4115	32.36	5.088	6.360	0.4987 −01	0.2750 −01
5.29	0.1356 −02	0.8947 −02	0.1516	31.41	79.511	10.90	0.4114	32.48	5.090	6.381	0.4949 −01	0.2740 −01
5.30	0.1341 −02	0.8875 −02	0.1511	31.65	79.596	10.88	0.4113	32.61	5.093	6.401	0.4912 −01	0.2730 −01
5.31	0.1326 −02	0.8804 −02	0.1506	31.89	79.681	10.85	0.4112	32.73	5.096	6.422	0.4876 −01	0.2720 −01
5.32	0.1311 −02	0.8734 −02	0.1501	32.14	79.766	10.83	0.4110	32.85	5.099	6.443	0.4840 −01	0.2710 −01
5.33	0.1297 −02	0.8665 −02	0.1497	32.39	79.850	10.81	0.4109	32.98	5.102	6.464	0.4804 −01	0.2700 −01
5.34	0.1282 −02	0.8596 −02	0.1492	32.64	79.934	10.79	0.4108	33.10	5.105	6.484	0.4768 −01	0.2690 −01
5.35	0.1268 −02	0.8528 −02	0.1487	32.89	80.018	10.77	0.4107	33.23	5.108	6.505	0.4732 −01	0.2680 −01
5.36	0.1254 −02	0.8460 −02	0.1482	33.15	80.102	10.75	0.4106	33.35	5.111	6.526	0.4697 −01	0.2670 −01
5.37	0.1240 −02	0.8393 −02	0.1478	33.40	80.185	10.73	0.4104	33.48	5.113	6.547	0.4663 −01	0.2660 −01
5.38	0.1227 −02	0.8327 −02	0.1473	33.66	80.268	10.71	0.4103	33.60	5.116	6.568	0.4628 −01	0.2650 −01
5.39	0.1213 −02	0.8261 −02	0.1468	33.92	80.351	10.69	0.4102	33.73	5.119	6.589	0.4594 −01	0.2641 −01

Table 2.14: Isentropic and normal shock properties for air, $\gamma = 1.4$ (continued).

$\begin{array}{c}M\\or\\M_1\end{array}$	$\dfrac{P}{P_t}$	$\dfrac{\rho}{\rho_t}$	$\dfrac{T}{T_t}$	$\dfrac{A}{A^*}$	ν	μ	M_2	$\dfrac{P_2}{P_1}$	$\dfrac{\rho_2}{\rho_1}$	$\dfrac{T_2}{T_1}$	$\dfrac{P_{t_2}}{P_{t_1}}$	$\dfrac{P_1}{P_{t_2}}$
5.40	0.1200 −02	0.8196 −02	0.1464	34.18	80.433	10.67	0.4101	33.85	5.122	6.610	0.4560 −01	0.2631 −01
5.41	0.1186 −02	0.8132 −02	0.1459	34.44	80.516	10.65	0.4100	33.98	5.125	6.631	0.4526 −01	0.2621 −01
5.42	0.1173 −02	0.8068 −02	0.1454	34.70	80.598	10.63	0.4099	34.11	5.127	6.652	0.4493 −01	0.2612 −01
5.43	0.1161 −02	0.8005 −02	0.1450	34.97	80.680	10.61	0.4098	34.23	5.130	6.673	0.4460 −01	0.2602 −01
5.44	0.1148 −02	0.7942 −02	0.1445	35.23	80.761	10.59	0.4096	34.36	5.133	6.694	0.4427 −01	0.2593 −01
5.45	0.1135 −02	0.7880 −02	0.1441	35.50	80.842	10.57	0.4095	34.49	5.136	6.715	0.4395 −01	0.2583 −01
5.46	0.1123 −02	0.7818 −02	0.1436	35.77	80.923	10.55	0.4094	34.61	5.138	6.737	0.4362 −01	0.2574 −01
5.47	0.1111 −02	0.7757 −02	0.1432	36.04	81.004	10.53	0.4093	34.74	5.141	6.758	0.4330 −01	0.2565 −01
5.48	0.1099 −02	0.7697 −02	0.1427	36.32	81.085	10.51	0.4092	34.87	5.144	6.779	0.4299 −01	0.2556 −01
5.49	0.1087 −02	0.7637 −02	0.1423	36.59	81.165	10.49	0.4091	35.00	5.146	6.800	0.4267 −01	0.2546 −01
5.50	0.1075 −02	0.7577 −02	0.1418	36.87	81.245	10.48	0.4090	35.13	5.149	6.822	0.4236 −01	0.2537 −01
5.51	0.1063 −02	0.7518 −02	0.1414	37.15	81.325	10.46	0.4089	35.25	5.152	6.843	0.4205 −01	0.2528 −01
5.52	0.1052 −02	0.7460 −02	0.1410	37.43	81.404	10.44	0.4088	35.38	5.154	6.865	0.4174 −01	0.2519 −01
5.53	0.1040 −02	0.7402 −02	0.1405	37.71	81.484	10.42	0.4086	35.51	5.157	6.886	0.4144 −01	0.2510 −01
5.54	0.1029 −02	0.7345 −02	0.1401	38.00	81.563	10.40	0.4085	35.64	5.159	6.908	0.4114 −01	0.2501 −01
5.55	0.1018 −02	0.7288 −02	0.1397	38.28	81.642	10.38	0.4084	35.77	5.162	6.929	0.4084 −01	0.2492 −01
5.56	0.1007 −02	0.7232 −02	0.1392	38.57	81.720	10.36	0.4083	35.90	5.165	6.951	0.4054 −01	0.2483 −01
5.57	0.9960 −03	0.7176 −02	0.1388	38.86	81.799	10.34	0.4082	36.03	5.167	6.973	0.4025 −01	0.2475 −01
5.58	0.9853 −03	0.7121 −02	0.1384	39.15	81.877	10.32	0.4081	36.16	5.170	6.994	0.3996 −01	0.2466 −01
5.59	0.9747 −03	0.7066 −02	0.1379	39.45	81.955	10.31	0.4080	36.29	5.172	7.016	0.3967 −01	0.2457 −01
5.60	0.9643 −03	0.7012 −02	0.1375	39.74	82.032	10.29	0.4079	36.42	5.175	7.038	0.3938 −01	0.2448 −01
5.61	0.9539 −03	0.6958 −02	0.1371	40.04	82.110	10.27	0.4078	36.55	5.177	7.060	0.3910 −01	0.2440 −01
5.62	0.9437 −03	0.6905 −02	0.1367	40.34	82.187	10.25	0.4077	36.68	5.180	7.082	0.3882 −01	0.2431 −01
5.63	0.9336 −03	0.6852 −02	0.1362	40.64	82.264	10.23	0.4076	36.81	5.183	7.103	0.3854 −01	0.2423 −01
5.64	0.9237 −03	0.6800 −02	0.1358	40.94	82.340	10.21	0.4075	36.94	5.185	7.125	0.3826 −01	0.2414 −01
5.65	0.9138 −03	0.6748 −02	0.1354	41.25	82.417	10.19	0.4074	37.08	5.187	7.147	0.3798 −01	0.2406 −01
5.66	0.9041 −03	0.6697 −02	0.1350	41.55	82.493	10.18	0.4073	37.21	5.190	7.169	0.3771 −01	0.2397 −01
5.67	0.8945 −03	0.6646 −02	0.1346	41.86	82.569	10.16	0.4072	37.34	5.192	7.191	0.3744 −01	0.2389 −01
5.68	0.8850 −03	0.6595 −02	0.1342	42.17	82.645	10.14	0.4071	37.47	5.195	7.213	0.3717 −01	0.2381 −01
5.69	0.8756 −03	0.6545 −02	0.1338	42.48	82.721	10.12	0.4070	37.61	5.197	7.236	0.3691 −01	0.2372 −01
5.70	0.8663 −03	0.6496 −02	0.1334	42.80	82.796	10.10	0.4069	37.74	5.200	7.258	0.3664 −01	0.2364 −01
5.71	0.8571 −03	0.6447 −02	0.1330	43.11	82.871	10.09	0.4068	37.87	5.202	7.280	0.3638 −01	0.2356 −01
5.72	0.8481 −03	0.6398 −02	0.1326	43.43	82.946	10.07	0.4067	38.01	5.205	7.302	0.3612 −01	0.2348 −01
5.73	0.8391 −03	0.6349 −02	0.1322	43.75	83.021	10.05	0.4066	38.14	5.207	7.324	0.3586 −01	0.2340 −01
5.74	0.8303 −03	0.6302 −02	0.1318	44.08	83.095	10.03	0.4065	38.27	5.209	7.347	0.3561 −01	0.2332 −01
5.75	0.8216 −03	0.6254 −02	0.1314	44.40	83.169	10.02	0.4064	38.41	5.212	7.369	0.3535 −01	0.2324 −01
5.76	0.8129 −03	0.6207 −02	0.1310	44.73	83.243	9.998	0.4063	38.54	5.214	7.392	0.3510 −01	0.2316 −01
5.77	0.8044 −03	0.6161 −02	0.1306	45.05	83.317	9.980	0.4062	38.68	5.217	7.414	0.3485 −01	0.2308 −01
5.78	0.7960 −03	0.6114 −02	0.1302	45.38	83.391	9.963	0.4061	38.81	5.219	7.436	0.3461 −01	0.2300 −01
5.79	0.7876 −03	0.6068 −02	0.1298	45.72	83.464	9.945	0.4060	38.95	5.221	7.459	0.3436 −01	0.2292 −01
5.80	0.7794 −03	0.6023 −02	0.1294	46.05	83.537	9.928	0.4059	39.08	5.224	7.482	0.3412 −01	0.2284 −01
5.81	0.7712 −03	0.5978 −02	0.1290	46.39	83.610	9.911	0.4058	39.22	5.226	7.504	0.3388 −01	0.2277 −01
5.82	0.7632 −03	0.5933 −02	0.1286	46.73	83.683	9.894	0.4058	39.35	5.228	7.527	0.3364 −01	0.2269 −01
5.83	0.7552 −03	0.5889 −02	0.1282	47.07	83.755	9.877	0.4057	39.49	5.231	7.549	0.3340 −01	0.2261 −01
5.84	0.7474 −03	0.5845 −02	0.1279	47.41	83.827	9.859	0.4056	39.62	5.233	7.572	0.3317 −01	0.2253 −01
5.85	0.7396 −03	0.5802 −02	0.1275	47.75	83.900	9.842	0.4055	39.76	5.235	7.595	0.3293 −01	0.2246 −01
5.86	0.7319 −03	0.5759 −02	0.1271	48.10	83.971	9.825	0.4054	39.90	5.237	7.618	0.3270 −01	0.2238 −01
5.87	0.7243 −03	0.5716 −02	0.1267	48.45	84.043	9.809	0.4053	40.03	5.240	7.640	0.3247 −01	0.2231 −01
5.88	0.7168 −03	0.5674 −02	0.1263	48.80	84.115	9.792	0.4052	40.17	5.242	7.663	0.3224 −01	0.2223 −01
5.89	0.7094 −03	0.5632 −02	0.1260	49.15	84.186	9.775	0.4051	40.31	5.244	7.686	0.3202 −01	0.2216 −01
5.90	0.7021 −03	0.5590 −02	0.1256	49.51	84.257	9.758	0.4050	40.45	5.246	7.709	0.3179 −01	0.2208 −01
5.91	0.6949 −03	0.5549 −02	0.1252	49.87	84.328	9.742	0.4049	40.58	5.249	7.732	0.3157 −01	0.2201 −01
5.92	0.6877 −03	0.5508 −02	0.1249	50.23	84.398	9.725	0.4049	40.72	5.251	7.755	0.3135 −01	0.2194 −01
5.93	0.6806 −03	0.5468 −02	0.1245	50.59	84.469	9.708	0.4048	40.86	5.253	7.778	0.3113 −01	0.2186 −01
5.94	0.6736 −03	0.5427 −02	0.1241	50.95	84.539	9.692	0.4047	41.00	5.255	7.801	0.3092 −01	0.2179 −01

Table 2.14: Isentropic and normal shock properties for air, $\gamma = 1.4$ (continued).

$\begin{matrix}M\\ or\\ M_1\end{matrix}$	$\frac{P}{P_t}$	$\frac{\rho}{\rho_t}$	$\frac{T}{T_t}$	$\frac{A}{A^*}$	ν	μ	M_2	$\frac{P_2}{P_1}$	$\frac{\rho_2}{\rho_1}$	$\frac{T_2}{T_1}$	$\frac{P_{t_2}}{P_{t_1}}$	$\frac{P_1}{P_{t_2}}$
5.95	0.6667 −03	0.5388 −02	0.1238	51.32	84.609	9.675	0.4046	41.14	5.257	7.824	0.3070 −01	0.2172 −01
5.96	0.6599 −03	0.5348 −02	0.1234	51.69	84.679	9.659	0.4045	41.28	5.260	7.848	0.3049 −01	0.2165 −01
5.97	0.6531 −03	0.5309 −02	0.1230	52.06	84.748	9.643	0.4044	41.41	5.262	7.871	0.3028 −01	0.2157 −01
5.98	0.6465 −03	0.5270 −02	0.1227	52.43	84.818	9.626	0.4043	41.55	5.264	7.894	0.3007 −01	0.2150 −01
5.99	0.6399 −03	0.5232 −02	0.1223	52.80	84.887	9.610	0.4042	41.69	5.266	7.917	0.2986 −01	0.2143 −01
6.00	0.6333 −03	0.5193 −02	0.1219	53.18	84.956	9.594	0.4042	41.83	5.268	7.941	0.2965 −01	0.2136 −01
6.01	0.6269 −03	0.5156 −02	0.1216	53.56	85.025	9.578	0.4041	41.97	5.270	7.964	0.2944 −01	0.2129 −01
6.02	0.6205 −03	0.5118 −02	0.1212	53.94	85.093	9.562	0.4040	42.11	5.273	7.987	0.2924 −01	0.2122 −01
6.03	0.6142 −03	0.5081 −02	0.1209	54.33	85.162	9.546	0.4039	42.25	5.275	8.011	0.2904 −01	0.2115 −01
6.04	0.6080 −03	0.5044 −02	0.1205	54.71	85.230	9.530	0.4038	42.40	5.277	8.034	0.2884 −01	0.2108 −01
6.05	0.6018 −03	0.5007 −02	0.1202	55.10	85.298	9.514	0.4037	42.54	5.279	8.058	0.2864 −01	0.2101 −01
6.06	0.5957 −03	0.4971 −02	0.1198	55.49	85.366	9.498	0.4037	42.68	5.281	8.081	0.2844 −01	0.2094 −01
6.07	0.5897 −03	0.4935 −02	0.1195	55.89	85.433	9.482	0.4036	42.82	5.283	8.105	0.2825 −01	0.2088 −01
6.08	0.5837 −03	0.4900 −02	0.1191	56.28	85.501	9.467	0.4035	42.96	5.285	8.129	0.2805 −01	0.2081 −01
6.09	0.5779 −03	0.4864 −02	0.1188	56.68	85.568	9.451	0.4034	43.10	5.287	8.152	0.2786 −01	0.2074 −01
6.10	0.5720 −03	0.4829 −02	0.1185	57.08	85.635	9.435	0.4033	43.25	5.289	8.176	0.2767 −01	0.2067 −01
6.11	0.5663 −03	0.4794 −02	0.1181	57.48	85.702	9.420	0.4033	43.39	5.291	8.200	0.2748 −01	0.2061 −01
6.12	0.5606 −03	0.4760 −02	0.1178	57.89	85.768	9.404	0.4032	43.53	5.293	8.224	0.2729 −01	0.2054 −01
6.13	0.5550 −03	0.4726 −02	0.1174	58.29	85.835	9.389	0.4031	43.67	5.295	8.247	0.2711 −01	0.2047 −01
6.14	0.5494 −03	0.4692 −02	0.1171	58.70	85.901	9.373	0.4030	43.82	5.297	8.271	0.2692 −01	0.2041 −01
6.15	0.5439 −03	0.4658 −02	0.1168	59.12	85.967	9.358	0.4029	43.96	5.299	8.295	0.2674 −01	0.2034 −01
6.16	0.5385 −03	0.4625 −02	0.1164	59.53	86.033	9.343	0.4029	44.10	5.301	8.319	0.2656 −01	0.2028 −01
6.17	0.5331 −03	0.4592 −02	0.1161	59.95	86.099	9.327	0.4028	44.25	5.303	8.343	0.2638 −01	0.2021 −01
6.18	0.5278 −03	0.4559 −02	0.1158	60.37	86.165	9.312	0.4027	44.39	5.305	8.367	0.2620 −01	0.2015 −01
6.19	0.5225 −03	0.4527 −02	0.1154	60.79	86.230	9.297	0.4026	44.54	5.307	8.391	0.2602 −01	0.2008 −01
6.20	0.5173 −03	0.4495 −02	0.1151	61.21	86.295	9.282	0.4025	44.68	5.309	8.415	0.2584 −01	0.2002 −01
6.21	0.5122 −03	0.4463 −02	0.1148	61.64	86.360	9.267	0.4025	44.83	5.311	8.440	0.2567 −01	0.1995 −01
6.22	0.5071 −03	0.4431 −02	0.1144	62.07	86.425	9.252	0.4024	44.97	5.313	8.464	0.2550 −01	0.1989 −01
6.23	0.5021 −03	0.4400 −02	0.1141	62.50	86.490	9.237	0.4023	45.12	5.315	8.488	0.2532 −01	0.1983 −01
6.24	0.4971 −03	0.4368 −02	0.1138	62.93	86.554	9.222	0.4022	45.26	5.317	8.512	0.2515 −01	0.1976 −01
6.25	0.4922 −03	0.4337 −02	0.1135	63.37	86.618	9.207	0.4022	45.41	5.319	8.536	0.2498 −01	0.1970 −01
6.26	0.4873 −03	0.4307 −02	0.1132	63.81	86.682	9.192	0.4021	45.55	5.321	8.561	0.2481 −01	0.1964 −01
6.27	0.4825 −03	0.4276 −02	0.1128	64.25	86.746	9.177	0.4020	45.70	5.323	8.585	0.2465 −01	0.1958 −01
6.28	0.4778 −03	0.4246 −02	0.1125	64.70	86.810	9.162	0.4019	45.85	5.325	8.610	0.2448 −01	0.1951 −01
6.29	0.4731 −03	0.4216 −02	0.1122	65.14	86.874	9.148	0.4019	45.99	5.327	8.634	0.2432 −01	0.1945 −01
6.30	0.4684 −03	0.4187 −02	0.1119	65.59	86.937	9.133	0.4018	46.14	5.329	8.659	0.2416 −01	0.1939 −01
6.31	0.4638 −03	0.4157 −02	0.1116	66.04	87.000	9.119	0.4017	46.29	5.331	8.683	0.2399 −01	0.1933 −01
6.32	0.4593 −03	0.4128 −02	0.1113	66.50	87.063	9.104	0.4016	46.43	5.332	8.708	0.2383 −01	0.1927 −01
6.33	0.4548 −03	0.4099 −02	0.1109	66.96	87.126	9.089	0.4016	46.58	5.334	8.732	0.2367 −01	0.1921 −01
6.34	0.4503 −03	0.4071 −02	0.1106	67.42	87.189	9.075	0.4015	46.73	5.336	8.757	0.2352 −01	0.1915 −01
6.35	0.4459 −03	0.4042 −02	0.1103	67.88	87.251	9.061	0.4014	46.88	5.338	8.782	0.2336 −01	0.1909 −01
6.36	0.4416 −03	0.4014 −02	0.1100	68.34	87.314	9.046	0.4014	47.03	5.340	8.806	0.2320 −01	0.1903 −01
6.37	0.4373 −03	0.3986 −02	0.1097	68.81	87.376	9.032	0.4013	47.17	5.342	8.831	0.2305 −01	0.1897 −01
6.38	0.4330 −03	0.3958 −02	0.1094	69.28	87.438	9.018	0.4012	47.32	5.344	8.856	0.2290 −01	0.1891 −01
6.39	0.4288 −03	0.3931 −02	0.1091	69.75	87.500	9.003	0.4011	47.47	5.345	8.881	0.2274 −01	0.1885 −01
6.40	0.4247 −03	0.3904 −02	0.1088	70.23	87.561	8.989	0.4011	47.62	5.347	8.906	0.2259 −01	0.1880 −01
6.41	0.4205 −03	0.3876 −02	0.1085	70.71	87.623	8.975	0.4010	47.77	5.349	8.931	0.2244 −01	0.1874 −01
6.42	0.4165 −03	0.3850 −02	0.1082	71.19	87.684	8.961	0.4009	47.92	5.351	8.956	0.2229 −01	0.1868 −01
6.43	0.4124 −03	0.3823 −02	0.1079	71.67	87.745	8.947	0.4009	48.07	5.353	8.981	0.2215 −01	0.1862 −01
6.44	0.4085 −03	0.3797 −02	0.1076	72.16	87.806	8.933	0.4008	48.22	5.354	9.006	0.2200 −01	0.1857 −01
6.45	0.4045 −03	0.3770 −02	0.1073	72.65	87.867	8.919	0.4007	48.37	5.356	9.031	0.2186 −01	0.1851 −01
6.46	0.4006 −03	0.3744 −02	0.1070	73.14	87.928	8.905	0.4007	48.52	5.358	9.056	0.2171 −01	0.1845 −01
6.47	0.3968 −03	0.3719 −02	0.1067	73.64	87.988	8.891	0.4006	48.67	5.360	9.081	0.2157 −01	0.1840 −01
6.48	0.3930 −03	0.3693 −02	0.1064	74.13	88.048	8.877	0.4005	48.82	5.362	9.106	0.2143 −01	0.1834 −01
6.49	0.3892 −03	0.3668 −02	0.1061	74.63	88.108	8.864	0.4004	48.97	5.363	9.131	0.2129 −01	0.1828 −01

Table 2.14: Isentropic and normal shock properties for air, $\gamma = 1.4$ (continued).

M or M_1	$\dfrac{P}{P_t}$	$\dfrac{\rho}{\rho_t}$	$\dfrac{T}{T_t}$	$\dfrac{A}{A^*}$	ν	μ	M_2	$\dfrac{P_2}{P_1}$	$\dfrac{\rho_2}{\rho_1}$	$\dfrac{T_2}{T_1}$	$\dfrac{P_{t_2}}{P_{t_1}}$	$\dfrac{P_1}{P_{t_2}}$
6.50	0.3854 −03	0.3643 −02	0.1058	75.14	88.168	8.850	0.4004	49.13	5.365	9.157	0.2115 −01	0.1823 −01
6.51	0.3818 −03	0.3618 −02	0.1055	75.64	88.228	8.836	0.4003	49.28	5.367	9.182	0.2101 −01	0.1817 −01
6.52	0.3781 −03	0.3593 −02	0.1052	76.15	88.288	8.822	0.4002	49.43	5.369	9.207	0.2087 −01	0.1812 −01
6.53	0.3745 −03	0.3568 −02	0.1050	76.66	88.347	8.809	0.4002	49.58	5.370	9.233	0.2073 −01	0.1806 −01
6.54	0.3709 −03	0.3544 −02	0.1047	77.18	88.407	8.795	0.4001	49.73	5.372	9.258	0.2060 −01	0.1801 −01
6.55	0.3674 −03	0.3520 −02	0.1044	77.70	88.466	8.782	0.4000	49.89	5.374	9.283	0.2046 −01	0.1795 −01
6.56	0.3639 −03	0.3496 −02	0.1041	78.22	88.525	8.768	0.4000	50.04	5.375	9.309	0.2033 −01	0.1790 −01
6.57	0.3604 −03	0.3472 −02	0.1038	78.74	88.584	8.755	0.3999	50.19	5.377	9.335	0.2020 −01	0.1784 −01
6.58	0.3570 −03	0.3448 −02	0.1035	79.26	88.643	8.741	0.3999	50.35	5.379	9.360	0.2007 −01	0.1779 −01
6.59	0.3536 −03	0.3425 −02	0.1032	79.79	88.701	8.728	0.3998	50.50	5.381	9.386	0.1994 −01	0.1774 −01
6.60	0.3503 −03	0.3402 −02	0.1030	80.33	88.760	8.715	0.3997	50.65	5.382	9.411	0.1981 −01	0.1768 −01
6.61	0.3469 −03	0.3379 −02	0.1027	80.86	88.818	8.701	0.3997	50.81	5.384	9.437	0.1968 −01	0.1763 −01
6.62	0.3437 −03	0.3356 −02	0.1024	81.40	88.876	8.688	0.3996	50.96	5.386	9.463	0.1955 −01	0.1758 −01
6.63	0.3404 −03	0.3333 −02	0.1021	81.94	88.934	8.675	0.3995	51.12	5.387	9.489	0.1942 −01	0.1753 −01
6.64	0.3372 −03	0.3311 −02	0.1019	82.48	88.992	8.662	0.3995	51.27	5.389	9.514	0.1930 −01	0.1747 −01
6.65	0.3340 −03	0.3289 −02	0.1016	83.03	89.049	8.649	0.3994	51.43	5.391	9.540	0.1917 −01	0.1742 −01
6.66	0.3309 −03	0.3266 −02	0.1013	83.58	89.107	8.636	0.3993	51.58	5.392	9.566	0.1905 −01	0.1737 −01
6.67	0.3278 −03	0.3244 −02	0.1010	84.13	89.164	8.623	0.3993	51.74	5.394	9.592	0.1893 −01	0.1732 −01
6.68	0.3247 −03	0.3223 −02	0.1008	84.68	89.221	8.610	0.3992	51.89	5.395	9.618	0.1881 −01	0.1727 −01
6.69	0.3217 −03	0.3201 −02	0.1005	85.24	89.278	8.597	0.3992	52.05	5.397	9.644	0.1869 −01	0.1721 −01
6.70	0.3187 −03	0.3180 −02	0.1002	85.80	89.335	8.584	0.3991	52.21	9.399	9.670	0.1857 −01	0.1716 −01
6.71	0.3157 −03	0.3158 −02	0.9995 −01	86.37	89.392	8.571	0.3990	52.36	5.400	9.696	0.1845 −01	0.1711 −01
6.72	0.3127 −03	0.3137 −02	0.9968 −01	86.94	89.448	8.558	0.3990	52.52	5.402	9.722	0.1833 −01	0.1706 −01
6.73	0.3098 −03	0.3116 −02	0.9942 −01	87.51	89.505	8.545	0.3989	52.68	5.404	9.748	0.1821 −01	0.1701 −01
6.74	0.3069 −03	0.3096 −02	0.9915 −01	88.08	89.561	8.532	0.3988	52.83	5.405	9.775	0.1809 −01	0.1696 −01
6.75	0.3041 −03	0.3075 −02	0.9889 −01	88.66	89.617	8.520	0.3988	52.99	5.407	9.801	0.1798 −01	0.1691 −01
6.76	0.3012 −03	0.3054 −02	0.9862 −01	89.24	89.673	8.507	0.3987	53.15	5.408	9.827	0.1786 −01	0.1686 −01
6.77	0.2984 −03	0.3034 −02	0.9836 −01	89.82	89.729	8.494	0.3987	53.31	5.410	9.854	0.1775 −01	0.1681 −01
6.78	0.2957 −03	0.3014 −02	0.9810 −01	90.41	89.784	8.482	0.3986	53.46	5.411	9.880	0.1764 −01	0.1676 −01
6.79	0.2929 −03	0.2994 −02	0.9784 −01	91.00	89.840	8.469	0.3985	53.62	5.413	9.906	0.1753 −01	0.1672 −01
6.80	0.2902 −03	0.2974 −02	0.9758 −01	91.59	89.895	8.456	0.3985	53.78	5.415	9.933	0.1741 −01	0.1667 −01
6.81	0.2875 −03	0.2955 −02	0.9732 −01	92.19	89.951	8.444	0.3984	53.94	5.416	9.959	0.1730 −01	0.1662 −01
6.82	0.2849 −03	0.2935 −02	0.9706 −01	92.79	90.006	8.431	0.3984	54.10	5.418	9.986	0.1719 −01	0.1657 −01
6.83	0.2823 −03	0.2916 −02	0.9681 −01	93.39	90.061	8.419	0.3983	54.26	5.419	10.01	0.1708 −01	0.1652 −01
6.84	0.2797 −03	0.2897 −02	0.9655 −01	94.00	90.115	8.407	0.3983	54.42	5.421	10.04	0.1698 −01	0.1647 −01
6.85	0.2771 −03	0.2878 −02	0.9630 −01	94.61	90.170	8.394	0.3982	54.58	5.422	10.07	0.1687 −01	0.1643 −01
6.86	0.2745 −03	0.2859 −02	0.9604 −01	95.22	90.225	8.382	0.3981	54.74	5.424	10.09	0.1676 −01	0.1638 −01
6.87	0.2720 −03	0.2840 −02	0.9579 −01	95.83	90.279	8.370	0.3981	54.90	5.425	10.12	0.1666 −01	0.1633 −01
6.88	0.2695 −03	0.2821 −02	0.9554 −01	96.45	90.333	8.357	0.3980	55.06	5.427	10.15	0.1655 −01	0.1628 −01
6.89	0.2671 −03	0.2803 −02	0.9529 −01	97.08	90.387	8.345	0.3980	55.22	5.428	10.17	0.1645 −01	0.1624 −01
6.90	0.2646 −03	0.2784 −02	0.9504 −01	97.70	90.441	8.333	0.3979	55.38	5.430	10.20	0.1634 −01	0.1619 −01
6.91	0.2622 −03	0.2766 −02	0.9479 −01	98.33	90.495	8.321	0.3979	55.54	5.431	10.23	0.1624 −01	0.1614 −01
6.92	0.2598 −03	0.2748 −02	0.9454 −01	98.96	90.549	8.309	0.3978	55.70	5.433	10.25	0.1614 −01	0.1610 −01
6.93	0.2574 −03	0.2730 −02	0.9429 −01	99.60	90.602	8.297	0.3977	55.86	5.434	10.28	0.1604 −01	0.1605 −01
6.94	0.2551 −03	0.2713 −02	0.9405 −01	100.2	90.656	8.285	0.3977	56.02	5.436	10.31	0.1594 −01	0.1601 −01
6.95	0.2528 −03	0.2695 −02	0.9380 −01	100.9	90.709	8.273	0.3976	56.19	5.437	10.33	0.1584 −01	0.1596 −01
6.96	0.2505 −03	0.2677 −02	0.9356 −01	101.5	90.762	8.261	0.3976	56.35	5.439	10.36	0.1574 −01	0.1591 −01
6.97	0.2482 −03	0.2660 −02	0.9332 −01	102.2	90.815	8.249	0.3975	56.51	5.440	10.39	0.1564 −01	0.1587 −01
6.98	0.2460 −03	0.2643 −02	0.9307 −01	102.8	90.868	8.237	0.3975	56.67	5.442	10.42	0.1554 −01	0.1582 −01
6.99	0.2437 −03	0.2626 −02	0.9283 −01	103.5	90.920	8.225	0.3974	56.84	5.443	10.44	0.1545 −01	0.1578 −01
7.00	0.2415 −03	0.2609 −02	0.9259 −01	104.1	90.973	8.213	0.3974	57.00	5.444	10.47	0.1535 −01	0.1573 −01
7.01	0.2394 −03	0.2592 −02	0.9235 −01	104.8	91.026	8.201	0.3973	57.16	5.446	10.50	0.1526 −01	0.1569 −01
7.02	0.2372 −03	0.2575 −02	0.9211 −01	105.5	91.078	8.190	0.3973	57.33	5.447	10.52	0.1516 −01	0.1565 −01
7.03	0.2351 −03	0.2559 −02	0.9187 −01	106.1	91.130	8.178	0.3972	57.49	5.449	10.55	0.1507 −01	0.1560 −01
7.04	0.2330 −03	0.2542 −02	0.9164 −01	106.8	91.182	8.166	0.3971	57.66	5.450	10.58	0.1497 −01	0.1556 −01

Table 2.14: Isentropic and normal shock properties for air, $\gamma = 1.4$ (continued).

M or M_1	$\dfrac{P}{P_t}$	$\dfrac{\rho}{\rho_t}$	$\dfrac{T}{T_t}$	$\dfrac{A}{A^*}$	ν	μ	M_2	$\dfrac{P_2}{P_1}$	$\dfrac{\rho_2}{\rho_1}$	$\dfrac{T_2}{T_1}$	$\dfrac{P_{t_2}}{P_{t_1}}$	$\dfrac{P_1}{P_{t_2}}$
7.05	0.2309 −03	0.2526 −02	0.9140 −01	107.5	91.234	8.154	0.3971	57.82	5.452	10.61	0.1488 −01	0.1551 −01
7.06	0.2288 −03	0.2510 −02	0.9117 −01	108.2	91.286	8.143	0.3970	57.98	5.453	10.63	0.1479 −01	0.1547 −01
7.07	0.2267 −03	0.2493 −02	0.9093 −01	108.9	91.337	8.131	0.3970	58.15	5.454	10.66	0.1470 −01	0.1543 −01
7.08	0.2247 −03	0.2477 −02	0.9070 −01	109.5	91.389	8.120	0.3969	58.31	5.456	10.69	0.1461 −01	0.1538 −01
7.09	0.2227 −03	0.2462 −02	0.9047 −01	110.2	91.440	8.108	0.3969	58.48	5.457	10.72	0.1452 −01	0.1534 −01
7.10	0.2207 −03	0.2446 −02	0.9023 −01	110.9	91.492	8.097	0.3968	58.65	5.459	10.74	0.1443 −01	0.1530 −01
7.11	0.2187 −03	0.2430 −02	0.9000 −01	111.6	91.543	8.085	0.3968	58.81	5.460	10.77	0.1434 −01	0.1526 −01
7.12	0.2168 −03	0.2415 −02	0.8977 −01	112.3	91.594	8.074	0.3967	58.98	5.461	10.80	0.1425 −01	0.1521 −01
7.13	0.2149 −03	0.2399 −02	0.8954 −01	113.0	91.644	8.062	0.3967	59.14	5.463	10.83	0.1416 −01	0.1517 −01
7.14	0.2129 −03	0.2384 −02	0.8932 −01	113.8	91.695	8.051	0.3966	59.31	5.464	10.85	0.1408 −01	0.1513 −01
7.15	0.2111 −03	0.2369 −02	0.8909 −01	114.5	91.746	8.040	0.3966	59.48	5.465	10.88	0.1399 −01	0.1509 −01
7.16	0.2092 −03	0.2354 −02	0.8886 −01	115.2	91.796	8.028	0.3965	59.64	5.467	10.91	0.1390 −01	0.1504 −01
7.17	0.2073 −03	0.2339 −02	0.8864 −01	115.9	91.847	8.017	0.3965	59.81	5.468	10.94	0.1382 −01	0.1500 −01
7.18	0.2055 −03	0.2324 −02	0.8841 −01	116.6	91.897	8.006	0.3964	59.98	5.470	10.97	0.1373 −01	0.1496 −01
7.19	0.2037 −03	0.2310 −02	0.8819 −01	117.4	91.947	7.995	0.3964	60.15	5.471	10.99	0.1365 −01	0.1492 −01
7.20	0.2019 −03	0.2295 −02	0.8796 −01	118.1	91.997	7.983	0.3963	60.31	5.472	11.02	0.1357 −01	0.1488 −01
7.21	0.2001 −03	0.2280 −02	0.8774 −01	118.8	92.047	7.972	0.3963	60.48	5.474	11.05	0.1349 −01	0.1484 −01
7.22	0.1983 −03	0.2266 −02	0.8752 −01	119.6	92.097	7.961	0.3962	60.65	5.475	11.08	0.1340 −01	0.1480 −01
7.23	0.1966 −03	0.2252 −02	0.8730 −01	120.3	92.146	7.950	0.3962	60.82	5.476	11.11	0.1332 −01	0.1476 −01
7.24	0.1949 −03	0.2238 −02	0.8708 −01	121.0	92.196	7.939	0.3961	60.99	5.478	11.13	0.1324 −01	0.1472 −01
7.25	0.1931 −03	0.2224 −02	0.8686 −01	121.8	92.245	7.928	0.3961	61.16	5.479	11.16	0.1316 −01	0.1468 −01
7.26	0.1914 −03	0.2210 −02	0.8664 −01	122.6	92.294	7.917	0.3960	61.33	5.480	11.19	0.1308 −01	0.1464 −01
7.27	0.1898 −03	0.2196 −02	0.8642 −01	123.3	92.343	7.906	0.3960	61.50	5.481	11.22	0.1300 −01	0.1460 −01
7.28	0.1881 −03	0.2182 −02	0.8621 −01	124.1	92.392	7.895	0.3959	61.66	5.483	11.25	0.1292 −01	0.1456 −01
7.29	0.1865 −03	0.2168 −02	0.8599 −01	124.8	92.441	7.884	0.3959	61.83	5.484	11.28	0.1285 −01	0.1452 −01
7.30	0.1848 −03	0.2155 −02	0.8578 −01	125.6	92.490	7.873	0.3958	62.01	5.485	11.30	0.1277 −01	0.1448 −01
7.31	0.1832 −03	0.2141 −02	0.8556 −01	126.4	92.539	7.863	0.3958	62.18	5.487	11.33	0.1269 −01	0.1444 −01
7.32	0.1816 −03	0.2128 −02	0.8535 −01	127.2	92.587	7.852	0.3957	62.35	5.488	11.36	0.1261 −01	0.1440 −01
7.33	0.1800 −03	0.2115 −02	0.8514 −01	127.9	92.635	7.841	0.3957	62.52	5.489	11.39	0.1254 −01	0.1436 −01
7.34	0.1785 −03	0.2102 −02	0.8492 −01	128.7	92.684	7.830	0.3956	62.69	5.490	11.42	0.1246 −01	0.1432 −01
7.35	0.1769 −03	0.2089 −02	0.8471 −01	129.5	92.732	7.820	0.3956	62.86	5.492	11.45	0.1239 −01	0.1428 −01
7.36	0.1754 −03	0.2076 −02	0.8450 −01	130.3	92.780	7.809	0.3955	63.03	5.493	11.48	0.1231 −01	0.1424 −01
7.37	0.1739 −03	0.2063 −02	0.8429 −01	131.1	92.828	7.798	0.3955	63.20	5.494	11.50	0.1224 −01	0.1420 −01
7.38	0.1724 −03	0.2050 −02	0.8408 −01	131.9	92.876	7.788	0.3955	63.38	5.496	11.53	0.1217 −01	0.1417 −01
7.39	0.1709 −03	0.2037 −02	0.8387 −01	132.7	92.923	7.777	0.3954	63.55	5.497	11.56	0.1209 −01	0.1413 −01
7.40	0.1694 −03	0.2025 −02	0.8367 −01	133.5	92.971	7.766	0.3954	63.72	5.498	11.59	0.1202 −01	0.1409 −01
7.41	0.1679 −03	0.2012 −02	0.8346 −01	134.3	93.018	7.756	0.3953	63.89	5.499	11.62	0.1195 −01	0.1405 −01
7.42	0.1665 −03	0.2000 −02	0.8325 −01	135.2	93.066	7.745	0.3953	64.07	5.500	11.65	0.1188 −01	0.1402 −01
7.43	0.1651 −03	0.1988 −02	0.8305 −01	136.0	93.113	7.735	0.3952	64.24	5.502	11.68	0.1181 −01	0.1398 −01
7.44	0.1636 −03	0.1975 −02	0.8284 −01	136.8	93.160	7.724	0.3952	64.41	5.503	11.71	0.1174 −01	0.1394 −01
7.45	0.1622 −03	0.1963 −02	0.8264 −01	137.6	93.207	7.714	0.3951	64.59	5.504	11.73	0.1167 −01	0.1390 −01
7.46	0.1608 −03	0.1951 −02	0.8244 −01	138.5	93.254	7.704	0.3951	64.76	5.505	11.76	0.1160 −01	0.1387 −01
7.47	0.1595 −03	0.1939 −02	0.8223 −01	139.3	93.300	7.693	0.3950	64.93	5.507	11.79	0.1153 −01	0.1383 −01
7.48	0.1581 −03	0.1927 −02	0.8203 −01	140.2	93.347	7.683	0.3950	65.11	5.508	11.82	0.1146 −01	0.1379 −01
7.49	0.1568 −03	0.1916 −02	0.8183 −01	141.0	93.394	7.672	0.3950	65.28	5.509	11.85	0.1140 −01	0.1376 −01
7.50	0.1554 −03	0.1904 −02	0.8163 −01	141.8	93.440	7.662	0.3949	65.46	5.510	11.88	0.1133 −01	0.1372 −01
7.51	0.1541 −03	0.1892 −02	0.8143 −01	142.7	93.486	7.652	0.3949	65.63	5.511	11.91	0.1126 −01	0.1368 −01
7.52	0.1528 −03	0.1881 −02	0.8123 −01	143.6	93.533	7.642	0.3948	65.81	5.513	11.94	0.1119 −01	0.1365 −01
7.53	0.1515 −03	0.1869 −02	0.8103 −01	144.4	93.579	7.631	0.3948	65.98	5.514	11.97	0.1113 −01	0.1361 −01
7.54	0.1502 −03	0.1858 −02	0.8084 −01	145.3	93.625	7.621	0.3947	66.16	5.515	12.00	0.1106 −01	0.1358 −01
7.55	0.1489 −03	0.1847 −02	0.8064 −01	146.2	93.670	7.611	0.3947	66.34	5.516	12.03	0.1100 −01	0.1354 −01
7.56	0.1477 −03	0.1835 −02	0.8044 −01	147.0	93.716	7.601	0.3946	66.51	5.517	12.06	0.1093 −01	0.1350 −01
7.57	0.1464 −03	0.1824 −02	0.8025 −01	147.9	93.762	7.591	0.3946	66.69	5.519	12.08	0.1087 −01	0.1347 −01
7.58	0.1452 −03	0.1813 −02	0.8005 −01	148.8	93.807	7.581	0.3946	66.87	5.520	12.11	0.1081 −01	0.1343 −01
7.59	0.1439 −03	0.1802 −02	0.7986 −01	149.7	93.853	7.571	0.3945	67.04	5.521	12.14	0.1074 −01	0.1340 −01

Table 2.14: Isentropic and normal shock properties for air, $\gamma = 1.4$ (continued).

M or M_1	$\dfrac{P}{P_t}$	$\dfrac{\rho}{\rho_t}$	$\dfrac{T}{T_t}$	$\dfrac{A}{A^*}$	ν	μ	M_2	$\dfrac{P_2}{P_1}$	$\dfrac{\rho_2}{\rho_1}$	$\dfrac{T_2}{T_1}$	$\dfrac{P_{t_2}}{P_{t_1}}$	$\dfrac{P_1}{P_{t_2}}$
7.60	0.1427 −03	0.1791 −02	0.7967 −01	150.6	93.898	7.561	0.3945	67.22	5.522	12.17	0.1068 −01	0.1336 −01
7.61	0.1415 −03	0.1781 −02	0.7947 −01	151.5	93.943	7.551	0.3944	67.40	5.523	12.20	0.1062 −01	0.1333 −01
7.62	0.1403 −03	0.1770 −02	0.7928 −01	152.4	93.988	7.541	0.3944	67.58	5.524	12.23	0.1056 −01	0.1329 −01
7.63	0.1391 −03	0.1759 −02	0.7909 −01	153.3	94.033	7.531	0.3943	67.75	5.525	12.26	0.1049 −01	0.1326 −01
7.64	0.1380 −03	0.1749 −02	0.7890 −01	154.2	94.078	7.521	0.3943	67.93	5.527	12.29	0.1043 −01	0.1322 −01
7.65	0.1368 −03	0.1738 −02	0.7871 −01	155.1	94.123	7.511	0.3943	68.11	5.528	12.32	0.1037 −01	0.1319 −01
7.66	0.1357 −03	0.1728 −02	0.7852 −01	156.0	94.168	7.501	0.3942	68.29	5.529	12.35	0.1031 −01	0.1316 −01
7.67	0.1345 −03	0.1717 −02	0.7833 −01	157.0	94.212	7.491	0.3942	68.47	5.530	12.38	0.1025 −01	0.1312 −01
7.68	0.1334 −03	0.1707 −02	0.7814 −01	157.9	94.257	7.482	0.3941	68.65	5.531	12.41	0.1019 −01	0.1309 −01
7.69	0.1323 −03	0.1697 −02	0.7796 −01	158.8	94.301	7.472	0.3941	68.83	5.532	12.44	0.1013 −01	0.1305 −01
7.70	0.1312 −03	0.1687 −02	0.7777 −01	159.8	94.345	7.462	0.3941	69.01	5.533	12.47	0.1007 −01	0.1302 −01
7.71	0.1301 −03	0.1677 −02	0.7759 −01	160.7	94.389	7.452	0.3940	69.18	5.534	12.50	0.1002 −01	0.1299 −01
7.72	0.1290 −03	0.1667 −02	0.7740 −01	161.7	94.433	7.443	0.3940	69.36	5.536	12.53	0.9959 −02	0.1295 −01
7.73	0.1279 −03	0.1657 −02	0.7722 −01	162.6	94.477	7.433	0.3939	69.55	5.537	12.56	0.9901 −02	0.1292 −01
7.74	0.1269 −03	0.1647 −02	0.7703 −01	163.6	94.521	7.423	0.3939	69.73	5.538	12.59	0.9844 −02	0.1289 −01
7.75	0.1258 −03	0.1637 −02	0.7685 −01	164.5	94.565	7.414	0.3939	69.91	5.539	12.62	0.9788 −02	0.1285 −01
7.76	0.1248 −03	0.1627 −02	0.7666 −01	165.5	94.608	7.404	0.3938	70.09	5.540	12.65	0.9731 −02	0.1282 −01
7.77	0.1237 −03	0.1618 −02	0.7648 −01	166.5	94.652	7.394	0.3938	70.27	5.541	12.68	0.9675 −02	0.1279 −01
7.78	0.1227 −03	0.1608 −02	0.7630 −01	167.4	94.695	7.385	0.3937	70.45	5.542	12.71	0.9620 −02	0.1276 −01
7.79	0.1217 −03	0.1599 −02	0.7612 −01	168.4	94.739	7.375	0.3937	70.63	5.543	12.74	0.9565 −02	0.1272 −01
7.80	0.1207 −03	0.1589 −02	0.7594 −01	169.4	94.782	7.366	0.3937	70.81	5.544	12.77	0.9510 −02	0.1269 −01
7.81	0.1197 −03	0.1580 −02	0.7576 −01	170.4	94.825	7.356	0.3936	71.00	5.545	12.80	0.9455 −02	0.1266 −01
7.82	0.1187 −03	0.1571 −02	0.7558 −01	171.4	94.868	7.347	0.3936	71.18	5.547	12.83	0.9401 −02	0.1263 −01
7.83	0.1177 −03	0.1561 −02	0.7540 −01	172.4	94.911	7.337	0.3935	71.36	5.548	12.86	0.9348 −02	0.1259 −01
7.84	0.1168 −03	0.1552 −02	0.7523 −01	173.4	94.954	7.328	0.3935	71.54	5.549	12.89	0.9294 −02	0.1256 −01
7.85	0.1158 −03	0.1543 −02	0.7505 −01	174.4	94.996	7.319	0.3935	71.73	5.550	12.92	0.9241 −02	0.1253 −01
7.86	0.1148 −03	0.1534 −02	0.7487 −01	175.4	95.039	7.309	0.3934	71.91	5.551	12.96	0.9189 −02	0.1250 −01
7.87	0.1139 −03	0.1525 −02	0.7470 −01	176.4	95.081	7.300	0.3934	72.09	5.552	12.99	0.9136 −02	0.1247 −01
7.88	0.1130 −03	0.1516 −02	0.7452 −01	177.5	95.124	7.291	0.3933	72.28	5.553	13.02	0.9084 −02	0.1244 −01
7.89	0.1120 −03	0.1507 −02	0.7435 −01	178.5	95.166	7.281	0.3933	72.46	5.554	13.05	0.9033 −02	0.1240 −01
7.90	0.1111 −03	0.1498 −02	0.7417 −01	179.5	95.208	7.272	0.3933	72.65	5.555	13.08	0.8981 −02	0.1237 −01
7.91	0.1102 −03	0.1490 −02	0.7400 −01	180.6	95.250	7.263	0.3932	72.83	5.556	13.11	0.8931 −02	0.1234 −01
7.92	0.1093 −03	0.1481 −02	0.7382 −01	181.6	95.292	7.254	0.3932	73.01	5.557	13.14	0.8880 −02	0.1231 −01
7.93	0.1084 −03	0.1472 −02	0.7365 −01	182.6	95.334	7.244	0.3931	73.20	5.558	13.17	0.8830 −02	0.1228 −01
7.94	0.1075 −03	0.1464 −02	0.7348 −01	183.7	95.376	7.235	0.3931	73.38	5.559	13.20	0.8780 −02	0.1225 −01
7.95	0.1067 −03	0.1455 −02	0.7331 −01	184.8	95.418	7.226	0.3931	73.57	5.560	13.23	0.8730 −02	0.1222 −01
7.96	0.1058 −03	0.1447 −02	0.7314 −01	185.8	95.460	7.217	0.3930	73.76	5.561	13.26	0.8681 −02	0.1219 −01
7.97	0.1050 −03	0.1438 −02	0.7297 −01	186.9	95.501	7.208	0.3930	73.94	5.562	13.29	0.8632 −02	0.1216 −01
7.98	0.1041 −03	0.1430 −02	0.7280 −01	188.0	95.542	7.199	0.3930	74.13	5.563	13.32	0.8584 −02	0.1213 −01
7.99	0.1033 −03	0.1422 −02	0.7263 −01	189.0	95.584	7.190	0.3929	74.31	5.564	13.36	0.8535 −02	0.1210 −01
8.00	0.1024 −03	0.1413 −02	0.7246 −01	190.1	95.625	7.181	0.3929	74.50	5.565	13.39	0.8487 −02	0.1207 −01
8.01	0.1016 −03	0.1405 −02	0.7229 −01	191.2	95.666	7.172	0.3929	74.69	5.566	13.42	0.8440 −02	0.1204 −01
8.02	0.1008 −03	0.1397 −02	0.7213 −01	192.3	95.707	7.163	0.3928	74.87	5.567	13.45	0.8392 −02	0.1201 −01
8.03	0.9996 −04	0.1389 −02	0.7196 −01	193.4	95.748	7.154	0.3928	75.06	5.568	13.48	0.8345 −02	0.1198 −01
8.04	0.9916 −04	0.1381 −02	0.7179 −01	194.5	95.789	7.145	0.3927	75.25	5.569	13.51	0.8299 −02	0.1195 −01
8.05	0.9836 −04	0.1373 −02	0.7163 −01	195.6	95.830	7.136	0.3927	75.44	5.570	13.54	0.8252 −02	0.1192 −01
8.06	0.9757 −04	0.1365 −02	0.7146 −01	196.7	95.871	7.127	0.3927	75.62	5.571	13.57	0.8206 −02	0.1189 −01
8.07	0.9679 −04	0.1357 −02	0.7130 −01	197.8	95.911	7.118	0.3926	75.81	5.572	13.61	0.8161 −02	0.1186 −01
8.08	0.9601 −04	0.1350 −02	0.7114 −01	199.0	95.952	7.109	0.3926	76.00	5.573	13.64	0.8115 −02	0.1183 −01
8.09	0.9524 −04	0.1342 −02	0.7097 −01	200.1	95.992	7.100	0.3926	76.19	5.574	13.67	0.8070 −02	0.1180 −01
8.10	0.9448 −04	0.1334 −02	0.7081 −01	201.2	96.032	7.092	0.3925	76.38	5.575	13.70	0.8025 −02	0.1177 −01
8.11	0.9372 −04	0.1327 −02	0.7065 −01	202.4	96.073	7.083	0.3925	76.57	5.576	13.73	0.7980 −02	0.1174 −01
8.12	0.9297 −04	0.1319 −02	0.7049 −01	203.5	96.113	7.074	0.3925	76.76	5.577	13.76	0.7936 −02	0.1172 −01
8.13	0.9223 −04	0.1312 −02	0.7033 −01	204.7	96.153	7.065	0.3924	76.95	5.578	13.79	0.7892 −02	0.1169 −01
8.14	0.9150 −04	0.1304 −02	0.7016 −01	205.8	96.193	7.057	0.3924	77.14	5.579	13.83	0.7848 −02	0.1166 −01

Table 2.14: Isentropic and normal shock properties for air, $\gamma = 1.4$ (continued).

M or M_1	$\dfrac{P}{P_t}$	$\dfrac{\rho}{\rho_t}$	$\dfrac{T}{T_t}$	$\dfrac{A}{A^*}$	ν	μ	M_2	$\dfrac{P_2}{P_1}$	$\dfrac{\rho_2}{\rho_1}$	$\dfrac{T_2}{T_1}$	$\dfrac{P_{t_2}}{P_{t_1}}$	$\dfrac{P_1}{P_{t_2}}$
8.15	0.9077 −04	0.1297 −02	0.7000 −01	207.0	96.232	7.048	0.3924	77.33	5.580	13.86	0.7805 −02	0.1163 −01
8.16	0.9005 −04	0.1289 −02	0.6984 −01	208.1	96.272	7.039	0.3923	77.52	5.581	13.89	0.7762 −02	0.1160 −01
8.17	0.8933 −04	0.1282 −02	0.6969 −01	209.3	96.312	7.030	0.3923	77.71	5.582	13.92	0.7719 −02	0.1157 −01
8.18	0.8862 −04	0.1275 −02	0.6953 −01	210.5	96.351	7.022	0.3923	77.90	5.583	13.95	0.7676 −02	0.1155 −01
8.19	0.8792 −04	0.1267 −02	0.6937 −01	211.7	96.391	7.013	0.3922	78.09	5.584	13.99	0.7634 −02	0.1152 −01
8.20	0.8722 −04	0.1260 −02	0.6921 −01	212.9	96.430	7.005	0.3922	78.28	5.585	14.02	0.7592 −02	0.1149 −01
8.21	0.8654 −04	0.1253 −02	0.6906 −01	214.0	96.470	6.996	0.3921	78.47	5.586	14.05	0.7550 −02	0.1146 −01
8.22	0.8585 −04	0.1246 −02	0.6890 −01	215.2	96.509	6.988	0.3921	78.66	5.587	14.08	0.7509 −02	0.1143 −01
8.23	0.8517 −04	0.1239 −02	0.6874 −01	216.5	96.548	6.979	0.3921	78.86	5.588	14.11	0.7467 −02	0.1141 −01
8.24	0.8450 −04	0.1232 −02	0.6859 −01	217.7	96.587	6.970	0.3920	79.05	5.588	14.15	0.7427 −02	0.1138 −01
8.25	0.8384 −04	0.1225 −02	0.6843 −01	218.9	96.626	6.962	0.3920	79.24	5.589	14.18	0.7386 −02	0.1135 −01
8.26	0.8318 −04	0.1218 −02	0.6828 −01	220.1	96.665	6.954	0.3920	79.43	5.590	14.21	0.7345 −02	0.1132 −01
8.27	0.8252 −04	0.1211 −02	0.6813 −01	221.3	96.704	6.945	0.3919	79.63	5.591	14.24	0.7305 −02	0.1130 −01
8.28	0.8187 −04	0.1205 −02	0.6797 −01	222.6	96.742	6.937	0.3919	79.82	5.592	14.27	0.7265 −02	0.1127 −01
8.29	0.8123 −04	0.1198 −02	0.6782 −01	223.8	96.781	6.928	0.3919	80.01	5.593	14.31	0.7226 −02	0.1124 −01
8.30	0.8060 −04	0.1191 −02	0.6767 −01	225.0	96.820	6.920	0.3918	80.21	5.594	14.34	0.7186 −02	0.1122 −01
8.31	0.7996 −04	0.1184 −02	0.6751 −01	226.3	96.858	6.911	0.3918	80.40	5.595	14.37	0.7147 −02	0.1119 −01
8.32	0.7934 −04	0.1178 −02	0.6736 −01	227.5	96.896	6.903	0.3918	80.59	5.596	14.40	0.7108 −02	0.1116 −01
8.33	0.7872 −04	0.1171 −02	0.6721 −01	228.8	96.935	6.895	0.3917	80.79	5.597	14.44	0.7070 −02	0.1114 −01
8.34	0.7811 −04	0.1165 −02	0.6706 −01	230.1	96.973	6.886	0.3917	80.98	5.598	14.47	0.7031 −02	0.1111 −01
8.35	0.7750 −04	0.1158 −02	0.6691 −01	231.3	97.011	6.878	0.3917	81.18	5.599	14.50	0.6993 −02	0.1108 −01
8.36	0.7689 −04	0.1152 −02	0.6676 −01	232.6	97.049	6.870	0.3917	81.37	5.599	14.53	0.6955 −02	0.1106 −01
8.37	0.7629 −04	0.1145 −02	0.6661 −01	233.9	97.087	6.862	0.3916	81.57	5.600	14.56	0.6917 −02	0.1103 −01
8.38	0.7570 −04	0.1139 −02	0.6647 −01	235.2	97.125	6.853	0.3916	81.76	5.601	14.60	0.6880 −02	0.1100 −01
8.39	0.7511 −04	0.1133 −02	0.6632 −01	236.5	97.163	6.845	0.3916	81.96	5.602	14.63	0.6843 −02	0.1098 −01
8.40	0.7453 −04	0.1126 −02	0.6617 −01	237.8	97.200	6.837	0.3915	82.15	5.603	14.66	0.6806 −02	0.1095 −01
8.41	0.7395 −04	0.1120 −02	0.6602 −01	239.1	97.238	6.829	0.3915	82.35	5.604	14.70	0.6769 −02	0.1093 −01
8.42	0.7338 −04	0.1114 −02	0.6588 −01	240.4	97.275	6.821	0.3915	82.55	5.605	14.73	0.6733 −02	0.1090 −01
8.43	0.7281 −04	0.1108 −02	0.6573 −01	241.7	97.313	6.813	0.3914	82.74	5.606	14.76	0.6696 −02	0.1087 −01
8.44	0.7225 −04	0.1102 −02	0.6559 −01	243.0	97.350	6.805	0.3914	82.94	5.606	14.79	0.6660 −02	0.1085 −01
8.45	0.7169 −04	0.1096 −02	0.6544 −01	244.4	97.387	6.796	0.3914	83.14	5.607	14.83	0.6624 −02	0.1082 −01
8.46	0.7114 −04	0.1090 −02	0.6530 −01	245.7	97.425	6.788	0.3913	83.33	5.608	14.86	0.6589 −02	0.1080 −01
8.47	0.7059 −04	0.1084 −02	0.6515 −01	247.0	97.462	6.780	0.3913	83.53	5.609	14.89	0.6554 −02	0.1077 −01
8.48	0.7005 −04	0.1078 −02	0.6501 −01	248.4	97.499	6.772	0.3913	83.73	5.610	14.93	0.6518 −02	0.1075 −01
8.49	0.6951 −04	0.1072 −02	0.6487 −01	249.7	97.536	6.764	0.3912	83.93	5.611	14.96	0.6484 −02	0.1072 −01
8.50	0.6898 −04	0.1066 −02	0.6472 −01	251.1	97.573	6.756	0.3912	84.13	5.612	14.99	0.6449 −02	0.1070 −01
8.51	0.6845 −04	0.1060 −02	0.6458 −01	252.5	97.609	6.748	0.3912	84.32	5.613	15.02	0.6414 −02	0.1067 −01
8.52	0.6793 −04	0.1054 −02	0.6444 −01	253.8	97.646	6.740	0.3911	84.52	5.613	15.06	0.6380 −02	0.1065 −01
8.53	0.6741 −04	0.1048 −02	0.6430 −01	255.2	97.683	6.732	0.3911	84.72	5.614	15.09	0.6346 −02	0.1062 −01
8.54	0.6689 −04	0.1043 −02	0.6416 −01	256.6	97.719	6.724	0.3911	84.92	5.615	15.12	0.6312 −02	0.1060 −01
8.55	0.6638 −04	0.1037 −02	0.6402 −01	258.0	97.756	6.717	0.3911	85.12	5.616	15.16	0.6279 −02	0.1057 −01
8.56	0.6587 −04	0.1031 −02	0.6388 −01	259.4	97.792	6.709	0.3910	85.32	5.617	15.19	0.6245 −02	0.1055 −01
8.57	0.6537 −04	0.1026 −02	0.6374 −01	260.8	97.828	6.701	0.3910	85.52	5.618	15.22	0.6212 −02	0.1052 −01
8.58	0.6487 −04	0.1020 −02	0.6360 −01	262.2	97.865	6.693	0.3910	85.72	5.618	15.26	0.6179 −02	0.1050 −01
8.59	0.6438 −04	0.1014 −02	0.6346 −01	263.6	97.901	6.685	0.3909	85.92	5.619	15.29	0.6146 −02	0.1047 −01
8.60	0.6389 −04	0.1009 −02	0.6332 −01	265.0	97.937	6.677	0.3909	86.12	5.620	15.32	0.6114 −02	0.1045 −01
8.61	0.6341 −04	0.1004 −02	0.6318 −01	266.5	97.973	6.670	0.3909	86.32	5.621	15.36	0.6081 −02	0.1043 −01
8.62	0.6292 −04	0.9981 −03	0.6305 −01	267.9	98.009	6.662	0.3908	86.52	5.622	15.39	0.6049 −02	0.1040 −01
8.63	0.6245 −04	0.9927 −03	0.6291 −01	269.3	98.045	6.654	0.3908	86.77	5.623	15.42	0.6017 −02	0.1038 −01
8.64	0.6198 −04	0.9873 −03	0.6277 −01	270.8	98.080	6.646	0.3908	86.92	5.623	15.46	0.5986 −02	0.1035 −01
8.65	0.6151 −04	0.9819 −03	0.6264 −01	272.2	98.116	6.639	0.3908	87.13	5.624	15.49	0.5954 −02	0.1033 −01
8.66	0.6104 −04	0.9766 −03	0.6250 −01	273.7	98.152	6.631	0.3907	87.33	5.625	15.53	0.5923 −02	0.1031 −01
8.67	0.6058 −04	0.9714 −03	0.6237 −01	275.1	98.187	6.623	0.3907	87.53	5.626	15.56	0.5891 −02	0.1028 −01
8.68	0.6012 −04	0.9661 −03	0.6223 −01	276.6	98.223	6.616	0.3907	87.73	5.627	15.59	0.5860 −02	0.1026 −01
8.69	0.5967 −04	0.9609 −03	0.6210 −01	278.1	98.258	6.608	0.3906	87.94	5.627	15.63	0.5830 −02	0.1024 −01

Table 2.14: Isentropic and normal shock properties for air, $\gamma = 1.4$ (continued).

$\begin{matrix}M\\ or\\ M_1\end{matrix}$	$\dfrac{P}{P_t}$	$\dfrac{\rho}{\rho_t}$	$\dfrac{T}{T_t}$	$\dfrac{A}{A^*}$	ν	μ	M_2	$\dfrac{P_2}{P_1}$	$\dfrac{\rho_2}{\rho_1}$	$\dfrac{T_2}{T_1}$	$\dfrac{P_{t_2}}{P_{t_1}}$	$\dfrac{P_1}{P_{t_2}}$
8.70	0.5922 −04	0.9558 −03	0.6196 −01	279.6	98.293	6.600	0.3906	88.14	5.628	15.66	0.5799 −02	0.1021 −01
8.71	0.5878 −04	0.9506 −03	0.6183 −01	281.1	98.329	6.593	0.3906	88.34	5.629	15.69	0.5769 −02	0.1019 −01
8.72	0.5834 −04	0.9455 −03	0.6170 −01	282.6	98.364	6.585	0.3906	88.54	5.630	15.73	0.5738 −02	0.1017 −01
8.73	0.5790 −04	0.9405 −03	0.6157 −01	284.1	98.399	6.577	0.3905	88.75	5.631	15.76	0.5708 −02	0.1014 −01
8.74	0.5747 −04	0.9354 −03	0.6143 −01	285.6	98.434	6.570	0.3905	88.95	5.631	15.80	0.5679 −02	0.1012 −01
8.75	0.5704 −04	0.9304 −03	0.6130 −01	287.1	98.469	6.562	0.3905	89.16	5.632	15.83	0.5649 −02	0.1010 −01
8.76	0.5661 −04	0.9254 −03	0.6117 −01	288.6	98.504	6.555	0.3904	89.37	5.633	15.86	0.5619 −02	0.1007 −01
8.77	0.5619 −04	0.9205 −03	0.6104 −01	290.2	98.538	6.547	0.3904	89.57	5.634	15.90	0.5590 −02	0.1005 −01
8.78	0.5577 −04	0.9156 −03	0.6091 −01	291.7	98.573	6.540	0.3904	89.77	5.635	15.93	0.5561 −02	0.1003 −01
8.79	0.5535 −04	0.9107 −03	0.6078 −01	293.2	98.608	6.532	0.3904	89.97	5.635	15.97	0.5532 −02	0.1001 −01
8.80	0.5494 −04	0.9058 −03	0.6065 −01	294.8	98.642	6.525	0.3903	90.18	5.636	16.00	0.5503 −02	0.9983 −02
8.81	0.5453 −04	0.9010 −03	0.6052 −01	296.3	98.677	6.517	0.3903	90.39	5.637	16.04	0.5475 −02	0.9960 −02
8.82	0.5412 −04	0.8962 −03	0.6039 −01	297.9	98.711	6.510	0.3903	90.59	5.638	16.07	0.5446 −02	0.9938 −02
8.83	0.5372 −04	0.8915 −03	0.6026 −01	299.5	98.746	6.503	0.3903	90.80	5.638	16.10	0.5418 −02	0.9915 −02
8.84	0.5332 −04	0.8868 −03	0.6013 −01	301.0	98.780	6.495	0.3902	91.00	5.639	16.14	0.5390 −02	0.9893 −02
8.85	0.5293 −04	0.8821 −03	0.6001 −01	302.6	98.814	6.488	0.3902	91.21	5.640	16.17	0.5362 −02	0.9871 −02
8.86	0.5254 −04	0.8774 −03	0.5988 −01	304.2	98.848	6.481	0.3902	91.42	5.641	16.21	0.5334 −02	0.9849 −02
8.87	0.5215 −04	0.8727 −03	0.5975 −01	305.8	98.882	6.473	0.3901	91.62	5.641	16.24	0.5307 −02	0.9827 −02
8.88	0.5176 −04	0.8681 −03	0.5963 −01	307.4	98.916	6.466	0.3901	91.83	5.642	16.28	0.5279 −02	0.9805 −02
8.89	0.5138 −04	0.8635 −03	0.5950 −01	309.0	98.950	6.459	0.3901	92.04	5.643	16.31	0.5252 −02	0.9783 −02
8.90	0.5100 −04	0.8590 −03	0.5937 −01	310.7	98.984	6.451	0.3901	92.25	5.644	16.35	0.5225 −02	0.9761 −02
8.91	0.5063 −04	0.8545 −03	0.5925 −01	312.3	99.018	6.444	0.3900	92.45	5.645	16.38	0.5198 −02	0.9739 −02
8.92	0.5025 −04	0.8500 −03	0.5912 −01	313.9	99.051	6.437	0.3900	92.66	5.645	16.41	0.5172 −02	0.9717 −02
8.93	0.4988 −04	0.8455 −03	0.5900 −01	315.5	99.085	6.430	0.3900	92.87	5.646	16.45	0.5145 −02	0.9696 −02
8.94	0.4952 −04	0.8411 −03	0.5888 −01	317.2	99.119	6.422	0.3900	93.08	5.647	16.48	0.5119 −02	0.9674 −02
8.95	0.4915 −04	0.8367 −03	0.5875 −01	318.8	99.152	6.415	0.3899	93.29	5.647	16.52	0.5092 −02	0.9652 −02
8.96	0.4879 −04	0.8323 −03	0.5863 −01	320.5	99.185	6.408	0.3899	93.50	5.648	16.55	0.5066 −02	0.9631 −02
8.97	0.4844 −04	0.8279 −03	0.5850 −01	322.2	99.219	6.401	0.3899	93.70	5.649	16.59	0.5040 −02	0.9610 −02
8.98	0.4808 −04	0.8236 −03	0.5838 −01	323.8	99.252	6.394	0.3899	93.91	5.650	16.62	0.5015 −02	0.9588 −02
8.99	0.4773 −04	0.8193 −03	0.5826 −01	325.5	99.285	6.386	0.3898	94.12	5.650	16.66	0.4989 −02	0.9567 −02
9.00	0.4738 −04	0.8150 −03	0.5814 −01	327.2	99.318	6.379	0.3898	94.33	5.651	16.69	0.4964 −02	0.9546 −02
9.01	0.4704 −04	0.8107 −03	0.5802 −01	328.9	99.352	6.372	0.3898	94.54	5.652	16.73	0.4938 −02	0.9525 −02
9.02	0.4669 −04	0.8065 −03	0.5790 −01	330.6	99.385	6.365	0.3897	94.75	5.653	16.76	0.4913 −02	0.9504 −02
9.03	0.4635 −04	0.8023 −03	0.5777 −01	332.3	99.417	6.358	0.3897	94.96	5.653	16.80	0.4888 −02	0.9483 −02
9.04	0.4602 −04	0.7981 −03	0.5765 −01	334.0	99.450	6.351	0.3897	95.18	5.654	16.83	0.4863 −02	0.9462 −02
9.05	0.4568 −04	0.7940 −03	0.5753 −01	335.8	99.483	6.344	0.3897	95.39	5.655	16.87	0.4839 −02	0.9441 −02
9.06	0.4535 −04	0.7899 −03	0.5741 −01	337.5	99.516	6.337	0.3896	95.60	5.656	16.90	0.4814 −02	0.9421 −02
9.07	0.4502 −04	0.7858 −03	0.5730 −01	339.2	99.549	6.330	0.3896	95.81	5.656	16.94	0.4790 −02	0.9400 −02
9.08	0.4470 −04	0.7817 −03	0.5718 −01	341.0	99.581	6.323	0.3896	96.02	5.657	16.97	0.4765 −02	0.9379 −02
9.09	0.4437 −04	0.7777 −03	0.5706 −01	342.7	99.614	6.316	0.3896	96.23	5.658	17.01	0.4741 −02	0.9359 −02
9.10	0.4405 −04	0.7736 −03	0.5694 −01	344.5	99.646	6.309	0.3895	96.45	5.658	17.05	0.4717 −02	0.9338 −02
9.11	0.4373 −04	0.7696 −03	0.5682 −01	346.2	99.678	6.302	0.3895	96.66	5.659	17.08	0.4693 −02	0.9318 −02
9.12	0.4342 −04	0.7657 −03	0.5670 −01	348.0	99.711	6.295	0.3895	96.87	5.660	17.12	0.4670 −02	0.9297 −02
9.13	0.4310 −04	0.7617 −03	0.5659 −01	349.8	99.743	6.288	0.3895	97.08	5.660	17.15	0.4646 −02	0.9277 −02
9.14	0.4279 −04	0.7578 −03	0.5647 −01	351.6	99.775	6.281	0.3894	97.30	5.661	17.19	0.4623 −02	0.9257 −02
9.15	0.4249 −04	0.7539 −03	0.5635 −01	353.4	99.807	6.274	0.3894	97.51	5.662	17.22	0.4600 −02	0.9237 −02
9.16	0.4218 −04	0.7500 −03	0.5624 −01	355.2	99.839	6.267	0.3894	97.72	5.663	17.26	0.4576 −02	0.9217 −02
9.17	0.4188 −04	0.7462 −03	0.5612 −01	357.0	99.871	6.261	0.3894	97.94	5.663	17.29	0.4553 −02	0.9197 −02
9.18	0.4158 −04	0.7423 −03	0.5601 −01	358.8	99.903	6.254	0.3893	98.15	5.664	17.33	0.4531 −02	0.9177 −02
9.19	0.4128 −04	0.7385 −03	0.5589 −01	360.7	99.935	6.247	0.3893	98.37	5.665	17.37	0.4508 −02	0.9157 −02
9.20	0.4098 −04	0.7348 −03	0.5578 −01	362.5	99.967	6.240	0.3893	98.58	5.665	17.40	0.4485 −02	0.9137 −02
9.21	0.4069 −04	0.7310 −03	0.5566 −01	364.3	99.999	6.233	0.3893	98.79	5.666	17.44	0.4463 −02	0.9117 −02
9.22	0.4040 −04	0.7273 −03	0.5555 −01	366.2	100.030	6.226	0.3892	99.01	5.667	17.47	0.4441 −02	0.9098 −02
9.23	0.4011 −04	0.7235 −03	0.5544 −01	368.0	100.062	6.220	0.3892	99.23	5.667	17.51	0.4418 −02	0.9078 −02
9.24	0.3982 −04	0.7199 −03	0.5532 −01	369.9	100.094	6.213	0.3892	99.44	5.668	17.54	0.4396 −02	0.9059 −02

Table 2.14: Isentropic and normal shock properties for air, $\gamma = 1.4$ (continued).

M or M_1	$\dfrac{P}{P_t}$	$\dfrac{\rho}{\rho_t}$	$\dfrac{T}{T_t}$	$\dfrac{A}{A^*}$	ν	μ	M_2	$\dfrac{P_2}{P_1}$	$\dfrac{\rho_2}{\rho_1}$	$\dfrac{T_2}{T_1}$	$\dfrac{P_{t_2}}{P_{t_1}}$	$\dfrac{P_1}{P_{t_2}}$
9.25	0.3954 −04	0.7162 −03	0.5521 −01	371.8	100.125	6.206	0.3892	99.66	5.669	17.58	0.4374 −02	0.9039 −02
9.26	0.3926 −04	0.7125 −03	0.5510 −01	373.7	100.157	6.199	0.3892	99.87	5.669	17.62	0.4353 −02	0.9020 −02
9.27	0.3898 −04	0.7089 −03	0.5498 −01	375.5	100.188	6.193	0.3891	100.1	5.670	17.65	0.4331 −02	0.9000 −02
9.28	0.3870 −04	0.7053 −03	0.5487 −01	377.4	100.219	6.186	0.3891	100.3	5.671	17.69	0.4309 −02	0.8981 −02
9.29	0.3843 −04	0.7017 −03	0.5476 −01	379.3	100.250	6.179	0.3891	100.5	5.671	17.72	0.4288 −02	0.8962 −02
9.30	0.3815 −04	0.6982 −03	0.5465 −01	381.3	100.282	6.173	0.3891	100.7	5.672	17.76	0.4267 −02	0.8943 −02
9.31	0.3788 −04	0.6946 −03	0.5454 −01	383.2	100.313	6.166	0.3890	101.0	5.673	17.80	0.4246 −02	0.8923 −02
9.32	0.3762 −04	0.6911 −03	0.5443 −01	385.1	100.344	6.159	0.3890	101.2	5.673	17.83	0.4224 −02	0.8904 −02
9.33	0.3735 −04	0.6876 −03	0.5432 −01	387.0	100.375	6.153	0.3890	101.4	5.674	17.87	0.4204 −02	0.8885 −02
9.34	0.3709 −04	0.6842 −03	0.5421 −01	389.0	100.406	6.146	0.3890	101.6	5.675	17.91	0.4183 −02	0.8866 −02
9.35	0.3682 −04	0.6807 −03	0.5410 −01	390.9	100.436	6.140	0.3889	101.8	5.675	17.94	0.4162 −02	0.8848 −02
9.36	0.3656 −04	0.6773 −03	0.5399 −01	392.9	100.467	6.133	0.3889	102.0	5.676	17.98	0.4142 −02	0.8829 −02
9.37	0.3631 −04	0.6739 −03	0.5388 −01	394.9	100.498	6.126	0.3889	102.3	5.677	18.01	0.4121 −02	0.8810 −02
9.38	0.3605 −04	0.6705 −03	0.5377 −01	396.8	100.529	6.120	0.3889	102.5	5.677	18.05	0.4101 −02	0.8791 −02
9.39	0.3580 −04	0.6671 −03	0.5366 −01	398.8	100.559	6.113	0.3888	102.7	5.678	18.09	0.4081 −02	0.8773 −02
9.40	0.3555 −04	0.6637 −03	0.5355 −01	400.8	100.590	6.107	0.3888	102.9	5.679	18.12	0.4061 −02	0.8754 −02
9.41	0.3530 −04	0.6604 −03	0.5345 −01	402.8	100.620	6.100	0.3888	103.1	5.679	18.16	0.4041 −02	0.8736 −02
9.42	0.3505 −04	0.6571 −03	0.5334 −01	404.8	100.651	6.094	0.3888	103.4	5.680	18.20	0.4021 −02	0.8717 −02
9.43	0.3480 −04	0.6538 −03	0.5323 −01	406.8	100.681	6.087	0.3888	103.6	5.681	18.23	0.4001 −02	0.8699 −02
9.44	0.3456 −04	0.6505 −03	0.5313 −01	408.8	100.711	6.081	0.3887	103.8	5.681	18.27	0.3981 −02	0.8680 −02
9.45	0.3432 −04	0.6473 −03	0.5302 −01	410.9	100.742	6.074	0.3887	104.0	5.682	18.31	0.3962 −02	0.8662 −02
9.46	0.3408 −04	0.6440 −03	0.5291 −01	412.9	100.772	6.068	0.3887	104.2	5.683	18.34	0.3943 −02	0.8644 −02
9.47	0.3384 −04	0.6408 −03	0.5281 −01	415.0	100.802	6.061	0.3887	104.5	5.683	18.38	0.3923 −02	0.8626 −02
9.48	0.3360 −04	0.6376 −03	0.5270 −01	417.0	100.832	6.055	0.3886	104.7	5.684	18.42	0.3904 −02	0.8608 −02
9.49	0.3337 −04	0.6345 −03	0.5260 −01	419.1	100.862	6.049	0.3886	104.9	5.684	18.46	0.3885 −02	0.8589 −02
9.50	0.3314 −04	0.6313 −03	0.5249 −01	421.2	100.892	6.042	0.3886	105.1	5.685	18.49	0.3866 −02	0.8571 −02
9.51	0.3291 −04	0.6282 −03	0.5239 −01	423.2	100.922	6.036	0.3886	105.3	5.686	18.53	0.3847 −02	0.8554 −02
9.52	0.3268 −04	0.6250 −03	0.5228 −01	425.3	100.952	6.030	0.3886	105.6	5.686	18.57	0.3829 −02	0.8536 −02
9.53	0.3245 −04	0.6219 −03	0.5218 −01	427.4	100.981	6.023	0.3885	105.8	5.687	18.60	0.3810 −02	0.8518 −02
9.54	0.3223 −04	0.6189 −03	0.5208 −01	429.5	101.011	6.017	0.3885	106.0	5.688	18.64	0.3791 −02	0.8500 −02
9.55	0.3200 −04	0.6158 −03	0.5197 −01	431.6	101.041	6.010	0.3885	106.2	5.688	18.68	0.3773 −02	0.8482 −02
9.56	0.3178 −04	0.6127 −03	0.5187 −01	433.8	101.070	6.004	0.3885	106.5	5.689	18.71	0.3755 −02	0.8465 −02
9.57	0.3156 −04	0.6097 −03	0.5177 −01	435.9	101.100	5.998	0.3884	106.7	5.689	18.75	0.3737 −02	0.8447 −02
9.58	0.3134 −04	0.6067 −03	0.5166 −01	438.0	101.129	5.992	0.3884	106.9	5.690	18.79	0.3718 −02	0.8429 −02
9.59	0.3113 −04	0.6037 −03	0.5156 −01	440.2	101.159	5.985	0.3884	107.1	5.691	18.83	0.3700 −02	0.8412 −02
9.60	0.3091 −04	0.6007 −03	0.5146 −01	442.3	101.188	5.979	0.3884	107.4	5.691	18.86	0.3683 −02	0.8394 −02
9.61	0.3070 −04	0.5978 −03	0.5136 −01	444.5	101.217	5.973	0.3884	107.6	5.692	18.90	0.3665 −02	0.8377 −02
9.62	0.3049 −04	0.5948 −03	0.5126 −01	446.7	101.247	5.967	0.3883	107.8	5.692	18.94	0.3647 −02	0.8360 −02
9.63	0.3028 −04	0.5919 −03	0.5116 −01	448.9	101.276	5.960	0.3883	108.0	5.693	18.98	0.3630 −02	0.8342 −02
9.64	0.3007 −04	0.5890 −03	0.5106 −01	451.1	101.305	5.954	0.3883	108.3	5.694	19.01	0.3612 −02	0.8325 −02
9.65	0.2987 −04	0.5861 −03	0.5096 −01	453.3	101.334	5.948	0.3883	108.5	5.694	19.05	0.3595 −02	0.8308 −02
9.66	0.2966 −04	0.5832 −03	0.5086 −01	455.5	101.363	5.942	0.3883	108.7	5.695	19.09	0.3577 −02	0.8291 −02
9.67	0.2946 −04	0.5804 −03	0.5076 −01	457.7	101.392	5.936	0.3882	108.9	5.695	19.13	0.3560 −02	0.8274 −02
9.68	0.2926 −04	0.5775 −03	0.5066 −01	459.9	101.421	5.929	0.3882	109.2	5.696	19.16	0.3543 −02	0.8257 −02
9.69	0.2906 −04	0.5747 −03	0.5056 −01	462.2	101.450	5.923	0.3882	109.4	5.697	19.20	0.3526 −02	0.8240 −02
9.70	0.2886 −04	0.5719 −03	0.5046 −01	464.4	101.478	5.917	0.3882	109.6	5.697	19.24	0.3509 −02	0.8223 −02
9.71	0.2866 −04	0.5691 −03	0.5036 −01	466.7	101.507	5.911	0.3882	109.8	5.698	19.28	0.3493 −02	0.8206 −02
9.72	0.2846 −04	0.5663 −03	0.5026 −01	468.9	101.536	5.905	0.3881	110.1	5.698	19.31	0.3476 −02	0.8189 −02
9.73	0.2827 −04	0.5636 −03	0.5016 −01	471.2	101.564	5.899	0.3881	110.3	5.699	19.35	0.3459 −02	0.8173 −02
9.74	0.2808 −04	0.5608 −03	0.5007 −01	473.5	101.593	5.893	0.3881	110.5	5.700	19.39	0.3443 −02	0.8156 −02
9.75	0.2789 −04	0.5581 −03	0.4997 −01	475.8	101.621	5.887	0.3881	110.7	5.700	19.43	0.3426 −02	0.8139 −02
9.76	0.2770 −04	0.5554 −03	0.4987 −01	478.1	101.650	5.881	0.3880	111.0	5.701	19.47	0.3410 −02	0.8123 −02
9.77	0.2751 −04	0.5527 −03	0.4977 −01	480.4	101.678	5.875	0.3880	111.2	5.701	19.50	0.3394 −02	0.8106 −02
9.78	0.2732 −04	0.5500 −03	0.4968 −01	482.7	101.707	5.869	0.3880	111.4	5.702	19.54	0.3378 −02	0.8089 −02
9.79	0.2714 −04	0.5474 −03	0.4958 −01	485.0	101.735	5.863	0.3880	111.7	5.703	19.58	0.3362 −02	0.8073 −02

Table 2.14: Isentropic and normal shock properties for air, $\gamma = 1.4$ (continued).

M or M_1	$\dfrac{p}{p_t}$	$\dfrac{\rho}{\rho_t}$	$\dfrac{T}{T_t}$	$\dfrac{A}{A^*}$	ν	μ	M_2	$\dfrac{P_2}{P_1}$	$\dfrac{\rho_2}{\rho_1}$	$\dfrac{T_2}{T_1}$	$\dfrac{P_{t_2}}{P_{t_1}}$	$\dfrac{P_1}{P_{t_2}}$
9.80	0.2695 −04	0.5447 −03	0.4948 −01	487.3	101.763	5.857	0.3880	111.9	5.703	19.62	0.3346 −02	0.8057 −02
9.81	0.2677 −04	0.5421 −03	0.4939 −01	489.7	101.791	5.851	0.3879	112.1	5.704	19.66	0.3330 −02	0.8040 −02
9.82	0.2659 −04	0.5395 −03	0.4929 −01	492.0	101.819	5.845	0.3879	112.3	5.704	19.69	0.3314 −02	0.8024 −02
9.83	0.2641 −04	0.5369 −03	0.4920 −01	494.4	101.848	5.839	0.3879	112.6	5.705	19.73	0.3298 −02	0.8008 −02
9.84	0.2623 −04	0.5343 −03	0.4910 −01	496.8	101.876	5.833	0.3879	112.8	5.705	19.77	0.3283 −02	0.7991 −02
9.85	0.2606 −04	0.5317 −03	0.4901 −01	499.1	101.904	5.827	0.3879	113.0	5.706	19.81	0.3267 −02	0.7975 −02
9.86	0.2588 −04	0.5291 −03	0.4891 −01	501.5	101.931	5.821	0.3878	113.3	5.707	19.85	0.3252 −02	0.7959 −02
9.87	0.2571 −04	0.5266 −03	0.4882 −01	503.9	101.959	5.815	0.3878	113.5	5.707	19.89	0.3236 −02	0.7943 −02
9.88	0.2553 −04	0.5241 −03	0.4872 −01	506.3	101.987	5.809	0.3878	113.7	5.708	19.92	0.3221 −02	0.7927 −02
9.89	0.2536 −04	0.5215 −03	0.4863 −01	508.8	102.015	5.803	0.3878	114.0	5.708	19.96	0.3206 −02	0.7911 −02
9.90	0.2519 −04	0.5190 −03	0.4854 −01	511.2	102.043	5.797	0.3878	114.2	5.709	20.00	0.3191 −02	0.7895 −02
9.91	0.2502 −04	0.5166 −03	0.4844 −01	513.6	102.070	5.791	0.3877	114.4	5.709	20.04	0.3176 −02	0.7879 −02
9.92	0.2486 −04	0.5141 −03	0.4835 −01	516.1	102.098	5.786	0.3877	114.6	5.710	20.08	0.3161 −02	0.7864 −02
9.93	0.2469 −04	0.5116 −03	0.4826 −01	518.5	102.125	5.780	0.3877	114.9	5.710	20.12	0.3146 −02	0.7848 −02
9.94	0.2453 −04	0.5092 −03	0.4817 −01	521.0	102.153	5.774	0.3877	115.1	5.711	20.16	0.3131 −02	0.7832 −02
9.95	0.2436 −04	0.5067 −03	0.4807 −01	523.5	102.180	5.768	0.3877	115.3	5.712	20.19	0.3117 −02	0.7816 −02
9.96	0.2420 −04	0.5043 −03	0.4798 −01	525.9	102.208	5.762	0.3877	115.6	5.712	20.23	0.3102 −02	0.7801 −02
9.97	0.2404 −04	0.5019 −03	0.4789 −01	528.4	102.235	5.756	0.3876	115.8	5.713	20.27	0.3088 −02	0.7785 −02
9.98	0.2388 −04	0.4995 −03	0.4780 −01	530.9	102.262	5.751	0.3876	116.0	5.713	20.31	0.3073 −02	0.7770 −02
9.99	0.2372 −04	0.4972 −03	0.4771 −01	533.4	102.289	5.745	0.3876	116.3	5.714	20.35	0.3059 −02	0.7754 −02
10.00	0.2356 −04	0.4948 −03	0.4762 −01	536.0	102.317	5.739	0.3876	116.5	5.714	20.39	0.3045 −02	0.7739 −02
10.02	0.2325 −04	0.4901 −03	0.4744 −01	541.0	102.371	5.728	0.3875	117.0	5.715	20.47	0.3016 −02	0.7708 −02
10.04	0.2294 −04	0.4855 −03	0.4726 −01	546.2	102.425	5.716	0.3875	117.4	5.716	20.54	0.2988 −02	0.7677 −02
10.06	0.2264 −04	0.4809 −03	0.4708 −01	551.3	102.479	5.705	0.3875	117.9	5.718	20.62	0.2961 −02	0.7647 −02
10.08	0.2234 −04	0.4764 −03	0.4690 −01	556.5	102.532	5.693	0.3874	118.4	5.719	20.70	0.2933 −02	0.7617 −02
10.10	0.2205 −04	0.4719 −03	0.4672 −01	561.7	102.586	5.682	0.3874	118.8	5.720	20.78	0.2906 −02	0.7587 −02
10.12	0.2176 −04	0.4675 −03	0.4655 −01	567.0	102.639	5.671	0.3874	119.3	5.721	20.86	0.2879 −02	0.7557 −02
10.14	0.2147 −04	0.4631 −03	0.4637 −01	572.3	102.692	5.660	0.3873	119.8	5.722	20.94	0.2853 −02	0.7527 −02
10.16	0.2119 −04	0.4587 −03	0.4620 −01	577.7	102.744	5.648	0.3873	120.3	5.723	21.02	0.2827 −02	0.7498 −02
10.18	0.2092 −04	0.4545 −03	0.4603 −01	583.0	102.797	5.637	0.3872	120.7	5.724	21.09	0.2801 −02	0.7468 −02
10.20	0.2064 −04	0.4502 −03	0.4585 −01	588.5	102.849	5.626	0.3872	121.2	5.725	21.17	0.2775 −02	0.7439 −02
10.22	0.2038 −04	0.4460 −03	0.4568 −01	594.0	102.902	5.615	0.3872	121.7	5.726	21.25	0.2750 −02	0.7410 −02
10.24	0.2011 −04	0.4419 −03	0.4551 −01	599.5	102.954	5.604	0.3871	122.2	5.727	21.33	0.2725 −02	0.7381 −02
10.26	0.1985 −04	0.4378 −03	0.4534 −01	605.0	103.005	5.593	0.3871	122.6	5.728	21.41	0.2700 −02	0.7353 −02
10.28	0.1959 −04	0.4337 −03	0.4517 −01	610.6	103.057	5.582	0.3871	123.1	5.729	21.49	0.2675 −02	0.7324 −02
10.30	0.1934 −04	0.4297 −03	0.4501 −01	616.3	103.108	5.571	0.3870	123.6	5.730	21.57	0.2651 −02	0.7296 −02
10.32	0.1909 −04	0.4258 −03	0.4484 −01	621.9	103.160	5.561	0.3870	124.1	5.731	21.65	0.2627 −02	0.7268 −02
10.34	0.1885 −04	0.4219 −03	0.4468 −01	627.7	103.211	5.550	0.3870	124.6	5.732	21.73	0.2603 −02	0.7240 −02
10.36	0.1860 −04	0.4180 −03	0.4451 −01	633.4	103.262	5.539	0.3869	125.1	5.733	21.81	0.2580 −02	0.7212 −02
10.38	0.1837 −04	0.4142 −03	0.4435 −01	639.2	103.312	5.528	0.3869	125.5	5.734	21.89	0.2556 −02	0.7184 −02
10.40	0.1813 −04	0.4104 −03	0.4418 −01	645.1	103.363	5.518	0.3869	126.0	5.735	21.97	0.2533 −02	0.7157 −02
10.42	0.1790 −04	0.4066 −03	0.4402 −01	651.0	103.413	5.507	0.3868	126.5	5.736	22.06	0.2511 −02	0.7129 −02
10.44	0.1767 −04	0.4029 −03	0.4386 −01	656.9	103.463	5.496	0.3868	127.0	5.737	22.14	0.2488 −02	0.7102 −02
10.46	0.1745 −04	0.3992 −03	0.4370 −01	662.9	103.513	5.486	0.3868	127.5	5.738	22.22	0.2466 −02	0.7075 −02
10.48	0.1722 −04	0.3956 −03	0.4354 −01	668.9	103.563	5.475	0.3867	128.0	5.739	22.30	0.2444 −02	0.7048 −02
10.50	0.1701 −04	0.3920 −03	0.4338 −01	675.0	103.613	5.465	0.3867	128.5	5.740	22.38	0.2422 −02	0.7022 −02
10.52	0.1679 −04	0.3885 −03	0.4323 −01	681.1	103.662	5.455	0.3867	129.0	5.741	22.46	0.2400 −02	0.6995 −02
10.54	0.1658 −04	0.3849 −03	0.4307 −01	687.3	103.711	5.444	0.3866	129.4	5.742	22.54	0.2379 −02	0.6969 −02
10.56	0.1637 −04	0.3815 −03	0.4291 −01	693.5	103.760	5.434	0.3866	129.9	5.743	22.63	0.2358 −02	0.6942 −02
10.58	0.1616 −04	0.3780 −03	0.4276 −01	699.7	103.809	5.424	0.3866	130.4	5.743	22.71	0.2337 −02	0.6916 −02
10.60	0.1596 −04	0.3746 −03	0.4260 −01	706.0	103.858	5.413	0.3865	130.9	5.744	22.79	0.2316 −02	0.6890 −02
10.62	0.1576 −04	0.3713 −03	0.4245 −01	712.4	103.906	5.403	0.3865	131.4	5.745	22.87	0.2296 −02	0.6864 −02
10.64	0.1556 −04	0.3679 −03	0.4230 −01	718.8	103.955	5.393	0.3865	131.9	5.746	22.96	0.2276 −02	0.6839 −02
10.66	0.1537 −04	0.3646 −03	0.4214 −01	725.2	104.003	5.383	0.3864	132.4	5.747	23.04	0.2256 −02	0.6813 −02
10.68	0.1518 −04	0.3614 −03	0.4199 −01	731.7	104.051	5.373	0.3864	132.9	5.748	23.12	0.2236 −02	0.6788 −02

Table 2.14: Isentropic and normal shock properties for air, $\gamma = 1.4$ (continued).

M or M_1	$\frac{P}{P_t}$	$\frac{\rho}{\rho_t}$	$\frac{T}{T_t}$	$\frac{A}{A^*}$	ν	μ	M_2	$\frac{P_2}{P_1}$	$\frac{\rho_2}{\rho_1}$	$\frac{T_2}{T_1}$	$\frac{P_{t_2}}{P_{t_1}}$	$\frac{P_1}{P_{t_2}}$
10.70	0.1499 −04	0.3582 −03	0.4184 −01	738.2	104.099	5.362	0.3864	133.4	5.749	23.21	0.2216 −02	0.6762 −02
10.72	0.1480 −04	0.3550 −03	0.4169 −01	744.8	104.146	5.352	0.3863	133.9	5.750	23.29	0.2197 −02	0.6737 −02
10.74	0.1462 −04	0.3518 −03	0.4155 −01	751.4	104.194	5.342	0.3863	134.4	5.751	23.37	0.2177 −02	0.6712 −02
10.76	0.1443 −04	0.3487 −03	0.4140 −01	758.1	104.241	5.333	0.3863	134.9	5.752	23.46	0.2158 −02	0.6687 −02
10.78	0.1426 −04	0.3456 −03	0.4125 −01	764.8	104.288	5.323	0.3862	135.4	5.752	23.54	0.2140 −02	0.6663 −02
10.80	0.1408 −04	0.3425 −03	0.4110 −01	771.6	104.335	5.313	0.3862	135.9	5.753	23.62	0.2121 −02	0.6638 −02
10.82	0.1391 −04	0.3395 −03	0.4096 −01	778.4	104.382	5.303	0.3862	136.4	5.754	23.71	0.2103 −02	0.6614 −02
10.84	0.1373 −04	0.3365 −03	0.4081 −01	785.3	104.429	5.293	0.3862	136.9	5.755	23.79	0.2084 −02	0.6589 −02
10.86	0.1357 −04	0.3336 −03	0.4067 −01	792.2	104.475	5.283	0.3861	137.4	5.756	23.88	0.2066 −02	0.6565 −02
10.88	0.1340 −04	0.3306 −03	0.4053 −01	799.1	104.521	5.274	0.3861	137.9	5.757	23.96	0.2048 −02	0.6541 −02
10.90	0.1323 −04	0.3277 −03	0.4038 −01	806.2	104.568	5.264	0.3861	138.4	5.758	24.05	0.2031 −02	0.6517 −02
10.92	0.1307 −04	0.3249 −03	0.4024 −01	813.2	104.614	5.254	0.3860	139.0	5.759	24.13	0.2013 −02	0.6493 −02
10.94	0.1291 −04	0.3220 −03	0.4010 −01	820.3	104.659	5.245	0.3860	139.5	5.759	24.22	0.1996 −02	0.6470 −02
10.96	0.1276 −04	0.3192 −03	0.3996 −01	827.5	104.705	5.235	0.3860	140.0	5.760	24.30	0.1979 −02	0.6446 −02
10.98	0.1260 −04	0.3164 −03	0.3982 −01	834.7	104.751	5.225	0.3860	140.5	5.761	24.39	0.1962 −02	0.6423 −02
11.00	0.1245 −04	0.3137 −03	0.3968 −01	842.0	104.796	5.216	0.3859	141.0	5.762	24.47	0.1945 −02	0.6400 −02
11.02	0.1230 −04	0.3109 −03	0.3954 −01	849.3	104.841	5.206	0.3859	141.5	5.763	24.56	0.1928 −02	0.6376 −02
11.04	0.1215 −04	0.3082 −03	0.3941 −01	856.6	104.886	5.197	0.3859	142.0	5.764	24.64	0.1912 −02	0.6353 −02
11.06	0.1200 −04	0.3056 −03	0.3927 −01	864.1	104.931	5.187	0.3858	142.5	5.764	24.73	0.1896 −02	0.6331 −02
11.08	0.1185 −04	0.3029 −03	0.3913 −01	871.5	104.976	5.178	0.3858	143.1	5.765	24.82	0.1879 −02	0.6308 −02
11.10	0.1171 −04	0.3003 −03	0.3900 −01	879.1	105.021	5.169	0.3858	143.6	5.766	24.90	0.1863 −02	0.6285 −02
11.12	0.1157 −04	0.2977 −03	0.3886 −01	886.6	105.065	5.159	0.3858	144.1	5.767	24.99	0.1848 −02	0.6263 −02
11.14	0.1143 −04	0.2952 −03	0.3873 −01	894.3	105.109	5.150	0.3857	144.6	5.768	25.07	0.1832 −02	0.6240 −02
11.16	0.1129 −04	0.2926 −03	0.3860 −01	901.9	105.153	5.141	0.3857	145.1	5.768	25.16	0.1816 −02	0.6218 −02
11.18	0.1116 −04	0.2901 −03	0.3846 −01	909.7	105.197	5.132	0.3857	145.7	5.769	25.25	0.1801 −02	0.6196 −02
11.20	0.1103 −04	0.2877 −03	0.3833 −01	917.5	105.241	5.122	0.3856	146.2	5.770	25.34	0.1786 −02	0.6174 −02
11.22	0.1089 −04	0.2852 −03	0.3820 −01	925.3	105.285	5.113	0.3856	146.7	5.771	25.42	0.1771 −02	0.6152 −02
11.24	0.1076 −04	0.2828 −03	0.3807 −01	933.2	105.328	5.104	0.3856	147.2	5.772	25.51	0.1756 −02	0.6130 −02
11.26	0.1064 −04	0.2804 −03	0.3794 −01	941.2	105.372	5.095	0.3856	147.8	5.772	25.60	0.1741 −02	0.6108 −02
11.28	0.1051 −04	0.2780 −03	0.3781 −01	949.2	105.415	5.086	0.3855	148.3	5.773	25.68	0.1727 −02	0.6087 −02
11.30	0.1039 −04	0.2756 −03	0.3768 −01	957.2	105.458	5.077	0.3855	148.8	5.774	25.77	0.1712 −02	0.6065 −02
11.32	0.1026 −04	0.2733 −03	0.3755 −01	965.3	105.501	5.068	0.3855	149.3	5.775	25.86	0.1698 −02	0.6044 −02
11.34	0.1014 −04	0.2710 −03	0.3743 −01	973.5	105.544	5.059	0.3855	149.9	5.775	25.95	0.1684 −02	0.6023 −02
11.36	0.1002 −04	0.2687 −03	0.3730 −01	981.7	105.586	5.050	0.3854	150.4	5.776	26.04	0.1670 −02	0.6002 −02
11.38	0.9903 −05	0.2664 −03	0.3717 −01	990.0	105.629	5.041	0.3854	150.9	5.777	26.13	0.1656 −02	0.5980 −02
11.40	0.9787 −05	0.2642 −03	0.3705 −01	998.4	105.671	5.032	0.3854	151.5	5.778	26.21	0.1642 −02	0.5960 −02
11.42	0.9672 −05	0.2619 −03	0.3692 −01	1007	105.714	5.024	0.3854	152.0	5.778	26.30	0.1629 −02	0.5939 −02
11.44	0.9558 −05	0.2598 −03	0.3680 −01	1015	105.756	5.015	0.3853	152.5	5.779	26.39	0.1615 −02	0.5918 −02
11.46	0.9446 −05	0.2576 −03	0.3667 −01	1024	105.798	5.006	0.3853	153.1	5.780	26.48	0.1602 −02	0.5898 −02
11.48	0.9336 −05	0.2554 −03	0.3655 −01	1032	105.839	4.997	0.3853	153.6	5.781	26.57	0.1589 −02	0.5877 −02
11.50	0.9227 −05	0.2533 −03	0.3643 −01	1041	105.881	4.988	0.3853	154.1	5.781	26.66	0.1575 −02	0.5857 −02
11.52	0.9119 −05	0.2512 −03	0.3631 −01	1050	105.922	4.980	0.3852	154.7	5.782	26.75	0.1563 −02	0.5836 −02
11.54	0.9013 −05	0.2491 −03	0.3619 −01	1058	105.964	4.971	0.3852	155.2	5.783	26.84	0.1550 −02	0.5816 −02
11.56	0.8909 −05	0.2470 −03	0.3607 −01	1067	106.005	4.963	0.3852	155.7	5.784	26.93	0.1537 −02	0.5796 −02
11.58	0.8805 −05	0.2450 −03	0.3595 −01	1076	106.046	4.954	0.3852	156.3	5.784	27.02	0.1524 −02	0.5776 −02
11.60	0.8703 −05	0.2429 −03	0.3583 −01	1085	106.087	4.945	0.3851	156.8	5.785	27.11	0.1512 −02	0.5756 −02
11.62	0.8603 −05	0.2409 −03	0.3571 −01	1094	106.128	4.937	0.3851	157.4	5.786	27.20	0.1500 −02	0.5737 −02
11.64	0.8503 −05	0.2389 −03	0.3559 −01	1103	106.169	4.928	0.3851	157.9	5.786	27.29	0.1487 −02	0.5717 −02
11.66	0.8405 −05	0.2370 −03	0.3547 −01	1112	106.209	4.920	0.3851	158.5	5.787	27.38	0.1475 −02	0.5697 −02
11.68	0.8309 −05	0.2350 −03	0.3535 −01	1121	106.250	4.911	0.3850	159.0	5.788	27.47	0.1463 −02	0.5678 −02
11.70	0.8213 −05	0.2331 −03	0.3524 −01	1130	106.290	4.903	0.3850	159.5	5.789	27.56	0.1451 −02	0.5659 −02
11.72	0.8119 −05	0.2312 −03	0.3512 −01	1140	106.330	4.895	0.3850	160.1	5.789	27.65	0.1440 −02	0.5639 −02
11.74	0.8026 −05	0.2293 −03	0.3501 −01	1149	106.370	4.886	0.3850	160.6	5.790	27.74	0.1428 −02	0.5620 −02
11.76	0.7934 −05	0.2274 −03	0.3489 −01	1158	106.410	4.878	0.3849	161.2	5.791	27.84	0.1417 −02	0.5601 −02
11.78	0.7844 −05	0.2255 −03	0.3478 −01	1168	106.450	4.870	0.3849	161.7	5.791	27.93	0.1405 −02	0.5582 −02

Table 2.14: Isentropic and normal shock properties for air, $\gamma = 1.4$ (continued).

M or M_1	$\dfrac{P}{P_t}$	$\dfrac{\rho}{\rho_t}$	$\dfrac{T}{T_t}$	$\dfrac{A}{A^*}$	ν	μ	M_2	$\dfrac{P_2}{P_1}$	$\dfrac{\rho_2}{\rho_1}$	$\dfrac{T_2}{T_1}$	$\dfrac{P_{t_2}}{P_{t_1}}$	$\dfrac{P_1}{P_{t_2}}$
11.80	0.7754 −05	0.2237 −03	0.3466 −01	1177	106.490	4.861	0.3849	162.3	5.792	28.02	0.1394 −02	0.5563 −02
11.82	0.7666 −05	0.2219 −03	0.3455 −01	1187	106.529	4.853	0.3849	162.8	5.793	28.11	0.1383 −02	0.5545 −02
11.84	0.7579 −05	0.2201 −03	0.3444 −01	1197	106.568	4.845	0.3848	163.4	5.793	28.20	0.1372 −02	0.5526 −02
11.86	0.7493 −05	0.2183 −03	0.3433 −01	1206	106.608	4.837	0.3848	163.9	5.794	28.29	0.1361 −02	0.5507 −02
11.88	0.7408 −05	0.2165 −03	0.3421 −01	1216	106.647	4.829	0.3848	164.5	5.795	28.39	0.1350 −02	0.5489 −02
11.90	0.7325 −05	0.2148 −03	0.3410 −01	1226	106.686	4.820	0.3848	165.0	5.795	28.48	0.1339 −02	0.5471 −02
11.92	0.7242 −05	0.2130 −03	0.3399 −01	1236	106.725	4.812	0.3848	165.6	5.796	28.57	0.1328 −02	0.5452 −02
11.94	0.7160 −05	0.2113 −03	0.3388 −01	1246	106.763	4.804	0.3847	166.2	5.797	28.66	0.1318 −02	0.5434 −02
11.96	0.7080 −05	0.2096 −03	0.3377 −01	1256	106.802	4.796	0.3847	166.7	5.797	28.76	0.1307 −02	0.5416 −02
11.98	0.7000 −05	0.2079 −03	0.3366 −01	1266	106.841	4.788	0.3847	167.3	5.798	28.85	0.1297 −02	0.5398 −02
12.00	0.6921 −05	0.2063 −03	0.3356 −01	1276	106.879	4.780	0.3847	167.8	5.799	28.94	0.1287 −02	0.5380 −02
12.02	0.6844 −05	0.2046 −03	0.3345 −01	1287	106.917	4.772	0.3846	168.4	5.799	29.04	0.1276 −02	0.5362 −02
12.04	0.6767 −05	0.2030 −03	0.3334 −01	1297	106.955	4.764	0.3846	169.0	5.800	29.13	0.1266 −02	0.5344 −02
12.06	0.6692 −05	0.2014 −03	0.3323 −01	1307	106.993	4.756	0.3846	169.5	5.801	29.23	0.1256 −02	0.5327 −02
12.08	0.6617 −05	0.1997 −03	0.3313 −01	1318	107.031	4.748	0.3846	170.1	5.801	29.32	0.1246 −02	0.5309 −02
12.10	0.6543 −05	0.1982 −03	0.3302 −01	1328	107.069	4.741	0.3846	170.7	5.802	29.41	0.1237 −02	0.5292 −02
12.12	0.6471 −05	0.1966 −03	0.3292 −01	1339	107.107	4.733	0.3845	171.2	5.803	29.51	0.1227 −02	0.5274 −02
12.14	0.6399 −05	0.1950 −03	0.3281 −01	1349	107.144	4.725	0.3845	171.8	5.803	29.60	0.1217 −02	0.5257 −02
12.16	0.6328 −05	0.1935 −03	0.3271 −01	1360	107.182	4.717	0.3845	172.3	5.804	29.70	0.1208 −02	0.5240 −02
12.18	0.6258 −05	0.1919 −03	0.3260 −01	1371	107.219	4.709	0.3845	172.9	5.804	29.79	0.1198 −02	0.5223 −02
12.20	0.6189 −05	0.1904 −03	0.3250 −01	1382	107.256	4.702	0.3844	173.5	5.805	29.89	0.1189 −02	0.5205 −02
12.22	0.6121 −05	0.1889 −03	0.3240 −01	1393	107.293	4.694	0.3844	174.1	5.806	29.98	0.1180 −02	0.5188 −02
12.24	0.6053 −05	0.1874 −03	0.3230 −01	1404	107.330	4.686	0.3844	174.6	5.806	30.08	0.1170 −02	0.5172 −02
12.26	0.5987 −05	0.1860 −03	0.3219 −01	1415	107.367	4.679	0.3844	175.2	5.807	30.17	0.1161 −02	0.5155 −02
12.28	0.5921 −05	0.1845 −03	0.3209 −01	1426	107.404	4.671	0.3844	175.8	5.807	30.27	0.1152 −02	0.5138 −02
12.30	0.5856 −05	0.1830 −03	0.3199 −01	1437	107.440	4.663	0.3843	176.3	5.808	30.36	0.1143 −02	0.5121 −02
12.32	0.5792 −05	0.1816 −03	0.3189 −01	1448	107.477	4.656	0.3843	176.9	5.809	30.46	0.1135 −02	0.5105 −02
12.34	0.5728 −05	0.1802 −03	0.3179 −01	1460	107.513	4.648	0.3843	177.5	5.809	30.55	0.1126 −02	0.5088 −02
12.36	0.5666 −05	0.1788 −03	0.3169 −01	1471	107.549	4.641	0.3843	178.1	5.810	30.65	0.1117 −02	0.5072 −02
12.38	0.5604 −05	0.1774 −03	0.3159 −01	1483	107.585	4.633	0.3843	178.6	5.810	30.75	0.1109 −02	0.5056 −02
12.40	0.5543 −05	0.1760 −03	0.3149 −01	1494	107.622	4.626	0.3842	179.2	5.811	30.84	0.1100 −02	0.5039 −02
12.42	0.5483 −05	0.1746 −03	0.3139 −01	1506	107.657	4.618	0.3842	179.8	5.812	30.94	0.1092 −02	0.5023 −02
12.44	0.5423 −05	0.1733 −03	0.3130 −01	1517	107.693	4.611	0.3842	180.4	5.812	31.04	0.1083 −02	0.5007 −02
12.46	0.5365 −05	0.1719 −03	0.3120 −01	1529	107.729	4.603	0.3842	181.0	5.813	31.13	0.1075 −02	0.4991 −02
12.48	0.5307 −05	0.1706 −03	0.3110 −01	1541	107.765	4.596	0.3842	181.5	5.813	31.23	0.1067 −02	0.4975 −02
12.50	0.5249 −05	0.1693 −03	0.3101 −01	1553	107.800	4.588	0.3841	182.1	5.814	31.33	0.1059 −02	0.4959 −02
12.52	0.5193 −05	0.1680 −03	0.3091 −01	1565	107.835	4.581	0.3841	182.7	5.815	31.42	0.1050 −02	0.4943 −02
12.54	0.5137 −05	0.1667 −03	0.3082 −01	1577	107.871	4.574	0.3841	183.3	5.815	31.52	0.1042 −02	0.4928 −02
12.56	0.5082 −05	0.1654 −03	0.3072 −01	1589	107.906	4.567	0.3841	183.9	5.816	31.62	0.1035 −02	0.4912 −02
12.58	0.5027 −05	0.1641 −03	0.3063 −01	1601	107.941	4.559	0.3841	184.5	5.816	31.72	0.1027 −02	0.4896 −02
12.60	0.4973 −05	0.1629 −03	0.3053 −01	1614	107.976	4.552	0.3840	185.1	5.817	31.81	0.1019 −02	0.4881 −02
12.62	0.4920 −05	0.1616 −03	0.3044 −01	1626	108.011	4.545	0.3840	185.6	5.817	31.91	0.1011 −02	0.4865 −02
12.64	0.4867 −05	0.1604 −03	0.3034 −01	1639	108.045	4.538	0.3840	186.2	5.818	32.01	0.1004 −02	0.4850 −02
12.66	0.4815 −05	0.1592 −03	0.3025 −01	1651	108.080	4.530	0.3840	186.8	5.818	32.11	0.9959 −03	0.4835 −02
12.68	0.4764 −05	0.1580 −03	0.3016 −01	1664	108.114	4.523	0.3840	187.4	5.819	32.21	0.9884 −03	0.4820 −02
12.70	0.4713 −05	0.1568 −03	0.3007 −01	1676	108.149	4.516	0.3839	188.0	5.820	32.31	0.9810 −03	0.4805 −02
12.72	0.4663 −05	0.1556 −03	0.2998 −01	1689	108.183	4.509	0.3839	188.6	5.820	32.41	0.9736 −03	0.4789 −02
12.74	0.4614 −05	0.1544 −03	0.2988 −01	1702	108.217	4.502	0.3839	189.2	5.821	32.50	0.9663 −03	0.4774 −02
12.76	0.4565 −05	0.1532 −03	0.2979 −01	1715	108.251	4.495	0.3839	189.8	5.821	32.60	0.9591 −03	0.4760 −02
12.78	0.4516 −05	0.1521 −03	0.2970 −01	1728	108.285	4.488	0.3839	190.4	5.822	32.70	0.9519 −03	0.4745 −02
12.80	0.4469 −05	0.1509 −03	0.2961 −01	1741	108.319	4.481	0.3839	191.0	5.822	32.80	0.9448 −03	0.4730 −02
12.82	0.4422 −05	0.1498 −03	0.2952 −01	1754	108.353	4.474	0.3838	191.6	5.823	32.90	0.9377 −03	0.4715 −02
12.84	0.4375 −05	0.1486 −03	0.2943 −01	1767	108.387	4.467	0.3838	192.2	5.823	33.00	0.9307 −03	0.4701 −02
12.86	0.4329 −05	0.1475 −03	0.2935 −01	1781	108.420	4.460	0.3838	192.8	5.824	33.10	0.9238 −03	0.4686 −02
12.88	0.4284 −05	0.1464 −03	0.2926 −01	1794	108.454	4.453	0.3838	193.4	5.824	33.20	0.9170 −03	0.4671 −02

Table 2.14: Isentropic and normal shock properties for air, $\gamma = 1.4$ (continued).

M or M_1	$\dfrac{P}{P_t}$	$\dfrac{\rho}{\rho_t}$	$\dfrac{T}{T_t}$	$\dfrac{A}{A^*}$	ν	μ	M_2	$\dfrac{P_2}{P_1}$	$\dfrac{\rho_2}{\rho_1}$	$\dfrac{T_2}{T_1}$	$\dfrac{P_{t_2}}{P_{t_1}}$	$\dfrac{P_1}{P_{t_2}}$
12.90	0.4239 −05	0.1453 −03	0.2917 −01	1808	108.487	4.446	0.3838	194.0	5.825	33.30	0.9102 −03	0.4657 −02
12.92	0.4194 −05	0.1442 −03	0.2908 −01	1821	108.520	4.439	0.3837	194.6	5.826	33.40	0.9034 −03	0.4643 −02
12.94	0.4150 −05	0.1431 −03	0.2899 −01	1835	108.554	4.432	0.3837	195.2	5.826	33.50	0.8967 −03	0.4628 −02
12.96	0.4107 −05	0.1421 −03	0.2891 −01	1849	108.587	4.425	0.3837	195.8	5.827	33.60	0.8901 −03	0.4614 −02
12.98	0.4064 −05	0.1410 −03	0.2882 −01	1862	108.620	4.418	0.3837	196.4	5.827	33.70	0.8835 −03	0.4600 −02
13.00	0.4022 −05	0.1400 −03	0.2873 −01	1876	108.652	4.412	0.3837	197.0	5.828	33.81	0.8770 −03	0.4586 −02
13.02	0.3980 −05	0.1389 −03	0.2865 −01	1890	108.685	4.405	0.3837	197.6	5.828	33.91	0.8706 −03	0.4572 −02
13.04	0.3939 −05	0.1379 −03	0.2856 −01	1904	108.718	4.398	0.3836	198.2	5.829	34.01	0.8642 −03	0.4558 −02
13.06	0.3898 −05	0.1369 −03	0.2848 −01	1918	108.751	4.391	0.3836	198.8	5.829	34.11	0.8578 −03	0.4544 −02
13.08	0.3858 −05	0.1359 −03	0.2839 −01	1933	108.783	4.385	0.3836	199.4	5.830	34.21	0.8516 −03	0.4530 −02
13.10	0.3818 −05	0.1349 −03	0.2831 −01	1947	108.815	4.378	0.3836	200.1	5.830	34.31	0.8453 −03	0.4516 −02
13.12	0.3778 −05	0.1339 −03	0.2823 −01	1961	108.848	4.371	0.3836	200.7	5.831	34.42	0.8391 −03	0.4502 −02
13.14	0.3739 −05	0.1329 −03	0.2814 −01	1976	108.880	4.365	0.3836	201.3	5.831	34.52	0.8330 −03	0.4489 −02
13.16	0.3701 −05	0.1319 −03	0.2806 −01	1990	108.912	4.358	0.3835	201.9	5.832	34.62	0.8270 −03	0.4475 −02
13.18	0.3663 −05	0.1309 −03	0.2798 −01	2005	108.944	4.351	0.3835	202.5	5.832	34.72	0.8209 −03	0.4462 −02
13.20	0.3625 −05	0.1300 −03	0.2789 −01	2020	108.976	4.345	0.3835	203.1	5.833	34.82	0.8150 −03	0.4448 −02
13.22	0.3588 −05	0.1290 −03	0.2781 −01	2035	109.008	4.338	0.3835	203.7	5.833	34.93	0.8091 −03	0.4435 −02
13.24	0.3551 −05	0.1281 −03	0.2773 −01	2050	109.040	4.332	0.3835	204.4	5.834	35.03	0.8032 −03	0.4421 −02
13.26	0.3515 −05	0.1271 −03	0.2765 −01	2065	109.071	4.325	0.3835	205.0	5.834	35.13	0.7974 −03	0.4408 −02
13.28	0.3479 −05	0.1262 −03	0.2757 −01	2080	109.103	4.318	0.3834	205.6	5.835	35.24	0.7916 −03	0.4395 −02
13.30	0.3444 −05	0.1253 −03	0.2749 −01	2095	109.134	4.312	0.3834	206.2	5.835	35.34	0.7859 −03	0.4382 −02
13.32	0.3409 −05	0.1244 −03	0.2741 −01	2110	109.166	4.305	0.3834	206.8	5.836	35.44	0.7803 −03	0.4369 −02
13.34	0.3374 −05	0.1235 −03	0.2733 −01	2125	109.197	4.299	0.3834	207.5	5.836	35.55	0.7746 −03	0.4355 −02
13.36	0.3340 −05	0.1226 −03	0.2725 −01	2141	109.228	4.293	0.3834	208.1	5.837	35.65	0.7691 −03	0.4342 −02
13.38	0.3306 −05	0.1217 −03	0.2717 −01	2157	109.259	4.286	0.3834	208.7	5.837	35.76	0.7636 −03	0.4330 −02
13.40	0.3272 −05	0.1208 −03	0.2709 −01	2172	109.290	4.280	0.3833	209.3	5.837	35.86	0.7581 −03	0.4317 −02
13.42	0.3239 −05	0.1199 −03	0.2701 −01	2188	109.321	4.273	0.3833	210.0	5.838	35.96	0.7527 −03	0.4304 −02
13.44	0.3207 −05	0.1191 −03	0.2693 −01	2204	109.352	4.267	0.3833	210.6	5.838	36.07	0.7473 −03	0.4291 −02
13.46	0.3174 −05	0.1182 −03	0.2686 −01	2220	109.383	4.261	0.3833	211.2	5.839	36.17	0.7420 −03	0.4278 −02
13.48	0.3142 −05	0.1173 −03	0.2678 −01	2236	109.413	4.254	0.3833	211.8	5.839	36.28	0.7367 −03	0.4266 −02
13.50	0.3111 −05	0.1165 −03	0.2670 −01	2252	109.444	4.248	0.3833	212.5	5.840	36.38	0.7314 −03	0.4253 −02
13.52	0.3080 −05	0.1157 −03	0.2662 −01	2268	109.474	4.242	0.3832	213.1	5.840	36.49	0.7262 −03	0.4240 −02
13.54	0.3049 −05	0.1148 −03	0.2655 −01	2284	109.505	4.235	0.3832	213.7	5.841	36.59	0.7211 −03	0.4228 −02
13.56	0.3018 −05	0.1140 −03	0.2647 −01	2301	109.535	4.229	0.3832	214.4	5.841	36.70	0.7160 −03	0.4216 −02
13.58	0.2988 −05	0.1132 −03	0.2640 −01	2317	109.565	4.223	0.3832	215.0	5.842	36.80	0.7109 −03	0.4203 −02
13.60	0.2958 −05	0.1124 −03	0.2632 −01	2334	109.595	4.217	0.3832	215.6	5.842	36.91	0.7059 −03	0.4191 −02
13.62	0.2929 −05	0.1116 −03	0.2625 −01	2350	109.625	4.210	0.3832	216.3	5.843	37.02	0.7009 −03	0.4179 −02
13.64	0.2900 −05	0.1108 −03	0.2617 −01	2367	109.655	4.204	0.3832	216.9	5.843	37.12	0.6960 −03	0.4166 −02
13.66	0.2871 −05	0.1100 −03	0.2610 −01	2384	109.685	4.198	0.3831	217.5	5.843	37.23	0.6911 −03	0.4154 −02
13.68	0.2842 −05	0.1092 −03	0.2602 −01	2401	109.715	4.192	0.3831	218.2	5.844	37.33	0.6862 −03	0.4142 −02
13.70	0.2814 −05	0.1085 −03	0.2595 −01	2418	109.745	4.186	0.3831	218.8	5.844	37.44	0.6814 −03	0.4130 −02
13.72	0.2786 −05	0.1077 −03	0.2587 −01	2435	109.774	4.180	0.3831	219.5	5.845	37.55	0.6766 −03	0.4118 −02
13.74	0.2759 −05	0.1069 −03	0.2580 −01	2452	109.804	4.174	0.3831	220.1	5.845	37.65	0.6719 −03	0.4106 −02
13.76	0.2731 −05	0.1062 −03	0.2573 −01	2470	109.833	4.168	0.3831	220.7	5.846	37.76	0.6672 −03	0.4094 −02
13.78	0.2705 −05	0.1054 −03	0.2565 −01	2487	109.863	4.161	0.3831	221.4	5.846	37.87	0.6625 −03	0.4082 −02
13.80	0.2678 −05	0.1047 −03	0.2558 −01	2505	109.892	4.155	0.3830	222.0	5.847	37.97	0.6579 −03	0.4070 −02
13.82	0.2652 −05	0.1039 −03	0.2551 −01	2522	109.921	4.149	0.3830	222.7	5.847	38.08	0.6533 −03	0.4059 −02
13.84	0.2626 −05	0.1032 −03	0.2544 −01	2540	109.950	4.143	0.3830	223.3	5.847	38.19	0.6488 −03	0.4047 −02
13.86	0.2600 −05	0.1025 −03	0.2537 −01	2558	109.979	4.137	0.3830	224.0	5.848	38.30	0.6443 −03	0.4035 −02
13.88	0.2574 −05	0.1018 −03	0.2530 −01	2576	110.008	4.131	0.3830	224.6	5.848	38.41	0.6398 −03	0.4024 −02
13.90	0.2549 −05	0.1011 −03	0.2522 −01	2594	110.037	4.125	0.3830	225.3	5.849	38.51	0.6354 −03	0.4012 −02
13.92	0.2524 −05	0.1004 −03	0.2515 −01	2612	110.066	4.120	0.3830	225.9	5.849	38.62	0.6310 −03	0.4001 −02
13.94	0.2500 −05	0.9965 −04	0.2508 −01	2630	110.094	4.114	0.3829	226.6	5.849	38.73	0.6266 −03	0.3989 −02
13.96	0.2475 −05	0.9896 −04	0.2501 −01	2649	110.123	4.108	0.3829	227.2	5.850	38.84	0.6223 −03	0.3978 −02
13.98	0.2451 −05	0.9827 −04	0.2494 −01	2667	110.152	4.102	0.3829	227.9	5.850	38.95	0.6180 −03	0.3967 −02

Table 2.14: Isentropic and normal shock properties for air, $\gamma = 1.4$ (continued).

M or M_1	$\dfrac{P}{P_t}$	$\dfrac{\rho}{\rho_t}$	$\dfrac{T}{T_t}$	$\dfrac{A}{A^*}$	ν	μ	M_2	$\dfrac{P_2}{P_1}$	$\dfrac{\rho_2}{\rho_1}$	$\dfrac{T_2}{T_1}$	$\dfrac{P_{t_2}}{P_{t_1}}$	$\dfrac{P_1}{P_{t_2}}$
14.00	0.2428 −05	0.9759 −04	0.2487 −01	2686	110.180	4.096	0.3829	228.5	5.851	39.06	0.6137 −03	0.3955 −02
14.02	0.2404 −05	0.9691 −04	0.2481 −01	2704	110.209	4.090	0.3829	229.2	5.851	39.17	0.6095 −03	0.3944 −02
14.04	0.2381 −05	0.9624 −04	0.2474 −01	2723	110.237	4.084	0.3829	229.8	5.852	39.27	0.6053 −03	0.3933 −02
14.06	0.2358 −05	0.9557 −04	0.2467 −01	2742	110.265	4.078	0.3829	230.5	5.852	39.38	0.6012 −03	0.3922 −02
14.08	0.2335 −05	0.9491 −04	0.2460 −01	2761	110.293	4.073	0.3828	231.1	5.852	39.49	0.5971 −03	0.3910 −02
14.10	0.2312 −05	0.9426 −04	0.2453 −01	2780	110.321	4.067	0.3828	231.8	5.853	39.60	0.5930 −03	0.3899 −02
14.12	0.2290 −05	0.9361 −04	0.2446 −01	2799	110.349	4.061	0.3828	232.4	5.853	39.71	0.5890 −03	0.3888 −02
14.14	0.2268 −05	0.9297 −04	0.2440 −01	2818	110.377	4.055	0.3828	233.1	5.854	39.82	0.5849 −03	0.3877 −02
14.16	0.2246 −05	0.9233 −04	0.2433 −01	2838	110.405	4.050	0.3828	233.8	5.854	39.93	0.5810 −03	0.3867 −02
14.18	0.2225 −05	0.9169 −04	0.2426 −01	2857	110.433	4.044	0.3828	234.4	5.854	40.04	0.5770 −03	0.3856 −02
14.20	0.2203 −05	0.9107 −04	0.2420 −01	2877	110.461	4.038	0.3828	235.1	5.855	40.15	0.5731 −03	0.3845 −02
14.22	0.2182 −05	0.9044 −04	0.2413 −01	2897	110.488	4.033	0.3827	235.8	5.855	40.26	0.5692 −03	0.3834 −02
14.24	0.2161 −05	0.8982 −04	0.2406 −01	2917	110.516	4.027	0.3827	236.4	5.856	40.37	0.5654 −03	0.3823 −02
14.26	0.2141 −05	0.8921 −04	0.2400 −01	2936	110.543	4.021	0.3827	237.1	5.856	40.48	0.5615 −03	0.3813 −02
14.28	0.2120 −05	0.8860 −04	0.2393 −01	2957	110.571	4.016	0.3827	237.7	5.856	40.60	0.5577 −03	0.3802 −02
14.30	0.2100 −05	0.8800 −04	0.2387 −01	2977	110.598	4.010	0.3827	238.4	5.857	40.71	0.5540 −03	0.3791 −02
14.32	0.2080 −05	0.8740 −04	0.2380 −01	2997	110.625	4.004	0.3827	239.1	5.857	40.82	0.5502 −03	0.3781 −02
14.34	0.2061 −05	0.8681 −04	0.2374 −01	3017	110.652	3.999	0.3827	239.7	5.858	40.93	0.5465 −03	0.3770 −02
14.36	0.2041 −05	0.8622 −04	0.2367 −01	3038	110.679	3.993	0.3827	240.4	5.858	41.04	0.5429 −03	0.3760 −02
14.38	0.2022 −05	0.8564 −04	0.2361 −01	3058	110.706	3.988	0.3826	241.1	5.858	41.15	0.5392 −03	0.3749 −02
14.40	0.2003 −05	0.8506 −04	0.2354 −01	3079	110.733	3.982	0.3826	241.8	5.859	41.27	0.5356 −03	0.3739 −02
14.42	0.1984 −05	0.8448 −04	0.2348 −01	3100	110.760	3.976	0.3826	242.4	5.859	41.38	0.5320 −03	0.3729 −02
14.44	0.1965 −05	0.8391 −04	0.2342 −01	3121	110.787	3.971	0.3826	243.1	5.859	41.49	0.5285 −03	0.3718 −02
14.46	0.1946 −05	0.8335 −04	0.2335 −01	3142	110.814	3.965	0.3826	243.8	5.860	41.60	0.5249 −03	0.3708 −02
14.48	0.1928 −05	0.8279 −04	0.2329 −01	3163	110.840	3.960	0.3826	244.5	5.860	41.71	0.5214 −03	0.3698 −02
14.50	0.1910 −05	0.8223 −04	0.2323 −01	3185	110.867	3.955	0.3826	245.1	5.861	41.83	0.5180 −03	0.3688 −02
14.52	0.1892 −05	0.8168 −04	0.2317 −01	3206	110.894	3.949	0.3825	245.8	5.861	41.94	0.5145 −03	0.3677 −02
14.54	0.1874 −05	0.8113 −04	0.2310 −01	3227	110.920	3.944	0.3825	246.5	5.861	42.05	0.5111 −03	0.3667 −02
14.56	0.1857 −05	0.8059 −04	0.2304 −01	3249	110.946	3.938	0.3825	247.2	5.862	42.17	0.5077 −03	0.3657 −02
14.58	0.1840 −05	0.8005 −04	0.2298 −01	3271	110.973	3.933	0.3825	247.8	5.862	42.28	0.5043 −03	0.3647 −02
14.60	0.1822 −05	0.7952 −04	0.2292 −01	3293	110.999	3.927	0.3825	248.5	5.862	42.39	0.5010 −03	0.3637 −02
14.62	0.1805 −05	0.7899 −04	0.2286 −01	3315	111.025	3.922	0.3825	249.2	5.863	42.51	0.4977 −03	0.3627 −02
14.64	0.1789 −05	0.7846 −04	0.2280 −01	3337	111.051	3.917	0.3825	249.9	5.863	42.62	0.4944 −03	0.3618 −02
14.66	0.1772 −05	0.7794 −04	0.2274 −01	3359	111.077	3.911	0.3825	250.6	5.864	42.73	0.4912 −03	0.3608 −02
14.68	0.1755 −05	0.7742 −04	0.2267 −01	3381	111.103	3.906	0.3825	251.3	5.864	42.85	0.4879 −03	0.3598 −02
14.70	0.1739 −05	0.7691 −04	0.2261 −01	3404	111.129	3.901	0.3824	251.9	5.864	42.96	0.4847 −03	0.3588 −02
14.72	0.1723 −05	0.7640 −04	0.2255 −01	3426	111.155	3.895	0.3824	252.6	5.865	43.08	0.4815 −03	0.3578 −02
14.74	0.1707 −05	0.7589 −04	0.2249 −01	3449	111.181	3.890	0.3824	253.3	5.865	43.19	0.4784 −03	0.3569 −02
14.76	0.1691 −05	0.7539 −04	0.2244 −01	3472	111.206	3.885	0.3824	254.0	5.865	43.31	0.4752 −03	0.3559 −02
14.78	0.1676 −05	0.7489 −04	0.2238 −01	3495	111.232	3.879	0.3824	254.7	5.866	43.42	0.4721 −03	0.3549 −02
14.80	0.1660 −05	0.7440 −04	0.2232 −01	3518	111.258	3.874	0.3824	255.4	5.866	43.54	0.4691 −03	0.3540 −02
14.82	0.1645 −05	0.7391 −04	0.2226 −01	3541	111.283	3.869	0.3824	256.1	5.866	43.65	0.4660 −03	0.3530 −02
14.84	0.1630 −05	0.7343 −04	0.2220 −01	3565	111.308	3.864	0.3824	256.8	5.867	43.77	0.4630 −03	0.3521 −02
14.86	0.1615 −05	0.7294 −04	0.2214 −01	3588	111.334	3.859	0.3823	257.5	5.867	43.88	0.4599 −03	0.3511 −02
14.88	0.1600 −05	0.7247 −04	0.2208 −01	3612	111.359	3.853	0.3823	258.2	5.868	44.00	0.4570 −03	0.3502 −02
14.90	0.1586 −05	0.7199 −04	0.2202 −01	3635	111.384	3.848	0.3823	258.9	5.868	44.11	0.4540 −03	0.3493 −02
14.92	0.1571 −05	0.7152 −04	0.2197 −01	3659	111.409	3.843	0.3823	259.5	5.868	44.23	0.4510 −03	0.3483 −02
14.94	0.1557 −05	0.7105 −04	0.2191 −01	3683	111.435	3.838	0.3823	260.2	5.869	44.35	0.4481 −03	0.3474 −02
14.96	0.1543 −05	0.7059 −04	0.2185 −01	3707	111.460	3.833	0.3823	260.9	5.869	44.46	0.4452 −03	0.3465 −02
14.98	0.1529 −05	0.7013 −04	0.2180 −01	3731	111.484	3.828	0.3823	261.6	5.869	44.58	0.4423 −03	0.3455 −02
15.00	0.1515 −05	0.6967 −04	0.2174 −01	3756	111.509	3.822	0.3823	262.3	5.870	44.70	0.4395 −03	0.3446 −02
15.02	0.1501 −05	0.6922 −04	0.2168 −01	3780	111.534	3.817	0.3823	263.0	5.870	44.81	0.4367 −03	0.3437 −02
15.04	0.1487 −05	0.6877 −04	0.2163 −01	3805	111.559	3.812	0.3822	263.7	5.870	44.93	0.4339 −03	0.3428 −02
15.06	0.1474 −05	0.6833 −04	0.2157 −01	3829	111.584	3.807	0.3822	264.4	5.871	45.05	0.4311 −03	0.3419 −02
15.08	0.1460 −05	0.6788 −04	0.2151 −01	3854	111.608	3.802	0.3822	265.1	5.871	45.16	0.4283 −03	0.3410 −02

Table 2.14: Isentropic and normal shock properties for air, $\gamma = 1.4$ (continued).

M or M_1	$\dfrac{P}{P_t}$	$\dfrac{\rho}{\rho_t}$	$\dfrac{T}{T_t}$	$\dfrac{A}{A^*}$	ν	μ	M_2	$\dfrac{P_2}{P_1}$	$\dfrac{\rho_2}{\rho_1}$	$\dfrac{T_2}{T_1}$	$\dfrac{P_{t_2}}{P_{t_1}}$	$\dfrac{P_1}{P_{t_2}}$
15.10	0.1447 −05	0.6745 −04	0.2146 −01	3879	111.633	3.797	0.3822	265.9	5.871	45.28	0.4255 −03	0.3401 −02
15.12	0.1434 −05	0.6701 −04	0.2140 −01	3904	111.657	3.792	0.3822	266.6	5.872	45.40	0.4228 −03	0.3392 −02
15.14	0.1421 −05	0.6658 −04	0.2135 −01	3929	111.682	3.787	0.3822	267.3	5.872	45.52	0.4201 −03	0.3383 −02
15.16	0.1408 −05	0.6615 −04	0.2129 −01	3955	111.706	3.782	0.3822	268.0	5.872	45.63	0.4174 −03	0.3374 −02
15.18	0.1396 −05	0.6572 −04	0.2124 −01	3980	111.730	3.777	0.3822	268.7	5.873	45.75	0.4148 −03	0.3365 −02
15.20	0.1383 −05	0.6530 −04	0.2118 −01	4006	111.755	3.772	0.3822	269.4	5.873	45.87	0.4121 −03	0.3356 −02
15.22	0.1371 −05	0.6488 −04	0.2113 −01	4032	111.779	3.767	0.3821	270.1	5.873	45.99	0.4095 −03	0.3347 −02
15.24	0.1359 −05	0.6447 −04	0.2107 −01	4057	111.803	3.762	0.3821	270.8	5.874	46.11	0.4069 −03	0.3339 −02
15.26	0.1346 −05	0.6405 −04	0.2102 −01	4084	111.827	3.757	0.3821	271.5	5.874	46.23	0.4043 −03	0.3330 −02
15.28	0.1334 −05	0.6365 −04	0.2097 −01	4110	111.851	3.752	0.3821	272.2	5.874	46.34	0.4018 −03	0.3321 −02
15.30	0.1322 −05	0.6324 −04	0.2091 −01	4136	111.875	3.747	0.3821	272.9	5.875	46.46	0.3992 −03	0.3313 −02
15.32	0.1311 −05	0.6284 −04	0.2086 −01	4162	111.899	3.743	0.3821	273.7	5.875	46.58	0.3967 −03	0.3304 −02
15.34	0.1299 −05	0.6244 −04	0.2081 −01	4189	111.923	3.738	0.3821	274.4	5.875	46.70	0.3942 −03	0.3295 −02
15.36	0.1287 −05	0.6204 −04	0.2075 −01	4216	111.947	3.733	0.3821	275.1	5.875	46.82	0.3917 −03	0.3287 −02
15.38	0.1276 −05	0.6164 −04	0.2070 −01	4242	111.970	3.728	0.3821	275.8	5.876	46.94	0.3892 −03	0.3278 −02
15.40	0.1265 −05	0.6125 −04	0.2065 −01	4269	111.994	3.723	0.3820	276.5	5.876	47.06	0.3868 −03	0.3270 −02
15.42	0.1253 −05	0.6087 −04	0.2059 −01	4297	112.017	3.718	0.3820	277.2	5.876	47.18	0.3843 −03	0.3261 −02
15.44	0.1242 −05	0.6048 −04	0.2054 −01	4324	112.041	3.713	0.3820	278.0	5.877	47.30	0.3819 −03	0.3253 −02
15.46	0.1231 −05	0.6010 −04	0.2049 −01	4351	112.064	3.709	0.3820	278.7	5.877	47.42	0.3795 −03	0.3245 −02
15.48	0.1221 −05	0.5972 −04	0.2044 −01	4379	112.088	3.704	0.3820	279.4	5.877	47.54	0.3772 −03	0.3236 −02
15.50	0.1210 −05	0.5934 −04	0.2039 −01	4406	112.111	3.699	0.3820	280.1	5.878	47.66	0.3748 −03	0.3228 −02
15.52	0.1199 −05	0.5897 −04	0.2034 −01	4434	112.134	3.694	0.3820	280.9	5.878	47.78	0.3725 −03	0.3220 −02
15.54	0.1189 −05	0.5860 −04	0.2028 −01	4462	112.158	3.689	0.3820	281.6	5.878	47.90	0.3701 −03	0.3211 −02
15.56	0.1178 −05	0.5823 −04	0.2023 −01	4490	112.181	3.685	0.3820	282.3	5.879	48.02	0.3678 −03	0.3203 −02
15.58	0.1168 −05	0.5786 −04	0.2018 −01	4518	112.204	3.680	0.3819	283.0	5.879	48.14	0.3655 −03	0.3195 −02
15.60	0.1158 −05	0.5750 −04	0.2013 −01	4547	112.227	3.675	0.3819	283.8	5.879	48.27	0.3633 −03	0.3187 −02
15.62	0.1147 −05	0.5714 −04	0.2008 −01	4575	112.250	3.671	0.3819	284.5	5.880	48.39	0.3610 −03	0.3178 −02
15.64	0.1137 −05	0.5679 −04	0.2003 −01	4604	112.273	3.666	0.3819	285.2	5.880	48.51	0.3588 −03	0.3170 −02
15.66	0.1128 −05	0.5643 −04	0.1998 −01	4633	112.296	3.661	0.3819	286.0	5.880	48.63	0.3566 −03	0.3162 −02
15.68	0.1118 −05	0.5608 −04	0.1993 −01	4662	112.319	3.656	0.3819	286.7	5.880	48.75	0.3543 −03	0.3154 −02
15.70	0.1108 −05	0.5573 −04	0.1988 −01	4691	112.341	3.652	0.3819	287.4	5.881	48.87	0.3522 −03	0.3146 −02
15.72	0.1098 −05	0.5538 −04	0.1983 −01	4720	112.364	3.647	0.3819	288.1	5.881	49.00	0.3500 −03	0.3138 −02
15.74	0.1089 −05	0.5504 −04	0.1978 −01	4749	112.387	3.643	0.3819	288.9	5.881	49.12	0.3478 −03	0.3130 −02
15.76	0.1079 −05	0.5470 −04	0.1973 −01	4779	112.409	3.638	0.3819	289.6	5.882	49.24	0.3457 −03	0.3122 −02
15.78	0.1070 −05	0.5436 −04	0.1968 −01	4809	112.432	3.633	0.3818	290.4	5.882	49.36	0.3436 −03	0.3114 −02
15.80	0.1061 −05	0.5402 −04	0.1963 −01	4838	112.454	3.629	0.3818	291.1	5.882	49.49	0.3414 −03	0.3107 −02
15.82	0.1052 −05	0.5369 −04	0.1959 −01	4868	112.477	3.624	0.3818	291.8	5.882	49.61	0.3393 −03	0.3099 −02
15.84	0.1042 −05	0.5336 −04	0.1954 −01	4899	112.499	3.620	0.3818	292.6	5.883	49.73	0.3373 −03	0.3091 −02
15.86	0.1033 −05	0.5303 −04	0.1949 −01	4929	112.522	3.615	0.3818	293.3	5.883	49.86	0.3352 −03	0.3083 −02
15.88	0.1025 −05	0.5270 −04	0.1944 −01	4959	112.544	3.610	0.3818	294.0	5.883	49.98	0.3332 −03	0.3075 −02
15.90	0.1016 −05	0.5238 −04	0.1939 −01	4990	112.566	3.606	0.3818	294.8	5.884	50.10	0.3311 −03	0.3068 −02
15.92	0.1007 −05	0.5205 −04	0.1935 −01	5021	112.588	3.601	0.3818	295.5	5.884	50.23	0.3291 −03	0.3060 −02
15.94	0.9984 −06	0.5174 −04	0.1930 −01	5051	112.610	3.597	0.3818	296.3	5.884	50.35	0.3271 −03	0.3052 −02
15.96	0.9898 −06	0.5142 −04	0.1925 −01	5082	112.632	3.592	0.3818	297.0	5.884	50.47	0.3251 −03	0.3045 −02
15.98	0.9814 −06	0.5110 −04	0.1920 −01	5114	112.654	3.588	0.3818	297.8	5.885	50.60	0.3231 −03	0.3037 −02
16.00	0.9730 −06	0.5079 −04	0.1916 −01	5145	112.676	3.583	0.3817	298.5	5.885	50.72	0.3212 −03	0.3030 −02
16.02	0.9647 −06	0.5048 −04	0.1911 −01	5176	112.698	3.579	0.3817	299.3	5.885	50.85	0.3192 −03	0.3022 −02
16.04	0.9564 −06	0.5017 −04	0.1906 −01	5208	112.720	3.574	0.3817	300.0	5.886	50.97	0.3173 −03	0.3014 −02
16.06	0.9483 −06	0.4987 −04	0.1902 −01	5240	112.742	3.570	0.3817	300.8	5.886	51.10	0.3154 −03	0.3007 −02
16.08	0.9402 −06	0.4956 −04	0.1897 −01	5272	112.763	3.565	0.3817	301.5	5.886	51.22	0.3135 −03	0.2999 −02
16.10	0.9322 −06	0.4926 −04	0.1892 −01	5304	112.785	3.561	0.3817	302.3	5.886	51.35	0.3116 −03	0.2992 −02
16.12	0.9243 −06	0.4896 −04	0.1888 −01	5336	112.807	3.557	0.3817	303.0	5.887	51.47	0.3097 −03	0.2985 −02
16.14	0.9165 −06	0.4867 −04	0.1883 −01	5369	112.828	3.552	0.3817	303.8	5.887	51.60	0.3078 −03	0.2977 −02
16.16	0.9087 −06	0.4837 −04	0.1879 −01	5401	112.850	3.548	0.3817	304.5	5.887	51.72	0.3060 −03	0.2970 −02
16.18	0.9010 −06	0.4808 −04	0.1874 −01	5434	112.871	3.543	0.3817	305.3	5.888	51.85	0.3041 −03	0.2963 −02

Table 2.14: Isentropic and normal shock properties for air, $\gamma = 1.4$ (continued).

M or M_1	$\dfrac{p}{p_t}$	$\dfrac{\rho}{\rho_t}$	$\dfrac{T}{T_t}$	$\dfrac{A}{A^*}$	ν	μ	M_2	$\dfrac{p_2}{p_1}$	$\dfrac{\rho_2}{\rho_1}$	$\dfrac{T_2}{T_1}$	$\dfrac{p_{t_2}}{p_{t_1}}$	$\dfrac{p_1}{p_{t_2}}$
16.20	0.8934 −06	0.4779 −04	0.1870 −01	5467	112.893	3.539	0.3817	306.0	5.888	51.98	0.3023 −03	0.2955 −02
16.22	0.8859 −06	0.4750 −04	0.1865 −01	5500	112.914	3.535	0.3816	306.8	5.888	52.10	0.3005 −03	0.2948 −02
16.24	0.8784 −06	0.4721 −04	0.1860 −01	5533	112.935	3.530	0.3816	307.5	5.888	52.23	0.2987 −03	0.2941 −02
16.26	0.8710 −06	0.4693 −04	0.1856 −01	5567	112.957	3.526	0.3816	308.3	5.889	52.35	0.2969 −03	0.2934 −02
16.28	0.8637 −06	0.4665 −04	0.1852 −01	5600	112.978	3.522	0.3816	309.1	5.889	52.48	0.2951 −03	0.2926 −02
16.30	0.8564 −06	0.4637 −04	0.1847 −01	5634	112.999	3.517	0.3816	309.8	5.889	52.61	0.2934 −03	0.2919 −02
16.32	0.8492 −06	0.4609 −04	0.1843 −01	5668	113.020	3.513	0.3816	310.6	5.889	52.73	0.2916 −03	0.2912 −02
16.34	0.8421 −06	0.4581 −04	0.1838 −01	5702	113.041	3.509	0.3816	311.3	5.890	52.86	0.2899 −03	0.2905 −02
16.36	0.8351 −06	0.4554 −04	0.1834 −01	5736	113.062	3.504	0.3816	312.1	5.890	52.99	0.2882 −03	0.2898 −02
16.38	0.8281 −06	0.4527 −04	0.1829 −01	5770	113.083	3.500	0.3816	312.9	5.890	53.12	0.2865 −03	0.2891 −02
16.40	0.8212 −06	0.4500 −04	0.1825 −01	5805	113.104	3.496	0.3816	313.6	5.890	53.24	0.2848 −03	0.2884 −02
16.42	0.8143 −06	0.4473 −04	0.1821 −01	5840	113.125	3.491	0.3816	314.4	5.891	53.37	0.2831 −03	0.2877 −02
16.44	0.8075 −06	0.4446 −04	0.1816 −01	5875	113.146	3.487	0.3815	315.2	5.891	53.50	0.2814 −03	0.2870 −02
16.46	0.8008 −06	0.4420 −04	0.1812 −01	5910	113.166	3.483	0.3815	315.9	5.891	53.63	0.2797 −03	0.2863 −02
16.48	0.7942 −06	0.4393 −04	0.1808 −01	5945	113.187	3.479	0.3815	316.7	5.892	53.75	0.2781 −03	0.2856 −02
16.50	0.7876 −06	0.4367 −04	0.1803 −01	5980	113.208	3.475	0.3815	317.5	5.892	53.88	0.2764 −03	0.2849 −02
16.52	0.7810 −06	0.4341 −04	0.1799 −01	6016	113.228	3.470	0.3815	318.2	5.892	54.01	0.2748 −03	0.2842 −02
16.54	0.7746 −06	0.4316 −04	0.1795 −01	6051	113.249	3.466	0.3815	319.0	5.892	54.14	0.2732 −03	0.2835 −02
16.56	0.7682 −06	0.4290 −04	0.1791 −01	6087	113.269	3.462	0.3815	319.8	5.893	54.27	0.2716 −03	0.2828 −02
16.58	0.7618 −06	0.4265 −04	0.1786 −01	6123	113.290	3.458	0.3815	320.6	5.893	54.40	0.2700 −03	0.2822 −02
16.60	0.7555 −06	0.4240 −04	0.1782 −01	6160	113.310	3.454	0.3815	321.3	5.893	54.53	0.2684 −03	0.2815 −02
16.62	0.7493 −06	0.4215 −04	0.1778 −01	6196	113.331	3.449	0.3815	322.1	5.893	54.66	0.2668 −03	0.2808 −02
16.64	0.7431 −06	0.4190 −04	0.1774 −01	6233	113.351	3.445	0.3815	322.9	5.894	54.79	0.2653 −03	0.2801 −02
16.66	0.7370 −06	0.4165 −04	0.1770 −01	6269	113.371	3.441	0.3815	323.7	5.894	54.91	0.2637 −03	0.2795 −02
16.68	0.7310 −06	0.4141 −04	0.1765 −01	6306	113.391	3.437	0.3814	324.4	5.894	55.04	0.2622 −03	0.2788 −02
16.70	0.7250 −06	0.4116 −04	0.1761 −01	6343	113.412	3.433	0.3814	325.2	5.894	55.17	0.2607 −03	0.2781 −02
16.72	0.7190 −06	0.4092 −04	0.1757 −01	6381	113.432	3.429	0.3814	326.0	5.895	55.30	0.2592 −03	0.2775 −02
16.74	0.7131 −06	0.4068 −04	0.1753 −01	6418	113.452	3.425	0.3814	326.8	5.895	55.43	0.2576 −03	0.2768 −02
16.76	0.7073 −06	0.4044 −04	0.1749 −01	6456	113.472	3.421	0.3814	327.6	5.895	55.56	0.2561 −03	0.2761 −02
16.78	0.7015 −06	0.4021 −04	0.1745 −01	6493	113.492	3.416	0.3814	328.3	5.895	55.70	0.2547 −03	0.2755 −02
16.80	0.6958 −06	0.3997 −04	0.1741 −01	6531	113.512	3.412	0.3814	329.1	5.896	55.83	0.2532 −03	0.2748 −02
16.82	0.6901 −06	0.3974 −04	0.1737 −01	6570	113.532	3.408	0.3814	329.9	5.896	55.96	0.2517 −03	0.2742 −02
16.84	0.6845 −06	0.3951 −04	0.1733 −01	6608	113.551	3.404	0.3814	330.7	5.896	56.09	0.2503 −03	0.2735 −02
16.86	0.6789 −06	0.3928 −04	0.1728 −01	6646	113.571	3.400	0.3814	331.5	5.896	56.22	0.2488 −03	0.2729 −02
16.88	0.6734 −06	0.3905 −04	0.1724 −01	6685	113.591	3.396	0.3814	332.3	5.897	56.35	0.2474 −03	0.2722 −02
16.90	0.6680 −06	0.3883 −04	0.1720 −01	6724	113.611	3.392	0.3814	333.1	5.897	56.48	0.2460 −03	0.2716 −02
16.92	0.6626 −06	0.3860 −04	0.1716 −01	6763	113.630	3.388	0.3813	333.8	5.897	56.61	0.2445 −03	0.2709 −02
16.94	0.6572 −06	0.3838 −04	0.1712 −01	6802	113.650	3.384	0.3813	334.6	5.897	56.74	0.2431 −03	0.2703 −02
16.96	0.6519 −06	0.3815 −04	0.1709 −01	6842	113.669	3.380	0.3813	335.4	5.897	56.88	0.2417 −03	0.2697 −02
16.98	0.6466 −06	0.3793 −04	0.1705 −01	6881	113.689	3.376	0.3813	336.2	5.898	57.01	0.2403 −03	0.2690 −02
17.00	0.6414 −06	0.3772 −04	0.1701 −01	6921	113.708	3.372	0.3813	337.0	5.898	57.14	0.2390 −03	0.2684 −02
17.02	0.6362 −06	0.3750 −04	0.1697 −01	6961	113.728	3.368	0.3813	337.8	5.898	57.27	0.2376 −03	0.2678 −02
17.04	0.6311 −06	0.3728 −04	0.1693 −01	7001	113.747	3.364	0.3813	338.6	5.898	57.41	0.2362 −03	0.2671 −02
17.06	0.6260 −06	0.3707 −04	0.1689 −01	7042	113.767	3.360	0.3813	339.4	5.899	57.54	0.2349 −03	0.2665 −02
17.08	0.6210 −06	0.3686 −04	0.1685 −01	7082	113.786	3.356	0.3813	340.2	5.899	57.67	0.2336 −03	0.2659 −02
17.10	0.6160 −06	0.3664 −04	0.1681 −01	7123	113.805	3.352	0.3813	341.0	5.899	57.80	0.2322 −03	0.2653 −02
17.12	0.6111 −06	0.3643 −04	0.1677 −01	7164	113.824	3.349	0.3813	341.8	5.899	57.94	0.2309 −03	0.2647 −02
17.14	0.6062 −06	0.3623 −04	0.1673 −01	7205	113.844	3.345	0.3813	342.6	5.900	58.07	0.2296 −03	0.2640 −02
17.16	0.6014 −06	0.3602 −04	0.1670 −01	7246	113.863	3.341	0.3813	343.4	5.900	58.20	0.2283 −03	0.2634 −02
17.18	0.5965 −06	0.3581 −04	0.1666 −01	7288	113.882	3.337	0.3812	344.2	5.900	58.34	0.2270 −03	0.2628 −02
17.20	0.5918 −06	0.3561 −04	0.1662 −01	7329	113.901	3.333	0.3812	345.0	5.900	58.47	0.2257 −03	0.2622 −02
17.22	0.5871 −06	0.3541 −04	0.1658 −01	7371	113.920	3.329	0.3812	345.8	5.901	58.60	0.2244 −03	0.2616 −02
17.24	0.5824 −06	0.3520 −04	0.1654 −01	7413	113.939	3.325	0.3812	346.6	5.901	58.74	0.2232 −03	0.2610 −02
17.26	0.5778 −06	0.3500 −04	0.1651 −01	7455	113.958	3.321	0.3812	347.4	5.901	58.87	0.2219 −03	0.2604 −02
17.28	0.5732 −06	0.3480 −04	0.1647 −01	7498	113.977	3.318	0.3812	348.2	5.901	59.01	0.2206 −03	0.2598 −02

Table 2.14: Isentropic and normal shock properties for air, $\gamma = 1.4$ (continued).

M or M_1	$\frac{P}{P_t}$	$\frac{\rho}{\rho_t}$	$\frac{T}{T_t}$	$\frac{A}{A^*}$	ν	μ	M_2	$\frac{P_2}{P_1}$	$\frac{\rho_2}{\rho_1}$	$\frac{T_2}{T_1}$	$\frac{P_{t_2}}{P_{t_1}}$	$\frac{P_1}{P_{t_2}}$
17.30	0.5686 −06	0.3461 −04	0.1643 −01	7541	113.995	3.314	0.3812	349.0	5.901	59.14	0.2194 −03	0.2592 −02
17.32	0.5641 −06	0.3441 −04	0.1639 −01	7583	114.014	3.310	0.3812	349.8	5.902	59.28	0.2182 −03	0.2586 −02
17.34	0.5597 −06	0.3422 −04	0.1636 −01	7626	114.033	3.306	0.3812	350.6	5.902	59.41	0.2169 −03	0.2580 −02
17.36	0.5552 −06	0.3402 −04	0.1632 −01	7670	114.052	3.302	0.3812	351.4	5.902	59.55	0.2157 −03	0.2574 −02
17.38	0.5509 −06	0.3383 −04	0.1628 −01	7713	114.070	3.298	0.3812	352.3	5.902	59.68	0.2145 −03	0.2568 −02
17.40	0.5465 −06	0.3364 −04	0.1625 −01	7757	114.089	3.295	0.3812	353.1	5.903	59.82	0.2133 −03	0.2562 −02
17.42	0.5422 −06	0.3345 −04	0.1621 −01	7800	114.107	3.291	0.3812	353.9	5.903	59.95	0.2121 −03	0.2556 −02
17.44	0.5379 −06	0.3326 −04	0.1617 −01	7844	114.126	3.287	0.3811	354.7	5.903	60.09	0.2109 −03	0.2550 −02
17.46	0.5337 −06	0.3308 −04	0.1614 −01	7889	114.144	3.283	0.3811	355.5	5.903	60.22	0.2097 −03	0.2545 −02
17.48	0.5295 −06	0.3289 −04	0.1610 −01	7933	114.163	3.280	0.3811	356.3	5.903	60.36	0.2086 −03	0.2539 −02
17.50	0.5254 −06	0.3270 −04	0.1606 −01	7978	114.181	3.276	0.3811	357.1	5.904	60.49	0.2074 −03	0.2533 −02
17.52	0.5212 −06	0.3252 −04	0.1603 −01	8022	114.200	3.272	0.3811	358.0	5.904	60.63	0.2063 −03	0.2527 −02
17.54	0.5172 −06	0.3234 −04	0.1599 −01	8067	114.218	3.268	0.3811	358.8	5.904	60.77	0.2051 −03	0.2521 −02
17.56	0.5131 −06	0.3216 −04	0.1596 −01	8113	114.236	3.265	0.3811	359.6	5.904	60.90	0.2040 −03	0.2516 −02
17.58	0.5091 −06	0.3198 −04	0.1592 −01	8158	114.254	3.261	0.3811	360.4	5.904	61.04	0.2028 −03	0.2510 −02
17.60	0.5051 −06	0.3180 −04	0.1588 −01	8204	114.273	3.257	0.3811	361.2	5.905	61.18	0.2017 −03	0.2504 −02
17.62	0.5012 −06	0.3162 −04	0.1585 −01	8250	114.291	3.253	0.3811	362.1	5.905	61.31	0.2006 −03	0.2499 −02
17.64	0.4973 −06	0.3145 −04	0.1581 −01	8296	114.309	3.250	0.3811	362.9	5.905	61.45	0.1995 −03	0.2493 −02
17.66	0.4934 −06	0.3127 −04	0.1578 −01	8342	114.327	3.246	0.3811	363.7	5.905	61.59	0.1984 −03	0.2487 −02
17.68	0.4896 −06	0.3110 −04	0.1574 −01	8388	114.345	3.242	0.3811	364.5	5.906	61.73	0.1973 −03	0.2482 −02
17.70	0.4858 −06	0.3093 −04	0.1571 −01	8435	114.363	3.239	0.3811	365.4	5.906	61.86	0.1962 −03	0.2476 −02
17.72	0.4820 −06	0.3076 −04	0.1567 −01	8482	114.381	3.235	0.3810	366.2	5.906	62.00	0.1951 −03	0.2471 −02
17.74	0.4783 −06	0.3058 −04	0.1564 −01	8529	114.399	3.231	0.3810	367.0	5.906	62.14	0.1940 −03	0.2465 −02
17.76	0.4746 −06	0.3042 −04	0.1560 −01	8576	114.417	3.228	0.3810	367.8	5.906	62.28	0.1930 −03	0.2459 −02
17.78	0.4709 −06	0.3025 −04	0.1557 −01	8624	114.434	3.224	0.3810	368.7	5.907	62.42	0.1919 −03	0.2454 −02
17.80	0.4673 −06	0.3008 −04	0.1554 −01	8671	114.452	3.221	0.3810	369.5	5.907	62.55	0.1909 −03	0.2448 −02
17.82	0.4637 −06	0.2991 −04	0.1550 −01	8719	114.470	3.217	0.3810	370.3	5.907	62.69	0.1898 −03	0.2443 −02
17.84	0.4601 −06	0.2975 −04	0.1547 −01	8767	114.488	3.213	0.3810	371.2	5.907	62.83	0.1888 −03	0.2437 −02
17.86	0.4566 −06	0.2959 −04	0.1543 −01	8816	114.505	3.210	0.3810	372.0	5.907	62.97	0.1877 −03	0.2432 −02
17.88	0.4531 −06	0.2942 −04	0.1540 −01	8864	114.523	3.206	0.3810	372.8	5.908	63.11	0.1867 −03	0.2427 −02
17.90	0.4496 −06	0.2926 −04	0.1536 −01	8913	114.541	3.202	0.3810	373.7	5.908	63.25	0.1857 −03	0.2421 −02
17.92	0.4462 −06	0.2910 −04	0.1533 −01	8962	114.558	3.199	0.3810	374.5	5.908	63.39	0.1847 −03	0.2416 −02
17.94	0.4427 −06	0.2894 −04	0.1530 −01	9011	114.576	3.195	0.3810	375.3	5.908	63.53	0.1837 −03	0.2410 −02
17.96	0.4394 −06	0.2878 −04	0.1526 −01	9061	114.593	3.192	0.3810	376.2	5.908	63.67	0.1827 −03	0.2405 −02
17.98	0.4360 −06	0.2863 −04	0.1523 −01	9110	114.611	3.188	0.3810	377.0	5.909	63.81	0.1817 −03	0.2400 −02
18.00	0.4327 −06	0.2847 −04	0.1520 −01	9160	114.628	3.185	0.3810	377.8	5.909	63.95	0.1807 −03	0.2394 −02
18.02	0.4294 −06	0.2832 −04	0.1516 −01	9210	114.645	3.181	0.3809	378.7	5.909	64.09	0.1797 −03	0.2389 −02
18.04	0.4261 −06	0.2816 −04	0.1513 −01	9260	114.663	3.178	0.3809	379.5	5.909	64.23	0.1787 −03	0.2384 −02
18.06	0.4229 −06	0.2801 −04	0.1510 −01	9311	114.680	3.174	0.3809	380.4	5.909	64.37	0.1778 −03	0.2379 −02
18.08	0.4196 −06	0.2786 −04	0.1506 −01	9362	114.697	3.171	0.3809	381.2	5.910	64.51	0.1768 −03	0.2373 −02
18.10	0.4165 −06	0.2770 −04	0.1503 −01	9413	114.715	3.167	0.3809	382.1	5.910	64.65	0.1759 −03	0.2368 −02
18.12	0.4133 −06	0.2755 −04	0.1500 −01	9464	114.732	3.164	0.3809	382.9	5.910	64.79	0.1749 −03	0.2363 −02
18.14	0.4102 −06	0.2740 −04	0.1497 −01	9515	114.749	3.160	0.3809	383.7	5.910	64.93	0.1740 −03	0.2358 −02
18.16	0.4071 −06	0.2726 −04	0.1493 −01	9567	114.766	3.157	0.3809	384.6	5.910	65.07	0.1730 −03	0.2352 −02
18.18	0.4040 −06	0.2711 −04	0.1490 −01	9619	114.783	3.153	0.3809	385.4	5.911	65.21	0.1721 −03	0.2347 −02
18.20	0.4009 −06	0.2696 −04	0.1487 −01	9671	114.800	3.150	0.3809	386.3	5.911	65.35	0.1712 −03	0.2342 −02
18.22	0.3979 −06	0.2682 −04	0.1484 −01	9723	114.817	3.146	0.3809	387.1	5.911	65.50	0.1703 −03	0.2337 −02
18.24	0.3949 −06	0.2667 −04	0.1481 −01	9776	114.834	3.143	0.3809	388.0	5.911	65.64	0.1694 −03	0.2332 −02
18.26	0.3919 −06	0.2653 −04	0.1477 −01	9828	114.851	3.139	0.3809	388.8	5.911	65.78	0.1684 −03	0.2327 −02
18.28	0.3890 −06	0.2639 −04	0.1474 −01	9881	114.868	3.136	0.3809	389.7	5.912	65.92	0.1675 −03	0.2322 −02
18.30	0.3861 −06	0.2624 −04	0.1471 −01	9935	114.885	3.132	0.3809	390.6	5.912	66.06	0.1667 −03	0.2317 −02
18.32	0.3832 −06	0.2610 −04	0.1468 −01	9988	114.902	3.129	0.3809	391.4	5.912	66.21	0.1658 −03	0.2312 −02
18.34	0.3803 −06	0.2596 −04	0.1465 −01	0.1004 +05	114.918	3.126	0.3808	392.3	5.912	66.35	0.1649 −03	0.2307 −02
18.36	0.3774 −06	0.2582 −04	0.1462 −01	0.1010 +05	114.935	3.122	0.3808	393.1	5.912	66.49	0.1640 −03	0.2302 −02
18.38	0.3746 −06	0.2569 −04	0.1458 −01	0.1015 +05	114.952	3.119	0.3808	394.0	5.912	66.63	0.1631 −03	0.2297 −02

Table 2.14: Isentropic and normal shock properties for air, $\gamma = 1.4$ (continued).

M or M_1	$\dfrac{P}{P_t}$	$\dfrac{\rho}{\rho_t}$	$\dfrac{T}{T_t}$	$\dfrac{A}{A^*}$	ν	μ	M_2	$\dfrac{P_2}{P_1}$	$\dfrac{\rho_2}{\rho_1}$	$\dfrac{T_2}{T_1}$	$\dfrac{P_{t_2}}{P_{t_1}}$	$\dfrac{P_1}{P_{t_2}}$
18.40	0.3718 −06	0.2555 −04	0.1455 −01	0.1020 +05	114.968	3.115	0.3808	394.8	5.913	66.78	0.1623 −03	0.2292 −02
18.42	0.3690 −06	0.2541 −04	0.1452 −01	0.1026 +05	114.985	3.112	0.3808	395.7	5.913	66.92	0.1614 −03	0.2287 −02
18.44	0.3663 −06	0.2528 −04	0.1449 −01	0.1031 +05	115.002	3.109	0.3808	396.6	5.913	67.06	0.1605 −03	0.2282 −02
18.46	0.3636 −06	0.2514 −04	0.1446 −01	0.1037 +05	115.018	3.105	0.3808	397.4	5.913	67.21	0.1597 −03	0.2277 −02
18.48	0.3609 −06	0.2501 −04	0.1443 −01	0.1042 +05	115.035	3.102	0.3808	398.3	5.913	67.35	0.1588 −03	0.2272 −02
18.50	0.3582 −06	0.2488 −04	0.1440 −01	0.1048 +05	115.051	3.099	0.3808	399.1	5.914	67.50	0.1580 −03	0.2267 −02
18.52	0.3555 −06	0.2474 −04	0.1437 −01	0.1054 +05	115.068	3.095	0.3808	400.0	5.914	67.64	0.1572 −03	0.2262 −02
18.54	0.3529 −06	0.2461 −04	0.1434 −01	0.1059 +05	115.084	3.092	0.3808	400.9	5.914	67.78	0.1563 −03	0.2257 −02
18.56	0.3503 −06	0.2448 −04	0.1431 −01	0.1065 +05	115.101	3.088	0.3808	401.7	5.914	67.93	0.1555 −03	0.2252 −02
18.58	0.3477 −06	0.2435 −04	0.1428 −01	0.1070 +05	115.117	3.085	0.3808	402.6	5.914	68.07	0.1547 −03	0.2247 −02
18.60	0.3451 −06	0.2422 −04	0.1425 −01	0.1076 +05	115.133	3.082	0.3808	403.5	5.915	68.22	0.1539 −03	0.2243 −02
18.62	0.3425 −06	0.2410 −04	0.1422 −01	0.1082 +05	115.149	3.079	0.3808	404.3	5.915	68.36	0.1531 −03	0.2238 −02
18.64	0.3400 −06	0.2397 −04	0.1419 −01	0.1088 +05	115.166	3.075	0.3808	405.2	5.915	68.51	0.1523 −03	0.2233 −02
18.66	0.3375 −06	0.2384 −04	0.1416 −01	0.1093 +05	115.182	3.072	0.3807	406.1	5.915	68.65	0.1515 −03	0.2228 −02
18.68	0.3350 −06	0.2372 −04	0.1413 −01	0.1099 +05	115.198	3.069	0.3807	406.9	5.915	68.80	0.1507 −03	0.2223 −02
18.70	0.3326 −06	0.2359 −04	0.1410 −01	0.1105 +05	115.214	3.065	0.3807	407.8	5.915	68.94	0.1499 −03	0.2219 −02
18.72	0.3301 −06	0.2347 −04	0.1407 −01	0.1111 +05	115.230	3.062	0.3807	408.7	5.916	69.09	0.1491 −03	0.2214 −02
18.74	0.3277 −06	0.2334 −04	0.1404 −01	0.1116 +05	115.246	3.059	0.3807	409.6	5.916	69.23	0.1483 −03	0.2209 −02
18.76	0.3253 −06	0.2322 −04	0.1401 −01	0.1122 +05	115.262	3.056	0.3807	410.4	5.916	69.38	0.1476 −03	0.2205 −02
18.78	0.3229 −06	0.2310 −04	0.1398 −01	0.1128 +05	115.279	3.052	0.3807	411.3	5.916	69.52	0.1468 −03	0.2200 −02
18.80	0.3205 −06	0.2298 −04	0.1395 −01	0.1134 +05	115.294	3.049	0.3807	412.2	5.916	69.67	0.1460 −03	0.2195 −02
18.82	0.3182 −06	0.2286 −04	0.1392 −01	0.1140 +05	115.310	3.046	0.3807	413.1	5.916	69.82	0.1453 −03	0.2190 −02
18.84	0.3159 −06	0.2274 −04	0.1389 −01	0.1146 +05	115.326	3.043	0.3807	414.0	5.917	69.96	0.1445 −03	0.2186 −02
18.86	0.3136 −06	0.2262 −04	0.1386 −01	0.1152 +05	115.342	3.039	0.3807	414.8	5.917	70.11	0.1438 −03	0.2181 −02
18.88	0.3113 −06	0.2250 −04	0.1383 −01	0.1158 +05	115.358	3.036	0.3807	415.7	5.917	70.26	0.1430 −03	0.2177 −02
18.90	0.3090 −06	0.2239 −04	0.1380 −01	0.1164 +05	115.374	3.033	0.3807	416.6	5.917	70.40	0.1423 −03	0.2172 −02
18.92	0.3068 −06	0.2227 −04	0.1377 −01	0.1170 +05	115.390	3.030	0.3807	417.5	5.917	70.55	0.1415 −03	0.2167 −02
18.94	0.3045 −06	0.2215 −04	0.1375 −01	0.1176 +05	115.405	3.026	0.3807	418.4	5.918	70.70	0.1408 −03	0.2163 −02
18.96	0.3023 −06	0.2204 −04	0.1372 −01	0.1182 +05	115.421	3.023	0.3807	419.2	5.918	70.85	0.1401 −03	0.2158 −02
18.98	0.3001 −06	0.2192 −04	0.1369 −01	0.1189 +05	115.437	3.020	0.3807	420.1	5.918	70.99	0.1394 −03	0.2154 −02
19.00	0.2980 −06	0.2181 −04	0.1366 −01	0.1195 +05	115.452	3.017	0.3806	421.0	5.918	71.14	0.1386 −03	0.2149 −02
19.02	0.2958 −06	0.2170 −04	0.1363 −01	0.1201 +05	115.468	3.014	0.3806	421.9	5.918	71.29	0.1379 −03	0.2145 −02
19.04	0.2937 −06	0.2159 −04	0.1360 −01	0.1207 +05	115.484	3.011	0.3806	422.8	5.918	71.44	0.1372 −03	0.2140 −02
19.06	0.2915 −06	0.2147 −04	0.1358 −01	0.1213 +05	115.499	3.007	0.3806	423.7	5.919	71.59	0.1365 −03	0.2136 −02
19.08	0.2894 −06	0.2136 −04	0.1355 −01	0.1220 +05	115.515	3.004	0.3806	424.6	5.919	71.73	0.1358 −03	0.2131 −02
19.10	0.2874 −06	0.2125 −04	0.1352 −01	0.1226 +05	115.530	3.001	0.3806	425.5	5.919	71.88	0.1351 −03	0.2127 −02
19.12	0.2853 −06	0.2114 −04	0.1349 −01	0.1232 +05	115.546	2.998	0.3806	426.4	5.919	72.03	0.1344 −03	0.2122 −02
19.14	0.2832 −06	0.2104 −04	0.1346 −01	0.1239 +05	115.561	2.995	0.3806	427.2	5.919	72.18	0.1337 −03	0.2118 −02
19.16	0.2812 −06	0.2093 −04	0.1344 −01	0.1245 +05	115.576	2.992	0.3806	428.1	5.919	72.33	0.1330 −03	0.2114 −02
19.18	0.2792 −06	0.2082 −04	0.1341 −01	0.1251 +05	115.592	2.989	0.3806	429.0	5.920	72.48	0.1324 −03	0.2109 −02
19.20	0.2772 −06	0.2071 −04	0.1338 −01	0.1258 +05	115.607	2.985	0.3806	429.9	5.920	72.63	0.1317 −03	0.2105 −02
19.22	0.2752 −06	0.2061 −04	0.1335 −01	0.1264 +05	115.622	2.982	0.3806	430.8	5.920	72.78	0.1310 −03	0.2100 −02
19.24	0.2732 −06	0.2050 −04	0.1333 −01	0.1271 +05	115.638	2.979	0.3806	431.7	5.920	72.93	0.1304 −03	0.2096 −02
19.26	0.2713 −06	0.2040 −04	0.1330 −01	0.1277 +05	115.653	2.976	0.3806	432.6	5.920	73.08	0.1297 −03	0.2092 −02
19.28	0.2693 −06	0.2029 −04	0.1327 −01	0.1284 +05	115.668	2.973	0.3806	433.5	5.920	73.23	0.1290 −03	0.2087 −02
19.30	0.2674 −06	0.2019 −04	0.1324 −01	0.1290 +05	115.683	2.970	0.3806	434.4	5.921	73.38	0.1284 −03	0.2083 −02
19.32	0.2655 −06	0.2009 −04	0.1322 −01	0.1297 +05	115.699	2.967	0.3806	435.3	5.921	73.53	0.1277 −03	0.2079 −02
19.34	0.2636 −06	0.1998 −04	0.1319 −01	0.1304 +05	115.714	2.964	0.3806	436.2	5.921	73.68	0.1271 −03	0.2074 −02
19.36	0.2617 −06	0.1988 −04	0.1316 −01	0.1310 +05	115.729	2.961	0.3805	437.1	5.921	73.83	0.1264 −03	0.2070 −02
19.38	0.2599 −06	0.1978 −04	0.1314 −01	0.1317 +05	115.744	2.958	0.3805	438.0	5.921	73.98	0.1258 −03	0.2066 −02
19.40	0.2580 −06	0.1968 −04	0.1311 −01	0.1324 +05	115.759	2.955	0.3805	438.9	5.921	74.13	0.1252 −03	0.2062 −02
19.42	0.2562 −06	0.1958 −04	0.1308 −01	0.1330 +05	115.774	2.952	0.3805	439.8	5.921	74.28	0.1245 −03	0.2057 −02
19.44	0.2544 −06	0.1948 −04	0.1306 −01	0.1337 +05	115.789	2.949	0.3805	440.7	5.922	74.43	0.1239 −03	0.2053 −02
19.46	0.2526 −06	0.1938 −04	0.1303 −01	0.1344 +05	115.804	2.946	0.3805	441.7	5.922	74.58	0.1233 −03	0.2049 −02
19.48	0.2508 −06	0.1929 −04	0.1300 −01	0.1351 +05	115.819	2.943	0.3805	442.6	5.922	74.73	0.1227 −03	0.2045 −02

Table 2.14: Isentropic and normal shock properties for air, $\gamma = 1.4$ (continued).

M or M_1	$\dfrac{P}{P_t}$	$\dfrac{\rho}{\rho_t}$	$\dfrac{T}{T_t}$	$\dfrac{A}{A^*}$	ν	μ	M_2	$\dfrac{P_2}{P_1}$	$\dfrac{\rho_2}{\rho_1}$	$\dfrac{T_2}{T_1}$	$\dfrac{P_{t_2}}{P_{t_1}}$	$\dfrac{P_1}{P_{t_2}}$
19.50	0.2490 −06	0.1919 −04	0.1298 −01	0.1358 +05	115.833	2.939	0.3805	443.5	5.922	74.88	0.1220 −03	0.2041 −02
19.52	0.2473 −06	0.1909 −04	0.1295 −01	0.1364 +05	115.848	2.936	0.3805	444.4	5.922	75.04	0.1214 −03	0.2036 −02
19.54	0.2455 −06	0.1899 −04	0.1293 −01	0.1371 +05	115.863	2.933	0.3805	445.3	5.922	75.19	0.1208 −03	0.2032 −02
19.56	0.2438 −06	0.1890 −04	0.1290 −01	0.1378 +05	115.878	2.930	0.3805	446.2	5.923	75.34	0.1202 −03	0.2028 −02
19.58	0.2421 −06	0.1880 −04	0.1287 −01	0.1385 +05	115.893	2.927	0.3805	447.1	5.923	75.49	0.1196 −03	0.2024 −02
19.60	0.2404 −06	0.1871 −04	0.1285 −01	0.1392 +05	115.907	2.924	0.3805	448.0	5.923	75.64	0.1190 −03	0.2020 −02
19.62	0.2387 −06	0.1862 −04	0.1282 −01	0.1399 +05	115.922	2.921	0.3805	449.0	5.923	75.80	0.1184 −03	0.2016 −02
19.64	0.2370 −06	0.1852 −04	0.1280 −01	0.1406 +05	115.937	2.919	0.3805	449.9	5.923	75.95	0.1178 −03	0.2012 −02
19.66	0.2354 −06	0.1843 −04	0.1277 −01	0.1413 +05	115.951	2.916	0.3805	450.8	5.923	76.10	0.1172 −03	0.2007 −02
19.68	0.2337 −06	0.1834 −04	0.1274 −01	0.1420 +05	115.966	2.913	0.3805	451.7	5.924	76.26	0.1167 −03	0.2003 −02
19.70	0.2321 −06	0.1825 −04	0.1272 −01	0.1428 +05	115.980	2.910	0.3805	452.6	5.924	76.41	0.1161 −03	0.1999 −02
19.72	0.2304 −06	0.1815 −04	0.1269 −01	0.1435 +05	115.995	2.907	0.3805	453.5	5.924	76.56	0.1155 −03	0.1995 −02
19.74	0.2288 −06	0.1806 −04	0.1267 −01	0.1442 +05	116.009	2.904	0.3805	454.5	5.924	76.72	0.1149 −03	0.1991 −02
19.76	0.2272 −06	0.1797 −04	0.1264 −01	0.1449 +05	116.024	2.901	0.3804	455.4	5.924	76.87	0.1144 −03	0.1987 −02
19.78	0.2257 −06	0.1788 −04	0.1262 −01	0.1456 +05	116.038	2.898	0.3804	456.3	5.924	77.02	0.1138 −03	0.1983 −02
19.80	0.2241 −06	0.1780 −04	0.1259 −01	0.1464 +05	116.053	2.895	0.3804	457.2	5.924	77.18	0.1132 −03	0.1979 −02
19.82	0.2225 −06	0.1771 −04	0.1257 −01	0.1471 +05	116.067	2.892	0.3804	458.2	5.925	77.33	0.1127 −03	0.1975 −02
19.84	0.2210 −06	0.1762 −04	0.1254 −01	0.1478 +05	116.082	2.889	0.3804	459.1	5.925	77.49	0.1121 −03	0.1971 −02
19.86	0.2195 −06	0.1753 −04	0.1252 −01	0.1486 +05	116.096	2.886	0.3804	460.0	5.925	77.64	0.1116 −03	0.1967 −02
19.88	0.2179 −06	0.1744 −04	0.1249 −01	0.1493 +05	116.110	2.883	0.3804	460.9	5.925	77.79	0.1110 −03	0.1963 −02
19.90	0.2164 −06	0.1736 −04	0.1247 −01	0.1500 +05	116.125	2.880	0.3804	461.9	5.925	77.95	0.1105 −03	0.1959 −02
19.92	0.2149 −06	0.1727 −04	0.1244 −01	0.1508 +05	116.139	2.877	0.3804	462.8	5.925	78.10	0.1099 −03	0.1955 −02
19.94	0.2134 −06	0.1719 −04	0.1242 −01	0.1515 +05	116.153	2.875	0.3804	463.7	5.925	78.26	0.1094 −03	0.1952 −02
19.96	0.2120 −06	0.1710 −04	0.1239 −01	0.1523 +05	116.167	2.872	0.3804	464.7	5.926	78.41	0.1088 −03	0.1948 −02
19.98	0.2105 −06	0.1702 −04	0.1237 −01	0.1530 +05	116.181	2.869	0.3804	465.6	5.926	78.57	0.1083 −03	0.1944 −02
20.00	0.2090 −06	0.1693 −04	0.1235 −01	0.1538 +05	116.196	2.866	0.3804	466.5	5.926	78.72	0.1078 −03	0.1940 −02
20.20	0.1951 −06	0.1612 −04	0.1210 −01	0.1615 +05	116.335	2.838	0.3803	475.9	5.927	80.29	0.1026 −03	0.1902 −02
20.40	0.1823 −06	0.1536 −04	0.1187 −01	0.1695 +05	116.473	2.810	0.3803	485.4	5.929	81.87	0.9777 −04	0.1865 −02
20.60	0.1704 −06	0.1463 −04	0.1164 −01	0.1779 +05	116.607	2.782	0.3802	494.9	5.930	83.46	0.9319 −04	0.1829 −02
20.80	0.1594 −06	0.1395 −04	0.1142 −01	0.1866 +05	116.739	2.756	0.3802	504.6	5.931	85.07	0.8886 −04	0.1794 −02
21.00	0.1492 −06	0.1331 −04	0.1121 −01	0.1956 +05	116.869	2.729	0.3802	514.4	5.933	86.70	0.8477 −04	0.1760 −02
21.20	0.1397 −06	0.1270 −04	0.1100 −01	0.2050 +05	116.996	2.704	0.3801	524.2	5.934	88.34	0.8090 −04	0.1727 −02
21.40	0.1309 −06	0.1212 −04	0.1080 −01	0.2147 +05	117.121	2.678	0.3801	534.1	5.935	89.99	0.7725 −04	0.1695 −02
21.60	0.1227 −06	0.1158 −04	0.1060 −01	0.2248 +05	117.243	2.653	0.3800	544.2	5.936	91.67	0.7379 −04	0.1663 −02
21.80	0.1151 −06	0.1106 −04	0.1041 −01	0.2352 +05	117.363	2.629	0.3800	554.3	5.938	93.36	0.7051 −04	0.1633 −02
22.00	0.1081 −06	0.1057 −04	0.1022 −01	0.2461 +05	117.482	2.605	0.3800	564.5	5.939	95.06	0.6741 −04	0.1603 −02
22.20	0.1015 −06	0.1011 −04	0.1004 −01	0.2573 +05	117.598	2.582	0.3799	574.8	5.940	96.78	0.6447 −04	0.1575 −02
22.40	0.9540 −07	0.9669 −05	0.9866 −02	0.2690 +05	117.711	2.559	0.3799	585.2	5.941	98.51	0.6168 −04	0.1547 −02
22.60	0.8970 −07	0.9253 −05	0.9694 −02	0.2811 +05	117.823	2.536	0.3799	595.7	5.942	100.3	0.5903 −04	0.1519 −02
22.80	0.8438 −07	0.8858 −05	0.9526 −02	0.2936 +05	117.933	2.514	0.3798	606.3	5.943	102.0	0.5652 −04	0.1493 −02
23.00	0.7942 −07	0.8483 −05	0.9363 −02	0.3065 +05	118.042	2.492	0.3798	617.0	5.944	103.8	0.5414 −04	0.1467 −02
23.20	0.7480 −07	0.8127 −05	0.9204 −02	0.3199 +05	118.148	2.470	0.3798	627.8	5.945	105.6	0.5187 −04	0.1442 −02
23.40	0.7047 −07	0.7788 −05	0.9049 −02	0.3338 +05	118.252	2.449	0.3797	638.7	5.946	107.4	0.4972 −04	0.1417 −02
23.60	0.6643 −07	0.7467 −05	0.8897 −02	0.3482 +05	118.355	2.428	0.3797	649.6	5.947	109.2	0.4767 −04	0.1394 −02
23.80	0.6265 −07	0.7161 −05	0.8750 −02	0.3630 +05	118.456	2.408	0.3797	660.7	5.948	111.1	0.4573 −04	0.1370 −02
24.00	0.5912 −07	0.6870 −05	0.8606 −02	0.3784 +05	118.555	2.388	0.3796	671.9	5.948	112.9	0.4387 −04	0.1347 −02
24.20	0.5581 −07	0.6593 −05	0.8465 −02	0.3942 +05	118.653	2.368	0.3796	683.1	5.949	114.8	0.4211 −04	0.1325 −02
24.40	0.5271 −07	0.6329 −05	0.8328 −02	0.4106 +05	118.749	2.349	0.3796	694.4	5.950	116.7	0.4043 −04	0.1304 −02
24.60	0.4981 −07	0.6078 −05	0.8194 −02	0.4275 +05	118.844	2.330	0.3796	705.9	5.951	118.6	0.3883 −04	0.1283 −02
24.80	0.4708 −07	0.5839 −05	0.8064 −02	0.4450 +05	118.937	2.311	0.3795	717.4	5.952	120.5	0.3731 −04	0.1262 −02
25.00	0.4453 −07	0.5611 −05	0.7936 −02	0.4631 +05	119.029	2.292	0.3795	729.0	5.952	122.5	0.3586 −04	0.1242 −02
25.20	0.4213 −07	0.5394 −05	0.7812 −02	0.4817 +05	119.119	2.274	0.3795	740.7	5.953	124.4	0.3447 −04	0.1222 −02
25.40	0.3988 −07	0.5186 −05	0.7690 −02	0.5010 +05	119.207	2.256	0.3795	752.5	5.954	126.4	0.3315 −04	0.1203 −02
25.60	0.3777 −07	0.4988 −05	0.7571 −02	0.5208 +05	119.295	2.239	0.3794	764.4	5.955	128.4	0.3189 −04	0.1184 −02
25.80	0.3578 −07	0.4799 −05	0.7455 −02	0.5413 +05	119.381	2.221	0.3794	776.4	5.955	130.4	0.3068 −04	0.1166 −02

Table 2.14: Isentropic and normal shock properties for air, $\gamma = 1.4$ (continued).

M or M_1	$\dfrac{P}{P_t}$	$\dfrac{\rho}{\rho_t}$	$\dfrac{T}{T_t}$	$\dfrac{A}{A^*}$	ν	μ	M_2	$\dfrac{P_2}{P_1}$	$\dfrac{\rho_2}{\rho_1}$	$\dfrac{T_2}{T_1}$	$\dfrac{P_{t_2}}{P_{t_1}}$	$\dfrac{P_1}{P_{t_2}}$
26.00	0.3391 −07	0.4619 −05	0.7342 −02	0.5624 +05	119.466	2.204	0.3794	788.5	5.956	132.4	0.2953 −04	0.1148 −02
26.20	0.3215 −07	0.4446 −05	0.7231 −02	0.5842 +05	119.549	2.187	0.3794	800.7	5.957	134.4	0.2843 −04	0.1131 −02
26.40	0.3050 −07	0.4282 −05	0.7123 −02	0.6066 +05	119.631	2.171	0.3794	813.0	5.957	136.5	0.2738 −04	0.1114 −02
26.60	0.2894 −07	0.4124 −05	0.7017 −02	0.6297 +05	119.712	2.154	0.3793	825.3	5.958	138.5	0.2638 −04	0.1097 −02
26.80	0.2747 −07	0.3974 −05	0.6913 −02	0.6536 +05	119.792	2.138	0.3793	837.8	5.959	140.6	0.2542 −04	0.1081 −02
27.00	0.2609 −07	0.3830 −05	0.6812 −02	0.6781 +05	119.870	2.123	0.3793	850.4	5.959	142.7	0.2450 −04	0.1065 −02
27.20	0.2478 −07	0.3692 −05	0.6713 −02	0.7034 +05	119.948	2.107	0.3793	863.0	5.960	144.8	0.2362 −04	0.1049 −02
27.40	0.2355 −07	0.3560 −05	0.6616 −02	0.7294 +05	120.024	2.092	0.3793	875.7	5.960	146.9	0.2278 −04	0.1034 −02
27.60	0.2239 −07	0.3434 −05	0.6521 −02	0.7562 +05	120.099	2.076	0.3792	888.6	5.961	149.1	0.2197 −04	0.1019 −02
27.80	0.2129 −07	0.3313 −05	0.6428 −02	0.7838 +05	120.174	2.061	0.3792	901.5	5.961	151.2	0.2120 −04	0.1004 −02
28.00	0.2026 −07	0.3197 −05	0.6337 −02	0.8122 +05	120.247	2.047	0.3792	914.5	5.962	153.4	0.2046 −04	0.9902 −03
28.20	0.1928 −07	0.3086 −05	0.6248 −02	0.8414 +05	120.319	2.032	0.3792	927.6	5.963	155.6	0.1975 −04	0.9762 −03
28.40	0.1835 −07	0.2979 −05	0.6161 −02	0.8714 +05	120.390	2.018	0.3792	940.8	5.963	157.8	0.1907 −04	0.9625 −03
28.60	0.1748 −07	0.2877 −05	0.6075 −02	0.9023 +05	120.460	2.004	0.3792	954.1	5.964	160.0	0.1842 −04	0.9491 −03
28.80	0.1665 −07	0.2779 −05	0.5992 −02	0.9340 +05	120.529	1.990	0.3791	967.5	5.964	162.2	0.1779 −04	0.9359 −03
29.00	0.1587 −07	0.2685 −05	0.5910 −02	0.9667 +05	120.597	1.976	0.3791	981.0	5.965	164.5	0.1719 −04	0.9231 −03
29.20	0.1513 −07	0.2595 −05	0.5830 −02	0.1000 +06	120.664	1.963	0.3791	994.6	5.965	166.7	0.1662 −04	0.9105 −03
29.40	0.1443 −07	0.2508 −05	0.5751 −02	0.1035 +06	120.731	1.949	0.3791	1008	5.965	169.0	0.1606 −04	0.8981 −03
29.60	0.1376 −07	0.2425 −05	0.5674 −02	0.1070 +06	120.796	1.936	0.3791	1022	5.966	171.3	0.1553 −04	0.8861 −03
29.80	0.1313 −07	0.2345 −05	0.5599 −02	0.1107 +06	120.861	1.923	0.3791	1036	5.966	173.6	0.1502 −04	0.8742 −03
30.00	0.1253 −07	0.2269 −05	0.5525 −02	0.1144 +06	120.924	1.910	0.3790	1050	5.967	175.9	0.1453 −04	0.8626 −03
30.20	0.1197 −07	0.2195 −05	0.5452 −02	0.1182 +06	120.987	1.898	0.3790	1064	5.967	178.3	0.1406 −04	0.8512 −03
30.40	0.1143 −07	0.2124 −05	0.5381 −02	0.1222 +06	121.049	1.885	0.3790	1078	5.968	180.6	0.1361 −04	0.8401 −03
30.60	0.1092 −07	0.2056 −05	0.5311 −02	0.1262 +06	121.110	1.873	0.3790	1092	5.968	183.0	0.1317 −04	0.8291 −03
30.80	0.1044 −07	0.1990 −05	0.5243 −02	0.1304 +06	121.171	1.861	0.3790	1107	5.969	185.4	0.1275 −04	0.8184 −03
31.00	0.9976 −08	0.1927 −05	0.5176 −02	0.1346 +06	121.231	1.849	0.3790	1121	5.969	187.8	0.1235 −04	0.8079 −03
31.20	0.9539 −08	0.1867 −05	0.5110 −02	0.1390 +06	121.289	1.837	0.3790	1136	5.969	190.2	0.1196 −04	0.7975 −03
31.40	0.9123 −08	0.1808 −05	0.5045 −02	0.1435 +06	121.348	1.825	0.3789	1150	5.970	192.7	0.1159 −04	0.7874 −03
31.60	0.8729 −08	0.1752 −05	0.4982 −02	0.1481 +06	121.405	1.813	0.3789	1165	5.970	195.1	0.1123 −04	0.7775 −03
31.80	0.8354 −08	0.1698 −05	0.4920 −02	0.1528 +06	121.462	1.802	0.3789	1180	5.970	197.6	0.1088 −04	0.7677 −03
32.00	0.7997 −08	0.1646 −05	0.4859 −02	0.1576 +06	121.518	1.791	0.3789	1195	5.971	200.1	0.1055 −04	0.7582 −03
32.20	0.7657 −08	0.1595 −05	0.4799 −02	0.1626 +06	121.573	1.780	0.3789	1210	5.971	202.6	0.1023 −04	0.7488 −03
32.40	0.7334 −08	0.1547 −05	0.4740 −02	0.1677 +06	121.628	1.769	0.3789	1225	5.972	205.1	0.9916 −05	0.7396 −03
32.60	0.7026 −08	0.1500 −05	0.4683 −02	0.1729 +06	121.682	1.758	0.3789	1240	5.972	207.6	0.9617 −05	0.7305 −03
32.80	0.6733 −08	0.1455 −05	0.4626 −02	0.1782 +06	121.735	1.747	0.3789	1255	5.972	210.1	0.9330 −05	0.7217 −03
33.00	0.6454 −08	0.1412 −05	0.4570 −02	0.1837 +06	121.788	1.736	0.3789	1270	5.973	212.7	0.9052 −05	0.7129 −03
33.20	0.6187 −08	0.1370 −05	0.4516 −02	0.1893 +06	121.840	1.726	0.3788	1286	5.973	215.3	0.8784 −05	0.7044 −03
33.40	0.5934 −08	0.1330 −05	0.4462 −02	0.1950 +06	121.891	1.716	0.3788	1301	5.973	217.9	0.8526 −05	0.6960 −03
33.60	0.5692 −08	0.1291 −05	0.4409 −02	0.2009 +06	121.942	1.705	0.3788	1317	5.974	220.5	0.8277 −05	0.6877 −03
33.80	0.5461 −08	0.1253 −05	0.4357 −02	0.2069 +06	121.992	1.695	0.3788	1333	5.974	223.1	0.8036 −05	0.6796 −03
34.00	0.5241 −08	0.1217 −05	0.4307 −02	0.2131 +06	122.042	1.685	0.3788	1349	5.974	225.7	0.7804 −05	0.6716 −03
34.20	0.5031 −08	0.1182 −05	0.4257 −02	0.2194 +06	122.091	1.676	0.3788	1364	5.974	228.4	0.7580 −05	0.6638 −03
34.40	0.4831 −08	0.1148 −05	0.4207 −02	0.2259 +06	122.139	1.666	0.3788	1380	5.975	231.0	0.7363 −05	0.6561 −03
34.60	0.4640 −08	0.1116 −05	0.4159 −02	0.2325 +06	122.187	1.656	0.3788	1397	5.975	233.7	0.7154 −05	0.6485 −03
34.80	0.4457 −08	0.1084 −05	0.4112 −02	0.2392 +06	122.234	1.647	0.3788	1413	5.975	236.4	0.6952 −05	0.6411 −03
35.00	0.4282 −08	0.1054 −05	0.4065 −02	0.2462 +06	122.281	1.637	0.3788	1429	5.976	239.1	0.6757 −05	0.6338 −03
35.20	0.4116 −08	0.1024 −05	0.4019 −02	0.2532 +06	122.328	1.628	0.3787	1445	5.976	241.9	0.6568 −05	0.6266 −03
35.40	0.3956 −08	0.9955 −06	0.3974 −02	0.2605 +06	122.373	1.619	0.3787	1462	5.976	244.6	0.6385 −05	0.6196 −03
35.60	0.3804 −08	0.9680 −06	0.3930 −02	0.2679 +06	122.419	1.610	0.3787	1478	5.976	247.4	0.6209 −05	0.6126 −03
35.80	0.3658 −08	0.9414 −06	0.3886 −02	0.2755 +06	122.463	1.601	0.3787	1495	5.977	250.2	0.6038 −05	0.6058 −03
36.00	0.3519 −08	0.9156 −06	0.3843 −02	0.2832 +06	122.508	1.592	0.3787	1512	5.977	253.0	0.5873 −05	0.5991 −03
36.20	0.3385 −08	0.8907 −06	0.3801 −02	0.2911 +06	122.551	1.583	0.3787	1529	5.977	255.8	0.5714 −05	0.5925 −03
36.40	0.3258 −08	0.8666 −06	0.3759 −02	0.2992 +06	122.595	1.574	0.3787	1546	5.977	258.6	0.5559 −05	0.5860 −03
36.60	0.3136 −08	0.8432 −06	0.3719 −02	0.3075 +06	122.638	1.566	0.3787	1563	5.978	261.4	0.5410 −05	0.5796 −03
36.80	0.3019 −08	0.8207 −06	0.3678 −02	0.3159 +06	122.680	1.557	0.3787	1580	5.978	264.3	0.5265 −05	0.5733 −03

Table 2.14: Isentropic and normal shock properties for air, $\gamma = 1.4$ (continued).

M or M_1	$\dfrac{P}{P_t}$	$\dfrac{\rho}{\rho_t}$	$\dfrac{T}{T_t}$	$\dfrac{A}{A^*}$	ν	μ	M_2	$\dfrac{P_2}{P_1}$	$\dfrac{\rho_2}{\rho_1}$	$\dfrac{T_2}{T_1}$	$\dfrac{P_{t_2}}{P_{t_1}}$	$\dfrac{P_1}{P_{t_2}}$
37.00	0.2907 −08	0.7988 −06	0.3639 −02	0.3246 +06	122.722	1.549	0.3787	1597	5.978	267.1	0.5125 −05	0.5672 −03
37.20	0.2799 −08	0.7776 −06	0.3600 −02	0.3334 +06	122.763	1.540	0.3787	1614	5.978	270.0	0.4989 −05	0.5611 −03
37.40	0.2697 −08	0.7571 −06	0.3562 −02	0.3424 +06	122.804	1.532	0.3787	1632	5.979	272.9	0.4858 −05	0.5551 −03
37.60	0.2598 −08	0.7373 −06	0.3524 −02	0.3516 +06	122.845	1.524	0.3787	1649	5.979	275.8	0.4731 −05	0.5492 −03
37.80	0.2504 −08	0.7180 −06	0.3487 −02	0.3611 +06	122.885	1.516	0.3786	1667	5.979	278.8	0.4608 −05	0.5434 −03
38.00	0.2413 −08	0.6994 −06	0.3451 −02	0.3707 +06	122.925	1.508	0.3786	1685	5.979	281.7	0.4488 −05	0.5377 −03
38.20	0.2327 −08	0.6813 −06	0.3415 −02	0.3805 +06	122.964	1.500	0.3786	1702	5.980	284.7	0.4372 −05	0.5321 −03
38.40	0.2243 −08	0.6638 −06	0.3379 −02	0.3905 +06	123.003	1.492	0.3786	1720	5.980	287.7	0.4260 −05	0.5266 −03
38.60	0.2164 −08	0.6469 −06	0.3344 −02	0.4007 +06	123.042	1.484	0.3786	1738	5.980	290.7	0.4152 −05	0.5211 −03
38.80	0.2087 −08	0.6304 −06	0.3310 −02	0.4112 +06	123.080	1.477	0.3786	1756	5.980	293.7	0.4046 −05	0.5158 −03
39.00	0.2013 −08	0.6145 −06	0.3276 −02	0.4219 +06	123.117	1.469	0.3786	1774	5.980	296.7	0.3944 −05	0.5105 −03
39.20	0.1943 −08	0.5990 −06	0.3243 −02	0.4327 +06	123.155	1.462	0.3786	1793	5.981	299.7	0.3845 −05	0.5053 −03
39.40	0.1875 −08	0.5840 −06	0.3210 −02	0.4439 +06	123.192	1.454	0.3786	1811	5.981	302.8	0.3749 −05	0.5002 −03
39.60	0.1810 −08	0.5695 −06	0.3178 −02	0.4552 +06	123.228	1.447	0.3786	1829	5.981	305.9	0.3655 −05	0.4951 −03
39.80	0.1747 −08	0.5553 −06	0.3146 −02	0.4668 +06	123.265	1.440	0.3786	1848	5.981	309.0	0.3565 −05	0.4902 −03
40.00	0.1687 −08	0.5416 −06	0.3115 −02	0.4786 +06	123.300	1.433	0.3786	1867	5.981	312.1	0.3477 −05	0.4853 −03
40.20	0.1630 −08	0.5283 −06	0.3084 −02	0.4906 +06	123.336	1.425	0.3786	1885	5.981	315.2	0.3392 −05	0.4805 −03
40.40	0.1574 −08	0.5154 −06	0.3054 −02	0.5029 +06	123.371	1.418	0.3786	1904	5.982	318.3	0.3309 −05	0.4757 −03
40.60	0.1521 −08	0.5029 −06	0.3024 −02	0.5154 +06	123.406	1.411	0.3786	1923	5.982	321.5	0.3228 −05	0.4711 −03
40.80	0.1470 −08	0.4907 −06	0.2995 −02	0.5282 +06	123.440	1.404	0.3785	1942	5.982	324.6	0.3150 −05	0.4665 −03
41.00	0.1420 −08	0.4789 −06	0.2966 −02	0.5412 +06	123.474	1.398	0.3785	1961	5.982	327.8	0.3075 −05	0.4619 −03
41.20	0.1373 −08	0.4674 −06	0.2937 −02	0.5545 +06	123.508	1.391	0.3785	1980	5.982	331.0	0.3001 −05	0.4574 −03
41.40	0.1327 −08	0.4563 −06	0.2909 −02	0.5680 +06	123.542	1.384	0.3785	1999	5.983	334.2	0.2930 −05	0.4530 −03
41.60	0.1283 −08	0.4455 −06	0.2881 −02	0.5818 +06	123.575	1.377	0.3785	2019	5.983	337.4	0.2860 −05	0.4487 −03
41.80	0.1241 −08	0.4349 −06	0.2853 −02	0.5959 +06	123.608	1.371	0.3785	2038	5.983	340.7	0.2793 −05	0.4444 −03
42.00	0.1200 −08	0.4247 −06	0.2826 −02	0.6102 +06	123.640	1.364	0.3785	2058	5.983	344.0	0.2727 −05	0.4402 −03
42.20	0.1161 −08	0.4148 −06	0.2800 −02	0.6249 +06	123.673	1.358	0.3785	2078	5.983	347.2	0.2663 −05	0.4360 −03
42.40	0.1124 −08	0.4051 −06	0.2773 −02	0.6398 +06	123.704	1.351	0.3785	2097	5.983	350.5	0.2601 −05	0.4319 −03
42.60	0.1087 −08	0.3957 −06	0.2748 −02	0.6549 +06	123.736	1.345	0.3785	2117	5.984	353.8	0.2541 −05	0.4279 −03
42.80	0.1052 −08	0.3866 −06	0.2722 −02	0.6704 +06	123.767	1.339	0.3785	2137	5.984	357.1	0.2482 −05	0.4239 −03
43.00	0.1019 −08	0.3777 −06	0.2697 −02	0.6862 +06	123.798	1.333	0.3785	2157	5.984	360.5	0.2425 −05	0.4200 −03
43.20	0.9861 −09	0.3690 −06	0.2672 −02	0.7022 +06	123.829	1.326	0.3785	2177	5.984	363.8	0.2370 −05	0.4161 −03
43.40	0.9548 −09	0.3606 −06	0.2647 −02	0.7186 +06	123.860	1.320	0.3785	2197	5.984	367.2	0.2316 −05	0.4123 −03
43.60	0.9246 −09	0.3525 −06	0.2623 −02	0.7352 +06	123.890	1.314	0.3785	2218	5.984	370.6	0.2264 −05	0.4085 −03
43.80	0.8955 −09	0.3445 −06	0.2599 −02	0.7522 +06	123.920	1.308	0.3785	2238	5.984	374.0	0.2213 −05	0.4048 −03
44.00	0.8675 −09	0.3368 −06	0.2576 −02	0.7695 +06	123.949	1.302	0.3785	2259	5.985	377.4	0.2163 −05	0.4011 −03
44.20	0.8405 −09	0.3292 −06	0.2553 −02	0.7871 +06	123.979	1.296	0.3785	2279	5.985	380.8	0.2115 −05	0.3975 −03
44.40	0.8144 −09	0.3219 −06	0.2530 −02	0.8050 +06	124.008	1.291	0.3785	2300	5.985	384.3	0.2068 −05	0.3939 −03
44.60	0.7892 −09	0.3148 −06	0.2507 −02	0.8232 +06	124.037	1.285	0.3785	2321	5.985	387.7	0.2022 −05	0.3904 −03
44.80	0.7650 −09	0.3078 −06	0.2485 −02	0.8418 +06	124.065	1.279	0.3784	2341	5.985	391.2	0.1977 −05	0.3869 −03
45.00	0.7415 −09	0.3011 −06	0.2463 −02	0.8607 +06	124.094	1.273	0.3784	2362	5.985	394.7	0.1934 −05	0.3835 −03
45.20	0.7189 −09	0.2945 −06	0.2441 −02	0.8799 +06	124.122	1.268	0.3784	2383	5.985	398.2	0.1892 −05	0.3801 −03
45.40	0.6971 −09	0.2881 −06	0.2420 −02	0.8995 +06	124.150	1.262	0.3784	2405	5.985	401.7	0.1850 −05	0.3767 −03
45.60	0.6760 −09	0.2818 −06	0.2399 −02	0.9194 +06	124.177	1.257	0.3784	2426	5.986	405.3	0.1810 −05	0.3734 −03
45.80	0.6557 −09	0.2757 −06	0.2378 −02	0.9397 +06	124.205	1.251	0.3784	2447	5.986	408.8	0.1771 −05	0.3702 −03
46.00	0.6360 −09	0.2698 −06	0.2357 −02	0.9604 +06	124.232	1.246	0.3784	2469	5.986	412.4	0.1733 −05	0.3670 −03
46.20	0.6170 −09	0.2640 −06	0.2337 −02	0.9814 +06	124.259	1.240	0.3784	2490	5.986	416.0	0.1696 −05	0.3638 −03
46.40	0.5987 −09	0.2584 −06	0.2317 −02	0.1003 +07	124.285	1.235	0.3784	2512	5.986	419.6	0.1660 −05	0.3607 −03
46.60	0.5810 −09	0.2529 −06	0.2297 −02	0.1024 +07	124.312	1.230	0.3784	2533	5.986	423.2	0.1625 −05	0.3576 −03
46.80	0.5639 −09	0.2476 −06	0.2278 −02	0.1047 +07	124.338	1.224	0.3784	2555	5.986	426.8	0.1590 −05	0.3545 −03
47.00	0.5473 −09	0.2424 −06	0.2258 −02	0.1069 +07	124.364	1.219	0.3784	2577	5.986	430.5	0.1557 −05	0.3515 −03
47.20	0.5313 −09	0.2373 −06	0.2239 −02	0.1092 +07	124.390	1.214	0.3784	2599	5.987	434.1	0.1524 −05	0.3486 −03
47.40	0.5159 −09	0.2323 −06	0.2220 −02	0.1115 +07	124.415	1.209	0.3784	2621	5.987	437.8	0.1493 −05	0.3456 −03
47.60	0.5009 −09	0.2275 −06	0.2202 −02	0.1139 +07	124.441	1.204	0.3784	2643	5.987	441.5	0.1462 −05	0.3427 −03
47.80	0.4865 −09	0.2228 −06	0.2184 −02	0.1163 +07	124.466	1.199	0.3784	2666	5.987	445.2	0.1431 −05	0.3399 −03

Table 2.14: Isentropic and normal shock properties for air, $\gamma = 1.4$ (continued).

M or M_1	$\dfrac{P}{P_t}$	$\dfrac{\rho}{\rho_t}$	$\dfrac{T}{T_t}$	$\dfrac{A}{A^*}$	ν	μ	M_2	$\dfrac{P_2}{P_1}$	$\dfrac{\rho_2}{\rho_1}$	$\dfrac{T_2}{T_1}$	$\dfrac{P_{t_2}}{P_{t_1}}$	$\dfrac{P_1}{P_{t_2}}$
48.00	0.4725 −09	0.2182 −06	0.2165 −02	0.1187 +07	124.491	1.194	0.3784	2688	5.987	449.0	0.1402 −05	0.3370 −03
48.20	0.4590 −09	0.2137 −06	0.2148 −02	0.1212 +07	124.515	1.189	0.3784	2710	5.987	452.7	0.1373 −05	0.3342 −03
48.40	0.4459 −09	0.2093 −06	0.2130 −02	0.1238 +07	124.540	1.184	0.3784	2733	5.987	456.5	0.1345 −05	0.3315 −03
48.60	0.4332 −09	0.2051 −06	0.2112 −02	0.1263 +07	124.564	1.179	0.3784	2756	5.987	460.2	0.1318 −05	0.3288 −03
48.80	0.4210 −09	0.2009 −06	0.2095 −02	0.1289 +07	124.588	1.174	0.3784	2778	5.987	464.0	0.1291 −05	0.3261 −03
49.00	0.4091 −09	0.1969 −06	0.2078 −02	0.1316 +07	124.612	1.169	0.3784	2801	5.988	467.8	0.1265 −05	0.3234 −03
49.20	0.3976 −09	0.1929 −06	0.2061 −02	0.1343 +07	124.636	1.165	0.3784	2824	5.988	471.6	0.1239 −05	0.3208 −03
49.40	0.3865 −09	0.1890 −06	0.2045 −02	0.1370 +07	124.659	1.160	0.3784	2847	5.988	475.5	0.1215 −05	0.3182 −03
49.60	0.3758 −09	0.1853 −06	0.2028 −02	0.1398 +07	124.683	1.155	0.3784	2870	5.988	479.3	0.1190 −05	0.3156 −03
49.80	0.3653 −09	0.1816 −06	0.2012 −02	0.1427 +07	124.706	1.151	0.3784	2893	5.988	483.2	0.1167 −05	0.3131 −03
50.00	0.3553 −09	0.1780 −06	0.1996 −02	0.1456 +07	124.729	1.146	0.3784	2917	5.988	487.1	0.1144 −05	0.3106 −03
51.00	0.3094 −09	0.1612 −06	0.1919 −02	0.1607 +07	124.841	1.124	0.3783	3034	5.988	506.7	0.1036 −05	0.2986 −03
52.00	0.2701 −09	0.1463 −06	0.1846 −02	0.1770 +07	124.949	1.102	0.3783	3155	5.989	526.7	0.9405 −06	0.2872 −03
53.00	0.2364 −09	0.1331 −06	0.1777 −02	0.1947 +07	125.052	1.081	0.3783	3277	5.989	547.1	0.8553 −06	0.2765 −03
54.00	0.2075 −09	0.1212 −06	0.1712 −02	0.2137 +07	125.152	1.061	0.3783	3402	5.990	568.0	0.7792 −06	0.2663 −03
55.00	0.1825 −09	0.1106 −06	0.1650 −02	0.2342 +07	125.249	1.042	0.3783	3529	5.990	589.1	0.7110 −06	0.2567 −03
56.00	0.1609 −09	0.1011 −06	0.1592 −02	0.2562 +07	125.341	1.023	0.3783	3659	5.990	610.7	0.6499 −06	0.2476 −03
57.00	0.1422 −09	0.9255 −07	0.1537 −02	0.2799 +07	125.431	1.005	0.3783	3790	5.991	632.7	0.5949 −06	0.2390 −03
58.00	0.1259 −09	0.8485 −07	0.1484 −02	0.3052 +07	125.518	0.9879	0.3783	3925	5.991	655.1	0.5455 −06	0.2308 −03
59.00	0.1117 −09	0.7791 −07	0.1434 −02	0.3324 +07	125.601	0.9712	0.3782	4061	5.991	677.8	0.5009 −06	0.2231 −03
60.00	0.9936 −10	0.7164 −07	0.1387 −02	0.3615 +07	125.682	0.9550	0.3782	4200	5.992	701.0	0.4606 −06	0.2157 −03
61.00	0.8852 −10	0.6596 −07	0.1342 −02	0.3926 +07	125.760	0.9393	0.3782	4341	5.992	724.5	0.4241 −06	0.2087 −03
62.00	0.7900 −10	0.6082 −07	0.1299 −02	0.4258 +07	125.836	0.9242	0.3782	4485	5.992	748.4	0.3911 −06	0.2020 −03
63.00	0.7064 −10	0.5615 −07	0.1258 −02	0.4612 +07	125.909	0.9095	0.3782	4630	5.992	772.7	0.3611 −06	0.1957 −03
64.00	0.6328 −10	0.5190 −07	0.1219 −02	0.4989 +07	125.980	0.8953	0.3782	4779	5.993	797.4	0.3338 −06	0.1896 −03
65.00	0.5678 −10	0.4803 −07	0.1182 −02	0.5391 +07	126.049	0.8815	0.3782	4929	5.993	822.5	0.3089 −06	0.1838 −03
66.00	0.5103 −10	0.4451 −07	0.1147 −02	0.5818 +07	126.115	0.8681	0.3782	5082	5.993	848.0	0.2862 −06	0.1783 −03
67.00	0.4594 −10	0.4129 −07	0.1113 −02	0.6272 +07	126.180	0.8552	0.3782	5237	5.993	873.8	0.2655 −06	0.1730 −03
68.00	0.4142 −10	0.3834 −07	0.1080 −02	0.6753 +07	126.243	0.8426	0.3782	5395	5.994	900.1	0.2466 −06	0.1679 −03
69.00	0.3740 −10	0.3565 −07	0.1049 −02	0.7264 +07	126.304	0.8304	0.3782	5554	5.994	926.7	0.2293 −06	0.1631 −03
70.00	0.3382 −10	0.3318 −07	0.1019 −02	0.7805 +07	126.363	0.8185	0.3782	5717	5.994	953.7	0.2134 −06	0.1585 −03
71.00	0.3062 −10	0.3091 −07	0.9909 −03	0.8378 +07	126.421	0.8070	0.3782	5881	5.994	981.2	0.1988 −06	0.1541 −03
72.00	0.2777 −10	0.2882 −07	0.9636 −03	0.8984 +07	126.477	0.7958	0.3782	6048	5.994	1009	0.1854 −06	0.1498 −03
73.00	0.2522 −10	0.2690 −07	0.9374 −03	0.9625 +07	126.531	0.7849	0.3781	6217	5.994	1037	0.1730 −06	0.1457 −03
74.00	0.2293 −10	0.2513 −07	0.9122 −03	0.1030 +08	126.584	0.7743	0.3781	6389	5.995	1066	0.1617 −06	0.1418 −03
75.00	0.2087 −10	0.2350 −07	0.8881 −03	0.1102 +08	126.636	0.7640	0.3781	6562	5.995	1095	0.1512 −06	0.1381 −03
76.00	0.1903 −10	0.2200 −07	0.8649 −03	0.1177 +08	126.686	0.7539	0.3781	6739	5.995	1124	0.1415 −06	0.1345 −03
77.00	0.1736 −10	0.2061 −07	0.8426 −03	0.1256 +08	126.735	0.7441	0.3781	6917	5.995	1154	0.1326 −06	0.1310 −03
78.00	0.1587 −10	0.1932 −07	0.8211 −03	0.1340 +08	126.782	0.7346	0.3781	7098	5.995	1184	0.1243 −06	0.1276 −03
79.00	0.1451 −10	0.1813 −07	0.8005 −03	0.1428 +08	126.829	0.7253	0.3781	7281	5.995	1214	0.1166 −06	0.1244 −03
80.00	0.1329 −10	0.1703 −07	0.7806 −03	0.1521 +08	126.874	0.7162	0.3781	7467	5.995	1245	0.1095 −06	0.1213 −03
81.00	0.1218 −10	0.1600 −07	0.7615 −03	0.1618 +08	126.918	0.7074	0.3781	7654	5.995	1277	0.1029 −06	0.1184 −03
82.00	0.1118 −10	0.1505 −07	0.7430 −03	0.1720 +08	126.961	0.6987	0.3781	7845	5.996	1308	0.9682 −07	0.1155 −03
83.00	0.1027 −10	0.1417 −07	0.7253 −03	0.1828 +08	127.003	0.6903	0.3781	8037	5.996	1340	0.9113 −07	0.1127 −03
84.00	0.9448 −11	0.1334 −07	0.7081 −03	0.1940 +08	127.045	0.6821	0.3781	8232	5.996	1373	0.8584 −07	0.1101 −03
85.00	0.8697 −11	0.1258 −07	0.6916 −03	0.2059 +08	127.085	0.6741	0.3781	8429	5.996	1406	0.8091 −07	0.1075 −03
86.00	0.8014 −11	0.1186 −07	0.6756 −03	0.2182 +08	127.124	0.6662	0.3781	8629	5.996	1439	0.7632 −07	0.1050 −03
87.00	0.7392 −11	0.1120 −07	0.6601 −03	0.2312 +08	127.162	0.6586	0.3781	8830	5.996	1473	0.7204 −07	0.1026 −03
88.00	0.6824 −11	0.1058 −07	0.6452 −03	0.2448 +08	127.199	0.6511	0.3781	9035	5.996	1507	0.6804 −07	0.1003 −03
89.00	0.6305 −11	0.9995 −08	0.6308 −03	0.2590 +08	127.236	0.6438	0.3781	9241	5.996	1541	0.6431 −07	0.9805 −04
90.00	0.5831 −11	0.9452 −08	0.6169 −03	0.2739 +08	127.272	0.6366	0.3781	9450	5.996	1576	0.6082 −07	0.9588 −04
91.00	0.5397 −11	0.8944 −08	0.6034 −03	0.2894 +08	127.307	0.6296	0.3781	9661	5.996	1611	0.5755 −07	0.9378 −04
92.00	0.5000 −11	0.8469 −08	0.5904 −03	0.3057 +08	127.341	0.6228	0.3781	9875	5.996	1647	0.5449 −07	0.9176 −04
93.00	0.4636 −11	0.8024 −08	0.5778 −03	0.3226 +08	127.374	0.6161	0.3781	10090	5.997	1683	0.5163 −07	0.8979 −04
94.00	0.4302 −11	0.7606 −08	0.5655 −03	0.3404 +08	127.407	0.6095	0.3781	10310	5.997	1719	0.4894 −07	0.8789 −04

Table 2.14: Isentropic and normal shock properties for air, $\gamma = 1.4$ (continued).

$\begin{array}{c}M\\ or\\ M_1\end{array}$	$\frac{P}{P_t}$	$\frac{\rho}{\rho_t}$	$\frac{T}{T_t}$	$\frac{A}{A^*}$	ν	μ	M_2	$\frac{P_2}{P_1}$	$\frac{\rho_2}{\rho_1}$	$\frac{T_2}{T_1}$	$\frac{P_{t_2}}{P_{t_1}}$	$\frac{P_1}{P_{t_2}}$
95.00	0.3995 −11	0.7214 −08	0.5537 −03	0.3588 +08	127.439	0.6031	0.3781	10530	5.997	1756	0.4642 −07	0.8605 −04
96.00	0.3712 −11	0.6846 −08	0.5422 −03	0.3781 +08	127.471	0.5968	0.3781	10750	5.997	1793	0.4405 −07	0.8427 −04
97.00	0.3453 −11	0.6501 −08	0.5311 −03	0.3982 +08	127.501	0.5907	0.3781	10980	5.997	1830	0.4183 −07	0.8254 −04
98.00	0.3214 −11	0.6176 −08	0.5203 −03	0.4191 +08	127.531	0.5847	0.3781	11200	5.997	1868	0.3974 −07	0.8086 −04
99.00	0.2993 −11	0.5871 −08	0.5099 −03	0.4410 +08	127.561	0.5788	0.3781	11430	5.997	1907	0.3778 −07	0.7924 −04
100.00	0.2790 −11	0.5583 −08	0.4997 −03	0.4637 +08	127.590	0.5730	0.3781	11670	5.997	1945	0.3593 −07	0.7766 −04

Figure 2.5: Variation of shock angle with flow deflection angle for various upstream Mach numbers. Perfect gas, $\gamma = 1.4$.

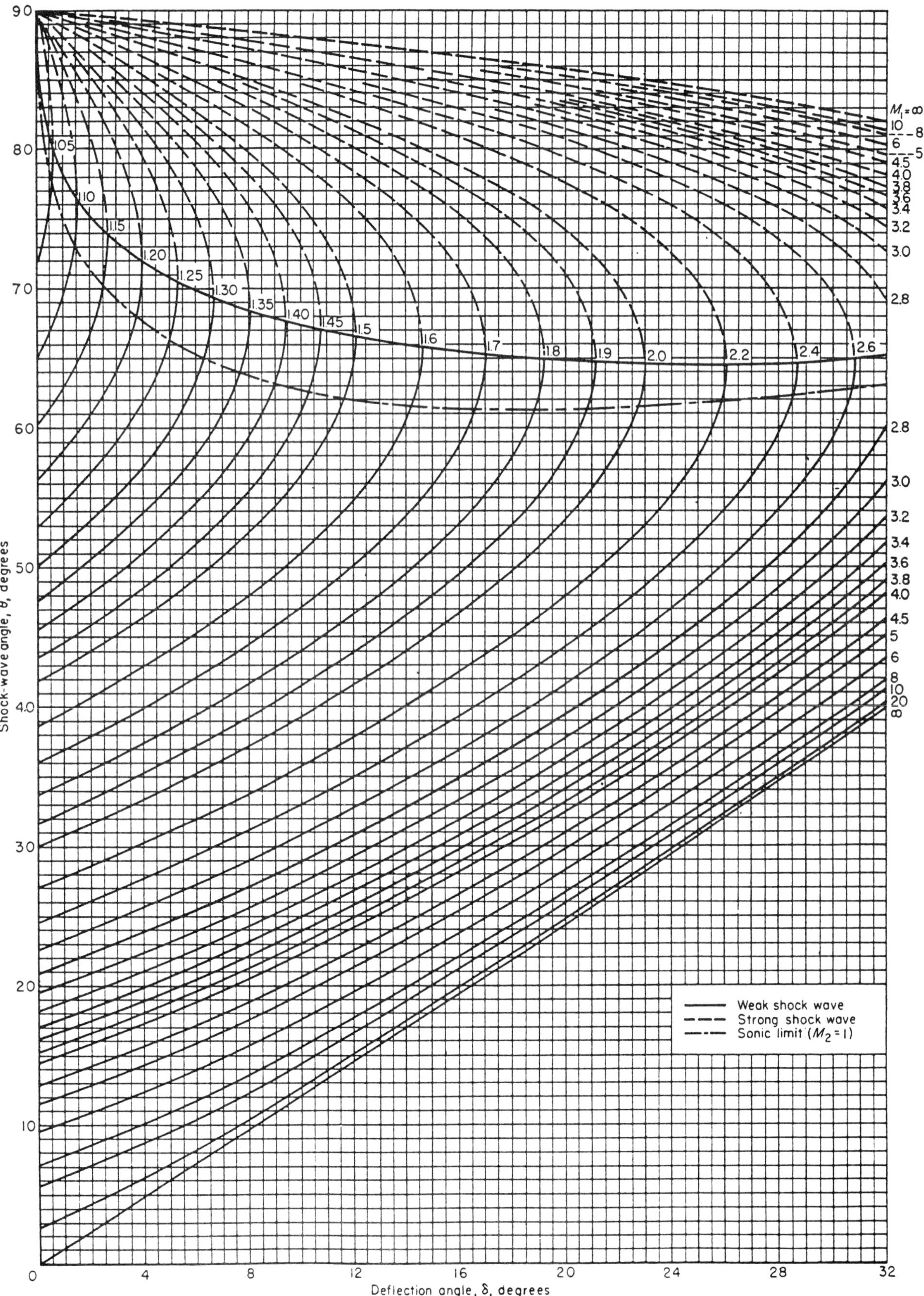

Figure 2.5: Concluded.

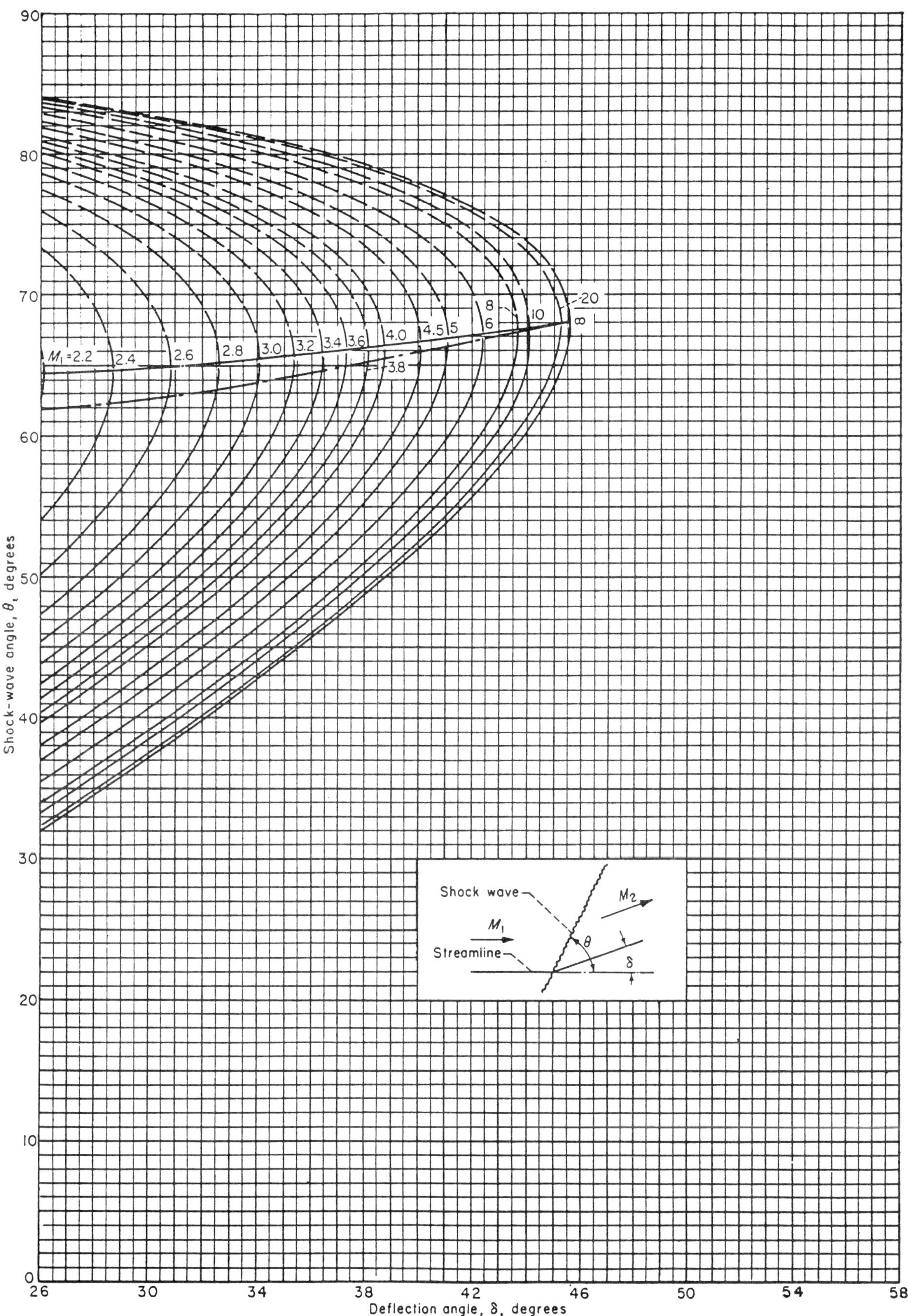

Figure 2.6: Variation of pressure coefficient across shock waves with flow deflection angle for various upstream Mach numbers. Perfect gas, $\gamma = 1.4$.

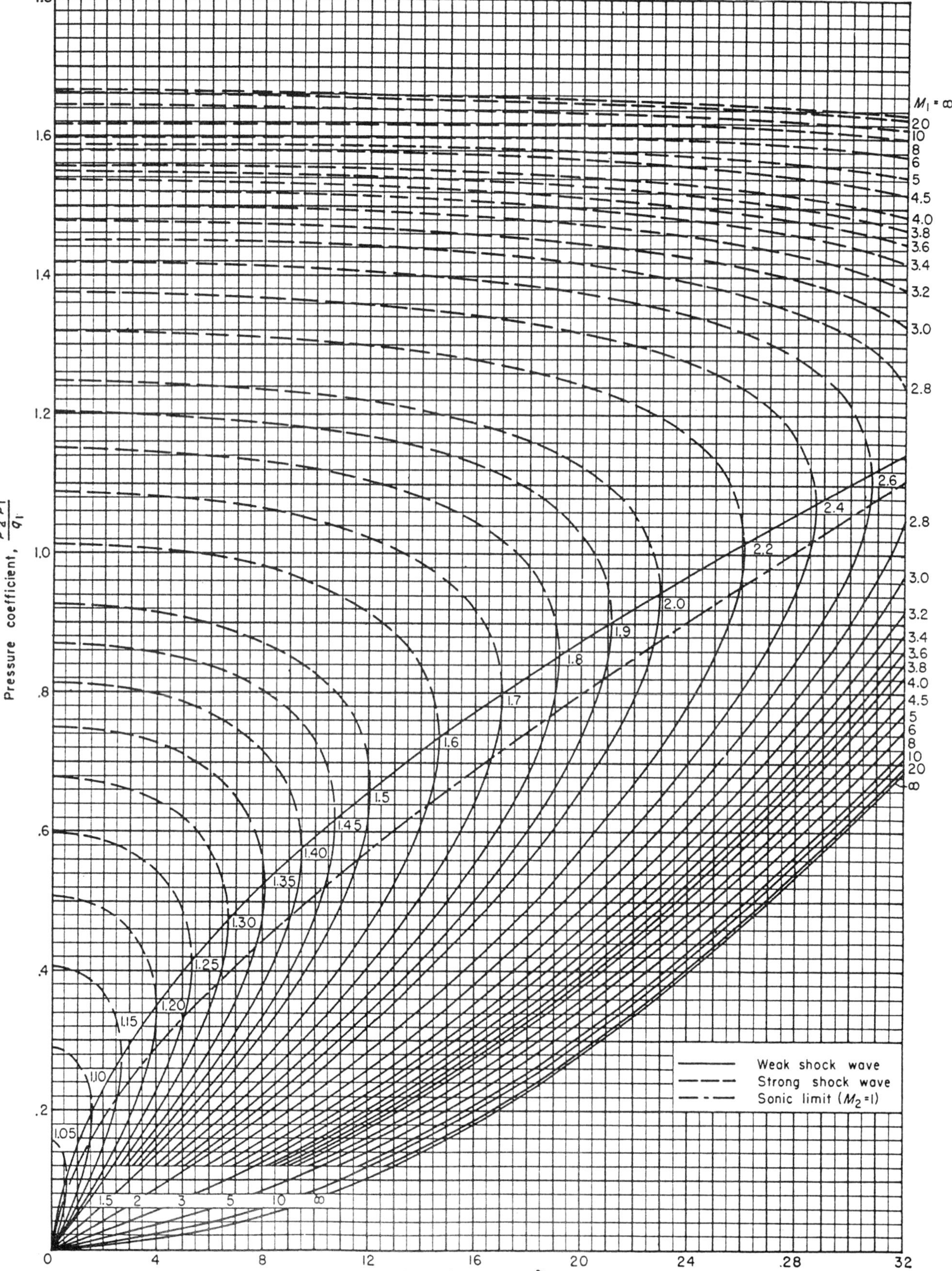

Figure 2.6: Concluded.

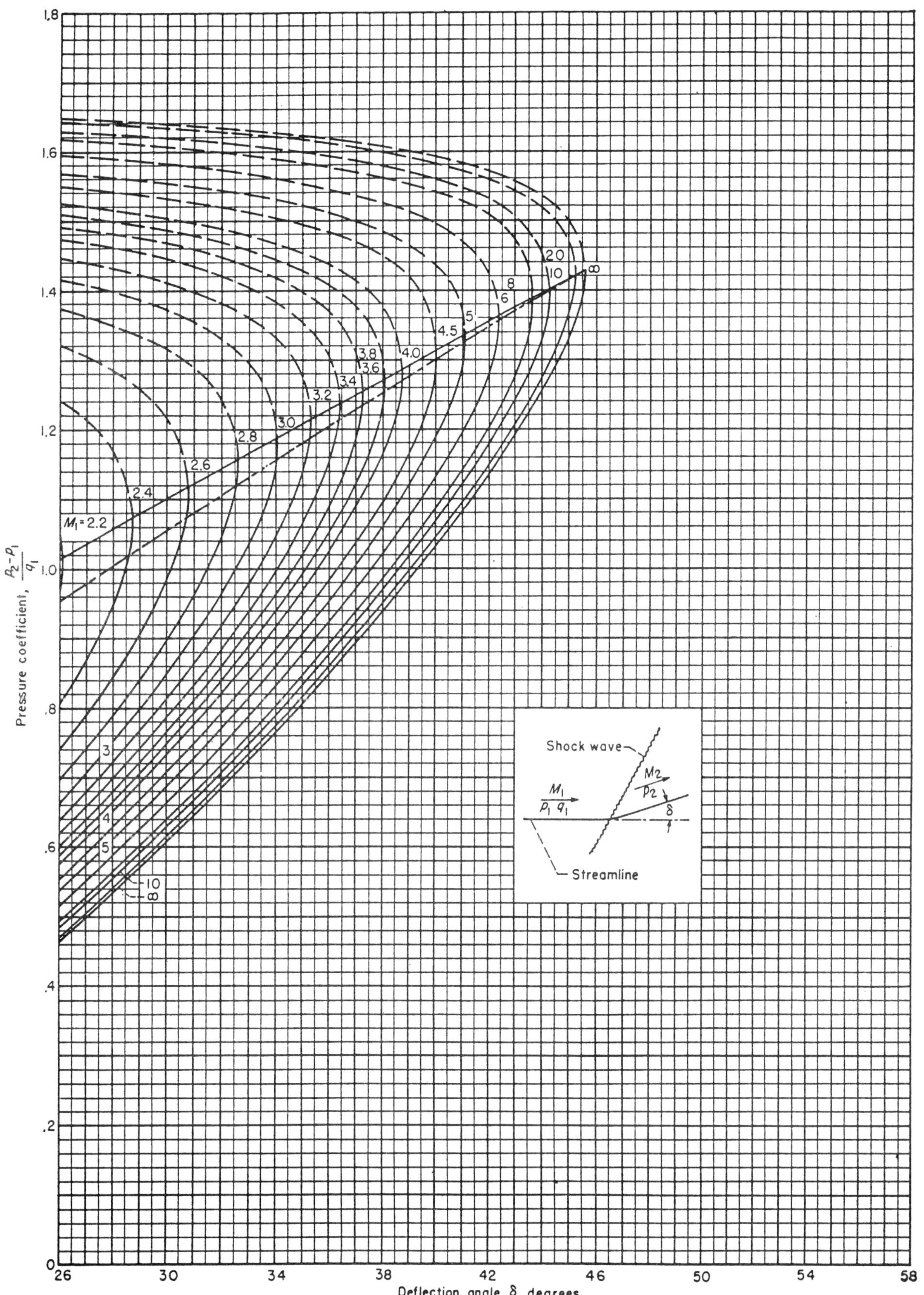

Pressure coefficient, $\frac{p_2 - p_1}{q_1}$

Deflection angle, δ, degrees

Figure 2.7: Variation of Mach number downstream of a shock wave with flow deflection angle for various upstream Mach numbers. Perfect gas, $\gamma = 1.4$.

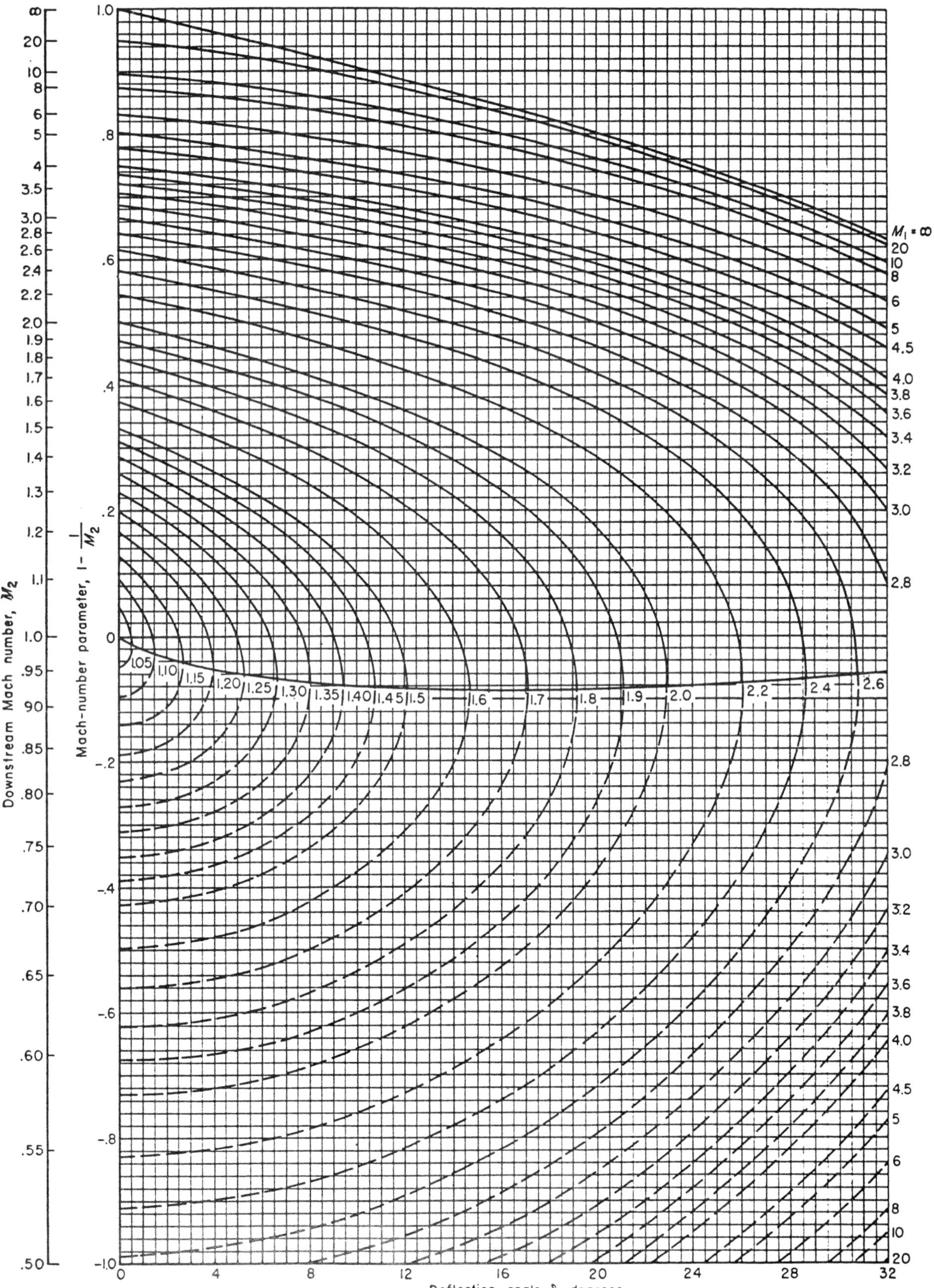

Figure 2.7: Concluded.

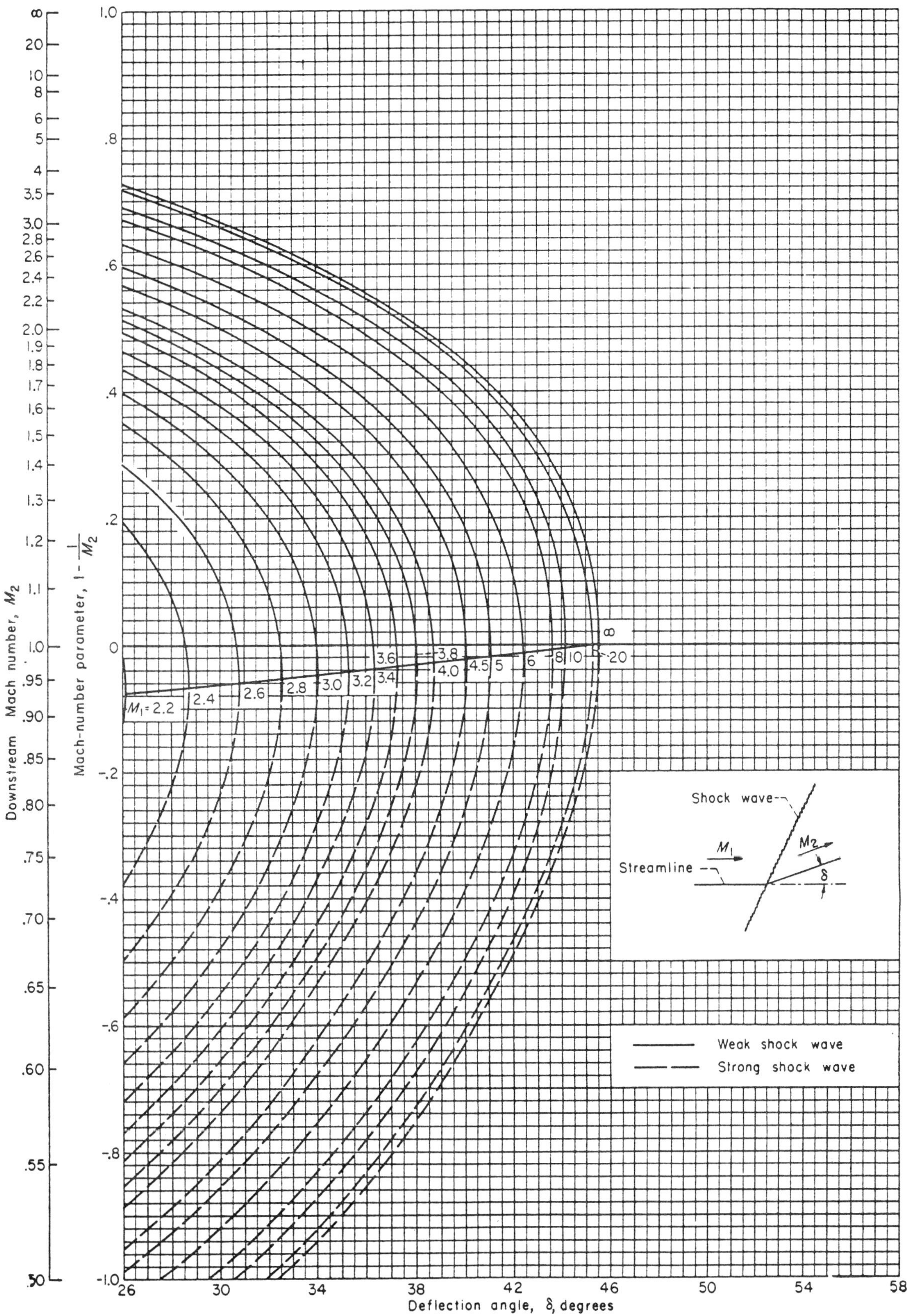

Figure 2.8: Variation of shock wave angle with cone semivertex angle for various upstream Mach numbers. Perfect gas, $\gamma = 1.4$.

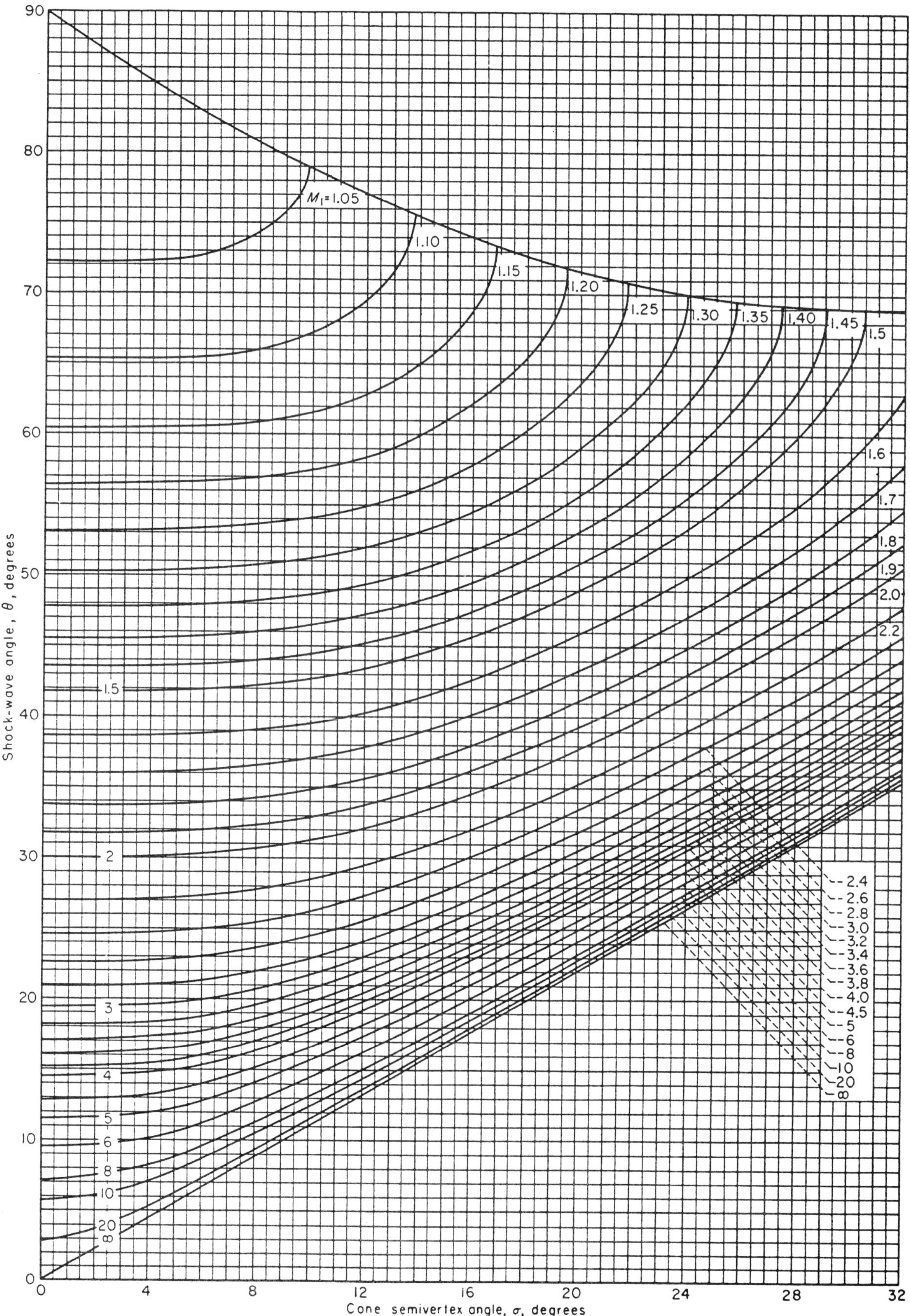

Figure 2.8: Concluded.

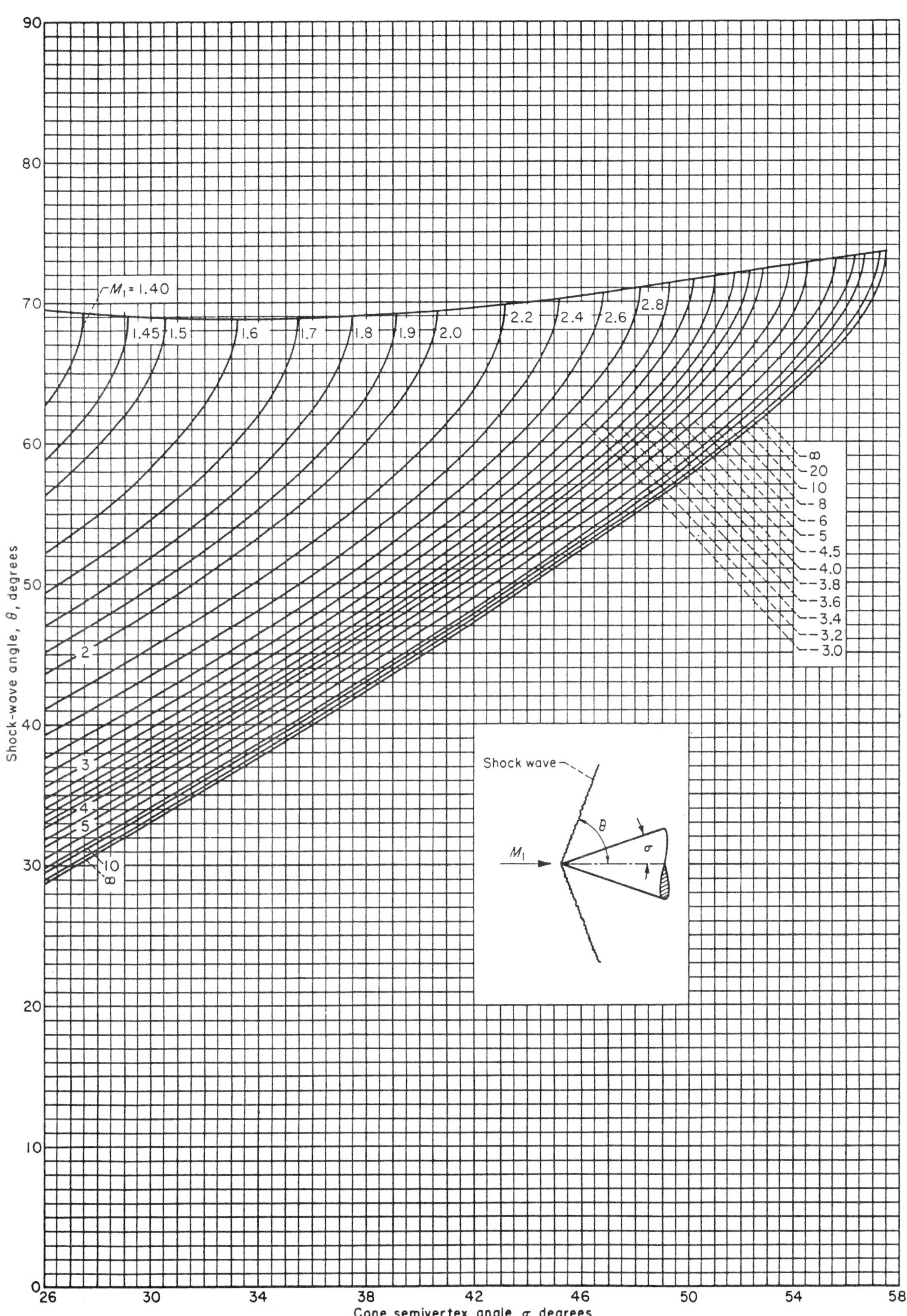

Figure 2.9: Variation of surface pressure coefficient with cone semivertex angle for various upstream Mach numbers. Perfect gas, $\gamma = 1.4$.

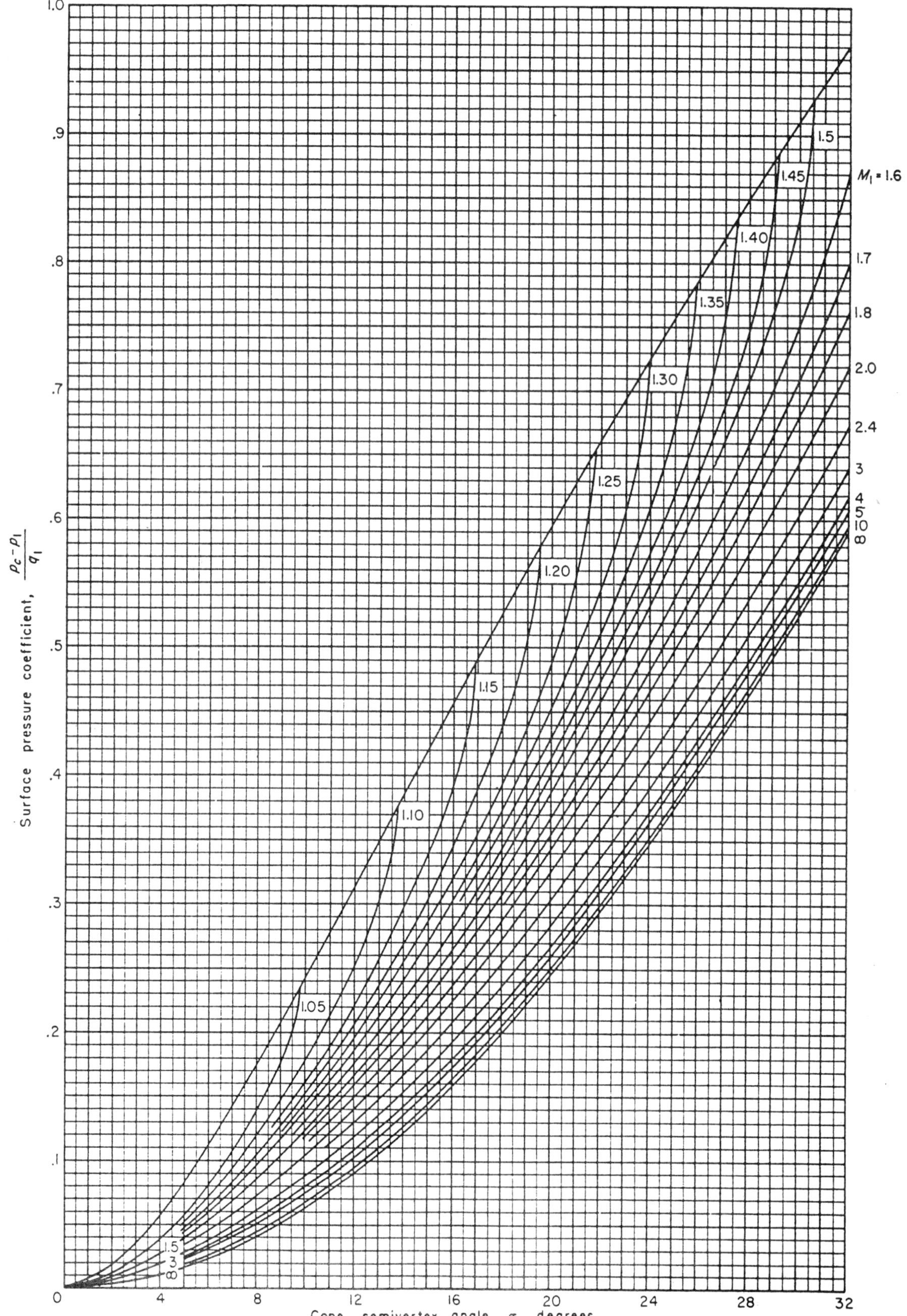

Figure 2.9: Concluded.

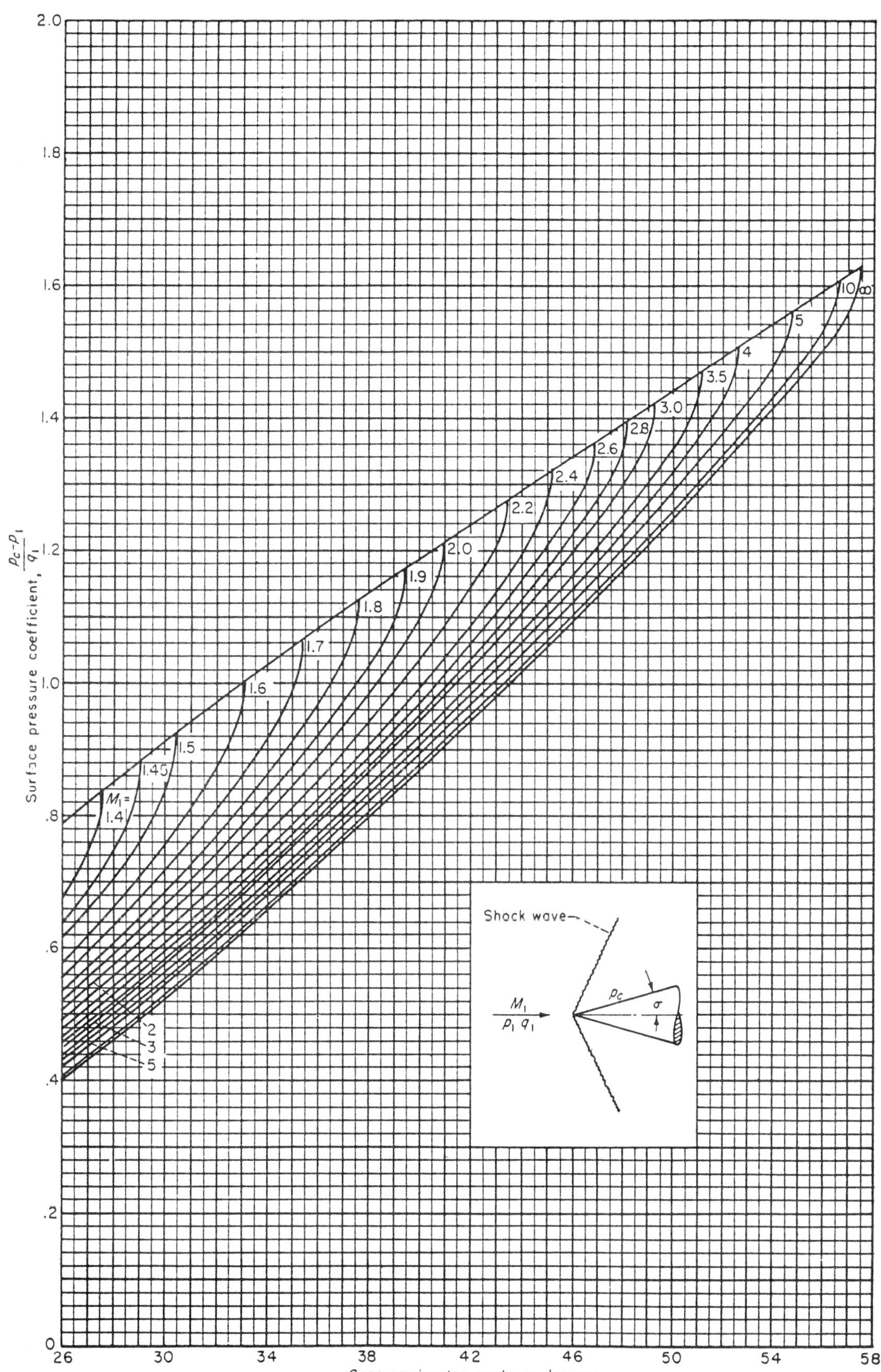

Surface pressure coefficient, $\frac{p_c - p_1}{q_1}$

Cone semivertex angle, σ, degrees

Figure 2.10: Variation of Mach number at the surface of a cone with cone semivertex angle for various upstream Mach numbers. Perfect gas, $\gamma = 1.4$.

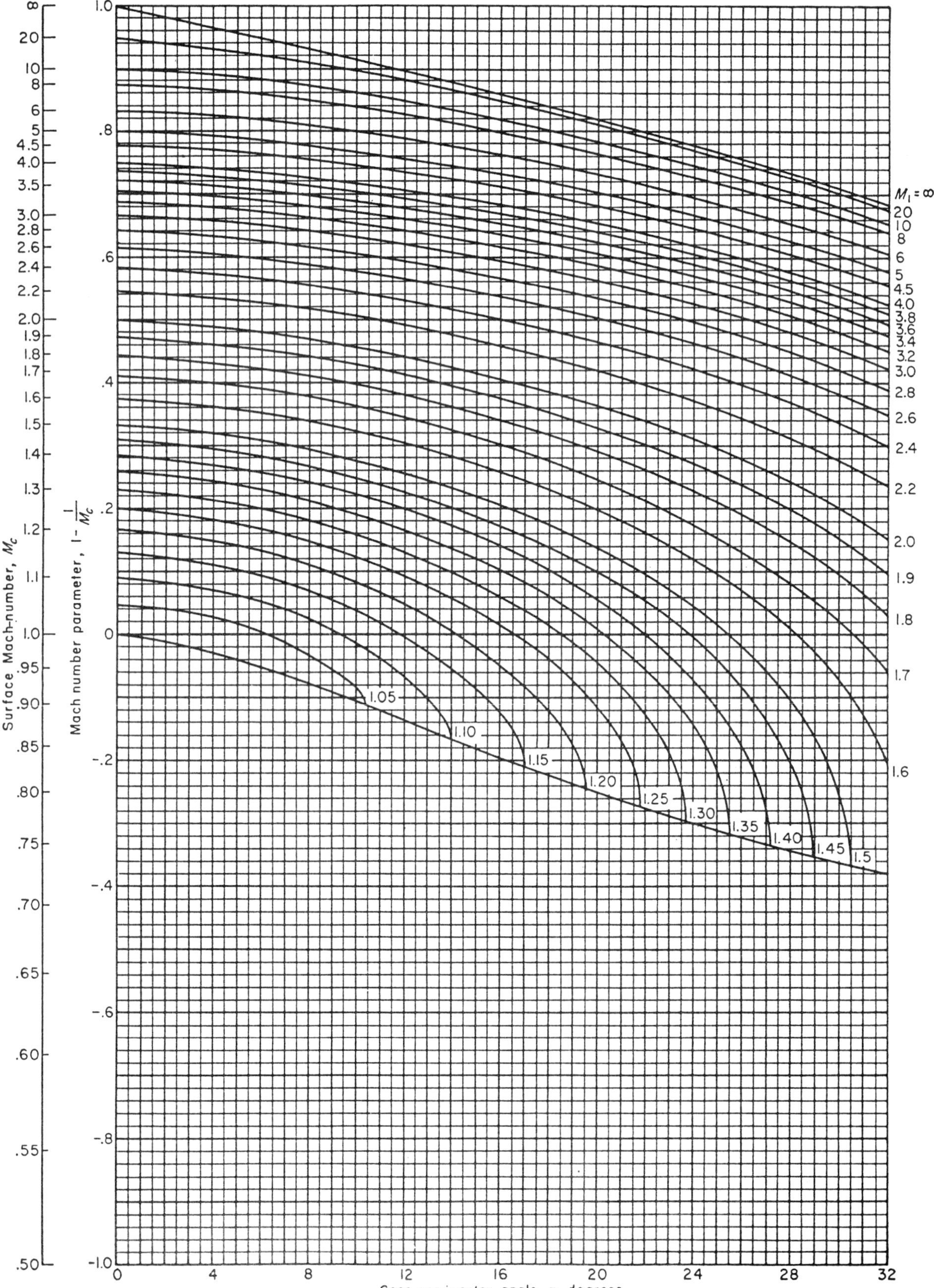

Figure 2.10: Concluded.

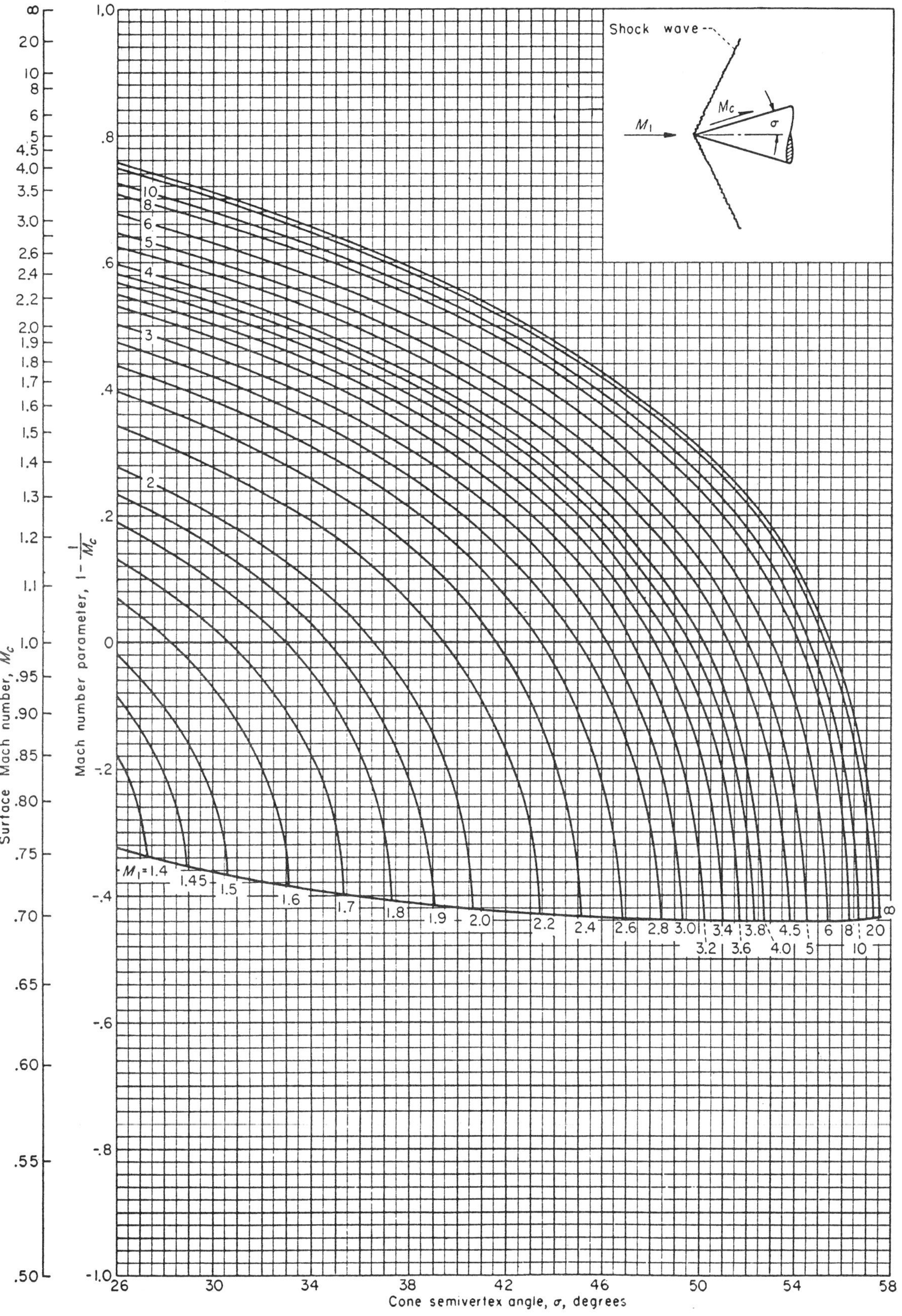

2.19 References

[2.1] Shapiro, A. H., "The Dynamics and Thermodynamics of Compressible Fluid Flow," Vol. 1, The Ronald Press Co., 1953.

[2.2] Shapiro, A. H., and Kline, S. J., "On the Thickness of Normal Shock Waves in Air," Eighth International Congress on Theoretical and Applied Mechanics, 1952.

[2.3] Staff, "Equations, Tables, and Charts for Compressible Flow," Report 1135, NACA, 1953.

Chapter 3
Boundary Layer Equations

3.1 Introductory Remarks

In many high Reynolds number flows, the effect of viscous forces is dominant in the regions adjacent to the surface. This region is known as the boundary layer. From elementary fluid mechanics, it is well known that a boundary layer usually starts with well-behaved streamlines, and the fluid mixing is at a microscopic level. This type of boundary layer flow is identified as a laminar boundary layer. Due to conditions imposed by the geometry and flowfield, such as surface roughness, surface temperature, surface injection, and pressure gradient, the mixing of the fluid increases and takes place at the macroscopic level; streamlines are no longer well-behaved. This type of boundary layer is known as a turbulent boundary layer. There is a transitional region between laminar and turbulent boundary layers known as the *transition region*. The various boundary layer regions are shown in Figure 3.1.

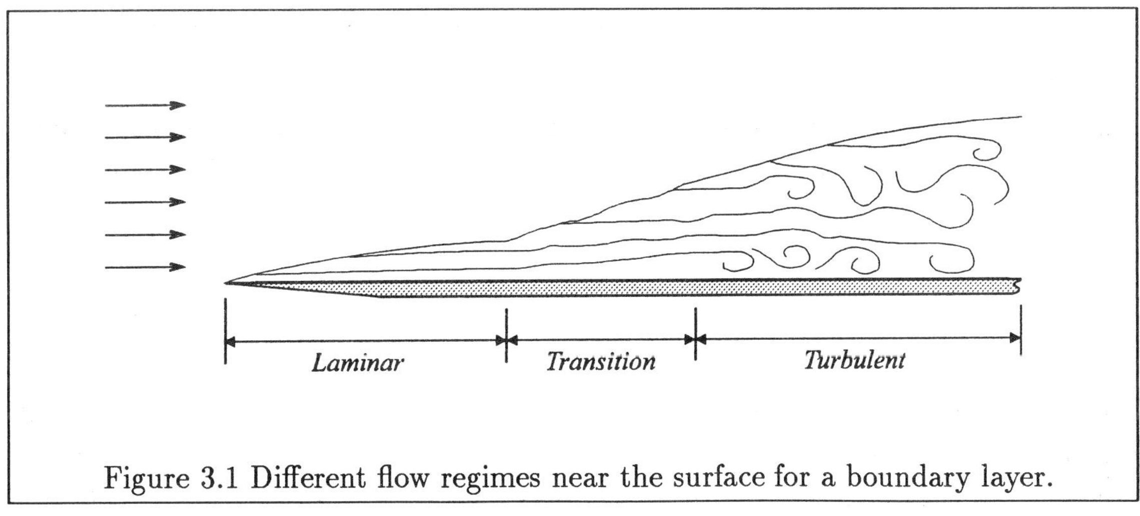

Figure 3.1 Different flow regimes near the surface for a boundary layer.

As a result of the heavy mixing of the fluid in a turbulent boundary layer and the associated large momentum flux, the velocity profile becomes fuller in a turbulent boundary layer as compared to a laminar boundary layer; that is, the velocity gradient near the wall for a turbulent boundary layer is larger than the velocity gradient of a laminar boundary layer. This results in larger values of the wall shear stress for turbulent flows, as compared to the corresponding laminar flows. Typical velocity profiles for laminar and turbulent boundary layers are shown in Figure 3.2.

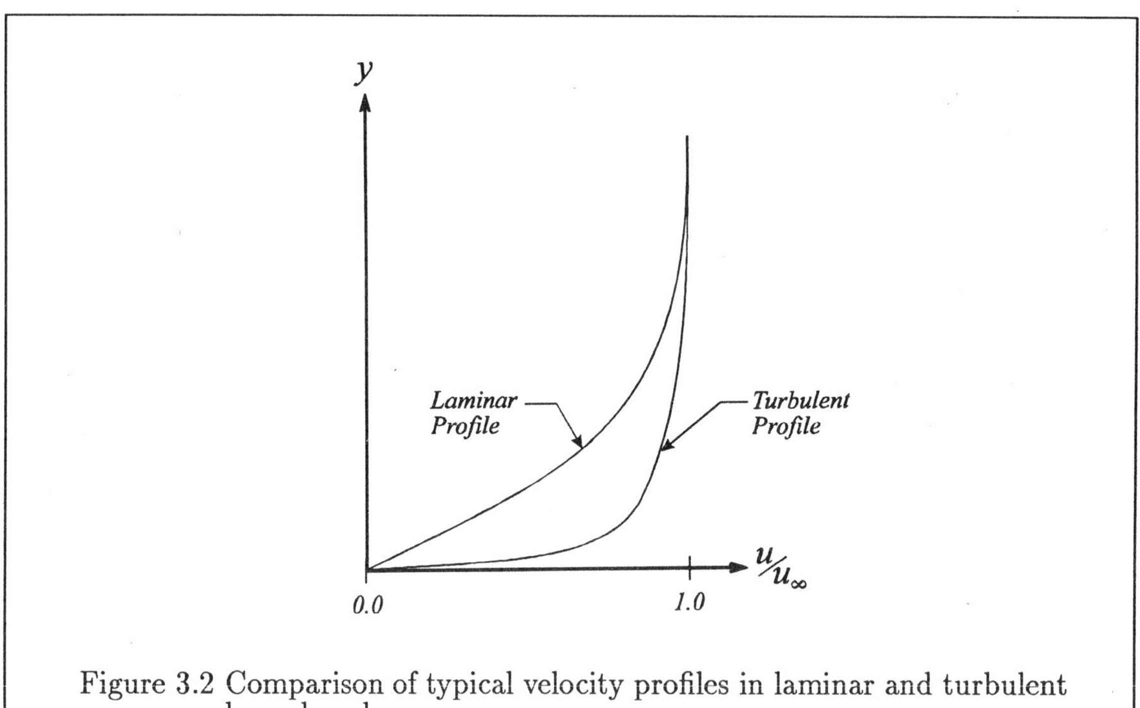

Figure 3.2 Comparison of typical velocity profiles in laminar and turbulent boundary layers.

Boundary layers can also be classified as either velocity boundary layers or thermal boundary layers. For most problems, velocity boundary layer thicknesses and thermal boundary layer thicknesses are not identical. Typical velocity and thermal boundary layers are shown in Figure 3.3.

Recall that a nondimensional parameter was previously defined as the Prandtl number, and expressed as

$$Pr = \frac{\nu}{\alpha} = \frac{\mu c_p}{k}$$

This parameter represents the ratio of momentum transfer in the flow to that of heat transfer. It may therefore be used to obtain a measure of the relative velocity and thermal boundary layer thicknesses. This relationship is illustrated in Figure 3.4.

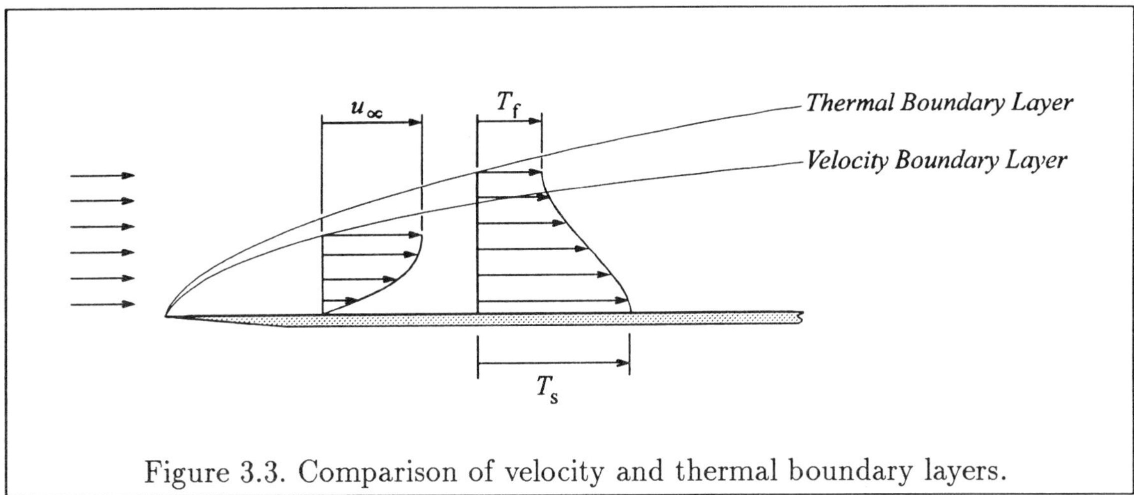

Figure 3.3. Comparison of velocity and thermal boundary layers.

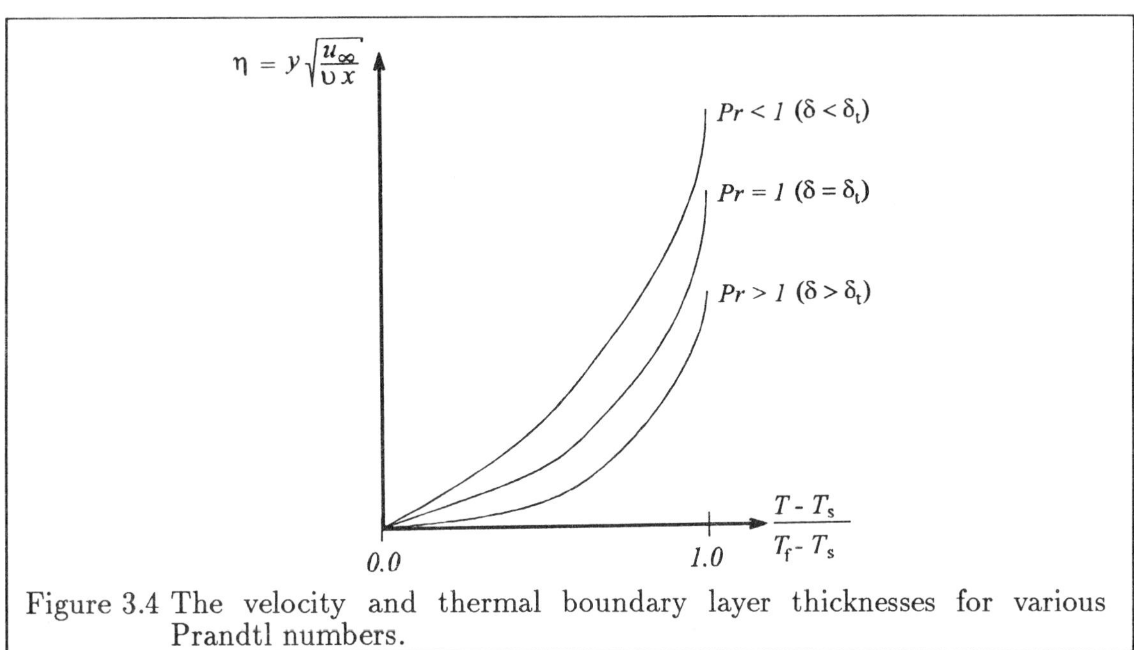

Figure 3.4 The velocity and thermal boundary layer thicknesses for various Prandtl numbers.

If δ and δ_t are used to denote the velocity and thermal boundary layer thicknesses, respectively, then,

$$
\begin{array}{llll}
\text{for} & Pr < 1 & \rightarrow & \delta_t > \delta \\
& Pr = 1 & \rightarrow & \delta_t = \delta \\
\text{and for} & Pr > 1 & \rightarrow & \delta_t < \delta
\end{array}
$$

Generally speaking, the equations of fluid motion may be reduced to the laminar boundary equations for high Reynolds number flows. Subsequently, the laminar boundary layer equations may be modified by introduction of fluctuating terms to obtain equations for turbulent boundary layers.

3.2   Laminar Boundary Layer Equations

The reduction of the equations of fluid motion is accomplished by an order of magnitude analysis. As a result, for an incompressible, steady, two-dimensional flow in Cartesian coordinate system, the equations of laminar boundary layer are obtained as:

Continuity

$$\frac{\partial u}{\partial x} + \frac{\partial v}{\partial y} = 0 \tag{3.1}$$

x-Component of momentum equation

$$u\frac{\partial u}{\partial x} + v\frac{\partial u}{\partial y} = -\frac{1}{\rho}\frac{\partial p}{\partial x} + \nu\frac{\partial^2 u}{\partial y^2} \tag{3.2}$$

y-Component of momentum equation

$$\frac{\partial p}{\partial y} = 0 \tag{3.3}$$

Energy

$$u\frac{\partial T}{\partial x} + v\frac{\partial T}{\partial y} = \frac{k}{\rho c_p}\left(\frac{\partial^2 T}{\partial y^2}\right) + \frac{\mu}{\rho c_p}\left(\frac{\partial u}{\partial y}\right)^2 \tag{3.4}$$

Equation (3.3) indicates that the pressure is constant across the boundary layer. Consequently, the streamwise pressure gradient is computed from the inviscid solution and imposed on the boundary layer which reduces the unknowns to three, that is, u, v, and T. Note that under the stated assumptions, the energy equation is decoupled from the continuity and momentum equations. Therefore, Equations (3.1) and (3.2) are solved initially for the velocity components, and subsequently the energy equation is solved for the temperature distribution.

The boundary layer equations are parabolic and, therefore, an initial and two boundary conditions must be provided. The boundary condition at the surface is the no-slip condition and the outer boundary condition is imposed at the edge of the boundary layer by an inviscid solution. It is customary to define the edge of the boundary layer at a location where the velocity is about 99% of the free stream velocity. Problems involving injection or suction at the surface require suitable modifications of these boundary conditions.

For applications where temperature (and pressure) dependency of the viscosity and thermal conductivity must be included, the energy equation is coupled with the continuity and momentum equations and, therefore, the boundary layer equations must all be solved simultaneously. General forms of the laminar boundary layer equations for incompressible and compressible flows are given in Tables 3.1 and 3.2, respectively.

3.2.1 Axisymmetric Flows

The boundary layer equations for axisymmetric flows are provided in Tables 3.1 and 3.2. These equations are expressed in terms of the x,y coordinate system where x is defined along the body surface, and y is perpendicular to the body surface. Typically, the cross-section radius of axisymmetric body is denoted by r_0, whereas r represents the radial coordinate as illustrated in Figure 3.5. One can easily establish the following relation

$$r(x,y) = r_0(x) + y \cos \beta \qquad (3.5)$$

Now, for axisymmetric flows where the body radius r_0 is much larger than the boundary layer thickness, the variation of r_0 through the boundary layer can be neglected. Therefore,

$$\frac{\partial r_0}{\partial y} \approx 0$$

and, since $r_0 \geq \delta$, according to (3.5), then $r \approx r_0$. Note that some of the axisymmetric boundary layer equations presented in Tables 3.1 and 3.2 are expressed in terms of r_0, and, therefore, the assumption stated above is imposed. Also note that, for many applications where $r_0 \geq \delta$, r may be replaced by r_0 in the axisymmetric boundary layer equations.

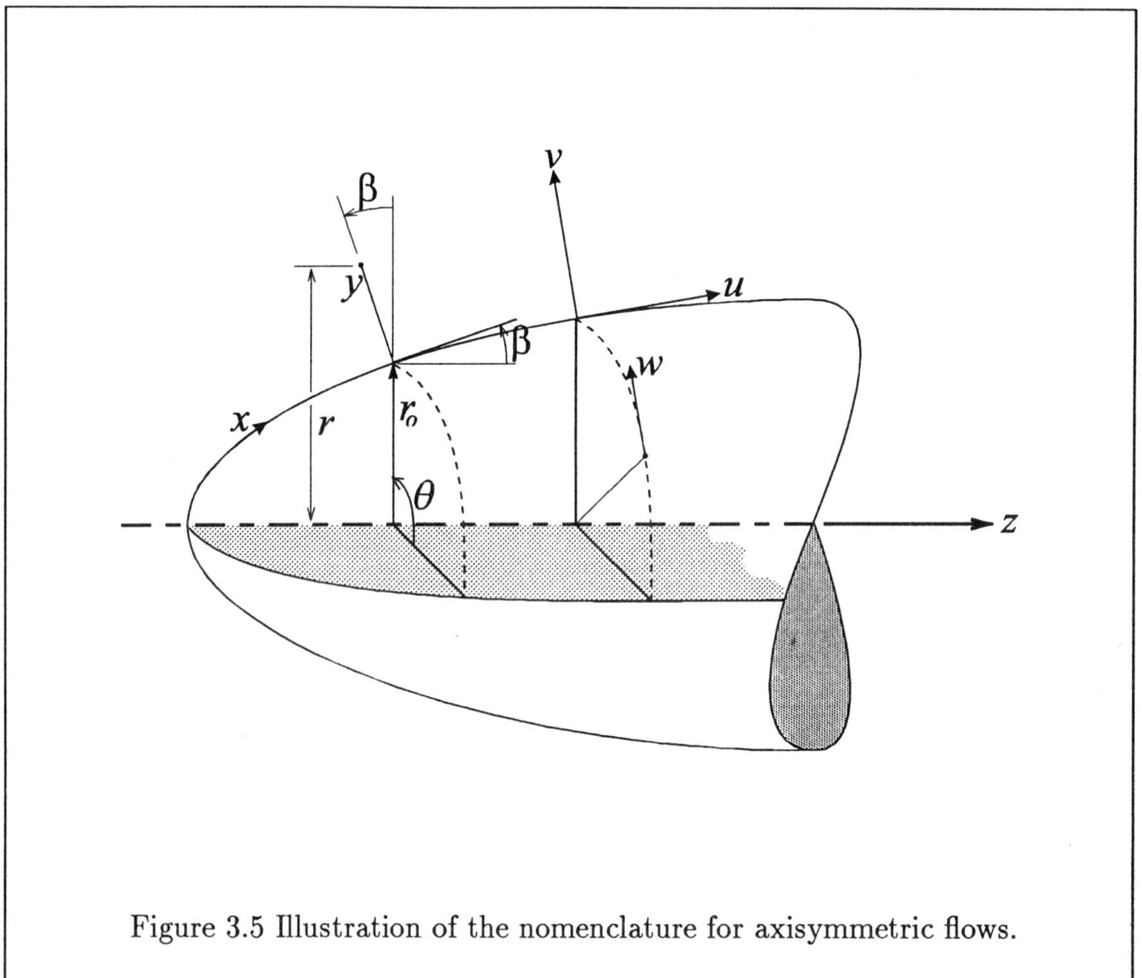

Figure 3.5 Illustration of the nomenclature for axisymmetric flows.

Table 3.1: Laminar boundary layer equations for an incompressible constant property flow

Two-dimensional Cartesian Coordinates

Mass:
$$\frac{\partial u}{\partial x} + \frac{\partial v}{\partial y} = 0 \tag{3.6}$$

x-Momentum:
$$\frac{\partial u}{\partial t} + u\frac{\partial u}{\partial x} + v\frac{\partial u}{\partial y} = -\frac{1}{\rho}\frac{\partial p}{\partial x} + \nu\frac{\partial^2 u}{\partial y^2} \tag{3.7}$$

y-Momentum:
$$\frac{\partial p}{\partial y} = 0 \tag{3.8}$$

Energy:
$$\frac{\partial T}{\partial t} + u\frac{\partial T}{\partial x} + v\frac{\partial T}{\partial y} = \frac{k}{\rho c_p}\frac{\partial^2 T}{\partial y^2} + \frac{\mu}{\rho c_p}\left(\frac{\partial u}{\partial y}\right)^2 \tag{3.9}$$

Three-dimensional Cartesian Coordinates

Mass:
$$\frac{\partial u}{\partial x} + \frac{\partial v}{\partial y} + \frac{\partial w}{\partial z} = 0 \tag{3.10}$$

x-Momentum:
$$\frac{\partial u}{\partial t} + u\frac{\partial u}{\partial x} + v\frac{\partial u}{\partial y} + w\frac{\partial u}{\partial z} = -\frac{1}{\rho}\frac{\partial p}{\partial x} + \nu\frac{\partial^2 u}{\partial y^2} \tag{3.11}$$

y-Momentum:
$$\frac{\partial p}{\partial y} = 0 \tag{3.12}$$

z-Momentum:
$$\frac{\partial w}{\partial t} + u\frac{\partial w}{\partial x} + v\frac{\partial w}{\partial y} + w\frac{\partial w}{\partial y} = -\frac{1}{\rho}\frac{\partial p}{\partial z} + \nu\frac{\partial^2 w}{\partial y^2} \tag{3.13}$$

Energy: $\dfrac{\partial T}{\partial t} + u\dfrac{\partial T}{\partial x} + v\dfrac{\partial T}{\partial y} + w\dfrac{\partial T}{\partial z} = \dfrac{k}{\rho c_p}\left(\dfrac{\partial^2 T}{\partial y^2}\right)$

$$+\dfrac{\mu}{\rho c_p}\left[\left(\dfrac{\partial u}{\partial y}\right)^2 + \left(\dfrac{\partial w}{\partial y}\right)^2\right] \tag{3.14}$$

Rotationally Symmetric Boundary Layers

Mass: $\quad \dfrac{\partial}{\partial x}(r_0 u) + r_0\dfrac{\partial v}{\partial y} = 0 \tag{3.15}$

x-Momentum: $\dfrac{\partial u}{\partial t} + u\dfrac{\partial u}{\partial x} + v\dfrac{\partial u}{\partial y} - \dfrac{w^2}{r_0}\dfrac{dr_0}{dx} = -\dfrac{1}{\rho}\dfrac{\partial p}{\partial x} + \nu\dfrac{\partial^2 u}{\partial y^2} \tag{3.16}$

y-Momentum: $\dfrac{\partial p}{\partial y} = 0 \tag{3.17}$

θ-Momentum: $\dfrac{\partial w}{\partial t} + u\dfrac{\partial w}{\partial x} + v\dfrac{\partial w}{\partial y} + \dfrac{uw}{r_0}\dfrac{dr_0}{dx} = \nu\dfrac{\partial^2 w}{\partial y^2} \tag{3.18}$

Axisymmetric Boundary Layers ($w = 0$)

Mass: $\quad \dfrac{\partial}{\partial x}(r_o u) + r_0\dfrac{\partial v}{\partial y} = 0 \tag{3.19}$

x-Momentum: $\dfrac{\partial u}{\partial t} + u\dfrac{\partial u}{\partial x} + v\dfrac{\partial u}{\partial y} = -\dfrac{1}{\rho}\dfrac{\partial p}{\partial x} + \nu\dfrac{\partial^2 u}{\partial y^2} \tag{3.20}$

y-Momentum: $\dfrac{\partial p}{\partial y} = 0 \tag{3.21}$

General Orthogonal Curvilinear Coordinate System

Mass:
$$\frac{1}{h_1 h_2}\left[\frac{\partial}{\partial x_1}(h_2 u_1) + \frac{\partial}{\partial x_2}(h_1 u_2) + \frac{\partial}{\partial x_3}(h_1 h_2 u_3)\right] = 0 \qquad (3.22)$$

x_1-Momentum:
$$\frac{du_1}{dt} + \frac{u_1 u_2}{h_1 h_2}\frac{\partial h_1}{\partial x_2} - \frac{u_2^2}{h_1 h_2}\frac{\partial h_2}{\partial x_1} = -\frac{1}{\rho h_1}\frac{\partial p}{\partial x_1} + \nu\frac{\partial^2 u_1}{\partial x_3^2} \qquad (3.23)$$

x_2-Momentum:
$$\frac{du_2}{dt} - \frac{u_1^2}{h_1 h_2}\frac{\partial h_1}{\partial x_2} + \frac{u_1 u_2}{h_1 h_2}\frac{\partial h_2}{\partial x_1} = -\frac{1}{\rho h_2}\frac{\partial p}{\partial x_2} + \nu\frac{\partial^2 u_2}{\partial x_3^2} \qquad (3.24)$$

x_3-Momentum:
$$\frac{\partial p}{\partial x_3} = 0 \qquad (3.25)$$

Energy:
$$\frac{dh_t}{dt} = \frac{1}{\rho}\frac{\partial p}{\partial t} + \alpha\frac{\partial^2}{\partial x_3^2}\left[h_t + \frac{(Pr - 1)}{2}(u_1^2 + u_2^2)\right] \qquad (3.26)$$

where
$$\frac{d}{dt} = \frac{\partial}{\partial t} + \frac{u_1}{h_1}\frac{\partial}{\partial x_1} + \frac{u_2}{h_2}\frac{\partial}{\partial x_2} + u_3\frac{\partial}{\partial x_3}$$

$$\alpha = \frac{k}{\rho c_p} \quad \text{and} \quad h_t = c_p T + \frac{1}{2}(u_1^2 + u_2^2) = \text{Total enthalpy}$$

Note: It is assumed that x_3 is everywhere normal to the surface, and therefore, $h_3 = 1$. This restriction was not applied to the formulations of the governing equations of fluid motion expressed in the general orthogonal system in Chapter One. The orthogonal coordinate system for the boundary layer equations above is shown in Figure 3.6.

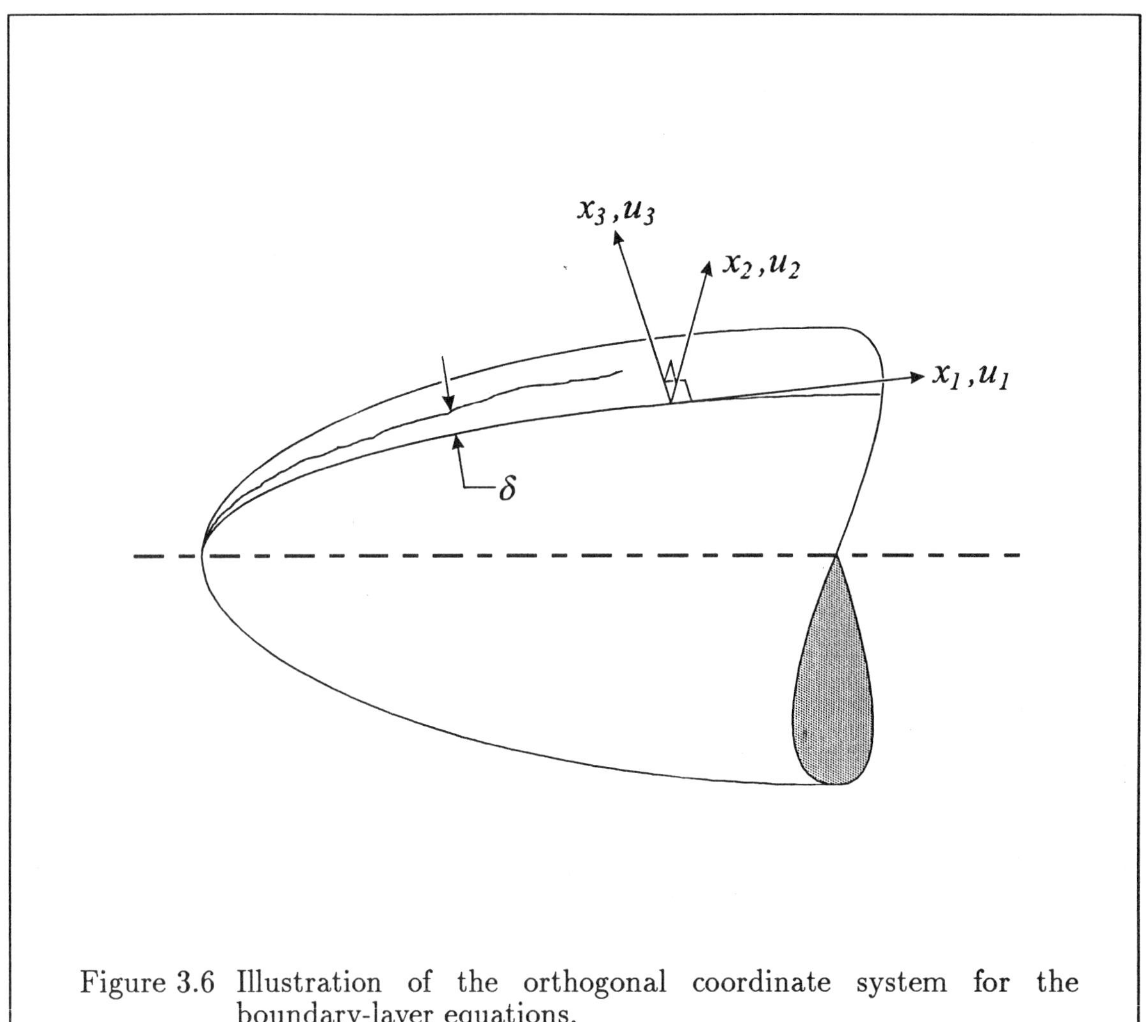

Figure 3.6 Illustration of the orthogonal coordinate system for the boundary-layer equations.

Table 3.2: Laminar boundary layer equations for a compressible flow

Two-dimensional Cartesian Coordinates

Mass: $\dfrac{\partial \rho}{\partial t} + \dfrac{\partial}{\partial x}(\rho u) + \dfrac{\partial}{\partial y}(\rho v) = 0$ \hfill (3.27)

x-Momentum: $\rho\left(\dfrac{\partial u}{\partial t} + u\dfrac{\partial u}{\partial x} + v\dfrac{\partial u}{\partial y}\right) = -\dfrac{\partial p}{\partial x} + \dfrac{\partial}{\partial y}\left(\mu\dfrac{\partial u}{\partial y}\right)$ \hfill (3.28)

y-Momentum: $\dfrac{\partial p}{\partial y} = 0$ \hfill (3.29)

Energy: $\rho\left(\dfrac{\partial h}{\partial t} + u\dfrac{\partial h}{\partial x} + v\dfrac{\partial h}{\partial y}\right) = \dfrac{\partial p}{\partial t} + u\dfrac{\partial p}{\partial x} + \dfrac{\partial}{\partial y}\left(k\dfrac{\partial T}{\partial y}\right) + \mu\left(\dfrac{\partial u}{\partial y}\right)^2$ \hfill (3.30)

The energy equation (3.30) may be expressed in terms of total enthalpy as

$$\rho\left(\frac{\partial h_t}{\partial t} + u\frac{\partial h_t}{\partial x} + v\frac{\partial h_t}{\partial y}\right) = \frac{\partial p}{\partial t}$$

$$+ \frac{\partial}{\partial y}\left(\frac{\mu}{Pr}\frac{\partial h_t}{\partial y}\right) + \frac{\partial}{\partial y}\left[\left(1 - \frac{1}{Pr}\right)\mu\frac{\partial}{\partial y}\left(\frac{u^2}{2}\right)\right] \qquad (3.31)$$

Note that $h_t = h + \dfrac{1}{2}u^2$ and $h = e + \dfrac{p}{\rho}$

Three-dimensional Cartesian Coordinates

Mass: $\dfrac{\partial \rho}{\partial t} + \dfrac{\partial}{\partial x}(\rho u) + \dfrac{\partial}{\partial y}(\rho v) + \dfrac{\partial}{\partial z}(\rho w) = 0$ \hfill (3.32)

x-Momentum: $\rho\left(\dfrac{\partial u}{\partial t} + u\dfrac{\partial u}{\partial x} + v\dfrac{\partial u}{\partial y} + w\dfrac{\partial u}{\partial z}\right) = -\dfrac{\partial p}{\partial x} + \dfrac{\partial}{\partial y}\left(\mu\dfrac{\partial u}{\partial y}\right)$ \hfill (3.33)

y-Momentum: $\dfrac{\partial p}{\partial y} = 0$ \hfill (3.34)

z-Momentum: $\rho\left(\dfrac{\partial w}{\partial t} + u\dfrac{\partial w}{\partial x} + v\dfrac{\partial w}{\partial y} + w\dfrac{\partial w}{\partial z}\right) = -\dfrac{\partial p}{\partial z} + \dfrac{\partial}{\partial y}\left(\mu\dfrac{\partial w}{\partial y}\right)$ \hfill (3.35)

Energy:

$$\rho\left(\frac{\partial h_t}{\partial t} + u\frac{\partial h_t}{\partial x} + v\frac{\partial h_t}{\partial y} + w\frac{\partial h_t}{\partial z}\right) = \frac{\partial p}{\partial t} +$$

$$\frac{\partial}{\partial y}\left(\frac{\mu}{Pr}\frac{\partial h_t}{\partial y}\right) + \frac{\partial}{\partial y}\left[\mu\left(1 - \frac{1}{Pr}\right)\left(u\frac{\partial u}{\partial y} + w\frac{\partial w}{\partial y}\right)\right] \tag{3.36}$$

Three-Dimensional Cartesian Coordinates in Conservative Form

Mass:

$$\frac{\partial \rho}{\partial t} + \frac{\partial}{\partial x}(\rho u) + \frac{\partial}{\partial y}(\rho v) + \frac{\partial}{\partial z}(\rho w) = 0 \tag{3.37}$$

x-Momentum:

$$\frac{\partial}{\partial t}(\rho u) + \frac{\partial}{\partial x}(\rho u^2) + \frac{\partial}{\partial y}(\rho uv) + \frac{\partial}{\partial z}(\rho uw) = -\frac{\partial p}{\partial x} + \frac{\partial}{\partial y}\left(\mu\frac{\partial u}{\partial y}\right) \tag{3.38}$$

y-Momentum:

$$\frac{\partial p}{\partial y} = 0 \tag{3.39}$$

z-Momentum:

$$\frac{\partial}{\partial t}(\rho w) + \frac{\partial}{\partial x}(\rho uw) + \frac{\partial}{\partial y}(\rho vw) + \frac{\partial}{\partial z}(\rho w^2) =$$

$$-\frac{\partial p}{\partial z} + \frac{\partial}{\partial y}\left(\mu\frac{\partial w}{\partial y}\right) \tag{3.40}$$

Energy in terms of e_t:

$$\frac{\partial}{\partial t}(\rho e_t) + \frac{\partial}{\partial x}(\rho u e_t) + \frac{\partial}{\partial y}(\rho v e_t) + \frac{\partial}{\partial z}(\rho w e_t) = -\left[\frac{\partial}{\partial x}(pu) + \right.$$

$$\frac{\partial}{\partial y}(pv) + \frac{\partial}{\partial z}(pw)\bigg] + \frac{\partial}{\partial y}\left(k\frac{\partial T}{\partial y}\right) + \frac{\partial}{\partial y}\left[\mu\left(u\frac{\partial u}{\partial y} + w\frac{\partial w}{\partial y}\right)\right] \tag{3.41}$$

Energy in terms of H:

$$\frac{\partial}{\partial t}(\rho h_t) + \frac{\partial}{\partial x}(\rho u h_t) + \frac{\partial}{\partial y}(\rho v h_t) + \frac{\partial}{\partial z}(\rho w h_t) = \frac{\partial p}{\partial t} +$$

$$\frac{\partial}{\partial y}\left(\frac{\mu}{Pr}\frac{\partial h_t}{\partial y}\right) + \frac{\partial}{\partial y}\left[\mu\left(1 - \frac{1}{Pr}\right)\left(u\frac{\partial u}{\partial y} + w\frac{\partial w}{\partial y}\right)\right] \tag{3.42}$$

Flux Vector Form

$$\frac{\partial Q}{\partial t} + \frac{\partial E_B}{\partial x} + \frac{\partial F_B}{\partial y} + \frac{\partial G_B}{\partial z} = \frac{\partial F_{VB}}{\partial y} \tag{3.43}$$

where

$$Q = \begin{bmatrix} \rho \\ \rho u \\ \rho v \\ \rho w \\ \rho e_t \end{bmatrix} \quad E_B = \begin{bmatrix} \rho u \\ p + \rho u^2 \\ 0 \\ \rho u w \\ (p + \rho e_t)u \end{bmatrix} \quad F_B = \begin{bmatrix} \rho v \\ \rho u v \\ 0 \\ \rho u w \\ (p + \rho e_t)v \end{bmatrix}$$

$$G_B = \begin{bmatrix} \rho w \\ \rho u w \\ 0 \\ p + \rho w^2 \\ (p + \rho e_t)w \end{bmatrix} \quad F_{VB} = \begin{bmatrix} 0 \\ \mu \dfrac{\partial u}{\partial y} \\ 0 \\ \mu \dfrac{\partial w}{\partial y} \\ k \dfrac{\partial T}{\partial y} + \mu \left(u \dfrac{\partial u}{\partial y} + w \dfrac{\partial w}{\partial y} \right) \end{bmatrix}$$

Two-dimensional Axisymmetric

Mass: $\dfrac{\partial \rho}{\partial t} + \dfrac{\partial}{\partial x}(r\rho u) + \dfrac{\partial}{\partial y}(\rho r v) = 0 \tag{3.44}$

x-Momentum: $\rho \left(\dfrac{\partial u}{\partial t} + u \dfrac{\partial u}{\partial x} + v \dfrac{\partial u}{\partial y} \right) = -\dfrac{\partial p}{\partial x} + \dfrac{1}{r} \dfrac{\partial}{\partial y} \left(r \mu \dfrac{\partial u}{\partial y} \right) \tag{3.45}$

y-Momentum: $\dfrac{\partial p}{\partial y} = 0 \tag{3.46}$

Energy: $\rho \left(\dfrac{\partial h_t}{\partial t} + u \dfrac{\partial h_t}{\partial x} + v \dfrac{\partial h_t}{\partial y} \right) = \dfrac{1}{r} \dfrac{\partial}{\partial y} \left[r \dfrac{\mu}{Pr} \dfrac{\partial h_t}{\partial y} \right] + \dfrac{1}{r} \dfrac{\partial}{\partial y} \left[r u \mu \left(1 - \dfrac{1}{Pr} \right) \dfrac{\partial u}{\partial y} \right]$

$$\tag{3.47}$$

General Orthogonal Coordinate System

Mass: $\dfrac{\partial \rho}{\partial t} + \dfrac{1}{h_1}\dfrac{\partial(\rho u_1)}{\partial x_1} + \dfrac{1}{h_2}\dfrac{\partial(\rho u_2)}{\partial x_2} + \dfrac{\partial(\rho u_3)}{\partial x_3} + \dfrac{\rho u_1}{h_1 h_2}\dfrac{\partial h_2}{\partial x_1} + \dfrac{\rho u_2}{h_1 h_2}\dfrac{\partial h_1}{\partial x_2} = 0$ (3.48)

x_1-Momentum: $\dfrac{\partial u_1}{\partial t} + \dfrac{u_1}{h_1}\dfrac{\partial u_1}{\partial x_1} + \dfrac{u_2}{h_2}\dfrac{\partial u_1}{\partial x_2} + u_3\dfrac{\partial u_1}{\partial x_3} + \dfrac{u_1 u_2}{h_1 h_2}\dfrac{\partial h_1}{\partial x_2} - \dfrac{u_2^2}{h_1 h_2}\dfrac{\partial h_2}{\partial x_1} =$

$$-\dfrac{1}{\rho h_1}\dfrac{\partial p}{\partial x_1} + \dfrac{\partial}{\partial x_3}\left(\nu\dfrac{\partial u_1}{\partial x_3}\right) \qquad (3.49)$$

x_2-Momentum: $\dfrac{\partial u_2}{\partial t} + \dfrac{u_2}{h_1}\dfrac{\partial u_2}{\partial x_1} + \dfrac{u_2}{h_2}\dfrac{\partial u_2}{\partial x_2} + u_3\dfrac{\partial u_2}{\partial x_3} - \dfrac{u_1^2}{h_1 h_2}\dfrac{\partial h_1}{\partial x_2} + \dfrac{u_1 u_2}{h_1 h_2}\dfrac{\partial h_2}{\partial x_1} =$

$$-\dfrac{1}{\rho h_2}\dfrac{\partial p}{\partial x_2} + \dfrac{\partial}{\partial x_3}\left(\nu\dfrac{\partial u_2}{\partial x_3}\right) \qquad (3.50)$$

x_3-Momentum: $\dfrac{\partial p}{\partial x_3} = 0$ (3.51)

Energy: $\dfrac{\rho u_1}{h_1}\dfrac{\partial h_t}{\partial x_1} + \dfrac{\rho u_2}{h_2}\dfrac{\partial h_t}{\partial x_2} + \dfrac{\rho u_3}{h_3}\dfrac{\partial h_t}{\partial x_3} = \dfrac{1}{h_3}\dfrac{\partial}{\partial x_3}\left\{\dfrac{\mu}{Pr}\dfrac{1}{h_3}\dfrac{\partial h_t}{\partial x_3}\right.$

$$\left.+\dfrac{1}{h_3}\left[\mu\left(1-\dfrac{1}{Pr}\right)\right]\dfrac{\partial}{\partial x_3}\left(\dfrac{u_1^2+u_2^2}{2}\right)\right\}$$

3.3   Integral Equations of the Boundary Layer

Integration of the differential form of the boundary layer equations across the boundary layer provides the integral form of the boundary layer equations. The limits of integration are typically from the surface ($y = 0$) to the free stream ($y \rightarrow \infty$) or to the edge of the boundary layer ($y \rightarrow \delta$). These equations are usually expressed in terms of shear stress τ and heat transfer q and are, thus, generally valid for turbulent flows as well. The integral equations of boundary layer are commonly expressed in compact form in terms of the displacement thickness, momentum thickness, etc. These quantities for an incompressible flow and compressible flow are defined as follows:

Incompressible flow

Displacement thickness:
$$\delta^* = \int_0^\infty \left(1 - \frac{u}{U_e}\right) dy \tag{3.52}$$

Momentum thickness:
$$\theta = \int_0^\infty \frac{u}{U_e}\left(1 - \frac{u}{U_e}\right) dy \tag{3.53}$$

Energy thickness:
$$\delta_3 = \int_0^\infty \frac{u}{U_e}\left(1 - \frac{u^2}{U_e^2}\right) dy \tag{3.54}$$

Enthalpy thickness:
$$\delta_h = \int_0^\infty \frac{u}{U_e}\left(\frac{T - T_e}{T_w - T_e}\right) dy \tag{3.55}$$

Compressible flow

Displacement thickness:
$$\delta^* = \int_0^\infty \left(1 - \frac{\rho u}{\rho_e U_e}\right) dy \tag{3.56}$$

Momentum thickness:
$$\theta = \int_0^\infty \frac{\rho u}{\rho_e U_e}\left(1 - \frac{u}{U_e}\right) dy \tag{3.57}$$

Energy thickness:
$$\delta_3 = \int_0^\infty \frac{\rho u}{\rho_e U_e}\left(1 - \frac{u^2}{U_e^2}\right) dy \tag{3.58}$$

Enthalpy thickness:
$$\delta_h = \int_0^\infty \frac{\rho u}{\rho_e U_e}\left(\frac{h - h_e}{h_w - h_e}\right) dy \tag{3.59}$$

The ratio of displacement thickness to momentum thickness is known as the *shape factor*, $H = \dfrac{\delta^*}{\theta}$.

The integral equations of boundary layer are summarized in Table 3.3 for an incompressible flow and in Table 3.4 for a compressible flow.

Table 3.3: **Integral boundary layer equations for two-dimensional, incompressible flows**

Momentum equation/ von Karman integral equation

$$\frac{\tau_w}{\rho} = \frac{\partial}{\partial t}\int_0^\infty (U_e - u)dy + \frac{\partial}{\partial x}\int_0^\infty u(U_e - u)dy + \frac{\partial U_e}{\partial x}\int_0^\infty (U_e - u)dy - U_e v_w$$

$$(3.60)$$

where v_w = normal velocity at porous wall ($+v_w$ for injection)

Equation above can be written as

$$\frac{\tau_w}{\rho U_e^2} = \frac{C_f}{2} = \frac{1}{U_e^2}\frac{\partial}{\partial t}(U_e \delta^*) + \frac{\partial \theta}{\partial x} + (2\theta + \delta^*)\frac{1}{U_e}\frac{\partial U_e}{\partial x} - \frac{v_w}{U_e} \qquad (3.61)$$

For a steady flow with an impermeable wall

$$\frac{\tau_w}{\rho} = \frac{d}{dx}(U_e^2 \theta) + \delta^* U_e \frac{dU_e}{dx} \qquad (3.62)$$

or

$$\frac{C_f}{2} = \frac{d\theta}{dx} + (2 + H)\frac{\theta}{U_e}\frac{dU_e}{dx} \qquad (3.63)$$

Energy equation

$$\frac{dQ}{dt} = \rho U_e c_p (T_w - T_e)(\delta_h - Ec\delta_3) \qquad (3.64)$$

where

$$Ec = \frac{U_e^2}{2c_p(T_w - T_e)}, \quad \frac{dQ}{dt} \text{ is the integrated value of local } q_w, \text{ i.e., total heat transfer rate.}$$

or

$$\frac{q_w}{\rho U_e c_p(T_w - T_e)} = \frac{d\delta_h}{dx} - Ec\frac{d\delta_3}{dx} \qquad (3.65)$$

Table 3.4: Integral boundary layer equations for two-dimensional, compressible flows

Momentum equation

$$\frac{\tau_w}{\rho_e U_e^2} = \frac{C_f}{2} = \frac{d\theta}{dx} + \left(2 + H + \frac{U_e}{\rho_e}\frac{d\rho_e}{dU_e}\right)\frac{\theta}{U_e}\frac{dU_e}{dx} + \frac{1}{\rho_e U_e^2}\frac{d}{dx}\left(p_e\delta - \int_0^\delta p\,dy\right) \quad (3.66)$$

For an adiabatic free stream $\dfrac{U_e}{\rho_e}\dfrac{d\rho_e}{dU_e} = -M_e^2$

If the pressure term in Equation (3.66) above is neglected, then it is reduced to

$$\frac{C_f}{2} = \frac{d\theta}{dx} + (2 + H - M_e^2)\frac{\theta}{U_e}\frac{dU_e}{dx} \quad (3.67)$$

Energy equation

$$\frac{2}{\rho_e U_e^3}\int_0^\infty \tau\frac{\partial u}{\partial y}dy = \frac{d\delta_3}{dx} + \left[3 - (2-\gamma)M_e^2\right]\frac{\delta_3}{U_e}\frac{dU_e}{dx} \quad (3.68)$$

or

$$q_w = \rho\frac{\partial}{\partial t}\int_0^\infty c_p T\,dy + \rho\frac{\partial}{\partial x}\int_0^\infty u(h_t - h_{te})dy - \rho v_w c_p T_w \quad (3.69)$$

The momentum and energy integral equations for compressible axisymmetric bodies may be expressed as

$$\frac{\tau_w}{\rho_e U_e^2} + \frac{\rho_w V_w}{\rho_e U_e} = \frac{d\theta}{dx} + \theta\left[\left(2 + \frac{\delta^*}{\theta}\right)\frac{1}{U_e}\frac{dU_e}{dx} + \frac{1}{\rho_e}\frac{d\rho_e}{dx} + \frac{1}{r_o}\frac{dr_o}{dx}\right] \quad (3.70)$$

$$\frac{q_w}{\rho_e U_e c_p(T_w - T_e)} = \frac{d\delta_h}{dx} + \delta_h\left[\frac{1}{U_e}\frac{dU_e}{dx} + \frac{1}{(T_w - T_e)}\frac{d}{dx}(T_w - T_e)\right.$$
$$\left. + \frac{1}{\rho_e}\frac{d\rho_e}{dx} + \frac{1}{r_o}\frac{dr_o}{dx}\right]$$

3.4 Turbulent Boundary Layer Equations

In order to include and account for the effect of turbulence in a flowfield, the laminar boundary layer equations discussed previously must be modified. Traditionally this modification is accomplished by representing the instantaneous flow quantities by the sum of a mean value (denoted by a bar over the variable) and a time-dependent fluctuating value (denoted by a prime). Mathematically, it is expressed as

$$f = \bar{f} + f' \tag{3.71}$$

where

$$\bar{f} = \frac{1}{\Delta t} \int_{t_0}^{t_0+\Delta t} f\,dt$$

These quantities are shown graphically in Figure 3.7. The time averaging of a fluctuating quantity over a time interval Δt results in

$$\overline{f'} = \frac{1}{\Delta t} \int_{t_0}^{t_0+\Delta t} f\,dt = 0$$

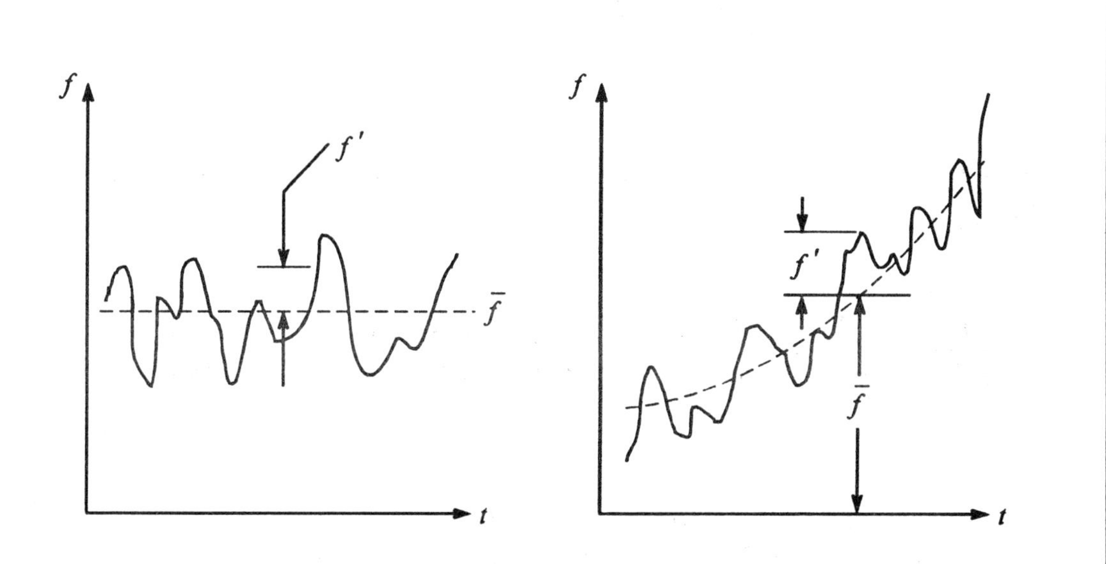

Figure 3.7 Illustration of the average and fluctuating quantities for steady and unsteady flows.

The time interval, Δt, used in the definitions above must be larger than the period of fluctuating quantities but smaller than the time interval associated with the unsteady flow. Thus the time interval is problem dependent, that is, it depends on the geometry and physics of the flowfield being investigated. Note that for a steady flow the time-averaged mean value is constant while for an unsteady flow it is a function of time.

Following the substitution of (3.71) into the laminar boundary layer equations, they are time averaged, and the terms involving the average of a single fluctuating term are dropped. The incompressible, steady, turbulent boundary layer equations obtained from Equations (3.1) through (3.4) are

$$\frac{\partial \overline{u}}{\partial x} + \frac{\partial \overline{v}}{\partial y} = 0 \tag{3.72}$$

$$\overline{u}\frac{\partial \overline{u}}{\partial x} + \overline{v}\frac{\partial \overline{u}}{\partial y} = -\frac{1}{\rho}\frac{\partial p}{\partial x} + \nu\frac{\partial^2 \overline{u}}{\partial y^2} - \frac{\partial}{\partial y}\left(\overline{u'v'}\right) \tag{3.73}$$

$$\overline{u}\frac{\partial \overline{T}}{\partial x} + \overline{v}\frac{\partial \overline{T}}{\partial y} = \frac{\nu}{c_p}\left(\frac{\partial \overline{u}}{\partial y}\right)^2 - \frac{1}{c_p}\left(\overline{u'v'}\right)\frac{\partial \overline{u}}{\partial y} + \alpha\frac{\partial^2 \overline{T}}{\partial y^2} - \frac{\partial}{\partial y}\left(\overline{v'T'}\right) \tag{3.74}$$

Additional assumptions are made in the process of obtaining Equations (3.73) and (3.74), for example,

$$\frac{\partial}{\partial y}\left(\overline{u'v'}\right) \gg \frac{\partial}{\partial x}\left(\overline{u'^2}\right)$$

Equations (3.72) through (3.74) reveal that in addition to the three unknowns, $\overline{u}, \overline{v}$ and $\overline{T}$ (which were previously encountered in the laminar boundary layer), two additional unknowns, namely $\overline{u'v'}$, and $\overline{v'T'}$, have been introduced. In order to close the system, additional relations must be introduced. These relations are known as turbulence models, which in general relate the fluctuating correlations to the mean flow properties by means of empirical constants. Turbulence models vary in degree of sophistication from simple algebraic equations to systems of partial differential equations. Surprisingly enough, in many instances simple algebraic models provide as good a result as sophisticated models, and with less computation time.

Complete review of all turbulence models is beyond the scope of this text; however, review of some simple and most commonly used models is in order. The time-averaged turbulent boundary layer equations are summarized in Table 3.5.

Table 3.5: Turbulent boundary layer equations

Steady, two-dimensional, incompressible

$$\text{Mass:}\quad \frac{\partial \bar{u}}{\partial x} + \frac{\partial \bar{v}}{\partial y} = 0 \tag{3.75}$$

$$x\text{-momentum}\quad \bar{u}\frac{\partial \bar{u}}{\partial x} + \bar{v}\frac{\partial \bar{u}}{\partial y} = -\frac{1}{\rho}\frac{\partial \bar{p}}{\partial x} + \nu\frac{\partial^2 \bar{u}}{\partial y^2} - \frac{\partial}{\partial y}(\overline{u'v'}) \tag{3.76}$$

$$\text{Energy:}\quad \bar{u}\frac{\partial \overline{T}}{\partial x} + \bar{v}\frac{\partial \overline{T}}{\partial y} = \frac{\nu}{c_p}\left(\frac{\partial \bar{u}}{\partial y}\right)^2 - \frac{1}{c_p}\left(\overline{u'v'}\right)\frac{\partial \bar{u}}{\partial y} + \alpha\frac{\partial^2 \overline{T}}{\partial y^2} - \frac{\partial}{\partial y}(\overline{v'T'}) \tag{3.77}$$

In terms of the eddy viscosity and eddy diffusivity

$$\text{Mass:}\quad \frac{\partial \bar{u}}{\partial x} + \frac{\partial \bar{v}}{\partial y} = 0 \tag{3.78}$$

$$x\text{-Momentum:}\quad \bar{u}\frac{\partial \bar{u}}{\partial x} + \bar{v}\frac{\partial \bar{u}}{\partial y} = -\frac{1}{\rho}\frac{\partial \bar{p}}{\partial x} + (\nu + \nu_t)\frac{\partial \bar{u}}{\partial y} \tag{3.79}$$

$$\text{Energy:}\quad \bar{u}\frac{\partial \overline{T}}{\partial x} + \bar{v}\frac{\partial \overline{T}}{\partial y} = \frac{1}{c_p}(\nu + \nu_t)\left(\frac{\partial \bar{u}}{\partial y}\right)^2 + (\alpha + \alpha_t)\frac{\partial \overline{T}}{\partial y} \tag{3.80}$$

Two-dimensional planar or axisymmetric compressible

$$\text{Mass:}\quad \frac{\partial \bar{\rho}}{\partial t} + \frac{\partial}{\partial x}(\bar{\rho}\bar{u}r^k) + \frac{\partial}{\partial y}(\bar{\rho}Vr^k) = 0 \tag{3.81}$$

$$\text{where}\quad V = \bar{v} + \frac{\overline{\rho'v'}}{\bar{\rho}}$$

$$x\text{-Momentum:}\quad \bar{\rho}\frac{\partial \bar{u}}{\partial t} + \bar{\rho}\bar{u}\frac{\partial \bar{u}}{\partial x} + \bar{\rho}V\frac{\partial \bar{u}}{\partial y} = -\frac{\partial \bar{p}}{\partial x} + \frac{1}{r^k}\frac{\partial}{\partial y}\left[r^k\left(\mu\frac{\partial \bar{u}}{\partial y} - \bar{\rho}\overline{u'v'}\right)\right] \tag{3.82}$$

Energy in terms of the total enthalpy h_t:

$$\bar{\rho}\frac{\partial \bar{h}_t}{\partial t} + \bar{\rho}\bar{u}\frac{\partial \bar{h}_t}{\partial x} + \bar{\rho}V\frac{\partial \bar{h}_t}{\partial y} = \frac{1}{r^k}\frac{\partial}{\partial y}\left\{ r^k\left[\bar{\mu}\left(1 - \frac{1}{Pr}\right)\bar{u}\frac{\partial \bar{u}}{\partial y}. + \frac{\bar{\mu}}{Pr}\frac{\partial \bar{h}_t}{\partial y} - \overline{\rho v'h_t'}\right]\right\} \quad (3.83)$$

where $k = 0 \Longrightarrow$ 2-D planar and $k = 1 \Longrightarrow$ 2-D axisymmetric.

Equations above for a steady flow are commonly written as

Mass: $\dfrac{\partial}{\partial x}(\bar{\rho}\bar{u}r^k) + \dfrac{\partial}{\partial y}(\bar{\rho}Vr^k) = 0$

x-momentum $\ \bar{\rho}\bar{u}\dfrac{\partial \bar{u}}{\partial x} + \bar{\rho}V\dfrac{\partial \bar{u}}{\partial y} + \dfrac{\partial}{\partial y}\left(\overline{\rho u'v'}\right) = -\dfrac{\partial \bar{p}}{\partial x} + \dfrac{\partial}{\partial y}\left(\bar{\mu}\dfrac{\partial \bar{u}}{\partial y}\right)$ $\qquad (3.84)$

Energy: $\ \bar{\rho}\bar{u}\dfrac{\partial \bar{h}}{\partial x} + \bar{\rho}V\dfrac{\partial \bar{h}}{\partial y} = \dfrac{\partial}{\partial y}\left(\bar{k}\dfrac{\partial \overline{T}}{\partial y}\right) - \dfrac{\partial}{\partial y}\left(\overline{\rho v'h'}\right)$

$$+ \bar{u}\frac{\partial \bar{p}}{\partial x} + \bar{\mu}\left(\frac{\partial \bar{u}}{\partial y}\right)^2 - \overline{\rho u'v'}\frac{\partial \bar{u}}{\partial y} \qquad (3.85)$$

In terms of the eddy viscosity and eddy diffusivity

Mass: $\dfrac{\partial}{\partial x}(\bar{\rho}\bar{u}r^k) + \dfrac{\partial}{\partial y}(\bar{\rho}Vr^k) = 0$ $\qquad\qquad (3.86)$

x-Momentum: $\ \bar{\rho}\bar{u}\dfrac{\partial \bar{u}}{\partial x} + \bar{\rho}V\dfrac{\partial \bar{u}}{\partial y} = -\dfrac{\partial \bar{p}}{\partial x} + \dfrac{\partial}{\partial y}(\bar{\mu} + \mu_t)\dfrac{\partial \bar{u}}{\partial y}$ $\qquad (3.87)$

Energy: $\ \bar{\rho}\bar{u}\dfrac{\partial \bar{h}}{\partial x} + \bar{\rho}V\dfrac{\partial \bar{h}}{\partial y} = \bar{u}\dfrac{\partial \bar{p}}{\partial x} + (\bar{\mu} + \mu_t)\left(\dfrac{\partial \bar{u}}{\partial y}\right)^2 + \dfrac{\partial}{\partial y}(k + k_t)\dfrac{\partial \overline{T}}{\partial y}$ $\qquad (3.88)$

The energy equation can be written in terms of the total enthalpy as follows:

$$\bar{\rho}\bar{u}\frac{\partial \bar{h}_t}{\partial x} + \bar{\rho}V\frac{\partial \bar{h}_t}{\partial y} = \frac{\partial}{\partial y}\left[\left(\frac{\mu}{Pr} + \frac{\mu_t}{Pr_t}\right)\frac{\partial \bar{h}_t}{\partial y}\right]$$

$$+ \frac{\partial}{\partial y}\left\{\left[\mu\left(1 - \frac{1}{Pr}\right) + \mu_t\left(1 - \frac{1}{Pr_t}\right)\right]\frac{\partial}{\partial y}\left(\frac{1}{2}\bar{u}^2\right)\right\} \tag{3.89}$$

where $\quad Pr_t = \dfrac{\mu_t c_p}{k_t}$

Three-Dimensional Cartesian Coordinate

Mass: $\quad \dfrac{\partial \bar{\rho}}{\partial t} + \dfrac{\partial}{\partial x}(\bar{\rho}\bar{u}) + \dfrac{\partial}{\partial y}(\bar{\rho}V) + \dfrac{\partial}{\partial z}(\bar{\rho}\bar{w}) = 0 \tag{3.90}$

x-Momentum: $\quad \dfrac{\partial \bar{u}}{\partial t} + \bar{\rho}\bar{u}\dfrac{\partial \bar{u}}{\partial x} + \bar{\rho}V\dfrac{\partial \bar{u}}{\partial y} + \bar{\rho}\bar{w}\dfrac{\partial \bar{u}}{\partial z} = -\dfrac{\partial \bar{p}}{\partial x} + \dfrac{\partial}{\partial y}\left(\bar{\mu}\dfrac{\partial \bar{u}}{\partial y} - \overline{\rho u'v'}\right) \tag{3.91}$

y-Momentum: $\quad \dfrac{\partial \bar{p}}{\partial y} = 0 \tag{3.92}$

z-Momentum: $\quad \dfrac{\partial \bar{w}}{\partial t} + \bar{\rho}\bar{u}\dfrac{\partial \bar{w}}{\partial x} + \bar{\rho}V\dfrac{\partial \bar{w}}{\partial y} + \bar{\rho}\bar{w}\dfrac{\partial \bar{w}}{\partial z} = -\dfrac{\partial p}{\partial z} + \dfrac{\partial}{\partial y}\left(\bar{\mu}\dfrac{\partial \bar{w}}{\partial y} - \overline{\rho w'v'}\right) \tag{3.93}$

Energy: $\quad \dfrac{\partial \bar{h}_t}{\partial t} + \bar{\rho}\bar{u}\dfrac{\partial \bar{h}_t}{\partial x} + \bar{\rho}V\dfrac{\partial \bar{h}_t}{\partial y} + \bar{\rho}\bar{w}\dfrac{\partial \bar{h}_t}{\partial z} =$

$$\frac{\partial}{\partial y}\left[\frac{\mu}{Pr}\frac{\partial \bar{h}_t}{\partial y} - \bar{\rho}c_p\overline{v'T'} + \bar{\mu}\left(1 - \frac{1}{Pr}\right)\frac{\partial}{\partial y}\left(\frac{u^2 + w^2}{2}\right) - \bar{\rho}\bar{u}\overline{u'v'} - \bar{\rho}\bar{w}\overline{v'w'}\right] \tag{3.94}$$

3.5　Turbulence Models

3.5.1 General Remarks

A turbulence model is a semi-empirical equation relating the fluctuating correlation to mean flow variables with various constants provided from experimental investigations. When this equation is expressed as an algebraic equation, it is referred to as the *zero-equation model*. When partial differential equations are used, they are referred to as *one-equation* or *two-equation models*, depending on the number of PDEs utilized. Some models employ ordinary differential equations, in which case they are referred to as *half-equation models*. Finally, it is possible to write a partial differential equation directly for each of the turbulence correlations such as $\overline{u'v'}$, $\overline{u'^2}$, etc., in which case they compose a system of PDEs known as the Reynolds stress equations.

Before various turbulence models are introduced, the commonly defined zones within a turbulent boundary layer are identified. A very small layer next to the surface is defined as the *laminar sublayer* where viscous shear is dominant. The outer region where turbulent shear (terms such as $\overline{u'v'}$) is dominant is called the *fully turbulent zone*. A region which connects these two zones is called the *buffer zone*, where both viscous shear and turbulent shear are of equal importance.

A typical velocity profile for a turbulent flow over a flat surface, indicating various zones, is shown in Figure 3.8.

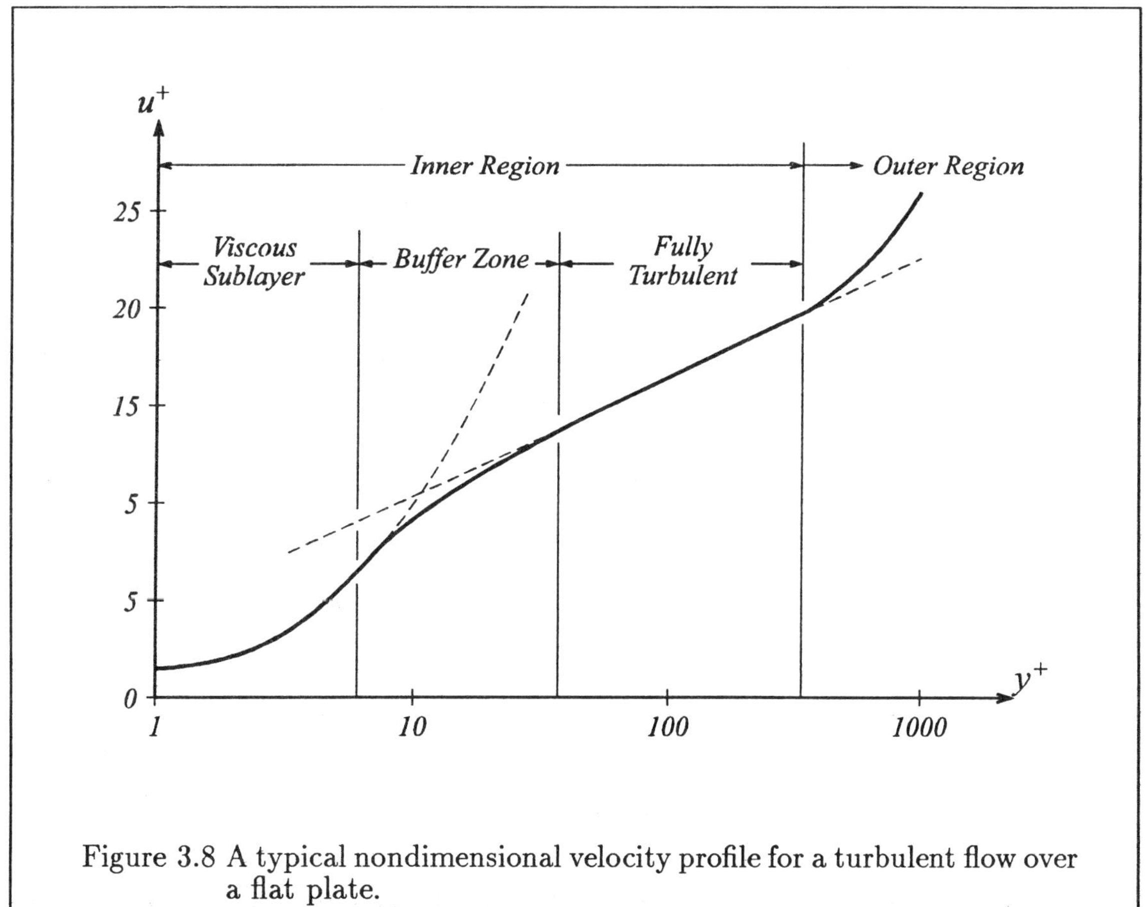

Figure 3.8 A typical nondimensional velocity profile for a turbulent flow over a flat plate.

The nondimensional velocity u^+ and space coordinate y^+ are defined as

$$u^+ = \frac{u}{u^*} \tag{3.95}$$

and

$$y^+ = y\frac{u^*}{\nu} \tag{3.96}$$

where u^* is known as the friction velocity given by

$$u^* = \left(\frac{\tau_w}{\rho}\right)^{\frac{1}{2}} \tag{3.97}$$

The velocity profile across a turbulent boundary layer (as shown in Figure 3.8) can be represented by the universal distributions, given as:

Laminar sublayer	$0 < y^+ < 5$	$u^+ = y^+$
Buffer layer	$5 < y^+ < 30$	$u^+ = 5\ln y^+ - 3.05$
Fully turbulent layer	$y^+ > 30$	$u^+ = 2.5\ln y^+ + 5.5$

3.5.2 Turbulent Shear Stress and Heat Flux

In this section, some of the fundamental concepts introduced by the early investigating pioneers are explored. For the sake of simplicity, the steady incompressible boundary layer equations are considered. The governing equations are expressed as

$$\frac{\partial \overline{u}}{\partial x} + \frac{\partial \overline{v}}{\partial y} = 0 \tag{3.98}$$

$$\overline{u}\frac{\partial \overline{u}}{\partial x} + \overline{v}\frac{\partial \overline{u}}{\partial y} = -\frac{1}{\rho}\frac{\partial \overline{p}}{\partial x} + \nu\frac{\partial^2 \overline{u}}{\partial y^2} - \frac{\partial}{\partial y}\left(\overline{u'v'}\right) \tag{3.99}$$

$$\overline{u}\frac{\partial \overline{T}}{\partial x} + \overline{v}\frac{\partial \overline{T}}{\partial y} = \frac{\nu}{c_p}\left(\frac{\partial \overline{u}}{\partial y}\right)^2 - \frac{1}{c_p}\left(\overline{u'v'}\right)\frac{\partial \overline{u}}{\partial y} + \alpha\frac{\partial^2 \overline{T}}{\partial y^2} - \frac{\partial}{\partial y}\left(\overline{v'T'}\right) \tag{3.100}$$

In the equation above, α is the thermal diffusivity.

It can be shown that the macroscopic momentum exchange due to turbulence is represented by $-\rho\overline{u'v'}$, which is referred to as the *turbulent shear stress* (or *Reynolds stress*), denoted herein by τ_t. It is customary to combine the laminar and turbulent shear stress terms in Equation (3.99) as

$$\nu\frac{\partial^2 \overline{u}}{\partial y^2} - \frac{\partial}{\partial y}\left(\overline{u'v'}\right) = \frac{1}{\rho}\frac{\partial}{\partial y}(\tau_l + \tau_t) \tag{3.101}$$

For a laminar flow under the assumptions stated, one may write

$$\tau_l = \mu \frac{\partial \overline{u}}{\partial y} = \rho \nu \frac{\partial \overline{u}}{\partial y} \tag{3.102}$$

In order to express the turbulent shear in a similar form, consider the concept of eddy viscosity, where the turbulent shear stress is related to the gradient of the mean flow velocity. This analogy, introduced by Boussinesq, is referred to as the *Boussinesq assumption*. Thus, one writes

$$\tau_t = \rho \nu_t \frac{\partial \overline{u}}{\partial y} \tag{3.103}$$

where ν_t is known as the eddy viscosity. Hence, Equation (3.99) may be expressed as

$$\overline{u}\frac{\partial \overline{u}}{\partial x} + \overline{v}\frac{\partial \overline{u}}{\partial y} = -\frac{1}{\rho}\frac{\partial \overline{p}}{\partial x} + (\nu + \nu_t)\frac{\partial^2 \overline{u}}{\partial y^2} \tag{3.104}$$

In a similar fashion, eddy diffusivity is defined to combine laminar and turbulent heat fluxes. For a laminar flow, the Fourier heat conduction law is expressed as

$$\left(\frac{q}{A}\right)_l = -k\frac{\partial \overline{T}}{\partial y} = \rho c_p \alpha \frac{\partial \overline{T}}{\partial y} \tag{3.105}$$

For turbulent heat flux, the following is written,

$$\left(\frac{q}{A}\right)_t = \rho c_p \overline{v'T'} = -\rho c_p \alpha_t \frac{\partial \overline{T}}{\partial y} \tag{3.106}$$

where α_t is the eddy diffusivity. Now, the energy equation may be rearranged as

$$\overline{u}\frac{\partial \overline{T}}{\partial x} + \overline{v}\frac{\partial \overline{T}}{\partial y} = \frac{1}{\rho c_p}(\tau_l + \tau_t)\frac{\partial \overline{u}}{\partial y} - \frac{1}{\rho c_p}\frac{\partial}{\partial y}\left[\left(\frac{q}{A}\right)_l + \left(\frac{q}{A}\right)_t\right] \tag{3.107}$$

or in terms of eddy parameters,

$$\overline{u}\frac{\partial \overline{T}}{\partial x} + \overline{v}\frac{\partial \overline{T}}{\partial y} = \frac{1}{c_p}(\nu + \nu_t)\left(\frac{\partial \overline{u}}{\partial y}\right)^2 + (\alpha + \alpha_t)\frac{\partial^2 \overline{T}}{\partial y^2} \tag{3.108}$$

Note that, up to this point, the problem of closure has not been addressed, that is, two additional unknowns still remain. All that has been accomplished is the rewriting of the equations by introducing the Boussinesq assumption.

It is well known that in a laminar flow the mixing of fluid is at the molecular level, and the viscous stresses and heat fluxes are due to momentum and energy transfer by molecules travelling a distance of mean-free path before collision. A similar concept is extended to turbulent flows, where it is assumed that lumps of fluid travel a finite distance before collision and before losing their identity. The resulting momentum and energy exchange give rise to what was defined as turbulent shear stress and turbulent heat flux. This finite distance is defined as the mixing length, l. This concept, introduced by Prandtl, is known as the *Prandtl mixing length*. The Prandtl hypothesis is expressed as

$$\tau_t = -\rho\overline{u'v'} = \rho l^2 \left(\frac{\partial \overline{u}}{\partial y}\right)^2 \tag{3.109}$$

Similarly

$$\left(\frac{q}{A}\right)_t = -\rho c_p l_e^2 \left(\frac{\partial \overline{T}}{\partial y}\right)^2 \tag{3.110}$$

In terms of eddy viscosity and eddy diffusivity, the following may be written:

$$\nu_t = l^2 \left(\frac{\partial \overline{u}}{\partial y}\right) \tag{3.111}$$

$$\alpha_t = l_e^2 \left(\frac{\partial \overline{T}}{\partial y}\right) \tag{3.112}$$

where l_e is the thermal mixing length.

The concept of eddy viscosity and eddy diffusivity can be easily extended to general flowfields governed by the Navier-Stokes equations. For example, for a two-dimensional problem,

$$\nu_t = l^2 \left[\left(\frac{\partial \overline{u}}{\partial y}\right)^2 + \left(\frac{\partial \overline{v}}{\partial x}\right)^2\right]^{\frac{1}{2}} \tag{3.113}$$

What has been accomplished up to this point is the introduction of new parameters, and the central issue of closure is not complete as yet since there are still two additional unknowns, now in the form of l and l_e.

In order to write expressions for mixing lengths, experimental investigations are heavily relied upon, and l and l_e are modelled for the various flow regimes, namely the inner and outer regions.

As discussed previously, these semi-empirical relations are known as turbulence models and may be expressed as algebraic equations or partial differential equations. In the next three subsections some commonly used zero-equation, one-equation, and two-equation turbulence models, as a sample of the wide range of models, are investigated. Several recently published survey and assessment papers on turbulence models [3.1–3.14] are excellent sources for those interested in various turbulence models.

3.5.3 Zero-Equation Turbulence Models

The zero-equation models are equations in which the turbulent fluctuating correlations are related to the mean flowfield quantities by algebraic relations. The underlying assumption in zero-equation models is that the local rate of production of turbulence and the rate of dissipation of turbulence are approximately equal. However, they do not include the convection of turbulence. Obviously, it is contrary to the physics of most flowfields, since the past history of the flow must be accounted for. Nevertheless, these models are mathematically simple and their incorporation into a numerical code can be accomplished with relative ease.

3.5.3.1 *Inner Region/Outer Region Model:* Generally, most models employ an inner region/outer region formulation to represent mixing length. A commonly used model utilizes an exponential function (van Driest damping function) for the inner region, whereas the outer region is proportional to the boundary layer thickness. Mathematically they are expressed as

$$l_i = \kappa(1 - e^{-y^+/A^+})y \tag{3.114}$$

and

$$l_0 = C_0\delta \tag{3.115}$$

where κ is the von Karman constant (~ 0.41), and A^+ is a parameter which depends on the streamwise pressure gradient. For a zero-pressure gradient flow, it has a value of 26. The constant C_o in Equation (3.115) is usually assigned a value of $0.08 \sim 0.09$, whereas δ is the boundary layer thickness.

A detailed description of the inner/outer law of the wall follows.

3.5.3.1.1 *Inner law or the law of the wall:* Very close to the wall the velocity profile is linear, and the following relation holds

$$u^+ = y^+ \tag{3.116}$$

The Spalding law [3.15] is a simple and quite accurate composite formula which is valid near the wall as well as in the overlap region.

$$y^+ = u^+ + e^{-\kappa B}\left[e^{\kappa u^+} - 1 - \kappa u^+ - \frac{(\kappa u^+)^2}{2} - \frac{(\kappa u^+)^3}{6}\right] \tag{3.117}$$

where $\kappa =$ von Karman constant $= 0.4$ and $B = 5.5$.

The expression for eddy viscosity used in the derivation of the above expression is

$$\mu_{t_i} = \mu \kappa e^{-\kappa B} \left[e^{\kappa u^+} - 1 - \kappa u^+ - \frac{(\kappa u^+)^2}{2} \right] \tag{3.118}$$

Other empirical expressions for the inner law eddy viscosity [3.16] are:

Rotta:

$$\mu_{t_i} = \begin{cases} e\kappa^2 \left(y - \dfrac{5\nu}{u^*} \right)^2 \left| \dfrac{\partial \bar{u}}{\partial y} \right| & \text{for} \quad y > \dfrac{5\nu}{u^*} \\[2ex] 0 & \text{for} \quad y < \dfrac{5\nu}{u^*} \end{cases}$$

Reichardt:

$$\mu_{t_i} = \mu \kappa \left(y^+ - 5 \tanh \frac{y^+}{5} \right)$$

Deissler:

$$\mu_{t_i} = 0.012 \mu u^+ y^+ \left[1 - \exp(-0.012 u^+ y^+) \right]$$

Mellor:

$$\mu_{t_i} = \frac{\mu (\kappa y^+)^4}{(\kappa y^+)^3 + 328.5}$$

3.5.3.1.2 Overlap law: The overlap layer is generally assumed to exist in the region $35 \le y^+ \le 350$. The profiles of velocity all collapse into a single curve represented by a logarithmic expression

$$u^+ = \frac{1}{\kappa} \ln y^+ + B \tag{3.119}$$

For the constants κ and B, Nikuradse suggests values of 0.4 and 5.5, while Cole uses values of 0.41 and 5.0, respectively.

3.5.3.1.3 Outer law or the law of the wake: In the analysis of the outer layer or the wake region, it is necessary to take into account the pressure gradient. The parameter which accounts for this is generally referred to as *Clauser's equilibrium parameter,*

$$\beta = \frac{\delta^*}{\tau_w} \frac{\partial p_e}{\partial x}$$

Clauser proposed the following relationship for eddy viscosity in the outer layer

$$\mu_{t_o} = 0.016 \rho U_e \delta^*$$

Coles' law of the wake is perhaps the most well known formulation

$$u^+ = \frac{1}{\kappa} \ln y^+ + B + \frac{\Pi}{\kappa} W(\frac{y}{\delta})$$

where W is the wake function, normalized to zero at the wall, $W = 2$ at $y = \delta$. Coles proposed the following curve fit approximation to the wake function.

$$W\left(\frac{y}{\delta}\right) = 2 \sin^2 \left(\frac{\pi}{2} \frac{y}{\delta} \right)$$

Π is called the wake parameter and is assumed to be a constant in an equilibrium boundary layer.

$$\Pi = \frac{\kappa A}{2} \qquad A = 2.35 \text{ for flat plates}$$

$$\text{or} \ \ \Pi \cong 0.5$$

In general, $\Pi = \Pi(B)$. A curve fit is given by $\Pi = 0.8(\beta + 0.5)^{0.75}$

3.5.3.2 Cebeci/Smith Model:

Another formulation commonly used for the outer region turbulent viscosity is the Cebeci/Smith model [3.17] expressed as

$$\nu_t = \alpha U_e \delta^* \tag{3.120}$$

where α is usually assigned a value of 0.0168 for flows where Re_θ (momentum thickness Reynolds number) is greater than 5000 and δ^* is the displacement thickness (incompressible flow) defined as

$$\delta^* = \int_0^\infty \left(1 - \frac{u}{U_e}\right) dy \tag{3.121}$$

Recall that the Reynolds number based on momentum thickness is defined as

$$Re_\theta = \frac{\rho_e u_e \theta}{\mu_e} \tag{3.122}$$

where the momentum thickness θ is

$$\theta = \int_0^\infty \frac{u}{U_e} \left(1 - \frac{u}{U_e}\right) dy \tag{3.123}$$

The algebraic model described above requires information regarding the boundary layer thickness and flow properties at the boundary layer edge. When the Navier-Stokes equations are being solved, it may be a difficult task to determine the boundary layer thickness and the required properties at the edge. That is especially the case when flow separation exists within the domain. However, when it is necessary to determine the extent of the viscous region within the domain, the total enthalpy is usually used.

3.5.3.3 Baldwin-Lomax Model:

A turbulence model which is not written in terms of the boundary layer quantities was introduced by Baldwin and Lomax [3.18]. The inner region is approximated by

$$\mu_t = \rho l^2 \mid \omega_z \mid \tag{3.124}$$

where ω_z is the vorticity defined as

$$\omega_z = \frac{\partial v}{\partial x} - \frac{\partial u}{\partial y} \tag{3.125}$$

The outer region is approximated by

$$\mu_t = \alpha \bar{\rho} U_0 L_0 \tag{3.126}$$

where α is assigned a value of 0.0168 (as in the Cebeci/Smith model) and

$$U_0 = \min\left[G_{\text{max}}, \frac{(\Delta V)^2}{4 G_{\text{max}}} \right] \tag{3.127}$$

and

$$L_0 = 1.6 I_0 y_{\text{max}} \tag{3.128}$$

In Equations (3.127) and (3.128), the following definitions are employed,

$$G_{\text{max}} = \max\left(\frac{l}{\kappa} \, |\, \omega_z \,| \right) \tag{3.129}$$

where κ is the von Karman constant, and l, the mixing length, is determined by the van Driest function given by (3.114). The difference between the absolute values of the maximum and minimum velocities is denoted by ΔV. I_o is the intermittency factor defined as

$$I_o = \left[l + 5.5 \left(\frac{0.3y}{y_{\text{max}}} \right)^6 \right]^{-1} \tag{3.130}$$

and y_{max} is the y location where G_{max} occurs.

To model the eddy diffusivity, the Reynolds analogy may be used. Recall that the Reynolds analogy assumes a similarity between the momentum transfer and heat transfer. Therefore, a turbulent Prandtl number is defined as

$$Pr_t = \frac{\nu_t}{\alpha_t} = \frac{\mu_t c_p}{k_t}$$

For most flows, it is assumed that the turbulent Prandtl number remains constant across the boundary layer. For air, $Pr_t = 0.9$. Thus, the turbulent conductivity is determined as

$$k_t = \frac{\mu_t c_p}{Pr_t}$$

where μ_t is provided by the turbulence models just described.

3.5.4 One-Equation Turbulence Models

As seen in the previous section, zero-equation models employ an algebraic relation for the eddy viscosity. These models specify length and velocity scales in terms of the mean flow, thus implying an equilibrium between mean motion and turbulence.

One-equation models employ a partial differential equation for velocity scale, whereas the length scale is specified algebraically. The velocity scale is typically written in terms of turbulent kinetic energy k defined as

$$k = \frac{1}{2}(u'^2 + v'^2 + w'^2) \tag{3.131}$$

The turbulent viscosity μ_t is now written as

$$\mu_t = \rho k^{\frac{1}{2}} \ell \tag{3.132}$$

In the following section, two recently developed one-equation turbulence models are reviewed.

3.5.4.1 Baldwin-Barth One-Equation Turbulence Model: The Baldwin-Barth one-equation model is obtained from the standard k-ϵ two-equation model [3.19]. The two equations are combined to provide a single equation for the turbulence Reynolds number Re_T, where $Re_T = k^2/(\nu\epsilon)$. To account for the near wall regions, the turbulence Reynolds number is split into two parts as $Re_T = \overline{Re}_T f(\overline{Re}_T)$, where f is a damping function such that $Re_T \cong \overline{Re}_T$ for large Re_T.

The transport equation for turbulence Reynolds number $\overline{Re}_T$ is written as

$$\frac{d}{dt}(\nu\overline{Re}_T) = (c_{\epsilon_2}f_2 - c_{\epsilon_1}\sqrt{\nu Re_T P} + \left(\nu + \frac{\nu_t}{\sigma_\epsilon}\right)\nabla^2(\nu\overline{Re}_T)$$

$$- \frac{1}{\sigma_\epsilon}(\nabla\nu_t) \cdot \nabla(\nu\overline{Re}_T) \tag{3.133}$$

The turbulence production based on Boussinesq assumption is provided from

$$P = \nu_t\left(\frac{\partial U_i}{\partial x_j} + \frac{\partial U_j}{\partial x_i}\right)\frac{\partial U_i}{\partial x_j} - \frac{2}{3}\nu_t\left(\frac{\partial U_k}{\partial x_k}\right)^2 \tag{3.134}$$

The kinematic eddy viscosity ν_t is determined from

$$\nu_t = c_\mu(\nu\overline{Re}_T)D_1 D_2$$

where D_1 and D_2 are damping functions which extend the validity of the model to near wall regions and are given by

$$D_1 = 1 - \exp(-y^+/A^+) \tag{3.135}$$

$$D_2 = 1 - \exp(-y^+/A_2^+) \tag{3.136}$$

Various functions and constants appearing in Equation (3.133) are as follows.

$$\frac{1}{\sigma_\epsilon} = (c_{\epsilon_2} - c_{\epsilon_1})\sqrt{c_\mu}/\kappa^2$$

$$\mu_t = \rho\nu_t$$

$$f_2(y^+) = \frac{c_{\epsilon_1}}{c_{\epsilon_2}} + \left(1 - \frac{c_{\epsilon_1}}{c_{\epsilon_2}}\right)\left(\frac{1}{\kappa y^+} + D_1 D_2\right)$$

$$\left[\sqrt{D_1 D_2} + \frac{y^+}{\sqrt{D_1 D_2}}\left(\frac{1}{A^+}\exp(-y^+/A^+)D_2\right.\right.$$

$$\left.\left. + \frac{1}{A_2^+}\exp(-y^+/A_2^+)D_1\right)\right]$$

$$\kappa = 0.41, \qquad c_{\epsilon_1} = 1.2, \qquad c_{\epsilon_2} = 2.0$$

$$c_\mu = 0.09, \qquad A^+ = 26, \qquad A_2^+ = 10$$

3.5.4.2 Spalart-Allmaras One-Equation Turbulence Model:

The Spalart-Allmaras model solves a transport equation for a working variable $\bar{\nu}$ which is related to the eddy viscosity. The governing equation is derived by using empiricism, dimensional analysis, Galilean invariance, and selected dependence on the molecular viscosity [3.20]. The transport equation is expressed as

$$\frac{d\bar{\nu}}{dt} = c_{b1}\bar{S}\bar{\nu}(1 - f_{t2}) + \frac{1}{\sigma}\left[\nabla\cdot\left((\nu + \bar{\nu})\nabla\bar{\nu}\right) + c_{b2}(\nabla\bar{\nu})^2\right]$$

$$- \left[c_{w1}f_w - \frac{c_{b1}}{\kappa^2}f_{t2}\right]\left[\frac{\bar{\nu}}{d}\right]^2 + f_{t1}(\Delta U)^2 \qquad (3.137)$$

where the eddy viscosity is given by

$$\nu_t = \bar{\nu}f_{v1}$$

and

$$f_{v1} = \frac{\chi^3}{\chi^3 + c_{v1}^3}$$

$$\chi = \frac{\bar{\nu}}{\nu}$$

Various functions and constants appearing in Equation (3.137) are defined as

$$\bar{S} = S + \frac{\bar{\nu}}{\kappa^2 d^2}f_{v2}$$

where d is the distance to the wall, κ is the von Karman constant, S is the magnitude of the vorticity, and

$$f_{v2} = 1 - \frac{\chi}{1 + \chi f_{v1}}$$

The function f_w is

$$f_w(r) = g \left(\frac{1 + c_{w3}^6}{g^6 + c_{w3}^6} \right)^{\frac{1}{6}}$$

where

$$g = r + c_{w2}(r^6 - r)$$

and

$$r = \frac{\bar{\nu}}{\bar{S}\kappa^2 d^2}$$

Large values of r should be truncated to a value of about 10. The function f_{t2} is given by

$$f_{t2} = c_{t3} \exp(-c_{t4}\chi^2)$$

and the trip function f_{t1} is

$$f_{t1} = c_{t1} g_t \exp\left[-c_{t2} \left(\frac{\omega_t}{\Delta U} \right)^2 (d^2 + f_t^2 d_t^2) \right] \tag{3.138}$$

The following are used in Equation (3.138):

d_t: The distance from the field point to the trip which is located on the surface.

ω_t: The wall vorticity at the trip.

ΔU:The difference between the velocity at the field point and trip.

g_t: $g_t = \min[1.0, \Delta U/\omega_t \Delta x]$, where ΔX is the grid spacing along the wall at the trip.

The constants used in the equations above are

$$\sigma = \frac{2}{3} \qquad c_{b1} = 0.1355 \qquad c_{b2} = 0.622$$

$$c_{w1} = \frac{c_{b1}}{\kappa^2} + (1 + c_{b2})/\sigma \qquad c_{w2} = 0.3 \qquad c_{w3} = 2 \qquad \kappa = 0.41$$

$$c_{v1} = 7.1 \qquad c_{t1} = 1.0 \qquad c_{t2} = 2.0 \qquad c_{t3} = 1.1 \qquad c_{t4} = 2.0$$

3.5.5 Two-Equation Models

It was pointed out previously that the convection of turbulence is not modeled in zero-equation models. Therefore, the physical effect of past history of the flow is not included in simple algebraic models. In order to account for this physical effect, a transport equation based on the Navier-Stokes equation may be derived. When one such equation is employed, it is referred to as a one-equation model which was discussed in the previous section. When two transport equations are used, it is known as a *two-equation model.*

Complex flowfields which include massively separated flows, unsteadiness, and flows involving multiple-length scales occur frequently in fluid mechanics applications. In these types of complex flows, the lower order turbulence models, that is, zero-, half-, or one-equation models, become very complicated and often ambiguous. Two-equation models are developed to better represent the physics of turbulence in these types of complex flowfields. In this section, k-ϵ and k-ω two-equation turbulence models are reviewed.

3.5.5.1 k-ϵ Two-Equation Turbulence Model:

A commonly used two-equation turbulence model is the k-ϵ model. The first low Reynolds number k-ϵ model was developed by Jones and Launder [3.21] and has subsequently been modified by several investigators.

The partial differential equations are derived for kinetic energy of turbulence (k), and the dissipation of turbulence (ϵ), where

$$k = \frac{1}{2}\left[\overline{u'^2} + \overline{v'^2} + \overline{w'^2}\right]$$

and

$$\epsilon = \nu_t \overline{\left(\frac{\partial u_i'}{\partial x_j}\right)\left(\frac{\partial u_i'}{\partial x_j}\right)}$$

The standard k-ϵ two-equation model may be expressed by the turbulent kinetic equation

$$\frac{d}{dt}(\rho k) = \Gamma_{ij}\frac{\partial u_i}{\partial x_j} - \rho\epsilon + \frac{\partial}{\partial x_j}\left[\left(\mu + \frac{\mu_t}{Pr_\kappa}\right)\frac{\partial k}{\partial x_j}\right] \tag{3.139}$$

and the dissipation rate equation

$$\frac{d}{dt}(\rho\epsilon) = c_1\frac{\epsilon}{k}\tau_{ij}\frac{\partial u_i}{\partial x_j} - c_2\rho\frac{\epsilon^2}{k} + \frac{\partial}{\partial x_j}\left[\left(\mu + \frac{\mu_t}{Pr_\epsilon}\right)\frac{\partial\epsilon}{\partial x_j}\right] \tag{3.140}$$

The turbulent kinetic equation and the dissipation rate equation can be written in an expanded form as

$$\frac{\partial}{\partial t}(\rho k) + \frac{\partial}{\partial x}(\rho u k) + \frac{\partial}{\partial y}(\rho v k) =$$

$$\frac{\partial}{\partial x}\left[\left(\mu + \frac{\mu_t}{Pr_k}\right)\frac{\partial k}{\partial x}\right] + \frac{\partial}{\partial y}\left[\left(\mu + \frac{\mu_t}{Pr_k}\right)\frac{\partial k}{\partial y}\right]$$

$$+\mu_t\left[2\left(\frac{\partial u}{\partial x}\right)^2 + 2\left(\frac{\partial v}{\partial y}\right)^2 + \left(\frac{\partial u}{\partial y} + \frac{\partial v}{\partial x}\right)^2\right] - \rho\epsilon \tag{3.141}$$

and

$$\frac{\partial}{\partial t}(\rho\epsilon) + \frac{\partial}{\partial x}(\rho u\epsilon) + \frac{\partial}{\partial y}(\rho v\epsilon) =$$

$$\frac{\partial}{\partial x}\left[\left(\mu + \frac{\mu_t}{Pr_\epsilon}\right)\frac{\partial\epsilon}{\partial x}\right] + \frac{\partial}{\partial y}\left[\left(\mu + \frac{\mu_t}{Pr_\epsilon}\right)\frac{\partial\epsilon}{\partial y}\right]$$

$$+c_1\frac{\epsilon}{k}\mu_t\left[2\left(\frac{\partial u}{\partial x}\right)^2 + 2\left(\frac{\partial v}{\partial y}\right)^2 + \left(\frac{\partial u}{\partial y} + \frac{\partial v}{\partial x}\right)^2\right] - \rho c_2\frac{\epsilon^2}{k} \qquad (3.142)$$

where typical constants are

$$Pr_k = 1.0 \quad Pr_\epsilon = 1.3 \quad c_1 = 1.44 \quad c_2 = 1.92$$

and the turbulent viscosity is related to ϵ by

$$\mu_t = \rho c_\mu \frac{k^2}{\epsilon} \qquad (3.143)$$

where $c_\mu = 0.09$.

As for the zero-equation models, a two-layer approach is incorporated. That is, Equations (3.141) and (3.142) are used in the outer region down to the vicinity of the viscous sublayer. For a typical wall flow, that corresponds to a y^+ range of 30–50. Adjacent to the surface, wall functions such as (3.114) are employed. The values of k and ϵ must be provided at the interface as the inner boundary conditions for Equations (3.141) and (3.142). These values may be obtained by the following: if y_i denotes the interface between the outer region and the viscous sublayer, then

$$k_{y=y_i} = \frac{\tau_i}{\rho c_D^{\frac{3}{2}}}$$

where c_D is typically 0.164, and τ_i is the shear stress at y_i location. Similarly,

$$\epsilon_{y=y_i} = \frac{c_D k_{y=y_i}^{\frac{3}{2}}}{l}$$

where the mixing length may be evaluated by the simple linear relation for the viscous sublayer given by

$$l = \kappa y$$

Once the local values of the kinetic energy of turbulence and dissipation of turbulence have been computed, turbulent viscosity given by (3.143) may be determined. It is noted that correlations such as $\overline{u'^2}$, $\overline{v'^2}$, and $\overline{w'^2}$ can be determined by semi-empirical relations. For example,

$$\overline{u'^2} = 2a_2 k$$

$$\overline{v'^2} = 2a_3 k$$

$$\overline{w'^2} = 2(1 - a_2 - a_3)k$$

where a_2 and a_3 are structural scales usually assigned values of 0.556 and 0.15, re-spectively.

3.5.5.2 k-ω Two-Equation Turbulence Model:

This two-equation model includes one equation for the turbulent kinetic energy k, and a second equation for the specific turbulent dissipation rate (or turbulent frequency) ω. As in the case of the k-ϵ model, there are several versions of the k-ω model. The k-ω model of Wilcox [3.22] is provided below.

The turbulent kinetic energy equation is given by

$$\frac{d\rho k}{dt} = \tau_{ij}\frac{\partial u_i}{\partial x_j} - \beta^* \rho \omega k + \frac{\partial}{\partial x_j}\left((\mu + \sigma^* \mu_t)\frac{\partial k}{\partial x_j}\right) \tag{3.144}$$

and the specific dissipation rate equation is

$$\frac{d\rho\omega}{dt} = \alpha\frac{\omega}{k}\tau_{ij}\frac{\partial u_i}{\partial x_j} - \beta\rho\omega^2 + \frac{\partial}{\partial x_j}\left((\mu + \sigma \mu_t)\frac{\partial \omega}{\partial x_j}\right) \tag{3.145}$$

The eddy viscosity is determined from

$$\mu_t = \rho\frac{k}{\omega}$$

and the auxiliary relations are

$$\epsilon = \beta^* \omega k$$

$$\ell = \frac{k^{\frac{1}{2}}}{\omega}$$

The constants used in Equations (3.144) and (3.145) are

$$\alpha = \frac{5}{9} \qquad \beta = \frac{3}{40} \qquad \beta^* = \frac{9}{100} \qquad \sigma = \frac{1}{2} \qquad \sigma^* = \frac{1}{2}$$

3.6 Analytical Solutions of the Boundary Layer Equations

3.6.1 Laminar Incompressible Flows Past Flat Surfaces

The governing equations and the associated boundary conditions are:

$$\frac{\partial u}{\partial x} + \frac{\partial v}{\partial y} = 0 \tag{3.146}$$

$$u\frac{\partial u}{\partial x} + v\frac{\partial u}{\partial y} = \nu\frac{\partial^2 u}{\partial y^2} \tag{3.147}$$

$$u\frac{\partial T}{\partial x} + v\frac{\partial T}{\partial y} = \alpha\frac{\partial^2 T}{\partial y^2} \tag{3.148}$$

and

$$u = v = 0 \quad \text{and} \quad T = T_s \ \text{ at } \ y = 0$$

$$u = U_e \quad \text{and} \quad T = T_e \ \text{ at } \ y = \delta$$

One class of solutions to the system of equations above is referred to as the "Similarity Solutions." Blasius solved the governing equations by introducing a "similarity parameter" in the form

$$\eta = y\sqrt{\frac{U_e}{\nu x}} \tag{3.149}$$

A nondimensional velocity is defined as a derivative of $f(\eta)$, an arbitrary function of η, that is,

$$\frac{u}{U_e} = \frac{\partial f}{\partial \eta} = f'(\eta) \tag{3.150}$$

Subsequently, from continuity,

$$\frac{v}{U_e} = \frac{1}{2}\sqrt{\frac{\nu}{U_e x}}(\eta f' - f)$$

With the substitution of expressions above, the momentum equation is obtained as

$$f''' + \frac{1}{2}ff'' = 0 \tag{3.151}$$

and the boundary conditions are

$$f'(0) = 0 \ , \quad f'(\infty) = 1 \ , \quad \text{and } f(0) = 0$$

A numerical solution of (3.151) yields the values presented in Table (3.6). Some simple expressions obtained from the solution above are provided in Table (3.7).

Table 3.6. Solution of laminar boundary layer over a flat plate (Blasius solution)

η	f	f'	f''
0	0	0	0.33206
0.2	0.00664	0.06641	0.33199
0.4	0.02656	0.13277	0.33147
0.6	0.05974	0.19894	0.33008
0.8	0.10611	0.26471	0.32739
1.0	0.16557	0.32979	0.32301
1.2	0.23795	0.39378	0.31659
1.4	0.32298	0.45627	0.30787
1.6	0.42032	0.51676	0.29667
1.8	0.52952	0.57477	0.28293
2.0	0.65003	0.62977	0.26675
2.2	0.78120	0.68132	0.24835
2.4	0.92230	0.72899	0.22809
2.6	1.07252	0.77246	0.20646
2.8	1.23099	0.81152	0.18401
3.0	1.39682	0.84605	0.16136
3.2	1.56911	0.87609	0.13913
3.4	1.74696	0.90177	0.11788
3.6	1.92954	0.92333	0.09809
3.8	2.11605	0.94112	0.08013
4.0	2.30576	0.95552	0.06424
4.2	2.49806	0.96696	0.05052
4.4	2.69238	0.97587	0.03897
4.6	2.88826	0.98269	0.02948
4.8	3.08534	0.98779	0.02187
5.0	3.28329	0.99155	0.01591
5.2	3.48189	0.99425	0.01134
5.4	3.68094	0.99616	0.00793
5.6	3.88031	0.99748	0.00543
5.8	4.07990	0.99838	0.00365
6.0	4.27964	0.99898	0.00240
7.0	5.27926	0.99992	0.00022
8.0	6.27923	1.00000	0.00001

> ### Table 3.7. Incompressible laminar boundary layer.
>
> Boundary layer thickness
> $$\frac{\delta}{x} = \frac{5}{\sqrt{Re_x}} \tag{3.152}$$
>
> Displacement thickness
> $$\frac{\delta^*}{x} = \frac{1.72}{\sqrt{Re_x}} \tag{3.153}$$
>
> Momentum thickness
> $$\frac{\theta}{x} = \frac{0.664}{\sqrt{Re_x}} \tag{3.154}$$
>
> Skin friction coefficient (local)
> $$C_f = \frac{0.664}{\sqrt{Re_x}} \tag{3.155}$$
>
> Skin friction coefficient (total)
> $$C_D = \frac{1.328}{\sqrt{Re_L}} \tag{3.156}$$
> over a plate of length L

The energy equation is solved in a similar manner to quantize the thermal boundary layer. A nondimensional temperature ζ is defined as

$$\zeta = \frac{T - T_s}{T_f - T_s} \tag{3.157}$$

which reduces the energy equation to

$$u\frac{\partial \zeta}{\partial x} + v\frac{\partial \zeta}{\partial y} = \alpha\frac{\partial^2 \zeta}{\partial y^2} \tag{3.158}$$

with the boundary conditions specified by

$$\zeta = 0 \quad \text{at} \quad y = 0$$

$$\zeta = 1 \quad \text{at} \quad y = \infty \quad \text{and at} \quad x = 0$$

Assuming $\zeta = \zeta(\eta)$, the partial differential equation is reduced to a linear first order differential equation.

$$\zeta'' + \frac{Pr}{2}f(\eta)\zeta' = 0 \tag{3.159}$$

with

$$\zeta(0) = 0 \quad \text{and} \quad \zeta'(\infty) = 1$$

The solution of Equation (3.159) can be expressed in terms of the Nusselt number as

$$Nu_x = \zeta_0'(Pr)\sqrt{Re_x} \tag{3.160}$$

$$Nu_L = 2\zeta_0'(Pr)\sqrt{Re_L} \tag{3.161}$$

The function $\zeta_0'(Pr)$ is provided in Table 3.8 and can be approximated by

$$\zeta_0'(Pr) = 0.332 Pr^{\frac{1}{3}}$$

Therefore, expressions (3.160) and (3.161) can be written as

$$Nu_x = 0.332 Pr^{\frac{1}{3}} Re_x^{\frac{1}{2}} \tag{3.162}$$

$$Nu_L = 0.664 Pr^{\frac{1}{3}} Re_L^{\frac{1}{2}} \tag{3.163}$$

Recall that the Nusselt number is defined as $Nu_x = (hx)/k$ where h is the heat transfer coefficient and k is the thermal conductivity.

Table 3.8

Pr	0.6	0.7	0.8	0.9	1.0	1.1	7.0	10.0	15.0
$\zeta_0'(Pr)$	0.276	0.293	0.307	0.320	0.332	0.344	0.645	0.730	0.835

3.6.2 Turbulent Incompressible Flows Past Flat Surfaces

The transition from laminar to turbulent flow takes place at a typical Reynolds number of $Re_{tr} = 5 \times 10^5$ for flows over a flat plate. In a fully developed flow, the boundary layer near the surface is similar to that of a laminar flow. This is called the laminar sublayer and has been discussed in Section 3.5.1. Away from the wall, a fully turbulent profile exists. One of the common representations of a turbulent boundary layer velocity profile is the one-seventh power law:

$$\frac{u}{U_e} = \left(\frac{y}{\delta}\right)^{\frac{1}{7}} \tag{3.164}$$

The use of this velocity profile in the momentum integral equation provides the results presented in Table 3.9.

Table 3.9. Incompressible turbulent boundary layer.

Boundary layer thickness	$\dfrac{\delta}{x} = \dfrac{0.371}{Re_x^{\frac{1}{5}}}$	(3.165)
Displacement thickness	$\dfrac{\delta^*}{x} = \dfrac{0.046}{Re_x^{\frac{1}{5}}}$	(3.166)
Momentum thickness	$\dfrac{\theta}{x} = \dfrac{0.036}{Re_x^{\frac{1}{5}}}$	(3.167)
Skin friction coefficient (local)	$C_f = \dfrac{0.0577}{Re_x^{\frac{1}{5}}}$	(3.168)

The momentum thickness and displacement thickness are related to the boundary layer thickness by the following expressions, valid only for a one-seventh power law velocity distribution

$$\delta^* = \frac{\delta}{8} \tag{3.169}$$

and

$$\theta = \frac{7\delta}{72} \tag{3.170}$$

For a general power law, $\frac{u}{U_e} = \left(\frac{y}{\delta}\right)^{1/n}$

$$\delta^* = \frac{\delta}{1+n} \tag{3.171}$$

and

$$\frac{\theta}{\delta} = \frac{n}{(n+1)(n+2)} \tag{3.172}$$

The heat transfer aspects in turbulent flow are examined through the similarity of momentum and of heat transfer in laminar flows. When $Pr = 1$, it turns out that the ratio of heat flux to shear stress is a constant value within a laminar boundary layer. Reynolds' analogy extends this fact to turbulent boundary layers. In this case, the turbulent Prandtl number $Pr_t = \nu_t/\alpha_t$ is also assumed to be close to unity. The ratio of the heat flux to shear stress yields

$$Nu_x = \frac{C_f}{2} Re_x Pr \tag{3.173}$$

Using the experimentally modified value of $C_f = 0.0596/Re_x^{1/5}$, the Nusselt number is written as

$$Nu_x = 0.0296 Re_x^{0.8} Pr \qquad (3.174)$$

The correlations presented in this section have been modified extensively by various researchers, based on experimental investigation. Additional relations for the Nusselt number are provided in Table 3.10.

3.7 Simple Formulas for Forced Convection

Computation of skin friction and heat transfer for complex geometries typically requires solution of the boundary layer equations and, in many instances, the Navier-Stokes equations. However, for relatively simple geometries, useful relations have been developed to provide an estimate of skin friction and heat transfer. Some of these relations are developed based on theoretical investigations, while others have been developed based on correlations of experimental data. In particular, relations involving turbulent flows have heavily relied upon experimental data. It is important to recognize that various approximations have been introduced in the theoretical analyses to obtain these simple expressions. That is also the case for the experimentally obtained expressions, as experimental errors introduce inaccuracies into the relations. In fact, it is not uncommon to encounter different expressions for, say, skin friction coefficient for similar flow conditions. Typically, a deviation of 5–10% would not be unreasonable. The deviation for complex geometries and for turbulent flows would be even higher. Therefore, it is emphasized that these simple expressions should be used to obtain only an estimate of the desired quantities within their range of applicability. It is also noted that most of these expressions are based on constant surface temperature boundary conditions.

The expressions for skin friction coefficient or heat transfer are typically given in terms of the nondimensional parameters such as the Reynolds number, the Prandtl number, etc. Since the fluid properties such as ρ, μ, c_p, and k are typically functions of temperature, it is obvious that these quantities will vary within the boundary layer. However, for most applications, the temperature variation within the boundary layer is not known; instead, the surface temperature and the free stream temperature are given. Thus, a question arises: What value of the temperature should it be used to determine the fluid properties which are required in the calculation of various parameters, such as the Reynolds number? For most applications involving external flows, a mean film temperature is defined:

$$T_m = \frac{1}{2}(T_s + T_e) \qquad (3.175)$$

where T_s is the surface temperature and T_e is the fluid temperature at the boundary layer edge, typically free stream temperature. Subsequently, the mean film temperature is used to determine the fluid properties. For internal flows, such as flow in a pipe of radius R, a bulk fluid temperature is defined as

$$T_b = \frac{\int_0^R \rho c_p 2\pi r T \, dr}{\int_0^R \rho c_p 2\pi r \, dr} \tag{3.176}$$

Now this temperature is used to determine the fluid properties. It is also common to define a mean film temperature as

$$T_m = \frac{1}{2}(T_s + T_b) \tag{3.177}$$

which is then used to evaluate the fluid properties.

Many of the simple expressions developed for incompressible flows can be used in high speed flows if a factor is included to account for the effect of compressibility. A simple scheme was introduced by Eckert [3.23], where the flow properties are evaluated at a reference temperature T^*, defined as

$$T^* = \frac{1}{2}(T_s + T_e) + 0.22(T_r - T_e) \tag{3.178}$$

where T_r is the recovery temperature.

Recovery temperature T_r is the temperature of the wall corresponding to zero heat transfer. Recovery temperature is also called *adiabatic wall temperature* T_{aw}. The recovery temperature can be calculated from either one of the following relations.

$$T_r = T_e + r\frac{V_e^2}{2c_p} \tag{3.179}$$

or

$$T_r = T_e \left(1 + r\frac{\gamma - 1}{2}M_e^2\right) \tag{3.180}$$

where e denotes the property at the boundary layer edge, and r is the recovery factor. The recovery factor is defined as

$$r = \frac{T_r - T_e}{\dfrac{V_e^2}{2c_p}} = \frac{T_r - T_e}{T_{te} - T_e} \tag{3.181}$$

For a laminar flow

$$r = Pr^{1/2} \tag{3.182}$$

and for a turbulent flow

$$r = Pr^{1/3} \tag{3.183}$$

Note that for low speed flows, T_r approaches T_e, and therefore the reference temperature becomes the mean film temperature given by Equation (3.172), $T_m = \frac{1}{2}(T_e + T_s)$.

Some commonly used expressions for the skin friction coefficient and drag coefficient are given in Table 3.11. The correlations for heat transfer are typically provided in terms of the nondimensional heat transfer coefficient known as the Nusselt number. These relations are provided in Table 3.10.

Table 3.10: Correlation equations for heat transfer coefficient for forced convection

Geometry	Equation(s)	Restrictions/Remarks	Ref(s)
Flat Plate	$Nu_x = 0.332\,Re_x^{\frac{1}{2}} Pr^{\frac{1}{3}}$ (3.184)	Incompressible laminar flow, $0.6 \le Pr \le 50$, properties at T_m	[3.24]
	$Nu_L = 0.664\,Re_L^{\frac{1}{2}} Pr^{\frac{1}{3}}$ (3.185)		[3.24]
	$Nu_x = 0.565\,(Re_x Pr)^{\frac{1}{2}}$ (3.186)	Incompressible laminar flow, $Pr < 0.05$, properties at T_m	[3.35]
	$Nu_L = 1.130\,(Re_L Pr)^{\frac{1}{2}}$ (3.187)		
	$Nu_x = \dfrac{0.3387\,Re_x^{\frac{1}{2}} Pr^{\frac{1}{3}}}{\left[1 + (0.0468/Pr)^{\frac{2}{3}}\right]^{\frac{1}{4}}}$ (3.188)	Incompressible laminar flow, $RePr > 100$, properties at T_m	[3.36]
	$Nu_L = \dfrac{0.6774\,Re_L^{\frac{1}{2}} Pr^{\frac{1}{3}}}{\left[1 + (0.0468/Pr)^{\frac{2}{3}}\right]^{\frac{1}{4}}}$ (3.189)		
	$Nu_x = 0.0296\,Re_x^{0.8} Pr^{\frac{1}{3}}$ (3.190)	Incompressible turbulent flow, $5 \times 10^5 < Re_x < 10^7$, Colburn analogy	[3.25]
	$Nu_x = \dfrac{0.185\,Re_x Pr^{\frac{1}{3}}}{(\log_{10} Re_x)^{2.584}}$ (3.191)	Incompressible turbulent flow, $Re_x > 10^7$, $0.6 \le Pr \le 60$, properties at T_m	

(Continued on next page)

Table 3.10: Continued.

Geometry	Equation(s)		Restrictions/Remarks	Ref(s)
Flat Plate	$Nu_x = \dfrac{0.0296 Re_x^{0.8} Pr}{1 + 0.86 Re_x^{-0.1}(Pr - 1)}$	(3.192)	Incompressible turbulent flow, includes effect of laminar sublayer $5 \times 10^5 < Re_x < 10^7$	[3.24]
	$Nu_x = \dfrac{0.185(\log_{10} Re_x)^{-2.584} Re_x Pr}{1 + 2.15(\log_{10} Re_x)^{-1.292}(Pr - 1)}$	(3.193)	$Re_x > 10^7$	[3.24]
	$Nu_x = (0.0296 Re_x^{0.8} Pr) \div$ $\left(1 + 0.86 Re_x^{-0.1}\left\{(Pr - 1) + \ln\left[1 + \dfrac{5}{6}(Pr - 1)\right]\right\}\right)$	(3.194)	Incompressible turbulent flow, includes effect of buffer zone $5 \times 10^5 < Re_x < 10^7$	[3.24]
	$Nu_x = (0.185[\log_{10} Re_x]^{-2.584} Re_x Pr) \div$ $\left(1 + 2.15(\log_{10} Re_x)^{-1.292}\left\{(Pr - 1) + \ln\left[1 + \dfrac{5}{6}(Pr - 1)\right]\right\}\right)$	(3.195)	$Re_x > 10^7$	
	$Nu_L = \left(0.037 Re_L^{0.8} - \dfrac{A}{2}\right) Pr^{\frac{1}{3}}$	(3.196)	Average heat transfer which accounts for the initial laminar portion of the boundary layer, $0.6 \leq Pr \leq 60$, properties at T_m , $5 \times 10^5 < Re_L < 10^7$	[3.25]

(Continued on next page)

Table 3.10: Continued.

Geometry	Equation(s)	Restrictions/Remarks	Ref(s)
Flat Plate	$Nu_L = \left[0.228\,Re_L(\log_{10} Re_L)^{-2.584} - \dfrac{A}{2}\right]Pr^{\frac{1}{3}}$ $A = 0.074\,Re_{tr}^{\frac{4}{5}} - 1.328\,Re_{tr}^{\frac{1}{2}}$ (3.197)	$10^7 < Re_L < 10^9$	
	$Nu_x^* = 0.332\,Re_x^{*\frac{1}{2}}Pr^{*\frac{1}{3}}$ (3.198)	Laminar flow with compressibility effects included, properties at T^*, $0.6 \le Pr^* \le 50$, $r = Pr^{*\frac{1}{2}}$	
	$Nu_L^* = 0.664\,Re_L^{*\frac{1}{2}}Pr^{*\frac{1}{3}}$ (3.199)		
	$Nu_x^* = 0.0296\,Re_x^{*0.8}Pr^{*\frac{1}{3}}$ (3.200)	Turbulent flow with compressibility effects included, properties at T^*, $0.6 \le Pr^* \le 60$, $5 \times 10^5 < Re_x^* < 10^7$, $r = Pr^{*\frac{1}{3}}$	
	$Nu_x^* = 0.185\,Re_x^*(\log_{10} Re_x^*)^{-2.584}Pr^{*\frac{1}{3}}$ (3.201)	$Re_x^* > 10^7$	
Circular Cylinder in crossflow	$Nu_\theta = 1.14(Re_D)^{\frac{1}{2}}Pr^{0.4}\left[1 - \left(\dfrac{\theta}{90}\right)^3\right]$ (3.202)	Incompressible, properties at T_m, forward section of cylinder $0 < \theta < 80°$, $Nu_\theta = \dfrac{h_\theta D}{k}$	

(Continued on next page)

Table 3.10: Continued.

Geometry	Equation(s)		Restrictions/Remarks	Ref(s)
Circular Cylinder in crossflow	$Nu_D = C(Re_D)^m Pr^n \left(\dfrac{Pr_\infty}{Pr_s}\right)^{\frac{1}{4}}$ where	(3.203)	$Nu_D = \dfrac{hD}{k}$: Average Nusselt number. Pr_s evaluated at T_s, Pr_∞ at T_∞, Remaining properties at T_m. $0.7 < Pr < 500$, $1 < Re_D < 10^6$	[3.27]

Re_D	C	m
1–40	0.75	0.4
40–1000	0.51	0.5
1000–200,000	0.26	0.6
200,000–1,000,000	0.076	0.7

and

$$n = 0.36 \quad \text{for} \quad Pr > 10$$
$$n = 0.37 \quad \text{for} \quad Pr \le 10$$

Geometry	Equation(s)		Restrictions/Remarks	Ref(s)
	$Nu_D = 0.3 + \dfrac{0.62\,Re_D^{\frac{1}{2}} Pr^{\frac{1}{3}}}{\left[1 + (0.4/Pr)^{\frac{2}{3}}\right]^{\frac{1}{4}}} \left[1 + \left(\dfrac{Re_D}{2.82 \times 10^5}\right)^{\frac{5}{8}}\right]^{\frac{4}{5}}$	(3.204)	The Churchill and Bernstein relation. $Re_D Pr > 0.2$, properties at T_m	[3.38]
	$Nu_D = 0.123 Re_D^{0.651} + \left(\dfrac{D}{L}\right)^{0.85} Re_D^{0.792}$	(3.205)	Short cylinder in a gas, $\dfrac{L}{D} < 4$, $7 \times 10^4 < Re_D < 1.1 \times 10^5$	[3.39]
Sphere	$Nu_D = 0.37(Re_D)^{0.6}$	(3.206)	Gases, $25 < Re_D < 100,000$	
	$Nu_D = 2 + \left(0.4 Re_D^{\frac{1}{2}} + 0.06 Re_D^{\frac{2}{3}}\right) Pr^{0.4} \left(\dfrac{\mu_\infty}{\mu_s}\right)^{\frac{1}{4}}$	(3.207)	Liquids or gases, $3.5 < Re_D < 7.6 \times 10^4$, $0.15 < Pr < 380$	[3.40]

(Continued on next page)

Table 3.10: Continued.

Geometry	Equation(s)	Restrictions/Remarks	Ref(s)
	$Nu_D = 2 + \left(\dfrac{Re_D}{4} + 3 \times 10^4 Re_D^{1.6} \right)^{\frac{1}{2}}$ (3.208)	Air, constant surface temperature, $100 < Re_D < 2 \times 10^5$	[3.41]
	$Nu_D = 430 + 5 \times 10^{-3} Re_D + 0.25 \times 10^{-9} Re_D^2 - 3.1 \times 10^{-17} Re_D^3$ (3.209)	Air, $4 \times 10^5 < Re_D < 5 \times 10^6$	
Noncircular cylinders in crossflow	$Nu_D = C\, Re_D^m\, Pr^{\frac{1}{3}}$ (3.210)	Gases, properties at T_m	[3.42]

Re_D	C	m
5000–100000	0.246	0.588
2500–7500	0.289	0.624
5000–100000	0.102	0.675
2500–8000	0.145	0.699
5000–19500	0.159	0.638
19500–100000	0.039	0.782
5000–100000	0.153	0.638
2500–15000	0.245	0.612
3000–15000	0.094	0.804
4000–15000	0.227	0.731

Table 3.11: Correlations for skin friction coefficient and drag coefficient for forced convection

Geometry	Equation(s)		Restrictions/Remarks	Reference(s)
Flat Plate	$C_f = 0.664\, Re_x^{-\frac{1}{2}}$	(3.211)	Incompressible laminar flow	[3.24]
	$C_D = 1.328\, Re_L^{-\frac{1}{2}}$	(3.212)	Incompressible laminar flow	[3.24]
	$C_f = 0.0592\, Re_x^{-\frac{1}{5}}$	(3.213)	Incompressible turbulent flow $5 \times 10^5 < Re_x < 10^7$	[3.24]
	$C_D = 0.074\, Re_L^{-\frac{1}{5}}$	(3.214)	Incompressible turbulent flow $5 \times 10^5 < Re_L < 10^7$	[3.24]
	$C_f = \dfrac{0.455}{\ln^2(0.06\,Re_x)}$	(3.215)	Incompressible turbulent flow entire range	[3.16]
	$C_D = \dfrac{0.523}{\ln^2(0.06\,Re_L)}$	(3.216)	Incompressible turbulent flow	[3.16]
	$C_f = (2.0 \log Re_x - 0.65)^{-2.3}$	(3.217)	Incompressible turbulent flow entire range	[3.25]
	$C_D = \dfrac{0.455}{(\log_{10} Re_L)^{2.58}}$	(3.218)	Incompressible turbulent flow	[3.25]
	$\dfrac{1}{\sqrt{C_f}} = 4.15 \log_{10} Re_x\, C_f + 1.7$	(3.219)	Incompressible turbulent flow entire range	[3.26, 3.27]
	$\dfrac{1}{\sqrt{C_D}} = 4.13 \log_{10} Re_L\, C_D$	(3.220)	Incompressible turbulent flow	[3.26, 3.27]

(Continued on next page)

Table 3.11: Continued.

Geometry	Equation(s)		Restrictions/Remarks	Ref(s)
Flat Plate	$C_f = \dfrac{0.37}{(\log Re_x)^{2.584}}$	(3.221)	Incompressible turbulent flow $Re_x > 10^7$	[3.28]
	$C_D = \dfrac{0.427}{(\log Re_L - 0.407)^{2.64}}$	(3.222)		[3.28]
	$C_D = 0.074 Re_L^{-\frac{1}{5}} - Re_L^{-1}(0.074 Re_{tr}^{\frac{4}{5}} - 1.328 Re_{tr}^{\frac{1}{2}})$	(3.223)	Averaged combined laminar and turbulent flow, smooth surface, $Re_{tr} < Re_L < 10^7$	[3.25]
	$C_D = 0.455(\log_{10} Re_L)^{-2.584} - Re_L^{-1}(0.074 Re_{tr}^{\frac{4}{5}} - 1.328 Re_{tr}^{\frac{1}{2}})$	(3.224)	Averaged combined laminar and turbulent flow, smooth surface $Re_{tr} < Re_L < 10^9$	[3.25]
	$C_D = 0.455(\log_{10} Re_L)^{-2.584} - Re_L^{-1}(1700)$	(3.225)	Averaged combined laminar and turbulent flow, smooth surface, $Re_{tr} < Re_L < 10^9$, $Re_{tr} = 5 \times 10^5$	
	$Re_x = 1.73125(1 + 0.3k^+)e^{0.4\lambda}\,[A]$	(3.226)	Incompressible turbulent flow, Rough plate, with average roughness height $= k$,	[3.16]
	$[A] = z^2 - 4z + 6 - \dfrac{0.3k^+}{1 + 0.3k^+}(z - 1)$	(3.227)	$u^* = \sqrt{\dfrac{\tau_w}{\rho}} \equiv$ The friction velocity	
	where $z = 0.4\lambda$, $\lambda = \sqrt{\dfrac{2}{C_f}}$, $k^+ = \dfrac{ku^*}{\nu} = \dfrac{Re_x(k/x)}{\lambda}$		λ is obtained from the relation for the local skin friction coefficient	

(Continued on next page)

Table 3.11: Continued.

Geometry	Equation(s)	Restrictions/Remarks	Ref(s)
Flat Plate	$C_D = \dfrac{0.554}{Re_L} e^{0.4\lambda_L} \left[1 + \dfrac{0.3 Re_L(k/L)}{\lambda_L} \right] \left(1 - \dfrac{5}{\lambda_L} \right)$ where $\lambda_L = \sqrt{\dfrac{2}{C_f(L)}}$ (3.228)		[3.16]
	$C_f = \left(1.4 + 3.7 \log \dfrac{x}{k} \right)^{-2}$ (3.229)	Fully rough $\dfrac{x}{k} > 100$	[3.16]
	$C_D = 0.024 \left(\dfrac{k}{L} \right)^{\frac{1}{6}}$ (3.230)	Fully rough	[3.16]
	$C_f = \left(2.87 + 1.58 \log \dfrac{x}{k} \right)^{-2.5}$ (3.231)	Fully rough	[3.29]
	$C_D = \left(1.89 + 1.62 \log \dfrac{L}{k} \right)^{-2.5}$ (3.232)	Fully rough	[3.29]
	$\dfrac{C_D}{C_{D_i}} = (1 + 0.144 M_\infty^2)^{-0.65}$ (3.233)	Compressibility correction, Turbulent flow C_{D_i}: incompressible drag coefficient	[3.30]
	$C_f \sqrt{Re} = 0.45 \cos \alpha + \dfrac{4.6 U_\infty}{10000} \sin \alpha \cos^{2.2} \alpha$ (3.234)	Laminar hypersonic flow, α is the angle of attack, U_∞ in fps, results are typically $\pm 20\%$ for low altitudes at high α or $\pm 10\%$ at high altitude	[3.31]

(Continued on next page)

Table 3.11: Continued.

Geometry	Equation(s)		Restrictions/Remarks	Ref(s)
	$C_f (Re)^{0.2} = 0.48 \sin(4.5\alpha) + 0.7 \dfrac{U_\infty}{10000} \cos^{2.2}\alpha \sin^{1.5}\alpha$	(3.235)	Turbulent hypersonic flow,	[3.32]
Cone	$C_{f_{\text{cone}}} = \sqrt{3}\, C_{f_{\text{flat plate}}}$	(3.236)	Laminar flow, identical boundary layer edge conditions as the flat plate	
	$C_{D_{\text{cone}}} = \dfrac{2}{\sqrt{3}}\, C_{D_{\text{flat plate}}}$	(3.237)	Laminar flow, identical boundary layer edge conditions as the flat plate	
	$C_{f_{\text{cone}}} \cong 1.15\, C_{f_{\text{flat plate}}}$	(3.238)	Turbulent flow, identical boundary layer edge conditions as the flat plate	
Cylinder parallel to freestream	$C_f \sqrt{Re_x} = \zeta \left[\dfrac{2}{\chi} + \dfrac{0.571}{\chi^2} - \dfrac{0.835}{\chi^3} + \dots \right]$	(3.239)	Incompressible, laminar flow, long cylinder of constant radius a, $\zeta > 30$ $\chi = \ln 2\zeta$, $\zeta = \dfrac{x/a}{\sqrt{Re_x}}$	[3.33]
	$C_f \sqrt{Re_x} = 0.664 + 1.392\zeta - 1.594\zeta^2 + \dots$	(3.240)	Incompressible, laminar flow, short cylinder, $\zeta < 0.1$	[3.34]
	$C_D = \dfrac{4}{Re_a \ln\left[(4L/a)/Re_a\right]}$	(3.241)	Incompressible, laminar flow, $\zeta > 100$	[3.16]
	$C_f = 0.0015 + \left[0.2 + 0.016 \left(\dfrac{x}{a}\right)^{0.4} \right] Re_x^{-\frac{1}{3}}$	(3.243)	Incompressible, turbulent flow, very long cylinder	[3.16]
	$C_D = 0.0015 + \left[0.3 + 0.015 \left(\dfrac{L}{a}\right)^{0.4} \right] Re_L^{-\frac{1}{3}}$	(3.244)	Incompressible, turbulent flow $\dfrac{L}{a} < 10^6$, $10^6 < Re_L < 10^9$	[3.16]

(Continued on next page)

Table 3.11: Continued.

Geometry	Equation(s)		Restrictions/Remarks	Ref(s)
Cylinder Parallel to free stream	$C_f = \dfrac{4}{Re_a}\left[\dfrac{1}{\zeta} + \dfrac{0.5772}{\zeta^2} + \dots\right]$		Extremely thick turbulent boundary layer, $Re_a < 20$, $\zeta = \ln\left[\dfrac{4Re_x}{Re_a^2}\right]$	[3.16]
	$C_D = \dfrac{4}{Re_a}\left[\dfrac{1}{\zeta_L} + \dfrac{1.5772}{\zeta_L^2} + \dots\right]$	(3.245)	Same as above, $\zeta_L = \ln\left[\dfrac{4Re_L}{Re_a^2}\right]$	[3.16]

3.8   Flow Over Tube Bundles

The design and analysis of many types of heat exchangers require knowledge of heat transfer of bundles of circular tubes in a crossflow. The flow pattern of such an arrangement is quite complex and includes turbulence, flow separation and attachment, and flow unsteadiness. Most available equations for analysis of flow over tube bundles are based almost entirely on experimental data. The typical **arrangement** of tubes is either in-line or staggered. The geometrical arrangement and the nomenclature are shown in Figures 3.9 and 3.10.

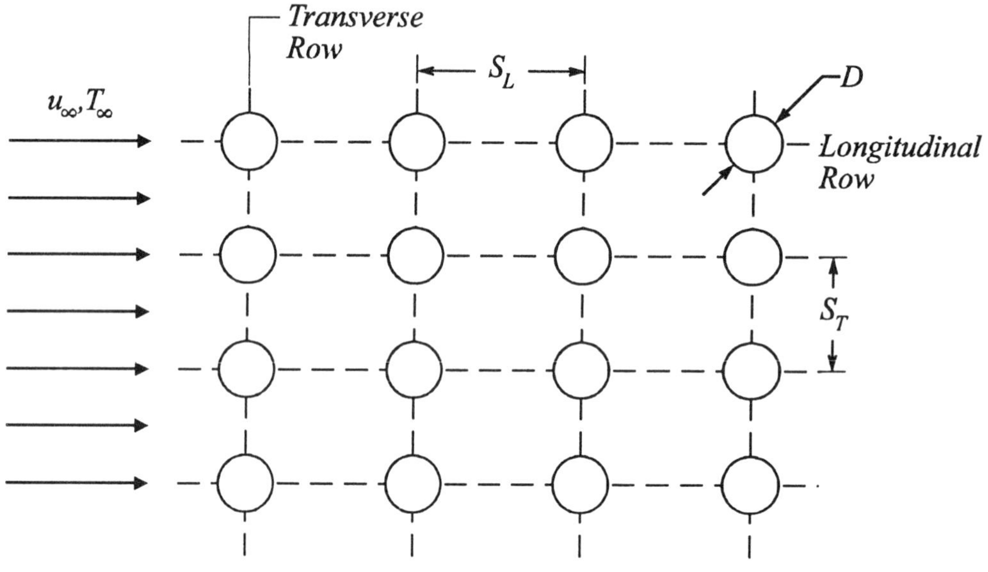

Figure 3.9. Nomenclature for tube bundles, in line arrangement.

The heat transfer coefficient for the first row is similar to that of a single cylinder, while the heat transfer coefficient for the inner rows is generally larger. In fact, once the flow has passed several rows of tubes, the heat transfer coefficient becomes independent of location. An average Nusselt number may be determined from the following relations developed by Zukauskas [3.37]. The Reynolds number is determined based on maximum average velocity u_m, which occurs at minimum free area. Thus,

for in-line arrangement: $$u_m = u_\infty \frac{S_T}{S_T - D} \qquad (3.246)$$

for staggered arrangement if $\quad 2(S_D - D) > S_T - D \qquad (3.247)$

$$u_m = u_\infty \frac{S_T}{S_T - D}$$

or for staggered arrangement if $\quad 2(S_D - D) < S_T - D \qquad (3.248)$

$$u_m = u_\infty \frac{S_T}{2(S_D - D)}$$

where

$$S_D = \left[S_L^2 + \left(\frac{S_T}{2} \right)^2 \right]^{\frac{1}{2}}$$

The average heat transfer for the inner rows [3.37] is given in Table 3.12 where the Prandtl number range is $0.7 < Pr < 500$.

Pr_s is evaluated at the tube surface temperature, and Pr_∞ is evaluated at T_∞. Fluid properties are evaluated at T_m.

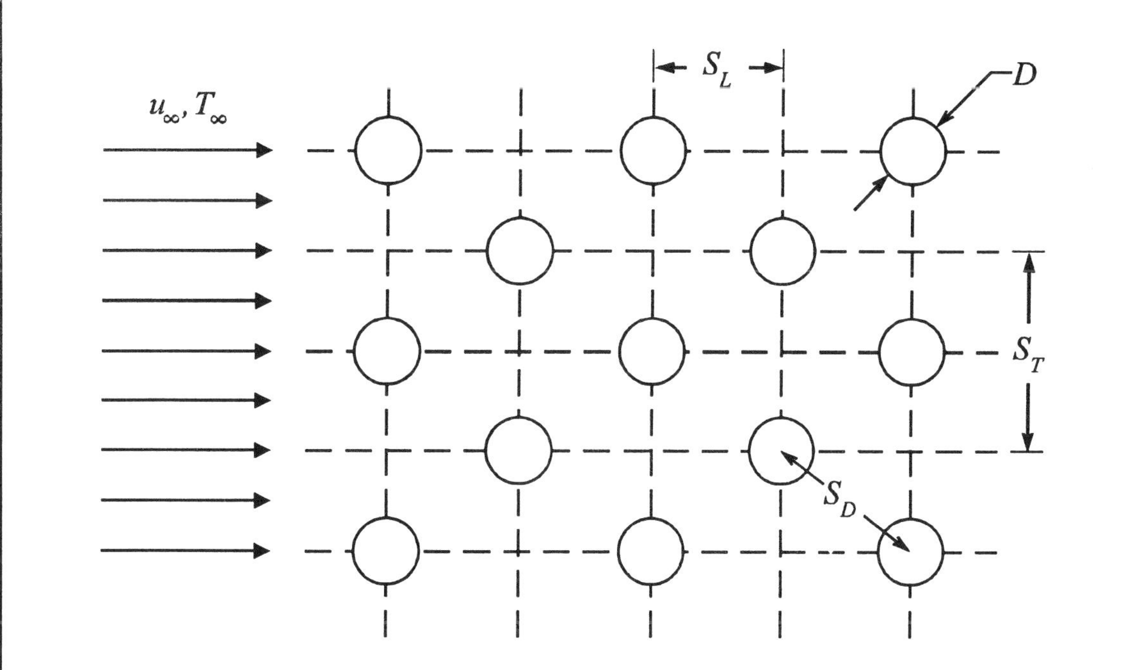

Figure 3.10. Nomenclature for tube bundles, staggered arrangement.

Table 3.12: Average Nusselt number for flow over tube bundles

Arrangement	Average Nusselt Number		Restrictions/ Remarks
In-line tubes	$\overline{Nu}_D = 0.8\,Re_D^{0.4}Pr^{0.36}\left(\dfrac{Pr_\infty}{Pr_s}\right)^{0.25}$	(3.249)	Laminar flow, $10 < Re_D < 100$
Staggered tubes	$\overline{Nu}_D = 0.9\,Re_D^{0.4}Pr^{0.36}\left(\dfrac{Pr_\infty}{Pr_s}\right)^{0.25}$	(3.250)	Laminar flow, $10 < Re_D < 100$
In-line tubes	$\overline{Nu}_D = 0.27\,Re_D^{0.63}Pr^{0.36}\left(\dfrac{Pr_\infty}{Pr_s}\right)^{0.25}$	(3.251)	Transition regime, $10^3 < Re_D < 2\times10^5$ $\dfrac{S_T}{S_L} \geq 0.7$
Staggered tubes	$\overline{Nu}_D = 0.35\left(\dfrac{S_T}{S_L}\right)^{0.2}Re_D^{0.60}Pr^{0.36}\left(\dfrac{Pr_\infty}{Pr_s}\right)^{0.25}$	(3.252)	$\dfrac{S_T}{S_L} < 2$
Staggered tubes	$\overline{Nu}_D = 0.40\,Re_D^{0.60}Pr^{0.36}\left(\dfrac{Pr_\infty}{Pr_s}\right)^{0.25}$	(3.253)	$\dfrac{S_T}{S_L} \geq 2$
In-line tubes	$\overline{Nu}_D = 0.021\,Re_D^{0.84}Pr^{0.36}\left(\dfrac{Pr_\infty}{Pr_s}\right)^{0.25}$	(3.254)	Turbulent regime, $Re_D > 2\times10^5$
Staggered tubes	$\overline{Nu}_D = 0.022\,Re_D^{0.84}Pr^{0.36}\left(\dfrac{Pr_\infty}{Pr_s}\right)^{0.25}$	(3.255)	$Pr > 1$
	$\overline{Nu}_D = 0.019\,Re_D^{0.84}$	(3.256)	$Pr = 0.7$

3.9 Hydrodynamic Entrance Length

When a uniform, essentially potential flow enters a domain between two parallel plates, boundary layers begin to develop at the upper and lower surfaces. As the boundary layers grow in the streamwise direction, they could eventually merge. The streamwise distance to the location where the boundary layer flow fills the domain is called *the entrance length* and is denoted by X_L, as shown in Figure 3.11.

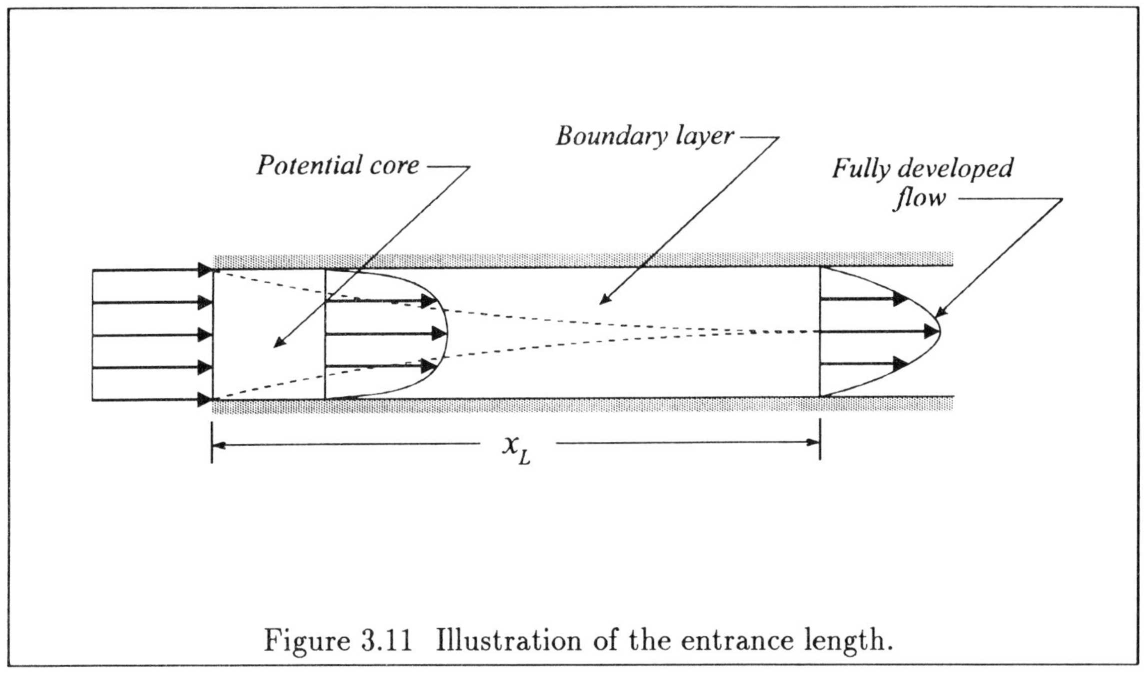

Figure 3.11 Illustration of the entrance length.

The flow beyond X_L is called a *fully developed flow* and becomes essentially independent of streamwise direction. A similar flow pattern occurs when flow enters ducts or pipes. Application of the integral momentum equation provides an expression for the pressure drop over a specified distance such as

$$\frac{2(p_i - p_x)}{\rho \bar{u}^2} = \lambda \frac{x}{D} + K \tag{3.257}$$

where D is the diameter of the pipe or the distance between plates, $\bar{u}$ is the mean velocity, and p_i is the pressure at the entrance, that is, $x = 0.0$. Furthermore, the nondimensional parameter K is

$$K = \frac{2}{3} + \int_0^{x/a} \frac{4(\tau - \tau_w)}{\rho \bar{u}^2} \frac{dx}{a} \tag{3.258}$$

where $a = D/2$. An approximate value for K is given by

$$K = 1.31 \quad \text{for pipe entrance}$$

$$K = 0.67 \quad \text{for channel entrance}$$

The value of λ defined by $\lambda = (8\tau_w)/(\rho \bar{u}^2)$ is shown to be

$$\lambda = \frac{64}{Re_D} \quad \text{for pipe flow} \tag{3.259}$$

$$\lambda = \frac{96}{Re_D} \quad \text{for channel flow} \tag{3.260}$$

Since $\tau_w = C_f(\rho \bar{u}^2/2)$, the above relations can be written as $C_f = 24/Re_D$ for channel flow and as $C_f = 16/Re_D$ for pipe flow. An estimate of entrance length is

$$\frac{X_L}{D} = 0.08 Re_D + 0.7 \quad \text{for pipe entrance} \tag{3.261}$$

$$\frac{X_L}{D} = 0.04 Re_D + 0.5 \quad \text{for channel entrance} \tag{3.262}$$

3.10 Thermal Entrance Length—Laminar Flow

Similar to that of the establishment of hydrodynamic entrance length, a merging of thermal boundary layers will result in thermal entrance length. A limited number of solutions are reviewed in the following for a circular pipe with uniform wall temperature.

The governing energy equation is reduced to

$$\frac{\partial^2 T}{\partial r^2} + \frac{1}{r}\frac{\partial T}{\partial r} = \frac{u}{\alpha}\frac{\partial T}{\partial x} - \frac{\partial^2 T}{\partial x^2} \tag{3.263}$$

which is nondimensionalized by

$$r^* = \frac{r}{R}, \quad u^* = \frac{u}{\bar{u}}, \quad x^* = \frac{x}{R\,Pr\,Re}, \quad \theta = \frac{T_w - T}{T_w - T_i}$$

where T_i is the inlet temperature and R is the radius. Equation (3.263) is hence expressed in a nondimensional form as

$$\frac{\partial^2 \theta}{\partial r^{*2}} + \frac{1}{r^*}\frac{\partial \theta}{\partial r^*} = \frac{u^*}{2}\frac{\partial \theta}{\partial x^*} - \frac{1}{(RePr)^2}\frac{\partial^2 \theta}{\partial x^{*2}} \tag{3.264}$$

For applications where $RePr > 100$, the last term is neglected, and the following results include this requirement. Furthermore, utilizing a parabolic velocity profile and assuming hydrodynamically fully developed laminar flow, Equation (3.264) is now written as

$$\frac{\partial^2 \theta}{\partial r^{*2}} + \frac{1}{r^*}\frac{\partial \theta}{\partial r^*} = (1 - r^{*2})\frac{\partial \theta}{\partial x^*} \tag{3.265}$$

Subject to the boundary conditions of

$$\theta(0, r^*) = 1 \tag{3.266}$$

and

$$\theta(x^*, 1) = 0 \tag{3.267}$$

The solution is given by

$$\theta(x^*,\, r^*) = \sum_{n=0}^{\infty} C_n \, \boldsymbol{R}_n \, (r^*) \, \exp(-\lambda_n^2 x^*) \tag{3.268}$$

where λ_n's are the eigenvalues for the eigenfunctions $\boldsymbol{R}_n$, and C_n's are the associated constants. The solution above yields the following quantities of interest.

Heat flux:
$$q_w = \frac{2k}{R} \sum_{n=0}^{\infty} G_n \exp(-\lambda_n^2 x^*)(T_w - T_i) \qquad (3.269)$$

Mixed mean temperature:
$$\theta_m = 8 \sum_{n=0}^{\infty} \frac{G_n}{\lambda_n^2} \exp(-\lambda_n^2 x^*) \qquad (3.270)$$

Local Nusselt number:
$$Nu_x = \frac{\sum_{n=0}^{\infty} G_n \exp\left(-\lambda_n^2 x^*\right)}{2 \sum_{n=0}^{\infty} (G_n/\lambda_n^2) \exp(-\lambda_n^2 x^*)} \qquad (3.271)$$

Mean Nusselt number:
$$Nu_m = \frac{1}{2x^*} \ln \left[\frac{1}{8 \sum_{n=0}^{\infty} (G_n/\lambda_n^2) \exp(-\lambda_n^2 x^*)} \right] \qquad (3.272)$$

where
$$G_n = -\frac{1}{2} C_n \boldsymbol{R}_n'(1)$$

The eigenvalues and constants in the expressions above are provided in Table 3.13 as presented in Ref. [3.43]. For $n > 2$, $\lambda_n = 4n + (8/3)$, and $G_n = 1.01276\lambda_n^{-\frac{1}{3}}$.

Table 3.13: Eigenvalues λ_n and function G_n for laminar flow in circular pipe subject to uniform wall temperature

n	G_n	λ_n^2
0	0.749	7.313
1	0.544	44.61
2	0.463	113.9
3	0.415	215.2
4	0.383	348.6

For $x^* > 0.1$, only the first term makes a contribution to Equation (3.271), thus yielding $Nu_x = 3.658$, which is the Nusselt number for a fully developed temperature profile. Therefore, it is reasonable to assume that the thermal entrance length is at

$$x^* = 0.1$$

or

$$\frac{X_L}{D} = 0.05 Re_D \, Pr \qquad (3.273)$$

An empirical relation for short tubes has been developed by Seider and Tate [3.44].

$$Nu_D = 1.86 \left[\left(\frac{D}{L}\right) Re_D Pr \right]^{\frac{1}{3}} \left(\frac{\mu}{\mu_s}\right)^{0.14} \qquad (3.274)$$

where $0.48 < Pr < 16,700$ and $(\frac{D}{L})Re_D Pr > 10$. The fluid properties are all evaluated at T_b, except μ_s which is evaluated at T_s.

3.10.1 Constant Heat Flux

The governing equation is the same as for the previous section except that the boundary condition is specified by a constant temperature gradient. Furthermore, the nondimensional temperature is defined as

$$\theta = \frac{T_i - T}{\dot{q}_w \dfrac{D}{k}}$$

where T_i is the inlet temperature, and k is the thermal conductivity. The solution of governing equation yields the local Nusselt number as

$$Nu_x = \left[\frac{1}{Nu_\infty} - \frac{1}{2} \sum_{m=1}^{\infty} \frac{\exp(-\gamma_m^2 x^*)}{A_m \gamma_m^4} \right]^{-1} \tag{3.275}$$

where the eigenvalues and constants as provided in Ref. [3.44] are presented in Table 3.14. For $m > 5$, $\gamma_m = 4m + (4/3)$, and $A_m = 0.428\gamma_m^{-\frac{7}{3}}$.

Table 3.14: Eigenvalues and constants for circular pipe subject to uniform heat flux.

m	γ_m^2	A_m
1	25.68	7.630×10^{-3}
2	83.86	2.053×10^{-3}
3	174.20	0.903×10^{-3}
4	296.50	0.491×10^{-3}
5	450.90	0.307×10^{-3}

The thermal entry lengths and the corresponding local Nusselt numbers are given in Table 3.15. Recall that $Nu_\infty = 4.36$ for fully developed flow (Sec. 1.11.4).

Table 3.15: Thermal entry length for circular tube with constant heat flux.

x^+	0	0.002	0.004	0.010	0.020	0.04	0.10	∞
Nu_x	∞	12.00	9.93	7.49	6.14	5.19	4.51	4.36

3.10.2 Thermal Entry Lengths for Noncircular Tubes and Ducts in Laminar Flow

3.10.2.1 Flow Between Parallel Planes: The expressions for the Nusselt number obtained as Equations (3.271) and (3.272) can be used if x^* is defined as

$$x^* = \frac{2x/D_h}{RePr} \qquad (3.276)$$

where D_h is the hydraulic diameter defined by $4 \times$ cross-sectional area/perimeter. For the flow between parallel plates $D_h = 2h$, where h is the spacing between plates. The solution corresponds to the same constant temperature at both surfaces. The eigenvalues and constants required for the equation are given in Table 3.16 [3.35]. For $n > 2$, $\lambda_n = \dfrac{16n}{\sqrt{3}} + \dfrac{20}{3\sqrt{3}}$, $G_n = 2.68\lambda_n^{-\frac{1}{3}}$.

Table 3.16: Eigenvalues and constants for parallel planes configuration

n	λ_n^2	G_n
0	15.09	1.717
1	171.30	1.139
2	498.00	0.952

3.10.2.2 Rectangular Ducts: These solutions are not very well developed; however, numerical solutions for uniform surface temperature have been obtained [3.46]. The eigenvalues and constants for a square tube are given in Table 3.17 [3.47].

Table 3.17: Eigenvalues and constants for square tube

n	λ_n^2	G_n
0	5.96	0.598
1	35.64	0.462
2	78.9	0.138

The thermal entry length for rectangular tube for uniform temperature and constant axial heat flux and constant surface temperature are given in Tables 3.18 and 3.19, respectively. These values are due to numerical solution presented in Ref. [3.46]. The quantity b/a in Tables 3.18 and 3.19 is the ratio of the long side of the rectangle to its short side.

Table 3.18: Thermal entry length – Nusselt numbers for rectangular tubes at constant wall temperature

| | Nu_x | | | | | Nu_m | | | | |
| | b/a | | | | | b/a | | | | |
x^*	1	2	4	6	∞	1	2	4	6	∞
0	∞	∞	∞	∞	∞	∞	∞	∞	∞	∞
0.01	4.55	5.72	6.57	7.02	8.52	8.63	8.58	9.47	10.01	11.63
0.02	4.12	4.72	5.55	6.07	7.75	6.48	6.84	7.71	8.17	9.83
0.05	3.46	3.85	4.87	5.48	7.55	4.83	5.24	6.16	6.70	8.48
0.10	3.10	3.54	4.65	5.34	7.55	4.04	4.46	5.44	6.04	8.02
0.20	2.99	3.43	4.53	5.24	7.55	3.53	3.95	5.00	5.66	7.78
∞	2.98	3.39	4.51	5.22	7.55	2.98	3.39	4.51	5.22	7.55

Table 3.19: Thermal entry length – Nusselt number for rectangular tubes with constant axial heat flux

| | Nu_x | | | | |
| | b/a | | | | |
x^*	1	2	0.333	4	∞
0	∞	∞	∞	∞	∞
0.01	7.10	7.46	8.02	8.44	
0.02	5.69	6.05	6.57	7.00	8.80
0.05	4.45	4.84	5.39	5.87	
0.10	3.91	4.38	5.00	5.62	8.25
0.20	3.71	4.22	4.85	5.45	
∞	3.60	4.11	4.77	5.35	8.235

3.11   Turbulent Flow Inside Tubes

Experiments have shown that below a Reynolds number of 2000, the flow through smooth pipes and tubes remains laminar. Beyond a Reynolds number of 2000, the flow becomes increasingly unstable until a turbulent flow is established. In commercial pipes, this occurs around a Reynolds number of 4000. The following relations may be established for a fully developed turbulent flow through circular pipes of constant cross section. The nomenclature is shown in Figure 3.12.

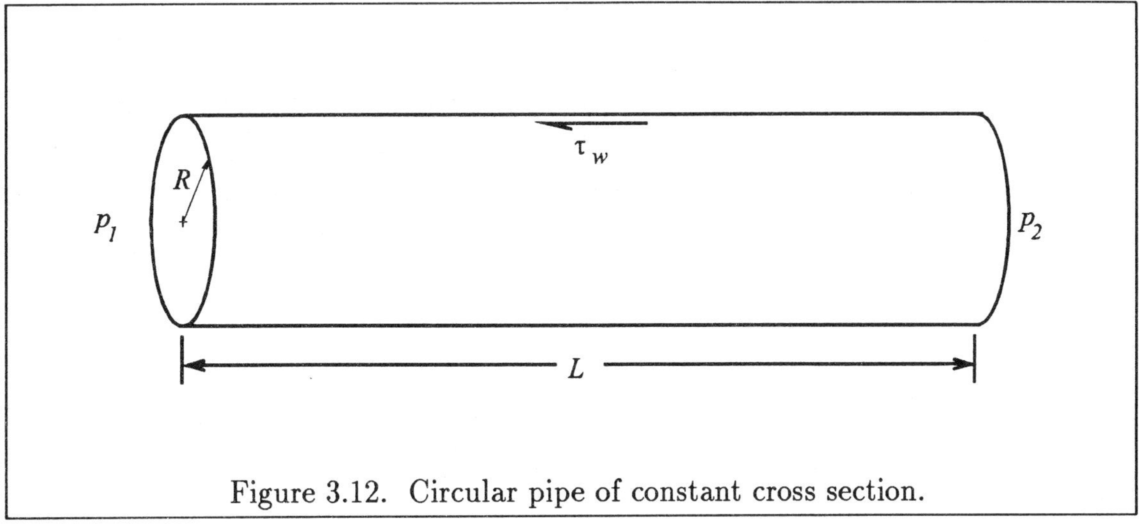

Figure 3.12. Circular pipe of constant cross section.

Shear stress $\quad \tau_w = \dfrac{p_1 - p_2}{L}\dfrac{R}{2}$

$\quad$ since $\quad \dfrac{p_1 - p_2}{L} = \dfrac{\lambda}{4R}\rho\bar{u}^2$

$\quad$ then $\quad \tau_w = \dfrac{\lambda}{8}\rho\bar{u}^2$

where $\bar{u}$ is the average velocity and $\lambda =$ Darcy's friction factor.

3.11.1 Universal Velocity Distribution Laws

It is common to introduce the following nondimensional variables

$$u^+ = \frac{u}{u^*} \tag{3.277}$$

$$y^+ = \frac{yu^*}{\nu} \tag{3.278}$$

where $u^* = \sqrt{\tau_w/\rho}$ is known as the friction velocity. The ordinate y^+ is measured from the wall and is used, generally, to define three distinct regions of turbulent flow within a pipe. In the immediate vicinity of the wall for $y^+ < 5$, the turbulent shear stresses are negligible and laminar stresses predominate. This region is called the *laminar sublayer*, and, within this region, the velocity is expressed as

$$u^+ = y^+ \tag{3.279}$$

In the *buffer zone*, defined by $5 < y^+ < 30$, the laminar as well as turbulent stresses are of the same order of magnitude, and the velocity profile is given by

$$u^+ = 5.0\ln y^+ - 3.05 \tag{3.280}$$

Beyond $y^+ > 30$, the fully turbulent core is reached, in which the turbulent stresses are dominant. The velocity profile is given by

$$u^+ = 2.5 \ln y^+ + 5.5 \tag{3.281}$$

Expression (3.281) is the most familiar form of the logarithmic law of the wall. Other expressions for the velocity profile and turbulent viscosity ν_t are provided below.

Deisslers' law

$$y^+ = \sqrt{\frac{\pi}{2m}} \exp\left(\frac{1}{2}m\, u^{+^2}\right) erf\left(\sqrt{\frac{m}{2}}u^+\right) \qquad y^+ \le 26$$

$$u^+ = 3.8 + 2.78 \ln y^+ \quad \text{for the fully turbulent core}$$

$$m = \text{experimentally determined constant} = 0.0119$$

$$\frac{\nu_t}{\nu} = mu^+y^+\left[1 - \exp\left(-mu^+y^+\right)\right] \qquad 0 \le y^+ \le 26$$

with $m = 0.0154$.

Spalding's law

For the entire region

$$y^+ = u^+ + \frac{1}{E}\left[e^{\kappa u^+} - 1 - \kappa u^+ - \frac{(\kappa u^+)^2}{2!} - \frac{(\kappa u^+)^3}{3!} - \frac{(\kappa u^+)^4}{4!}\right]$$

$$\frac{\nu_t}{\nu} = \frac{\kappa}{E}\left[e^{\kappa u^+} - 1 - \kappa u^+ - \frac{(\kappa u^+)^2}{2!} - \frac{(\kappa u^+)^3}{3!}\right]$$

where $\kappa = 0.407$ and $E = 10$

Reichardts' law

For the turbulent core

$$u^+ = 5.5 + 2.5 \ln\left[y^+ \frac{1.5\left(1 + \dfrac{r}{R}\right)}{1 + 2\left(\dfrac{r}{R}\right)^2}\right]$$

For moderate Reynolds numbers, the expression above can be approximated by

$$u^+ = 8.6(y^+)^{\frac{1}{7}}$$

$$\frac{\nu_t}{\nu} = \frac{\kappa\, y^+}{6}\left[1 + \left(\frac{r}{R}\right)\right]\left[1 + 2\left(\frac{r}{R}\right)^2\right]$$

where $\kappa = 0.4$ and $r = R - y$

3.11.2 Friction Coefficient in Fully Developed Turbulent Flow

Defining a Reynolds number based on the tube diameter and the mean velocity, $Re = \frac{\bar{u}D}{\nu}$, the following relation may be established

$$C_f = 0.078 Re^{-0.25} \tag{3.282}$$

or

$$\lambda = 0.312 Re^{-0.25} \tag{3.283}$$

where $\lambda = 4C_f$ is the Darcy's friction factor. The expressions above approximate experimental data quite well within the range of $10^4 < Re < 5 \times 10^4$. For higher Reynolds numbers, the logarithmic law is used to provide the Karman-Nikuradse equation where the coefficients are adjusted slightly to provide a better match with experimental data.

$$\frac{1}{\sqrt{C_f/2}} = 2.46 \ln\left(Re\sqrt{C_f/2}\right) + 0.30 \tag{3.284}$$

or, in terms of the friction factor λ,

$$\frac{1}{\sqrt{\lambda}} = 0.87 \ln\left(Re\sqrt{\lambda}\right) - 0.8 \tag{3.285}$$

An empirical approximation to the equation above in the range of $3 \times 10^4 < Re < 10^6$ is obtained as

$$C_f = 0.046 Re^{-0.2} \tag{3.286}$$

or

$$\lambda = 0.184 Re^{-0.2} \tag{3.287}$$

A second relation over the range of $10^4 < Re < 5 \times 10^6$ is the Petukov equation [3.48]

$$\frac{C_f}{2} = (2.236 \ln Re - 4.639)^{-2} \tag{3.288}$$

For Reynolds numbers below 100,000, Blasius obtained the following experimental relation

$$\lambda = 0.3164 Re^{-\frac{1}{4}} \tag{3.289}$$

This correlation can be used to obtain the wall shear stress in terms of other flow parameters

$$\tau_w = 0.03325 \rho \bar{u}^{\frac{7}{4}} \nu^{\frac{1}{4}} Re^{-\frac{1}{4}} \tag{3.290}$$

3.11.3 Turbulent Flow in Rough Pipes

The results presented in the preceding section have been for smooth pipes only. Surface roughness, though of negligible consequence in laminar flow, is an important consideration in turbulent flow. The surface roughness is measured in terms of $k^+ = (kv^*)/\nu$ where k is the average height of the roughness elements. As this height is increased, the viscous layer is gradually obscured, until around $k^+ = O(30)$, the flow near the wall is independent of viscosity. The outer flow, however, is unaffected by roughness. The overall effect of surface roughness is to shift the profiles downward by an amount ΔB which is determined by k^+. The law of the wall takes the form

$$u^+ = 2.5 \ln y^+ + 5.5 - \Delta B\, k^+ \tag{3.291}$$

ΔB is not a unique function of k^+, but depends on the type of roughness. For the Prandtl-Schlichting sand grain roughness, it has the following approximation

$$\Delta B = \frac{1}{\kappa} \ln 0.3 k^+ \qquad \text{for } k^+ > 60$$

The following three sand grain regimes are defined

$$k^+ < 4 \qquad \text{hydraulically smooth wall}$$

$$4 < k^+ < 60 \qquad \text{transitional-roughness regime}$$

$$k^+ > 60 \qquad \text{fully rough regime}$$

A friction factor for smooth, transitional, or fully rough pipes is given by

$$\frac{1}{\sqrt{\lambda}} = 0.87 \ln \frac{Re_D\sqrt{\lambda}}{1 + 0.1 \left(\dfrac{k}{D}\right) Re_D\sqrt{\lambda}} - 0.8 \tag{3.292}$$

where, for $\frac{k}{D} < \frac{10}{Re_D}$, the roughness is unimportant, while for $\frac{k}{D} \geq \frac{1000}{Re_D}$, the flow is considered to be fully rough, and the friction factor varies only with roughness. Other relations are

$$\textit{Karman's formula} \qquad \lambda = \left(0.87 \ln \frac{D}{k} + 1.2\right)^{-2} \tag{3.293}$$

$$\textit{Colbrooks' formula} \qquad \frac{1}{\sqrt{\lambda}} = 1.74 - 0.87 \ln \left(\frac{2k}{D} + \frac{18.7}{Re_D\sqrt{\lambda}}\right) \tag{3.294}$$

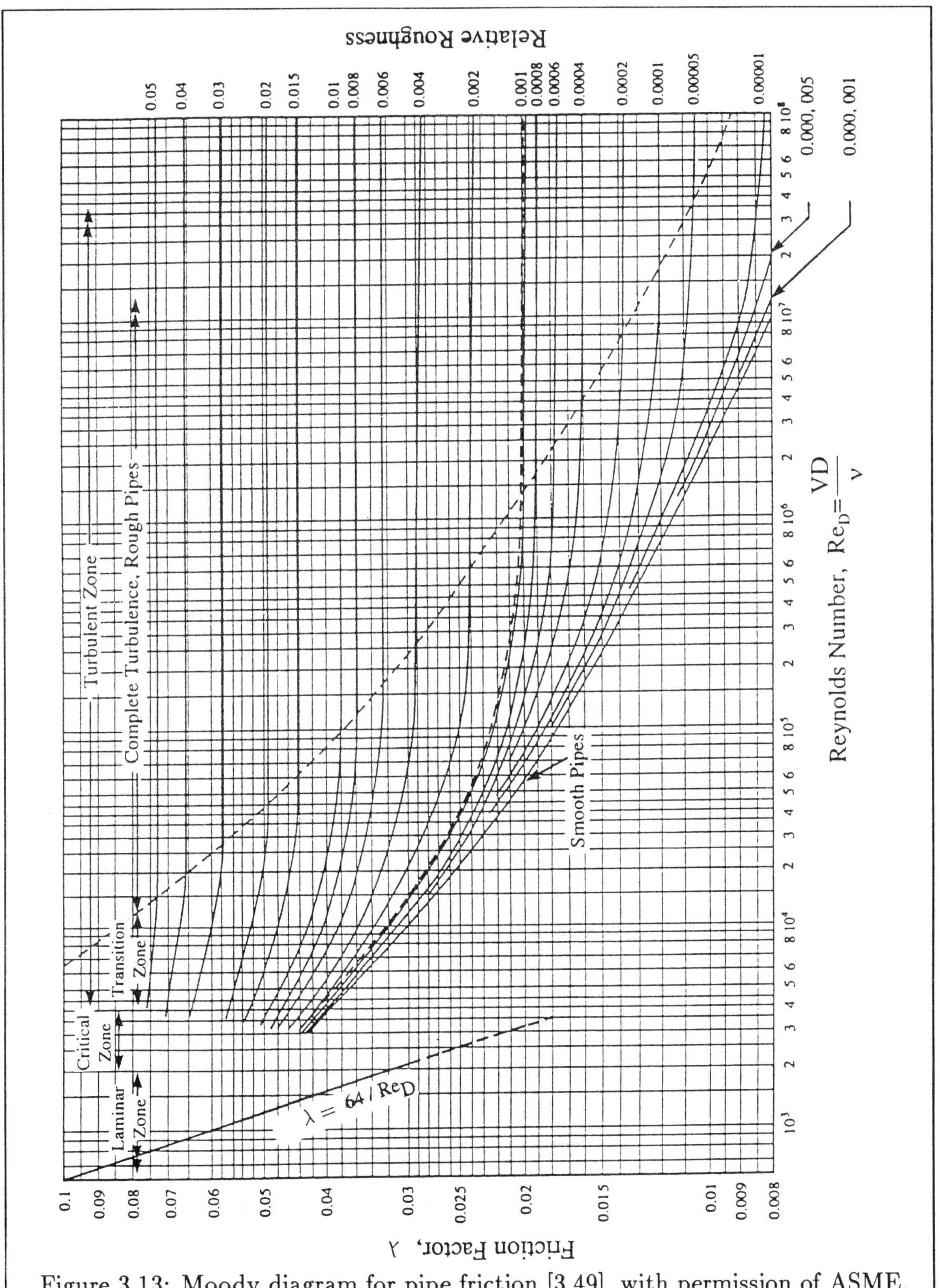

Figure 3.13: Moody diagram for pipe friction [3.49], with permission of ASME.

Friction factors in terms of relative roughness k/D and Re_D have been compiled by Moody [3.49] and are known as *the Moody diagram*. The Moody diagram is shown in Figure 3.13 and can be used to determine the friction factor.

Turbulent flow in a channel

Friction factor
$$\frac{1}{\sqrt{\lambda}} = 0.88 \ln \left(Re_h \sqrt{\lambda} \right) + 0.142 \tag{3.295}$$

where $Re_h = \dfrac{h\bar{u}}{\nu}$, $h =$ channel half width, and $\bar{u} =$ average velocity across the channel.

The following are approximations to the expression (3.295)

Curve fit $\qquad \lambda = 0.495(\log_{10} Re_h)^{-2.2} \qquad \pm 2\%$

Blasius power law $\qquad \lambda = 0.24 Re_h^{-\frac{1}{4}} \qquad$ for $Re_h < 500,000$

3.12   Turbulent Flow Inside Tubes – Heat Transfer

The general analysis of heat transfer in fully developed pipe flow is carried out by the extension of analogies similar to that of a flow over a flat plate. The expressions for the Nusselt number are summarized in Table 3.20. The Reynolds number is based on the pipe diameter, and the Nusselt number Nu represents the local heat transfer rate.

3.13   Thermal Entry Length for Turbulent Flow in Circular Tubes

The solutions for turbulent flow can be formulated in the same form as laminar flow, and, therefore, laminar flow solutions previously described can be used. These equations are given by Equations (3.271) and (3.272) for constant wall temperature and by Equation (3.275) for constant heat flux boundary. However, now the eigenvalues and constants used in these equations are dependent on the Reynolds number as well as on the Prandtl number. The eigenvalues and constants, as provided in Ref. [3.35], are given in Table 3.21.

Table 3.20: Heat transfer rate for fully developed pipe flow

Description	Equation	Restrictions/ Remarks	Refs.
Reynolds Analogy	$Nu = \dfrac{\lambda}{8} Pr\, Re_D \qquad (3.296)$	$\lambda = 0.316 Re_D^{-0.25}$ for $10^4 < Re_D < 5 \times 10^4$ or $\lambda = 0.184 Re^{-0.2}$ for $3 \times 10^4 < Re_D < 10^6$ or λ from the Moody chart	[3.24]
Colburn Analogy	$Nu = \dfrac{\lambda}{8} Re_D\, Pr^{1/3} \qquad (3.297)$ $Nu = 0.0395 Re_D^{0.75} Pr^{0.33} \qquad (3.298)$ $Nu = 0.023 Re_D^{0.80} Pr^{0.33} \qquad (3.299)$	$10^4 < Re_D < 5 \times 10^4$ $3 \times 10^4 < Re_D < 10^6$	[3.24]
Prandtl Analogy	$Nu = \dfrac{\left(\dfrac{\lambda}{8}\right) Re_D\, Pr}{1 + 5\sqrt{\dfrac{\lambda}{8}}(Pr - 1)} \qquad (3.300)$	$\lambda = 0.316 Re_D^{-0.25}$ for $10^4 < Re_D < 5 \times 10^4$ $\lambda = 0.184 Re_D^{-0.2}$ for $3 \times 10^4 < Re_D < 10^6$	[3.24]
Von Karman Analogy	$Nu = \left(\dfrac{\lambda}{8}\right) Re_D\, Pr \div$ $1 + 5\sqrt{\dfrac{\lambda}{8}}\left\{(Pr - 1) + \ln\left[1 + \dfrac{5}{6}(Pr - 1)\right]\right\} \qquad (3.301)$	Same as above	[3.24]

Table 3.21: Eigenvalues and constants for turbulent flow in a circular pipe

n,m	Pr	Re	λ_m^2	G_n	γ_m^2	A_m
0	0.002	50,000	10.61	0.975		
1			58.19	0.897	29.66	0.00773
2			144.6	0.862	98.51	0.01015
3			269.7	0.841	206.5	0.00105
4					353.4	0.000607
0	0.002	100,000	11.23	1.036		
1			61.74	0.943	31.61	0.00725
2			153.6	0.903	104.7	0.00209
3			286.7	0.88	219.2	0.000977
4					374.9	0.00056
0	0.002	500,000	15.91	1.527		
1			90.19	1.246	48.83	0.00437
2			227.4	1.137	158.0	0.001202
3			426.8	1.08	331.0	0.000547
4					564.8	0.000307
0	0.01	50,000	13.39	1.277		
1			75.23	1.079	40.12	0.00431
2			188.9	0.996	131.8	0.00152
3			353.9	0.949	275.0	0.000696
4					469.3	0.000395
0	0.01	100,000	16.75	1.642		
1			96.15	1.29	53.11	0.00394
2			243.8	1.154	172.7	0.001068
3			458.5	1.085	358.8	0.000479
4						
0	0.01	500,000	39.31	4.2		
1			252.9	2.42	159.8	0.001152
2			665.6	1.93	504.6	0.000273
3			1271.0	1.752	1032.0	0.000112

(Continued on next page)

Table 3.21: Continued.

n, m	Pr	Re	λ_m^2	G_n	γ_m^2	A_m
4					1737.0	0.0000591
0	0.03	50,000	21.28	2.17		
1			126.7	1.532	73.68	0.00272
2			325.2	1.32	237.8	0.000705
3			614.8	1.189	491.7	0.000306
4					835.3	0.000167
0	0.03	100,000	31.06	3.28		
1			194.2	2.03	119.4	0.0016
2			506.9	1.649	380.5	0.00039
3			964.8	1.488	781.4	0.000162
4					1,320.0	0.000087
0	0.03	500,000	91.5	10.38		
1			678.1	4.39	477.7	0.000362
2			1,844.0	3.22	1,490.0	0.000076
3			3,562.0	2.9	3,011.0	0.000029
4					5,027.0	0.000015
0	0.72	10,000	64.38	7.596		
1			646.8	1.829	519.5	0.000301
2			1,870.0	1.217	1,624.0	0.0000513
3					3,202.0	0.0000149
0	0.72	50,000	219.0	26.6		
1			2,350.0	5.63	1,952.0	0.0000886
2			6,808.0	3.32	6,154.0	0.0000179
3					12,480.0	0.000006
0	0.72	100,000	375.9	45.8		
1			4,183.0	9.25	3,510.0	0.0000494
2			12,130.0	5.48	11,030.0	0.0000101
3					22,340.0	0.0000035
0	8.0	50,000	685.6	85.4		
		100,000	1,232.0	154.0		
		500,000	5,020.0	625.0		

(Continued on next page)

Table 3.21: Continued.

n, m	Pr	Re	λ_m^2	G_n	γ_m^2	A_m
1	10.0	50,000			2.736×10^4	5.02×10^{-6}
2					7.316×10^4	1.21×10^{-6}
3					1.373×10^5	4.36×10^{-3}
4					2.196×10^5	1.87×10^{-7}
5					3.198×10^5	8.79×10^{-8}
1	10.0	100,000			5.040×10^4	2.78×10^{-6}
2					1.346×10^5	6.97×10^{-4}
3					2.528×10^5	2.68×10^{-7}
4					4.046×10^5	1.26×10^{-7}
5					5.904×10^5	6.56×10^{-8}
0	50.0	100,000	2,570.0	321.0		
		200,000	4,778.0	598.0		
		500,000	10,800.0	1,350.0		
0	100.0	100,000	3,317.0	415.0		
		200,000	6,129.0	766.0		
		500,000	14,040.0	1,750.0		

3.13.1 Thermal Entry Length for Turbulent Flow Between Parallel Plates

Once again the equations originally developed for laminar flow, that is, Equations (3.270) and (3.271) are applicable. The appropriate eigenvalues and constants have been obtained by Hatton and Quarmby [3.49] and are presented in Tables 3.22 and 3.23. The two cases considered are with one surface at either constant temperature or constant heat flux and the other surface insulated. For the latter case, the Nusselt numbers and the influence coefficients are listed directly.

These values can be used in Equations (1.244) and (1.245) for specified constant heat rate at one surface.

Table 3.22: Eigenvalues and constants for turbulent flow between parallel plates, one surface at a constant temperature and the other surface is insulated, thermal entry length

| | $Re = 7096$ | | $Re = 73,612$ | | $Re = 494,576$ | |
n	λ_n^2	G_n	λ_n^2	G_n	λ_n^2	G_n
			$Pr = 0.1$			
0	16.7	1.86	59.4	6.87	228.0	27.1
1	158.0	1.34	611.0	3.78	2,685.0	10.9
2	450.0	1.03	1,788.0	2.54	8,200.0	6.47
3	893.0	0.859	3,535.0	2.14	16,210.0	5.37
4	1,473.0	0.794	5,830.0	2.04	26,700.0	5.10
5	2,201.0	0.762	8,660.0	1.95	39,820.0	4.85
			$Pr = 1$			
0	52.1	6.28	296.0	36.1	1,303.0	160.0
1	656.0	2.17	4,400.0	9.44	2,244.0	33.0
2	2,067.0	1.16	14,100.0	4.85	7,360.0	16.3
3	4,175.0	0.939	22,150.0	3.73	14,890.0	12.6
4	6,775.0	0.936	46,850.0	3.27	24,530.0	10.9
5	9,945.0	0.981	70,100.0	2.87	36,660.0	9.55
			$Pr = 10$			
0	153.0	19.1	1,038.0	129.4	5,180.0	648.0
1	4,880.0	1.81	37,200.0	10.2	199,600.0	46.9
2	17,000.0	0.914	132,000.0	4.45	696,000.0	20.2
3	34,250.0	0.865	265,600.0	3.26	1,427,000.0	14.5
4	53,800.0	1.01	442,000.0	2.73	2,368,000.0	11.8
5	76,700.0	1.18	663,000.0	2.35	3,556,000.0	9.65

Table 3.23: Nusselt's number and influence coefficients for turbulent flow between parallel plates, one surface subject to constant heat flux and the other surface is insulated, thermal entry length solution

	$Re = 7096$		$Re = 73,612$		$Re = 494,576$	
x/D_h	Nu_{11}	θ_1^*	Nu_{11}	θ_1^*	Nu_{11}	θ_1^*
$Pr = 0.1$						
1	19.7	0.056	75.2	0.018	241.0	0.005
3	14.3	0.122	56.2	0.146	194.0	0.023
10	10.7	0.267	42.4	0.115	155.0	0.062
30	9.44	0.352	34.8	0.233	132.0	0.147
100	9.34	0.359	32.1	0.290	120.0	0.219
$Pr = 1.0$						
1	47.3	0.013	234.0	0.005	940.0	0.000
3	37.9	0.033	203.0	0.018	851.0	0.009
10	31.5	0.089	177.0	0.049	761.0	0.030
30	28.0	0.173	160.0	0.114	697.0	0.077
100	27.1	0.200	152.0	0.155	661.0	0.123
$Pr = 10$						
1	102.0	0.004	602.0	0.004	2925.0	0.000
3	88.6	0.012	575.0	0.008	2829.0	0.003
10	81.9	0.027	550.0	0.018	2724.0	0.010
30	78.6	0.057	532.0	0.041	2640.0	0.027
100	77.5	0.070	522.0	0.057	2590.0	0.045

3.14   Free Convection Boundary Layers

Natural or free convection flow is a flow in which the bouyant force, resulting from a density difference, provides the driving mechanism. The density gradient could be due to mass concentrations or temperature gradients within the fluid.

If the x-axis is selected along a vertical surface, then the governing boundary layer equations for a steady, two-dimensional, incompressible flow are as follows:

$$\frac{\partial u}{\partial x} + \frac{\partial v}{\partial y} = 0 \tag{3.302}$$

$$u\frac{\partial u}{\partial x} + v\frac{\partial u}{\partial y} = \nu\frac{\partial^2 u}{\partial y^2} + g\,\beta\,\Delta T \tag{3.303}$$

$$u\frac{\partial T}{\partial x} + v\frac{\partial T}{\partial y} = \alpha\frac{\partial^2 T}{\partial y^2} \tag{3.304}$$

where β is the coefficient of volume expansion and $\Delta T = T - T_\infty$.

The boundary conditions are prescribed as follows:

$$\text{At} \quad y = 0 \;, \quad u = v = 0 \;, \quad T = T_s$$

$$\text{At} \quad x = 0 \;, \quad u = 0 \quad\;, \quad T = T_\infty$$

$$\text{At} \quad y \to \infty \;, \quad u \to 0 \;, \quad T \to T_\infty$$

The coordinate system is illustrated in Figure 3.14. Several heat transfer expressions for free convection boundary layer flow are summarized in Table 3.24. The Grashof number and Rayleigh number are defined as

$$Gr_x = \frac{g\beta x^3 \Delta T}{\nu^2}$$

$$Ra = Gr\,Pr = \frac{\rho g\beta x^3 \Delta T}{\mu\alpha}$$

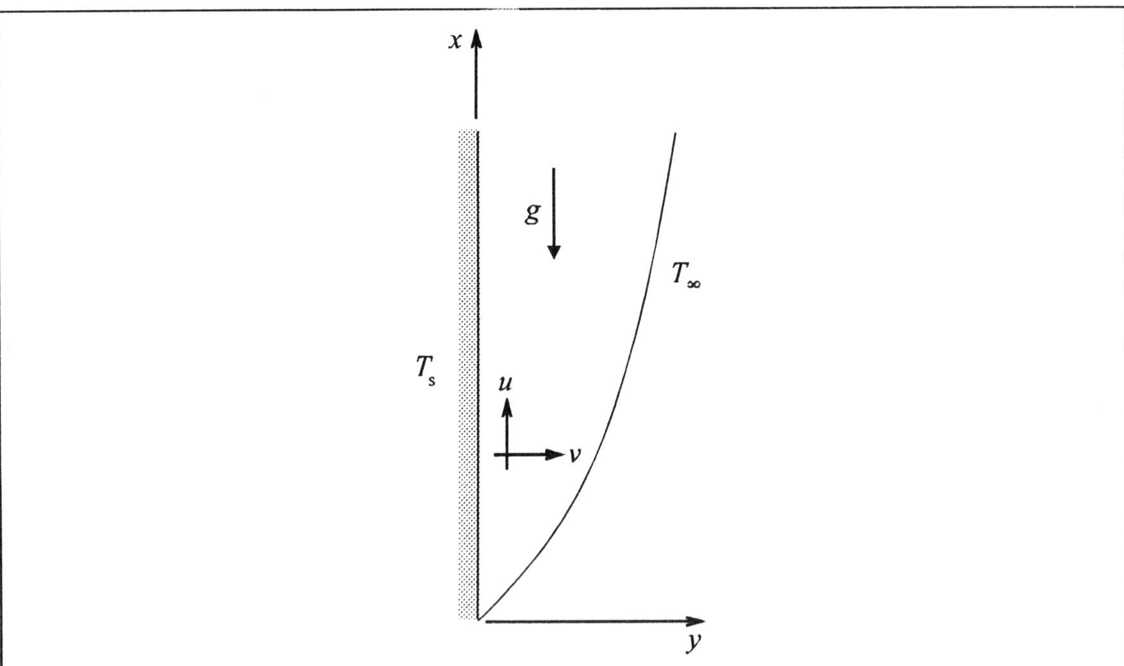

Figure 3.14 Schematic of coordinate system for free convection boundary layer.

Table 3.24: Heat transfer correlations for free convection flows

Description	Equation(s)		Restrictions/Remarks	Reference(s)
Vertical plate–laminar flow	$Nu_x = 0.478\,Gr_x^{\frac{1}{4}}Pr^{\frac{1}{2}}(0.861 + Pr)^{-\frac{1}{4}}$	(3.305)	Fluid properties at $T_m = \frac{1}{2}(T_f + T_s)$, β evaluated at T_m for liquids and at T_f for gases	[3.24]
	$Nu_L = 0.637\,Gr_L^{\frac{1}{4}}Pr^{\frac{1}{2}}(0.861 + Pr)^{-\frac{1}{4}}$	(3.306)		
In terms of Rayleigh number	$Nu_x = 0.478\,Ra_x^{\frac{1}{4}}\left(1 + \dfrac{0.861}{Pr}\right)^{-\frac{1}{4}}$	(3.307)		[3.24]
	$Nu_L = 0.637\,Ra_L^{\frac{1}{4}}\left(1 + \dfrac{0.861}{Pr}\right)^{-\frac{1}{4}}$	(3.308)		
An alternate correlation	$Nu_x = \dfrac{3}{4}\left[\dfrac{2Pr}{5(1 + 2Pr^{\frac{1}{2}} + 2Pr)}\right]^{\frac{1}{4}}(Gr_xPr)^{\frac{1}{4}}$	(3.309)		[3.35]
Very small Prandtl numbers	$Nu_x = 0.600Gr_x^{\frac{1}{4}}Pr^{\frac{1}{2}}$	(3.310)	$Pr \to 0$	[3.35]
Very large Prandtl numbers	$Nu_x = 0.503\,Gr_x^{\frac{1}{4}}Pr^{\frac{1}{4}}$	(3.311)	$Pr \to \infty$	[3.35]
Constant heat flux	$Nu_x = \left[\dfrac{Pr}{4 + 9Pr^{\frac{1}{2}} + 10Pr}\right](Gr_x^*Pr)^{\frac{1}{5}}$	(3.312)	$Gr_x^* = Gr_xNu_x = \dfrac{g\beta x^4\dot{q}_w}{k\nu^2}$	[3.52, 3.35]
A general correlation	$Nu_L = 0.68 + 0.670Ra_L^{\frac{1}{4}}\left[1 + \left(\dfrac{0.492}{Pr}\right)^{\frac{9}{16}}\right]^{-\frac{4}{9}}$	(3.313)	$0 < Ra_L < 10^9$, all Pr	[3.53]

(Continued on next page)

Table 3.24: Continued.

Description	Equation(s)	Restrictions/Remarks	Ref(s)
Integral solution	$Nu_x = 0.508 \left[\dfrac{Pr^2}{0.952 + Pr} \right]^{\frac{1}{4}} Gr_x^{\frac{1}{4}}$ (3.314)		[3.35]
Vertical plate–turbulent flow	$Nu_x = 0.0295 Gr_x^{\frac{2}{5}} Pr^{\frac{7}{15}} (1 + 0.494 Pr^{\frac{2}{3}})^{-\frac{2}{5}}$ (3.315)	Integral solution	[3.24]
	$Nu_L = 0.0246 Gr_L^{\frac{2}{5}} Pr^{\frac{7}{15}} (1 + 0.494 Pr^{\frac{2}{3}})^{-\frac{2}{5}}$ (3.316)	Based on $\frac{1}{7}$th power law	[3.24]
A general correlation for an isothermal plate	$Nu_L = \dfrac{0.15(Gr_L Pr)^{\frac{1}{3}}}{\left[1 + (0.492/Pr)^{\frac{9}{16}} \right]^{\frac{16}{27}}}$ (3.317)	All Pr, $Gr_x \leq 10^{12}$	[3.53]
Constant heat flux	$Nu_x = 0.0942(Gr_x Pr)^{\frac{1}{3}}$ (3.318)	For air	[3.54]
	$Nu_L = \dfrac{0.15(Gr_L Pr)^{\frac{1}{3}}}{\left[1 + (0.437/Pr)^{\frac{9}{16}} \right]^{\frac{16}{27}}}$ (3.319)	All Pr, $Gr_x \leq 10^{12}$	[3.53]
A general correlation	$Nu_L = \left\{ 0.825 + 0.387 Ra_L^{\frac{1}{6}} \left[1 + \left(\dfrac{0.492}{Pr} \right)^{\frac{9}{16}} \right]^{-\frac{8}{27}} \right\}^2$ (3.320)	$Ra_L > 10^9$	[3.51]
Inclined surfaces	Previous equations are valid if g in Grashof or Rayleigh numbers is replaced with $g \cos\theta$, where θ is measured from vertical position.	Range of validity is $-60° \leq \theta \leq +60°$	[3.24, 3.35]

3.15 References

[3-1] Papadakis, M., Lall, V., and Hoffmann, K. A., "Performance of Turbulence Models for Planar Flows: A Selected Review," AIAA 94-1873, June 1994.

[3-2] Purtell, L. P., "Turbulence in Complex Flows, A Selected Review," AIAA 92-0435, Jan. 1992.

[3-3] El Tahry, S. H., and Haworth, D. C., "A Critical Review of Turbulence Models for Applications in the Automotive Industry," AIAA 91-0516, Jan. 1991.

[3-4] Bushnell, D. M., "Turbulence Modeling in Aerodynamic Shear Flow – Status and Problems," AIAA 91-0214, Jan. 1991.

[3-5] Menter, F. R., "Performance of Popular Turbulence Models for Attached and Separated Adverse Pressure Gradient Flows," AIAA 91-1784, June 1991.

[3-6] Renze, K. J., Buning, P. G., and Rajagopalan, R. G., "A Comparative Study of Turbulence Models for Overset Grids," AIAA 92-0437, Jan. 1992.

[3-7] Rumsey, C. L., and Vatsa, V. N., "A Comparison of the Predictive Capabilities of Several Turbulence Models Using Upwind and Central-Difference Computer Codes," AIAA 93-0192, Jan. 1993.

[3-8] Coakley, T. J., Huang, P. G., "Turbulence Modeling for High Speed Flows," AIAA 92-0436, Jan. 1992.

[3-9] Coakley, T. J., "Turbulence Modeling Methods for the Compressible Navier-Stokes Equations," AIAA 83-1693, July 1983.

[3-10] Marvin, J. G., and Coakley, T. J., "Turbulence Modeling for Hypersonic Flows," NASA TM 101079, June 1979.

[3-11] Papadakis, M., Hoffmann, K. A., and Lall, V., "A Comparative Study of Numerical Schemes and Turbulence Models for Predicting Airfoil Flows at Various Mach Numbers," AIAA 94-0392, Jan. 1994.

[3-12] Chen, H. C., and Patel, V. C., "Near Wall Turbulence Models for Complex Flows Including Separation," *AIAA Journal,* Vol. 26, No. 6, pp. 641–648, June 1988.

[3-13] Shabbir, A., and Shih, T., "Critical Assessment of Reynolds Stress Turbulence Models Using Homogeneous Flows," AIAA 93-0082, Jan. 1993.

[3-14] Edwards, J. R., and Chandra, S., "Eddy Viscosity Transport Turbulence Models for High Speed, Two-Dimensional, Shock Separated Flowfields," AIAA 94-0310, Jan. 1994.

[3-15] Spalding, D. B., *Journal of Applied Mechanics,* Vol. 28, pp. 455–457, 1961.

[3-16] White, F. M., "Viscous Fluid Flow," McGraw-Hill, 1974.

[3-17] Cebeci, T., and Smith, A. M. O., "Analysis of Turbulent Boundary Layers," Academic Press, 1974.

[3-18] Baldwin, B. S., and Lomax, H., "Thin-Layer Approximation and Algebraic Model for Separated Turbulent Flows," AIAA-78-257, 1978.

[3-19] Baldwin, B., Barth, T. J., "A One-Equation Turbulence Transport Model for High Reynolds Number Wall-Bounded Flows," NASA TM-102847, Aug. 1990.

[3-20] Spalart, P. R., Allmaras, S. R., "A One-Equation Turbulence Model for Aerodynamic Flows," AIAA 92-0439, Jan. 1992.

[3-21] Jones, W. P., and Launder, B. E., "The Calculation of Low-Reynolds Number Phenomena with a Two-Equation Model of Turbulence," *International Journal of Heat and Mass Transfer,* Vol. 16, pp. 1119–1130, 1973.

[3-22] Wilcox, D. C., "Reassessment of Scale Determining Equations for Advanced Turbulence Models," *AIAA Journal,* Vol. 26, No. 11, pp. 1299–1310, Nov. 1988.

[3-23] Eckert, E. R. G., "Survey on Heat Transfer at High Speeds," Aeronaut. Res. Lab Rep. 189, Office of Aerospace Research, Wright-Patterson Air Force Base, OH, 1961.

[3-24] Chapman, A. J., "Heat Transfer," 4th ed., Macmillan, 1984.

[3-25] Schlichting, H., "Boundary Layer Theory," 6th ed., McGraw-Hill, 1968.

[3-26] von Karman, T., Mechanische Ähnlichkeit und Turbulenz. Proc. of the IIIrd Intern. Congr. of Appl. Mech., 85, Stockholm 1931. Also NACA TM 611, 1931.

[3-27] Schaenherr, K. E., "Resistance of Flat Surfaces Moving Through a Fluid," Trans. Soc. Nav. Arch. and Mar. Engr. 40, pp. 279, 1932.

[3-28] Schultz-Grunow, F., Luftfahrtforschung, Vol. 17, pp. 239–249, 1940. Also NACA TM 986.

[3-29] Prandtl, L., and Schlichting, H., Werft, Reederei, Hafen, Vol. 1934, pp. 1–4, 1934.

[3-30] Hayes, W. D., and Probstein, R. F., "Hypersonic Flow Theory," Academic, 1959.

[3-31] Schmidt, H., "Laminar Skin Friction and Heat Transfer Parameters for a Flat Plate at Hypersonic Speeds in Terms of Free-Stream Flow Properties," NACA TN D-8, Sept. 1959.

[3-32] Hankey, W. L., and Neumann, R. D., "Design Procedures for Computing Aerodynamic Heating at Hypersonic Speeds," WADC TR 59-610, 1960.

[3-33] Glauert, M. B., and Lighthill, M. J., Proc. R. Soc., ser. A, vol. 230, pp. 180–203, 1955.

[3-34] Seban, R. A., and Bond, R., *Journal of Aeronautical Sci.,* vol. 18, p. 671, 1951.

[3-35] Kays, W. M., and Crawford, M. E., "Convective Heat and Mass Transfer," 3rd ed., McGraw-Hill, 1993.

[3-36] Churchill, S. W., and Ozoe, H., "Correlations for Laminar Forced Convection in Flow over an Isothermal Flat Plate and in Developing and Fully Developed Flows in an Isothermal Tube," *J. Heat Transfer,* Trans. ASME, Vol. 95, pp. 416, 1973.

[3-37] Zukauskas, A. A., "Heat Transfer from Tubes in Cross Flow," *Adv. Heat Transfer,* Academic, Vol. 8, pp. 93–106, 1972.

[3-38] Churchill, S. W., and Bernstein, M., "A Correlating Equation for Forced Convection from Gases and Liquids to Circular Cylinder in Crossflow," *J. Heat Transfer,* Trans. ASME, Vol. 94, p. 300, 1977.

[3-39] Quarmly, A., and Al-Fakhii, A. A. M., "Effect of Finite ength on Forced Convection Heat Transfer from Cylinders," *Int. J. Heat Mass Transfer,* Vol. 23, pp. 463–469, 1980.

[3-40] Whitaker, S., "Forced Convection Heat Transfer Correlations for Flows in Pipes past Flat Plates, Single Cylinders, Single Spheres, and for Flows in Packed Beds and Tube Bundles," *AIChE J.,* Vol. 18, pp. 361–371, 1972.

[3-41] Achenbach, E., "Heat Transfer from Spheres up to Re = 6 x 10^6," Proc. 6th Int. Heat Transfer Conf., Vol. 5, Hemisphere, Washington DC, 1978.

[3-42] Jakob, M., "Heat Transfer," Vol. 1, 1949, and Vol. 2, 1957, Wiley.

[3-43] Sellars, J. R., Trilus, M., Klein, J. S., "Heat Transfer to Laminar Flow in a Round Tube or Flat Conduit – The Graetz Problem Extended," *Trans. ASME,* Vol. 78, pp. 441–448, 1956.

[3-44] Sieder, E. N., and Tate, E. G., "Heat Transfer and Pressure Drop of Liquids in Tubes," *Ind. Eng. Chem.,* Vol. 28, p. 1429, 1936.

[3-45] Seigel, R., Sparrow, E. M., and Hallman, T. M., "Steady Laminar Heat Transfer in a Circular Tube with Prescribed Wall Heat Flux," *Appl. Sci. Res.,* Sec. A., Vol. 7, pp. 386–392, 1958.

[3-46] Wibulswas, P., PhD Thesis, London University, 1966.

[3-47] Dennis, S. C. R., Mercer, A. McD., and Poots, G., *O. Appl. Math.,* Vol. 17, pp. 285–297, 1959.

[3-48] Petukov, B. S., "Advances in Heat Transfer," Vol. 6, pp. 503–504, Academic Press, 1970.

[3-49] Moody, L. F., "Friction Factors for Pipe Flow," *Trans. ASME,* Vol. 66, pp. 671–684, 1944.

[3-50] Hatton, A. P., and Quarmby, A., "The Effect of Axially Varying and Unsymmetrical Boundary Conditions on Heat Transfer with Turbulent Flow Between Parallel Plates," *Int. J. Heat Mass Transfer,* Vol. 6, pp. 903–914, 1963.

[3-51] Ede, A. J., "Advances in Heat Transfer," Vol. 4, pp. 1–64, Academic Press, 1964.

[3-52] Fujii, T., and Fujii, M., "The Dependence of Local Nusselt Number on Prandtl Number in the Case of Free Convection Along a Vertical Surface with Uniform Heat Flux," *Int. J. Heat Mass Transfer,* Vol. 19, pp. 121–122, 1976.

[3-53] Churchill, S. W., and Chu, H. H. S., "Correlating Equations for Laminar and Turbulent Free Convection from a Vertical Plate,"*Int. J. Heat Mass Transfer,* Vol. 18, pp. 1323–1329, 1975.

[3-54] Vliet, G. C., and Ross, D. C., "Turbulent Natural Convection on Upward and Downward Facing Inclined Constant Heat Flux Surfaces," *J. Heat Transfer,* Vol. 97, pp. 549–555, 1975.

[3-55] Kuehn, T. H., and Goldstein, R. J., "Correlating Equations for Natural Convection Heat Transfer Between Horizontal Circular Cylinders," *Int. J. Heat Mass Transfer,* Vol. 19, pp. 1127–1134, 1976.

Chapter 4

Hypersonics/Chemistry Equations

4.1 Introductory Remarks

For high speed flows, particularly those in the hypersonic regime, the assumption of a calorically perfect gas is no longer valid. That is due to the high temperatures associated with such flowfields. As a consequence of high temperatures, molecules will dissociate and may ionize. Therefore, the effect of chemistry must be accounted for if a reasonable result is to be obtained.

The underlying assumption of a calorically perfect gas results in constant specific heats and, hence, a constant ratio of specific heats, γ. For example, a value of $\gamma = 1.4$ is used for air. In addition, the internal energy of the system is expressed solely as a function of temperature. Indeed, it is assumed that the internal energy is composed of translational and rotational modes of energy only. The previous set of equations was primarily based on these assumptions. For a chemically reacting flow, the energy may be a function of temperature as well as pressure. The ratio of specific heats, γ, is no longer constant and, in general, a new definition for γ in the form of the ratio of enthalpy to internal energy is introduced.

4.2 Fundamental Concepts and Definitions

In the following subsections, various definitions from physical as well as mathematical points of views and some relevant formulations are provided. Furthermore, a nomenclature for this chapter is provided.

4.2.1 Nomenclature

a	measure of the intermolecular force used in the Van der Waals equation
a	speed of sound
A_s	element symbol of species s
b	effective volume of molecules used in the Van der Waals equation
$c_{v,s}$	specific heat at constant volume for species s, per unit mass
$c_{p,s}$	specific heat at constant pressure for species s, per unit mass
C_s	mass fraction of species s, $C_s = \rho_s/\rho$ and $\sum_{s=1}^{n} C_s = 1$
D	self diffusion coefficient ($= D_{ss}$)
Da	Damkohler number
D_{sr}	binary diffusion coefficient for s-r pair
e	internal energy of mixture, per unit mass $e = \sum_{s=1}^{n} C_s e_s$
e_s	internal energy of species s, per unit mass
$e_{t,s}$	translational (internal) energy of species s
$e_{e,s}$	electronic (internal) energy of species s
$e_{r,s}$	rotational (internal) energy of species s
$e_{v,s}$	vibrational (internal) energy of species s
$e_{0,s}$	zero-point energy of species s
E	Arrhenius activation energy (dissociation energy)
$g_0, g_1, g_2, \dots$	different levels of degenerate energy
h_s	enthalpy of species s, per unit mass
h	enthalpy of gas mixture
$\hat{k}$	Boltzmann constant (per particle), $\hat{k} = \mathcal{R}/\hat{N} = 1.38 \times 10^{-23}$ J/K $= 1.38 \times 10^{-16}$ erg/K $= 0.565 \times 10^{-23}$ ft-lb/°R
k	thermal conductivity of gas mixture
k_s	thermal conductivity of species s
k_b	backward rate constant
k_c	equilibrium rate constant
k_f	forward rate constant
Kn	Knudsen number $Kn = \lambda/L$
L	characteristic length

Le	Lewis number, $Le = \dfrac{\rho D C_p}{k}$
$\mathcal{M}$	total mass of mixture, $\mathcal{M} = \rho V$
MW	molecular weight of gas mixture, $MW = \left(\sum_{s=1}^{n} C_s/MW_s\right)^{-1}$
M	nonreacting heavy particles in a chemical reaction
MW_s	molecular (or atomic) weight of species s
$\bar{m}_{si}$	reduced molecular weight of s-i pair
n	number of different species in a gas mixture
$\mathcal{N}$	number of moles of mixture in system, $\mathcal{N} = \mathcal{M}/MW$
N	number of gas particles of mixture in system, $N = \mathcal{N}\hat{N}$
$\hat{N}$	Avogadro's number, $\hat{N} = 6.02 \times 10^{26}$ mole^{-1}
p	pressure of gas mixture
p_c	critical pressure
p_r	reduced pressure
p_s	partial pressure contributed from species s
q	heat flux
Q_e	electronic partition function
$\dot{Q}_{\text{rad}}$	radiation heat flux
Q_s^{T-V}	translational-vibrational energy exchange rate
R	gas constant of gas mixture, per unit mass $R = \sum_s C_s R_s = \sum_s \gamma_s \mathcal{R}$
R_s	gas constant of species s, per unit mass, $R_s = \mathcal{R}/MW_s$
$\mathcal{R}$	universal gas constant, per mole, $\mathcal{R} = 8314$ J/(kg$\cdot$mole$\cdot$K) $= 8.314 \times 10^7$ erg/(g$\cdot$mole$\cdot$K) $= 1.9873$ cal/(g$\cdot$mole$\cdot$K) $= 4.97 \times 10^4$ ft-lb/(slug$\cdot$mole$\cdot$°R)
Sc	Schmidt number, $Sc = \dfrac{\mu}{\rho D} = \dfrac{Pr}{Le}$
T	temperature (for equilibrium flow)
$T_t, T_r, T_e, T_{v,s}$	translational, rotational, electronic, vibrational (of species s) temperature
T_r	reduced temperature
T_c	critical temperature
u, v, w	velocity components in x, y, z coordinates system
v	specific volume of mixture (volume per unit mass)
v_s^j	diffusion velocity of species s in j direction

V	volume of gas mixture
$\dot{w}_s$	mass production rate of species s
Z	compressibility factor used in real gas, $Z = p/\rho RT$

GREEK SYMBOLS

$\beta, \tilde{\gamma}$	"effective" ratio of specific heats, $\beta, \tilde{\gamma} = h/e$
γ	ratio of specific heats, $\gamma = \frac{c_p}{c_v}\big\|_s$
γ_s	the number of moles of s per unit mass of gas mixture; mole-mass ratio of species s (mass concentration), $\gamma_s = C_s/M_s = \chi_s/\rho$
Γ	mole-mass ratio of gas mixture, the total number of moles per unit mass of gas mixture, $\Gamma = \sum_{s=1}^{n} \gamma_s$
δ_s	mole fraction of species s, the number of moles of s per mole of mixture, $\delta_s = p_s/p = \gamma_s/\Gamma$
$\theta_{v,s}$	vibrational characteristic temperature of species s
$\theta_{e,1}, \theta_{e,2}, \ldots$	first, second, $\ldots$ level of degeneracy factor used in determining Q_e
λ	mean free path of gas particle
μ	viscosity of gas mixture
μ_s	viscosity of species s
ν_s^R, ν_s^P	Stoichiometric number of species s in a chemical reaction, R : reactant, P : product
ρ	density of gas mixture
ρ_s	partial density contributed from species s
σ	collision section of gas particles
σ_{sr}	the average collision diameter of s-r pair
σ_v	vibrational excitation cross section
τ_c	characteristic time of chemical process
$<\tau_s>$	the average vibrational relaxation time of species s
τ_{si}	the vibrational relaxation time for collision pair s-i
τ_{ij}	stress in ij direction (tensor form)
χ_s	concentration of species s, the number of moles of s per unit volume of mixture, $\chi_s = \rho_s/M_s = \gamma_s\rho$
χ	concentration of gas mixture, the total number of moles per unit volume of mixture, $\chi = \sum_{s=1}^{n} \chi_s$
$\Omega^{(\cdot)*}$	Lennard-Jones 6–12 potential integrals

4.2.2 Real Gas

When the molecules are more compact, the intermolecular forces become important and the effective volume of the molecules cannot be ignored; the gas is then defined as a real gas. This translates into conditions where the pressure is extremely high and/or temperature is very low.

The equation of state for a real gas may be expressed as

$$\left(p + \frac{a}{v^2}\right)(v - b) = RT \tag{4.1}$$

which is known as the Van der Waals equation. Note that, in the equation above, a and b are gas-dependent constants. The coefficient "a" is a measure of the intermolecular force between the molecules, and "b" is the effective volume of the molecules.

Typical units used in Equation (4.1) are as follows:

$$p = \text{atm}, \quad v = \text{liters}, \quad R = 0.08206 \text{ liter-atm/mole-K}, \quad T = \text{K}$$

The values of coefficients a and b for various gases are included in Table 4.1 [4.1].

	$a \left(\dfrac{\text{liter}^2 \cdot \text{atm}}{\text{mole}^2}\right)$	$b \left(\dfrac{\text{liter}}{\text{mole}}\right)$
CO_2	3.592	0.04267
CO	1.485	0.03985
He	0.03412	0.02370
H_2	0.2444	0.02661
NO	1.340	0.2789
N_2	1.390	0.03913
O_2	1.360	0.03183
H_2O	5.464	0.03049

Table 4.1. Constants a and b for the Van der Waals equation of state. (Source: Ref. [4.1])

4.2.3 Perfect Gas

When the intermolecular forces and the effective volume of the molecules are negligible, the gas is defined as a perfect gas. For the majority of problems in aerodynamics, the assumption of perfect gas is a valid one and is utilized extensively. In addition, the perfect gas includes the following assumptions [4.2]:

(1) The gas particles may collide only with solid boundaries (i.e., particles cannot collide with each other);

(2) Upon collision with a solid boundary, a gas particle is reflected specularly (i.e., the angle of reflection is equal to the angle of incidence) with its speed unchanged (i.e., an elastic reflection).

The equation of state for a single gas species is given by

$$p = \rho R T \tag{4.2}$$

To be more specific, one may write

$$p_s = \rho_s R_s T$$

where subscript "s" denotes a specific species of gas, T represents the translational temperature and the gas constant $R_s = \dfrac{\mathcal{R}}{MW_s}$ where $\mathcal{R}$ is the universal gas constant, MW_s is the molecular weight of species s.

4.2.4 Partial Pressure

Partial pressure of a species (p_s) is defined as the pressure within a domain if the species s is the only matter within the region. Mathematically, the pressure of a mixture within a domain can be written as the sum of the partial pressures, i.e.,

$$p = \sum_{s=1}^{n} p_s \tag{4.3}$$

This relation is known as *Dalton's law of partial pressures*. Equation (4.3) states that the total pressure of an "n" species mixture is the sum of individual partial pressures. Now the equation of state for a mixture of perfect gases may be expressed as $p = \rho R T$ where

$$\rho = \sum_{s=1}^{n} \rho_s$$

$$R = \sum_{s=1}^{n} \left(\frac{\rho_s}{\rho}\right) R_s = \frac{\mathcal{R}}{\rho} \sum_{s=1}^{n} \frac{\rho_s}{MW_s} \tag{4.4}$$

When one defines the mass fraction of species s as $C_s = \dfrac{\rho_s}{\rho}$, then

$$p = \sum_{s=1}^{n} \rho_s R_s T = \sum_{s=1}^{n} \rho_s \frac{\mathcal{R}}{MW_s} T = \sum_{s=1}^{n} \rho C_s \frac{\mathcal{R}}{MW_s} T$$

$$= \rho T \sum_{s=1}^{n} \left(C_s \frac{\mathcal{R}}{MW_s}\right) = \rho R T \tag{4.5}$$

Note that the gas constant for the mixture R, in terms of mass fraction, is defined as

$$R = \sum_{s=1}^{n} C_s R_s \tag{4.6}$$

4.2.5 Various Modes of Energy

The internal energy (per unit of mass) for a species "s" may be written as

$$e_s = e_{t,s} + e_{e,s} + e_{o,s} \quad \text{for an atom}$$

and

$$e_s = e_{t,s} + e_{r,s} + e_{v,s} + e_{e,s} + e_{o,s} \quad \text{for a molecule}$$

The different modes of energies are as follow:

(1) The *translational mode* of energy "$e_{t,s}$" is associated with the translational motion of its center of gravity and is usually a linear function of the translational temperature T_t.

$$e_{t,s} = \frac{3}{2} \frac{\mathcal{R}}{MW_s} T_t \tag{4.7}$$

(2) The *rotational mode* of energy "$e_{r,s}$" is due to the rotation of molecules about their orthogonal axes in space. For a diatomic molecule (such as O_2, H_2, ...) or a linear polyatomic molecule (such as CO_2), it is expressed as

$$e_{r,s} = \left(\frac{\mathcal{R}}{MW_s} \right) T_r \tag{4.8}$$

However, for a nonlinear polyatomic molecule (such as H_2O),

$$e_{r,s} = \frac{3}{2} \left(\frac{\mathcal{R}}{MW_s} \right) T_r \tag{4.9}$$

where T_r is the rotational temperature.

(3) The *vibrational mode* of energy "$e_{v,s}$" is due to vibrational motion of atoms forming the molecule and is given by

$$e_{v,s} = \frac{\mathcal{R}}{MW_s} \cdot \left(\frac{\theta_{v,s}}{\exp\left(\theta_{v,s}/T_{v,s} \right) - 1} \right) \tag{4.10}$$

where $T_{v,s}$ is the vibrational temperature of species "s," and $\theta_{v,s}$ is the vibrational characteristic temperature of species "s."

(4) The *electronic mode* of energy "$e_{e,s}$" is composed of the kinetic energy of the electrons orbiting about the nucleus of the atom and a potential energy established by an electromagnetic field. An expression for the electronic energy in terms of the electronic temperature T_e is given by

$$e_{e,s} = \frac{\mathcal{R}}{MW_s} \left[\frac{\theta_{e,1} \left(g_1/g_0 \right) \exp\left(-\theta_{e,1}/T_e \right)}{1 + \left(g_1/g_0 \right) \exp\left(-\theta_{e,1}/T_e \right)} \right]_s \tag{4.11}$$

Species	O_2	N_2	NO	O	N	NO^+	e^-
MW_s (g/mole)	32	28	30	16	14	30	0.000549
$h_{o,s}$ (erg/g)	0	0	2.995E10	1.544E11	3.364E11	3.283E11	0
$\theta_{v,s}$ (K)	2275.2	3394.4	2740.7	–	–	2740.7	
$\theta_{e,1}$ (K)	11390.	100,000.	174.	228.	28000.	–	
g_0	3	–	2	5	4	–	
g_1	2	–	2	3	6	–	

Table 4.2. Molecular weight, heat formation, characteristic temperatures, and degeneracy factors for air species.

The expression above is based on the assumption that the electronic partition function Q_e is

$$Q_e = g_0 + g_1 \cdot \exp\left(-\theta_{e,1}/T_e\right) + \underset{\longmapsto \approx 0}{g_2 \cdot \exp\left(-\theta_{e,2}/T_e\right) + \cdots} \tag{4.12}$$

That is, the higher degenerate levels of energy (g_2, ...) are ignored. If the g_2 level is included, then

$$e_{e,s} = \frac{\mathcal{R}}{MW_s}\left[\frac{\theta_{e,1}g_1\exp\left(-\theta_{e,1}/T_e\right) + \theta_{e,2}g_2\exp\left(-\theta_{e,2}/T_e\right)}{g_0 + g_1\exp\left(-\theta_{e,1}/T_e\right) + g_2\exp\left(-\theta_{e,2}/T_e\right)}\right]_s \tag{4.13}$$

(5) The *zero-point energy* "$e_{o,s}$" is the energy of the molecule (or atom) at absolute zero degree of temperature. Note that the value of zero-point energy cannot be calculated or measured. Thus, an effective zero-point energy $h_{o,s}$, called the heat of formation, is generally used to replace $e_{o,s}$. The value of MW_s, $h_{o,s}$, $\theta_{e,1}$, g_0, and g_1 for various species of air are presented in Table 4.2.

The values of $\theta_{e,1}$ and g_0, g_1 ... may differ somewhat in different references, e.g., Refs. [4.3] and [4.4]. Furthermore, the value of $\theta_{e,1}$ for N_2 is about 100,000K, and therefore the corresponding $e_{e,s}$ is relatively small compared with the electronic energy of other species. Thus, the values of g_0 and g_1 can generally be ignored.

4.2.6 Compressibility Factor

Modification of Equation (4.2), i.e., the perfect gas equation of state to include real gas effects, may be accomplished by the introduction of a so-called *compressibility factor Z*, such that

$$p = \frac{ZRT}{v} = Z\rho RT \tag{4.14}$$

Substance		Molecular Weight g/mole	T_c (K)	p_c (kN/m^2)kPa
Air		28.96	132.4	3,770
Ammonia	NH$_3$	17.03	405.5	11,280
Argon	Ar	39.948	151.0	4,863
Butane	C$_4$H$_{10}$	58.124	425.2	3,800
Carbondioxide	CO$_2$	44.01	304.2	7,390
Ethane	C$_2$H$_6$	30.07	305.5	4,880
Freon-12	CCl$_2$F$_2$	120.91	384.7	4,010
Freon-22	CHClF$_2$	86.48	369.2	4,980
Helium	He	4.003	5.3	230
Hydrogen	H$_2$	2.016	33.3	1,296
Methane	CH$_4$	16.043	191.1	4,640
Neon	Ne	20.183	44.5	2,624
Nitrogen	N$_2$	28.013	126.2	3,390
Oxygen	O$_2$	31.999	154.8	5,080
Propane	C$_3$H$_8$	44.097	370.0	4,260
Xenon	Xe	131.30	289.81	5,865

Table 4.3. Critical values for various gases.

The compressibility factor Z is usually given as a function of reduced pressure and temperature. Recall that the reduced values are defined as the ratio of the property to its critical value. Thus,

$$p_r = \frac{p}{p_c} \quad \text{and} \quad T_r = \frac{T}{T_c} \tag{4.15}$$

The critical values for various gases are given in Table 4.3.

The variation of compressibility factor as a function of reduced pressure and reduced temperature can be found in thermodynamics texts such as [4.5]. Note that the real gas effects are prominant only for gases under high pressure and/or low temperature, i.e., the compressibility factor is away from one.

4.2.7 Thermodynamic Properties of a Mixture

Various definitions of thermodynamic properties for a mixture are presented in this section. Internal energy of a mixture is the sum of the internal energies of the species given by:

$$e = \sum_{s=1}^{n} C_s e_s \tag{4.16}$$

where $C_s = \dfrac{\rho_s}{\rho}$ is defined as the mass fraction of s.

Enthalpy:

$$h = \sum_{s=1}^{n} C_s h_s = \sum_{s=1}^{n} C_s e_s + \sum_{s=1}^{n} C_s R_s T_t$$

$$= e + RT_t = e + \frac{p}{\rho} \tag{4.17}$$

Specific heat at constant volume:

$$c_v = \sum_{s=1}^{n} C_s c_{v,s} \tag{4.18}$$

Specific heat at constant pressure:

$$c_p = \sum_{s=1}^{n} C_s c_{p,s} = \sum_{s=1}^{n} C_s (c_{v,s} + R_s) = c_v + R \tag{4.19}$$

4.2.8 Thermally Perfect Gas and Calorically Perfect Gas

When the internal energy of a gas is a function of temperature only, it is called a thermally perfect gas. If the ratio of specific heats "γ" is constant, then it is known as calorically perfect gas. A couple of comments is in order at this point. (1) For a system without chemical reaction, the gas can be assumed as thermally perfect. For example, air at temperatures corresponding to values where any dissocation can take place may be treated as thermally perfect. (2) For gases at relatively low temperatures, the vibrational mode of energy as well as the electronic mode of energy are rather small and can usually be ignored. Therefore,

$$e_s = \frac{3}{2} R_s T_t + e_{o,s} \qquad \text{for an atom}$$

$$e_s = \frac{5}{2} R_s T_t + e_{o,s} \qquad \text{for a molecule}$$

and

$$\gamma = \left.\frac{c_p}{c_v}\right|_s = \frac{5}{3} \qquad \text{for an atom}$$

$$\gamma = \frac{7}{5} \qquad\qquad \text{for a molecule}$$

Indeed, the value of $\gamma = 1.4$ for air at room temperature originates from the above argument.

4.2.9 Equation of State and the Speed of Sound for a Mixture

The equation of state for a mixture of calorically perfect gas can be expressed as:

$$p = (\gamma - 1)\rho e \qquad (4.20)$$

The speed of sound is given by

$$a = \sqrt{\gamma \frac{p}{\rho}} = \sqrt{\gamma R T} \qquad (4.21)$$

Note that, for a general case of a mixture with chemical reaction, parameter $\tilde{\gamma}$ is used to replace γ. Therefore,

$$p = (\tilde{\gamma} - 1)\rho e$$

and

$$a = \sqrt{\tilde{\gamma} R T_t} \qquad (4.22)$$

where $\tilde{\gamma}$ is defined as [4.3, 4.6, 4.7]

$$\tilde{\gamma} = \frac{h}{e} = 1 + \frac{p}{\rho e}$$

Another scheme for $\tilde{\gamma}$ is prescribed in Ref. [4.8].

4.3 Equilibrium, Nonequilibrium, and Frozen Flows

A generic and general definition of equilibrium implies that the properties of a system have no tendency to proceed with a change. Any change which can occur in a flowfield may be with regards to a change in the mass fraction of species or exchange of energy between various modes of energies; hence, there is a need to define chemical or thermal equilibrium and nonequilibrium flows.

4.3.1 Chemical Equilibrium

For flowfields where the chemical reaction rates are extremely high, the reactions take place instantaneously. Thus, reactions are completed before the fluid has a chance to move downstream. Note that chemical equilibrium flow is indeed an idealized situation. Any molecular process requires molecular collisions, therefore, a certain amount of time is required. However, for a chemical equilibrium flow, the characteristic time of the chemical process is much smaller than the translational time of the fluid. For a flow in an equilibrium state, the specific heats are functions of both pressure and temperature. Therefore, the ratio of specific heats is no longer constant and becomes a function of temperature and pressure as well. The gas constant is also a variable due to changes in the molecular weight of the mixture.

4.3.2 Chemical Nonequilibrium

In reality, chemical reactions occur as particles are moving within the domain. When the characteristic time of the chemical process and the characteristic time of fluid motion are about the same order of magnitude, the flow is defined to be a chemical nonequilibrium flow. For nonequilibrium flows, the perfect gas equation of state still holds, except the gas constant is now a variable because the molecular weight of the mixture is changing.

4.3.3 Frozen Flow

When the chemical reaction rates within the flowfield are extremely slow, such that fluid particles moving within the domain do not experience any change in the chemical composition, it is referred to as a *frozen flow.*

From a physical point of view, recall that the chemical reactions occur due to molecular collisions. Relating the number of molecular collisions to the chemical time scales, it is apparent that for a frozen flow the molecular collisions are sufficiently few such that no chemical processes take place.

A frozen flow model can be used in the following situations:

(1) If the temperature of the gas is relatively low, molecules will not have enough internal energy to produce the number of collisions required for a chemical process to occur. This is the case in most supersonic/subsonic flight.

(2) If the density of the gas is low and the temperature is very high, gas cannot produce any chemical reaction simply because there are not enough gas particles to collide with each other, i.e., the chance for collision is too small. This can occur at very high altitude (say above 80km) flights.

4.3.4 Thermal Equilibrium

For a flow in thermal equilibrium, there is no exchange of energy between various modes of energies. For a thermal equilibrium flow, all relaxation times are comparable and small compared to the flow characteristic time (the relaxation time is the time required for a particular internal energy mode to come to equilibrium with another energy mode; see Section 4.6.5 for a detailed description).

4.3.5 Thermal Nonequilibrium

For a thermal nonequilibrium flow, the relaxation times of various energy modes are vastly different, and some are in the same order of the flow characteristic time (the relaxation times are great compared to the time between collisions).

4.4 Damkohler Number and Knudsen Number

A set of nondimensional parameters commonly used in problems associated with the phenomenon of equilibrium and nonequilibrium flows are the Damkohler number and Knudsen number defined in this section.

4.4.1 Damkohler Number

The ratio of the characteristic time of fluid motion to that of the characteristic time of the chemical process is defined as the Damkohler number. The definition can also be interpreted as the ratio of available flow residence time to the time required for equilibrium.

Mathematically [4.9],

$$Da = \frac{l/u}{\tau_c} \tag{4.23}$$

where l is the characteristic length, u is the velocity of the fluid, and τ_c is the characteristic time of the chemical process. The characteristic time may be determined by the following relation [4.10, 4.11]

$$\tau_c = \left[\left(\prod_{s=1}^{n} \chi_s^R \right) \rho k_f \right]^{-1} \tag{4.24}$$

where χ_s^R represents the concentration of reacted species, and k_f represents the forward reaction rate at local value of the temperature. The values of forward reaction rates are provided in Section 4.7.2. Note that if $Da \gg 1$, the flow can be considered in chemical equilibrium state.

4.4.2 Knudsen Number

The Knudsen number is defined as the ratio of the mean free path of molecules to the characteristic length of the flow. It is primarily used to identify applicability of the continuum model.

Mathematically,

$$Kn = \frac{\lambda}{L} \tag{4.25}$$

where [4.2]

$$\lambda = \frac{(MW/\hat{N})}{\sqrt{2}\pi d^2 \rho}$$

is the mean free path, and

$$d = \left(\frac{MW}{\hat{N}\rho_L} \right)^{\frac{1}{3}}$$

Knudsen	Governing Equations	Chemistry Model
$Kn \leq 0.001$	Navier-Stokes	Equilibrium
$0.001 \leq Kn \leq 0.03$	Navier-Stokes	Nonequilibrium
$0.03 \leq Kn \leq 0.2$	Modified Navier-Stokes	Nonequilibrium/Frozen
$Kn \approx 1$	DSMC	
$Kn \gg 1$	Boltzmann Equation	

Table 4.4.

is a measure of the diameter of the gas particle, or the radius of a sphere of influence with $\hat{N} = 6.02 \times 10^{26}$ mole^{-1} is the Avogadro's number, ρ_L is the density of the gas at liquid state. For example, for air, $\rho_L = 0.86$ gm/cm^3, $MW = 28.9$ gm/mole, and $d = 3.8 \times 10^{-8}$ cm. L is the characteristic length of the body, for example the nose radius of a blunt body.

4.5 Navier-Stokes Equations for Nonequilibrium Flows

The Navier-Stokes equations can be applied to solve flowfields with embedded nonequilibrium regions. However, the fundamental assumption of continuum model must be satisfied. It is for this purpose that the Knudsen number is used. Generally speaking, for a Knudsen value of less than about 0.03, the Navier-Stokes equations may be used. For a Kn number range of $0.03 < Kn < 0.2$, some modification of the Navier-Stokes equation may be required. For a Kn number greater than 0.2, discrete particle model (noncontinuum model) must be employed. A commonly used scheme for this purpose is the Direct Simulation Monte Carlo (DSMC) technique [4.12, 4.13].

A general range of application of the Navier-Stokes equation is summarized in Table 4.4.

The modifications required for the Navier-Stokes equations for the Kn number range of $0.03 \leq Kn \leq 0.2$ may include the following considerations: shock slip, wall slip, jump effect, and non-zero bulk viscosity [4.14, 4.15, 4.16].

Before the Navier-Stokes equations are presented, the phenomenon of various temperature models are briefly reviewed.

4.5.1 Temperature Models

As described previously in Section 4.2.5, one may define a temperature associated with each energy mode. Thus, for a mixture of n different species of molecules, the number of different temperatures can be as many as $n + 3$ (T_t, T_r, $T_{v,s}$ for $s =$

$1, \ldots, n$, and T_e). Since the translational and rotational modes of energies equilibrate rather fast, the translational temperature and the rotational temperature are usually assumed to be equal. This assumption is used extensively to simplify the calculation process [4.17, 4.18].

In Refs. [4.17] and [4.18], a six-temperature model was used for a seven-species model of air (four vibrational temperatures, T_t, and T_e). The results indicate similarity in the four vibrational temperatures. Therefore, one may argue against implementation of a multi-vibrational temperature model [4.9]. Instead, the three-temperature model [4.19], representing T_t, T_v and T_e, should be sufficient for the majority of applications. In fact, even the three-temperature model is quite complex and requires many chemical rate parameters which are not yet known, as discussed in Ref. [4.20]. On the other hand, the one-temperature model, where one assumes $T_t = T_v = T_e = T_r$, in general will overestimate the rate of equilibrium. Therefore, as a compromise, a two-temperature model where $T_t = T_r$ and $T_v = T_e$ has been developed [4.20], which is widely used.

4.5.2 Navier-Stokes Equations for a Two-Temperature Model

In Tables 4.5a and 4.5b, the Navier-Stokes equations to be employed, along with a two-temperature model, are provided.

Table 4.5a. Navier-Stokes equations in tensor notation.

Global Conservation of Mass:

$$\frac{\partial \rho}{\partial t} + \frac{\partial (\rho u^j)}{\partial x^j} = 0 \tag{4.26}$$

Species Conservation:

$$\frac{\partial (\rho C_s)}{\partial t} + \frac{\partial (\rho u^j C_s)}{\partial x^j} + \frac{\partial (\rho v_s^j C_s)}{\partial x^j} = \dot{w}_s \tag{4.27}$$

where $s = 1, \ldots n$, for n species fluid flow. The diffusion velocity is v_s^j, see Section 4.6.1.

Conservation of Momentum:

$$\frac{\partial (\rho u^i)}{\partial t} + \frac{\partial (\rho u^i u^j)}{\partial x^j} = -\frac{\partial p}{\partial x^i} + \frac{\partial \tau^{ij}}{\partial x^j} \tag{4.28}$$

Vibrational Energy Equation:

$$\frac{\partial\left[\sum\limits_{s=m}(\rho C_s e_{v,s})\right]}{\partial t} + \frac{\partial\left[\sum\limits_{s=m}(\rho C_s e_{v,s})\right]u^j}{\partial x^j}$$

$$= \frac{\partial(-q_v^j)}{\partial x^j} - \frac{\partial}{\partial x^j}\left[\sum\limits_{s=m}(\rho C_s e_{v,s} v_s^j)\right] + \sum\limits_{s=m}\dot{w}_s e_{v,s} + \sum\limits_{s=m} Q_s^{T\text{-}V} \qquad (4.29)$$

where subscript m represents molecule species.

Global Conservation of Energy:

$$\frac{\partial(\rho e_t)}{\partial t} + \frac{\partial}{\partial x^j}\left[\rho u^j e_t\right] = \frac{\partial}{\partial x^j}\left[-q^j\right] - \frac{\partial}{\partial x^j}\left[u^j p\right] + \frac{\partial}{\partial x^j}\left[u^i \tau^{ij}\right] - \dot{Q}_{\mathrm{rad}} \qquad (4.30)$$

where $\dot{Q}_{\mathrm{rad}}$ is the radiation heat flux.

Table 4.5b. Navier-Stokes equations in Cartesian coordinates.

Global Conservation of Mass:

$$\frac{\partial\rho}{\partial t} + \frac{\partial(\rho u)}{\partial x} + \frac{\partial(\rho v)}{\partial y} + \frac{\partial(\rho w)}{\partial z} = 0 \qquad (4.31)$$

Species Conservation:

$$\frac{\partial\rho C_s}{\partial t} + \frac{\partial(\rho u C_s)}{\partial x} + \frac{\partial(\rho v C_s)}{\partial y} + \frac{\partial(\rho w C_s)}{\partial z} = -\left[\frac{\partial(\rho C_s v_s^x)}{\partial x} + \frac{\partial(\rho C_s v_s^y)}{\partial y} + \frac{\partial(\rho C_s v_s^z)}{\partial z}\right] + \dot{w}_s$$
$$(4.32)$$

for $s = 1, \ldots n$.

x-Component of the Conservation of Momentum:

$$\frac{\partial}{\partial t}(\rho u) + \frac{\partial}{\partial x}(\rho u^2) + \frac{\partial}{\partial y}(\rho uv) + \frac{\partial}{\partial z}(\rho uw) = \frac{-\partial p}{\partial x} + \frac{\partial \tau_{xx}}{\partial x} + \frac{\partial \tau_{xy}}{\partial y} + \frac{\partial \tau_{xz}}{\partial z} \qquad (4.33)$$

y-Component of the Conservation of Momentum:

$$\frac{\partial}{\partial t}(\rho v) + \frac{\partial}{\partial x}(\rho uv) + \frac{\partial}{\partial y}(\rho v^2) + \frac{\partial}{\partial z}(\rho vw) = \frac{\partial \tau_{xy}}{\partial x} + \frac{-\partial p}{\partial y} + \frac{\partial \tau_{yy}}{\partial y} + \frac{\partial \tau_{yz}}{\partial z} \qquad (4.34)$$

z-Component of the Conservation of Momentum:

$$\frac{\partial}{\partial t}(\rho w) + \frac{\partial}{\partial x}(\rho uw) + \frac{\partial}{\partial y}(\rho vw) + \frac{\partial}{\partial z}(\rho w^2) = \frac{\partial \tau_{xz}}{\partial x} + \frac{\partial \tau_{yz}}{\partial y} + \frac{\partial \tau_{zz}}{\partial z} + \frac{-\partial p}{\partial z} \qquad (4.35)$$

where

$$\tau_{xx} = \frac{2}{3}\mu \left(2\frac{\partial u}{\partial x} - \frac{\partial v}{\partial y} - \frac{\partial w}{\partial z} \right) \tag{4.36}$$

$$\tau_{yy} = \frac{2}{3}\mu \left(2\frac{\partial v}{\partial y} - \frac{\partial u}{\partial x} - \frac{\partial w}{\partial z} \right) \tag{4.37}$$

$$\tau_{zz} = \frac{2}{3}\mu \left(2\frac{\partial w}{\partial z} - \frac{\partial u}{\partial x} - \frac{\partial v}{\partial y} \right) \tag{4.38}$$

$$\tau_{xy} = \mu \left(\frac{\partial u}{\partial y} + \frac{\partial v}{\partial x} \right) = \tau_{yx} \tag{4.39}$$

$$\tau_{xz} = \mu \left(\frac{\partial w}{\partial x} + \frac{\partial u}{\partial z} \right) = \tau_{zx} \tag{4.40}$$

$$\tau_{yz} = \mu \left(\frac{\partial v}{\partial z} + \frac{\partial w}{\partial y} \right) = \tau_{zy} \tag{4.41}$$

Vibrational Energy Equation:

$$\frac{\partial \left[\sum_{s=m} (\rho C_s e_{v,s}) \right]}{\partial t} + \frac{\partial \left[\sum_{s=m} (\rho C_s e_{v,s}) \right] u}{\partial x} + \frac{\partial \left[\sum_{s=m} (\rho C_s e_{v,s}) \right] v}{\partial y} + \frac{\partial \left[\sum_{s=m} (\rho C_s e_{v,s}) \right] w}{\partial z}$$

$$= -\frac{\partial}{\partial x} \left[q_v^x + \sum_{s=m} \rho C_s e_{v,s} v_s^x \right] - \frac{\partial}{\partial y} \left[q_v^y + \sum_{s=m} \rho C_s e_{v,s} v_s^y \right]$$

$$- \frac{\partial}{\partial z} \left[q_v^z + \sum_{s=m} \rho C_s e_{v,s} v_s^z \right] + \sum_{s=m} \dot{w}_s e_{v,s} + \sum_{s=m} Q_s^{T\text{-}V} \tag{4.42}$$

Global Conservation of Energy:

$$\frac{\partial}{\partial t}(\rho e_t) + \frac{\partial}{\partial x}(\rho u e_t) + \frac{\partial}{\partial y}(\rho v e_t) + \frac{\partial}{\partial z}(\rho w e_t)$$

$$= \frac{\partial}{\partial x} \left[-q^x - up + \mu(u\tau_{xx} + v\tau_{xy} + w\tau_{xz}) \right]$$

$$+ \frac{\partial}{\partial y} \left[-q^y - vp + \mu(u\tau_{xy} + v\tau_{yy} + w\tau_{yz}) \right]$$

$$+ \frac{\partial}{\partial z} \left[-q^z - wp + \mu(u\tau_{xz} + v\tau_{yz} + w\tau_{zz}) \right] - \dot{Q}_{\text{rad}} \tag{4.43}$$

4.6 Transport Properties

The calculation for transport properties (viscosity, thermal conductivity, diffusion, ...) is based on the first approximation of the Chapman-Enskog expansion [4.2, 4.19] to the simple gases and the multi-temperature gas mixtures. The calculations involve determinants of certain collision integrals. These integrals have been reduced to standard forms that have been evaluated for a number of intermolecular potentials. For a detailed discussion of the evaluation of transport properties, Ref. [4.5] should be reviewed.

4.6.1 Diffusion Velocity of Species s

A complete form of the diffusion velocity of species denoted by v_s^j includes many terms and is given in Ref. [4.9]. However, for most applications, the expression may be reduced by neglecting some of the terms. For example, terms representing the thermal diffusion and pressure diffusion may in general be neglected. Nevertheless, there are a variety of simplified forms to represent v_s^j, three of which are provided here.

$$\text{(i) } v_s^j = \frac{1}{\gamma_s MW_s} D_s \frac{\partial \delta_s}{\partial x^j} \text{ and } D_s = \frac{\Gamma^2 MW_s(1 - C_s)}{\sum\limits_{r \neq s} \frac{\gamma_r}{D_{sr}}} \quad \text{(from Ref. [4.19])} \qquad (4.44)$$

$$\text{(ii) } v_s^j = \frac{-1}{C_s} \left[\frac{(1 - C_s)}{\sum\limits_{r \neq s} \delta_r / D_{sr}} \right] \frac{\partial C_s}{\partial x^j} \quad \text{(from Ref. [4.3])} \qquad (4.45)$$

$$\text{(iii) } v_s^j = \frac{-1}{C_s} \left[\frac{(1 - C_s)}{\sum\limits_{r} \delta_r / D_{sr}} \right] \frac{\partial C_s}{\partial x^j} \quad \text{(from Ref. [4.21])} \qquad (4.46)$$

where

$$D_{sr} = \frac{2.628 \times 10^{-3} T^{\frac{3}{2}} \sqrt{\dfrac{MW_s + MW_r}{2 MW_s \cdot MW_r}}}{p \sigma_{sr}^2 \Omega_{sr}^{(1,1)*}} \quad (\text{unit} : \text{cm}^2/\text{sec}) \qquad (4.47)$$

is the binary diffusion coefficient for the s-r pair, p is the total pressure (units of atm). The average collision diameter of particles s and r is $\sigma_{sr} = \frac{1}{2}(\sigma_s + \sigma_r)$. The values of σ for different particles are provided in Table 4.6a.

The values of $\Omega_{sr}^{(1,1)*}$ (Lennard-Jones 6–12 potential integrals) can be determined by numerous schemes, a couple of which are provided below.

(1) The values of $\Omega_{sr}^{(1,1)*}$ are provided in Table 4.6b [4.5] as a function of a nondimensional temperature T^* defined as:

$$T^* = \frac{T}{\epsilon_{sr}/\hat{k}} \quad \text{for a } s\text{-}r \text{ pair collision}$$

The values of

$$\frac{\epsilon_{sr}}{\hat{k}} = \sqrt{\left(\frac{\epsilon_s}{\hat{k}}\right)\left(\frac{\epsilon_r}{\hat{k}}\right)}$$

are provided in Table 4.6a.

(2) $\Omega_{sr}^{(1,1)*}$ can also be calculated by [4.22]

$$\Omega_{sr}^{(1,1)*} = \frac{\bar{\Omega}_{s,r}^{(1,1)}}{\sigma_{sr}^2}$$

and values of $\bar{\Omega}_{s,r}^{(1,1)}$ are provided in Table 4.7 as a function of temperature (K).

(3) Some simple expressions by curve-fit of data are also available to determine the value of D_{sr} directly, such that the calculation of $\Omega_{sr}^{(1,1)*}$ would not be necessary; they are

(i) $D_{sr} = \dfrac{\bar{D}_{sr}}{p}$ where $\bar{D}_{sr} = e^c \cdot T^{(A \ln T + B)}$, as given in Ref. [4.24], (unit of $\bar{D}_{sr}$: $\text{cm}^2 \cdot \text{atm/sec}$) (units of: $T(\text{K})$, p (atm)). Coefficients A, B, and C are listed in Table 4.8.

(ii) Or $D_{sr} = \dfrac{AT^B}{p}$ (p: atm), as provided in Ref. [4.3]. Coefficients A and B are listed in Table 4.9.

Substance	Molecular Weight MW	Lennard-Jones Parameters	
		σ (Å)**	$\epsilon/\hat{k}$ (K)
Light elements:			
H_2	2.016	2.915	38.0
He	4.003	2.576	10.2
Noble gases:			
Ne	20.183	2.789	35.7
Ar	39.944	3.418	124.
Kr	83.80	3.498	225.
Xe	131.3	4.055	229.
Simple polyatomic substances :			
Air	28.97	3.617	97.0
N_2	28.02	3.681 (3.798)*	91.5 (71.4)*
O_2	32.00	3.433	113. (106.7)*
O_3	48.00		
CO	28.01	3.590	110.
CO_2	44.01	3.996	190.
NO	30.01	3.470 (3.492)*	119. (116.7)*
N_2O	44.02	3.879	220.
SO_2	64.07	4.290	252.
F_2	38.00	3.653	112.
Cl_2	70.91	4.115	357.
Br_2	159.83	4.268	520.
I_2	252.82	4.982	550.
Atoms			
O	(16.00)*	(3.05)*	(106.7)*
N	(14.01)*	(3.298) *	(71.4)*

Table 4.6a. Lennard-Jones parameters for various gases.
           Note: * values in ( ) are from Ref. [4.23], others from
                 Ref. [4.21].
              ** Å = 10^{-8} cm

T^*	$\Omega^{(1,1)*}$	$\Omega^{(2,2)*}$	T^*	$\Omega^{(1,1)*}$	$\Omega^{(2,2)*}$
0.30	2.662	2.785	2.7	0.9770	1.069
0.35	2.476	2.628	2.8	0.9672	1.058
0.40	2.318	2.492	2.9	0.9576	1.048
0.45	2.184	2.368	3.0	0.9490	1.039
0.50	2.066	2.257	3.1	0.9406	1.030
0.55	1.966	2.156	3.2	0.9328	1.022
0.60	1.877	2.065	3.3	0.9256	1.014
0.65	1.798	1.982	3.4	0.9186	1.007
0.70	1.729	1.908	3.5	0.9120	0.9999
0.75	1.667	1.841	3.6	0.9058	0.9932
0.80	1.612	1.780	3.7	0.8998	0.9870
0.85	1.562	1.725	3.8	0.8942	0.9811
0.90	1.517	1.675	3.9	0.8888	0.9755
0.95	1.476	1.629	4.0	0.8836	0.9700
1.00	1.439	1.587	4.1	0.8788	0.9649
1.05	1.406	1.549	4.2	0.8740	0.9600
1.10	1.375	1.514	4.3	0.8694	0.9553
1.15	1.346	1.482	4.4	0.8652	0.9507
1.20	1.320	1.452	4.5	0.8610	0.9464
1.25	1.296	1.424	4.6	0.8568	0.9422
1.30	1.273	1.399	4.7	0.8530	0.9382
1.35	1.253	1.375	4.8	0.8492	0.9343
1.40	1.233	1.353	4.9	0.8456	0.9305
1.45	1.215	1.333	5	0.8422	0.9269
1.50	1.198	1.314	6	0.8124	0.8963
1.55	1.182	1.296	7	0.7896	0.8727
1.60	1.167	1.279	8	0.7712	0.8538
1.65	1.153	1.264	9	0.7556	0.8379
1.70	1.140	1.248	10	0.7424	0.8242
1.75	1.128	1.234	20	0.6640	0.7432
1.80	1.116	1.221	30	0.6232	0.7005
1.85	1.105	1.209	40	0.5960	0.6718
1.90	1.094	1.197	50	0.5756	0.6504
1.95	1.084	1.186	60	0.5596	0.6335
2.00	1.075	1.175	70	0.5464	0.6194
2.10	1.057	1.156	80	0.5352	0.6076
2.20	1.041	1.138	90	0.5256	0.5973
2.30	1.026	1.122	100	0.5130	0.5882
2.40	1.012	1.107	200	0.4644	0.5320
2.50	0.9996	1.093	300	0.4360	0.5016
2.60	0.9878	1.081	400	0.4170	0.4811

Table 4.6b. The integrals $\Omega^{(r,s)*}$ for calculating the transport coefficients for the Lennard-Jones (6–12) potentials. (Source: Ref. [4.5])

T (K)	O-O		N-N		N-O	
	$<\bar{\Omega}^{(1,1)}>$, A^2	$<\bar{\Omega}^{(2,2)}>$, A^2	$<\bar{\Omega}^{(1,1)}>$, A^2	$<\bar{\Omega}^{(2,2)}>$, A^2	$<\bar{\Omega}^{(1,1)}>$, A^2	$<\bar{\Omega}^{(2,2)}>$, A^2
1000	6.193	7.182	6.235	7.039	7.156	8.418
1500	5.643	6.564	5.548	6.292	6.139	7.227
2000	5.271	6.139	5.106	5.812	5.518	6.494
2500	4.981	5.821	4.786	5.464	5.088	5.984
3000	4.746	5.567	4.538	5.193	4.767	5.606
3500	4.551	5.355	4.336	4.972	4.531	5.310
4000	4.386	5.174	4.166	4.786	4.347	5.075
4500	4.243	5.016	4.022	4.628	4.190	4.899
5000	4.117	4.876	3.895	4.488	4.054	4.745
5500	4.007	4.752	3.783	4.365	3.935	4.610
6000	3.905	4.640	3.683	4.254	3.828	4.490
6500	3.813	4.537	3.592	4.154	3.731	4.381
7000	3.727	4.438	3.509	4.062	3.644	4.282
7500	3.648	4.352	3.435	3.979	3.564	4.191
8000	3.575	4.271	3.365	3.902	3.491	4.108
8500	3.509	4.194	3.300	3.830	3.423	4.031
9000	3.442	4.118	3.240	3.764	3.360	3.959
9500	3.389	4.058	3.184	3.701	3.301	3.892
10000	3.335	3.996	3.131	3.643	3.246	3.829
11000	3.236	3.883	3.035	3.536	3.145	3.714
12000	3.148	3.782	2.949	3.440	3.055	3.611
13000	3.068	3.692	2.872	3.353	2.974	3.518
14000	2.996	3.609	2.801	3.274	2.900	3.433
15000	2.931	3.534	2.737	3.202	2.832	3.356

Table 4.7 Collision Integrals (Source: Ref. [4.22])

T (K)	N-N$_2$		O-O$_2$		N$_2$-O$_2$	
	$<\bar{\Omega}^{(1,1)}>$, A^2	$<\bar{\Omega}^{(2,2)}>$, A^2	$<\bar{\Omega}^{(1,1)}>$, A^2	$<\bar{\Omega}^{(2,2)}>$, A^2	$<\bar{\Omega}^{(1,1)}>$, A^2	$<\bar{\Omega}^{(2,2)}>$, A^2
1000	8.214	9.824	7.509	8.778	9.258	10.48
1500	7.396	8.897	6.890	8.093	8.709	10.05
2000	6.842	8.272	6.467	7.624	8.371	9.802
2500	6.427	7.802	6.148	7.269	8.134	9.598
3000	6.098	7.427	5.894	6.985	7.889	9.412
3500	5.827	7.115	5.683	6.749	7.705	9.246
4000	5.596	6.850	5.503	6.548	7.535	9.080
4500	5.397	6.622	5.347	6.374	7.375	8.916
5000	5.223	6.419	5.210	6.219	7.233	8.766
5500	5.067	6.238	5.087	6.082	7.095	8.620
6000	4.927	6.080	4.976	5.956	6.971	8.484
6500	4.800	5.933	4.875	5.843	6.853	8.354
7000	4.684	5.799	4.783	5.739	6.742	8.232
7500	4.577	5.676	4.697	5.642	6.631	8.110
8000	4.479	5.558	4.618	5.553	6.529	7.991
8500	4.387	5.453	4.545	5.470	6.405	7.853
9000	4.301	5.355	4.476	5.392	6.295	7.724
9500	4.221	5.259	4.411	5.319	6.186	7.603
10000	4.146	5.170	4.350	5.250	6.088	7.488
11000	4.008	5.010	4.238	5.123	5.907	7.278
12000	3.884	4.867	4.137	5.008	5.745	7.089
13000	3.772	4.734	4.045	4.903	5.597	6.918
14000	3.669	4.612	3.961	4.808	5.460	6.760
15000	3.575	4.501	3.884	4.719	5.335	6.615

Table 4.7 Continued

T (K)	O-N$_2$		N$_2$-N$_2$		O$_2$-O$_2$	
	$<\bar{\Omega}^{(1,1)}>$, A^2	$<\bar{\Omega}^{(2,2)}>$, A^2	$<\bar{\Omega}^{(1,1)}>$, A^2	$<\bar{\Omega}^{(2,2)}>$, A^2	$<\bar{\Omega}^{(1,1)}>$, A^2	$<\bar{\Omega}^{(2,2)}>$, A^2
1000	8.867	10.61	10.07	11.17	8.875	9.896
1500	7.975	9.610	9.365	10.54	8.381	9.470
2000	7.371	8.926	8.794	10.01	8.044	9.226
2500	6.919	8.406	8.297	9.520	7.810	9.020
3000	6.561	7.998	7.845	9.071	7.608	8.840
3500	6.265	7.659	7.450	8.697	7.437	8.672
4000	6.015	7.374	7.137	8.350	7.279	8.516
4500	5.798	7.126	6.844	8.100	7.128	8.368
5000	5.605	6.899	6.599	7.804	7.012	8.246
5500	5.439	6.712	6.369	7.592	6.868	8.090
6000	5.287	6.535	6.187	7.394	6.748	7.963
6500	5.149	6.374	6.023	7.209	6.622	7.827
7000	5.023	6.228	5.863	7.035	6.505	7.702
7500	4.907	6.094	5.708	6.873	6.402	7.586
8000	4.799	5.971	5.595	6.721	6.306	7.479
8500	4.700	5.851	5.480	6.611	6.217	7.379
9000	4.607	5.744	5.373	6.474	6.132	7.285
9500	4.520	5.641	5.284	6.375	6.053	7.197
10000	4.438	5.548	5.199	6.279	5.978	7.114
11000	4.288	5.369	5.069	6.097	5.844	6.960
12000	4.153	5.213	4.907	5.952	5.718	6.822
13000	4.032	5.068	4.810	5.815	5.608	6.696
14000	3.920	4.940	4.727	5.705	5.506	6.580
15000	3.818	4.818	4.630	5.599	5.412	6.473

Table 4.7 Continued

Interaction	A	B	C
N-O	-0.0043383	1.9119177	-11.891342
N-N$_2$	0.0191055	1.4904448	-10.358828
N-O$_2$	0.0191055	1.4904448	-10.358828
N-NO	0.0191055	1.4904448	-10.358828
O-O$_2$	0.0216586	1.3875747	-9.7389971
O-N$_2$	0.0168907	1.5276702	-10.629306
N$_2$-O$_2$	0.0435927	0.9784219	-8.3354916
O-NO	0.0183441	1.4750189	-10.265935
O$_2$-NO	0.0410864	1.0124720	-8.4455480
N$_2$-NO	0.0315955	1.2225368	-9.4862934
O-NO$^+$	0.0003467	1.8941393	-12.978394
N-NO$^+$	0.0003467	1.8941393	-12.978394
O$_2$-NO$^+$	0.0003467	1.8941393	-12.978394
N$_2$-NO$^+$	0.0003467	1.8941393	-12.978394
NO-NO$^+$	0.0039930	1.5689336	-11.441502

Table 4.8. Diffusion curve fit constants for $\bar{D}_{sr}$. (Source: Ref. [4.24])

4.6.2 Viscosity

The viscosity of a pure gas s, may be expressed as [4.5]:

$$\mu_s = \frac{2.6693 \times 10^{-5}\sqrt{MW_s T}}{\sigma_s^2 \Omega_s^{(2,2)*}} \quad \text{(units of g/cm-sec)} \tag{4.48}$$

where the values of $\Omega_s^{(2,2)*}$ are provided in Table 4.6 as a function of T^* or the term $\sigma_s^2 \Omega_s^{(2,2)*}$ denoted by $\bar{\Omega}_s^{(2,2)}$ (defined as collision integrals) which may be found in Table 4.7 given as a function of temperature.

A second expression to determine viscosity is given by [4.24]

$$\mu_s = e^C \cdot T^{(A\ln T + B)} \quad \text{(units of g/cm-sec)} \tag{4.49}$$

where the coefficients A, B, and C are provided in Table 4.10. The viscosity of a mixture of gases following Wilke's mixing rule [4.24] is given by

$$\mu = \sum_{s=1}^{n} \frac{\delta_s \mu_s}{\sum_{j=1}^{n} \delta_j \phi_{sj}} \tag{4.50}$$

where

$$\phi_{sj} = \frac{\left[1 + (\mu_s/\mu_j)^{0.5} (MW_j/MW_s)^{0.25}\right]^2}{\sqrt{8}\left[1 + (MW_s/MW_j)\right]^{0.5}}$$

Species Pair	A	B
N_2-O_2	$1.338E-05$	1.682
N-O	$2.480E-05$	1.682
N-NO	$1.874E-05$	1.684
N-N_2	$1.979E-05$	1.673
N-O_2	$1.914E-05$	1.682
O-NO	$1.655E-05$	1.696
O-N_2	$1.866E-05$	1.678
O-O_2	$1.705E-05$	1.695
NO-NO^+	$1.129E-05$	1.705
NO-N_2	$1.321E-05$	1.684
NO-O_2	$1.176E-05$	1.700

Table 4.9. Diffusion coefficients for D_{sr}. (Source: Ref. [4.3])

Species	A	B	C
O_2	0.0449290	-0.0826158	-9.2019475
N_2	0.0268142	0.3177838	-11.3155513
O	0.0203144	0.4294404	-11.6031403
N	0.0115572	0.6031679	-12.4327495
NO	0.0436378	-0.0335511	-9.5767430
NO^+	0.3020141	-3.5039791	-3.7355157

Table 4.10 Viscosity curve fit constants (temperatures 1000 K to 30000 K).

Equation (4.49) can be used for air (at temperatures below dissociation temperature), in which case

$$A = -0.1045186 \quad B = 1.9790489 \quad C = -16.48024$$

4.6.3 Thermal Conductivity

The thermal conductivity for a monatomic gas and a polyatomic molecule gas [4.25] are given respectively by

$$k_s = \frac{15}{4}\left(\frac{\mu_s \mathcal{R}}{MW_s}\right) \quad \text{(for a monatomic gas)} \tag{4.51}$$

and

$$k_s = \left[\frac{15}{4} + \frac{1}{S_c}\left(\frac{c_{p,s}\cdot MW_s}{\mathcal{R}} - \frac{5}{2}\right)\right]\left(\frac{\mu_s \mathcal{R}}{MW_s}\right) \quad \text{(for a molecule)} \tag{4.52}$$

where

$$Sc = \frac{\mu}{\rho D} = \frac{Pr}{Le} \quad \text{is the Schmidt Number} \tag{4.53}$$

$$Le = \frac{\rho D c_p}{k} \quad \text{is the Lewis Number} \tag{4.54}$$

Note that $c_{p,s}$ used in the equation is per unit mass, and

$$D = D_{ss} = \frac{2.628 \times 10^{-3} T^{\frac{3}{2}} MW_s^{-\frac{1}{2}}}{P\sigma_s^2 \Omega_{ss}^{(1,1)*}}$$

is the self diffusion coefficient. (Recall D_{sr} in Section 4.6.1.) The inverse of the Schmidt number may be set to 1.32, as in Ref. [4.23], or set to 1, as in Refs. [4.24] and [4.26]. This is known as the Eucken correction form.

The thermal conductivity of a mixture of gases may be determined from the following relation [4.23, 4.24, 4.26].

$$k = \sum_{s=1}^{n} \frac{\delta_s k_s}{\displaystyle\sum_{j=1}^{n} \delta_j \phi_{sj}} \tag{4.55}$$

Another relation for the thermal conductivity, as provided in Ref. [4.3] is

$$k_s = \left(0.469 + 0.354\frac{c_{v,s} MW_s}{\mathcal{R}}\right)\left(3.75\frac{\mu_s \mathcal{R}}{MW_s}\right) \tag{4.56}$$

for a pure polyatomic gas and

$$k = \sum_{s=1}^{n} \frac{k_s}{1 + \dfrac{1.065}{\delta_s}\displaystyle\sum_{j\neq s}\delta_j \phi_{sj}} \tag{4.57}$$

for a mixture, where ϕ_{sj} is defined in Section 4.6.2.

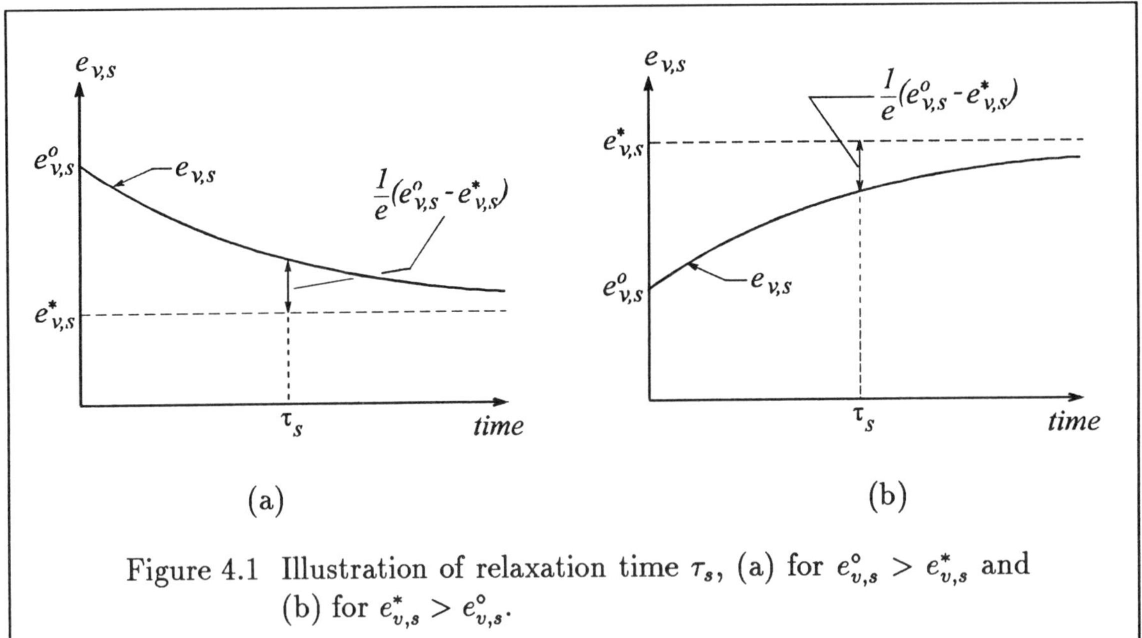

Figure 4.1 Illustration of relaxation time τ_s, (a) for $e_{v,s}^o > e_{v,s}^*$ and (b) for $e_{v,s}^* > e_{v,s}^o$.

4.6.4 Vibrational Heat Conduction Flux

The vibrational heat conduction flux, denoted by q_v^j, is given by [4.3]

$$q_v^j = -\mu \frac{\partial \bar{e}_v}{\partial x^j} \tag{4.58}$$

where

$$\bar{e}_v = \sum_{s=m} C_s e_{v,s}$$

(Subscript m denotes molecules)

4.6.5 *T-V* Energy-Exchange Process

The Landau-Teller equation provides the rate of energy exchange between the vibrational and translational modes of species s which may be expressed as [4.2]

$$Q^{T\text{-}V} = \frac{\rho C_s \left[e_{v,s}^*(T_t) - e_{v,s}(T_v) \right]}{<\tau_s>} \tag{4.59}$$

where $e_{v,s}^*$ is the equilibrium vibrational energy evaluated at T_t. $<\tau_s>$ is the average vibrational relaxation time of species s, which is the time required for the value of $\left| e_{v,s}^* - e_{v,s} \right|$ to fall to $1/e$ of its initial value (Fig. 4.1) [4.2]. The value of $<\tau_s>$ is given by

$$<\tau_s> = \sum_{i=1}^{n} \frac{\gamma_i}{\sum_{i=1}^{n} (\gamma_i/\tau_{si})} \quad \text{or,} \quad \frac{1}{<\tau_s>} = \sum_{i=1}^{n} \frac{\delta_i}{\tau_{si}} \tag{4.60}$$

where τ_{si} is the relaxation time for species s with the collision partner i [4.19].

The relaxation time τ_{si} may be calculated from the following expression [4.27]

$$p\tau_{si} = \exp\left[A_{si}\left(T^{-\frac{1}{3}} - 0.015\bar{m}_{si}^{\frac{1}{4}}\right) - 18.42\right] \tag{4.61}$$

where

$$A_{si} = 1.16 \times 10^{-3}\bar{m}_{si}^{\frac{1}{2}}\theta_{v,s}^{\frac{4}{3}}$$

and

$$\bar{m}_{si} = \frac{MW_s \cdot MW_i}{(MW_s + MW_i)}$$

is the reduced molecular weight of the s,i collision pair. p is the local pressure in units of atm, and $\theta_{v,s}$ is the vibrational characteristic temperature (Table 4.2).

For a temperature range higher than 8000K, the following modified expression is suggested [4.28]

$$\tau_{si} = \tau_{si}(\text{from the previous equations}) + \frac{1}{\sigma_v \bar{C}_s n_{si}} \tag{4.62}$$

where

$$\sigma_v = 10^{-17}\left(\frac{50,000}{T_t}\right)^2 \quad (\text{unit of cm}^2)$$

$$\bar{C}_s = [8\mathcal{R}T/(\pi MW_s)]^{\frac{1}{2}} \quad (\text{unit of cm/sec})$$

$$n_{si} = \rho\hat{N}\left(\frac{C_s}{MW_s} + \frac{C_i}{MW_i}\right) \quad (\text{unit of moleclues/cm}^3)$$

and the following definitions are used. The vibrational excitation cross section is σ_v, the average molecular speed is $\bar{C}_s$, and the number density of the colliding particles is n_{si}.

A nonlinear relation for $Q_s^{T\text{-}V}$ to replace the original Landau-Teller equation is suggested [4.20, 4.29] as follows:

$$Q_s^{T\text{-}V} = \frac{\rho C_s\left[e_{v,s}^*(T_t) - e_{v,s}(T_v)\right]}{<\tau_s>}\left|\frac{T_{\text{shock}} - T_v}{T_{\text{shock}} - T_{v,\text{shock}}}\right|^{\bar{S}-1} \tag{4.63}$$

where

$$\bar{S} = 3.5\exp\left(-5000/T_{\text{shock}}\right) \quad \text{for } T_t > 15,000\text{K}$$

$$\bar{S} = 1 \quad \text{for } T_t < 7,000\text{K}$$

For $7,000\text{K} < T_t < 15,000\text{K}$, values of $\bar{S}$ can be determined from Fig. 4.2 [4.29].

Note that T_{shock} and $T_{v,\text{shock}}$ are the translational temperature T_t and the vibrational temperature T_v immediately behind the shock wave. $T_{v,\text{shock}}$ can be taken as the freestream temperature, suggested by Ref. [4.3].

4.6.6 Heat Flux

The heat flux term denoted by q^j in the energy equation is given by

$$q^j = -k\frac{\partial T_t}{\partial x^j} + q_v^j + \sum_{s=1}^{n} \rho C_s h_s v_s^j \qquad (4.64)$$

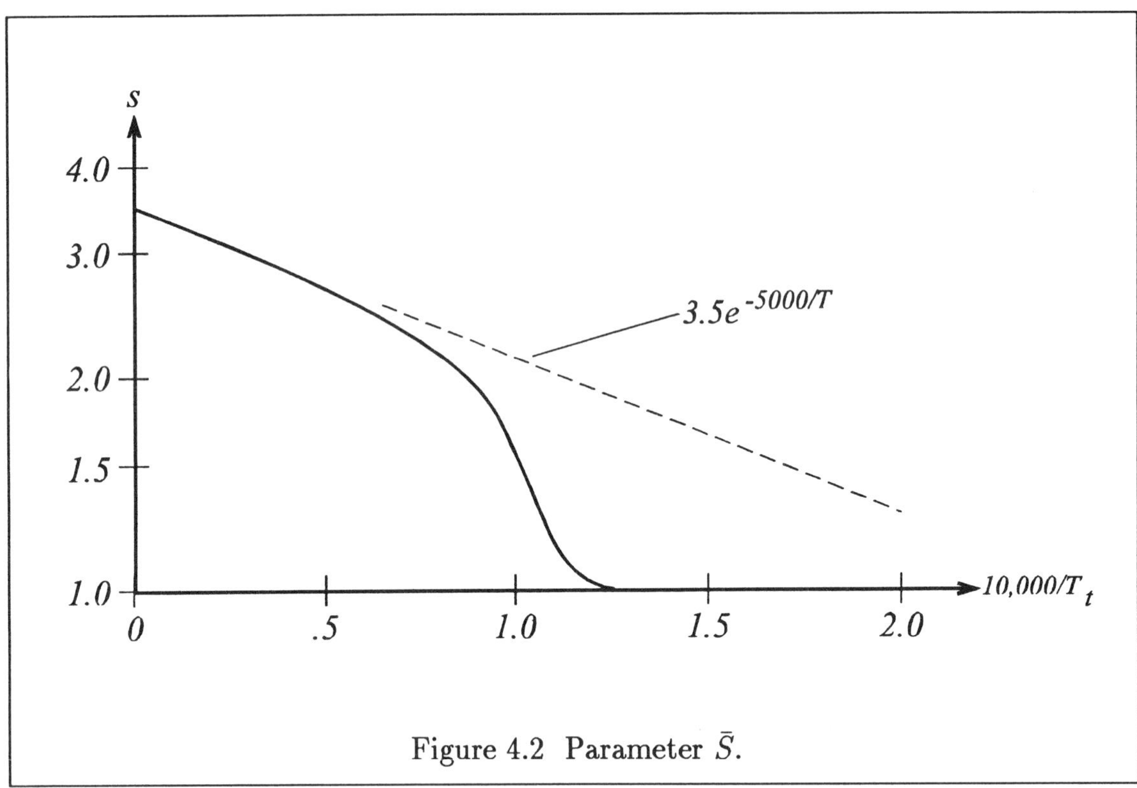

Figure 4.2 Parameter $\bar{S}$.

4.7   Chemical Reactions

In order to determine the mass production rate, $\dot{w}_s$ which appears in the species conservation equation and in the vibrational energy equation, the chemical reactions related to species s must be considered. The rate constants involved in these chemical reactions are either provided by experimental measurements or determined empirically and are typically presented in a curve-fit format.

4.7.1 Rate Equations

Any chemical reaction can be written as

$$\sum_{s=1}^{n} \nu_s^R A_s \; \underset{k_b}{\overset{k_f}{\rightleftharpoons}} \; \sum_{s=1}^{n} \nu_s^P A_s \qquad (4.65)$$

where ν_s^R represents the Stoichiometric mole number of the reactant A_s, and ν_s^P represents the Stoichiometric mole number of the product A_s. The reaction rate of A_s will be:

(i) Forward Rate

$$\left.\frac{d\chi_s}{dt}\right|_f = (\nu_s^P - \nu_s^R)\, k_f \prod_{i=1}^{n} \chi_i^{\nu_i^R} \tag{4.66}$$

(ii) Reverse (Backward) Rate

$$\left.\frac{d\chi_s}{dt}\right|_r = (\nu_s^R - \nu_s^P)\, k_b \prod_{i=1}^{n} \chi_i^{\nu_i^P} \tag{4.67}$$

(iii) Net forming rate of A_s

$$\left.\frac{d\chi_s}{dt}\right|_{\text{net}} = \left.\frac{d\chi_s}{dt}\right|_f + \left.\frac{d\chi_s}{dt}\right|_r = (\nu_s^P - \nu_s^R)\left(k_f \prod_{i=1}^{n} \chi_i^{\nu_i^R} - k_b \prod_{i=1}^{n} \chi_i^{\nu_i^P} \right) \tag{4.68}$$

[Note: if $\left.\dfrac{d\chi_s}{dt}\right|_{\text{net}} = 0$, then the reaction is in the equilibrium state.]

The mass production rate of A_s can be determined by

$$\dot{w}_s = MW_s \left.\frac{d\chi_s}{dt}\right|_{\text{net}} \tag{4.69}$$

4.7.2 Determination of the Rate Constants

4.7.2.1 *Rate constants for air, two-temperature model:*

4.7.2.1.1 Scheme presented by Park in Ref. [4.8]

The chemical reactions are divided into four groups where a different temperature is used in each group. The equilibrium rate constant is expressed by functional form as

$$k_c(T) = \exp(A_1 + A_2 \ln Z + A_3 Z + A_4 Z^2 + A_5 Z^3) \tag{4.70}$$

where $Z = 10,000/T$ and the values of the coefficients A_1 through A_5 for a temperature range of 500 to 50,000 K are given in Table 4.11.

Group 1: Reactions of the form $AB + M \underset{\leftarrow}{\rightarrow} A + B + M$

    (1a) If M is a heavy particle

$$k_f = CT_t^n \exp(-E/\hat{k}T_v - 1 + T_t/T_v) \tag{4.71}$$

$$k_c = k_c(T_t) \tag{4.72}$$

$$k_b = k_f^*/k_c \quad (4.73), \text{ where } k_f^* = CT_t^n \exp(-E/\hat{k}T_t)$$

    (1b) If M is an electron

$$k_f = CT_v^n \exp(-E/\hat{k}T_v) \tag{4.74}$$

$$k_c = k_c(\sqrt{T_t T_v}) \tag{4.75}$$

$$k_b = k_f^*/k_c \quad (4.76), \text{ where } k_f^* = C\left[\sqrt{T_t T_v}\right]^n \exp(-E/\hat{k}\sqrt{T_t T_v})$$

Group 2: Reactions of the form $AB + C \underset{\leftarrow}{\rightarrow} A + BC$

$$k_f = CT_t^n \exp(-E/\hat{k}T_v - 1 + T_t/T_v) \tag{4.77}$$

$$k_c = k_c(T_t) \tag{4.78}$$

$$k_b = k_f^*/k_c \quad (4.79) \text{ where } k_f^* = CT_t^n \exp(-E/\hat{k}T_t)$$

Group 3: Reactions of the form $A + B \underset{\leftarrow}{\rightarrow} AB^+ + e$

$$k_f = CT_t^n \exp(-E/\hat{k}T_t) \tag{4.80}$$

$$k_c = k_c(T_v) \tag{4.81}$$

$$k_b = k_f^*/k_c \quad (4.82) \text{ where } k_f^* = CT_v^n \exp(-E/\hat{k}T_v)$$

Group 4: Reactions of the form $A + e \underset{\leftarrow}{\rightarrow} A^+ + e + e$

$$k_f = CT_v^n \exp(-E/\hat{k}T_v) \tag{4.83}$$

$$k_c = k_c(T_v) \tag{4.84}$$

$$k_b = k_f^*/k_c \quad (4.85) \text{ where } k_f^* = CT_v^n \exp(-E/\hat{k}T_v)$$

The values of coefficients C, n, $E/\hat{k}$ are provided in Table 4.11.

Table 4.11. Reaction groups and rate constants of air for two-temperature model. (Source: Ref. [4.8])

j	Initial species	Final species	Reaction energy, kcal/mole	$A(i,j)$, $i=1$	$i=2$	$i=3$	$i=4$	$i=5$	C Rate constant, cm^3/mole	n Temperature exponent	$E/\bar{k}$ Activation temperature	Group
						Dissociation						
1	O_2 $\rightleftarrows$	$O+O$	-117.98	2.855	0.988	-6.181	-0.023	-0.001				
							$M = N$		$8.250E+19$	-1.000	$59{,}500$	
							$M = O$		$8.250E+19$	-1.000	$59{,}500$	
							$M = N_2$		$2.750E+19$	-1.000	$59{,}500$	
							$M = O_2$		$2.750E+19$	-1.000	$59{,}500$	
							$M = NO$		$2.750E+19$	-1.000	$59{,}500$	
							$M = N^+$		$8.250E+19$	-1.000	$59{,}500$	1a
							$M = O^+$		$8.250E+19$	-1.000	$59{,}500$	
							$M = N_2^+$		$2.750E+19$	-1.000	$59{,}500$	
							$M = O_2^+$		$2.750E+19$	-1.000	$59{,}500$	
							$M = NO^+$		$2.750E+19$	-1.000	$59{,}500$	
							$M = e^-$		$1.320E+22$	-1.000	$59{,}500$	1b
2	N_2 $\rightleftarrows$	$N+N$	-225.00	1.858	-1.325	-9.856	-0.174	0.008				
							$M = N$		$1.110E+22$	-1.600	$113{,}200$	
							$M = O$		$1.110E+22$	-1.600	$113{,}200$	
							$M = N_2$		$3.700E+21$	-1.600	$113{,}200$	
							$M = O_2$		$3.700E+21$	-1.600	$113{,}200$	
							$M = NO$		$3.700E+21$	-1.600	$113{,}200$	
							$M = N^+$		$1.110E+22$	-1.600	$113{,}200$	1a
							$M = O^+$		$1.110E+22$	-1.600	$113{,}200$	
							$M = N_2^+$		$3.700E+21$	-1.600	$113{,}200$	
							$M = O_2^+$		$3.700E+21$	-1.600	$113{,}200$	
							$M = NO^+$		$3.700E+21$	-1.600	$113{,}200$	
							$M = e^-$		$1.110E+24$	-1.600	$113{,}200$	1b
3	NO $\rightleftarrows$	$N+O$	-150.03	0.792	-0.492	-6.761	-0.091	0.004				
							$M = N$		$4.600E+17$	-0.500	$75{,}500$	
							$M = O$		$4.600E+17$	-0.500	$75{,}500$	
							$M = N_2$		$2.300E+17$	-0.500	$75{,}500$	
							$M = O_2$		$2.300E+17$	-0.500	$75{,}500$	
							$M = NO$		$2.300E+17$	-0.500	$75{,}500$	
							$M = N^+$		$4.600E+17$	-0.500	$75{,}500$	1a
							$M = O^+$		$4.600E+17$	-0.500	$75{,}000$	
							$M = N_2^+$		$2.300E+17$	-0.500	$75{,}000$	
							$M = O_2^+$		$2.300E+17$	-0.500	$75{,}500$	
							$M = NO^+$		$2.300E+17$	-0.500	$75{,}500$	
							$M = e^-$		$7.360E+19$	-0.500	$75{,}500$	1b

j	Initial species		Final species	Reaction energy, kcal/mole	Equilibrium constant coefficient $A(i,j)$					C Rate constant, cm^3/mole	n Temperature exponent	E/k Activation temperature	Group
					$i=1$	$i=2$	$i=3$	$i=4$	$i=5$				
					Exchange reactions								
4	NO+O	$\rightleftarrows$	$N+O_2$	−32.05	−2.063	−1.480	−0.580	−0.114	0.005	2.160E+08	1.290	19,220	
5	$O+N_2$	$\rightleftarrows$	N+NO	−74.97	1.066	−0.833	−3.095	−0.084	0.004	3.180E+13	0.100	37,700	
6	$O+O_2^+$	$\rightleftarrows$	O_2+O^+	−36.88	−0.276	0.888	−2.180	0.055	−0.003	6.850E+13	−0.520	18,600	
7	N_2+N^+	$\rightleftarrows$	$N+N_2^+$	−24.06	0.307	−1.076	−0.878	−0.004	−0.001	9.850E+12	−0.180	12,100	
8	$O+NO^+$	$\rightleftarrows$	$NO+O^-$	−101.34	0.148	−1.011	−4.121	−0.132	0.006	2.750E+13	0.010	51,000	2
9	N_2+O^+	$\rightleftarrows$	$O+N_2$	−44.23	2.979	0.382	−3.237	0.168	−0.009	6.330E+13	−0.210	22,200	
10	$N+NO^+$	$\rightleftarrows$	$NO+N^-$	−121.51	2.821	0.448	−6.481	0.040	−0.002	2.210E+15	−0.020	61,100	
11	O_2+NO^+	$\rightleftarrows$	$NO+O_2^+$	−64.46	0.424	−1.098	−1.941	−0.187	0.009	1.030E+16	−0.170	32,400	
12	NO^++N	$\rightleftarrows$	N_2^+ +O	−70.60	2.061	0.204	−4.263	0.119	−0.006	1.700E+13	0.400	35,500	
					Associative ionization								
13	O+N	$\rightleftarrows$	NO^++e^-	−63.69	−7.053	−0.532	−4.429	0.150	−0.007	1.530E+11	−0.370	32,000	
14	O+O	$\rightleftarrows$	$O_2^++e^-$	−160.20	−8.692	−3.110	−6.950	−0.151	0.007	3.850E+11	0.490	80,600	3
15	N+N	$\rightleftarrows$	$N_2^++e^-$	−134.29	−4.992	−0.328	−8.693	0.269	−0.013	1.790E+11	0.770	67,500	
					Electron impact ionization								
16	O	$\rightleftarrows$	O^++e^-	−315.06	−6.113	−2.035	−15.311	−0.073	0.004	3.900E+33	−3.780	158,500	
17	N	$\rightleftarrows$	N^++e^-	−335.23	−3.441	−0.577	−17.671	0.099	−0.005	2.500E+34	−3.820	168,600	4

Table 4.11 Continued.

4.7.2.1.2 Scheme presented by Park in Refs. [4.9] and [4.20].

$$k_f(T_a) = C(T_a)^n \exp(-E/\hat{k}T_a) \qquad (4.86)$$

where T_a is the rate-controlling temperature, which can be either T_t, T_v, or $T_v^q T_t^{1-q}$ $(0.3 < q < 0.5)$. The values of C, n, $E/\hat{k}$, and T_a are provided in Table 4.12, as presented in Ref. [4.9], or in Table 4.13, as given in Ref. [4.20].

The equilibrium rate is calculated based on the translational temperature T_t, i.e., $k_c = k_c(T_t)$. Once $k_c(T_t)$ and $k_f(T_a)$ are determined, then the backward rate k_b is calculated by

$$k_b = \frac{k_f(T_a)}{k_c(T_t)} \qquad (4.87)$$

The units of k_f and k_b are as follows:

$$k_f : \quad \text{cm}^3/\text{mole-sec}$$

$$k_b : \quad \text{cm}^6/\text{mole}^2\text{-sec}$$

No.	Reactions	Rate Expression	Remark
1	$O_2 + O_2 \rightarrow O + O + O_2$	$2 \times 10^{21} T_a^{-1.5} \exp(-59,500/T_a)$	
2	$O_2 + NO \rightarrow O + O + NO$	$2 \times 10^{21} T_a^{-1.5} \exp(-59,500/T_a)$	Estimated
3	$O_2 + N_2 \rightarrow O + O + N_2$	$2 \times 10^{21} T_a^{-1.5} \exp(-59,500/T_a)$	
4	$O_2 + O \rightarrow O + O + O$	$10^{22} T_a^{-1.5} \exp(-59,500/T_a)$	
5	$O_2 + N \rightarrow O + O + N$	$10^{22} T_a^{-1.5} \exp(-59,500/T_a)$	Estimated
6	$NO + O_2 \rightarrow N + O + NO$	$5 \times 10^{15} \exp(-75,500/T_a)$	Estimated
7	$NO + NO \rightarrow N + O + NO$	$1.1 \times 10^{17} \exp(-75,500/T_a)$	
8	$NO + N_2 \rightarrow N + O + N_2$	$5 \times 10^{15} \exp(-75,500/T_a)$	
9	$NO + O \rightarrow N + O + O$	$1.1 \times 10^{17} \exp(-75,500/T_a)$	Estimated
10	$NO + N \rightarrow N + O + N$	$1.1 \times 10^{17} \exp(-75,500/T_a)$	Estimated
11	$N_2 + O_2 \rightarrow N + N + O_2$	$7 \times 10^{21} T_a^{-1.6} \exp(-113,200/T_a)$	Estimated
12	$N_2 + NO \rightarrow N + N + NO$	$7 \times 10^{21} T_a^{-1.6} \exp(-113,200/T_a)$	Estimated
13	$N_2 + N_2 \rightarrow N + N + N_2$	$7 \times 10^{21} T_a^{-1.6} \exp(-113,200/T_a)$	
14	$N_2 + O \rightarrow N + N + O$	$3 \times 10^{22} T_a^{-1.6} \exp(-113,200/T_a)$	Estimated
15	$N_2 + N \rightarrow N + N + N$	$3 \times 10^{22} T_a^{-1.6} \exp(-113,200/T_a)$	
16	$N_2 + e \rightarrow N + N + e$	$3 \times 10^{24} T_e^{-1.6} \exp(-113,200/T_e)$	Estimated
17	$N_2 + O \rightarrow NO + N$	$6.4 \times 10^{17} T_a^{-1} \exp(-38,400/T_a)$	
18	$NO + O \rightarrow O_2 + N$	$8.4 \times 10^{12} \exp(-19,400/T_a)$	
19	$N + O \rightarrow NO^+ + e$	$5.3 \times 10^{12} \exp(-31,900/T_a)$	
20	$N + N \rightarrow N_2^+ + e$	$2 \times 10^{13} \exp(-67,500/T_a)$	
21	$O + O \rightarrow O_2^+ + e$	$1.1 \times 10^{13} \exp(-80,600/T_a)$	
22	$O + e \rightarrow O^+ + e + e$	$3.9 \times 10^{33} T_e^{-3.78} \exp(-158,500/T_e)$	Estimated
23	$N + e \rightarrow N^+ + e + e$	$2.5 \times 10^{33} T_e^{-3.82} \exp(-168,200/T_e)$	
24	$NO^+ + O \rightarrow N^+ + O_2$	$10^{12} T^{0.5} \exp(-77,200/T)$	
25	$O_2^+ + N \rightarrow N^+ + O_2$	$8.7 \times 10^{13} T^{0.14} \exp(-28,600/T)$	
26	$NO + O^+ \rightarrow N^+ + O_2$	$1.4 \times 10^5 T^{1.9} \exp(-15,300/T)$	
27	$O_2^+ + N_2 \rightarrow N_2^+ + O_2$	$9.9 \times 10^{12} \exp(-40,700/T)$	
28	$O_2^+ + O_2 \rightarrow O_2^+ + O_2$	$4 \times 10^{12} T^{-0.09} \exp(-18,000/T_a)$	
29	$NO^+ + N \rightarrow O^+ + N_2$	$3.4 \times 10^{13} T_a^{-1.08} \exp(-12,800/T)$	
30	$NO^+ + O_2 \rightarrow O_2^+ + NO$	$2.4 \times 10^{13} T^{0.41} \exp(-32,600/T)$	
31	$NO^+ + O \rightarrow O_2^+ + N$	$7.2 \times 10^{12} T^{0.29} \exp(-48,600/T)$	
32	$O^+ + N_2 \rightarrow N_2^+ + O$	$9 \times 10^{11} T^{0.36} \exp(-22,800/T)$	
33	$NO^+ + N \rightarrow N_2^+ + O$	$7.2 \times 10^{13} \exp(-35,500/T)$	

T is the translational-rotational temperature, T_e is the electron temperature, and T_a is the geometrical average between T and the vibrational temperature T_v, $T_a = T_v^q T^{1-q}$, where q is between 0.3 and 0.5.

Table 4.12 Rate constants for air, two-temperature model from Ref. [4.9].

Reaction	Rate-Controlling Temperature (T_a)	M	C	n	$E/\hat{k}$
$O_2 + M \rightarrow O + O + M$	$\sqrt{TT_v}$	N	2.90^{23}	-2.0	59750
		O	2.90^{23}		
		N_2	9.68^{22}		
		O_2	9.68^{22}		
		NO	9.68^{22}		
		e^-	9.68^{22}		
$N_2 + M \rightarrow N + N + M$	$\sqrt{TT_v}$	N	1.60^{22}	-1.6	113200
		O	4.98^{22}		
		N_2	3.70^{21}		
		O_2	3.70^{21}		
		NO	4.98^{21}		
		e^-	8.30^{24}		
$NO + M \rightarrow N + O + M$	$\sqrt{TT_v}$	N	7.95^{23}	-2.0	75500
		O	7.95^{23}		
		N_2	7.95^{23}		
		O_2	7.95^{23}		
		NO	7.95^{23}		
		e^-	7.95^{23}		
$NO + O \rightarrow N + O_2$	T		8.37^{12}	0	19450
$O + N_2 \rightarrow N + NO$			6.44^{17}	-1.0	38370
$O + O_2 \rightarrow O_2 + O^+$			6.85^{13}	-0.52	18600
$N_2 + N^+ \rightarrow N + N_2$			9.85^{12}	-0.18	12100
$O + NO^+ \rightarrow NO + O^+$			2.75^{13}	0.01	51000
$N_2 + O^+ \rightarrow O + N_2$			6.33^{13}	-0.21	22200
$O_2 + NO^+ \rightarrow NO + O_2^+$			1.03^{14}	-0.17	32400
$NO^+ + N \rightarrow N_2^+ + O$			1.70^{13}	0.40	35500
$O + N \rightarrow NO^+ + e^-$			1.53^{4}	0.37	32000
$O + O \rightarrow O_2^+ + e_-$			3.85^{4}	0.49	80600
$N + N \rightarrow N_2^+ + e$			1.79^{4}	0.77	67500
$O + e \rightarrow O^+ + e + e$	T_v		3.90^{33}	-3.78	158500
$N + e \rightarrow N^+ + e + e$	T_v		2.50^{33}	-3.82	168600

Table 4.13 Rate constants for air, two-temperature model from Ref. [4.20].

4.7.2.2 Rate constants for air, one-temperature model: A two-temperature model is considered to be useful to a very high temperature range; for example, the reaction rate of O_2 can be predicted up to 50,000 K [4.9]. However, in some cases (especially at "low" temperature ranges), a one-temperature model can be used for evaluating the reaction rates and simplifying the calculation, as noticed by Park [4.9]. Generally speaking, for $T_t < 7,000$ K, a one-temperature model will provide a reasonable result. The reaction rates of k_f and k_b for various reactions are listed in Table 4.14 from Ref. [4.30].

Both k_f and k_b are written in the following form:

$$K = CT^n \exp\left[-E/\hat{k}T\right] \tag{4.88}$$

	Reaction	M	Coefficients of k_f		
			C	n	$E/\hat{k}$
1a	$O_2 + M \leftrightarrow 2O + M$	N, NO	3.600E+18	−1.00	5.950E+04
1b	$O_2 + O \leftrightarrow 2O + O$		9.000E+19	−1.00	5.950E+04
1c	$O_2 + O_2 \leftrightarrow 2O + O_2$		3.240E+19	−1.00	5.950E+04
1d	$O_2 + N_2 \leftrightarrow 2O + N_2$		7.200E+18	−1.00	5.950E+04
2a	$N_2 + M \leftrightarrow 2N + M$	O, NO, O_2	1.900E+17	−0.50	1.130E+05
2b	$N_2 + N_2 \leftrightarrow 2N + N_2$		4.700E+17	−0.50	1.130E+05
3a	$NO+M \leftrightarrow N + O + M$	O_2, N_2	3.900E+20	−1.50	7.550E+04
3b	$NO+M \leftrightarrow N + O + M$	O, N, NO	7.800E+20	−1.50	7.550E+04
4	$O + NO \leftrightarrow N + O_2$		3.200E+09	1.00	1.970E+04
5	$O + N_2 \leftrightarrow N + NO$		7.000E+13	0.00	3.800E+04
6	$O + N \leftrightarrow NO^+ + e^-$		1.400E+06	1.50	3.190E+04
7	$N_2 + M \leftrightarrow N + N + M$	N	4.085E+22	−1.50	1.130E+05
8	$O_2 + N_2 \leftrightarrow NO + NO^+ + e^-$		1.380E+20	−1.84	1.410E+05
9	$NO + N_2 \leftrightarrow NO^+ + N_2 + e^-$		2.200E+15	−0.35	1.080E+05
10	$N + NO^+ \leftrightarrow NO + N^+$		1.000E+19	−0.93	6.100E+04
11	$O + NO^+ \leftrightarrow O_2 + N^+$		1.340E+13	0.31	7.727E+04
12	$N + e^- \leftrightarrow N^+ + e^- + e^-$		1.100E+32	−3.14	1.690E+05
13	$O + NO^+ \leftrightarrow NO + O^+$		3.360E+15	−0.60	5.080E+04
14	$O + e^- \leftrightarrow O^+ + e^- + e^-$		3.600E+31	−2.91	1.580E+05
15	$N_2 + O^+ \leftrightarrow O + N_2^+$		3.400E+19	−2.00	2.300E+04
16	$N_2 + N^+ \leftrightarrow N + N_2^+$		2.020E+11	0.81	1.300E+04
17	$N + N \leftrightarrow N_2^+ + e^-$		1.400E+13	0.00	6.780E+04
18	$O + O \leftrightarrow O_2^+ + e^-$		1.600E+17	−0.98	8.080E+04
19	$O_2 + NO^+ \leftrightarrow NO + O_2^+$		1.800E+15	0.17	3.300E+04
20	$NO + O_2^+ \leftrightarrow NO^+ + e^- + O_2$		8.800E+15	−0.35	1.080E+05
21	$O + O_2^+ \leftrightarrow O_2 + O^+$		2.920E+18	−1.11	2.800E+04

... continued on next page

Table 4.14. Reaction rate constants of a one-temperature model. (Source: Ref. [4.30])

Unit of k_f: cm^3/mole-sec

	Reaction	M	Coefficients of k_b		
			C	n	$E/\hat{k}$
1a	$O_2 + M \leftrightarrow 2O + M$	N, NO	3.000E+15	-0.50	0.00E+00
1b	$O_2 + O \leftrightarrow 2O + O$		7.500E+16	-0.50	0.00E+00
1c	$O_2 + O_2 \leftrightarrow 2O + O_2$		2.700E+16	-0.50	0.00E+00
1d	$O_2 + N_2 \leftrightarrow 2O + N_2$		6.000E+15	-0.50	0.00E+00
2a	$N_2 + M \leftrightarrow 2N + M$	O, NO, O_2	1.100E+16	-0.50	0.00E+00
2b	$N_2 + N_2 \leftrightarrow 2N + N_2$		2.720E+16	-0.50	0.00E+00
3a	$NO+M \leftrightarrow N + O + M$	O_2, N_2	1.000E+20	-1.50	0.00E+00
3b	$NO+M \leftrightarrow N + O + M$	O, N, NO	2.000E+20	-1.50	0.00E+00
4	$O + NO \leftrightarrow N + O_2$		1.300E+10	1.00	3.58E+03
5	$O + N_2 \leftrightarrow N + NO$		1.560E+13	0.00	0.00E+00
6	$O + N \leftrightarrow NO^+ + e^-$		6.700E+21	-1.50	0.00E+00
7	$N_2 + M \leftrightarrow N + N + M$	N	2.270E+21	-1.50	0.00E+00
8	$O_2 + N_2 \leftrightarrow NO + NO^+ + e^-$		1.000E+24	-2.50	0.00E+00
9	$NO + N_2 \leftrightarrow NO^+ + N_2 + e^-$		2.200E+26	-2.50	0.00E+00
10	$N + NO^+ \leftrightarrow NO + N^+$		4.800E+14	0.00	0.00E+00
11	$O + NO^+ \leftrightarrow O_2 + N^+$		1.000E+14	0.00	0.00E+00
12	$N + e^- \leftrightarrow N^+ + e^- + e^-$		2.200E+40	-4.50	0.00E+00
13	$O + NO^+ \leftrightarrow NO + O^+$		1.500E+13	0.00	0.00E+00
14	$O + e^- \leftrightarrow O^+ + e^- + e^-$		2.200E+40	-4.50	0.00E+00
15	$N_2 + O^+ \leftrightarrow O + N_2^+$		2.480E+19	-2.20	0.00E+00
16	$N_2 + N^+ \leftrightarrow N + N_2^+$		7.800E+11	0.50	0.00E+00
17	$N + N \leftrightarrow N_2^+ + e^-$		1.500E+22	-1.50	0.00E+00
18	$O + O \leftrightarrow O_2^+ + e^-$		8.000E+21	-1.50	0.00E+00
19	$O_2 + NO^+ \leftrightarrow NO + O_2^+$		1.800E+13	0.50	0.00E+00
20	$NO + O_2^+ \leftrightarrow NO^+ + e^- + O_2$		8.800E+26	-2.50	0.00E+00
21	$O + O_2^+ \leftrightarrow O_2 + O^+$		7.800E+11	0.50	0.00E+00

Table 4.14. Continued.

Unit of k_b: $cm^6/mole^2$-sec

4.7.2.3 Rate constants for air and H_2, one-temperature model: H_2/air mixture reactions are important for supersonic combustion processes, for example, for SCRAM jet engines at hypersonic speeds. Therefore, a 12-species/25-reactions chemical model is included, as shown in Table 4.15 [4.31]. Both k_f and k_b are written as

$$K = CT^n \exp\left[-E/\hat{k}T\right], \quad \text{with unit of} \quad \begin{array}{l} k_f : \quad \text{cm}^3/\text{mole-sec} \\ k_b : \quad \text{cm}^6/\text{mole}^2\text{-sec} \end{array}$$

Species		Reactions			
1	H	1	$HNO_2 + M \rightarrow NO + OH + M$	14	$OH + OH \rightarrow H + HO_2$
2	O	2	$NO_2 + M \rightarrow NO + O + M$	15	$H_2O + O \rightarrow H + HO_2$
3	H_2O	*3	$H_2 + M \rightarrow H + H + M$	16	$OH + O_2 \rightarrow O + HO_2$
4	OH	*4	$O_2 + M \rightarrow O + O + M$	17	$H_2O + O_2 \rightarrow OH + HO_2$
5	O_2	*5	$H_2O + M \rightarrow OH + H + M$	18	$H_2O + OH \rightarrow H_2 + HO_2$
6	H_2	*6	$OH + M \rightarrow O + H + M$	19	$O + N_2 \rightarrow N + NO$
7	N_2	7	$HO_2 + M \rightarrow H + O_2 + M$	20	$H + NO \rightarrow N + OH$
8	N	*8	$H_2O + O \rightarrow OH + OH$	21	$O + NO \rightarrow N + O_2$
9	NO	*9	$H_2O + H \rightarrow OH + H_2$	22	$NO + OH \rightarrow H + NO_2$
10	NO_2	*10	$O_2 + H \rightarrow OH + O$	23	$NO + O_2 \rightarrow O + NO_2$
11	NO_2	*11	$H_2 + O \rightarrow OH + H$	24	$NO_2 + H_2 \rightarrow H + HNO_2$
12	HNO_2	12	$H_2 + O_2 \rightarrow OH + OH$	25	$NO_2 + OH \rightarrow NO + HO_2$
		13	$H_2 + O_2 \rightarrow H + HO_2$		

	Reaction Rates					
	Forward Rate Constant k_f			Reverse Rate Constant k_b		
	C	n	$E/\hat{k}$	C	n	$E/\hat{k}$
1	5.0×10^{17}	-1.0	25000	8.0×10^{15}	0	-1000
2	1.1×10^{16}	0	32712	1.1×10^{15}	0	-941
3	5.5×10^{18}	-1.0	51987	1.8×10^{18}	-1.0	0
4	7.2×10^{18}	-1.0	59340	4.0×10^{17}	-1.0	0
5	5.2×10^{21}	-1.5	59386	4.4×10^{20}	-1.5	0
6	8.5×10^{18}	-1.0	50830	7.1×10^{18}	-1.0	0
7	1.7×10^{16}	0	23100	1.1×10^{16}	0	-400
8	5.8×10^{13}	0	9059	5.3×10^{12}	0	503
9	8.4×10^{13}	0	10116	2.0×10^{13}	0	2600
10	2.2×10^{14}	0	8455	1.5×10^{13}	0	0

...continued on next page

Table 4.15. Species and reactions of H_2/air. (Source: Ref. [4.31])

	Reaction Rates					
	Forward Rate Constant k_f			Reverse Rate Constant k_b		
	C	n	$E/\hat{k}$	C	n	$E/\hat{k}$
11	$.7.5 \times 10^{13}$	0	5586	3.0×10^{13}	0	4429
12	1.7×10^{13}	0	24232	5.7×10^{11}	0	14922
13	1.9×10^{13}	0	24100	1.3×10^{13}	0	0
14	1.7×10^{11}	0.5	21137	6.0×10^{13}	0	0
15	5.8×10^{11}	0.5	28686	3.0×10^{13}	0	0
16	3.7×10^{11}	0.64	27840	1.0×10^{13}	0	0
17	2.0×10^{11}	0.5	36296	1.2×10^{13}	0	0
18	1.2×10^{12}	0.21	39815	1.7×10^{13}	0	12582
19	5.0×10^{13}	0	37940	1.1×10^{13}	0	0
20	1.7×10^{14}	0	24500	4.5×10^{13}	0	0
21	2.4×10^{11}	0.5	19200	1.0×10^{12}	0.5	3120
22	2.0×10^{11}	0.5	15500	3.5×10^{14}	0	740
23	1.0×10^{12}	0	22800	1.0×10^{13}	0	302
24	2.4×10^{13}	0	14500	5.0×10^{11}	0.5	1500
25	1.0×10^{11}	0.5	6000	3.0×10^{12}	0.5	1200

Table 4.15. Continued.

4.7.3 Determination of the Mass Production Rate $\dot{w}_s$

The net mass production rate of species s is determined by the total reaction rate of species s from all the chemical reactions times the molecular weight of s, i.e.,

$$\dot{w}_s = MW_s \left(\sum_{\text{all reactions}} \frac{d\chi_s}{dt} \right) \tag{4.89}$$

One of the most popular models for air is the 7-species, 6-reactions chemistry model, for which the reactions are:

$$(1) \quad N + M \ \rightleftarrows \ 2N + M \tag{4.90}$$

$$(2) \quad O_2 + M \ \rightleftarrows \ 2O + M \tag{4.91}$$

$$(3) \quad NO + M \rightleftarrows N + O + M \tag{4.92}$$

$$(4) \quad NO + O \rightleftarrows N + O_2 \tag{4.93}$$

$$(5) \quad O + N_2 \ \rightleftarrows \ N + NO \tag{4.94}$$

$$(6) \quad N + O \ \rightleftarrows \ NO^+ + e^- \tag{4.95}$$

and

$$\dot{w}_N \ = \ M_N(-2R_1 - R_3 - R_4 - R_5 + R_6) \tag{4.96}$$

$$\dot{w}_{N_2} = MW_{N_2}(R_1 + R_5) \tag{4.97}$$

$$\dot{w}_{O} = MW_{O}(-2R_2 - R_3 + R_4 + R_5 + R_6) \tag{4.98}$$

$$\dot{w}_{O_2} = MW_{O_2}(R_2 - R_4) \tag{4.99}$$

$$\dot{w}_{NO} = MW_{NO}(R_3 + R_4 - R_5) \tag{4.100}$$

$$\dot{w}_{NO+} = MW_{NO+}(-R_6) \tag{4.101}$$

$$\dot{w}_{e-} = MW_{e-}(-R_6) \tag{4.102}$$

where

$$R_1 = \sum_M \left[-k_f^1(\chi_{N_2})(\chi_M) + k_b^1(\chi_N)^2(\chi_M) \right] \tag{4.103}$$

$$R_2 = \sum_M \left[-k_f^2(\chi_{O_2})(\chi_M) + k_b^2(\chi_O)^2(\chi_M) \right] \tag{4.104}$$

$$R_3 = \sum_M \left[-k_f^3(\chi_{NO})(\chi_M) + k_b^3(\chi_N)(\chi_O)(\chi_M) \right] \tag{4.105}$$

$$R_4 = -k_f^4(\chi_{NO})(\chi_O) + k_b^4(\chi_N)(\chi_{O_2}) \tag{4.106}$$

$$R_5 = -k_f^5(\chi_O)(\chi_{N_2}) + k_b^5(\chi_N)(\chi_{NO}) \tag{4.107}$$

$$R_6 = -k_f^6(\chi_N)(\chi_O) + k_b^6(\chi_{NO+})(\chi_{e-}) \tag{4.108}$$

The superscript appearing in k_f and k_b represents the number of the reaction given by (4.90) through (4.95). Note that $\sum_s \dot{w}_s = 0$, since the mass conservation law is still valid in a system with chemical reactions.

4.8 Equilibrium Flow

For flow problems with $Kn \leq 0.001$ (recall 4.4.2), the chemistry nonequilibrium model is still valid; however, it is a deficient method to be employed because the high reaction rates result in very "stiff" equations. Instead, the chemical equilibrium method is more appropriate, and the solution is rather efficient and relatively simple.

For a wide range of applications, the utilization of an equilibrium chemistry model will provide an accurate solution. That is particularly true for high density flows where a sufficient number of molecular collisions take place and where the reactions rates are high. For example, a flight regime below 50km can be adequately modeled by an equilibrium model. In general, specification of two thermodynamic state variables will provide the remaining thermodynamic properties.

The general approach to include an equilibrium chemistry effect is to consider the governing equations for partial pressures of the species. The relevant equations form a system of nonlinear algebraic equations. A typical system is composed of the Dalton's law of partial pressures, equilibrium constants (provided by statistical mechanics as a function of temperature), and equations for conservation of each nuclei. To solve this system of equations, two thermodynamic states, namely, pressure and temperature are required. Once the partial pressures of species are determined, the mass fraction of species may be evaluated. Subsequently, the density, enthalpy, and internal energy of the mixture are computed. A detailed description of the method is provided in the next section.

4.8.1 Equilibrium Composition

Chemical reactions which occur in a multicomponent gas mixture can be represented as:

$$\sum_{s=1}^{n} \nu_s^R A_s \underset{k_b}{\overset{k_f}{\rightleftharpoons}} \nu_s^P A_s \quad \text{(recall 4.7.1)} \tag{4.109}$$

When the gas is in chemical equilibrium, the net production of any species s is zero; therefore,

$$\frac{d\chi_s}{dt}\Big|_{\text{net}} = 0 \tag{4.110}$$

or

$$k_f \prod_{i=1}^{n} \chi_i^{\nu_i^R} - k_b \prod_{i=1}^{n} \chi_i^{\nu_i^P} = 0 \quad \text{(recall 4.7.1)} \tag{4.111}$$

Therefore,

$$(k_c)_{\text{eq}} = \left(\frac{k_f}{k_b}\right)_{\text{eq}} = \frac{\prod_{i=1}^{n} \chi_i^{\nu_i^P}}{\prod_{i=1}^{n} \chi_i^{\nu_i^R}} = \prod_{i=1}^{n} \chi_i^{\nu_i^P - \nu_i^R} \tag{4.112}$$

Replace $\chi_i = \rho \gamma_i$ in Equation (4.112); then

$$(k_c)_{\text{eq}} = \rho^{\sum_{i=1}^{n}\left(\nu_i^P - \nu_i^R\right)} \prod_{i=1}^{n} \gamma_i^{\left(\nu_i^P - \nu_i^R\right)} \tag{4.113}$$

In addition to the above equations, the equations for conservation of elements and charge are written as:

$$\gamma^j = \sum_{i=1}^{n} \alpha_i^j \gamma_i \tag{4.114}$$

where α_i^j = number of atoms of element j in species i. The equilibrium composition is obtained by solving the nonlinear algebraic equations described above. For the case of air, the following independent chemical reactions are used:

$$\text{Reaction} = 1 \qquad O_2 \rightleftharpoons 2O \tag{4.115}$$

$$= 2 \qquad N_2 \rightleftharpoons 2N \tag{4.116}$$

$$= 3 \qquad NO \rightleftharpoons N + O \tag{4.117}$$

$$= 4 \qquad N + O \rightleftharpoons NO^+ + e^- \tag{4.118}$$

Therefore,

$$(k_c^1)_{eq} = \rho(\gamma_O)^2/\gamma_{O_2} \tag{4.119}$$

$$(k_c^2)_{eq} = \rho(\gamma_N)^2/\gamma_{N_2} \tag{4.120}$$

$$(k_c^3)_{eq} = \rho\,\gamma_O\gamma_N/\gamma_{NO} \tag{4.121}$$

$$(k_c^4)_{eq} = (\gamma_{NO^+})\,(\gamma_{e^-})\,/\gamma_N\gamma_O \tag{4.122}$$

The equations for conservation of elements:

oxygen atom:
$$\gamma^O = \gamma_O + \gamma_{NO} + \gamma_{NO^+} + 2\gamma_{O_2} \tag{4.123}$$

nitrogen atom:
$$\gamma^N = \gamma_N + \gamma_{NO} + \gamma_{NO^+} + 2\gamma_{N_2} \tag{4.124}$$

where $\gamma^O = 0.01455$ and $\gamma^N = 0.05477$.

The equation for conservation of charge:

$$\gamma_{NO^+} = \gamma_{e^-} \tag{4.125}$$

The solution of Equations (4.119) through (4.125) will provide the composition of air.

The equilibrium constant is written in the form:

$$k_c = C(T)^n \exp(-E/\hat{k}T) \tag{4.126}$$

Values of C, n, and $E/\hat{k}$ are listed in Table 4.16 [4.24].

Reaction	C	n	$E/\hat{k}$
1	18.00	0	59,368
2	18.00	0	113,244
3	4.06	0	75,509
4	3.96×10^{-10}	1.5	32,041

Table 4.16. Coefficients for equilibrium constants for air reactions. (Source: Ref. [4.24])

It should be noted that the values provided in Table 4.16 are valid for the temperature range of 1000 K $< T <$ 6000 K. Therefore, one can use the curve-fitting form of k_c, as described in Section 4.7.2.1, along with the values provided in Table 4.11 for a wider range of applications.

For a typical application, the known thermodynamic states are usually pressure and temperature. Therefore, the density must be computed initially, for example, from the equation of state, i.e.,

$$\rho = p/(\mathcal{R}T \sum_{i=1}^{n} \gamma_i) \tag{4.127}$$

An iterative procedure is used to solve the system of nonlinear equations given by (4.119) to (4.125).

4.8.2 Thermodynamic and Transport Properties

4.8.2.1 Thermodynamic properties of air: Values for the thermodynamic properties as a function of temperature are obtained by using polynomial curve fits for each chemical species.

Specific heat:

$$\frac{c_{p,s}}{\mathcal{R}} = a_1 + a_2T + a_3T^2 + a_4T^3 + a_5T^4 \tag{4.128}$$

Enthalpy:

$$\frac{h_s}{\mathcal{R}T} = a_1 + \frac{a_2T}{2} + \frac{a_3T^2}{3} + \frac{a_4T^3}{4} + \frac{a_5T^4}{5} + \frac{a_6}{T} \tag{4.129}$$

Free energy:

$$\frac{F_s^0}{\mathcal{R}T} = a_1(1 - \log_e T) - \frac{a_2T}{2} - \frac{a_3T^2}{6} - \frac{a_4T^3}{12} - \frac{a_5T^4}{20} + \frac{a_6 - a_7}{T} \tag{4.130}$$

[Note: F_s^0 is the free energy of species s at 1 atm pressure (standard state)].

Values of a_1 to a_7 for different species of air are listed in Table 4.17 [4.23] (5 species, temperature range of 1,000 K to 15,000 K) and Table 4.18 (7 species, temperature range of 150 K to 44,000 K). The unit for $c_{p,s}$ in Equation (4.128) is cal/mole K, and the units for h_s and F_s^0 in Equations (4.129) and (4.130) are in cal/mole.

Once the values of $c_{p,s}$, h_s and γ_s are determined, the thermodynamic properties of the mixture can be calculated by the method described in Sec. 4.2.7, i.e.,

$$h = \sum_{s=1}^{n} C_s h_s = \sum_{s=1}^{n} \gamma_s MW_s h_s \tag{4.131}$$

$$c_p = \sum_{s=1}^{n} C_s c_{p,s} = \sum_{s=1}^{n} \gamma_s MW_s c_{p,c} \tag{4.132}$$

The units of $c_{p,s}$ and h_s in Equations (4.131) and (4.132) are based on per unit mass and not per unit mole.

4.8.2.2 Transport properties: The viscosity can be calculated by the method described in Sec. 4.6.2. However, curve-fitting of the viscosity of air based on the results of Hansen [4.32] can also be used. The result is

$$\mu = \mu_{\text{Suth}} \cdot A_1 \cdot A_2$$

where

$$A_1 = 1 + 0.023 \left(\frac{T}{1800}\right) \left\{ 1 + \text{Tanh}\left[\frac{\left(\frac{T}{1800}\right)(1 - 0.125\log_{10}p) - 6.5}{1.5 + 0.125\log_{10}p}\right]\right\} \quad (4.133)$$

and

$$A_2 = \left\{\exp\left[\frac{\frac{T}{1800} - 14.5 - 1.5\log_{10}p}{0.9 + 0.1\log_{10}p}\right] + 1\right\}^{-1}$$

where the units of T and p are °R and atm, respectively.

Furthermore,

$$\mu_{\text{Suth}} = \mu_0 \left(\frac{T}{T_O}\right)^{3/2}\left(\frac{T_0 + S_0}{T + S_0}\right)$$

is the viscosity evaluated by the Sutherland's equation, where

$$\mu_0 = 3.584 \times 10^{-7} \text{ lb}_\text{f}\cdot\text{s/ft}^2$$

$$T_0 = 491.6 \text{ °R}$$

$$S_0 = 199 \text{ °R}$$

The method described in Sec. 4.6.3 can be used to determine the thermal conductivity of a mixture of gases.

Species	Coefficients						
	a_1	a_2	a_3	a_4	a_5	a_6	a_7
1000 K to 6000 K							
O	2.670	-1.970×10^{-4}	7.193×10^{-8}	-8.901×10^{-12}	4.002×10^{-16}	2.915×10^{4}	4.504
O_2	3.316	1.151×10^{-3}	-3.726×10^{-7}	6.186×10^{-11}	-3.666×10^{-15}	-1.044×10^{3}	5.393
N	2.474	9.097×10^{-5}	-7.814×10^{-8}	2.218×10^{-11}	-1.489×10^{-15}	5.609×10^{4}	4.300
N_2	3.221	9.878×10^{-4}	-2.907×10^{-7}	3.938×10^{-11}	-2.000×10^{-15}	-1.043×10^{3}	4.326
NO	3.221	1.221×10^{-3}	-4.297×10^{-7}	6.559×10^{-11}	-3.451×10^{-15}	9.764×10^{3}	6.610
6000 K to 15,000 K							
O	2.548	-5.952×10^{-5}	2.701×10^{-8}	-2.798×10^{-12}	9.380×10^{-17}	2.915×10^{4}	5.049
O_2	3.721	4.254×10^{-4}	-2.835×10^{-8}	6.050×10^{-13}	-5.186×10^{-18}	-1.044×10^{3}	3.254
N	2.746	-3.909×10^{-4}	1.338×10^{-7}	-1.191×10^{-11}	3.365×10^{-14}	5.609×10^{4}	2.872
N_2	3.727	4.684×10^{-4}	-1.140×10^{-7}	1.154×10^{-11}	-3.293×10^{-16}	-1.043×10^{3}	1.294
NO	3.345	2.521×10^{-4}	-2.658×10^{-8}	2.162×10^{-12}	-6.381×10^{-17}	9.764×10^{3}	3.212

Table 4.17 Constants for polynomial approximations of thermodynamic properties of (5 species) air.

Species	Coefficients						
	a_1	a_2	a_3	a_4	a_5	a_6	a_7
150 K to 7250 K							
N	0.247732E1	0.82901E−4	−0.76069E−7	0.22462E−10	−0.15404E−14	0.56133E5	0.39988E1
NO	0.31968E1	0.11769E−2	−0.38778E−6	0.55731E−10	−0.28806E−14	0.98652E4	0.61134E1
N_2	0.32066E1	0.96095E−3	−0.26764E−6	0.33488E−10	−0.15440E−14	−0.99993E3	0.44602E1
O	0.26848E1	−0.24358E−3	0.96464E−7	−0.13271E−10	0.65179E−15	0.29177E5	0.41136E1
O_2	0.32206E1	0.13129E−2	−0.46651E−6	0.70960E−10	−0.38179E−14	−0.10143E4	0.59381E1
NO^+	0.32070E1	0.95464E−3	−0.26312E−6	0.32660E−10	−0.14946E−14	0.11809E6	0.52543E1
e^-	0.25000E1	0.00000	0.00000	0.00000	0.00000	−0.74537E3	−0.12427E2
7250 K to 44000 K							
N	0.29580E01	0.72397E−4	−0.19304E−8	0.65165E−14	−0.85711E−21	0.52845E5	−0.35782E0
NO	0.44678E01	0.22696E−4	−0.14025E−9	0.16592E−14	0.63494E−22	0.86956E4	−0.19150E1
N_2	0.44517E1	0.21996E−4	−0.20502E−9	0.24036E−14	0.82543E−22	−0.23784E4	−0.36415E1
O	0.25750E1	0.22027E−4	0.50533E−10	−0.11758E−13	−0.13901E−20	0.28692E5	0.44633E1
O_2	0.44763E1	0.33173E−4	−0.11257E−9	0.13687E−14	0.68543E−22	−0.20125E4	−0.18259E1
NO^+	0.44517E1	0.23791E−4	−0.20057E−9	0.23327E−14	0.73176E−22	0.11670E6	−0.28548E1
e^-	0.25000E1	0.00000	0.00000	0.00000	0.00000	−0.74537E3	−0.12427E2

Table 4.18 Constants for polynomial approximation of thermodynamic properties of (7 species) air.

4.8.3 Curve Fits for the Thermodynamic Properties of Equilibrium Air Suitable for Numerical Computation

The computations to determine the composition of air for many applications are not necessary nor are they required. What is essential is the computation of thermodynamic properties, as described in Secs. 4.8.1 and 4.8.2. For such applications, a simple curve-fit procedure [4.33] may be used. The procedure is as follows:

(i) $\tilde{\gamma}(= h/e) = \tilde{\gamma}(\rho, e)$ such that pressure p can be determined by $p = \rho e(\tilde{\gamma} - 1)$

(ii) $T = T(\rho, e)$

(iii) $\tilde{\gamma} = \tilde{\gamma}(\rho, p)$ such that the enthalpy can be calculated as $h = (p/\rho)\left(\frac{\tilde{\gamma}}{\tilde{\gamma}-1}\right)$

(iv) $T = T(\rho, p)$

Equations of Curve Fits

(i) $\tilde{\gamma} = a_1 + a_2 Y + a_3 Z + a_4 YZ + a_5 Y^2 + a_6 Z^2 + a_7 YZ^2 + a_8 Z^3$

$$+ \frac{a_9 + a_{10}Y + a_{11}Z + a_{12}YZ}{1 + \exp\left[(a_{13} + a_{14}Y)(Z + a_{15}Y + a_{16})\right]} \tag{4.134}$$

where $Y = \log_{10}(p/1.292)$ and $Z = \log_{10}(e/78408.4)$. The unit for ρ is kg/m^3, and the unit for e is m^2/sec^2. The coefficients a_1, a_2, ..., a_{16} are given in Table 4.19.

Once $\tilde{\gamma}$ is obtained, the pressure can be determined from $p = \rho e(\tilde{\gamma} - 1)$, and the speed of sound a can be determined by:

$$a = \left[e\left\{k_1 + (\tilde{\gamma} - 1)\left[\tilde{\gamma} + k_2\left(\frac{\partial\tilde{\gamma}}{\partial\log_e e}\right)_\rho\right] + k_3\left(\frac{\partial\tilde{\gamma}}{\partial\log_e \rho}\right)_e\right\}\right]^{1/2} \tag{4.135}$$

The coefficients k_1, k_2, and k_3 are also listed in Table 4.19.

(ii) In the calculation of $T = T(e, \rho)$, the pressure is initially determined from $p = \rho e(\tilde{\gamma} - 1)$, and subsequently the temperature is determined from:

$$\log_{10}(T/151.78) = b_1 + b_2 Y + b_3 Z + b_4 YZ + b_5 Z^2 + b_6 Y^2 + b_7 Y^2 Z$$

$$+ b_8 YZ^2 + \frac{b_9 + b_{10}Y + b_{11}Z + b_{12}YZ + b_{13}Z^2}{1 + \exp\left[(b_{14}Y + b_{15})(Z + b_{16})\right]} \tag{4.136}$$

where $Y = \log_{10}(\rho/1.225)$, $X = \log_{10}(p/1.0134 \times 10^5)$, and $Z = X - Y$. The unit for p is newtons/m^2, and the unit for T is K. The coefficients b_1, b_2, ..., b_{16} are given in Table 4.20.

(iii) The general form of the equation used for $\tilde{\gamma}$ was

$$\tilde{\gamma} = c_1 + c_2 Y + c_3 Z + c_4 YZ + \frac{c_5 + c_6 Y + c_7 Z + c_8 YZ}{1 + \exp\left[c_9\left(X + c_{10}Y + c_{11}\right)\right]} \tag{4.137}$$

where $Y = \log_{10}(\rho/1.292)$, $X = \log_{10}(p/1.013 \times 10^5)$, and $Z = X - Y$. The coefficients $c_1, c_2, \ldots, c_{11}$ are tabulated in Table 4.21.

Once $\tilde{\gamma}$ is obtained, then enthalpy $h = h(p, \rho)$ can be evaluated by

$$h = (p/\rho)\left(\frac{\tilde{\gamma}}{\tilde{\gamma} - 1}\right) \tag{4.138}$$

(iv) The general form of the equation used for the correlation $T = T(p, \rho)$ was

$$\log_{10}(T/T_{\mathrm{o}}) = d_1 + d_2 Y + d_3 Z + d_4 YZ + d_5 Z^2$$

$$+ \frac{d_6 + d_7 Y + d_8 Z + d_9 YZ + d_{10} Z^2}{1 + \exp\left[d_{11}\left(Z + d_{12}\right)\right]} \tag{4.139}$$

where $Y = \log_{10}(\rho/1.225)$, $X = \log_{10}(p/1.0134 \times 10^5)$, and $Z = X - Y$. The coefficients $d_1, d_2, \ldots, d_{12}$ are given in Table 4.22.

Density Range	Curve Range	a_1	a_2	a_3	a_4	a_5	a_6	a_7	a_8	a_9	a_{10}
$Y > -0.50$	$Z \leq 0.65$	1.40000	0	0	0	0	0	0	0	0	0
	$0.65 < Z \leq 1.68$	1.45510	-0.000102	-0.081537	0.000166	0	0	0	0	0.128647	-0.049454
	$1.68 < Z \leq 2.46$	1.59608	-0.042426	-0.192840	0.029353	0	0	0	0	-0.019430	0.005954
	$Z > 2.46$	1.54363	-0.049071	-0.153562	0.029209	0	0	0	0	-0.324907	-0.077599
$-4.5 < Y \leq -0.50$	$Z \leq 0.65$	1.40000	0	0	0	0	0	0	0	0	0
	$0.65 < Z \leq 1.54$	1.44813	0.001292	-0.073510	-0.001948	0	0	0	0	0.054745	-0.013705
	$1.54 < Z \leq 2.22$	1.73158	0.003902	-0.272846	0.006237	0	0	0	0	0.041419	0.037475
	$2.22 < Z \leq 2.90$	1.59350	0.075324	-0.176186	-0.026072	0	0	0	0	-0.200838	-0.058536
	$Z > 2.90$	1.12688	-0.025957	0.013602	0.013772	0	0	0	0	-0.127737	-0.087942
$-7 \leq Y \leq -4.5$	$Z \leq 0.65$	1.40000	0	0	0	0	0	0	0	0	0
	$0.65 < Z \leq 0.87$	1.46543	0.007625	-0.254500	-0.017244	0.000292	0.355907	0.015422	-0.163235	0	0
	$1.50 < Z \leq 2.20$	2.02636	0.058493	-0.454886	-0.027433	0	0	0	0	-0.165265	-0.014275
	$2.20 < Z \leq 3.05$	1.60804	0.034791	-0.188906	-0.010927	0	0	0	0	-0.124117	-0.007277
	$3.05 < Z \leq 3.38$	1.25672	0.007073	-0.039228	0.000491	0	0	0	0	0.721798	0.073753
	$Z > 3.38$	-84.0327	-0.831761	72.2066	0.491914	0.001153	-20.3559	-0.070617	1.90979	0	0

Density Range	Curve Range	a_{11}	a_{12}	a_{13}	a_{14}	a_{15}	a_{16}	k_1	k_2	k_3
$Y > -0.50$	$Z \leq 0.65$	0	0	0	0	0	0	0	0	0
	$0.65 < Z \leq 1.68$	-0.101036	0.033518	-15.0	0	0	-1.420	0.000450	0.203892	0.101797
	$1.68 < Z \leq 2.46$	0.026097	-0.006164	-15.0	0	0	-2.050	-0.006609	0.127637	0.297037
	$Z > 2.46$	0.142408	0.022071	-10.0	0	0	-2.708	-0.000081	0.226601	0.170922
$-4.5 < Y \leq -0.50$	$Z \leq 0.65$	0	0	0	0	0	0	0	0	0
	$0.65 < Z \leq 1.54$	-0.055473	-0.021874	-10.0	0	0	-1.420	-0.001973	0.185233	-0.059952
	$1.54 < Z \leq 2.22$	0.016984	-0.018038	-10.0	3.0	-0.025	-2.025	-0.013027	0.074270	0.012889
	$2.22 < Z \leq 2.90$	0.099687	0.025287	-10.0	5.0	0	-2.700	0.004342	0.212192	-0.001293
	$Z > 2.90$	0.043104	0.023547	-20.0	4.0	0	-3.30	0.006348	0.209716	-0.006001
$-7 \leq Y \leq -4.5$	$Z \leq 0.65$	0	0	0	0	0	0	0	0	0
	$0.65 < Z \leq 0.87$	0	0	0	0	0	0	-0.000954	0.171187	0.004567
	$1.50 < Z \leq 2.20$	0.136685	0.010071	-30.0	0	-0.0095	-1.947	0.008736	0.184842	-0.302441
	$2.20 < Z \leq 3.05$	0.069839	0.003985	-30.0	0	-0.007	-2.691	0.017884	0.153672	-0.030224
	$3.05 < Z \leq 3.38$	-0.198942	-0.021539	-50.0	0	-0.0085	-3.354	0.002379	0.217959	0.005943
	$Z > 3.38$	0	0	0	0	0	0	0.006572	0.183396	-0.135960

Table 4.19. Coefficients for curve fit $\tilde{\gamma} = \tilde{\gamma}(p, \rho)$ and $a = a(e, \rho)$.

Density Range	Curve Range	b_1	b_2	b_3	b_4	b_5	b_6	b_7	b_8
	$Z \leq 0.48$	$\mathrm{Log}_{10}(p/R\rho T_0)$	0	0	0	0	0	0	0
$Y > -0.50$	$0.48 < Z \leq 1.07$	0.279268	0	0.992172	0	0	0	0	0
	$Z > 1.07$	0.233261	-0.056383	1.19783	0.063121	-0.165985	0	0	0
	$Z \leq 0.48$	$\mathrm{Log}_{10}(p/R\rho T_0)$	0	0	0	0	0	0	0
$-4.5 < Y \leq -0.5$	$0.48 < Z \leq 0.9165$	0.284312	0.001644	0.987912	0	0	0	0	0
	$0.9165 < Z \leq 1.478$	0.502071	-0.012990	0.774818	0.025397	0	0	0	0
	$1.478 < Z \leq 2.176$	1.02294	0.021535	0.427213	0.006900	0	0	0	0
	$Z > 2.176$	1.47540	0.129620	0.254154	-0.046411	0	0	0	0
	$Z \leq 0.30$	$\mathrm{Log}_{10}(p/R\rho T_0)$	0	0	0	0	0	0	0
	$0.30 < Z \leq 1.00$	0.271800	0.000740	0.990136	-0.004947	0	0	0	0
$-7.0 \leq Y \leq -4.5$	$1.00 < Z \leq 1.35$	1.39925	0.167780	-0.143168	-0.159234	0	0	0	0
	$1.35 < Z \leq 1.79$	1.11401	0.002221	0.351875	0.017246	0	0	0	0
	$1.79 < Z \leq 2.47$	1.01722	-0.017918	0.473523	0.025456	0	0	0	0
	$Z > 2.47$	-45.0871	-9.00504	35.8685	6.79222	-6.77699	-0.064705	0.025325	-1.27370

Density Range	Curve Range	b_9	b_{10}	b_{11}	b_{12}	b_{13}	b_{14}	b_{15}	b_{16}
	$Z \leq 0.48$	0	0	0	0	0	0	0	0
$Y > -0.50$	$0.48 < Z \leq 1.07$	0	0	0	0	0	0	0	0
	$Z > 1.07$	-0.814535	0.099233	0.602385	-0.067428	-0.095991	5.0	-20.0	-1.78
	$Z \leq 0.48$	0	0	0	0	0	0	0	0
	$0.48 < Z \leq 0.9165$	0	0	0	0	0	0	0	0
$-4.5 < Y \leq -0.5$	$0.9165 < Z \leq 1.478$	0.009912	-0.150527	-0.000385	0.105734	0	0	-15.0	-1.28
	$1.478 < Z \leq 2.176$	-0.427823	-0.211991	0.257096	0.101192	0	0	-12.0	-1.778
	$Z > 2.176$	-0.221229	-0.057077	0.158116	0.030430	0	5.0	0	-2.40
	$Z \leq 0.30$	0	0	0	0	0	0	0	0
	$0.30 < Z \leq 1.00$	0.990717	0.175194	-0.982407	-0.159233	0	0	-20.0	-0.88
$-7.0 \leq Y \leq -4.5$	$1.00 < Z < 1.35$	-0.027614	-0.090761	0.307036	0.121621	0	0	-20.0	-1.17
	$1.35 < Z \leq 1.79$	-1.15099	-0.173555	0.673342	0.088399	0	0	-20.0	-1.56
	$1.79 < Z \leq 2.47$	-2.17978	-0.334716	0.898619	0.127386	0	0	-20.0	-2.22
	$Z > 2.47$	0	0	0	0	0	0	0	0

Table 4.20. Coefficients for curve fit $T = T(e, \rho)$.

Density Range	Curve Range	c_1	c_2	c_3	c_4	c_5	c_6
$Y > -0.50$	$z \leq 0.30$	1.40000	0	0	0	0	0
	$0.30 < Z \leq 1.15$	1.42598	0.000918	-0.092209	-0.002226	0.019772	-0.036600
	$1.15 < Z \leq 1.60$	1.64689	-0.062155	-0.334994	0.063612	-0.038332	-0.014468
	$Z > 1.60$	1.48558	-0.453562	-0.152096	0.303350	-0.459282	0.448395
$-4.50 < Y \leq 0.50$	$Z \leq 0.30$	1.40000	0	0	0	0	0
	$0.30 < Z \leq 0.98$	1.42176	-0.000366	-0.083614	0.000675	0.005272	-0.115853
	$0.98 < Z \leq 1.38$	1.74436	-0.035354	-0.415045	0.061921	0.018536	0.043582
	$1.38 < Z \leq 2.04$	1.49674	-0.021583	-0.197008	0.030886	-0.157738	-0.009158
	$Z > 2.04$	1.10421	-0.033664	0.031768	0.024335	-0.178802	-0.017456
$-7 \leq Y \leq -4.5$	$Z \leq 0.398$	1.40000	0	0	0	0	0
	$0.398 < Z \leq 0.87$	1.47003	0.007939	-0.244205	-0.025607	0.872248	0.049452
	$0.87 < Z \leq 1.27$	3.18652	0.137930	-1.89529	-0.103490	-2.14572	-0.272717
	$1.27 < Z \leq 1.863$	1.63963	-0.001004	-0.303549	0.016464	-0.852169	-0.101237
	$Z > 1.863$	1.55889	0.055932	-0.211764	-0.023548	-0.549041	-0.101758

Density Range	Curve Range	c_7	c_8	c_9	c_{10}	c_{11}
$Y > -0.50$	$z \leq 0.30$	0	0	0	0	0
	$0.30 < Z \leq 1.15$	-0.077469	0.043878	-15.0	-1.0	-1.040
	$1.15 < Z \leq 1.60$	0.073421	-0.002442	-15.0	-1.0	-1.360
	$Z > 1.60$	0.220546	-0.292293	-10.0	-1.0	-1.600
$-4.50 < Y \leq 0.50$	$Z \leq 0.30$	0	0	0	0	0
	$0.30 < Z \leq 0.98$	-0.007363	0.146179	-20.0	-1.0	-0.860
	$0.98 < Z \leq 1.38$	0.044353	-0.049750	-20.0	-1.04	-1.336
	$1.38 < Z \leq 2.04$	0.123213	-0.006553	-10.0	-1.05	-1.895
	$Z > 2.04$	0.080373	0.002511	-15.0	-1.08	-2.650
$-7 \leq Y \leq -4.5$	$Z \leq 0.398$	0	0	0	0	0
	$0.398 < Z \leq 0.87$	-0.764158	0.000147	-20.0	-1.0	-0.742
	$0.87 < Z \leq 1.27$	2.06586	0.223046	-15.0	-1.0	-1.041
	$1.27 < Z \leq 1.863$	0.503123	0.043580	-10.0	-1.0	-1.544
	$Z > 1.863$	0.276732	0.046031	-15.0	-1.0	2.250

Table 4.21. Coefficients for curve fit $h = h(p, \rho)$.

Density Range	Curve Range	d_1	d_2	d_3	d_4	d_5	d_6
$Y > -0.50$	$0.48 < Z \leq 0.90$	0.27407	0	1.00082	0	0	0
	$Z > 0.90$	0.235869	−0.043304	1.17619	0.046498	−0.143721	−1.37670
$-4.5 < Y \leq -0.5$	$0.48 < Z \leq 0.9165$	0.281611	0.001267	0.990406	0	0	0
	$0.9165 < Z \leq 1.478$	0.457643	−0.034272	0.819119	0.046471	0	−0.073233
	$1.478 < Z \leq 2.176$	1.04172	0.041961	0.412752	−0.009329	0	−0.434074
	$Z > 2.176$	0.418298	−0.252100	0.784048	0.144576	0	−2.00015
$-7 \leq Y \leq -4.5$	$0.30 < Z \leq 1.07$	2.72964	0.003725	0.938851	−0.011920	0	0.682406
	$1.07 < Z \leq 1.57$	2.50246	−0.042827	1.12924	0.041517	0	1.72067
	$1.57 < Z \leq 2.24$	2.44531	−0.047722	1.00488	0.034349	0	1.95893
	$Z > 2.24$	2.50342	0.026825	0.838860	−0.009819	0	3.58284

Density Range	Curve Range	d_7	d_8	d_9	d_{10}	d_{11}	d_{12}
$Y > -0.50$	$0.48 < Z \leq 0.90$	0	0	0	0	0	0
	$Z > 0.90$	0.160465	1.08988	−0.083489	−0.217748	−10.0	−1.78
$-4.5 < Y \leq -0.5$	$0.48 < Z \leq 0.9165$	0	0	0	0	0	0
	$0.9165 < Z \leq 1.478$	−0.169816	0.043264	0.111854	0	−15.0	−1.28
	$1.478 < Z \leq 2.176$	−0.196914	0.264883	0.100599	0	−15.0	−1.778
	$Z > 2.176$	−0.639022	0.716053	0.206457	0	−10.0	−2.40
$-7 \leq Y \leq -4.5$	$0.30 < Z \leq 1.07$	0.089153	−0.646541	−0.070769	0	−20.0	−0.82
	$1.07 < Z \leq 1.57$	0.268008	−1.25038	−0.179711	0	−20.0	−1.33
	$1.57 < Z \leq 2.24$	0.316244	−1.01200	−0.151561	0	−20.0	−1.88
	$Z > 2.24$	0.533853	−1.36147	−0.195436	0	−20.0	−2.47

Table 4.22. Coefficients for curve fit $T = T(p, \rho)$.

The calculation of p and T, as described in the previous procedure, requires the value of $\tilde{\gamma}$. Another procedure which provides the values of pressure and temperature without the need to determine $\tilde{\gamma}$, uses the following relations [4.34]:

(v)

$$p(\rho, e) = a_1 e + a_2 \rho e + a_3 + a_4 \rho e^2 + a_5 \rho e^3 + a_6 \rho e^4 + a_7 \rho^2 e^3 + a_8 \rho^2 e^4 +$$
$$a_9 \rho e^5 + a_{10} \rho^2 e^5 + a_{11} \rho^2 e^6 + a_{12} \rho e^6 + a_{13} \rho^2 e^7 + a_{14} \rho e^7 \qquad (4.140)$$

$$\frac{\partial p}{\partial \rho} = a_2 e + a_4 e^2 + a_5 e^3 + a_6 e^4 + 2a_7 \rho e^3 + 2a_8 \rho e^4 + a_9 e^5 + 2a_{10} \rho e^5 +$$
$$2a_{11} \rho e^6 + a_{12} e^6 + 2a_{13} \rho e^7 + a_{14} e^7 \qquad (4.141)$$

$$\frac{\partial p}{\partial e} = a_1 + a_2 \rho + 2a_4 \rho e + 3a_5 \rho e^2 + 4a_6 \rho e^3 + 3a_7 \rho^2 e^2 +$$
$$4a_8 \rho^2 e^3 + 5a_9 \rho e^4 + 5a_{10} \rho^2 e^4 + 6a_{11} \rho^2 e^5 + 6a_{12} \rho e^5 +$$
$$7a_{13} \rho^2 e^6 + 7a_{14} \rho e^6 \qquad (4.142)$$

(vi)

$$T(\rho, e) = b_1 e / \rho + b_2 e + b_3 / \rho + b_4 e^2 + b_5 e^3 + b_6 e^4 + b_7 \rho e^3 + b_8 \rho e^4 +$$
$$b_9 e^5 + b_{10} \rho e^5 + b_{11} \rho e^6 + b_{12} \rho e^6 + b_{13} \rho e^7 + b_{14} \rho e^7 \qquad (4.143)$$

$$\frac{\partial T}{\partial \rho} = -b_1 e / \rho^2 - b_3 / \rho^2 + b_7 e^3 + b_8 e^4 + b_{10} e^5 + b_{11} e^6 + b_{13} e^7 \qquad (4.144)$$

$$\frac{\partial T}{\partial e} = b_1 / \rho + b_2 + 2b_4 e + 3b_5 e^2 + 4b_6 e^3 + 3b_7 \rho e^2 + 4b_8 \rho e^3 +$$
$$5b_9 e^4 + 5b_{10} \rho e^4 + 6b_{11} \rho e^5 + 6b_{12} \rho e^5 + 7b_{13} \rho e^6 + 7b_{14} e^6 \qquad (4.145)$$

All the variables in the expressions above have been nondimensionalized by the following:

$$P_\infty = 101325.0 \ \text{N/m}^2$$

$$T_\infty = 300.0 \ K$$

$$\rho_\infty = 1.17196272 \ \text{Kg/m}^3$$

$$e_\infty = 8.64571924 \times 10^4 \ \text{m}^2/\text{s}^2$$

The coefficients a_n and b_n are provided in Table 4.23.

n	a_n	b_n
1	$-1.64184011 \times 10^{-08}$	$-4.66744619 \times 10^{-11}$
2	$+4.25555113 \times 10^{-01}$	$+4.11592537 \times 10^{-01}$
3	$+2.87728820 \times 10^{-06}$	$+9.27096790 \times 10^{-09}$
4	$-6.01056836 \times 10^{-03}$	$-4.76679827 \times 10^{-03}$
5	$+1.16745176 \times 10^{-04}$	$+8.29280665 \times 10^{-05}$
6	$-1.18673130 \times 10^{-06}$	$-7.78921374 \times 10^{-07}$
7	$+1.68428888 \times 10^{-07}$	$+1.35065409 \times 10^{-08}$
8	$-4.49056339 \times 10^{-09}$	$-3.13912759 \times 10^{-10}$
9	$+6.28302532 \times 10^{-09}$	$+3.84665192 \times 10^{-09}$
10	$+4.16636684 \times 10^{-11}$	$+2.65689892 \times 10^{-12}$
11	$-1.62288924 \times 10^{-13}$	$-9.73342645 \times 10^{-15}$
12	$-1.66714128 \times 10^{-11}$	$-9.60514023 \times 10^{-12}$
13	$+2.26513998 \times 10^{-16}$	$+1.30418711 \times 10^{-17}$
14	$+1.75501828 \times 10^{-14}$	$+9.61090540 \times 10^{-15}$

Table 4.23. Coefficients used in the curve fit equations.

Once the values of p, $\partial p/\partial \rho$, and $\partial p/\partial e$ are determined, the speed of sound can be calculated from

$$a^2 = \left(\frac{\partial p}{\partial \rho}\right)_s = \frac{\partial p}{\partial \rho} + \frac{p}{\rho^2}\frac{\partial p}{\partial e} \tag{4.146}$$

The range of applicability of Equations (4.140) through (4.145) is

$$1 \times 10^{-4} < p \text{ (atm)} < 100 \quad \text{and}$$

$$489 < T \text{ (K)} < 15000$$

$$\text{or for} \quad e > 350237.46 \text{ m}^2/\text{s}^2$$

For a temperature of less than 489K, the ideal gas model can be used.

4.9 Newtonian Flow

One of the frequently used expressions in predicting the surface pressure of simple configurations in hypersonic flow is based on Newtonian flow theory. The expression for the pressure coefficient is given by

$$C_p = 2\sin^2\theta \tag{4.147}$$

where the pressure coefficient is defined as

$$C_p = \frac{p - p_\infty}{\frac{1}{2}\rho_\infty u_\infty^2} \tag{4.148}$$

and θ is the surface inclination angle, as illustrated in Figure 4.3.

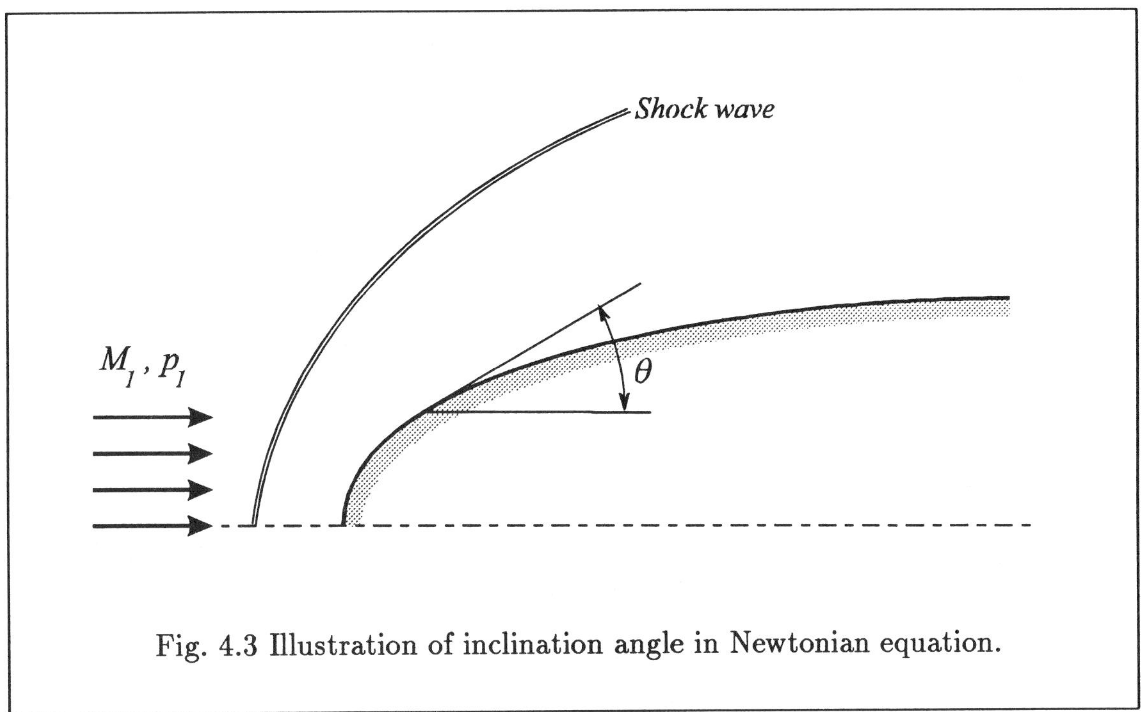

Fig. 4.3 Illustration of inclination angle in Newtonian equation.

Equation (4.147) is replaced by the following equation known as the *modified Newtonian equation*, which provides more accurate results.

$$C_p = C_{p_{\mathrm{max}}} \sin^2 \theta \tag{4.149}$$

where

$$C_{p_{\mathrm{max}}} = \frac{p_{t2} - p_\infty}{\frac{1}{2}\rho_\infty u_\infty^2} = \frac{2}{\gamma M_\infty^2} \left(\frac{p_{t2}}{p_\infty} - 1 \right) \tag{4.150}$$

Note that the notation of "∞" is the same as "1," denoting the quantities ahead of a normal shock wave, whereas "2" designates the quantities behind the shock. The surface pressure is determined subsequently from

$$p = p_\infty + \frac{1}{2}\rho_\infty u_\infty^2 C_p \tag{4.151}$$

Several aspects of flow/configuration must be considered before Newtonian or Modified Newtonian Theory is utilized. These considerations are as follows:

1. Newtonian flow theory can be used on the windward side of a body and not on the leeward side (shadow region).

2. Newtonian flow theory provides more accurate results at higher Mach numbers, typically at Mach numbers of 5 and higher.

3. Newtonian flow theory results in more accurate pressure distribution when applied to three-dimensional configurations such as a cone, than when applied to two-dimensional bodies such as a wedge.

4.10 Heat Transfer Rates

Computational procedures for the heat transfer rate of high speed flows may be categorized as (1) simple, yet reliable semi-empirical methods based on benchmark experimetnal data and numerical solution, or (2) advanced numerical schemes for the solution of the entire flowfield. The approximate algebraic equations can be used for simple geometries, whereas, for complex geometries, the numerical solution of the Navier-Stokes equation is usually required. In the following sections, expressions to predict stagnation point heat transfer, and laminar and turbulent heat transfer distributions for axisymmetric and/or three-dimensional configurations are provided.

4.10.1 Stagnation Point Heating

The stagnation point heat transfer rate may be computed by the approximate methods of Fay and Riddle [4.35] or Cohen [4.36]. These approximations are obtained by similarity transformation of the boundary layer equations and further reduction of the equations near the stagnation point. The relation proposed by Fay and Riddle has the following form.

$$\dot{q} = 0.763(Pr_w)^{-0.6} \left(\rho_w \mu_w\right)^{0.1} \left(\rho_{t2}\mu_{t2}\right)^{0.4} \left(h_{aw} - h_w\right) \left(\frac{du_e}{ds}\right)^{0.5} \qquad (4.152)$$

where the subscripts designate the following locations/conditions: w is the wall, $t2$ represents the properties behind the shock at the stagnation point, and aw is adiabatic wall condition. The velocity gradient appearing in Equation (4.152) is calculated from

$$\frac{du_e}{ds} = \frac{1}{R_n} \sqrt{\frac{2(p_{t2} - p_\infty)}{\rho_{t2}}} \qquad (4.153)$$

where R_n is the nose radius, and s is the distance along the surface. Equation (4.153) is a combination of Euler's equation and modified Newtonian in the vicinity of the stagnation point. For flow conditions where the freestream velocity exceeds 10,000 fps, the expression of Cohen may be used. Cohen's Expression is given by

$$\dot{q} = 0.767(Pr_w)^{-0.6} \left(\rho_w \mu_w\right)^{0.07} \left(\rho_{t2}\mu_{t2}\right)^{0.43} \left(h_{aw} - h_w\right) \left(\frac{du_e}{ds}\right)^{0.5} \qquad (4.154)$$

4.10.2 Surface Heat Flux Distribution

An approximate method for laminar heat transfer distribution is developed by Zoby et al. [4.37], and has the following form

$$\dot{q} = 0.22(\rho_e u_e) \left(\frac{\rho^*}{\rho_e}\right) \left(\frac{\mu^*}{\mu_e}\right) (h_{aw} - h_w)(Pr_w)^{-0.6}/Re_{\theta,e} \qquad (4.155)$$

where "e" designates properties at the boundary layer edge, and "$*$" requires properties evaluated at Eckert's reference temperature. The momentum thickness Reynolds number is defined as

$$Re_{\theta,e} = \frac{\rho_e u_e \theta_L}{\mu_e} \qquad (4.156)$$

where (the subscript L in θ_L represents laminar)

$$\theta_L = 0.664 \left[\int_0^s \rho^* \mu^* u_e r^2 ds\right]^{\frac{1}{2}} /(\rho_e u_e r) \qquad (4.157)$$

and r is the local body radius. Expression (4.155) may be applied to either nonreacting or reacting gas with either constant or variable entropy edge conditions.

For a turbulent flow, the expression for heat transfer rate is given by

$$\dot{q} = C_1 (Re_{\theta,e})^{-m} \left(\frac{\rho^*}{\rho_e}\right)^m \left(\frac{\mu^*}{\mu_e}\right)^m (\rho_e u_e)(h_{aw} - h_w)(Pr_w)^{-0.4} \qquad (4.158)$$

where now

$$Re_{\theta,e} = \frac{\rho_e u_e \theta_t}{\mu_e} \qquad (4.159)$$

and

$$\theta_t = \left[C_2 \int_0^s \rho^* u_e (\mu^*)^m r^{C_3} ds\right]^{C_4} /(\rho_e u_e r) \qquad (4.160)$$

The subscript "t" represents turbulent.

It is common to approximate a turbulent velocity profile by a power law formulation, i.e., $u/u_e = (y/\delta)^{1/n}$. An approximation widely used is the 1/7th power law. However, studies have shown that n can be expressed as a function of momentum thickness Reynolds number. This dependency has been modeled by Zoby, and a curve fit of experimental data yields,

$$n = 12.67 - 6.51 \log(Re_{\theta,e}) + 1.21(\log Re_{\theta,e})^2$$

The coefficients and exponents in Equations (4.158) and (4.160) are as follows.

$$m = \frac{2}{n+1} \qquad (4.161)$$

$$C_1 = \left(\frac{1}{C_5}\right)^{2n/(n+1)} \left[\frac{n}{(n+1)(n+2)}\right]^m \tag{4.162}$$

$$C_2 = (1+m)C_1 \tag{4.163}$$

$$C_3 = 1+m \tag{4.164}$$

$$C_4 = \frac{1}{C_3} \tag{4.165}$$

$$C_5 = 2.2433 + 0.93n \tag{4.166}$$

The edge velocity u_e may be determined from

$$u_e = \sqrt{2(h_t - h_e)} \tag{4.167}$$

Eckert reference condition is calculated according to

$$h^* = 0.5(h_w + h_e) + 0.22(h_{aw} - h_e) \tag{4.168}$$

where the adiabatic wall enthalpy is

$$h_{aw} = h_e + r\left(\frac{u_e^2}{2}\right) \tag{4.169}$$

where the recovery factor "r" is determined as

$$r = Pr^{\frac{1}{2}} \qquad \text{for a laminar flow}$$

or

$$r = Pr^{\frac{1}{3}} \qquad \text{for a turbulent flow}$$

4.11   References

[4.1] *CRC Handbook of Chemistry and Physics,* 68th edition, Editor-in-Chief Robert C. Weast.

[4.2] Vincenti, W. G., and Kruger, C. H., *Introduction to Physical Gas Dynamics,* Robert E. Krieger Publishing Co., Inc., 1965.

[4.3] Palmer, G., "The Development of an Explicit Thermochemical Nonequilibrium Algorithm and Its Application to Compute Three Dimensional AFE Flowfields," AIAA Paper 89-1701, AIAA 24th Thermophysics Conference, 1989.

[4.4] Stull, D. R., et al., *JANAF Thermochemical Tables,* National Bureau of Standards NSRDS-NBS 37, 1971.

[4.5] Cravalho, E. G., and Smith, J. L., "Engineering Thermodynamics," Pitman Publishing, 1981.

[4.6] Balakrishnan, A., "Application of a Flux-Split Algorithm to Chemically Relaxing, Hypervelocity Blunt-Body Flows," AIAA Paper 87-1578.

[4.7] Grossmann, B., and Cinnella, P., "The Computation of Nonequilibrium, Chemically-Reacting Flows," *Computers & Structures,* Vol. 30, No. 1/2, pp. 79–93, 1988.

[4.8] Park, C., "Convergence of Computation of Chemical Reacting Flows," AIAA Paper 85-0247.

[4.9] Park, C., *Nonequilibrium Hypersonic Aerothermodynamics,* John Wiley & Sons, Inc., 1990.

[4.10] Gnoffo, P. A., and McCandless, R. S., "Three-Dimensional AOTV Flowfields in Chemical Nonequilibrium," AIAA Paper 86-0230.

[4.11] *Turbulent Reacting Flows,* Topics in Applied Physics, Vol. 44, Editors: P. A. Libby & F. A. Williams, Springer-Verlag, 1980, p. 39.

[4.12] Bird, G. A., "Monte Carlo Simulation of Gas Flows," *Annual Review of Fluid Mechanics,* Vol. 10, 1978, pp. 11–32.

[4.13] Moss, J. N., and Bird, G. A., "Direct Simulation of Transitional Flow for Hypersonic Reentry Conditions," AIAA Paper 84-0223.

[4.14] Gupta, R. N., Scott, C. D., and Moss, J. N., "Surface Slip Equations for Low-Reynolds-Number Multicomponent Gas Flows," AIAA Paper 84-1732.

[4.15] Gupta, R. N., Simmonds, A. L., "Hypersonic Low-Density Solutions of the Navier-Stokes Equations with Chemical Nonequilibrium and Multicomponent Surface Slip," AIAA Paper 86-1349.

[4.16] Rutledge, W. H., *High Altitude Hypersonic Aerodynamics of Blunt Bodies,* Ph.D. Dissertation, 1990, The University of Texas at Austin.

[4.17] Candler, G., and MacCormack, R., "The Computation of Hypersonic Ionized Flows in Chemical and Thermal Nonequilibrium," AIAA Paper 88-0511.

[4.18] Candler, G., "On the Computation of Shock Shapes in Nonequilibrium Hypersonic Flows," AIAA Paper 89-0312.

[4.19] Lee, Jong-Hun, "Basic Governing Equations for the Flight Regimes of Aeroassisted Orbital Transfer Vehicles," AIAA Paper 84-1729.

[4.20] Park, C., "Assessment of Two-Temperature Kinetic Model for Ionizing Air," AIAA Paper 87-1574, 1987.

[4.21] Anderson, J. D., *Hypersonic and High Temperature Gas Dynamics,* McGraw-Hill, 1989.

[4.22] Yun, K. S., and Mason, E. A., "Collision Integrals for the Transport Properties of Dissociating Air at High Temperatures," *The Physics of Fluids,* Vol. 5, No. 4, p. 380, 1962.

[4.23] Moss, J. N., "Reacting Viscous-Shock-Layer Solutions with Multicomponent Diffusion and Mass Injection," NASA TR-R-411, June 1974.

[4.24] Blottner, F. G., "Chemically Reacting Viscous Flow Program for Multi-Component Gas Mixtures," Sandia Report SC-RR-70-754, 1971.

[4.25] Svehla, R. A., "Estimated Viscosities and Thermal Conductivities of Gases at High Temperatures," NASA TR-R-132, 1962.

[4.26] Prabhu, D. K., and Tannehill, J. C., "A New PNS Code for Chemical Nonequilibrium Flows," AIAA Paper 87-0284.

[4.27] Millikan, R. C., and White, D. R., "Systematics of Vibrational Relaxation," *The Journal of Chemical Physics,* Vol. 39, No. 12, 1963.

[4.28] Park, C., "Problems of Rate Chemistry in the Flight Regimes of Aeroassisted Orbital Transfer Vehicles," AIAA Paper 84-1730.

[4.29] Park, C., "Two-Temperature Interpretation of Dissociation Rate Data for N_2 and O_2," AIAA Paper 88-0458.

[4.30] Dunn, M. G., and Kang, S. W., "Theoretical and Experimental Studies of Reentry Plasmas," NASA CR-2232, 1973.

[4.31] Evans, J. S., and Schexnayder, C. J., Jr., "Influence of Chemical Kinetics and Unmixedness on Burning in Supersonic Hydrogen Flames," *AIAA Journal,* Vol. 18, No. 2, 1980, pp. 188–193.

[4.32] Hansen, F., "Approximations for the Thermodynamic and Transport Properties of High-Temperature Air," NASA TR-R50, 1959.

[4.33] Tannehill, J. C., and Mugge, P. H., "Improved Curve Fits for the Thermodynamic Properties of Equilibrium Air Suitable for Numerical Computation Using Time-Dependent or Shock Capturing Methods," NASA CR2470, 1974.

[4.34] Coirier, J. W. J., "Efficient Real Gas Navier-Stokes Computations of High Speed Flows Using an LU Scheme," AIAA Paper 90-0391, also, NASA TM-102429, 1990.

[4.35] Fay, J. A., and Riddell, F. R., "Theory of Stagnation Point Heat Transfer in Dissociated Air," *Journal of the Aeronautical Sciences,* Vol. 25, No. 2, 1958.

[4.36] Cohen, N. B., "Boundary Layer Similar Solutions and Correlation Equations for Laminar Heat-Transfer Distribution in Equilibrium Air at Velocities up to 41,000 Feet per Second," NASA TR R-118, 1961.

[4.37] Zoby, E. V., Moss, J. N., and Sutton, K., "Approximate Convective-Heating Equations for Hypersonic Flows," *Journal of Spacecraft and Rockets,* Vol. 18, No. 1, 1981.

Chapter 5

Transformation of the Equations of Fluid Motion from Physical Space to Computational Space

5.1 Introductory Remarks

To enhance the efficiency and accuracy of a numerical scheme and to simplify implementation of boundary conditions, a transformation from physical space to computational space is performed. This transformation allows clustering of grid points in regions where flow variables undergo high gradients and grid point motion when required. The computational domain in two dimensions is rectangular, which is divided into an equally spaced grid system. In order to solve the governing equations of motion in the computational space, a transformation of the equations from physical space into computational space is required. Any assumption on the simplification of the equations of motion is imposed on the transformed equations. For example, to reduce computational time and required storage, the full Navier-Stokes equations may be simplified by neglecting the circumferential and streamwise gradient of stresses, while retaining only the normal gradient of the stresses. The resulting equations are known as thin layer Navier-Stokes equations. The reduction of equations is performed on the transformed equations.

This chapter investigates generalized coordinate transformation of the governing equations of fluid motion expressed in the Cartesian coordinate system (x, y, z) from physical space to computational space (ξ, η, ζ). Various formulations of the equations which are discussed include full Navier-Stokes (NS), Euler, thin layer Navier-Stokes (TLNS), parabolized Navier-Stokes (PNS), and incompressible Navier-Stokes equations. A summary of the assumptions which are imposed in the reduction process of the equations is presented. The formulations include three-dimensional flows as well as two-dimensional flow fields.

In order to include the capability for shock capturing for compressible flows, the equations of motion are written in a conservative form. Any set of equations may then be approximated by finite difference formulation, which can be solved in the rectangular grid system. Furthermore, in the linearization process, Jacobian matrices are produced, which are included in this chapter.

This chapter summarizes the equations of fluid motion in computational space and presents them in concise and orderly manner. Thus, the objectives of this chapter are summarized as follows:

(1) Define the metrics and the Jacobian of transformation;

(2) Express the equations of fluid motion in a generalized coordinate system for NS, TLNS, Euler, PNS, and incompressible Navier-Stokes equations; and

(3) Derive the Jacobian matrices for each set of equations which are used in various numerical algorithms.

5.2 Generalized Coordinate Transformation

The equations of motion are transformed from the physical space (x, y, z) to the computational space (ξ, η, ζ) by the following relations:

$$\tau = t \tag{5.1}$$

$$\xi = \xi(t, x, y, z) \tag{5.2}$$

$$\eta = \eta(t, x, y, z) \tag{5.3}$$

$$\zeta = \zeta(t, x, y, z) \tag{5.4}$$

The chain rule of partial differentiation provides the following expressions for the Cartesian derivatives:

$$\frac{\partial}{\partial t} = \frac{\partial}{\partial \tau} + \xi_t \frac{\partial}{\partial \xi} + \eta_t \frac{\partial}{\partial \eta} + \zeta_t \frac{\partial}{\partial \zeta} \tag{5.5}$$

$$\frac{\partial}{\partial x} = \xi_x \frac{\partial}{\partial \xi} + \eta_x \frac{\partial}{\partial \eta} + \zeta_x \frac{\partial}{\partial \zeta} \tag{5.6}$$

$$\frac{\partial}{\partial y} = \xi_y \frac{\partial}{\partial \xi} + \eta_y \frac{\partial}{\partial \eta} + \zeta_y \frac{\partial}{\partial \zeta} \tag{5.7}$$

$$\frac{\partial}{\partial z} = \xi_z \frac{\partial}{\partial \xi} + \eta_z \frac{\partial}{\partial \eta} + \zeta_z \frac{\partial}{\partial \zeta} \tag{5.8}$$

5.2.1 Equations for the Metrics

From Equations (5.5) through (5.8), it is obvious that the value of the metrics ξ_t, η_t, ζ_t, ξ_x, η_x, ζ_x, ξ_y, η_y, ζ_y, ξ_z, η_z, and ζ_z must be provided in some fashion. In most cases the analytical determination of the metrics is not possible and therefore must be computed numerically. Since the stepsizes in the computational domain are equally spaced, x_ξ, x_η, x_ζ, etc., can be computed by various finite difference approximations. Thus, if the metrics appearing in Equations (5.5) through (5.8) can be expressed in terms of these derivatives, the numerical computation of metrics is completed. To obtain such relations, the following differential expressions are considered:

$$dt = \frac{\partial t}{\partial \tau}\, d\tau + \frac{\partial t}{\partial \xi}\, d\xi + \frac{\partial t}{\partial \eta}\, d\eta + \frac{\partial t}{\partial \zeta}\, d\zeta$$

But according to (5.1),

$$\frac{\partial t}{\partial \tau} = 1 \quad \text{and}$$

$$\frac{\partial t}{\partial \xi} = \frac{\partial t}{\partial \eta} = \frac{\partial t}{\partial \zeta} = 0 \quad \text{thus}$$

$$dt = d\tau \tag{5.9}$$

Similarly,

$$dx = x_\tau d\tau + x_\xi d\xi + x_\eta d\eta + x_\zeta d\zeta \tag{5.10}$$

$$dy = y_\tau d\tau + y_\xi d\xi + y_\eta d\eta + y_\zeta d\zeta \tag{5.11}$$

$$dz = z_\tau d\tau + z_\xi d\xi + z_\eta d\eta + z_\zeta d\zeta \tag{5.12}$$

Equations (5.9) through (5.12) are expressed in a matrix form as

$$\begin{bmatrix} dt \\ dx \\ dy \\ dz \end{bmatrix} = \begin{bmatrix} 1 & 0 & 0 & 0 \\ x_\tau & x_\xi & x_\eta & x_\zeta \\ y_\tau & y_\xi & y_\eta & y_\zeta \\ z_\tau & z_\xi & z_\eta & z_\zeta \end{bmatrix} \begin{bmatrix} d\tau \\ d\xi \\ d\eta \\ d\zeta \end{bmatrix} \tag{5.13}$$

Reversing the role of the independent variables, we may write

$$d\tau = dt \tag{5.14}$$

$$d\xi = \xi_t dt + \xi_x dx + \xi_y dy + \xi_z dz \tag{5.15}$$

$$d\eta = \eta_t dt + \eta_x dx + \eta_y dy + \eta_z dz \qquad (5.16)$$

$$d\zeta = \zeta_t dt + \zeta_x dx + \zeta_y dy + \zeta_z dz \qquad (5.17)$$

which are expressed as

$$
\begin{bmatrix} d\tau \\ d\xi \\ d\eta \\ d\zeta \end{bmatrix}
=
\begin{bmatrix}
1 & 0 & 0 & 0 \\
\xi_t & \xi_x & \xi_y & \xi_z \\
\eta_t & \eta_x & \eta_y & \eta_z \\
\zeta_t & \zeta_x & \zeta_y & \zeta_z
\end{bmatrix}
\begin{bmatrix} dt \\ dx \\ dy \\ dz \end{bmatrix}
\qquad (5.18)
$$

Comparing Equations (5.13) and (5.18), one concludes that

$$
\begin{bmatrix}
1 & 0 & 0 & 0 \\
\xi_t & \xi_x & \xi_y & \xi_z \\
\eta_t & \eta_x & \eta_y & \eta_z \\
\zeta_t & \zeta_x & \zeta_y & \zeta_z
\end{bmatrix}
=
\begin{bmatrix}
1 & 0 & 0 & 0 \\
x_\tau & x_\xi & x_\eta & x_\zeta \\
y_\tau & y_\xi & y_\eta & y_\zeta \\
z_\tau & z_\xi & z_\eta & z_\zeta
\end{bmatrix}^{-1}
$$

From which,

$$\xi_x = J(y_\eta z_\zeta - y_\zeta z_\eta) \qquad (5.19)$$

$$\xi_y = J(x_\zeta z_\eta - x_\eta z_\zeta) \qquad (5.20)$$

$$\xi_z = J(x_\eta y_\zeta - x_\zeta y_\eta) \qquad (5.21)$$

$$\eta_x = J(y_\zeta z_\xi - y_\xi z_\zeta) \qquad (5.22)$$

$$\eta_y = J(x_\xi z_\zeta - x_\zeta z_\xi) \qquad (5.23)$$

$$\eta_z = J(x_\zeta y_\xi - x_\xi y_\zeta) \qquad (5.24)$$

$$\zeta_x = J(y_\xi z_\eta - y_\eta z_\xi) \qquad (5.25)$$

$$\zeta_y = J(x_\eta z_\xi - x_\xi z_\eta) \qquad (5.26)$$

$$\zeta_z = J(x_\xi y_\eta - x_\eta y_\xi) \qquad (5.27)$$

$$\xi_t = -(x_\tau \xi_x + y_\tau \xi_y + z_\tau \xi_z) \qquad (5.28)$$

$$\eta_t = -(x_\tau \eta_x + y_\tau \eta_y + z_\tau \eta_z) \qquad (5.29)$$

$$\zeta_t = -(x_\tau \zeta_x + y_\tau \zeta_y + z_\tau \zeta_z) \qquad (5.30)$$

After substitution of Equations (5.19) through (5.27) into Equations (5.28), (5.29), and (5.30),

$$\xi_t = J[x_\tau(y_\zeta z_\eta - y_\eta z_\zeta) + y_\tau(x_\eta z_\zeta - x_\zeta z_\eta) + z_\tau(x_\zeta y_\eta - x_\eta y_\zeta)] \qquad (5.31)$$

$$\eta_t = J[x_\tau(y_\xi z_\zeta - y_\zeta z_\xi) + y_\tau(x_\zeta z_\xi - x_\xi z_\zeta) + z_\tau(x_\xi y_\zeta - x_\zeta y_\xi)] \qquad (5.32)$$

$$\zeta_t = J[x_\tau(y_\eta z_\xi - y_\xi z_\eta) + y_\tau(x_\xi z_\eta - x_\eta z_\xi) + z_\tau(x_\eta y_\xi - x_\xi y_\eta)] \qquad (5.33)$$

where J is the Jacobian of transformation defined by

$$J = \frac{\partial(\xi, \eta, \zeta)}{\partial(x, y, z)} = \frac{1}{x_\xi(y_\eta z_\zeta - y_\zeta z_\eta) - x_\eta(y_\xi z_\zeta - y_\zeta z_\xi) + x_\zeta(y_\xi z_\eta - y_\eta z_\xi)} \qquad (5.34)$$

5.3 Navier-Stokes Equations

The equations of fluid motion in complete form which include the conservation of mass, conservation of momentum, and conservation of energy are referred to as the Navier-Stokes equations. The nondimensional Navier-Stokes equations given by (1.170) in the Cartesian coordinate system are now transformed to the computational space by the following. Due to similarity of the left- and right-hand sides of Equation (1.170), the mathematical details are carried out for the left-hand side (LHS) of Equation (1.170) only. Equation (1.170), which is repeated here for convenience, is

$$\frac{\partial Q}{\partial t} + \frac{\partial E}{\partial x} + \frac{\partial F}{\partial y} + \frac{\partial G}{\partial z} = \frac{\partial E_v}{\partial x} + \frac{\partial F_v}{\partial y} + \frac{\partial G_v}{\partial z} \qquad (5.35)$$

The Cartesian derivatives are replaced by Equations (5.5) through (5.8) to yield

$$LHS = \frac{\partial Q}{\partial \tau} + \xi_t \frac{\partial Q}{\partial \xi} + \eta_t \frac{\partial Q}{\partial \eta} + \zeta_t \frac{\partial Q}{\partial \zeta} + \xi_x \frac{\partial E}{\partial \xi} + \eta_x \frac{\partial E}{\partial \eta} +$$

$$\zeta_x \frac{\partial E}{\partial \zeta} + \xi_y \frac{\partial F}{\partial \xi} + \eta_y \frac{\partial F}{\partial \eta} + \zeta_y \frac{\partial F}{\partial \zeta} + \xi_z \frac{\partial G}{\partial \xi} + \eta_z \frac{\partial G}{\partial \eta} + \zeta_z \frac{\partial G}{\partial \zeta} \qquad (5.36)$$

It is recognized that this equation is no longer in a conservative form. To recast the equation in a conservative form, some manipulation must be performed. To do so, Equation (5.36) is first divided by J, and then a combination of terms which sums up to zero is added. In the following, only four terms are considered to show the required mathematical steps. The conclusion is then extended to the remaining terms. The first four terms of Equation (5.36) divided by J and with the added zeros shown as the brackets are

$$\frac{1}{J}\frac{\partial Q}{\partial \tau} + \frac{1}{J}\xi_t \frac{\partial Q}{\partial \xi} + \frac{1}{J}\eta_t \frac{\partial Q}{\partial \eta} + \frac{1}{J}\zeta_t \frac{\partial Q}{\partial \zeta} + \left[Q\frac{\partial}{\partial \tau}\left(\frac{1}{J}\right) - Q\frac{\partial}{\partial \tau}\left(\frac{1}{J}\right) \right]$$

$$+ \left[Q\frac{\partial}{\partial \xi}\left(\frac{\xi_t}{J}\right) - Q\frac{\partial}{\partial \xi}\left(\frac{\xi_t}{J}\right) \right] + \left[Q\frac{\partial}{\partial \eta}\left(\frac{\eta_t}{J}\right) - Q\frac{\partial}{\partial \eta}\left(\frac{\eta_t}{J}\right) \right]$$

$$+ \left[Q\frac{\partial}{\partial \zeta}\left(\frac{\zeta_t}{J}\right) - Q\frac{\partial}{\partial \zeta}\left(\frac{\zeta_t}{J}\right) \right]$$

which may be rearranged as

$$\left[\frac{1}{J}\frac{\partial Q}{\partial \tau} + Q\frac{\partial}{\partial \tau}\left(\frac{1}{J}\right)\right] + \left[\frac{\xi_t}{J}\frac{\partial Q}{\partial \xi} + Q\frac{\partial}{\partial \xi}\left(\frac{\xi_t}{J}\right)\right]$$

$$+ \left[\frac{\eta_t}{J}\frac{\partial Q}{\partial \eta} + Q\frac{\partial}{\partial \eta}\left(\frac{\eta_t}{J}\right)\right] + \left[\frac{\zeta_t}{J}\frac{\partial Q}{\partial \zeta} + Q\frac{\partial}{\partial \zeta}\left(\frac{\zeta_t}{J}\right)\right]$$

$$- Q\left[\frac{\partial}{\partial \tau}\left(\frac{1}{J}\right) + \frac{\partial}{\partial \xi}\left(\frac{\xi_t}{J}\right) + \frac{\partial}{\partial \eta}\left(\frac{\eta_t}{J}\right) + \frac{\partial}{\partial \zeta}\left(\frac{\zeta_t}{J}\right)\right] \qquad (5.37)$$

The terms in the first four brackets may be combined, and, by substitution of expressions (5.31) through (5.34) into the last bracket, it can be shown that it is zero. Therefore, expression (5.37) is combined as

$$\frac{\partial}{\partial \tau}\left(\frac{Q}{J}\right) + \frac{\partial}{\partial \xi}\left(\xi_t\frac{Q}{J}\right) + \frac{\partial}{\partial \eta}\left(\eta_t\frac{Q}{J}\right) + \frac{\partial}{\partial \zeta}\left(\zeta_t\frac{Q}{J}\right)$$

Extending this conclusion to the remaining terms of Equation (5.36) results in the following expression:

$$LHS = \frac{\partial}{\partial \tau}\left(\frac{Q}{J}\right) + \frac{\partial}{\partial \xi}\left(\xi_t\frac{Q}{J}\right) + \frac{\partial}{\partial \eta}\left(\eta_t\frac{Q}{J}\right) + \frac{\partial}{\partial \zeta}\left(\zeta_t\frac{Q}{J}\right) +$$

$$\frac{\partial}{\partial \xi}\left(\xi_x\frac{E}{J}\right) + \frac{\partial}{\partial \eta}\left(\eta_x\frac{E}{J}\right) + \frac{\partial}{\partial \zeta}\left(\zeta_x\frac{E}{J}\right) + \frac{\partial}{\partial \xi}\left(\xi_y\frac{F}{J}\right) + \frac{\partial}{\partial \eta}\left(\eta_y\frac{F}{J}\right) +$$

$$\frac{\partial}{\partial \zeta}\left(\zeta_y\frac{F}{J}\right) + \frac{\partial}{\partial \xi}\left(\xi_z\frac{G}{J}\right) + \frac{\partial}{\partial \eta}\left(\eta_z\frac{G}{J}\right) + \frac{\partial}{\partial \zeta}\left(\zeta_z\frac{G}{J}\right) \qquad (5.38)$$

The right-hand side of Equation (5.35) is reformulated in a similar fashion, and therefore (5.35) is written as

$$\frac{\partial}{\partial \tau}\left(\frac{Q}{J}\right) + \frac{\partial}{\partial \xi}\left[\frac{1}{J}\left(\xi_t Q + \xi_x E + \xi_y F + \xi_z G\right)\right] +$$

$$\frac{\partial}{\partial \eta}\left[\frac{1}{J}\left(\eta_t Q + \eta_x E + \eta_y F + \eta_z G\right)\right] + \frac{\partial}{\partial \zeta}\left[\frac{1}{J}\left(\zeta_t Q + \zeta_x E + \zeta_y F + \zeta_z G\right)\right]$$

$$= \frac{\partial}{\partial \xi}\left[\frac{1}{J}\left(\xi_x E_v + \xi_y F_v + \xi_z G_v\right)\right] + \frac{\partial}{\partial \eta}\left[\frac{1}{J}\left(\eta_x E_v + \eta_y F_v + \eta_z G_v\right)\right] +$$

$$\frac{\partial}{\partial \zeta}\left[\frac{1}{J}\left(\zeta_x E_v + \zeta_y F_v + \zeta_z G_v\right)\right]$$

The terms in this equation are redefined such that it may be written as

$$\frac{\partial \bar{Q}}{\partial \tau} + \frac{\partial \bar{E}}{\partial \xi} + \frac{\partial \bar{F}}{\partial \eta} + \frac{\partial \bar{G}}{\partial \zeta} = \frac{\partial \bar{E}_v}{\partial \xi} + \frac{\partial \bar{F}_v}{\partial \eta} + \frac{\partial \bar{G}_v}{\partial \zeta} \tag{5.39}$$

where

$$\bar{Q} = \frac{Q}{J} \tag{5.40}$$

$$\bar{E} = \frac{1}{J}\left(\xi_t Q + \xi_x E + \xi_y F + \xi_z G\right) \tag{5.41}$$

$$\bar{F} = \frac{1}{J}\left(\eta_t Q + \eta_x E + \eta_y F + \eta_z G\right) \tag{5.42}$$

$$\bar{G} = \frac{1}{J}\left(\zeta_t Q + \zeta_x E + \zeta_y F + \zeta_z G\right) \tag{5.43}$$

$$\bar{E}_v = \frac{1}{J}\left(\xi_x E_v + \xi_y F_v + \xi_z G_v\right) \tag{5.44}$$

$$\bar{F}_v = \frac{1}{J}\left(\eta_x E_v + \eta_y F_v + \eta_z G_v\right) \tag{5.45}$$

$$\bar{G}_v = \frac{1}{J}\left(\zeta_x E_v + \zeta_y F_v + \zeta_z G_v\right) \tag{5.46}$$

The viscous shear stresses given by Equations (1.158), (1.159), (1.160), (1.162), (1.163), and (1.165) with Stoke's hypothesis ($\lambda = -\frac{2}{3}\mu$) in the transformed computational space are

$$\tau_{xx} = \frac{\mu}{Re_\infty}\left[\frac{4}{3}(\xi_x u_\xi + \eta_x u_\eta + \zeta_x u_\zeta) - \frac{2}{3}(\xi_y v_\xi + \eta_y v_\eta + \zeta_y v_\zeta)\right. \tag{5.47}$$

$$\left. - \frac{2}{3}(\xi_z w_\xi + \eta_z w_\eta + \zeta_z w_\zeta)\right]$$

$$\tau_{yy} = \frac{\mu}{Re_\infty}\left[\frac{4}{3}(\xi_y v_\xi + \eta_y v_\eta + \zeta_y v_\zeta) - \frac{2}{3}(\xi_x u_\xi + \eta_x u_\eta + \zeta_x u_\zeta)\right. \tag{5.48}$$

$$\left. - \frac{2}{3}(\xi_z w_\xi + \eta_z w_\eta + \zeta_z w_\zeta)\right]$$

$$\tau_{zz} = \frac{\mu}{Re_\infty}\left[\frac{4}{3}(\xi_z w_\xi + \eta_z w_\eta + \zeta_z w_\zeta) - \frac{2}{3}(\xi_x u_\xi + \eta_x u_\eta + \zeta_x u_\zeta)\right. \tag{5.49}$$

$$\left. - \frac{2}{3}(\xi_y v_\xi + \eta_y v_\eta + \zeta_y v_\zeta)\right]$$

$$\tau_{xy} = \tau_{yx} = \frac{\mu}{Re_\infty} \left(\xi_y u_\xi + \eta_y u_\eta + \zeta_y u_\zeta + \xi_x v_\xi + \eta_x v_\eta + \zeta_x v_\zeta \right) \tag{5.50}$$

$$\tau_{xz} = \tau_{zx} = \frac{\mu}{Re_\infty} \left(\xi_z u_\xi + \eta_z u_\eta + \zeta_z u_\zeta + \xi_x w_\xi + \eta_x w_\eta + \zeta_x w_\zeta \right) \tag{5.51}$$

$$\tau_{yz} = \tau_{zy} = \frac{\mu}{Re_\infty} \left(\xi_z v_\xi + \eta_z v_\eta + \zeta_z v_\zeta + \xi_y w_\xi + \eta_y w_\eta + \zeta_y w_\zeta \right) \tag{5.52}$$

where the heat conduction terms in the computation space are

$$q_x = -\frac{\mu}{Pr\,Re_\infty\,(\gamma - 1)\,M_\infty^2} \left(\xi_x T_\xi + \eta_x T_\eta + \zeta_x T_\zeta \right) \tag{5.53}$$

$$q_y = -\frac{\mu}{Pr\,Re_\infty\,(\gamma - 1)\,M_\infty^2} \left(\xi_y T_\xi + \eta_y T_\eta + \zeta_y T_\zeta \right) \tag{5.54}$$

$$q_z = -\frac{\mu}{Pr\,Re_\infty\,(\gamma - 1)\,M_\infty^2} \left(\xi_z T_\xi + \eta_z T_\eta + \zeta_z T_\zeta \right) \tag{5.55}$$

5.3.1 Linearization

In order to numerically solve Equation (5.39), a linearization procedure is introduced, and all flux vectors are expressed in terms of the flux vector $\bar{Q}$. For example, using Taylor series expansion we may write

$$\bar{E}^{n+1} = \bar{E}^n + \frac{\partial \bar{E}}{\partial \tau} \Delta\tau + O(\Delta\tau)^2 \tag{5.56}$$

In order to rewrite $\frac{\partial \bar{E}}{\partial \tau}$ in terms of the gradient of flux vector $\bar{Q}$, recall that

$$\bar{E} = f(\bar{Q}, \xi_t, \xi_x, \xi_y, \xi_z)$$

The chain-rule of differentiation yields the following relation:

$$\frac{\partial \bar{E}}{\partial \tau} = \frac{\partial \bar{E}}{\partial \bar{Q}} \frac{\partial \bar{Q}}{\partial \tau} + \frac{\partial \bar{E}}{\partial \xi_t} \frac{\partial \xi_t}{\partial \tau} + \frac{\partial \bar{E}}{\partial \xi_x} \frac{\partial \xi_x}{\partial \tau} + \frac{\partial \bar{E}}{\partial \xi_y} \frac{\partial \xi_y}{\partial \tau} + \frac{\partial \bar{E}}{\partial \xi_z} \frac{\partial \xi_z}{\partial \tau} \tag{5.57}$$

For many applications, the grid is independent of time, and therefore time gradients of the metrics are zero. Hence, for a time independent grid system, Equation (5.57) is reduced to

$$\frac{\partial \bar{E}}{\partial \tau} = \frac{\partial \bar{E}}{\partial \bar{Q}} \frac{\partial \bar{Q}}{\partial \tau} \tag{5.58}$$

This equation is substituted into Equation (5.56) to yield

$$\bar{E}^{n+1} = \bar{E}^n + \frac{\partial \bar{E}}{\partial \bar{Q}} \frac{\partial \bar{Q}}{\partial \tau} \Delta\tau + O(\Delta\tau)^2 \tag{5.59}$$

The partial derivative $\frac{\partial \bar{Q}}{\partial \tau}$ is approximated by a first-order backward difference expression as

$$\frac{\partial \bar{Q}}{\partial \tau} = \frac{\bar{Q}^{n+1} - \bar{Q}^n}{\Delta\tau} + O(\Delta\tau) = \frac{\Delta\bar{Q}}{\Delta\tau} + O(\Delta\tau)$$

Substitution of this equation into Equation (5.80) yields

$$\bar{E}^{n+1} = \bar{E}^n + \frac{\partial \bar{E}}{\partial \bar{Q}} \left[\frac{\Delta\bar{Q}}{\Delta\tau} + O(\Delta\tau) \right] \Delta\tau + O(\Delta\tau)^2$$

or

$$\bar{E}^{n+1} = \bar{E}^n + \frac{\partial \bar{E}}{\partial \bar{Q}} \Delta\bar{Q} + O(\Delta\tau)^2 \tag{5.60}$$

If a time dependent grid system is used, the additional terms which appeared in Equation (5.57) would be included in Equation (5.60) as well. In either case, terms such as $\frac{\partial \bar{E}}{\partial \bar{Q}}$ and similar terms related to flux vectors $\bar{F}, \bar{G}, \bar{E}_v, \bar{F}_v$, and $\bar{G}_v$ will always appear in the linearization process. They are defined as the Jacobian matrices, which are the focus of the next section.

For steady-state problems such as parabolized Navier-Stokes equations, the linearization process is similar. In that case, the Taylor series is expanded with respect to the streamwise space coordinate (marching direction).

5.3.2 Inviscid and Viscous Jacobian Matrices

To linearize Equation (5.39), the following and similar approximations are employed:

$$\bar{E}^{n+1} = \bar{E}^n + \frac{\partial \bar{E}}{\partial \bar{Q}} \Delta\bar{Q} + O(\Delta\tau)^2$$

where $\frac{\partial \bar{E}}{\partial \bar{Q}}$ is defined as the flux Jacobian matrix. The remaining flux Jacobian matrices are $\frac{\partial \bar{F}}{\partial \bar{Q}}$ and $\frac{\partial \bar{G}}{\partial \bar{Q}}$. The viscous Jacobian matrices are $\frac{\partial \bar{E}_v}{\partial \bar{Q}}$, $\frac{\partial \bar{F}_v}{\partial \bar{Q}}$, and $\frac{\partial \bar{G}_v}{\partial \bar{Q}}$. Since flux vectors $\bar{Q}$ and $\bar{E}$ are 5×1 vectors, each one of the Jacobian matrices are 5×5 (for three-dimensional problems). The derivations of the Jacobians are considered next. In the following derivations we will assume perfect gas; therefore, the equation of state is expressed as

$$p = \rho e (\gamma - 1).$$

The inviscid Jacobian $\frac{\partial \bar{E}}{\partial \bar{Q}}$ is

$$\frac{\partial \bar{E}}{\partial \bar{Q}} = \frac{\partial(\bar{E}_1, \bar{E}_2, \bar{E}_3, \bar{E}_4, \bar{E}_5)}{\partial(\bar{Q}_1, \bar{Q}_2, \bar{Q}_3, \bar{Q}_4, \bar{Q}_5)}$$

$$= \begin{bmatrix} \dfrac{\partial \bar{E}_1}{\partial \bar{Q}_1} & \dfrac{\partial \bar{E}_1}{\partial \bar{Q}_2} & \dfrac{\partial \bar{E}_1}{\partial \bar{Q}_3} & \dfrac{\partial \bar{E}_1}{\partial \bar{Q}_4} & \dfrac{\partial \bar{E}_1}{\partial \bar{Q}_5} \\[2mm] \dfrac{\partial \bar{E}_2}{\partial \bar{Q}_1} & \dfrac{\partial \bar{E}_2}{\partial \bar{Q}_2} & \dfrac{\partial \bar{E}_2}{\partial \bar{Q}_3} & \dfrac{\partial \bar{E}_2}{\partial \bar{Q}_4} & \dfrac{\partial \bar{E}_2}{\partial \bar{Q}_5} \\[2mm] \dfrac{\partial \bar{E}_3}{\partial \bar{Q}_1} & \dfrac{\partial \bar{E}_3}{\partial \bar{Q}_2} & \dfrac{\partial \bar{E}_3}{\partial \bar{Q}_3} & \dfrac{\partial \bar{E}_3}{\partial \bar{Q}_4} & \dfrac{\partial \bar{E}_3}{\partial \bar{Q}_5} \\[2mm] \dfrac{\partial \bar{E}_4}{\partial \bar{Q}_1} & \dfrac{\partial \bar{E}_4}{\partial \bar{Q}_2} & \dfrac{\partial \bar{E}_4}{\partial \bar{Q}_3} & \dfrac{\partial \bar{E}_4}{\partial \bar{Q}_4} & \dfrac{\partial \bar{E}_4}{\partial \bar{Q}_5} \\[2mm] \dfrac{\partial \bar{E}_5}{\partial \bar{Q}_1} & \dfrac{\partial \bar{E}_5}{\partial \bar{Q}_2} & \dfrac{\partial \bar{E}_5}{\partial \bar{Q}_3} & \dfrac{\partial \bar{E}_5}{\partial \bar{Q}_4} & \dfrac{\partial \bar{E}_5}{\partial \bar{Q}_5} \end{bmatrix} \tag{5.61}$$

In order to determine the elements of matrix (5.61), the flux vectors E, F, and G are expressed in terms of the components of vector Q according to

$$E = \begin{bmatrix} \rho u \\[1mm] \rho u^2 + p \\[1mm] \rho u v \\[1mm] \rho u w \\[1mm] (\rho e_t + p)u \end{bmatrix} = \begin{bmatrix} Q_2 \\[2mm] \dfrac{Q_2^2}{Q_1} + (\gamma - 1)\left[Q_5 - \dfrac{1}{2}\left(\dfrac{Q_2^2}{Q_1} + \dfrac{Q_3^2}{Q_1} + \dfrac{Q_4^2}{Q_1}\right)\right] \\[3mm] \dfrac{Q_2 Q_3}{Q_1} \\[3mm] \dfrac{Q_2 Q_4}{Q_1} \\[3mm] \left[\gamma Q_5 - \dfrac{\gamma - 1}{2}\left(\dfrac{Q_2^2}{Q_1} + \dfrac{Q_3^2}{Q_1} + \dfrac{Q_4^2}{Q_1}\right)\right]\dfrac{Q_2}{Q_1} \end{bmatrix} \tag{5.62}$$

$$F = \begin{bmatrix} \rho v \\[1mm] \rho u v \\[1mm] \rho v^2 + p \\[1mm] \rho v w \\[1mm] (\rho e_t + p)v \end{bmatrix} = \begin{bmatrix} Q_3 \\[2mm] \dfrac{Q_2 Q_3}{Q_1} \\[3mm] \dfrac{Q_3^2}{Q_1} + (\gamma - 1)\left[Q_5 - \dfrac{1}{2}\left(\dfrac{Q_2^2}{Q_1} + \dfrac{Q_3^2}{Q_1} + \dfrac{Q_4^2}{Q_1}\right)\right] \\[3mm] \dfrac{Q_3 Q_4}{Q_1} \\[3mm] \left[\gamma Q_5 - \dfrac{\gamma - 1}{2}\left(\dfrac{Q_2^2}{Q_1} + \dfrac{Q_3^2}{Q_1} + \dfrac{Q_4^2}{Q_1}\right)\right]\dfrac{Q_3}{Q_1} \end{bmatrix} \tag{5.63}$$

$$G = \begin{bmatrix} \rho w \\ \rho w u \\ \rho w v \\ \rho w^2 + p \\ (\rho e_t + p)w \end{bmatrix} = \begin{bmatrix} Q_4 \\ \dfrac{Q_2 Q_4}{Q_1} \\ \dfrac{Q_3 Q_4}{Q_1} \\ \dfrac{Q_4^2}{Q_1} + (\gamma - 1)\left[Q_5 - \dfrac{1}{2}\left(\dfrac{Q_2^2}{Q_1} + \dfrac{Q_3^2}{Q_1} + \dfrac{Q_4^2}{Q_1}\right)\right] \\ \left[\gamma Q_5 - \dfrac{\gamma - 1}{2}\left(\dfrac{Q_2^2}{Q_1} + \dfrac{Q_3^2}{Q_1} + \dfrac{Q_4^2}{Q_1}\right)\right]\dfrac{Q_4}{Q_1} \end{bmatrix} \tag{5.64}$$

Recall that, from Equation (5.41), we may write

$$\bar{E}_1 = \frac{1}{J}[\xi_t Q_1 + \xi_x E_1 + \xi_y F_1 + \xi_z G_1] \tag{5.65}$$

With the components E_1, F_1, and G_1 from Equations (5.62), (5.63), and (5.64), Equation (5.65) can be written as

$$\bar{E}_1 = \frac{1}{J}[\xi_t Q_1 + \xi_x Q_2 + \xi_y Q_3 + \xi_z Q_4] \tag{5.66}$$

Now the first element of (5.82) is determined as

$$\frac{\partial \bar{E}_1}{\partial \bar{Q}_1} = \frac{\partial\{\frac{1}{J}[\xi_t Q_1 + \xi_x Q_2 + \xi_y Q_3 + \xi_z Q_4]\}}{\partial\{\frac{Q_1}{J}\}}$$

$$= \frac{\partial(\xi_t Q_1 + \xi_x Q_2 + \xi_y Q_3 + \xi_z Q_4)}{\partial Q_1} = \xi_t \tag{5.67}$$

The remaining elements of the first row are

$$\frac{\partial \bar{E}_1}{\partial \bar{Q}_2} = \xi_x \ , \qquad \frac{\partial \bar{E}_1}{\partial \bar{Q}_3} = \xi_y \ , \qquad \frac{\partial \bar{E}_1}{\partial \bar{Q}_4} = \xi_z \ ,$$

and $\dfrac{\partial \bar{E}_1}{\partial \bar{Q}_5} = 0$.

The elements of the second, third, fourth, and fifth rows are determined in a similar fashion. The resulting inviscid Jacobians are:

$$\frac{\partial \bar{E}}{\partial \bar{Q}} =
\begin{bmatrix}
\xi_t & \xi_x & \xi_y & \xi_z & 0 \\[2ex]
\begin{aligned}
&-u(\xi_x u + \xi_y v + \xi_z w) \\
&+ \xi_x\left[\tfrac{1}{2}(\gamma-1)(u^2+v^2+w^2)\right]
\end{aligned}
&
\begin{aligned}
&\xi_t + \xi_x(2-\gamma)u \\
&+ (\xi_x u + \xi_y v + \xi_z w)
\end{aligned}
& \xi_y u - (\gamma-1)\xi_x v
& \xi_z u - (\gamma-1)\xi_x w
& (\gamma-1)\xi_x \\[3ex]
\begin{aligned}
&-v(\xi_x u + \xi_y v + \xi_z w) \\
&+ \xi_y\left[\tfrac{1}{2}(\gamma-1)(u^2+v^2+w^2)\right]
\end{aligned}
& \xi_x v - (\gamma-1)\xi_y u
&
\begin{aligned}
&\xi_t + \xi_y(2-\gamma)v \\
&+ (\xi_x u + \xi_y v + \xi_z w)
\end{aligned}
& \xi_z v - (\gamma-1)\xi_y w
& (\gamma-1)\xi_y \\[3ex]
\begin{aligned}
&-w(\xi_x u + \xi_y v + \xi_z w) \\
&+ \xi_z\left[\tfrac{1}{2}(\gamma-1)(u^2+v^2+w^2)\right]
\end{aligned}
& \xi_x w - (\gamma-1)\xi_z u
& \xi_y w - (\gamma-1)\xi_z v
&
\begin{aligned}
&\xi_t + \xi_z(2-\gamma)w \\
&+ (\xi_x u + \xi_y v + \xi_z w)
\end{aligned}
& (\gamma-1)\xi_z \\[3ex]
\begin{aligned}
&(\xi_x u + \xi_y v + \xi_z w)\left[-\gamma e_t\right. \\
&\left.+ (\gamma-1)(u^2+v^2+w^2)\right]
\end{aligned}
&
\begin{aligned}
&\xi_x\left[\gamma e_t - \tfrac{1}{2}(\gamma-1)\cdot\right. \\
&\left.(u^2+v^2+w^2)\right] \\
&- (\gamma-1)\left[\xi_x u + \xi_y v\right. \\
&\left.+ \xi_z w\right]u
\end{aligned}
&
\begin{aligned}
&\xi_y\left[\gamma e_t - \tfrac{1}{2}(\gamma-1)\cdot\right. \\
&\left.(u^2+v^2+w^2)\right] \\
&- (\gamma-1)\left[\xi_x u + \xi_y v\right. \\
&\left.+ \xi_z w\right]v
\end{aligned}
&
\begin{aligned}
&\xi_z\left[\gamma e_t - \tfrac{1}{2}(\gamma-1)\cdot\right. \\
&\left.(u^2+v^2+w^2)\right] \\
&- (\gamma-1)\left[\xi_x u + \xi_y v\right. \\
&\left.+ \xi_z w\right]w
\end{aligned}
&
\begin{aligned}
&\xi_t + \\
&\gamma\left[\xi_x u + \xi_y v + \xi_z w\right]
\end{aligned}
\end{bmatrix}$$

$$(5.68)$$

$$\frac{\partial \bar{F}}{\partial \bar{Q}} =$$

η_t	η_x	η_y	η_z	0
$-u(\eta_x u + \eta_y v + \eta_z w)$ $+ \eta_x\left[\frac{1}{2}(\gamma-1)(u^2+v^2+w^2)\right]$	$\eta_t + \eta_x(2-\gamma)u$ $+ (\eta_x u + \eta_y v + \eta_z w)$	$\eta_y u - (\gamma-1)\eta_x v$	$\eta_z u - (\gamma-1)\eta_x w$	$(\gamma-1)\eta_x$
$-v(\eta_x u + \eta_y v + \eta_z w)$ $+ \eta_y\left[\frac{1}{2}(\gamma-1)(u^2+v^2+w^2)\right]$	$\eta_x v - (\gamma-1)\eta_y u$	$\eta_t + \eta_y(2-\gamma)v$ $+ (\eta_x u + \eta_y v + \eta_z w)$	$\eta_z v - (\gamma-1)\eta_y w$	$(\gamma-1)\eta_y$
$-w(\eta_x u + \eta_y v + \eta_z w)$ $+ \eta_z\left[\frac{1}{2}(\gamma-1)(u^2+v^2+w^2)\right]$	$\eta_x w - (\gamma-1)\eta_z u$	$\eta_y w - (\gamma-1)\eta_z v$	$\eta_t + \eta_z(2-\gamma)w$ $+ (\eta_x u + \eta_y v + \eta_z w)$	$(\gamma-1)\eta_z$
$(\eta_x u + \eta_y v + \eta_z w)\left[-\gamma e_t + (\gamma-1)(u^2+v^2+w^2)\right]$	$\eta_x\left[\gamma e_t - \frac{1}{2}(\gamma-1)\cdot(u^2+v^2+w^2)\right]$ $- (\gamma-1)\left[\eta_x u + \eta_y v + \eta_z w\right]u$	$\eta_y\left[\gamma e_t - \frac{1}{2}(\gamma-1)\cdot(u^2+v^2+w^2)\right]$ $- (\gamma-1)\left[\eta_x u + \eta_y v + \eta_z w\right]v$	$\eta_z\left[\gamma e_t - \frac{1}{2}(\gamma-1)\cdot(u^2+v^2+w^2)\right]$ $- (\gamma-1)\left[\eta_x u + \eta_y v + \eta_z w\right]w$	$\eta_t + \gamma\left[\eta_x u + \eta_y v + \eta_z w\right]$

(5.69)

$$\frac{\partial \bar{G}}{\partial \bar{Q}} =
\begin{bmatrix}
\zeta_t & \zeta_x & \zeta_y & \zeta_z & 0 \\[2ex]
\begin{aligned}&-u(\zeta_x u + \zeta_y v + \zeta_z w) \\ &+ \zeta_x\left[\tfrac{1}{2}(\gamma-1)(u^2+v^2+w^2)\right]\end{aligned}
& \begin{aligned}&\zeta_t + \zeta_x(2-\gamma)u \\ &+ (\zeta_x u + \zeta_y v + \zeta_z w)\end{aligned}
& \zeta_y u - (\gamma-1)\zeta_x v
& \zeta_z u - (\gamma-1)\zeta_x w
& (\gamma-1)\zeta_x \\[3ex]
\begin{aligned}&-v(\zeta_x u + \zeta_y v + \zeta_z w) \\ &+ \zeta_y\left[\tfrac{1}{2}(\gamma-1)(u^2+v^2+w^2)\right]\end{aligned}
& \zeta_x v - (\gamma-1)\zeta_y u
& \begin{aligned}&\zeta_t + \zeta_y(2-\gamma)v \\ &+ (\zeta_x u + \zeta_y v + \zeta_z w)\end{aligned}
& \zeta_z v - (\gamma-1)\zeta_y w
& (\gamma-1)\zeta_y \\[3ex]
\begin{aligned}&-w(\zeta_x u + \zeta_y v + \zeta_z w) \\ &+ \zeta_z\left[\tfrac{1}{2}(\gamma-1)(u^2+v^2+w^2)\right]\end{aligned}
& \zeta_x w - (\gamma-1)\zeta_z u
& \zeta_y w - (\gamma-1)\zeta_z v
& \begin{aligned}&\zeta_t + \zeta_z(2-\gamma)w \\ &+ (\zeta_x u + \zeta_y v + \zeta_z w)\end{aligned}
& (\gamma-1)\zeta_z \\[3ex]
\begin{aligned}&(\zeta_x u + \zeta_y v + \zeta_z w)[-\gamma e_t \\ &+ (\gamma-1)(u^2+v^2+w^2)]\end{aligned}
& \begin{aligned}&\zeta_x\left[\gamma e_t - \tfrac{1}{2}(\gamma-1)\cdot\right. \\ &\left.(u^2+v^2+w^2)\right] \\ &- (\gamma-1)[\zeta_x u + \zeta_y v \\ &+ \zeta_z w]\,u\end{aligned}
& \begin{aligned}&\zeta_y\left[\gamma e_t - \tfrac{1}{2}(\gamma-1)\cdot\right. \\ &\left.(u^2+v^2+w^2)\right] \\ &- (\gamma-1)[\zeta_x u + \zeta_y v \\ &+ \zeta_z w]\,v\end{aligned}
& \begin{aligned}&\zeta_z\left[\gamma e_t - \tfrac{1}{2}(\gamma-1)\cdot\right. \\ &\left.(u^2+v^2+w^2)\right] \\ &- (\gamma-1)[\zeta_x u + \zeta_y v \\ &+ \zeta_z w]\,w\end{aligned}
& \begin{aligned}&\zeta_t + \\ &\gamma[\zeta_x u + \zeta_y v + \zeta_z w]\end{aligned}
\end{bmatrix}$$

$$(5.70)$$

Some of the terms appearing in the Jacobian matrices may be defined such that they are expressed in a compact form. For example,

$$U = \xi_t + \xi_x u + \xi_y v + \xi_z w \tag{5.71}$$

$$V = \eta_t + \eta_x u + \eta_y v + \eta_z w \tag{5.72}$$

$$W = \zeta_t + \zeta_x u + \zeta_y v + \zeta_z w \tag{5.73}$$

are known as the contravariant velocity components. They represent velocity components which are perpendicular to planes of constant ξ, η, and ζ, respectively. Furthermore, define

$$q^2 = u^2 + v^2 + w^2 \tag{5.74}$$

Before attempting to determine the viscous Jacobians, the viscous flux vectors must be rearranged by substituting expressions for shear stress and heat conduction terms. For example, consider the second component of $\bar{E}_v$, which from Equation (5.44) is given by

$$\bar{E}_{v_2} = \frac{1}{J} \left(\xi_x E_{v_2} + \xi_y F_{v_2} + \xi_z G_{v_2} \right) \tag{5.75}$$

From Equations (1.173), (1.175) and (1.177), the following is observed:

$$E_{v_2} = \tau_{xx} \quad , \quad F_{v_2} = \tau_{xy} \quad , \quad \text{and} \quad G_{v_2} = \tau_{xz}$$

Hence, Equation (5.75) may be expressed as

$$\bar{E}_{v_2} = \frac{1}{J} \left(\xi_x \tau_{xx} + \xi_y \tau_{xy} + \xi_z \tau_{xz} \right) \tag{5.76}$$

Now, using the expressions for the shear stresses given by Equations (5.47), (5.50) and (5.51), Equation (5.76) is rearranged as

$$\bar{E}_{v_2} = \frac{1}{Re_\infty J} \left\{ \xi_x \left[\frac{4}{3}\mu(\xi_x u_\xi + \eta_x u_\eta + \zeta_x u_\zeta) - \frac{2}{3}\mu(\xi_y v_\xi + \eta_y v_\eta \right. \right.$$

$$\left. + \zeta_y v_\zeta) - \frac{2}{3}\mu(\xi_z w_\xi + \eta_z w_\eta + \zeta_z w_\zeta) \right] + \xi_y \left[\mu(\xi_y u_\xi + \eta_y u_\eta + \right.$$

$$\left. + \zeta_y u_\zeta + \xi_x v_\xi + \eta_x v_\eta + \zeta_x v_\zeta) \right] + \xi_z \left[\mu(\xi_z u_\xi + \eta_z u_\eta + \right.$$

$$\left. \left. + \zeta_z u_\zeta + \xi_x w_\xi + \eta_x w_\eta + \zeta_x w_\zeta) \right] \right\}$$

which is factored as

$$\bar{E}_{v_2} = \frac{\mu}{Re_\infty J}\left\{(\frac{4}{3}\xi_x^2 + \xi_y^2 + \xi_z^2)u_\xi + (\frac{1}{3}\xi_x\xi_y)v_\xi + \right.$$

$$(\frac{1}{3}\xi_x\xi_z)w_\xi + (\frac{4}{3}\xi_x\eta_x + \xi_y\eta_y + \xi_z\eta_z)u_\eta + (\xi_y\eta_x - \frac{2}{3}\xi_x\eta_y)v_\eta +$$

$$(\xi_z\eta_x - \frac{2}{3}\xi_x\eta_z)w_\eta + (\frac{4}{3}\xi_x\zeta_x + \xi_y\zeta_y + \xi_z\zeta_z)u_\zeta +$$

$$\left.(\xi_y\zeta_x - \frac{2}{3}\xi_x\zeta_y)v_\zeta + (\xi_z\zeta_x - \frac{2}{3}\xi_x\zeta_z)w_\zeta\right\}$$

All the components of flux vector $\bar{E}_v$ are reformulated accordingly. In addition, viscous flux vectors $\bar{F}_v$ and $\bar{G}_v$ are modified in a similar fashion, resulting in the following form of the viscous flux vectors:

$$
\bar{E}_v = \frac{\mu}{Re_\infty J}
\begin{bmatrix}
0 \\[4pt]
\hline
a_1 u_\xi + a_5 v_\xi + a_7 w_\xi + d_1 u_\eta + d_7 v_\eta + d_9 w_\eta + e_1 u_\zeta + e_7 v_\zeta + e_9 w_\zeta \\[4pt]
\hline
a_5 u_\xi + a_2 v_\xi + a_6 w_\xi + d_5 u_\eta + d_2 v_\eta + d_{10} w_\eta + e_5 u_\zeta + e_2 v_\zeta + e_{10} w_\zeta \\[4pt]
\hline
a_7 u_\xi + a_6 v_\xi + a_3 w_\xi + d_6 u_\eta + d_8 v_\eta + d_3 w_\eta + e_6 u_\zeta + e_8 v_\zeta + e_3 w_\zeta \\[4pt]
\hline
\frac{1}{2} a_1 (u^2)_\xi + \frac{1}{2} a_2 (v^2)_\xi + \frac{1}{2} a_3 (w^2)_\xi + a_5 (uv)_\xi + a_6 (vw)_\xi + a_7 (uw)_\xi \\[6pt]
+ \; \frac{1}{Pr(\gamma-1)M_\infty^2} a_4 T_\xi + \frac{1}{2} d_1 (u^2)_\eta + \frac{1}{2} d_2 (v^2)_\eta + \frac{1}{2} d_3 (w^2)_\eta + d_5 v u_\eta \\[6pt]
+ \; d_6 w u_\eta + d_7 u v_\eta + d_8 w v_\eta + d_9 u w_\eta + d_{10} v w_\eta + \frac{1}{Pr(\gamma-1)M_\infty^2} d_4 T_\eta \\[6pt]
+ \; \frac{1}{2} e_1 (u^2)_\zeta + \frac{1}{2} e_2 (v^2)_\zeta + \frac{1}{2} e_3 (w^2)_\zeta + e_5 v u_\zeta + e_6 w u_\zeta + e_7 u v_\zeta \\[6pt]
+ \; e_8 w v_\zeta + e_9 u w_\zeta + e_{10} v w_\zeta + \frac{1}{Pr(\gamma-1)M_\infty^2} e_4 T_\zeta
\end{bmatrix}
\tag{5.77}
$$

where for a perfect gas

$$
\frac{1}{Pr(\gamma-1)M_\infty^2} T = \frac{\gamma}{Pr} \left[e_t - \frac{1}{2}(u^2 + v^2 + w^2) \right]
\tag{5.78}
$$

$$\bar{F}_v = \frac{\mu}{Re_\infty J} \begin{bmatrix} 0 \\ \\ d_1 u_\xi + d_5 v_\xi + d_6 w_\xi + b_1 u_\eta + b_5 v_\eta + b_7 w_\eta + f_1 u_\zeta + f_7 v_\zeta + f_9 w_\zeta \\ \\ d_7 u_\xi + d_2 v_\xi + d_8 w_\xi + b_5 u_\eta + b_2 v_\eta + b_6 w_\eta + f_5 u_\zeta + f_2 v_\zeta + f_{10} w_\zeta \\ \\ d_9 u_\xi + d_{10} v_\xi + d_3 w_\xi + b_7 u_\eta + b_6 v_\eta + b_3 w_\eta + f_6 u_\zeta + f_8 v_\zeta + f_3 w_\zeta \\ \\ \frac{1}{2} d_1 (u^2)_\xi + \frac{1}{2} d_2 (v^2)_\xi + \frac{1}{2} d_3 (w^2)_\xi + d_5 uv_\xi + d_6 uw_\xi + d_7 vu_\xi \\ \\ + \ d_8 vw_\xi + d_9 wu_\xi + d_{10} wv_\xi + \frac{1}{Pr(\gamma-1)M_\infty^2} d_4 T_\xi \\ \\ + \ \frac{1}{2} b_1 (u^2)_\eta + \frac{1}{2} b_2 (v^2)_\eta + \frac{1}{2} b_3 (w^2)_\eta + b_5 (uv)_\eta + b_6 (vw)_\eta + b_7 (uw)_\eta \\ \\ + \ \frac{1}{Pr(\gamma-1)M_\infty^2} b_4 T_\eta + \frac{1}{2} f_1 (u^2)_\zeta + \frac{1}{2} f_2 (v^2)_\zeta + \frac{1}{2} f_3 (w^2)_\zeta \\ \\ + \ f_5 vu_\zeta + f_6 wu_\zeta + f_7 uv_\zeta + f_8 wv_\zeta + f_9 uw_\zeta + f_{10} vw_\zeta \\ \\ + \ \frac{1}{Pr(\gamma-1)M_\infty^2} f_4 T_\zeta \end{bmatrix}$$

$$(5.79)$$

$$\bar{G}_v = \frac{\mu}{Re_\infty J}
\begin{bmatrix}
0 \\[4pt]
\hline
e_1 u_\xi + e_5 v_\xi + e_6 w_\xi + f_1 u_\eta + f_5 v_\eta + f_6 w_\eta + c_1 u_\zeta + c_5 v_\zeta + c_7 w_\zeta \\[4pt]
\hline
e_7 u_\xi + e_2 v_\xi + e_8 w_\xi + f_7 u_\eta + f_2 v_\eta + f_8 w_\eta + c_5 u_\zeta + c_2 v_\zeta + c_6 w_\zeta \\[4pt]
\hline
e_9 u_\xi + e_{10} v_\xi + e_3 w_\xi + f_9 u_\eta + f_{10} v_\eta + f_3 w_\eta + c_7 u_\zeta + c_6 v_\zeta + c_3 w_\zeta \\[4pt]
\hline
\frac{1}{2} e_1 (u^2)_\xi + \frac{1}{2} e_2 (v^2)_\xi + \frac{1}{2} e_3 (w^2)_\xi + e_5 u v_\xi + e_6 u w_\xi + e_7 v u_\xi \\[6pt]
+\ e_8 v w_\xi + e_9 w u_\xi + e_{10} w v_\xi + \frac{1}{Pr(\gamma-1)M_\infty^2} e_4 T_\xi + \frac{1}{2} f_1 (u^2)_\eta \\[6pt]
+\ \frac{1}{2} f_2 (v^2)_\eta + \frac{1}{2} f_3 (w^2)_\eta + f_5 u v_\eta + f_6 u w_\eta + f_7 v u_\eta + f_8 v w_\eta \\[6pt]
+\ f_9 w u_\eta + f_{10} w v_\eta + \frac{1}{Pr(\gamma-1)M_\infty^2} f_4 T_\eta + \frac{1}{2} c_1 (u^2)_\zeta + \frac{1}{2} c_2 (v^2)_\zeta \\[6pt]
+\ \frac{1}{2} c_3 (w^2)_\zeta + c_5 (uv)_\zeta + c_6 (vw)_\zeta + c_7 (uw)_\zeta + \frac{1}{Pr(\gamma-1)M_\infty^2} c_4 T_\zeta
\end{bmatrix}$$

$$(5.80)$$

where

$$a_1 = \tfrac{4}{3}\xi_x^2 + \xi_y^2 + \xi_z^2 \tag{5.81}$$

$$a_2 = \xi_x^2 + \tfrac{4}{3}\xi_y^2 + \xi_z^2 \tag{5.82}$$

$$a_3 = \xi_x^2 + \xi_y^2 + \tfrac{4}{3}\xi_z^2 \tag{5.83}$$

$$a_4 = \xi_x^2 + \xi_y^2 + \xi_z^2 \tag{5.84}$$

$$a_5 = \tfrac{1}{3}\xi_x\xi_y \tag{5.85}$$

$$a_6 = \tfrac{1}{3}\xi_y\xi_z \tag{5.86}$$

$$a_7 = \tfrac{1}{3}\xi_x\xi_z \tag{5.87}$$

$$b_1 = \tfrac{4}{3}\eta_x^2 + \eta_y^2 + \eta_z^2 \tag{5.88}$$

$$b_2 = \eta_x^2 + \tfrac{4}{3}\eta_y^2 + \eta_z^2 \tag{5.89}$$

$$b_3 = \eta_x^2 + \eta_y^2 + \tfrac{4}{3}\eta_z^2 \tag{5.90}$$

$$b_4 = \eta_x^2 + \eta_y^2 + \eta_z^2 \tag{5.91}$$

$$b_5 = \tfrac{1}{3}\eta_x\eta_y \tag{5.92}$$

$$b_6 = \tfrac{1}{3}\eta_y\eta_z \tag{5.93}$$

$$b_7 = \tfrac{1}{3}\eta_x\eta_z \tag{5.94}$$

$$c_1 = \tfrac{4}{3}\zeta_x^2 + \zeta_y^2 + \zeta_z^2 \tag{5.95}$$

$$c_2 = \zeta_x^2 + \tfrac{4}{3}\zeta_y^2 + \zeta_z^2 \tag{5.96}$$

$$c_3 = \zeta_x^2 + \zeta_y^2 + \tfrac{4}{3}\zeta_z^2 \tag{5.97}$$

$$c_4 = \zeta_x^2 + \zeta_y^2 + \zeta_z^2 \tag{5.98}$$

$$c_5 = \tfrac{1}{3}\zeta_x\zeta_y \tag{5.99}$$

$$c_6 = \tfrac{1}{3}\zeta_y\zeta_z \tag{5.100}$$

$$c_7 = \tfrac{1}{3}\zeta_x\zeta_z \tag{5.101}$$

$$d_1 = \tfrac{4}{3}\xi_x\eta_x + \xi_y\eta_y + \xi_z\eta_z \tag{5.102}$$

$$d_2 = \xi_x\eta_x + \tfrac{4}{3}\xi_y\eta_y + \xi_z\eta_z \tag{5.103}$$

$$d_3 = \xi_x\eta_x + \xi_y\eta_y + \tfrac{4}{3}\xi_z\eta_z \tag{5.104}$$

$$d_4 = \xi_x\eta_x + \xi_y\eta_y + \xi_z\eta_z \tag{5.105}$$

$$d_5 = \xi_x\eta_y - \tfrac{2}{3}\xi_y\eta_x \tag{5.106}$$

$$d_6 = \xi_x\eta_z - \tfrac{2}{3}\xi_z\eta_x \tag{5.107}$$

$$d_7 = \xi_y\eta_x - \tfrac{2}{3}\xi_x\eta_y \tag{5.108}$$

$$d_8 = \xi_y\eta_z - \tfrac{2}{3}\xi_z\eta_y \tag{5.109}$$

$$d_9 = \xi_z\eta_x - \tfrac{2}{3}\xi_x\eta_z \tag{5.110}$$

$$d_{10} = \xi_z\eta_y - \tfrac{2}{3}\xi_y\eta_z \tag{5.111}$$

$$e_1 = \tfrac{4}{3}\xi_x\zeta_x + \xi_y\zeta_y + \xi_z\zeta_z \quad (5.112) \qquad e_2 = \xi_x\zeta_x + \tfrac{4}{3}\xi_y\zeta_y + \xi_z\zeta_z \quad (5.113)$$

$$e_3 = \xi_x\zeta_x + \xi_y\zeta_y + \tfrac{4}{3}\xi_z\zeta_z \quad (5.114) \qquad e_4 = \xi_x\zeta_x + \xi_y\zeta_y + \xi_z\zeta_z \quad (5.115)$$

$$e_5 = \xi_x\zeta_y - \tfrac{2}{3}\xi_y\zeta_x \quad (5.116) \qquad e_6 = \xi_x\zeta_z - \tfrac{2}{3}\xi_z\zeta_x \quad (5.117)$$

$$e_7 = \xi_y\zeta_x - \tfrac{2}{3}\xi_x\zeta_y \quad (5.118) \qquad e_8 = \xi_y\zeta_z - \tfrac{2}{3}\xi_z\zeta_y \quad (5.119)$$

$$e_9 = \xi_z\zeta_x - \tfrac{2}{3}\xi_x\zeta_z \quad (5.120) \qquad e_{10} = \xi_z\zeta_y - \tfrac{2}{3}\xi_y\zeta_z \quad (5.121)$$

$$f_1 = \tfrac{4}{3}\eta_x\zeta_x + \eta_y\zeta_y + \eta_z\zeta_z \quad (5.122) \qquad f_2 = \eta_x\zeta_x + \tfrac{4}{3}\eta_y\zeta_y + \eta_z\zeta_z \quad (5.123)$$

$$f_3 = \eta_x\zeta_x + \eta_y\zeta_y + \tfrac{4}{3}\eta_z\zeta_z \quad (5.124) \qquad f_4 = \eta_x\zeta_x + \eta_y\zeta_y + \eta_z\zeta_z \quad (5.125)$$

$$f_5 = \eta_x\zeta_y - \tfrac{2}{3}\eta_y\zeta_x \quad (5.126) \qquad f_6 = \eta_x\zeta_z - \tfrac{2}{3}\eta_z\zeta_x \quad (5.127)$$

$$f_7 = \eta_y\zeta_x - \tfrac{2}{3}\eta_x\zeta_y \quad (5.128) \qquad f_8 = \eta_y\zeta_z - \tfrac{2}{3}\eta_z\zeta_y \quad (5.129)$$

$$f_9 = \eta_z\zeta_x - \tfrac{2}{3}\eta_x\zeta_z \quad (5.130) \qquad f_{10} = \eta_z\zeta_y - \tfrac{2}{3}\eta_y\zeta_z \quad (5.131)$$

The derivation of viscous Jacobians must be handled with special care. This is due to the fact that the components of the viscous flux vectors involve gradients of the dependent variable and the Jacobian of transformation, which is itself a function of the independent variables. Thus, J remains embedded inside the viscous Jacobians, unlike the inviscid Jacobians where the Js were cancelled. In order to generalize the mathematical procedure, it is recognized that all of the terms in the viscous flux vector may be expressed as

$$\bar{E}_t = (\text{Factor})\frac{1}{J}(\phi_\psi) , \qquad (5.132)$$

where ϕ represents the dependent variable such as u, or a combination of dependent variables such as u^2 and ψ represent an independent variable such as ξ, η, or ζ. For example, the first term of the second component of $\bar{E}_v$ (from Equation (5.77)) is

$$\bar{E}_{v2,\text{1st term}} = \frac{\mu}{Re_\infty}\frac{1}{J}a_1(u_\xi)$$

which may be expressed as

$$\bar{E}_{v2,1T} = (\frac{\mu}{Re_\infty}a_1)\frac{1}{J}(u_\xi)$$

or, in the general formulation of (5.132),

$$\bar{E}_{v2,1T} = (\text{Factor})\frac{1}{J}\phi_\psi$$

To obtain the viscous Jacobian $\frac{\partial \bar{E}_v}{\partial \bar{Q}}$, we need to compute $\frac{\partial(\bar{E}_{v1},\bar{E}_{v2},\bar{E}_{v3},\bar{E}_{v4},\bar{E}_{v5})}{\partial(\bar{Q}_1,\bar{Q}_2,\bar{Q}_3,\bar{Q}_4,\bar{Q}_5)}$. In a similar fashion, the Jacobians associated with flux vectors $\bar{F}_v$ and $\bar{G}_v$ may be determined. In order to perform this differentiation, the general expression (5.149) will be used to illustrate the procedure. Consider the determination of

$$\frac{\partial \bar{E}_t}{\partial \bar{Q}} = \frac{\partial}{\partial \bar{Q}}\left[(\text{Factor})\frac{1}{J}(\phi_\psi)\right] = (\text{Factor})(\frac{1}{J})\frac{\partial}{\partial \bar{Q}}(\phi_\psi)$$

$$= (\text{Factor})(\frac{1}{J})\frac{\partial}{\partial \bar{Q}}\frac{\partial \phi}{\partial \psi}$$

At this point the order of differentiation is interchanged to yield:

$$\frac{\partial \bar{E}_t}{\partial \bar{Q}} = (\text{Factor})(\frac{1}{J})\frac{\partial}{\partial \psi}(\frac{\partial \phi}{\partial \bar{Q}}) \tag{5.133}$$

Recall that $\bar{Q} = \frac{Q}{J}$. Hence, after substituting into (5.133), one obtains

$$\frac{\partial \bar{E}_t}{\partial \bar{Q}} = (\text{Factor})\left[\frac{1}{J}\frac{\partial}{\partial \psi}(J\frac{\partial \phi}{\partial Q})\right] = (\text{Factor})\left[\frac{1}{J}(J\frac{\partial \phi}{\partial Q})_\psi\right] \tag{5.134}$$

This formulation is used to obtain the viscous Jacobians given by Equations (5.135), (5.136), and (5.137):

$$\frac{\partial \bar{E}_v}{\partial \bar{Q}} = \frac{\mu}{Re_\infty J}\begin{bmatrix} 0 & 0 & 0 & 0 & 0 \\ EQ_{2,1} & EQ_{2,2} & EQ_{2,3} & EQ_{2,4} & 0 \\ EQ_{3,1} & EQ_{3,2} & EQ_{3,3} & EQ_{3,4} & 0 \\ EQ_{4,1} & EQ_{4,2} & EQ_{4,3} & EQ_{4,4} & 0 \\ EQ_{5,1} & EQ_{5,2} & EQ_{5,3} & EQ_{5,4} & EQ_{5,5} \end{bmatrix} \tag{5.135}$$

$$\frac{\partial \bar{F}_v}{\partial \bar{Q}} = \frac{\mu}{Re_\infty J} \begin{bmatrix} 0 & 0 & 0 & 0 & 0 \\ FQ_{2,1} & FQ_{2,2} & FQ_{2,3} & FQ_{2,4} & 0 \\ FQ_{3,1} & FQ_{3,2} & FQ_{3,3} & FQ_{3,4} & 0 \\ FQ_{4,1} & FQ_{4,2} & FQ_{4,3} & FQ_{4,4} & 0 \\ FQ_{5,1} & FQ_{5,2} & FQ_{5,3} & FQ_{5,4} & FQ_{5,5} \end{bmatrix} \tag{5.136}$$

$$\frac{\partial \bar{G}_v}{\partial \bar{Q}} = \frac{\mu}{Re_\infty J} \begin{bmatrix} 0 & 0 & 0 & 0 & 0 \\ GQ_{2,1} & GQ_{2,2} & GQ_{2,3} & GQ_{2,4} & 0 \\ GQ_{3,1} & GQ_{3,2} & GQ_{3,3} & GQ_{3,4} & 0 \\ GQ_{4,1} & GQ_{4,2} & GQ_{4,3} & GQ_{4,4} & 0 \\ GQ_{5,1} & GQ_{5,2} & GQ_{5,3} & GQ_{5,4} & GQ_{5,5} \end{bmatrix} \tag{5.137}$$

where

$$EQ_{2,1} = -\left[a_1 (J\frac{u}{\rho})_\xi + a_5 (J\frac{v}{\rho})_\xi + a_7 (J\frac{w}{\rho})_\xi \right]$$

$$-\left[d_1 (J\frac{u}{\rho})_\eta + d_7 (J\frac{v}{\rho})_\eta + d_9 (J\frac{w}{\rho})_\eta \right]$$

$$-\left[e_1 (J\frac{u}{\rho})_\zeta + e_7 (J\frac{v}{\rho})_\zeta + e_9 (J\frac{w}{\rho})_\zeta \right]$$

$$EQ_{2,2} = a_1 (\frac{J}{\rho})_\xi + d_1 (\frac{J}{\rho})_\eta + e_1 (\frac{J}{\rho})_\zeta$$

$$EQ_{2,3} \;=\; a_5(\frac{J}{\rho})_\xi + d_7(\frac{J}{\rho})_\eta + e_7(\frac{J}{\rho})_\zeta$$

$$EQ_{2,4} \;=\; a_7(\frac{J}{\rho})_\xi + d_9(\frac{J}{\rho})_\eta + e_9(\frac{J}{\rho})_\zeta$$

$$EQ_{3,1} \;=\; -\left[a_5(J\frac{u}{\rho})_\xi + a_2(J\frac{v}{\rho})_\xi + a_6(J\frac{w}{\rho})_\xi \right]$$

$$-\left[d_5(J\frac{u}{\rho})_\eta + d_2(J\frac{v}{\rho})_\eta + d_{10}(J\frac{w}{\rho})_\eta \right]$$

$$-\left[e_5(J\frac{u}{\rho})_\zeta + e_2(J\frac{v}{\rho})_\zeta + e_{10}(J\frac{w}{\rho})_\zeta \right]$$

$$EQ_{3,2} \;=\; a_5(\frac{J}{\rho})_\xi + d_5(\frac{J}{\rho})_\eta + e_5(\frac{J}{\rho})_\zeta$$

$$EQ_{3,3} \;=\; a_2(\frac{J}{\rho})_\xi + d_2(\frac{J}{\rho})_\eta + e_2(\frac{J}{\rho})_\zeta$$

$$EQ_{3,4} \;=\; a_6(\frac{J}{\rho})_\xi + d_{10}(\frac{J}{\rho})_\eta + e_{10}(\frac{J}{\rho})_\zeta$$

$$EQ_{4,1} \;=\; -\left[a_7(J\frac{u}{\rho})_\xi + a_6(J\frac{v}{\rho})_\xi + a_3(J\frac{w}{\rho})_\xi \right]$$

$$-\left[d_6(J\frac{u}{\rho})_\eta + d_8(J\frac{v}{\rho})_\eta + d_3(J\frac{w}{\rho})_\eta \right]$$

$$-\left[e_6(J\frac{u}{\rho})_\zeta + e_8(J\frac{v}{\rho})_\zeta + e_3(J\frac{w}{\rho})_\zeta \right]$$

$$EQ_{4,2} = a_7(\frac{J}{\rho})_\xi + d_6(\frac{J}{\rho})_\eta + e_6(\frac{J}{\rho})_\zeta$$

$$EQ_{4,3} = a_6(\frac{J}{\rho})_\xi + d_8(\frac{J}{\rho})_\eta + e_8(\frac{J}{\rho})_\zeta$$

$$EQ_{4,4} = a_3(\frac{J}{\rho})_\xi + d_3(\frac{J}{\rho})_\eta + e_3(\frac{J}{\rho})_\zeta$$

$$EQ_{5,1} = -\left\{ (a_1 - \frac{\gamma}{Pr}a_4)(J\frac{u^2}{\rho})_\xi + (a_2 - \frac{\gamma}{Pr}a_4)(J\frac{v^2}{\rho})_\xi + \right.$$

$$(a_3 - \frac{\gamma}{Pr}a_4)(J\frac{w^2}{\rho})_\xi + \frac{\gamma}{Pr}a_4(J\frac{e_t}{\rho})_\xi + 2a_5(J\frac{uv}{\rho})_\xi + 2a_6(J\frac{vw}{\rho})_\xi$$

$$\left. + \; 2a_7(J\frac{uw}{\rho})_\xi \right\} - \left\{ (d_1 - \frac{\gamma}{Pr}d_4)(J\frac{u^2}{\rho})_\eta + (d_2 - \frac{\gamma}{Pr}d_4)(J\frac{v^2}{\rho})_\eta \right.$$

$$+ \; (d_3 - \frac{\gamma}{Pr}a_4)(\gamma\frac{w^2}{\rho})_\eta + \frac{\gamma}{Pr}d_4(J\frac{e_t}{\rho})_\eta + (d_5v + d_6w)(J\frac{u}{\rho})_\eta$$

$$+ \; (d_7u + d_8w)(J\frac{v}{\rho})_\eta + (d_9u + d_{10}v)(J\frac{w}{\rho})_\eta + \frac{J}{\rho}(d_5v + d_6w)u_\eta$$

$$\left. + \; \frac{J}{\rho}(d_7u + d_8w)v_\eta + \frac{J}{\rho}(d_9u + d_{10}v)w_\eta \right\} - \left\{ (e_1 - \frac{\gamma e_4}{Pr})(J\frac{u^2}{\rho})_\zeta \right.$$

$$+ \; (e_2 - \frac{\gamma}{Pr}e_4)(J\frac{v^2}{\rho})_\zeta + (e_3 - \frac{\gamma}{Pr}e_4)(J\frac{w^2}{\rho})_\zeta + \frac{\gamma}{Pr}e_4(J\frac{e_t}{\rho})_\zeta$$

$$+ \; (e_5v + e_6w)(J\frac{u}{\rho})_\zeta + (e_7u + e_8w)(J\frac{v}{\rho})_\zeta + (e_9u + e_{10}v)(J\frac{w}{\rho})_\zeta$$

$$\left. + \; \frac{J}{\rho}(e_5v + e_6w)u_\zeta + \frac{J}{\rho}(e_7u + e_8w)v_\zeta + \frac{J}{\rho}(e_9u + e_{10}v)w_\zeta \right\}$$

$$EQ_{5,2} = \left[(a_1 - \frac{\gamma}{Pr}a_4)(J\frac{u}{\rho})_\xi + a_5(J\frac{v}{\rho})_\xi + a_7(J\frac{w}{\rho})_\xi\right]$$

$$+ \left[(d_1 - \frac{\gamma}{Pr}d_4)(J\frac{u}{\rho})_\eta + (d_5 v + d_6 w)(\frac{J}{\rho})_\eta + d_7(\frac{J}{\rho})v_\eta + d_9(\frac{J}{\rho})w_\eta\right]$$

$$+ \left[(e_1 - \frac{\gamma}{Pr}e_4)(J\frac{u}{\rho})_\zeta + (e_5 v + e_6 w)(\frac{J}{\rho})_\zeta + e_7(\frac{J}{\rho})v_\zeta + e_9(\frac{J}{\rho})w_\zeta\right]$$

$$EQ_{5,3} = \left[(a_2 - \frac{\gamma}{Pr}a_4)(J\frac{v}{\rho})_\xi + a_5(J\frac{u}{\rho})_\xi + a_6(J\frac{w}{\rho})_\xi\right]$$

$$+ \left[(d_2 - \frac{\gamma}{Pr}d_4)(J\frac{v}{\rho})_\eta + (d_7 u + d_{10} w)(\frac{J}{\rho})_\eta + d_5(\frac{J}{\rho})u_\eta + d_{10}(\frac{J}{\rho})w_\eta\right]$$

$$+ \left[(e_2 - \frac{\gamma}{Pr}e_4)(J\frac{v}{\rho})_\eta + (e_7 u + e_8 w)(\frac{J}{\rho})_\zeta + e_5(\frac{J}{\rho})u_\zeta + e_{10}(\frac{J}{\rho})w_\zeta\right]$$

$$EQ_{5,4} = \left[(a_3 - \frac{\gamma}{Pr}a_4)(J\frac{w}{\rho})_\xi + a_7(J\frac{u}{\rho})_\xi + a_6(J\frac{v}{\rho})_\xi\right]$$

$$+ \left[(d_3 - \frac{\gamma}{Pr}d_4)(J\frac{w}{\rho})_\eta + (d_9 u + d_{10} v)(\frac{J}{\rho})_\eta + d_6(\frac{J}{\rho})u_\eta + d_8(\frac{J}{\rho})v_\eta\right]$$

$$+ \left[(c_3 - \frac{\gamma}{Pr}c_4)(J\frac{w}{\rho})_\zeta + (e_9 u + e_{10} v)(\frac{J}{\rho})_\zeta + e_6(\frac{J}{\rho})u_\zeta + e_8(\frac{J}{\rho})v_\zeta\right]$$

$$EQ_{5,5} = \frac{\gamma}{Pr}\left[a_4(\frac{J}{\rho})_\xi + d_4(\frac{J}{\rho})_\eta + e_4(\frac{J}{\rho})_\zeta\right]$$

$$FQ_{2,1} = -\left[d_1(J\frac{u}{\rho})_\xi + d_5(J\frac{v}{\rho})_\xi + d_6(J\frac{w}{\rho})_\xi\right]$$

$$- \left[b_1(J\frac{u}{\rho})_\eta + b_5(J\frac{v}{\rho})_\eta + b_7(J\frac{w}{\rho})_\eta\right]$$

$$- \left[f_1(J\frac{u}{\rho})_\zeta + f_7(J\frac{v}{\rho})_\zeta + f_9(J\frac{w}{\rho})_\zeta \right]$$

$$FQ_{2,2} = d_1(\frac{J}{\rho})_\xi + b_1(\frac{J}{\rho})_\eta + f_1(\frac{J}{\rho})_\zeta$$

$$FQ_{2,3} = d_5(\frac{J}{\rho})_\xi + b_5(\frac{J}{\rho})_\eta + f_7(\frac{J}{\rho})_\zeta$$

$$FQ_{2,4} = d_6(\frac{J}{\rho})_\xi + b_7(\frac{J}{\rho})_\eta + f_9(\frac{J}{\rho})_\zeta$$

$$FQ_{3,1} = - \left[d_7(J\frac{u}{\rho})_\xi + d_2(J\frac{v}{\rho})_\xi + d_8(J\frac{w}{\rho})_\xi \right]$$

$$- \left[b_5(J\frac{u}{\rho})_\eta + b_2(J\frac{v}{\rho})_\eta + b_6(J\frac{w}{\rho})_\eta \right]$$

$$- \left[f_5(J\frac{u}{\rho})_\zeta + f_2(J\frac{v}{\rho})_\zeta + f_{10}(J\frac{w}{\rho})_\zeta \right]$$

$$FQ_{3,2} = d_7(\frac{J}{\rho})_\xi + b_5(\frac{J}{\rho})_\eta + f_5(\frac{J}{\rho})_\zeta$$

$$FQ_{3,3} = d_2(\frac{J}{\rho})_\xi + b_2(\frac{J}{\rho})_\eta + f_2(\frac{J}{\rho})_\zeta$$

$$FQ_{3,4} = d_8(\frac{J}{\rho})_\xi + b_6(\frac{J}{\rho})_\eta + f_{10}(\frac{J}{\rho})_\zeta$$

$$FQ_{4,1} = -\left[d_9(J\frac{u}{\rho})_\xi + d_{10}(J\frac{v}{\rho})_\xi + d_3(J\frac{w}{\rho})_\xi\right]$$

$$-\left[b_7(J\frac{u}{\rho})_\eta + b_6(J\frac{v}{\rho})_\eta + b_3(J\frac{w}{\rho})_\eta\right]$$

$$-\left[f_6(J\frac{u}{\rho})_\zeta + f_8(J\frac{v}{\rho})_\zeta + f_3(J\frac{w}{\rho})_\zeta\right]$$

$$FQ_{4,2} = d_9(\frac{J}{\rho})_\xi + b_7(\frac{J}{\rho})_\eta + f_6(\frac{J}{\rho})_\zeta$$

$$FQ_{4,3} = d_{10}(\frac{J}{\rho})_\xi + b_6(\frac{J}{\rho})_\eta + f_8(\frac{J}{\rho})_\zeta$$

$$FQ_{4,4} = d_3(\frac{J}{\rho})_\xi + b_3(\frac{J}{\rho})_\eta + f_3(\frac{J}{\rho})_\zeta$$

$$FQ_{5,1} = -\left\{(d_1 - \frac{\gamma}{Pr}d_4)(J\frac{u^2}{\rho})_\xi + (d_2 - \frac{\gamma}{Pr}d_4)(J\frac{v^2}{\rho})_\xi\right.$$

$$+ (d_3 - \frac{\gamma}{Pr}d_4)(J\frac{w^2}{\rho})_\xi + \frac{\gamma}{Pr}d_4(J\frac{e_t}{\rho})_\xi + (d_7v + d_9w)(J\frac{u}{\rho})_\xi$$

$$+ (d_5u + d_{10}w)(J\frac{v}{\rho})_\xi + (d_6u + d_8v)(J\frac{w}{\rho})_\xi + \frac{J}{\rho}(d_7v + d_9w)u_\xi$$

$$\left. + \frac{J}{\rho}(d_5u + d_{10}w)v_\xi + \frac{J}{\rho}(d_6u + d_8v)w_\xi\right\} - \left\{(b_1 - \frac{\gamma}{Pr}b_4)(J\frac{u^2}{\rho})_\eta\right.$$

$$+ (b_2 - \frac{\gamma}{Pr}b_4)(J\frac{v^2}{\rho})_\eta + (b_3 - \frac{\gamma}{Pr}b_4)(J\frac{w^2}{\rho})_\eta + \frac{\gamma}{Pr}b_4(J\frac{e_t}{\rho})_\eta$$

$$\left. + 2b_5(J\frac{uv}{\rho})_\eta + 2b_6(J\frac{vw}{\rho})_\eta + 2b_7(J\frac{uw}{\rho})_\eta\right\} - \left\{(f_1 - \frac{\gamma}{Pr}f_4)(J\frac{u^2}{\rho})_\zeta + \right.$$

$$+ (f_2 - \frac{\gamma}{Pr}f_4)(J\frac{v^2}{\rho})_\zeta + (f_3 - \frac{\gamma}{Pr}f_4)(J\frac{w^2}{\rho})_\zeta + \frac{\gamma}{Pr}f_4(J\frac{e_t}{\rho})_\zeta$$

$$+ (f_5 v + f_6 w)(J\frac{u}{\rho})_\zeta + (f_7 u + f_8 w)(J\frac{v}{\rho})_\zeta + (f_9 u + f_{10} v)(J\frac{w}{\rho})_\zeta$$

$$+ \frac{J}{\rho}(f_5 v + f_6 w)u_\zeta + \frac{J}{\rho}(f_7 u + f_8 w)v_\zeta + \frac{J}{\rho}(f_9 u + f_{10} v)w_\zeta \Big\}$$

$$FQ_{5,2} = \left[(d_1 - \frac{\gamma}{Pr}d_4)(J\frac{u}{\rho})_\xi + (d_7 v + d_9 w)(\frac{J}{\rho})_\xi + d_5(\frac{J}{\rho})v_\xi + d_6(\frac{J}{\rho})w_\xi \right]$$

$$+ \left[(b_1 - \frac{\gamma}{Pr}b_4)(J\frac{u}{\rho})_\eta + b_5(J\frac{v}{\rho})_\eta + b_7(J\frac{w}{\rho})_\eta \right] + \left[(f_1 - \frac{\gamma}{Pr}f_4)(J\frac{u}{\rho})_\zeta \right.$$

$$+ (f_5 v + f_6 w)(\frac{J}{\rho})_\xi + f_7(\frac{J}{\rho})v_\zeta + f_9(\frac{J}{\rho})w_\zeta \Big]$$

$$FQ_{5,3} = \left[(d_2 - \frac{\gamma}{Pr}d_4)(J\frac{v}{\rho})_\xi + (d_5 u + d_{10} w)(\frac{J}{\rho})_\xi + d_7(\frac{J}{\rho})u_\xi + d_8(\frac{J}{\rho})w_\xi \right]$$

$$+ \left[(b_2 - \frac{\gamma}{Pr}b_4)(J\frac{v}{\rho})_\eta + b_5(J\frac{u}{\rho})_\eta + b_6(J\frac{w}{\rho})_\eta \right]$$

$$+ \left[(f_2 - \frac{\gamma}{Pr}f_4)(J\frac{v}{\rho})_\zeta + (f_7 u + f_8 w)(\frac{J}{\rho})_\zeta + f_5(\frac{J}{\rho})u_\zeta + f_{10}(\frac{J}{\rho})w_\zeta \right]$$

$$FQ_{5,4} = \left[(d_3 - \frac{\gamma}{Pr}d_4)(J\frac{w}{\rho})_\xi + (d_6 u + d_8 v)(\frac{J}{\rho})_\xi + d_9(\frac{J}{\rho})u_\xi \right.$$

$$+ d_{10}(\frac{J}{\rho})v_\xi \Big] + \left[(b_3 - \frac{\gamma}{Pr}b_4)(J\frac{w}{\rho})_\eta + b_7(J\frac{u}{\rho})_\eta + b_6(J\frac{v}{\rho})_\eta \right]$$

$$+ \left[(f_3 - \frac{\gamma}{Pr}f_4)(J\frac{w}{\rho})_\zeta + (f_9 u + f_{10} v)(\frac{J}{\rho})_\zeta + f_6(\frac{J}{\rho})u_\zeta + f_8(\frac{J}{\rho})v_\zeta \right]$$

$$FQ_{5,5} = \frac{\gamma}{Pr}\left[d_4(\frac{J}{\rho})_\xi + b_4(\frac{J}{\rho})_\eta + f_4(\frac{J}{\rho})_\zeta\right]$$

$$GQ_{2,1} = -\left[e_1(J\frac{u}{\rho})_\xi + e_5(J\frac{v}{\rho})_\xi + e_6(J\frac{w}{\rho})_\xi\right]$$

$$-\left[f_1(J\frac{u}{\rho})_\eta + f_5(J\frac{v}{\rho})_\eta + f_6(J\frac{w}{\rho})_\eta\right]$$

$$-\left[c_1(J\frac{u}{\rho})_\zeta + c_5(J\frac{v}{\rho})_\zeta + c_7(J\frac{w}{\rho})_\zeta\right]$$

$$GQ_{2,2} = e_1(\frac{J}{\rho})_\xi + f_1(\frac{J}{\rho})_\eta + c_1(\frac{J}{\rho})_\zeta$$

$$GQ_{2,3} = e_5(\frac{J}{\rho})_\xi + f_5(\frac{J}{\rho})_\eta + c_5(\frac{J}{\rho})_\zeta$$

$$GQ_{2,4} = e_6(\frac{J}{\rho})_\xi + f_6(\frac{J}{\rho})_\eta + c_7(\frac{J}{\rho})_\zeta$$

$$GQ_{3,1} = -\left[e_7(J\frac{u}{\rho})_\xi + e_2(J\frac{v}{\rho})_\xi + e_8(J\frac{w}{\rho})_\xi\right]$$

$$-\left[f_7(J\frac{u}{\rho})_\eta + f_2(J\frac{v}{\rho})_\eta + f_8(J\frac{w}{\rho})_\eta\right]$$

$$-\left[c_5(J\frac{u}{\rho})_\zeta + c_2(J\frac{v}{\rho})_\zeta + c_6(J\frac{w}{\rho})_\zeta\right]$$

$$GQ_{3,2} = e_7(\frac{J}{\rho})_\xi + f_7(\frac{J}{\rho})_\eta + c_5(\frac{J}{\rho})_\zeta$$

$$GQ_{3,3} = e_2(\frac{J}{\rho})_\xi + f_2(\frac{J}{\rho})_\eta + c_2(\frac{J}{\rho})_\zeta$$

$$GQ_{3,4} = e_8(\frac{J}{\rho})_\xi + f_8(\frac{J}{\rho})_\eta + c_6(\frac{J}{\rho})_\zeta$$

$$GQ_{4,1} = -\left[e_9(J\frac{u}{\rho})_\xi + e_{10}(J\frac{v}{\rho})_\xi + e_3(J\frac{w}{\rho})_\xi\right]$$

$$-\left[f_9(J\frac{u}{\rho})_\eta + f_{10}(J\frac{v}{\rho})_\eta + f_3(J\frac{w}{\rho})_\eta\right]$$

$$-\left[c_7(J\frac{u}{\rho})_\zeta + c_6(J\frac{v}{\rho})_\zeta + c_3(J\frac{w}{\rho})_\zeta\right]$$

$$GQ_{4,2} = e_9(\frac{J}{\rho})_\xi + f_9(\frac{J}{\rho})_\eta + c_7(\frac{J}{\rho})_\zeta$$

$$GQ_{4,3} = e_{10}(\frac{J}{\rho})_\xi + f_{10}(\frac{J}{\rho})_\eta + c_6(\frac{J}{\rho})_\zeta$$

$$GQ_{4,4} = e_3(\frac{J}{\rho})_\xi + f_3(\frac{J}{\rho})_\eta + c_3(\frac{J}{\rho})_\zeta$$

$$GQ_{5,1} = -\left\{(e_1 - \frac{\gamma}{Pr}e_4)(J\frac{u^2}{\rho})_\xi + (e_2 - \frac{\gamma}{Pr}e_4)(J\frac{v^2}{\rho})_\xi\right.$$

$$+ (e_3 - \frac{\gamma}{Pr}e_4)(J\frac{w^2}{\rho})_\xi + \frac{\gamma}{Pr}e_4(J\frac{e_t}{\rho})_\xi + (e_7v + e_9w)(J\frac{u}{\rho})_\xi$$

$$+ (e_5u + e_{10}w)(J\frac{v}{\rho})_\xi + (e_6u + e_8v)(J\frac{w}{\rho})_\xi + \frac{J}{\rho}(e_7v + e_9w)u_\xi$$

$$+ \frac{J}{\rho}(e_5u + e_{10}w)v_\xi + \frac{J}{\rho}(e_6u + e_8v)w_\xi\left.\right\} - \left\{(f_1 - \frac{\gamma}{Pr}f_4)(J\frac{u^2}{\rho})_\eta\right.$$

$$+ (f_2 - \frac{\gamma}{Pr}f_4)(J\frac{v^2}{\rho})_\eta + (f_3 - \frac{\gamma}{Pr}f_4)(J\frac{w^2}{\rho})_\eta + \frac{\gamma}{Pr}f_4(J\frac{e_t}{\rho})_\eta +$$

$$+ (f_7 v + f_9 w)(J\frac{u}{\rho})_\eta + (f_5 u + f_{10} w)(J\frac{v}{\rho})_\eta + (f_6 u + f_8 v)(J\frac{w}{\rho})_\eta$$

$$+ \frac{J}{\rho}(f_7 v + f_9 w)u_\eta + \frac{J}{\rho}(f_5 u + f_{10} w)v_\eta + \frac{J}{\rho}(f_6 u + f_8 v)w_\eta \bigg\}$$

$$- \bigg\{ (c_1 - \frac{\gamma}{Pr}c_4)(J\frac{u^2}{\rho})_\zeta + (c_2 - \frac{\gamma}{Pr}c_4)(J\frac{v^2}{\rho})_\zeta + (c_3 - \frac{\gamma}{Pr}c_4)(J\frac{w^2}{\rho})_\zeta$$

$$+ \frac{\gamma}{Pr}c_4(J\frac{e_t}{\rho})_\zeta + 2c_5(J\frac{uv}{\rho})_\zeta + 2c_6(J\frac{vw}{\rho})_\zeta + 2c_7(J\frac{uw}{\rho})_\zeta \bigg\}$$

$$GQ_{5,2} = \bigg[(e_1 - \frac{\gamma}{Pr}e_4)(J\frac{u}{\rho})_\xi + (e_7 v + e_9 w)(\frac{J}{\rho})_\xi + e_5(\frac{J}{\rho})v_\xi$$

$$+ e_6(\frac{J}{\rho})w_\xi\bigg] + \bigg[(f_1 - \frac{\gamma}{Pr}f_4)(J\frac{u}{\rho})_\eta + (f_7 v + f_9 w)(\frac{J}{\rho})_\eta + f_5(\frac{J}{\rho})v_\eta$$

$$+ f_6(\frac{J}{\rho})w_\eta\bigg] + \bigg[(c_1 - \frac{\gamma}{Pr}c_4)(J\frac{u}{\rho})_\zeta + c_5(J\frac{v}{\rho})_\zeta + c_7(J\frac{w}{\rho})_\zeta\bigg]$$

$$GQ_{5,3} = \bigg[(e_2 - \frac{\gamma}{Pr}e_4)(J\frac{v}{\rho})_\xi + (e_5 u + e_{10} w)(\frac{J}{\rho})_\xi + e_7(\frac{J}{\rho})u_\xi + e_8(\frac{J}{\rho})w_\xi\bigg]$$

$$+ \bigg[(f_2 - \frac{\gamma}{Pr}f_4)(J\frac{v}{\rho})_\eta + (f_5 u + f_{10} w)(\frac{J}{\rho})_\eta + f_7(\frac{J}{\rho})u_\eta + f_8(\frac{J}{\rho})w_\eta\bigg]$$

$$+ \bigg[(c_2 - \frac{\gamma}{Pr}c_4)(J\frac{v}{\rho})_\zeta + c_5(J\frac{u}{\rho})_\zeta + c_6(J\frac{w}{\rho})_\zeta\bigg]$$

$$GQ_{5,4} = \bigg[(e_3 - \frac{\gamma}{Pr}e_4)(J\frac{w}{\rho})_\xi + (e_6 u + e_8 v)(\frac{J}{\rho})_\xi + e_9(\frac{J}{\rho})u_\xi + e_{10}(\frac{J}{\rho})v_\xi\bigg] +$$

$$+ \left[(f_3 - \frac{\gamma}{Pr} f_4)(J\frac{w}{\rho})_\eta + (f_6 u + f_8 v)(\frac{J}{\rho})_\eta + f_9(\frac{J}{\rho})u_\eta + f_{10}(\frac{J}{\rho})v_\eta \right]$$

$$+ \left[(c_3 - \frac{\gamma}{Pr} c_4)(J\frac{w}{\rho})_\zeta + c_7(J\frac{u}{\rho})_\zeta + c_6(J\frac{v}{\rho})_\zeta \right]$$

$$GQ_{5,5} = \frac{\gamma}{Pr} \left[e_4(\frac{J}{\rho})_\xi + f_4(\frac{J}{\rho})_\eta + c_4(\frac{J}{\rho})_\zeta \right]$$

5.4   Thin Layer Approximation

The Navier-Stokes equations are reduced rather drastically by assuming thin layer approximation. Under this assumption the gradients of the viscous stresses in the direction parallel to the surface (ξ and ζ directions) are neglected. The thin layer equations are expressed as

$$\frac{\partial \bar{Q}}{\partial \tau} + \frac{\partial \bar{E}}{\partial \xi} + \frac{\partial \bar{F}}{\partial \eta} + \frac{\partial \bar{G}}{\partial \zeta} = \frac{\partial \bar{F}_{vT}}{\partial \eta} \tag{5.138}$$

where flux vectors $\bar{Q}$, $\bar{E}$, $\bar{F}$, and $\bar{G}$ are given by (5.40), (5.41), (5.42), (5.43) and

$$\bar{F}_{vT} = \frac{\mu}{Re_\infty J} \begin{bmatrix} 0 \\ \hline b_1 u_\eta + b_5 v_\eta + b_7 w_\eta \\ \hline b_5 u_\eta + b_2 v_\eta + b_6 w_\eta \\ \hline b_7 u_\eta + b_6 v_\eta + b_3 w_\eta \\ \hline \frac{1}{2} b_1 (u^2)_\eta + \frac{1}{2} b_2 (v^2)_\eta + \frac{1}{2} b_3 (w^2)_\eta + b_5 (uv)_\eta \\ + b_6 (vw)_\eta + b_7 (uw)_\eta + \frac{1}{Pr(\gamma-1)M_\infty^2} b_4 T_\eta \end{bmatrix} \tag{5.139}$$

The viscous flux vector $\bar{F}_{vT}$ is obtained from the viscous flux vector $\bar{F}_v$ given as Equation (5.79) by omitting all the mixed partial derivatives. To illustrate this point, consider for example the second component of the flux vector $\bar{F}_v$, where

$$\bar{F}_{v_2} = d_1 u_\xi + d_5 v_\xi + d_6 w_\xi + b_1 u_\eta + b_5 v_\eta + b_7 w_\eta + f_1 u_\zeta + f_7 v_\zeta + f_9 w_\zeta$$

The partial differential equation (5.138) includes the η gradient of the viscous terms. Therefore, this gradient will include terms such as $\frac{\partial^2 u}{\partial\eta\partial\xi}$, $\frac{\partial^2 v}{\partial\eta\partial\xi}$, ... and so on. These and similar mixed partial derivatives are omitted. As a result, only η derivatives of the flow variables remain, which have been redefined as the flux vector $\bar{F}_{vT}$.

The inviscid Jacobian matrices are the same as those of the Navier-Stokes equations, and the Jacobian matrix $\dfrac{\partial \bar{F}_{vT}}{\partial \bar{Q}}$ is

$$\frac{\partial \bar{F}_{vT}}{\partial \bar{Q}} = \frac{\mu}{Re_\infty J}
\begin{bmatrix}
0 & 0 & 0 & 0 & 0 \\[2ex]
-\left[b_1(J\frac{u}{\rho})_\eta + b_5(J\frac{v}{\rho})_\eta + b_7(J\frac{w}{\rho})_\eta\right] & b_1(\frac{J}{\rho})_\eta & b_5(\frac{J}{\rho})_\eta & b_7(\frac{J}{\rho})_\eta & 0 \\[2ex]
-\left[b_5(J\frac{u}{\rho})_\eta + b_2(J\frac{v}{\rho})_\eta + b_6(J\frac{w}{\rho})_\eta\right] & b_5(\frac{J}{\rho})_\eta & b_2(\frac{J}{\rho})_\eta & b_6(\frac{J}{\rho})_\eta & 0 \\[2ex]
-\left[b_7(J\frac{u}{\rho})_\eta + b_6(J\frac{v}{\rho})_\eta + b_3(J\frac{w}{\rho})_\eta\right] & b_7(\frac{J}{\rho})_\eta & b_6(\frac{J}{\rho})_\eta & b_3(\frac{J}{\rho})_\eta & 0 \\[2ex]
\begin{aligned} -\Big\{ &(b_1 - \tfrac{\gamma}{Pr}b_4)(J\tfrac{u^2}{\rho})_\eta \\ &+(b_2 - \tfrac{\gamma}{Pr}b_4)(J\tfrac{v^2}{\rho})_\eta \\ &+(b_3 - \tfrac{\gamma}{Pr}b_4)(J\tfrac{w^2}{\rho})_\eta \\ &+\tfrac{\gamma}{Pr}b_4(J\tfrac{e_t}{\rho})_\eta \\ &+2b_5(J\tfrac{uv}{\rho})_\eta + 2b_6(J\tfrac{vw}{\rho})_\eta \\ &+2b_7(J\tfrac{uw}{\rho})_\eta \end{aligned} &
\begin{aligned} &(b_1 - \tfrac{\gamma}{Pr}b_4)(J\tfrac{u}{\rho})_\eta \\ &+b_5(J\tfrac{v}{\rho})_\eta \\ &+b_7(J\tfrac{w}{\rho})_\eta \end{aligned} &
\begin{aligned} &b_5(J\tfrac{u}{\rho})_\eta \\ &+(b_2 - \tfrac{\gamma}{Pr}b_4)(J\tfrac{v}{\rho})_\eta \\ &+b_6(J\tfrac{w}{\rho})_\eta \end{aligned} &
\begin{aligned} &b_7(J\tfrac{u}{\rho})_\eta \\ &+b_6(J\tfrac{v}{\rho})_\eta \\ &+(b_3 - \tfrac{\gamma}{Pr}b_4)(J\tfrac{w}{\rho})_\eta \end{aligned} &
\frac{\gamma}{Pr}b_4(\frac{J}{\rho})_\eta
\end{bmatrix}$$

5.5 Parabolized Navier-Stokes Equations

For steady supersonic flow fields, PNS equations provide an efficient method of solution. It is assumed that the streamwise gradient of viscous stress is small compared to the gradients of stresses in η and ζ directions. Therefore, the streamwise gradient of viscous stress is dropped. In addition, the streamwise pressure gradient within the subsonic portion of the boundary layer must be approximated to prevent a departure solution. This approximation may employ any one of the following methods:

a. Neglect the pressure gradient completely.
b. Treat the pressure gradient explicitly.
c. Impose the pressure from the first supersonic point. This is known as the sublayer approximation.
d. Retain a fraction ω of the pressure gradient.

The PNS approximations will exclude streamwise flow separation; however, cross flow separation is predicted. In the following formulation of the PNS equation, the streamwise pressure gradient within the subsonic portion is approximated by method d. The PNS equations are expressed as

$$\frac{\partial \bar{E}_P}{\partial \xi} + \frac{\partial \bar{E}_{PP}}{\partial \xi} + \frac{\partial \bar{F}}{\partial \eta} + \frac{\partial \bar{G}}{\partial \zeta} = \frac{\partial \bar{F}_{vP}}{\partial \eta} + \frac{\partial \bar{G}_{vP}}{\partial \zeta} \tag{5.140}$$

where

$$\bar{E}_P = \frac{1}{J}[\xi_x E_P + \xi_y F_P + \xi_z G_P] \tag{5.141}$$

$$\bar{E}_{PP} = \frac{1}{J}[\xi_x E_{PP} + \xi_y F_{PP} + \xi_z G_{PP}] \tag{5.142}$$

$$\bar{F} = \frac{1}{J}[\eta_x E + \eta_y F + \eta_z G] \tag{5.143}$$

$$\bar{G} = \frac{1}{J}[\zeta_x E + \zeta_y F + \zeta_z G] \tag{5.144}$$

$$\bar{F}_{vP} = \frac{1}{J}[\eta_x E_v + \eta_y F_v + \eta_z G_v] \tag{5.145}$$

$$\bar{G}_{vP} = \frac{1}{J}[\zeta_x E_v + \zeta_y F_v + \zeta_z G_v] \tag{5.146}$$

and flux vectors are defined as

$$
E_P = \begin{bmatrix} \rho u \\ \rho u^2 + \omega p \\ \rho u v \\ \rho u w \\ (\rho e_t + p)u \end{bmatrix} \quad (5.147) \qquad
F_P = \begin{bmatrix} \rho v \\ \rho v u \\ \rho v^2 + \omega p \\ \rho v w \\ (\rho e_t + p)v \end{bmatrix} \quad (5.148)
$$

$$
G_P = \begin{bmatrix} \rho w \\ \rho w u \\ \rho w v \\ \rho w^2 + \omega p \\ (\rho e_t + p)w \end{bmatrix} \quad (5.149) \qquad
E_{PP} = \begin{bmatrix} 0 \\ (1 - \omega)p \\ 0 \\ 0 \\ 0 \end{bmatrix} \quad (5.150)
$$

$$
F_{PP} = \begin{bmatrix} 0 \\ 0 \\ (1 - \omega)p \\ 0 \\ 0 \end{bmatrix} \quad (5.151) \qquad
G_{PP} = \begin{bmatrix} 0 \\ 0 \\ 0 \\ (1 - \omega)p \\ 0 \end{bmatrix} \quad (5.152)
$$

The flux vectors E, F, and G are the same as E_P, F_P, and G_P with $\omega = 1$. The parameter ω, introduced in Equations (5.147) through (5.152), is determined by stability analysis. The viscous flux vectors $\bar{F}_{vP}$ and $\bar{G}_{vP}$ are

$$
\bar{F}_{vP} = \frac{\mu}{Re_\infty J}
\begin{bmatrix}
0 \\[4pt]
\hline
b_1 u_\eta + b_5 v_\eta + b_7 w_\eta \\[4pt]
\hline
b_5 u_\eta + b_2 v_\eta + b_6 w_\eta \\[4pt]
\hline
b_7 u_\eta + b_6 v_\eta + b_3 w_\eta \\[4pt]
\hline
\frac{1}{2}b_1(u^2)_\eta + \frac{1}{2}b_2(v^2)_\eta + \frac{1}{2}b_3(w^2)_\eta + b_5(uv)_\eta + b_6(uw)_\eta \\[4pt]
+ b_7(uw)_\eta + \dfrac{1}{Pr(\gamma-1)M_\infty^2} b_4 T_\eta
\end{bmatrix}
\tag{5.153}
$$

$$
\bar{G}_{vP} = \frac{\mu}{Re_\infty J}
\begin{bmatrix}
0 \\[4pt]
\hline
c_1 u_\zeta + c_5 v_\zeta + c_7 w_\zeta \\[4pt]
\hline
c_5 u_\zeta + c_2 v_\zeta + c_6 w_\zeta \\[4pt]
\hline
c_7 u_\zeta + c_6 v_\zeta + c_3 w_\zeta \\[4pt]
\hline
\frac{1}{2}c_1(u^2)_\zeta + \frac{1}{2}c_2(v^2)_\zeta + \frac{1}{2}c_3(w^2)_\zeta + c_5(uv)_\zeta + c_6(uw)_\zeta \\[4pt]
+ c_7(uw)_\zeta + \frac{1}{Pr(\gamma-1)M_\infty^2}c_4 T_\zeta
\end{bmatrix}
\tag{5.154}
$$

where

$$b_1 = \tfrac{4}{3}\eta_x^2 + \eta_y^2 + \eta_z^2 \tag{5.155} \qquad b_2 = \eta_x^2 + \tfrac{4}{3}\eta_y^2 + \eta_z^2 \tag{5.156}$$

$$b_3 = \eta_x^2 + \eta_y^2 + \tfrac{4}{3}\eta_y^2 \tag{5.157} \qquad b_4 = \eta_x^2 + \eta_y^2 + \eta_z^2 \tag{5.158}$$

$$b_5 = \tfrac{1}{3}\eta_x\eta_y \tag{5.159} \qquad b_6 = \tfrac{1}{3}\eta_y\eta_z \tag{5.160}$$

$$b_7 = \tfrac{1}{3}\eta_x\eta_z \tag{5.161}$$

$$c_1 = \tfrac{4}{3}\zeta_x^2 + \zeta_y^2 + \zeta_z^2 \tag{5.162} \qquad c_2 = \zeta_x^2 + \tfrac{4}{3}\zeta_y^2 + \zeta_z^2 \tag{5.163}$$

$$c_3 = \zeta_x^2 + \zeta_y^2 + \tfrac{4}{3}\zeta_z^2 \tag{5.164} \qquad c_4 = \zeta_x^2 + \zeta_y^2 + \zeta_z^2 \tag{5.165}$$

$$c_5 = \tfrac{1}{3}\zeta_x\zeta_y \tag{5.166} \qquad c_6 = \tfrac{1}{3}\zeta_y\zeta_z \tag{5.167}$$

$$c_7 = \tfrac{1}{3}\zeta_x\zeta_z \tag{5.168}$$

Note that the viscous flux vectors $\bar{F}_{vP}$ and $\bar{G}_{vP}$ are obtained from relations (5.79) and (5.80) by omitting all the mixed partial derivatives. This was illustrated previously for the thin layer Navier-Stokes equations.

The inviscid and viscous Jacobian matrices for the PNS equations are as follows:

$$\frac{\partial \bar{E}_p}{\partial \bar{Q}} =
\begin{bmatrix}
0 & \xi_x & \xi_y & \xi_z & 0 \\[2ex]
\begin{aligned}&-u(\xi_x u + \xi_y v + \xi_z w)\\ &+ \xi_x\left[\tfrac{1}{2}\omega(\gamma-1)(u^2+v^2+w^2)\right]\end{aligned}
& \begin{aligned}&\xi_x[1-\omega(\gamma-1)]u\\ &+(\xi_x u + \xi_y v + \xi_z w)\end{aligned}
& \xi_y u - \omega(\gamma-1)\xi_x v
& \xi_z u - \omega(\gamma-1)\xi_x w
& (\gamma-1)\xi_x \\[2ex]
\begin{aligned}&-v(\xi_x u + \xi_y v + \xi_z w)\\ &+ \xi_y\left[\tfrac{1}{2}\omega(\gamma-1)(u^2+v^2+w^2)\right]\end{aligned}
& \xi_x v - \omega(\gamma-1)\xi_y u
& \begin{aligned}&\xi_y[1-\omega(\gamma-1)]v\\ &+(\xi_x u + \xi_y v + \xi_z w)\end{aligned}
& \xi_z v - \omega(\gamma-1)\xi_y w
& (\gamma-1)\xi_y \\[2ex]
\begin{aligned}&-w(\xi_x u + \xi_y v + \xi_z w)\\ &+ \xi_z\left[\tfrac{1}{2}\omega(\gamma-1)(u^2+v^2+w^2)\right]\end{aligned}
& \xi_x w - \omega(\gamma-1)\xi_z u
& \xi_y w - \omega(\gamma-1)\xi_z v
& \begin{aligned}&\xi_z[1-\omega(\gamma-1)]w\\ &+(\xi_x u + \xi_y v + \xi_z w)\end{aligned}
& (\gamma-1)\xi_z \\[2ex]
\begin{aligned}&(\xi_x u + \xi_y v + \xi_z w)[-\gamma e_t\\ &+ (\gamma-1)(u^2+v^2+w^2)]\end{aligned}
& \begin{aligned}&\xi_x\left[\gamma e_t - \tfrac{1}{2}(\gamma-1)\cdot\right.\\ &\left.(u^2+v^2+w^2)\right]\\ &-(\gamma-1)[\xi_x u + \xi_y v\\ &+ \xi_z w]u\end{aligned}
& \begin{aligned}&\xi_y\left[\gamma e_t - \tfrac{1}{2}(\gamma-1)\cdot\right.\\ &\left.(u^2+v^2+w^2)\right]\\ &-(\gamma-1)[\xi_x u + \xi_y v\\ &+ \xi_z w]v\end{aligned}
& \begin{aligned}&\xi_z\left[\gamma e_t - \tfrac{1}{2}(\gamma-1)\cdot\right.\\ &\left.(u^2+v^2+w^2)\right]\\ &-(\gamma-1)[\xi_x u + \xi_y v\\ &+ \xi_z w]w\end{aligned}
& \gamma[\xi_x u + \xi_y v + \xi_z w]
\end{bmatrix}$$

$$(5.169)$$

$$\frac{\partial \bar{F}}{\partial \bar{Q}} = \begin{bmatrix}
0 & \eta_x & \eta_y & \eta_z & 0 \\[2ex]
\begin{aligned} &-u(\eta_x u + \eta_y v + \eta_z w) \\ &+ \eta_x \left[\tfrac{1}{2}(\gamma-1)(u^2+v^2+w^2)\right] \end{aligned} & \begin{aligned} &\eta_x(2-\gamma)u \\ &+ (\eta_x u + \eta_y v + \eta_z w) \end{aligned} & \eta_y u - (\gamma-1)\eta_x v & \eta_z u - (\gamma-1)\eta_x w & (\gamma-1)\eta_x \\[3ex]
\begin{aligned} &-v(\eta_x u + \eta_y v + \eta_z w) \\ &+ \eta_y \left[\tfrac{1}{2}(\gamma-1)(u^2+v^2+w^2)\right] \end{aligned} & \eta_x v - (\gamma-1)\eta_y u & \begin{aligned} &\eta_y(2-\gamma)v \\ &+ (\eta_x u + \eta_y v + \eta_z w) \end{aligned} & \eta_z v - (\gamma-1)\eta_y w & (\gamma-1)\eta_y \\[3ex]
\begin{aligned} &-w(\eta_x u + \eta_y v + \eta_z w) \\ &+ \eta_z \left[\tfrac{1}{2}(\gamma-1)(u^2+v^2+w^2)\right] \end{aligned} & \eta_x w - (\gamma-1)\eta_z u & \eta_y w - (\gamma-1)\eta_z v & \begin{aligned} &\eta_z(2-\gamma)w \\ &+ (\eta_x u + \eta_y v + \eta_z w) \end{aligned} & (\gamma-1)\eta_z \\[3ex]
\begin{aligned} &(\eta_x u + \eta_y v + \eta_z w)\left[-\gamma e_t\right. \\ &\left.+ (\gamma-1)(u^2+v^2+w^2)\right] \end{aligned} & \begin{aligned} &\eta_x\left[\gamma e_t - \tfrac{1}{2}(\gamma-1)\cdot\right. \\ &\left.(u^2+v^2+w^2)\right] \\ &- (\gamma-1)[\eta_x u + \eta_y v \\ &+ \eta_z w]u \end{aligned} & \begin{aligned} &\eta_y\left[\gamma e_t - \tfrac{1}{2}(\gamma-1)\cdot\right. \\ &\left.(u^2+v^2+w^2)\right] \\ &- (\gamma-1)[\eta_x u + \eta_y v \\ &+ \eta_z w]v \end{aligned} & \begin{aligned} &\eta_z\left[\gamma e_t - \tfrac{1}{2}(\gamma-1)\cdot\right. \\ &\left.(u^2+v^2+w^2)\right] \\ &- (\gamma-1)[\eta_x u + \eta_y v \\ &+ \eta_z w]w \end{aligned} & \gamma[\eta_x u + \eta_y v + \eta_z w]
\end{bmatrix}$$

$$(5.170)$$

$$\frac{\partial \bar{G}}{\partial \bar{Q}} =
\begin{bmatrix}
0 & \zeta_x & \zeta_y & \zeta_z & 0 \\[2ex]
\begin{aligned}&-u(\zeta_x u + \zeta_y v + \zeta_z w)\\ &+\zeta_x\left[\tfrac{1}{2}(\gamma-1)(u^2+v^2+w^2)\right]\end{aligned}
& \begin{aligned}&\zeta_x(2-\gamma)u\\ &+(\zeta_x u + \zeta_y v + \zeta_z w)\end{aligned}
& \zeta_y u - (\gamma-1)\zeta_x v
& \zeta_z u - (\gamma-1)\zeta_x w
& (\gamma-1)\zeta_x \\[3ex]
\begin{aligned}&-v(\zeta_x u + \zeta_y v + \zeta_z w)\\ &+\zeta_y\left[\tfrac{1}{2}(\gamma-1)(u^2+v^2+w^2)\right]\end{aligned}
& \zeta_x v - (\gamma-1)\zeta_y u
& \begin{aligned}&\zeta_y(2-\gamma)v\\ &+(\zeta_x u + \zeta_y v + \zeta_z w)\end{aligned}
& \zeta_z v - (\gamma-1)\zeta_y w
& (\gamma-1)\zeta_y \\[3ex]
\begin{aligned}&-w(\zeta_x u + \zeta_y v + \zeta_z w)\\ &+\zeta_z\left[\tfrac{1}{2}(\gamma-1)(u^2+v^2+w^2)\right]\end{aligned}
& \zeta_x w - (\gamma-1)\zeta_z u
& \zeta_y w - (\gamma-1)\zeta_z v
& \begin{aligned}&\zeta_z(2-\gamma)w\\ &+(\zeta_x u + \zeta_y v + \zeta_z w)\end{aligned}
& (\gamma-1)\zeta_z \\[3ex]
\begin{aligned}&(\zeta_x u + \zeta_y v + \zeta_z w)\left[-\gamma e_t\right.\\ &\left.+\ (\gamma-1)(u^2+v^2+w^2)\right]\end{aligned}
& \begin{aligned}&\zeta_x\left[\gamma e_t - \tfrac{1}{2}(\gamma-1)\cdot\right.\\ &\left.(u^2+v^2+w^2)\right]\\ &-(\gamma-1)\left[\zeta_x u + \zeta_y v\right.\\ &\left.+\zeta_z w\right]u\end{aligned}
& \begin{aligned}&\zeta_y\left[\gamma e_t - \tfrac{1}{2}(\gamma-1)\cdot\right.\\ &\left.(u^2+v^2+w^2)\right]\\ &-(\gamma-1)\left[\zeta_x u + \zeta_y v\right.\\ &\left.+\zeta_z w\right]v\end{aligned}
& \begin{aligned}&\zeta_z\left[\gamma e_t - \tfrac{1}{2}(\gamma-1)\cdot\right.\\ &\left.(u^2+v^2+w^2)\right]\\ &-(\gamma-1)\left[\zeta_x u + \zeta_y v\right.\\ &\left.+\zeta_z w\right]w\end{aligned}
& \gamma\left[\zeta_x u + \zeta_y v + \zeta_z w\right]
\end{bmatrix}$$

$$(5.171)$$

$$
\frac{\partial \bar{F}_{vP}}{\partial \bar{Q}} = \frac{\mu}{Re_\infty J}
\begin{bmatrix}
0 & 0 & 0 & 0 & 0 \\[2ex]
-\left[b_1\left(J\frac{u}{\rho}\right)_\eta + b_5\left(J\frac{v}{\rho}\right)_\eta + b_7\left(J\frac{w}{\rho}\right)_\eta \right] & b_1\left(\frac{J}{\rho}\right)_\eta & b_5\left(\frac{J}{\rho}\right)_\eta & b_7\left(\frac{J}{\rho}\right)_\eta & 0 \\[2ex]
-\left[b_5\left(J\frac{u}{\rho}\right)_\eta + b_2\left(J\frac{v}{\rho}\right)_\eta + b_6\left(J\frac{w}{\rho}\right)_\eta \right] & b_5\left(\frac{J}{\rho}\right)_\eta & b_2\left(\frac{J}{\rho}\right)_\eta & b_6\left(\frac{J}{\rho}\right)_\eta & 0 \\[2ex]
-\left[b_7\left(J\frac{u}{\rho}\right)_\eta + b_6\left(J\frac{v}{\rho}\right)_\eta + b_3\left(J\frac{w}{\rho}\right)_\eta \right] & b_7\left(\frac{J}{\rho}\right)_\eta & b_6\left(\frac{J}{\rho}\right)_\eta & b_3\left(\frac{J}{\rho}\right)_\eta & 0 \\[2ex]
\begin{aligned} -&\left[\left(b_1 - \frac{\gamma}{Pr}b_4\right)\left(J\frac{u^2}{\rho}\right)_\eta \right. \\ +&\left(b_2 - \frac{\gamma}{Pr}b_4\right)\left(J\frac{v^2}{\rho}\right)_\eta \\ +&\left(b_3 - \frac{\gamma}{Pr}b_4\right)\left(J\frac{w^2}{\rho}\right)_\eta + \frac{\gamma}{Pr}b_4\left(J\frac{e_t}{\rho}\right)_\eta \\ +&2b_5\left(J\frac{uv}{\rho}\right)_\eta + 2b_6\left(J\frac{vw}{\rho}\right)_\eta \\ +&\left. 2b_7\left(J\frac{uw}{\rho}\right)_\eta \right] \end{aligned}
&
\begin{aligned} &\left(b_1 - \frac{\gamma}{Pr}b_4\right)\left(J\frac{u}{\rho}\right)_\eta \\ +&b_5\left(J\frac{v}{\rho}\right)_\eta \\ +&b_7\left(J\frac{w}{\rho}\right)_\eta \end{aligned}
&
\begin{aligned} &\left(b_2 - \frac{\gamma}{Pr}b_4\right)\left(J\frac{v}{\rho}\right)_\eta \\ +&b_5\left(J\frac{u}{\rho}\right)_\eta \\ +&b_6\left(J\frac{w}{\rho}\right)_\eta \end{aligned}
&
\begin{aligned} &\left(b_3 - \frac{\gamma}{Pr}b_4\right)\left(J\frac{w}{\rho}\right)_\eta \\ +&b_7\left(J\frac{u}{\rho}\right)_\eta \\ +&b_6\left(J\frac{v}{\rho}\right)_\eta \end{aligned}
&
\frac{\gamma}{Pr}b_4\left(\frac{J}{\rho}\right)_\eta
\end{bmatrix}
\tag{5.172}
$$

$$\frac{\partial \bar{G}_{vP}}{\partial \bar{Q}} = \frac{\mu}{Re_\infty J}
\begin{bmatrix}
0 & 0 & 0 & 0 & 0 \\[2ex]
-\left[c_1\left(J\frac{u}{\rho}\right)_\zeta + c_5\left(J\frac{v}{\rho}\right)_\zeta + c_7\left(J\frac{w}{\rho}\right)_\zeta\right] & c_1\left(\frac{J}{\rho}\right)_\zeta & c_5\left(\frac{J}{\rho}\right)_\zeta & c_7\left(\frac{J}{\rho}\right)_\zeta & 0 \\[2ex]
-\left[c_5\left(J\frac{u}{\rho}\right)_\zeta + c_2\left(J\frac{v}{\rho}\right)_\zeta + c_6\left(J\frac{w}{\rho}\right)_\zeta\right] & c_5\left(\frac{J}{\rho}\right)_\zeta & c_2\left(\frac{J}{\rho}\right)_\zeta & c_6\left(\frac{J}{\rho}\right)_\zeta & 0 \\[2ex]
-\left[c_7\left(J\frac{u}{\rho}\right)_\zeta + c_6\left(J\frac{v}{\rho}\right)_\zeta + c_3\left(J\frac{w}{\rho}\right)_\zeta\right] & c_7\left(\frac{J}{\rho}\right)_\zeta & c_6\left(\frac{J}{\rho}\right)_\zeta & c_3\left(\frac{J}{\rho}\right)_\zeta & 0 \\[2ex]
\begin{aligned} &-\left[\left(c_1 - \tfrac{\gamma}{Pr}c_4\right)\left(J\frac{u^2}{\rho}\right)_\zeta \right. \\ &+\left(c_2 - \tfrac{\gamma}{Pr}c_4\right)\left(J\frac{v^2}{\rho}\right)_\zeta \\ &+\left(c_3 - \tfrac{\gamma}{Pr}c_4\right)\left(J\frac{w^2}{\rho}\right)_\zeta + \tfrac{\gamma}{Pr}c_4\left(J\frac{e_t}{\rho}\right)_\zeta \\ &+2c_5\left(J\frac{uv}{\rho}\right)_\zeta + 2c_6\left(J\frac{vw}{\rho}\right)_\zeta \\ &\left. +2c_7\left(J\frac{uw}{\rho}\right)_\zeta\right] \end{aligned} & \begin{aligned} &\left(c_1 - \tfrac{\gamma}{Pr}c_4\right)\left(J\frac{u}{\rho}\right)_\zeta \\ &+c_5\left(J\frac{v}{\rho}\right)_\zeta \\ &+c_7\left(J\frac{w}{\rho}\right)_\zeta \end{aligned} & \begin{aligned} &\left(c_2 - \tfrac{\gamma}{Pr}c_4\right)\left(J\frac{v}{\rho}\right)_\zeta \\ &+c_5\left(J\frac{u}{\rho}\right)_\zeta \\ &+c_6\left(J\frac{w}{\rho}\right)_\zeta \end{aligned} & \begin{aligned} &\left(c_3 - \tfrac{\gamma}{Pr}c_4\right)\left(J\frac{w}{\rho}\right)_\zeta \\ &+c_7\left(J\frac{u}{\rho}\right)_\zeta \\ &+c_6\left(J\frac{v}{\rho}\right)_\zeta \end{aligned} & \tfrac{\gamma}{Pr}c_4\left(\frac{J}{\rho}\right)_\zeta
\end{bmatrix}$$

$$(5.173)$$

5.6 Two-Dimensional Planar or Axisymmetric Formulation

For a two-dimensional planar or axisymmetric flow field, the nondimensionalized Navier-Stokes equation may be expressed in a combined form as

$$\frac{\partial Q}{\partial t} + \frac{\partial E}{\partial x} + \frac{\partial F}{\partial y} + \alpha H = \frac{\partial E_v}{\partial x} + \frac{\partial F_v}{\partial y} + \alpha H_v \qquad (5.174)$$

where $\alpha = 0$ represents two-dimensional planar flow and $\alpha = 1$ represents two-dimensional axisymmetric flow. The flux vectors in Equation (5.174) are

$$Q = \begin{bmatrix} \rho \\ \rho u \\ \rho v \\ \rho e_t \end{bmatrix} \quad (5.175) \qquad E = \begin{bmatrix} \rho u \\ \rho u^2 + p \\ \rho u v \\ (\rho e_t + p)u \end{bmatrix} \quad (5.176)$$

$$F = \begin{bmatrix} \rho v \\ \rho u v \\ \rho v^2 + p \\ (\rho e_t + p)v \end{bmatrix} \quad (5.177) \qquad H = \frac{1}{y}\begin{bmatrix} \rho v \\ \rho u v \\ \rho v^2 \\ (\rho e_t + p)v \end{bmatrix} \quad (5.178)$$

$$E_v = \begin{bmatrix} 0 \\ \tau_{xxp} \\ \tau_{xy} \\ u\tau_{xxp} + v\tau_{xy} - q_x \end{bmatrix} \quad (5.179) \qquad F_v = \begin{bmatrix} 0 \\ \tau_{xy} \\ \tau_{yyp} \\ u\tau_{xy} + v\tau_{yyp} - q_y \end{bmatrix} \quad (5.180)$$

$$H_v = \frac{1}{y}\begin{bmatrix} 0 \\ \hline \tau_{xy} - \dfrac{2}{3}\dfrac{y}{Re_\infty}\dfrac{\partial}{\partial x}\left(\mu\dfrac{v}{y}\right) \\ \hline \tau_{yyp} - \tau_{\theta\theta} - \dfrac{2}{3}\dfrac{\mu}{Re_\infty}\left(\dfrac{v}{y}\right) - \dfrac{y}{Re_\infty}\dfrac{2}{3}\dfrac{\partial}{\partial y}\left(\mu\dfrac{v}{y}\right) \\ \hline u\tau_{xy} + v\tau_{yyp} - q_y - \dfrac{2}{3}\dfrac{\mu}{Re_\infty}\dfrac{v^2}{y} - \dfrac{y}{Re_\infty}\dfrac{\partial}{\partial y}\left(\dfrac{2}{3}\mu\dfrac{v^2}{y}\right) \\ - \dfrac{y}{Re_\infty}\dfrac{\partial}{\partial x}\left(\dfrac{2}{3}\mu\dfrac{uv}{y}\right) \end{bmatrix} \quad (5.181)$$

where

$$\tau_{xxp} = \frac{\mu}{Re_\infty}\left(\frac{4}{3}\frac{\partial u}{\partial x} - \frac{2}{3}\frac{\partial v}{\partial y}\right) \tag{5.182}$$

$$\tau_{yyp} = \frac{\mu}{Re_\infty}\left(\frac{4}{3}\frac{\partial v}{\partial y} - \frac{2}{3}\frac{\partial u}{\partial x}\right) \tag{5.183}$$

$$\tau_{xy} = \frac{\mu}{Re_\infty}\left(\frac{\partial u}{\partial y} + \frac{\partial v}{\partial x}\right) \tag{5.184}$$

$$\tau_{\theta\theta} = \frac{\mu}{Re_\infty}\left[-\frac{2}{3}(\frac{\partial u}{\partial x} + \frac{\partial v}{\partial y}) + \frac{4}{3}\frac{v}{y}\right] \tag{5.185}$$

$$q_x = -\frac{\mu}{Re_\infty Pr(\gamma - 1)M_\infty^2}\frac{\partial T}{\partial x} \tag{5.186}$$

$$q_y = -\frac{\mu}{Re_\infty Pr(\gamma - 1)M_\infty^2}\frac{\partial T}{\partial y} \tag{5.187}$$

The Navier-Stokes equations given by Equation (5.174) are transformed to the computational space by relations (5.6) and (5.7), resulting in

$$\frac{\partial \bar{Q}}{\partial \tau} + \frac{\partial \bar{E}}{\partial \xi} + \frac{\partial \bar{F}}{\partial \eta} + \alpha \bar{H} = \frac{\partial \bar{E}_v}{\partial \xi} + \frac{\partial \bar{F}_v}{\partial \eta} + \alpha \bar{H}_v \tag{5.188}$$

where

$$\bar{Q} = \frac{Q}{J} \tag{5.189}$$

$$\bar{E} = \frac{1}{J}[\xi_t Q + \xi_x E + \xi_y F] \tag{5.190}$$

$$\bar{F} = \frac{1}{J}[\eta_t Q + \eta_x E + \eta_y F] \tag{5.191}$$

$$\bar{H} = \frac{H}{J} \tag{5.192}$$

$$\bar{E}_v = \frac{1}{J}[\xi_x E_v + \xi_y F_v] \tag{5.193}$$

$$\bar{F}_v = \frac{1}{J}[\eta_x E_v + \eta_y F_v] \tag{5.194}$$

$$\bar{H}_v = \frac{H_v}{J} \tag{5.195}$$

Substitution of viscous shear stresses into expressions (5.193) and (5.194) provides

$$\bar{E}_v = \frac{\mu}{Re_\infty J}
\begin{bmatrix}
0 \\[2pt]
\hline
a_1 u_\xi + a_3 v_\xi + c_1 u_\eta + c_3 v_\eta \\[2pt]
\hline
a_3 u_\xi + a_2 v_\xi + c_4 u_\eta + c_2 v_\eta \\[2pt]
\hline
\frac{1}{2}a_1(u^2)_\xi + \frac{1}{2}a_2(v^2)_\xi + a_3(uv)_\xi + \frac{1}{Pr(\gamma-1)M_\infty^2}a_4 T_\xi \\[4pt]
+\frac{1}{2}c_1(u^2)_\eta + \frac{1}{2}c_2(v^2)_\eta + c_3 uv_\eta + c_4 vu_\eta + \frac{1}{Pr(\gamma-1)M_\infty^2}c_5 T_\eta
\end{bmatrix} \tag{5.196}$$

and

$$\bar{F}_v = \frac{\mu}{Re_\infty J}
\begin{bmatrix}
0 \\[2pt]
\hline
c_1 u_\xi + c_4 v_\xi + b_1 u_\eta + b_3 v_\eta \\[2pt]
\hline
c_3 u_\xi + c_2 v_\xi + b_3 u_\eta + b_2 v_\eta \\[2pt]
\hline
\frac{1}{2}c_1(u^2)_\xi + \frac{1}{2}c_2(v^2)_\xi + c_3 vu_\xi + c_4 uv_\xi + \frac{1}{Pr(\gamma-1)M_\infty^2}c_5 T_\xi \\[4pt]
+\frac{1}{2}b_1(u^2)_\eta + \frac{1}{2}b_2(v^2)_\eta + b_3(uv)_\eta + \frac{1}{Pr(\gamma-1)M_\infty^2}b_4 T_\eta
\end{bmatrix} \tag{5.197}$$

where

$$a_1 = \tfrac{4}{3}\xi_x^2 + \xi_y^2 \qquad (5.198)$$

$$a_2 = \xi_x^2 + \tfrac{4}{3}\xi_y^2 \qquad (5.199)$$

$$a_3 = \tfrac{1}{3}\xi_x\xi_y \qquad (5.200)$$

$$a_4 = \xi_x^2 + \xi_y^2 \qquad (5.201)$$

$$b_1 = \tfrac{4}{3}\eta_x^2 + \eta_y^2 \qquad (5.202)$$

$$b_2 = \eta_x^2 + \tfrac{4}{3}\eta_y^2 \qquad (5.203)$$

$$b_3 = \tfrac{1}{3}\eta_x\eta_y \qquad (5.204)$$

$$b_4 = \eta_x^2 + \eta_y^2 \qquad (5.205)$$

$$c_1 = \tfrac{4}{3}\eta_x\xi_x + \xi_y\eta_y \qquad (5.206)$$

$$c_2 = \xi_x\eta_x + \tfrac{4}{3}\xi_y\eta_y \qquad (5.207)$$

$$c_3 = \eta_x\xi_y - \tfrac{2}{3}\xi_x\eta_y \qquad (5.208)$$

$$c_4 = \xi_x\eta_y - \tfrac{2}{3}\xi_y\eta_x \qquad (5.209)$$

$$c_5 = \xi_x\eta_x + \xi_y\eta_y \qquad (5.210)$$

The inviscid and viscous Jacobian matrices are:

$$
\frac{\partial \bar{E}}{\partial \bar{Q}} =
\begin{bmatrix}
\xi_t & \xi_x & \xi_y & 0 \\[2ex]
\begin{array}{l} -u(\xi_x u + \xi_y v)+ \\ \xi_x[\tfrac{1}{2}(\gamma-1)(u^2+v^2)] \end{array} &
\begin{array}{l} \xi_t + \xi_x(2-\gamma)u \\ +\xi_x u + \xi_y v \end{array} &
-(\gamma-1)\xi_x v + \xi_y u &
(\gamma-1)\xi_x \\[3ex]
\begin{array}{l} -v(\xi_x u + \xi_y v)+ \\ \xi_y[\tfrac{1}{2}(\gamma-1)(u^2+v^2)] \end{array} &
-(\gamma-1)\xi_y u + \xi_x v &
\begin{array}{l} \xi_t + \xi_y(2-\gamma)v \\ +\xi_x u + \xi_y v \end{array} &
(\gamma-1)\xi_y \\[3ex]
\begin{array}{l} (\xi_x u + \xi_y v)[-\gamma e_t + \\ (\gamma-1)(u^2+v^2)] \end{array} &
\begin{array}{l} \xi_x[\gamma e_t - \tfrac{1}{2}(\gamma-1)(u^2+v^2)] \\ -(\gamma-1)(\xi_x u + \xi_y v)u \end{array} &
\begin{array}{l} \xi_y[\gamma e_t - \tfrac{1}{2}(\gamma-1)(u^2+v^2)] \\ -(\gamma-1)(\xi_x u + \xi_y v)v \end{array} &
\xi_t + \gamma(\xi_x u + \xi_y v)
\end{bmatrix}
$$

$$(5.211)$$

$$\frac{\partial \bar{F}}{\partial \bar{Q}} =
\begin{bmatrix}
\eta_t & \eta_x & \eta_y & 0 \\[2ex]
-u(\eta_x u + \eta_y v) + \eta_x[\tfrac{1}{2}(\gamma - 1)(u^2 + v^2)] & \eta_t + \eta_x(2 - \gamma)u + \eta_x u + \eta_y v & -(\gamma - 1)\eta_x v + \eta_y u & (\gamma - 1)\eta_x \\[2ex]
-v(\eta_x u + \eta_y v) + \eta_y[\tfrac{1}{2}(\gamma - 1)(u^2 + v^2)] & -(\gamma - 1)\eta_y u + \eta_x v & \eta_t + \eta_y(2 - \gamma)v + \eta_x u + \eta_y v & (\gamma - 1)\eta_y \\[2ex]
(\eta_x u + \eta_y v)[-\gamma e_t + (\gamma - 1)(u^2 + v^2)] & \eta_x[\gamma e_t - \tfrac{1}{2}(\gamma - 1)(u^2 + v^2)] - (\gamma - 1)(\eta_x u + \eta_y v)u & \eta_y[\gamma e_t - \tfrac{1}{2}(\gamma - 1)(u^2 + v^2)] - (\gamma - 1)(\eta_x u + \eta_y v)v & \eta_t + \gamma(\eta_x u + \eta_y v)
\end{bmatrix}$$

$$(5.212)$$

$$C = \frac{\partial \bar{H}}{\partial \bar{Q}} = \frac{1}{y}
\begin{bmatrix}
0 & 0 & 1 & 0 \\
-uv & v & u & 0 \\
-v^2 & 0 & 2v & 0 \\
v[-\gamma e_t + (\gamma - 1)(u^2 + v^2)] & -(\gamma - 1)uv & \gamma e_t - \left(\frac{\gamma - 1}{2}\right)(3v^2 + u^2) & \gamma v
\end{bmatrix}$$

$$(5.213)$$

$$
\frac{\partial \bar{E}_v}{\partial \bar{Q}} = \frac{\mu}{Re_\infty J}
\begin{bmatrix}
0 & 0 & 0 & 0 \\[2ex]
\begin{aligned}
&- \left[a_1 \left(J\frac{u}{\rho} \right)_\xi + a_3 \left(J\frac{v}{\rho} \right)_\xi \right] \\
&- \left[c_1 \left(J\frac{u}{\rho} \right)_\eta + c_3 \left(J\frac{v}{\rho} \right)_\eta \right]
\end{aligned}
&
a_1 \left(\frac{J}{\rho} \right)_\xi + c_1 \left(\frac{J}{\rho} \right)_\eta
&
a_3 \left(\frac{J}{\rho} \right)_\xi + c_3 \left(\frac{J}{\rho} \right)_\eta
&
0 \\[3ex]
\begin{aligned}
&- \left[a_2 \left(J\frac{v}{\rho} \right)_\xi + a_3 \left(J\frac{u}{\rho} \right)_\xi \right] \\
&- \left[c_2 \left(J\frac{v}{\rho} \right)_\eta + c_4 \left(J\frac{u}{\rho} \right)_\eta \right]
\end{aligned}
&
a_3 \left(\frac{J}{\rho} \right)_\xi + c_4 \left(\frac{J}{\rho} \right)_\eta
&
a_2 \left(\frac{J}{\rho} \right)_\xi + c_2 \left(\frac{J}{\rho} \right)_\eta
&
0 \\[3ex]
\begin{aligned}
&- \left\{ \left(a_1 - \frac{\gamma}{Pr}a_4 \right) \left(J\frac{u^2}{\rho} \right)_\xi \right. \\
&+ \left(a_2 - \frac{\gamma}{Pr}a_4 \right) \left(J\frac{v^2}{\rho} \right)_\xi + 2a_3 \left(J\frac{uv}{\rho} \right)_\xi \\
&\left. + \frac{\gamma}{Pr}a_4 \left(J\frac{e_t}{\rho} \right)_\xi \right\} \\
&- \left\{ \left(c_1 - \frac{\gamma}{Pr}c_5 \right) \left(J\frac{u^2}{\rho} \right)_\eta \right. \\
&+ \left(c_2 - \frac{\gamma}{Pr}c_5 \right) \left(J\frac{v^2}{\rho} \right)_\eta \\
&+ c_3 u \left(J\frac{v}{\rho} \right)_\eta + c_4 v \left(J\frac{u}{\rho} \right)_\eta \\
&+ c_3 \left(J\frac{u}{\rho} \right) v_\eta + c_4 \left(J\frac{v}{\rho} \right) u_\eta \\
&\left. + \frac{\gamma}{Pr}c_5 \left(J\frac{e_t}{\rho} \right)_\eta \right\}
\end{aligned}
&
\begin{aligned}
&\left[a_3 \left(J\frac{v}{\rho} \right)_\xi \right. \\
&+ \left(a_1 - \frac{\gamma}{Pr}a_4 \right) \left(J\frac{u}{\rho} \right)_\xi \right] \\
&+ \left[c_4 v \left(\frac{J}{\rho} \right)_\eta + c_3 \frac{J}{\rho} (v_\eta) \right. \\
&\left. + \left(c_1 - \frac{\gamma}{Pr}c_5 \right) \left(J\frac{u}{\rho} \right)_\eta \right]
\end{aligned}
&
\begin{aligned}
&\left[a_3 \left(J\frac{u}{\rho} \right)_\xi \right. \\
&+ \left(a_2 - \frac{\gamma}{Pr}a_4 \right) \left(J\frac{v}{\rho} \right)_\xi \right] \\
&+ \left[c_3 u \left(\frac{J}{\rho} \right)_\eta + c_4 \frac{J}{\rho} (u_\eta) \right. \\
&\left. + \left(c_2 - \frac{\gamma}{Pr}c_5 \right) \left(J\frac{v}{\rho} \right)_\eta \right]
\end{aligned}
&
\begin{aligned}
&\frac{\gamma}{Pr}a_4 \left(\frac{J}{\rho} \right)_\xi \\
&+ \frac{\gamma}{Pr}c_5 \left(\frac{J}{\rho} \right)_\eta
\end{aligned}
\end{bmatrix}
\tag{5.214}
$$

$$\frac{\partial \bar{F}_v}{\partial \bar{Q}} = \frac{\mu}{Re_\infty J}\begin{bmatrix}
0 & 0 & 0 & 0 \\[2ex]
\begin{aligned}&-\left[c_1\left(J\frac{u}{\rho}\right)_\xi + c_4\left(J\frac{v}{\rho}\right)_\xi\right]\\ &-\left[b_1\left(J\frac{u}{\rho}\right)_\eta + b_3\left(J\frac{v}{\rho}\right)_\eta\right]\end{aligned} & \begin{aligned}&c_1\left(\frac{J}{\rho}\right)_\xi\\ &+b_1\left(\frac{J}{\rho}\right)_\eta\end{aligned} & \begin{aligned}&c_4\left(\frac{J}{\rho}\right)_\xi\\ &+b_3\left(\frac{J}{\rho}\right)_\eta\end{aligned} & 0 \\[4ex]
\begin{aligned}&-\left[c_3\left(J\frac{u}{\rho}\right)_\xi + c_2\left(J\frac{v}{\rho}\right)_\xi\right]\\ &-\left[b_3\left(J\frac{u}{\rho}\right)_\eta + b_2\left(J\frac{v}{\rho}\right)_\eta\right]\end{aligned} & \begin{aligned}&c_3\left(\frac{J}{\rho}\right)_\xi\\ &+b_3\left(\frac{J}{\rho}\right)_\eta\end{aligned} & \begin{aligned}&c_2\left(\frac{J}{\rho}\right)_\xi\\ &+b_2\left(\frac{J}{\rho}\right)_\eta\end{aligned} & 0 \\[4ex]
\begin{aligned}&-\left\{\left(c_1-\frac{\gamma}{Pr}c_5\right)\left(J\frac{u^2}{\rho}\right)_\xi\right.\\ &+\left(c_2-\frac{\gamma}{Pr}c_5\right)\left(J\frac{v^2}{\rho}\right)_\xi + c_3v\left(J\frac{u}{\rho}\right)_\xi\\ &+c_4u\left(J\frac{v}{\rho}\right)_\xi + c_3\left(J\frac{v}{\rho}\right)u_\xi + c_4\left(J\frac{u}{\rho}\right)v_\xi\\ &\left.+\frac{\gamma}{Pr}c_5\left(J\frac{e_t}{\rho}\right)_\xi\right\}\\ &-\left\{\left(b_1-\frac{\gamma}{Pr}b_4\right)\left(J\frac{u^2}{\rho}\right)_\eta\right.\\ &+\left(b_2-\frac{\gamma}{Pr}b_4\right)\left(J\frac{v^2}{\rho}\right)_\eta + 2b_3\left(J\frac{uv}{\rho}\right)_\eta\\ &\left.+\frac{\gamma}{Pr}b_4\left(J\frac{e_t}{\rho}\right)_\eta\right\}\end{aligned} & \begin{aligned}&\left[c_3v\left(\frac{J}{\rho}\right)_\xi + c_4\frac{J}{\rho}v_\xi\right.\\ &+\left(c_1-\frac{\gamma}{Pr}c_5\right)\left(J\frac{u}{\rho}\right)_\xi\right]\\ &+\left[b_3\left(J\frac{v}{\rho}\right)_\eta\right.\\ &\left.+\left(b_1-\frac{\gamma}{Pr}b_4\right)\left(J\frac{u}{\rho}\right)_\eta\right]\end{aligned} & \begin{aligned}&\left[c_3\frac{J}{\rho}u_\xi + c_4u\left(\frac{J}{\rho}\right)_\xi\right.\\ &+\left(c_2-\frac{\gamma}{Pr}c_5\right)\left(J\frac{v}{\rho}\right)_\xi\right]\\ &+\left[b_3\left(J\frac{u}{\rho}\right)_\eta\right.\\ &\left.+\left(b_2-\frac{\gamma}{Pr}b_4\right)\left(J\frac{v}{\rho}\right)_\eta\right]\end{aligned} & \begin{aligned}&\frac{\gamma}{Pr}c_5\left(\frac{J}{\rho}\right)_\xi\\ &+\frac{\gamma}{Pr}b_4\left(\frac{J}{\rho}\right)_\eta\end{aligned}
\end{bmatrix} \tag{5.215}$$

$$C_v = \frac{1}{Re_\infty Jy} \begin{bmatrix} 0 & 0 & 0 & 0 \\ HQ_{2,1} & HQ_{2,2} & HQ_{2,3} & 0 \\ HQ_{3,1} & HQ_{3,2} & HQ_{3,3} & 0 \\ HQ_{4,1} & HQ_{4,2} & HQ_{4,3} & HQ_{4,4} \end{bmatrix} \tag{5.216}$$

where

$$HQ_{2,1} = -\mu\left[\xi_x(J\frac{v}{\rho})_\xi + \xi_y(J\frac{u}{\rho})_\xi\right] + \frac{2}{3}y\xi_x\left[(\frac{\mu}{y})(J\frac{v}{\rho})\right]_\xi$$

$$- \mu\left[\eta_x(J\frac{v}{\rho})_\eta + \eta_y(J\frac{u}{\rho})_\eta\right] + \frac{2}{3}y\eta_x\left[(\frac{\mu}{y})(J\frac{v}{\rho})_\eta\right]$$

$$HQ_{2,2} = \left[\mu\xi_y(\frac{J}{\rho})_\xi + \mu\eta_y(\frac{J}{\rho})_\eta\right]$$

$$HQ_{2,3} = \mu\xi_x(\frac{J}{\rho})_\xi - \frac{2}{3}y\xi_x\left[(\frac{J}{\rho})(\frac{\mu}{y})\right]_\xi + \mu\eta_x(\frac{J}{\rho})_\eta - \frac{2}{3}y\eta_x\left[(\frac{J}{\rho})(\frac{\mu}{y})\right]_\eta$$

$$HQ_{3,1} = 2\mu\left[-\xi_y(J\frac{v}{\rho})_\xi\right] + \frac{2}{3}y\xi_y\left[(\frac{\mu}{y})(J\frac{v}{\rho})\right]_\xi + 2\mu\left[-\eta_y(J\frac{v}{\rho})_\eta + \frac{J}{y}(\frac{v}{\rho})\right]$$

$$+ \frac{2}{3}y\eta_y\left[(\frac{\mu}{y})(J\frac{v}{\rho})\right]_\eta$$

$$HQ_{3,2} = 0$$

$$HQ_{3,3} = 2\mu\xi_y(\frac{J}{\rho})_\xi - \frac{2}{3}y\xi_y\left[(\frac{\mu}{y})(\frac{J}{\rho})\right]_\xi + 2\mu\left[\eta_y(\frac{J}{\rho})_\eta - \frac{1}{y}\frac{J}{\rho}\right] - \frac{2}{3}y\eta_y\left[(\frac{\mu}{y})(\frac{J}{\rho})\right]_\eta$$

$$HQ_{4,1} = -\mu(J\frac{u}{\rho})\Big[\xi_x v_\xi + \xi_y u_\xi\Big] - \mu u\Big[\xi_x(J\frac{v}{\rho})_\xi + \xi_y(J\frac{u}{\rho})_\xi\Big] - \frac{2}{3}\mu(J\frac{v}{\rho})\Big[2\xi_y v_\xi - \xi_x u_\xi\Big]$$

$$+ \frac{2}{3}\mu v\Big[-2\xi_y(J\frac{v}{\rho})_\xi + \xi_x(J\frac{u}{\rho})_\xi\Big] + \frac{4}{3}y\xi_y\Big[(\frac{\mu}{y})(J\frac{v^2}{\rho})\Big]_\xi + \frac{4}{3}y\xi_x\Big[(\frac{\mu}{y})(J\frac{uv}{\rho})\Big]_\xi$$

$$+ \frac{\gamma\mu}{Pr}\xi_y\Big[-J\frac{e_t}{\rho} + (\frac{J}{\rho}u^2 + \frac{J}{\rho}v^2)\Big]_\xi - \mu(J\frac{u}{\rho})\Big[\eta_x v_\eta + \eta_y u_\eta\Big]$$

$$- \mu u\Big[\eta_x(J\frac{v}{\rho})_\eta + \eta_y(J\frac{u}{\rho})_\eta\Big] - \frac{2}{3}\mu(J\frac{v}{\rho})\Big[2\eta_y v_\eta - \eta_x u_\eta\Big]$$

$$+ \frac{2}{3}\mu v\Big[-2\eta_y(J\frac{v}{\rho})_\eta + \eta_x(J\frac{u}{\rho})_\eta\Big] + \frac{4}{3}\frac{\mu}{y}(J\frac{v^2}{\rho}) + \frac{4}{3}y\eta_y\Big[(\frac{\mu}{y})(J\frac{v^2}{\rho})\Big]_\eta$$

$$+ \frac{4}{3}y\eta_x\Big[(\frac{\mu}{y})(J\frac{uv}{\rho})\Big]_\eta + \frac{\gamma\mu}{Pr}\eta_y\Big[-J\frac{e_t}{\rho} + (\frac{J}{\rho}u^2 + \frac{J}{\rho}v^2)\Big]_\eta$$

$$HQ_{4,2} = \mu(\frac{J}{\rho})(\xi_x v_\xi + \xi_y u_\xi) + \mu u\xi_y(\frac{J}{\rho})_\xi - \frac{2}{3}\mu v\xi_x(\frac{J}{\rho})_\xi - \frac{2}{3}y\xi_x\Big[(\frac{\mu}{y})(J\frac{v}{\rho})\Big]_\xi$$

$$+ \mu(\frac{J}{\rho})(\eta_x v_\eta + \eta_y u_\eta) + \mu u\eta_y(\frac{J}{\rho})_\eta - \frac{2}{3}\mu v\eta_x(\frac{J}{\rho})_\eta - \frac{\gamma\mu}{Pr}\xi_y(J\frac{u}{\rho})_\xi$$

$$- \frac{2}{3}y\eta_x\Big[(\frac{\mu}{y})(J\frac{v}{\rho})\Big]_\eta - \frac{\gamma\mu}{Pr}\eta_y(J\frac{u}{\rho})_\eta$$

$$HQ_{4,3} = \mu u\xi_x(\frac{J}{\rho})_\xi + \frac{2}{3}\mu\frac{J}{\rho}(2\xi_y v_\xi - \xi_x u_\xi) + \frac{4}{3}\mu v\xi_y(\frac{J}{\rho})_\xi$$

$$- \frac{4}{3}y\xi_y\Big[(\frac{\mu}{\rho})(J\frac{v}{\rho})\Big]_\xi - \frac{2}{3}y\xi_x\Big[(\frac{\mu}{y})(J\frac{u}{\rho})\Big]_\xi - \frac{\gamma\mu}{Pr}\xi_y(J\frac{v}{\rho})_\xi + \mu u\eta_x(\frac{J}{\rho})_\eta$$

$$+ \frac{2}{3}\mu\frac{J}{\rho}(2\eta_y v_\eta - \eta_x u_\eta) + \frac{4}{3}\mu v\eta_y(\frac{J}{\rho})_\eta - \frac{4}{3}\frac{\mu}{y}(J\frac{v}{\rho}) - \frac{4}{3}y\eta_y\Big[(\frac{\mu}{y})(J\frac{v}{\rho})\Big]_\eta +$$

$$-\frac{2}{3}y\eta_x\left[(\frac{\mu}{y})(J\frac{u}{\rho})\right]_\eta - \frac{\gamma\mu}{Pr}y_\eta(J\frac{v}{\rho})_\eta$$

$$HQ_{4,4} = \frac{\gamma\mu}{Pr}\left[\eta_y(\frac{J}{\rho})_\eta + \xi_y(\frac{J}{\rho})_\xi\right]$$

The thin layer Navier-Stokes equation for two-dimensional planar or axisymmetric flow is reduced to

$$\frac{\partial\bar{Q}}{\partial\tau} + \frac{\partial\bar{E}}{\partial\xi} + \frac{\partial\bar{F}}{\partial\eta} + \alpha\bar{H} = \frac{\partial\bar{F}_{vT}}{\partial\eta} + \alpha\bar{H}_v \tag{5.217}$$

where flux vectors $\bar{Q}$, $\bar{E}$, and $\bar{F}$ are given by (5.189), (5.190), and (5.191), and

$$\bar{F}_{vT} = \frac{\mu}{Re_\infty J}\begin{bmatrix} 0 \\[2ex] b_1 u_\eta + b_3 v_\eta \\[2ex] b_3 u_\eta + b_2 v_\eta \\[2ex] \frac{1}{2}b_1(u^2)_\eta + \frac{1}{2}b_2(v^2)_\eta + b_3(uv)_\eta + \frac{1}{Pr(\gamma-1)M_\infty^2}b_4 T_\eta \end{bmatrix} \tag{5.218}$$

The Jacobian matrices $\frac{\partial\bar{E}}{\partial\bar{Q}}$ and $\frac{\partial\bar{F}}{\partial\bar{Q}}$ are given by (5.211) and (5.212), and

$$\frac{\partial \bar{F}_{vT}}{\partial \bar{Q}} = \frac{\mu}{Re_\infty J}
\begin{bmatrix}
0 & 0 & 0 & 0 \\[2ex]
-\left[b_1 \left(J\frac{u}{\rho} \right)_\eta + b_3 \left(J\frac{v}{\rho} \right)_\eta \right] & b_1 \left(\frac{J}{\rho} \right)_\eta & b_3 \left(\frac{J}{\rho} \right)_\eta & 0 \\[2ex]
-\left[b_3 \left(J\frac{u}{\rho} \right)_\eta + b_2 \left(J\frac{v}{\rho} \right)_\eta \right] & b_3 \left(\frac{J}{\rho} \right)_\eta & b_2 \left(\frac{J}{\rho} \right)_\eta & 0 \\[2ex]
\begin{aligned}
-\Bigg\{ &\left(b_1 - \frac{\gamma}{Pr}b_4 \right) \left(J\frac{u^2}{\rho} \right)_\eta \\
&+ \left(b_2 - \frac{\gamma}{Pr}b_4 \right) \left(J\frac{v^2}{\rho} \right)_\eta \\
&+ \frac{\gamma}{Pr}b_4 \left(J\frac{e_t}{\rho} \right)_\eta \\
&+ 2b_3 \left(J\frac{uv}{\rho} \right)_\eta \Bigg\}
\end{aligned}
&
\begin{aligned}
&\left(b_1 - \frac{\gamma}{Pr}b_4 \right) \left(J\frac{u}{\rho} \right)_\eta \\
&+ b_3 \left(J\frac{v}{\rho} \right)_\eta
\end{aligned}
&
\begin{aligned}
&b_3 \left(J\frac{u}{\rho} \right)_\eta \\
&+ \left(b_2 - \frac{\gamma}{Pr}b_4 \right) \left(J\frac{v}{\rho} \right)_\eta
\end{aligned}
& \frac{\gamma}{Pr}b_4 \left(\frac{J}{\rho} \right)_\eta
\end{bmatrix}$$

$$(5.219)$$

The parabolized Navier-Stokes equations under the assumptions stated previously are formulated as

$$\frac{\partial \bar{E}_P}{\partial \xi} + \frac{\partial \bar{E}_{PP}}{\partial \xi} + \frac{\partial \bar{F}}{\partial \eta} + \alpha \bar{H} = \frac{\partial \bar{F}_{vP}}{\partial \eta} + \alpha \bar{H}_v \tag{5.220}$$

where

$$\bar{E}_P = \frac{1}{J}[\xi_x E_P + \xi_y F_P] \tag{5.221}$$

$$\bar{E}_{PP} = \frac{1}{J}[\xi_x E_{PP} + \xi_y F_{PP}] \tag{5.222}$$

and

$$E_P = \begin{bmatrix} \rho u \\ \rho u^2 + \omega p \\ \rho u v \\ (\rho e_t + p)u \end{bmatrix} \tag{5.223} \qquad F_P = \begin{bmatrix} \rho v \\ \rho u v \\ \rho v^2 + \omega p \\ (\rho e_t + p)v \end{bmatrix} \tag{5.224}$$

$$E_{PP} = \begin{bmatrix} 0 \\ (1-\omega)p \\ 0 \\ 0 \end{bmatrix} \tag{5.225} \qquad F_{PP} = \begin{bmatrix} 0 \\ 0 \\ (1-\omega)p \\ 0 \end{bmatrix} \tag{5.226}$$

and the viscous flux vector $\bar{F}_{vP}$ is

$$\bar{F}_{vP} = \frac{\mu}{Re_\infty J} \begin{bmatrix} 0 \\ b_1 u_\eta + b_3 v_\eta \\ b_3 u_\eta + b_2 v_\eta \\ \frac{1}{2}b_1(u^2)_\eta + \frac{1}{2}b_2(v^2)_\eta + b_3(uv)_\eta + \frac{1}{Pr(\gamma - 1)M_\infty^2}b_4 T_\eta \end{bmatrix} \tag{5.227}$$

The inviscid and viscous Jacobian matrices are:

$$\frac{\partial \bar{E}_P}{\partial \bar{Q}} =
\begin{bmatrix}
0 & \xi_x & \xi_y & 0 \\[2ex]
-u\left(\xi_x u + \xi_y v\right) + \xi_x\left[\frac{1}{2}\omega\left(\gamma-1\right)\left(u^2+v^2\right)\right] & \xi_x\left[1-\omega\left(\gamma-1\right)\right]u + \xi_x u + \xi_y v & -\omega\left(\gamma-1\right)\xi_x v + \xi_y u & \omega\left(\gamma-1\right)\xi_x \\[2ex]
-v\left(\xi_x u + \xi_y v\right) + \xi_y\left[\frac{1}{2}\omega\left(\gamma-1\right)\left(u^2+v^2\right)\right] & \xi_x v - \omega\left(\gamma-1\right)\xi_y u & \xi_y\left[1-\omega\left(\gamma-1\right)\right]v + \xi_x u + \xi_y v & \omega\left(\gamma-1\right)\xi_y \\[2ex]
\left(\xi_x u + \xi_y v\right)\left[-\gamma e_t + \left(\gamma-1\right)\left(u^2+v^2\right)\right] & \xi_x\left[\gamma e_t - \frac{1}{2}\left(\gamma-1\right)\cdot\left(u^2+v^2\right)\right] - \left(\gamma-1\right)\left[\xi_x u + \xi_y v\right]u & \xi_y\left[\gamma e_t - \frac{1}{2}\left(\gamma-1\right)\cdot\left(u^2+v^2\right)\right] - \left(\gamma-1\right)\left[\xi_x u + \xi_y v\right]v & \gamma\left[\xi_x u + \xi_y v\right]
\end{bmatrix}$$

$$(5.228)$$

$$\frac{\partial \bar{F}}{\partial \bar{Q}} =
\begin{bmatrix}
0 & \eta_x & \eta_y & 0 \\[2ex]
-u(\eta_x u + \eta_y v) + \eta_x[\tfrac{1}{2}(\gamma-1)(u^2+v^2)] & \eta_x(2-\gamma)u + \eta_x u + \eta_y v & -(\gamma-1)\eta_x v + \eta_y u & (\gamma-1)\eta_x \\[3ex]
-v(\eta_x u + \eta_y v) + \eta_y[\tfrac{1}{2}(\gamma-1)(u^2+v^2)] & -(\gamma-1)\eta_y u + \eta_x v & \eta_y(2-\gamma)v + \eta_x u + \eta_y v & (\gamma-1)\eta_y \\[3ex]
(\eta_x u + \eta_y v)[-\gamma e_t + (\gamma-1)(u^2+v^2)] & \eta_x[\gamma e_t - \tfrac{1}{2}(\gamma-1)(u^2+v^2)] - (\gamma-1)(\eta_x u + \eta_y v)u & \eta_y[\gamma e_t - \tfrac{1}{2}(\gamma-1)(u^2+v^2)] - (\gamma-1)(\eta_x u + \eta_y v)v & \gamma(\eta_x u + \eta_y v)
\end{bmatrix}$$

$$(5.229)$$

$$
\frac{\partial \bar{F}_{vP}}{\partial \bar{Q}} = \frac{\mu}{Re_\infty J}
\begin{bmatrix}
0 & 0 & 0 & 0 \\[2ex]
- \left[b_1 \left(J\frac{u}{\rho} \right)_\eta + b_3 \left(J\frac{v}{\rho} \right)_\eta \right] & b_1 \left(\frac{J}{\rho} \right)_\eta & b_3 \left(\frac{J}{\rho} \right)_\eta & 0 \\[2ex]
- \left[b_3 \left(J\frac{u}{\rho} \right)_\eta + b_2 \left(J\frac{v}{\rho} \right)_\eta \right] & b_3 \left(\frac{J}{\rho} \right)_\eta & b_2 \left(\frac{J}{\rho} \right)_\eta & 0 \\[2ex]
- \left\{ \left(b_1 - \frac{\gamma}{Pr} b_4 \right) \left(J\frac{u^2}{\rho} \right)_\eta + \left(b_2 - \frac{\gamma}{Pr} b_4 \right) \left(J\frac{v^2}{\rho} \right)_\eta + \frac{\gamma}{Pr} b_4 \left(J\frac{e_t}{\rho} \right)_\eta + 2b_3 \left(J\frac{uv}{\rho} \right)_\eta \right\} & \left(b_1 - \frac{\gamma}{Pr} b_4 \right) \left(J\frac{u}{\rho} \right)_\eta + b_3 \left(J\frac{v}{\rho} \right)_\eta & b_3 \left(J\frac{u}{\rho} \right)_\eta + \left(b_2 - \frac{\gamma}{Pr} b_4 \right) \left(J\frac{v}{\rho} \right)_\eta & \frac{\gamma}{Pr} b_4 \left(\frac{J}{\rho} \right)_\eta
\end{bmatrix}
$$

$$(5.230)$$

5.7 Incompressible Navier-Stokes Equations

A common scheme to solve numerically the incompressible Navier-Stokes equations expressed in primitive variables is the modification of the continuity equation to include an artificial compresibility term. If τ is used to denote this artificial compressibility and β is used to represent its inverse, then the incompressible Navier-Stokes equations are given by

$$\frac{\partial Q}{\partial t} + \frac{\partial E}{\partial x} + \frac{\partial F}{\partial y} + \frac{\partial G}{\partial z} = \frac{\partial E_v}{\partial x} + \frac{\partial F_v}{\partial y} + \frac{\partial G_v}{\partial z} \tag{5.231}$$

where

$$Q = \begin{bmatrix} p \\ u \\ v \\ w \end{bmatrix} \tag{5.232}$$

$$E = \begin{bmatrix} \beta u \\ u^2 + p \\ uv \\ uw \end{bmatrix} \tag{5.233} \qquad E_v = \begin{bmatrix} 0 \\ \tau_{xx} \\ \tau_{xy} \\ \tau_{xz} \end{bmatrix} \tag{5.234}$$

$$F = \begin{bmatrix} \beta v \\ vu \\ v^2 + p \\ vw \end{bmatrix} \tag{5.235} \qquad F_v = \begin{bmatrix} 0 \\ \tau_{yx} \\ \tau_{yy} \\ \tau_{yz} \end{bmatrix} \tag{5.236}$$

$$G = \begin{bmatrix} \beta w \\ wu \\ wv \\ w^2 + p \end{bmatrix} \tag{5.237} \qquad G_v = \begin{bmatrix} 0 \\ \tau_{zx} \\ \tau_{zy} \\ \tau_{zz} \end{bmatrix} \tag{5.238}$$

and

$$\tau_{xx} = 2\nu \frac{\partial u}{\partial x}$$

$$\tau_{xy} = \tau_{yx} = \nu \left(\frac{\partial u}{\partial y} + \frac{\partial v}{\partial x} \right)$$

$$\tau_{yy} = 2\nu \frac{\partial u}{\partial y}$$

$$\tau_{xz} = \tau_{zx} = \nu \left(\frac{\partial w}{\partial x} + \frac{\partial u}{\partial z} \right)$$

$$\tau_{zz} = 2\nu \frac{\partial w}{\partial z}$$

$$\tau_{yz} = \tau_{zy} = \nu \left(\frac{\partial w}{\partial y} + \frac{\partial v}{\partial z} \right)$$

Equation (5.231) may be expressed in a nondimensional form if the variables are nondimensionalized according to nondimensional terms defined previously in Section 5.3. The nondimensional form of the incompressible Navier-Stokes equation in a flux vector form is:

$$\frac{\partial Q^*}{\partial t^*} + \frac{\partial E^*}{\partial x^*} + \frac{\partial F^*}{\partial y^*} + \frac{\partial G^*}{\partial z^*} = \frac{\partial E_v^*}{\partial x^*} + \frac{\partial F_v^*}{\partial y^*} + \frac{\partial G_v^*}{\partial z^*} \tag{5.239}$$

where

$$Q^* = \begin{bmatrix} p^* \\ u^* \\ v^* \\ w^* \end{bmatrix}$$

$$E^* = \begin{bmatrix} \beta^* u^* \\ u^{*2} + p^* \\ u^* v^* \\ u^* w^* \end{bmatrix} \tag{5.240} \qquad E_v^* = \begin{bmatrix} 0 \\ \tau_{xx}^* \\ \tau_{xy}^* \\ \tau_{xz}^* \end{bmatrix} \tag{5.241}$$

$$F^* = \begin{bmatrix} \beta^* v^* \\ v^* u^* \\ v^* + p^* \\ v^* w^* \end{bmatrix} \tag{5.242} \qquad F_v^* = \begin{bmatrix} 0 \\ \tau_{yx}^* \\ \tau_{yy}^* \\ \tau_{yz}^* \end{bmatrix} \tag{5.243}$$

$$G^* = \begin{bmatrix} \beta^* w^* \\ w^* u^* \\ w^* v^* \\ w^{*2} + p^* \end{bmatrix} \tag{5.244} \qquad G_v^* = \begin{bmatrix} 0 \\ \tau_{zx}^* \\ \tau_{zy}^* \\ \tau_{zz}^* \end{bmatrix} \tag{5.245}$$

where

$$\tau_{xx}^* = \frac{2}{Re_\infty} \frac{\partial u^*}{\partial x^*} \tag{5.246}$$

$$\tau_{xy}^* = \tau_{yx}^* = \frac{1}{Re_\infty}\left(\frac{\partial u^*}{\partial y^*} + \frac{\partial v^*}{\partial x^*}\right) \tag{5.247}$$

$$\tau_{yy}^* = \frac{2}{Re_\infty}\frac{\partial v^*}{\partial y^*} \tag{5.248}$$

$$\tau_{xz}^* = \tau_{zx}^* = \frac{1}{Re_\infty}\left(\frac{\partial w^*}{\partial x^*} + \frac{\partial u^*}{\partial z^*}\right) \tag{5.249}$$

$$\tau_{zz}^* = \frac{2}{Re_\infty}\frac{\partial w^*}{\partial z^*} \tag{5.250}$$

$$\tau_{yz}^* = \tau_{zy}^* = \frac{1}{Re_\infty}\left(\frac{\partial w^*}{\partial y^*} + \frac{\partial v^*}{\partial z^*}\right) \tag{5.251}$$

and

$$Re_\infty = \frac{\rho_\infty u_\infty L}{\mu_\infty}$$

The asterisk which is used to denote the nondimensional quantities will be dropped for convenience. Therefore, all the expressions to follow are in nondimensional form unless specified otherwise. Note that the flux vector formulations in either dimensional or nondimensional forms are identical. Therefore, the transformed formulation applies to either one. However, appropriate nondimensional terms and expressions for shear stresses must be utilized.

Following the procedure of Section 5.3, the nondimensional incompressible Navier-Stokes equations in the computational space are expressed as

$$\frac{\partial \bar{Q}}{\partial t} + \frac{\partial \bar{E}}{\partial \xi} + \frac{\partial \bar{F}}{\partial \eta} + \frac{\partial \bar{G}}{\partial \zeta} = \frac{\partial \bar{E}_v}{\partial \xi} + \frac{\partial \bar{F}_v}{\partial \eta} + \frac{\partial \bar{G}_v}{\partial \zeta} \tag{5.252}$$

where

$$\bar{Q} = \frac{Q}{J}$$

and

$$\bar{E} = \frac{1}{J}\left(\xi_x E + \xi_y F + \xi_z G\right) \tag{5.253}$$

$$\bar{F} = \frac{1}{J}\left(\eta_x E + \eta_y F + \eta_z G\right) \tag{5.254}$$

$$\bar{G} = \frac{1}{J}\left(\zeta_x E + \zeta_y F + \zeta_z G\right) \tag{5.255}$$

$$\bar{E}_v = \frac{1}{J}\left(\xi_x E_v + \xi_y F_v + \xi_z G_v\right) \tag{5.256}$$

$$\bar{F}_v = \frac{1}{J}\left(\eta_x E_v + \eta_y F_v + \eta_z G_v\right) \tag{5.257}$$

$$\bar{G}_v = \frac{1}{J}\left(\zeta_x E_v + \zeta_y F_v + \zeta_z G_v\right) \tag{5.258}$$

The shear stresses given by Equations (5.246) through (5.251) expressed in the computational space are as follow:

$$\tau_{xx} = \frac{2}{Re_\infty}\left(\xi_x u_\xi + \eta_x u_\eta + \zeta_x u_\zeta\right) \tag{5.259}$$

$$\tau_{yy} = \frac{2}{Re_\infty}\left(\xi_y v_\xi + \eta_y v_\eta + \zeta_x v_\zeta\right) \tag{5.260}$$

$$\tau_{zz} = \frac{2}{Re_\infty}\left(\xi_z w_\xi + \eta_z w_\eta + \zeta_z w_\zeta\right) \tag{5.261}$$

$$\tau_{xy} = \tau_{yx} = \frac{1}{Re_\infty}\left(\xi_y u_\xi + \eta_z u_\eta + \zeta_y u_\zeta + \xi_x v_\xi + \eta_x v_\eta + \zeta_x v_\zeta\right) \tag{5.262}$$

$$\tau_{xz} = \tau_{zx} = \frac{1}{Re_\infty}\left(\xi_z u_\xi + \eta_z u_\eta + \zeta_z u_\zeta + \xi_x w_\xi + \eta_x w_\eta + \zeta_x w_\zeta\right) \tag{5.263}$$

$$\tau_{yz} = \tau_{zy} = \frac{1}{Re_\infty}\left(\xi_y w_\xi + \eta_y w_\eta + \zeta_y w_\zeta + \xi_z v_\xi + \eta_y v_\eta + \zeta_z v_\zeta\right) \tag{5.264}$$

The expressions for the shear stresses given by Equations (5.259) through (5.264) are substituted into the viscous flux vectors $\bar{E}_v$, $\bar{F}_v$, and $\bar{G}_v$ to provide the following:

$$\bar{E}_v = \frac{1}{JRe_\infty}\begin{bmatrix} 0 \\ a_1 u_\xi + b_1 u_\eta - c_1 v_\eta + c_2 w_\eta + b_2 u_\zeta - c_4 v_\zeta + c_5 w_\zeta \\ a_1 v_\xi + c_1 u_\eta + b_1 v_\eta - c_3 w_\eta + c_4 u_\zeta + b_2 v_\zeta - c_6 w_\zeta \\ a_1 w_\xi - c_2 u_\eta + c_3 v_\eta + b_1 w_\eta - c_5 u_\zeta + c_6 v_\zeta + b_2 w_\zeta \end{bmatrix} \tag{5.265}$$

$$\bar{F}_v = \frac{1}{JRe_\infty}\begin{bmatrix} 0 \\ a_2 u_\eta + b_1 u_\xi + c_1 v_\xi - c_2 w_\xi + b_2 u_\zeta - c_7 v_\zeta + c_8 w_\zeta \\ a_2 v_\eta - c_1 u_\xi + b_1 v_\xi + c_3 w_\xi + c_7 u_\zeta + b_3 v_\zeta - c_9 w_\zeta \\ a_2 w_\eta + c_2 u_\xi - c_3 v_\xi + b_1 w_\xi - c_8 u_\zeta + c_9 v_\zeta + b_3 w_\zeta \end{bmatrix} \tag{5.266}$$

$$\bar{G}_v = \frac{1}{JRe_\infty}\begin{bmatrix} 0 \\ a_3 u_\zeta + b_2 u_\xi + c_4 v_\xi - c_5 w_\xi + b_3 u_\eta + c_7 v_\eta - c_8 w_\eta \\ a_3 v_\zeta - c_4 u_\xi + b_2 v_\xi + c_6 w_\xi - c_7 u_\eta + b_3 v_\eta + c_9 w_\eta \\ a_3 w_\zeta + c_5 u_\xi - c_6 v_\xi + b_2 w_\xi + c_8 u_\eta - c_9 v_\eta + b_3 w_\eta \end{bmatrix} \tag{5.267}$$

where

$$a_1 = \xi_x^2 + \xi_y^2 + \xi_z^2 \tag{5.268}$$

$$a_2 = \eta_x^2 + \eta_y^2 + \eta_z^2 \tag{5.269}$$

$$a_3 = \zeta_x^2 + \zeta_y^2 + \zeta_z^2 \tag{5.270}$$

$$b_1 = \xi_x \eta_x + \xi_y \eta_y + \xi_z \eta_z \tag{5.271}$$

$$b_2 = \xi_x \zeta_x + \xi_y \zeta_y + \xi_z \zeta_z \tag{5.272}$$

$$b_3 = \zeta_x \eta_x + \zeta_y \eta_y + \zeta_z \eta_z \tag{5.273}$$

$$c_1 = \xi_x \eta_y - \eta_x \xi_y \quad (5.274) \qquad c_2 = \eta_x \xi_z - \xi_x \eta_z \quad (5.275) \qquad c_3 = \xi_y \eta_z - \eta_y \xi_z \quad (5.276)$$

$$c_4 = \xi_x \zeta_y - \zeta_x \xi_y \quad (5.277) \qquad c_5 = \zeta_x \xi_z - \xi_x \zeta_z \quad (5.278) \qquad c_6 = \xi_y \zeta_z - \zeta_y \xi_z \quad (5.279)$$

$$c_7 = \eta_x \zeta_y - \zeta_x \eta_y \quad (5.280) \qquad c_8 = \zeta_x \eta_z - \eta_x \zeta_z \quad (5.281) \qquad c_9 = \eta_y \zeta_z - \zeta_y \eta_z \quad (5.282)$$

5.7.1 Inviscid and Viscous Jacobian Matrices

The inviscid Jacobian matrices are determined according to

$$A = \frac{\partial \bar{E}}{\partial \bar{Q}}$$

$$B = \frac{\partial \bar{F}}{\partial \bar{Q}}$$

and

$$C = \frac{\partial \bar{G}}{\partial \bar{Q}}$$

Using the contravariant velocity components defined by

$$U = \xi_x u + \xi_y v + \xi_z w \tag{5.283}$$

$$V = \eta_x u + \eta_y v + \eta_z w \tag{5.284}$$

and

$$W = \zeta_x u + \zeta_y v + \zeta_z w \tag{5.285}$$

the resulting inviscid Jacobian matrices are:

$$A = \begin{bmatrix} 0 & \xi_x \beta & \xi_y \beta & \xi_z \beta \\ \xi_x & U + \xi_x u & \xi_y u & \xi_z u \\ \xi_y & \xi_x v & U + \xi_y v & \xi_z v \\ \xi_z & \xi_x w & \xi_y w & U + \xi_z w \end{bmatrix} \tag{5.286}$$

$$
B = \begin{bmatrix}
0 & \eta_x\beta & \eta_y\beta & \eta_z\beta \\
\eta_x & V+\eta_x u & \eta_y u & \eta_z u \\
\eta_y & \eta_x v & V+\eta_y v & \eta_z v \\
\eta_z & \eta_x w & \eta_y w & V+\eta_z w
\end{bmatrix}
\tag{5.287}
$$

$$
C = \begin{bmatrix}
0 & \zeta_x\beta & \zeta_y\beta & \zeta_z\beta \\
\zeta_x & W+\zeta_x u & \zeta_y u & \zeta_z u \\
\zeta_y & \zeta_x v & W+\zeta_y v & \zeta_z v \\
\zeta_z & \zeta_x w & \zeta_y w & W+\zeta_z w
\end{bmatrix}
\tag{5.288}
$$

Following the procedure outlined in Section 5.3.2 and given by the general expression (5.134), the viscous Jacobian matrices are determined and are provided by the following:

$$
A_v = \frac{\partial \bar{E}_v}{\partial \bar{Q}} = \frac{1}{Re_\infty J}
\begin{bmatrix}
0 & 0 & 0 & 0 \\
0 & A_1 & -B_1 & C_1 \\
0 & B_1 & A_1 & -D_1 \\
0 & -C_1 & D_1 & A_1
\end{bmatrix}
\tag{5.289}
$$

where

$$A_1 = a_1(J)_\xi + b_1(J)_\eta + b_2(J)_\zeta \tag{5.290}$$

$$B_1 = c_1(J)_\eta + c_4(J)_\zeta \tag{5.291}$$

$$C_1 = c_2(J)_\eta + c_5(J)_\zeta \tag{5.292}$$

$$D_1 = c_3(J)_\eta + c_6(J)_\zeta \tag{5.293}$$

$$
B_v = \frac{\partial \bar{F}_v}{\partial \bar{Q}} = \frac{1}{Re_\infty J}
\begin{bmatrix}
0 & 0 & 0 & 0 \\
0 & A_2 & -B_2 & C_2 \\
0 & B_2 & A_2 & -D_2 \\
0 & -C_2 & D_2 & A_2
\end{bmatrix}
\tag{5.294}
$$

where

$$A_2 = a_2(J)_\eta + b_1(J)_\xi + b_3(J)_\zeta \tag{5.295}$$

$$B_2 = -c_1(J)_\xi + c_7(J)_\zeta \tag{5.296}$$

$$C_2 = -c_2(J)_\xi + c_8(J)_\zeta \tag{5.297}$$

$$D_2 = -c_3(J)_\xi + c_9(J)_\zeta \tag{5.298}$$

$$C_v = \frac{\partial \bar{G}_v}{\partial \bar{Q}} = \frac{1}{Re_\infty J} \begin{bmatrix} 0 & 0 & 0 & 0 \\ 0 & A_3 & -B_3 & C_3 \\ 0 & B_3 & A_3 & -D_3 \\ 0 & -C_3 & D_3 & A_3 \end{bmatrix} \tag{5.299}$$

where

$$A_3 = a_3(J)_\zeta + b_2(J)_\xi + b_3(J)_\eta \tag{5.300}$$

$$B_3 = -c_4(J)_\xi - c_7(J)_\eta \tag{5.301}$$

$$C_3 = -c_5(J)_\xi - c_8(J)_\eta \tag{5.302}$$

$$D_3 = -c_6(J)_\xi - c_9(J)_\eta \tag{5.303}$$

5.7.2 Two-Dimensional Incompressible Navier-Stokes Equations

The nondimensional, incompressible Navier-Stokes equations in the computational domain ξ,η formulated in a flux vector form are expressed as

$$\frac{\partial \bar{Q}}{\partial t} + \frac{\partial \bar{E}}{\partial \xi} + \frac{\partial \bar{F}}{\partial \eta} = \frac{\partial \bar{E}_v}{\partial \xi} + \frac{\partial \bar{F}_v}{\partial \eta} \tag{5.304}$$

where the flux vectors are defined by the following:

$$\bar{Q} = \frac{Q}{J} \tag{5.305}$$

$$\bar{E} = \frac{1}{J}(\xi_x E + \xi_y F) \tag{5.306} \qquad\qquad \bar{F} = \frac{1}{J}(\eta_x E + \eta_y F) \tag{5.307}$$

$$\bar{E}_v = \frac{1}{J}(\xi_x E_v + \xi_y F_v) \tag{5.308} \qquad\qquad \bar{F}_v = \frac{1}{J}(\eta_x E_v + \eta_y F_v) \tag{5.309}$$

and

$$Q = \begin{bmatrix} p \\ u \\ v \end{bmatrix} \tag{5.310}$$

$$E = \begin{bmatrix} \beta u \\ u^2 + p \\ uv \end{bmatrix} \tag{5.311} \qquad\qquad F = \begin{bmatrix} \beta v \\ vu \\ v^2 + p \end{bmatrix} \tag{5.312}$$

$$E_v = \begin{bmatrix} 0 \\ \tau_{xx} \\ \tau_{xy} \end{bmatrix} \tag{5.313} \qquad\qquad F_v = \begin{bmatrix} 0 \\ \tau_{yx} \\ \tau_{yy} \end{bmatrix} \tag{5.314}$$

The shear stresses in the physical space are

$$\tau_{xx} = \frac{2}{Re_\infty} \left(\frac{\partial u}{\partial x} \right) \tag{5.315}$$

$$\tau_{xy} = \tau_{yx} = \frac{1}{Re_\infty} \left(\frac{\partial u}{\partial y} + \frac{\partial v}{\partial x} \right) \tag{5.316}$$

$$\tau_{yy} = \frac{2}{Re_\infty} \left(\frac{\partial v}{\partial y} \right) \tag{5.317}$$

The shear stresses in the computational space are

$$\tau_{xx} = \frac{2}{Re_\infty} \left(\xi_x u_\xi + \eta_x u_\eta \right) \tag{5.318}$$

$$\tau_{xy} = \tau_{yx} = \frac{1}{Re_\infty} \left(\xi_y u_\xi + \eta_y u_\eta + \xi_x v_\xi + \eta_x v_\eta \right) \tag{5.319}$$

$$\tau_{yy} = \frac{2}{Re_\infty} \left(\xi_y v_\xi + \eta_y v_\eta \right) \tag{5.320}$$

Upon substitution of shear stresses into the viscous flux vectors $\bar{E}_v$ and $\bar{F}_v$ given by Equations (5.308) and (5.309) and utilizing continuity, one obtains

$$\bar{E}_v = \frac{1}{JRe_\infty} \begin{bmatrix} 0 \\ au_\xi + bu_\eta - cv_\eta \\ av_\xi + cu_\eta + bv_\eta \end{bmatrix} \tag{5.321}$$

and

$$\bar{F}_v = \frac{1}{JRe_\infty} \begin{bmatrix} 0 \\ du_\eta + bu_\xi + cv_\xi \\ dv_\eta - cu_\xi + bv_\xi \end{bmatrix} \tag{5.322}$$

where

$$a = \xi_x^2 + \xi_y^2 \tag{5.323}$$

$$b = \xi_x \eta_x + \xi_y \eta_y \tag{5.324}$$

$$c = \xi_x \eta_y - \xi_y \eta_x \tag{5.325}$$

$$d = \eta_x^2 + \eta_y^2 \tag{5.326}$$

The inviscid and viscous Jacobian matrices are determined to be as follows

$$A = \frac{\partial \bar{E}}{\partial \bar{Q}} = \begin{bmatrix} 0 & \xi_x \beta & \xi_y \beta \\ \xi_x & U + \xi_x u & \xi_y u \\ \xi_y & \xi_x u & U + \xi_y v \end{bmatrix} \tag{5.327}$$

where the contravariant velocity U is

$$U = \xi_x u + \xi_y v \tag{5.328}$$

$$B = \frac{\partial \bar{F}}{\partial \bar{Q}} = \begin{bmatrix} 0 & \eta_x \beta & \eta_y \beta \\ \eta_x & V + \eta_x u & \eta_y u \\ \eta_y & \eta_x v & V + \eta_y v \end{bmatrix} \tag{5.329}$$

The contravariant velocity V is defined as

$$V = \eta_x u + \eta_y v \tag{5.330}$$

$$A_v = \frac{\partial \bar{E}_v}{\partial \bar{Q}} = \frac{1}{JRe_\infty} \begin{bmatrix} 0 & 0 & 0 \\ 0 & a(J)_\xi + b(J)_\eta & -c(J)_\eta \\ 0 & c(J)_\eta & a(J)_\xi + b(J)_\eta \end{bmatrix} \tag{5.331}$$

and

$$B_v = \frac{1}{JRe_\infty} \begin{bmatrix} 0 & 0 & 0 \\ 0 & b(J)_\xi + d(J)_\eta & c(J)_\xi \\ 0 & -c(J)_\xi & b(J)_\xi + d(J)_\eta \end{bmatrix} \tag{5.332}$$

Chapter Six:

Computational Fluid Dynamics

6.1   Introductory Remarks

An essential element in analysis and design procedures is the compilation of relevant data. The approach by which this goal is accomplished can be categorized into three groups. The first and perhaps the simplest is analytical solutions of the relevant equations. Unfortunately, analytical solutions are limited to the simplest of flows, configurations, and the imposed boundary conditions. However, they provide the cornerstone of many investigations. In fact, the solutions obtained by analytical methods are used routinely in the design and validation of numerical solutions as well as in baseline data for comparison with experimental data.

The second category by which information about a flowfield is obtained is by experiment. Typically in this approach, an instrumented model under appropriate flow conditions is used to obtain the required data. There are several difficulties and limitations with the experimental approach, among which are limitations on the hardware, such as the model and tunnel size, appropriate scaling between the model and the prototype, and the difficulty in adequately simulating the prototype flowfield. Furthermore, experimental setup and simulations can be very costly. Nevertheless, the flowfield information from an experiment is extremely valuable in analysis and design as well as in the validation of numerical solutions.

With improvements in computer technology resulting in increased memory, speed, and efficiency, a third category of solution schemes has evolved. This category is referred to as computational fluid dynamics (CFD). In this category, the governing equations are solved by a variety of numerical schemes. Unlike the first two categories, namely, analytical and experimental methods, the geometry and flow conditions in CFD schemes can be easily varied to simulate various design conditions.

In order to numerically solve the governing partial differential equations, the partial derivatives in the equations are approximated by algebraic expressions. Subsequently, the resulting algebraic equations are solved within the domain at specified discrete points.

The CFD approach is also limited in the scope of application. The first limitation is imposed by the hardware, which limits the number of grid points used. However, the ceiling of this limitation is being raised as computer technology advances. The second limitation is due to the numerical scheme itself. Typically, applications to relatively complex problems are subject to stability requirements, and therefore a substantial amount of computer time (and associated cost) may be required. Finally, user friendliness and robustness of computer codes need to be improved. Progress in all of the above-mentioned obstacles is being accomplished, and it is anticipated that the use of CFD in design and analysis will become routine practice in the near future.

Computational schemes are classified into three groups: finite difference schemes, finite volume schemes, and finite element schemes. In this chapter, a summary of finite difference schemes and relevant materials is presented.

6.2   Classification of Partial Differential Equations

Partial differential equations (PDEs) can be classified into different categories, where within each category they may be classified further into subcategories. The numerical procedure used to solve a partial differential equation very much depends on the classification of the governing equation. A brief review of the classification of partial differential equations is provided in the following subsections.

6.2.1 Linear and Nonlinear PDEs

(I) Linear PDE: There is no product of the dependent variable and/or product of its derivatives within the equation.

(II) Nonlinear PDE: The equation contains a product of the dependent variable and/or a product of the derivatives.

6.2.2 Classification Based on Characteristics

(I) First-order PDE: Almost all first-order PDEs have real characteristics, and therefore behave much like hyperbolic equations of second order.

(II) Second-order PDE: A second-order PDE with two independent variables, x and y, may be expressed in a general form as

$$A\frac{\partial^2 \phi}{\partial x^2} + B\frac{\partial^2 \phi}{\partial x \partial y} + C\frac{\partial^2 \phi}{\partial y^2} + D\frac{\partial \phi}{\partial x} + E\frac{\partial \phi}{\partial y} + F\phi + G = 0 \qquad (6.1)$$

The equation is classified according to the expression $(B^2 - 4AC)$ as follows:

$$\left(B^2 - 4AC\right) \begin{cases} < 0 \rightarrow \text{elliptic equation} \\ = 0 \rightarrow \text{parabolic equation} \\ > 0 \rightarrow \text{hyperbolic equation} \end{cases}$$

The following criteria may be stated with regard to each category defined above:

(a) Elliptic equations

- No real characteristic lines exist.
- A disturbance propagates in all directions.
- Domain of solution is a closed region.
- Boundary conditions must be specified on the boundaries of the domain.

(b) Parabolic equations

- Only one characteristic line exists.
- A disturbance propagates along the characteristic line.
- Domain of solution is an open region.
- An initial condition and two boundary conditions are required.

(c) Hyperbolic equations

- Two characteristic lines exist.
- A disturbance propagates along the characteristic lines.
- Domain of solution is an open region.
- Two initial conditions along with two boundary conditions are required.

(III) System of First-Order PDEs

A system of first-order PDEs may be expressed in a vector form as

$$\frac{\partial \boldsymbol{\Phi}}{\partial t} + [A]\frac{\partial \boldsymbol{\Phi}}{\partial x} + [B]\frac{\partial \boldsymbol{\Phi}}{\partial y} + \boldsymbol{\Psi} = 0 \qquad (6.2)$$

where the vector $\boldsymbol{\Phi}$ contains the dependent variables or the unknowns, and the vector $\boldsymbol{\Psi}$ is a function of x, y, and $\boldsymbol{\Phi}$. The system is classified according to the eigenvalues of coefficient matrices $[A]$ and $[B]$. If the eigenvalues of matrix $[A]$ are all real and distinct, the system is classified as hyperbolic in t and x. If the eigenvalues of $[A]$ are complex, the system is elliptic in t and x. Similarly, the system is classified with respect to the independent variables t and y based on the eigenvalues of matrix $[B]$.

For a steady state equivalent of (6.2), given by

$$[A]\frac{\partial \boldsymbol{\Phi}}{\partial x} + [B]\frac{\partial \boldsymbol{\Phi}}{\partial y} + \boldsymbol{\Psi} = 0 \qquad (6.3)$$

the classification is as follows:

$$H \begin{cases} < 0 \rightarrow \text{elliptic} \\ = 0 \rightarrow \text{parabolic} \\ > 0 \rightarrow \text{hyperbolic} \end{cases}$$

where

$$H = R^2 - 4PQ$$

and

$$P = |A|, \quad Q = |B|$$

For a system composed of two equations, R is given by

$$R = \begin{vmatrix} a_1 & a_4 \\ b_1 & b_4 \end{vmatrix} + \begin{vmatrix} a_3 & a_2 \\ b_3 & b_2 \end{vmatrix}$$

where

$$[A] = \begin{bmatrix} a_1 & a_2 \\ b_1 & b_2 \end{bmatrix} \quad \text{and} \quad [B] = \begin{bmatrix} u_3 & u_4 \\ b_3 & b_4 \end{bmatrix}$$

(IV) System of Second-Order PDEs

The classification of a system of second-order PDEs is facilitated if the second-order PDEs are reduced to their equivalent first-order PDEs. Subsequently, the system of first-order PDEs is classified, but the procedure could be cumbersome. For specific details and examples, References [6.1] and [6.2] should be consulted.

6.3   Boundary Conditions

A set of specific information with regard to the dependent variable and/or its derivative must be specified along the boundaries of the domain. This set of information is known as the *boundary condition* and may be categorized as follows.

(I) The Dirichlet boundary condition: The value of the dependent variable along the boundary is specified.

(II) The Neumann boundary condition: The normal gradient of the dependent variable along the boundary is specified.

(III) The mixed boundary condition: A combination of the Dirichlet and the Neumann type boundary conditions is specified.

6.4   Finite Difference Expressions

The partial derivatives appearing in PDEs must be approximated by algebraic expressions in order to numerically solve the resulting algebraic equations. A method often used to obtain these expressions is by Taylor series expansion. Various degrees of accuracy can be achieved depending on how many terms in the series are retained, that is, the series is truncated beyond certain terms. Hence, the error associated with dropping of higher order terms in the series is referred to as *truncation error.*

It should be noted that computers carry up to a certain number of digits in computations (e.g., 16 or 32) and that numbers are rounded off beyond the specified digit. This approximation of numbers introduces errors known as *round-off errors,* which are machine-dependent.

Finite difference expressions for forward, backward, and central difference approximations of first- and higher-order derivatives (up to fourth-order), with truncation errors of (Δx) and $(\Delta x)^2$ for the forward and backward differencing, and with truncation errors of $(\Delta x)^2$ and $(\Delta x)^4$ for central differencing, are provided in Tables 6.1 through 6.6.

For example, a second-order forward difference approximation of $\frac{\partial^2 f}{\partial x^2}$ is obtained from Table 6.2 as follows:

$$(\Delta x)^2 \frac{\partial^2 f}{\partial x^2} = 2f_i - 5f_{i+1} + 4f_{i+2} - f_{i+3}$$

or

$$\frac{\partial^2 f}{\partial x^2} = \frac{2f_i - 5f_{i+1} + 4f_{i+2} - f_{i+3}}{(\Delta x)^2} + O(\Delta x)^2$$

	f_i	f_{i+1}	f_{i+2}	f_{i+3}	f_{i+4}
$(\Delta x)\dfrac{\partial f}{\partial x}$	-1	1			
$(\Delta x)^2\dfrac{\partial^2 f}{\partial x^2}$	1	-2	1		
$(\Delta x)^3\dfrac{\partial^3 f}{\partial x^3}$	-1	3	-3	1	
$(\Delta x)^4\dfrac{\partial^4 f}{\partial x^4}$	1	-4	6	-4	1

Table 6.1 Forward difference representations of $O(\Delta x)$.

	f_i	f_{i+1}	f_{i+2}	f_{i+3}	f_{i+4}	f_{i+5}
$2(\Delta x)\dfrac{\partial f}{\partial x}$	-3	4	-1			
$(\Delta x)^2\dfrac{\partial^2 f}{\partial x^2}$	2	-5	4	-1		
$2(\Delta x)^3\dfrac{\partial^3 f}{\partial x^3}$	-5	18	-24	14	-3	
$(\Delta x)^4\dfrac{\partial^4 f}{\partial x^4}$	3	-14	26	-24	11	-2

Table 6.2 Forward difference representations of $O(\Delta x)^2$.

	f_{i-4}	f_{i-3}	f_{i-2}	f_{i-1}	f_i
$(\Delta x)\dfrac{\partial f}{\partial x}$				-1	1
$(\Delta x)^2\dfrac{\partial^2 f}{\partial x^2}$			1	-2	1
$(\Delta x)^3\dfrac{\partial^3 f}{\partial x^3}$		-1	3	-3	1
$(\Delta x)^4\dfrac{\partial^4 f}{\partial x^4}$	1	-4	6	-4	1

Table 6.3 Backward difference representations of $O(\Delta x)$.

	f_{i-5}	f_{i-4}	f_{i-3}	f_{i-2}	f_{i-1}	f_i
$2(\Delta x)\dfrac{\partial f}{\partial x}$				1	-4	3
$(\Delta x)^2\dfrac{\partial^2 f}{\partial x^2}$			-1	4	-5	2
$2(\Delta x)^3\dfrac{\partial^3 f}{\partial x^3}$		3	-14	24	-18	5
$(\Delta x)^4\dfrac{\partial^4 f}{\partial x^4}$	-2	11	-24	26	-14	3

Table 6.4 Backward difference representations of $O(\Delta x)^2$.

	f_{i-2}	f_{i-1}	f_i	f_{i+1}	f_{i+2}
$2(\Delta x)\dfrac{\partial f}{\partial x}$		-1	0	1	
$(\Delta x)^2\dfrac{\partial^2 f}{\partial x^2}$		1	-2	1	
$2(\Delta x)^3\dfrac{\partial^3 f}{\partial x^3}$	-1	2	0	-2	1
$(\Delta x)^4\dfrac{\partial^4 f}{\partial x^4}$	1	-4	6	-4	1

Table 6.5 Central difference representations of $O(\Delta x)^2$.

	f_{i-3}	f_{i-2}	f_{i-1}	f_i	f_{i+1}	f_{i+2}	f_{i+3}
$12(\Delta x)\dfrac{\partial f}{\partial x}$		1	-8	0	8	-1	
$12(\Delta x)^2\dfrac{\partial^2 f}{\partial x^2}$		-1	16	-30	16	-1	
$8(\Delta x)^3\dfrac{\partial^3 f}{\partial x^3}$	1	-8	13	0	-13	8	-1
$6(\Delta x)^4\dfrac{\partial^4 f}{\partial x^4}$	-1	12	-39	56	-39	12	-1

Table 6.6 Central difference representations of $O(\Delta x)^4$.

6.5   Finite Difference Equations

The partial derivatives appearing in the differential equations are replaced by approximate algebraic expressions to provide an equivalent algebraic equation known as the *finite difference equation*. Subsequently, the finite difference equation is solved within a domain which has been discretized into an equally spaced grid. Finite difference equations commonly used for the solution of parabolic, elliptic, and hyperbolic equations are reviewed in this section.

6.5.1   Explicit and Implicit Formulations

The finite difference equation (FDE) of a PDE may be expressed in such a way that only one unknown appears in the equation. This type of formulation is known as an *explicit formulation*. On the other hand, if the formulation is written such that more than one unknown appears in the FDE, then the formulation is called *implicit*. Any implicit formulation results in a system of equations which must be solved simultaneously. Typically, explicit formulations are subject to some stability requirements; whereas, most implicit formulations are unconditionally stable or have less stringent stability requirements.

6.5.2   Modified Equation

In order to determine the dominant error term of a finite difference equation, Taylor series expansions are substituted back into the finite difference equation and are rearranged. The resulting equation is known as *the modified equation*.

6.5.3   Parabolic Equations

Various finite difference formulations are reviewed for the one-dimensional parabolic equations initially and are subsequently extended to multi-dimensional problems.

6.5.3.1   One-space dimension: The simple diffusion equation is used in this section to review various finite difference equations. The model equation is given by

$$\frac{\partial u}{\partial t} = \alpha \frac{\partial^2 u}{\partial x^2} \tag{6.4}$$

where α is assumed to be a constant and hence a linear equation. To facilitate the review process, various aspects of each finite difference formulation such as the order of accuracy, amplification factor, stability requirement, and the corresponding modified equation are summarized. In the formulations to follow, the diffusion number is designated by d, which is

$$d = \alpha \frac{\Delta t}{(\Delta x)^2} \qquad (6.5)$$

and θ is the phase angle.

Scheme:	Forward Time–Central Space (FTCS) explicit
Formulation:	$u_i^{n+1} = u_i^n + d\left(u_{i+1}^n - 2u_i^n + u_{i-1}^n\right)$
Order:	$O\left[(\Delta t),\ (\Delta x)^2\right]$
Amplification Factor:	$G = 1 + 2d(\cos\theta - 1)$
Stability Requirement:	$d \leq \dfrac{1}{2}$
Modified Equation:	$\dfrac{\partial u}{\partial t} = \alpha\dfrac{\partial^2 u}{\partial x^2} + \left[-\dfrac{1}{2}\alpha^2(\Delta t) + \dfrac{1}{12}\alpha(\Delta x)^2\right]\dfrac{\partial^4 u}{\partial x^4}$
	$\quad + \left[\dfrac{1}{3}\alpha^3(\Delta t)^2\dfrac{\partial^4 u}{\partial x^4} - \dfrac{1}{12}\alpha^2(\Delta t)(\Delta x)^2\right.$
	$\quad \left. + \dfrac{1}{360}\alpha(\Delta x)^2\right]\dfrac{\partial^6 u}{\partial x^6} + \ldots$
Special Considerations:	None

Scheme:	DuFort-Frankel explicit
Formulation:	$u_i^{n+1} = \dfrac{1-2d}{1+2d}u_i^{n-1} + \dfrac{2d}{1+2d}\left(u_{i+1}^n + u_{i-1}^n\right)$
Order:	$O\left[(\Delta t)^2,\ (\Delta x)^2,\ \left(\dfrac{\Delta t}{\Delta x}\right)^2\right]$
Amplification Factor:	$G = \dfrac{1}{1+2d}\left[2d\,\cos\theta \pm (1-4d^2\sin^2\theta)^{\frac{1}{2}}\right]$
Stability Requirement:	Unconditionally stable
Modified Equation:	$\dfrac{\partial u}{\partial x} = \alpha\dfrac{\partial^2 u}{\partial x^2} + \left[\dfrac{1}{12}\alpha(\Delta x^2) - \alpha^3\left(\dfrac{\Delta t}{\Delta x}\right)^2\right]\dfrac{\partial^4 u}{\partial x^4}$
	$\qquad + \left[-\dfrac{1}{3}\alpha^3(\Delta t)^2 + \dfrac{1}{360}\alpha(\Delta x)^4\right.$
	$\qquad \left. + 2\alpha^5\left(\dfrac{\Delta t}{\Delta x}\right)^4\right]\dfrac{\partial^6 u}{\partial x^6} + \cdots$
Special Considerations:	Requires two sets of data to proceed, i.e., data at time levels $n-1$ and n

Scheme:	Laasonen implicit
Formulation:	$du_{i-1}^{n+1} - (1+2d)u_i^{n+1} + du_{i-1}^n = u_i^n$
Order:	$O\left[(\Delta t),\ (\Delta x)^2\right]$
Amplification Factor:	$G = \dfrac{1}{1+2d(1-\cos\theta)}$
Stability Requirement:	Unconditionally stable
Modified Equation:	$\dfrac{\partial u}{\partial t} = \alpha\dfrac{\partial^2 u}{\partial x^2} + \left[\dfrac{1}{2}\alpha^2(\Delta t) + \dfrac{1}{12}\alpha(\Delta x)^2\right]\dfrac{\partial^4 u}{\partial x^4}$
	$\qquad + \left[\dfrac{1}{3}\alpha^3(\Delta t)^2 + \dfrac{1}{12}\alpha^2(\Delta t)(\Delta x)^2\right.$
	$\qquad \left. + \dfrac{1}{360}\alpha(\Delta x)^4\right]\dfrac{\partial^6 u}{\partial x^6} + \cdots$
Special Considerations:	Requires solution of a tridiagonal system

Scheme:	Crank-Nicolson implicit
Formulation:	$\frac{1}{2}du_{i+1}^{n+1} - (1+d)u_i^{n+1} + \frac{1}{2}du_{i-1}^{n+1} = -\frac{1}{2}u_{i+1}^n$
	$+ (d-1)u_i^n - \frac{1}{2}du_{i-1}^n$
Order:	$O\left[(\Delta t)^2,\ (\Delta x)^2\right]$
Amplification Factor:	$G = \dfrac{1 - d(1 - \cos\theta)}{1 + d(1 - \cos\theta)}$
Stability Requirement:	Unconditionally stable
Modified Equation:	$\dfrac{\partial u}{\partial x} = \alpha\dfrac{\partial^2 u}{\partial x^2} + \left[\dfrac{1}{12}\alpha(\Delta x)^2\right]\dfrac{\partial^4 u}{\partial x^4}$
	$+ \left[\dfrac{1}{12}\alpha^3(\Delta t)^2 + \dfrac{1}{360}\alpha(\Delta x)^4\right]\dfrac{\partial^6 u}{\partial x^6} + \ldots$
Special Considerations:	Requires solution of a tridiagonal system

6.5.3.2 *Multi-space dimensions:*

6.5.3.2 Multi-space dimensions: The review of multi-dimensional problems will be limited to two-space dimensions. The procedure to three-space dimensions is similar. However, it is cautioned that the extension may not be trivial, and certain formulations which may have been unconditionally stable in two-space dimensions may become only conditionally stable in three-space dimensions. The model equation used is the diffusion equation in two-space dimensions given by

$$\frac{\partial u}{\partial t} = \alpha\left(\frac{\partial^2 u}{\partial x^2} + \frac{\partial^2 u}{\partial y^2}\right) \tag{6.6}$$

Scheme:	FTCS explicit
Formulation:	$u_{i,j}^{n+1} = u_{i,j}^n + d_x(u_{i+1,j}^n - 2u_{i,j}^n + u_{i-1,j}^n) + d_y\left(u_{i,j+1}^n\right.$
	$\left. -2u_{i,j}^n + u_{i,j-1}^n\right)$
Order:	$O\left[(\Delta t),\ (\Delta x)^2,\ (\Delta y)^2\right]$
Stability Requirement:	$(d_x + d_y) \leq \dfrac{1}{2}$

Scheme: ADI

Formulation:

$$-\left(\frac{1}{2}d_x\right) u_{i-1,j}^{n+\frac{1}{2}} + (1+d_x)u_{i,j}^{n+\frac{1}{2}} - \left(\frac{1}{2}d_x\right) u_{i+1,j}^{n+\frac{1}{2}}$$

$$= \left(\frac{1}{2}d_y\right) u_{i,j+1}^{n} + (1-d_y)u_{i,j}^{n} + \left(\frac{1}{2}d_y\right) u_{i,j-1}^{n}$$

and

$$-\left(\frac{1}{2}d_y\right) u_{i,j-1}^{n+1} + (1+d_y)u_{i,j}^{n+1} - \left(\frac{1}{2}d_y\right) u_{i,j+1}^{n+1}$$

$$= \left(\frac{1}{2}d_x\right) u_{i+1,j}^{n+\frac{1}{2}} + (1-d_x)u_{i,j}^{n+\frac{1}{2}} + \left(\frac{1}{2}d_x\right) u_{i-1,j}^{n+\frac{1}{2}}$$

Order: $O\left[(\Delta t)^2,\ (\Delta x)^2,\ (\Delta y)^2\right]$

Amplification Factor: $G = \dfrac{[1 - d_x(1 - \cos\theta_x)]\,[1 - d_y(1 - \cos\theta_y)]}{[1 + d_x(1 - \cos\theta_x)]\,[1 + d_y(1 - \cos\theta_y)]}$

Stability Requirement: Unconditionally stable

Scheme: Fractional step

Formulation:

$$d_x u_{i+1,j}^{n+\frac{1}{2}} - (1 + 2d_x)u_{i,j}^{n+\frac{1}{2}} + d_x u_{i-1,j}^{n+\frac{1}{2}}$$

$$= -d_x u_{i+1,j}^{n} + (2d_x - 1)u_{i,j}^{n} - d_x u_{i-1,j}^{n}$$

and

$$d_y u_{i,j+1}^{n+1} - (1 + 2d_y)u_{i,j}^{n+1} + d_y u_{i,j-1}^{n+1}$$

$$= -d_y u_{i,j+1}^{n+\frac{1}{2}} + (2d_y - 1)u_{i,j}^{n+\frac{1}{2}} - d_y u_{i,j-1}^{n+\frac{1}{2}}$$

Order: $O\left[(\Delta t)^2,\ (\Delta x)^2,\ (\Delta y)^2\right]$

Stability Requirement: Unconditionally stable

6.5.4 Elliptic Equations

The model elliptic equation utilized to represent various finite difference formulations is Laplace's equation expressed in two-space dimensions given by

$$\frac{\partial^2 u}{\partial x^2} + \frac{\partial^2 u}{\partial y^2} = 0 \qquad (6.7)$$

Only iterative schemes which are usually the most efficient schemes to solve elliptic equations are reviewed in this section. In the formulations to follow, the ratio of stepsizes is designated by β, that is, $\beta = \dfrac{\Delta x}{\Delta y}$.

Scheme: Point Gauss-Seidel (PGS)

Formulation: $u_{i,j}^{k+1} = \dfrac{1}{2(1+\beta^2)} \left[u_{i+1,j}^{k} + u_{i-1,j}^{k+1} + \beta^2 \left(u_{i,j+1}^{k} + u_{i,j-1}^{k+1} \right) \right]$

Order: $O\left[(\Delta x)^2 , (\Delta y)^2 \right]$

Scheme: Line Gauss-Seidel (LGS)

Formulation:
(x direction) $u_{i-1,j}^{k+1} - 2(1+\beta^2) u_{i,j}^{k+1} + u_{i+1,j}^{k+1} = -\beta^2 u_{i,j+1}^{k} - \beta^2 u_{i,j-1}^{k+1}$

Order: $O\left[(\Delta x)^2 , (\Delta y)^2 \right]$

Modified Equation: $\dfrac{\partial^2 u}{\partial x^2} + \dfrac{\partial^2 u}{\partial y^2} = -\dfrac{1}{12}(\Delta x)^2 \dfrac{\partial^4 u}{\partial x^4} - \dfrac{1}{12}(\Delta y)^2 \dfrac{\partial^4 u}{\partial y^4} + \cdots$

Scheme: Point Successive Over-Relaxation (PSOR)

Formulation: $u_{i,j}^{k+1} = (1-\omega) u_{i,j}^{k} + \dfrac{\omega}{2(1+\beta^2)} \left[u_{i+1,j}^{k} + u_{i-1,j}^{k+1} \right.$

$\qquad \left. + \beta^2 \left(u_{i,j+1}^{k} + u_{i,j-1}^{k+1} \right) \right]$

Order: $O\left[(\Delta x)^2 , (\Delta y)^2 \right]$

Special Considerations: The range of relaxation parameter is $1 \leq \omega < 2$

Scheme: Line Successive Over-Relaxation (LSOR)

Formulation:
(x direction)

$$\omega u_{i-1,j}^{k+1} - 2\left(1 + \beta^2\right) u_{i,j}^{k+1} + \omega u_{i+1,j}^{k+1}$$

$$= -(1 - \omega)\left[2(1 + \beta^2)\right] u_{i,j}^k - \omega \beta^2 \left(u_{i,j+1}^k + u_{i,j-1}^{k+1}\right)$$

Special Considerations: The range of relaxation parameter is $1 \leq \omega < 2$

Scheme: ADI

Formulation: $u_{i-1,j}^{k+\frac{1}{2}} - 2(1 + \beta^2)u_{i,j}^{k+\frac{1}{2}} + u_{i+1,j}^{k+\frac{1}{2}} = -\beta^2 \left(u_{i,j+1}^k + u_{i,j-1}^{k+\frac{1}{2}}\right)$

and $\beta^2 u_{i,j-1}^{k+1} - 2(1 + \beta^2)u_{i,j}^{k+1} + \beta^2 u_{i,j+1}^{k+1} = -u_{i+1,j}^{k+\frac{1}{2}} - u_{i-1,j}^{k+1}$

Scheme: ADI with relaxation parameter

Formulation: $\omega\, u_{i-1,j}^{k+\frac{1}{2}} - 2(1 + \beta^2)u_{i,j}^{k+\frac{1}{2}} + \omega\, u_{i+1,j}^{k+\frac{1}{2}}$

$$= -(1 - \omega)\left[2\left(1 + \beta^2\right)\right] u_{i,j}^k - \omega\,\beta^2 \left(u_{i,j+1}^k + u_{i,j-1}^{k+\frac{1}{2}}\right)$$

and $\omega\,\beta^2 u_{i,j-1}^{k+1} - 2(1 + \beta^2)u_{i,j}^{k+1} + \omega\,\beta^2 u_{i,j+1}^{k+1}$

$$= -\left(1 - \omega\right)\left[2\left(1 + \beta^2\right)\right] u_{i,j}^{k+\frac{1}{2}} - \omega \left(u_{i+1,j}^{k+\frac{1}{2}} + u_{i-1,j}^{k+1}\right)$$

6.5.5 Hyperbolic Equations

Investigation of various finite difference equations is easily accomplished with regard to linear hyperbolic equations. Subsequently, the conclusions may be extended to nonlinear hyperbolic equations. With that in mind, the review of the formulations is performed sequentially in two parts.

6.5.5.1 *Linear equations:* The wave equation given by

$$\frac{\partial u}{\partial t} = -a\frac{\partial u}{\partial x}, \quad a > 0$$

is used to review various numerical schemes for the solution of linear hyperbolic equations. Note that the speed of sound, a in the equation above, is assumed to be a constant, and, hence, the equation is linear. The parameter, $a\frac{\Delta t}{\Delta x}$, defined as the Courant number and designated by c, will be used in the formulations to follow.

Scheme:	First upwind differencing
Formulation:	$u_i^{n+1} = u_i^n - c\left(u_i^n - u_{i-1}^n\right)$
Order:	$O\left[(\Delta t),\ (\Delta x)\right]$
Amplification Factor:	$G = 1 - c(1 - \cos\theta) - i(c\sin\theta)$
Stability Requirement:	$c \leq 1$

$$\frac{\partial u}{\partial t} = -a\frac{\partial u}{\partial x} + \frac{1}{2}a\Delta x(1-c)\frac{\partial^2 u}{\partial x^2}$$

$$-\frac{1}{6}a(\Delta x)^2(2c^2 - 3c + 1)\frac{\partial^3 u}{\partial x^3} + \ldots$$

(Modified Equation)

Scheme:	Lax
Formulation:	$u_i^{n+1} = \frac{1}{2}\left(u_{i+1}^n + u_{i-1}^n\right) - \frac{1}{2}c\left(u_{i+1}^n - u_{i-1}^n\right)$
Order:	$O\left[(\Delta t),\ \frac{(\Delta x^2)}{(\Delta t)}\right]$
Amplification Factor:	$G = \cos\theta - i(c\sin\theta)$
Stability Requirement:	$c \leq 1$

$$\frac{\partial u}{\partial t} = a\frac{\partial u}{\partial x} + \frac{1}{2}a(\Delta x)\left(\frac{1}{c} - c\right)\frac{\partial^2 u}{\partial x^2}$$

$$+\frac{1}{3}a(\Delta x)^2(1 - c^2)\frac{\partial^3 u}{\partial x^3} + \ldots$$

(Modified Equation)

Scheme: Midpoint Leapfrog

Formulation: $u_i^{n+1} = u_i^{n-1} - c\left(u_{i+1}^n - u_{i-1}^n\right)$

Order: $O\left[(\Delta t)^2,\ (\Delta x)^2\right]$

Amplification Factor: $G = \pm\left[1 - c^2 \sin^2\theta\right]^{\frac{1}{2}} - i(c\sin\theta)$

Stability Requirement: $c \leq 1$

Modified Equation: $\dfrac{\partial u}{\partial t} = -a\dfrac{\partial u}{\partial x} - \dfrac{1}{6}a(\Delta x)^2\left(1 - c^2\right)\dfrac{\partial^3 u}{\partial x^3} + \ldots$

Special Considerations: Requires two sets of data for the solution
                                  to proceed

Scheme: Lax-Wendroff

Formulation: $u_i^{n+1} = u_i^n - \dfrac{1}{2}c\left(u_{i+1}^n - u_{i-1}^n\right)$

$\qquad\qquad\qquad + \dfrac{1}{2}c^2\left(u_{i+1}^n - 2u_i^n + u_{i-1}^n\right)$

Order: $O\left[(\Delta t)^2,\ (\Delta x)^2\right]$

Amplification Factor: $G = 1 - c^2\left(1 - \cos\theta\right) - i(c\sin\theta)$

Stability Requirement: $c \leq 1$

Modified Equation: $\dfrac{\partial u}{\partial t} = -a\dfrac{\partial u}{\partial x} - \dfrac{1}{6}a(\Delta x)^2\left(1 - c^2\right)\dfrac{\partial^3 u}{\partial x^3}$

$\qquad\qquad\qquad - \dfrac{1}{8}a(\Delta x)^3 c(1 - c^2)\dfrac{\partial^4 u}{\partial x^4} + \ldots$

Scheme: BTCS implicit

Formulation:
$$\frac{1}{2}c\,u_{i-1}^{n+1} - u_i^{n+1} - \frac{1}{2}c\,u_{i+1}^{n+1} = -u_i^n$$

Order:
$$O\left[(\Delta t)\,,\,(\Delta x)^2\right]$$

Amplification Factor:
$$G = \frac{1 - i(c\sin\theta)}{1 + c^2\left(\sin^2\theta\right)}$$

Stability Requirement: *None*

Modified Equation:
$$\frac{\partial u}{\partial t} = -a\frac{\partial u}{\partial x} + \frac{1}{2}a^2(\Delta t)\frac{\partial^2 u}{\partial x^2}$$
$$- \left[\frac{1}{6}a(\Delta x)^2 + \frac{1}{3}a^3(\Delta t)^2\right]\frac{\partial^3 u}{\partial x^3} + \cdots$$

Scheme: Crank-Nicolson

Formulation:
$$\frac{1}{4}c\,u_{i+1}^{n+1} - u_i^{n+1} - \frac{1}{4}c\,u_{i-1}^{n+1} = u_i^n - \frac{1}{4}c\left(u_{i+1}^n - u_{i-1}^n\right)$$

Order:
$$O\left[(\Delta t)^2\,,\,(\Delta x)^2\right]$$

Amplification Factor:
$$G = \frac{1 - 0.5i(c\sin\theta)}{1 + 0.5i\left(c\sin\theta\right)}$$

Stability Requirement: None

Modified Equation:
$$\frac{\partial u}{\partial t} = -a\frac{\partial u}{\partial x} - \frac{1}{6}a\left(\Delta x\right)^2\left(1 + \frac{1}{2}c^2\right)\frac{\partial^3 u}{\partial x^3}\cdots$$

Scheme: Lax-Wendroff

Formulation:
$$u_{i+\frac{1}{2}}^{n+\frac{1}{2}} = \frac{1}{2}\left(u_{i+1}^n + u_i^n\right) - \frac{1}{2}c\left(u_{i+1}^n - u_i^{n}\right)$$

and

$$u_i^{n+1} = u_i^n - c\left(u_{i+\frac{1}{2}}^{n+\frac{1}{2}} - u_{i-\frac{1}{2}}^{n+\frac{1}{2}}\right)$$

Order:
$$O\left[(\Delta t)^2\,,\,(\Delta x)^2\right]$$

Stability Requirement: $c \leq 1$

Scheme: MacCormack

Formulation: $u_i^* = u_i^n - c\left(u_{i+1}^n - u_i^n\right)$

     and $u_i^{n+1} = \dfrac{1}{2}\left[u_i^n + u_i^* - c\left(u_i^* - u_i^*\right)\right]$

Order: $O\left[(\Delta t)^2\,,\,(\Delta x)^2\right]$

Stability Requirement: $c \leq 1$

6.5.5.2 Nonlinear equations: The inviscid Burgers equation is used to review various schemes for the solution of hyperbolic equations. Recall that the model equation is given by

$$\frac{\partial u}{\partial t} = -u\frac{\partial u}{\partial x} = -\frac{\partial E}{\partial x} \tag{6.8}$$

where $E = \frac{1}{2}u^2$. Now, the Courant number is defined as $c = u\frac{\Delta t}{\Delta x}$.

Scheme: Lax

Formulation: $u_i^{n+1} = \dfrac{1}{2}\left(u_{i+1}^n + u_{i-1}^n\right) - \dfrac{1}{2}\dfrac{\Delta t}{\Delta x}\left(E_{i+1}^n - E_{i-1}^n\right)$

Order: $O\left[(\Delta t)\,,\,(\Delta x)^2\right]$

Amplification Factor: $G = \cos\theta - i(c\sin\theta)$

Stability Requirement: $\left|u_{\mathrm{max}}\dfrac{\Delta t}{\Delta x}\right| \leq 1$

Scheme: Lax-Wendroff

Formulation: $u_i^{n+1} = u_i^n - \dfrac{1}{2}\dfrac{\Delta t}{\Delta x}\left(E_{i+1}^n - E_{i-1}^n\right) + \dfrac{1}{4}\left(\dfrac{\Delta t}{\Delta x}\right)^2$

$\left[\left(u_{i+1}^n + u_i^n\right)\left(E_{i+1}^n - E_i^n\right) - \left(u_i^n + u_{i-1}^n\right)\left(E_i^n - E_{i-1}^n\right)\right]$

Amplification Factor: $G = 1 - 2c^2(1 - \cos\theta) - 2i(c\sin\theta)$

Stability Requirement: $\left|u_{\mathrm{max}}\dfrac{\Delta t}{\Delta x}\right| \leq 1$

Scheme: MacCormack

Formulation: $$u_i^* = u_i^n - \frac{\Delta t}{\Delta x}\left(E_{i+1}^n - E_i^n\right)$$

and $$u_i^{n+1} = \frac{1}{2}\left[u_i^n + u_i^* - \frac{\Delta t}{\Delta x}\left(E_i^* - E_{i-1}^*\right)\right]$$

Order: $O\left[(\Delta t)^2,\ (\Delta x)^2\right]$

6.6 Stability Analysis

The error introduced in the finite difference equations due to the truncation of the higher order terms in the Taylor series expansion may grow unbounded, producing an unstable solution. The control of errors within the solution is of primary concern for any numerical scheme. To establish the necessary requirements, a stability analysis must be performed. Among various methods available for stability analysis are: (1) the discrete perturbation stability analysis, (2) the von Neumann (or Fourier) stability analysis, and (3) the matrix method. It should be noted that direct stability analysis of a nonlinear, multi-dimensional, coupled system of equations is usually cumbersome. In most cases, expressions are proposed which are based on the stability analysis of simple model equations complemented and reinforced by numerical experimentation. Thus, one encounters the suggested stability requirement for a particular scheme which resembles those of simple model equations, but includes some modifications based on numerical experimentations.

A summary of the limitations and conclusions provided from the von Neumann stability analysis is provided below.

1. The von Neumann stability analysis can be applied to linear equations only.

2. The influence of the boundary conditions on the stability of the solution is not included.

3. For a scalar PDE which is approximated by a two-level FDE, the mathematical requirement is imposed on the amplification factor G as follows:

   (a) if G is real, then $|G| \leq 1$

   (b) if G is complex, then $|G|^2 \leq 1$, where $|G|^2 = G\overline{G}$

4. For a scalar PDE which is approximated by a three-level FDE, the amplification factor is a matrix. In this case, the requirement is imposed on the eigenvalues, λ, of G as follows:

(a) if λ is real, then $|\lambda| \leq 1$

(b) if λ is complex, then $|\lambda|^2 \leq 1$

5. The method can be easily extended to multi-dimensional problems.

6. The procedure can be used for stability analysis of a system of linear PDEs. The requirement is imposed on the largest eigenvalue of the amplification matrix.

7. Benchmark values for the stability of unsteady one-dimensional problems may be established as follows:

   (a) For most explicit formulations:

      I. Courant number, $c \leq 1$

      II. Diffusion number, $d \leq \dfrac{1}{2}$

      III. Cell Reynolds number, $Re_c \leq (2/c)$

   (b) For implicit formulation, most are unconditionally stable.

8. For multi-dimensional problems with equal grid spacing in all spatial directions, the stated benchmark values are adjusted usually by dividing them by the number of spatial dimensions.

9. On occasions where the amplification factor is a difficult expression to analyze, a graphical solution along with some numerical experimentation will facilitate the analysis.

6.7 Error Analysis

The truncation of terms in the approximation of a partial derivative could begin from an odd-order or an even-order derivative term. For example, one may approximate a first-order derivative by either

$$\frac{\partial u}{\partial x} = \frac{u_{i+1} - u_i}{\Delta x} + \frac{(\Delta x)}{2!}\frac{\partial^2 u}{\partial x^2} + \frac{(\Delta x)^2}{3!}\frac{\partial^3 u}{\partial x^3} + \cdots \qquad (6.9)$$

or

$$\frac{\partial u}{\partial x} = \frac{u_{i+1} - u_{i-1}}{2\Delta x} + \frac{(\Delta x)^2}{3!}\frac{\partial^3 u}{\partial x^3} + \cdots \qquad (6.10)$$

The approximation (6.9) may be truncated and expressed as

$$\frac{\partial u}{\partial x} = \frac{u_{i+1} - u_i}{\Delta x} + O(\Delta x)$$

where the dominant (or leading) error term includes a second-order derivative, that is, even. The second expression (6.10) is written as

$$\frac{\partial u}{\partial x} = \frac{u_{i+1} - u_{i-1}}{2\Delta x} + O(\Delta x)^2$$

where the dominant error term now includes an odd derivative. The behavior of error associated with finite difference equations is strongly influenced by the dominant error term. To clarify the types of error introduced to the finite difference equations, a convection dominated equation, where physical viscosity is absent, will be used. Thus, consider the wave equation and two different finite difference equations given by

$$u_i^{n+1} = u_i^n - c(u_i^n - u_{i-1}^n) \tag{6.11}$$

and

$$u_i^{n+1} = u_i^{n-1} - c(u_{i+1}^n - u_{i-1}^n) \tag{6.12}$$

The FDEs are recognized as the first upwind differencing scheme and the midpoint leapfrog method. To identify the dominant error term of an FDE, the modified equation must be investigated. The corresponding modified equations for the FDEs given by (6.11) and (6.12) are, respectively:

$$\frac{\partial u}{\partial t} = -a\frac{\partial u}{\partial x} + \frac{1}{2}\alpha(\Delta x)(1-c)\frac{\partial^2 u}{\partial x^2} - \frac{1}{6}\alpha(\Delta x)^2 \left(2c^2 - 3c + 1\right)\frac{\partial^3 u}{\partial x^3} + \dots \tag{6.13}$$

and

$$\frac{\partial u}{\partial t} = -a\frac{\partial u}{\partial x} + \frac{1}{6}\alpha(\Delta x)^2 \left(c^2 - 1\right)\frac{\partial^3 u}{\partial x^3} + \dots \tag{6.14}$$

Observe that the dominant error terms in Equations (6.13) and (6.14) include second-order derivative and third-order derivative, respectively. Recall that from a physical point of view a second-order derivative is associated with diffusion. Indeed, the coefficient of the second-order derivative in Equation (6.13) is known as the *numerical viscosity*. Thus, it is obvious that the error associated with Equation (6.13) is dissipative and, hence, it is called *dissipation error*. On the other hand, an FDE scheme, where its corresponding modified equation possesses an odd-order derivative as the lead term in error, is associated with oscillations within the solution. Such an error is called *dispersion error*.

6.8   Grid Generation-Structured Grids

Finite difference equations are most efficiently solved in a rectangular domain for two-dimensional applications (an equivalent hexahedral domain for three-dimensional applications) with equal grid spacings. Unfortunately, the majority of physical domains encountered are nonrectangular in shape. Thus, it is necessary to transform the nonrectangular physical domain to a rectangular computational domain where grid points are distributed at equal spacings. It is also important to note that the transformation allows the alignment of one of the coordinates along the body, thus facilitating the implementation of the boundary conditions. The objective of grid generation is then to identify the location of the grid points in the computational domain and the location of the corresponding grid points in the physical space. Furthermore, the metrics and the Jacobian of transformation which are required for the solution of flow equations are computed within the grid generation routine.

Typically, grid generation schemes may be categorized as algebraic methods or differential methods. In the latter case, the scheme is based on the solution of a set of PDEs and may be subcategorized as either an elliptic, parabolic, or hyperbolic grid generation. Either category of grid generation scheme should include the following considerations.

1. A mapping which guarantees one-to-one correspondence ensuring grid lines of the same family do not cross each other;

2. Smoothness of the grid distribution;

3. Orthogonality or near orthogonality of the grid lines;

4. Options for grid clustering.

A brief summary of the advantages and disadvantages of each method is provided below.

1. Algebraic grids

   The advantages of this category of grid generators are:

   (a) They are very fast computationally;

   (b) Metrics may be evaluated analytically, thus avoiding numerical errors;

   (c) The ability to cluster grid points in different regions can be easily implemented.

   The disadvantages are:

   (a) Discontinuities at a boundary may propagate into the interior region which could lead to errors due to sudden changes in the metrics;

   (b) Smoothness and skewness may be difficult to control.

2. Elliptic grids

   The advantages of this class of grid generators are:

(a) Will provide smooth grid point distribution, that is, if a boundary discontinuity point exists, it will be smoothed out in the interior domain;

(b) Numerous options for grid clustering and surface orthogonality are available;

(c) Method can be extended to three-dimensional problems.

The disadvantages of the method are:

(a) Computation time is large (compared to algebraic methods or hyperbolic grid generators);

(b) Specification of the forcing functions (or the constants used in these functions) is not easy;

(c) Metrics must be computed numerically.

3. Hyperbolic grids

The advantages of hyperbolic grid generators are:

(a) The grid system is orthogonal in two dimensions;

(b) Since a marching scheme is used for the solution of the system, computationally they are much faster than elliptic systems;

(c) Grid line spacing may be controlled by the cell area or arc-length functions.

The disadvantages are:

(a) Boundary discontinuity may be propagated into the interior domain;

(b) Specifying the cell-area or arc-length functions must be handled carefully. A bad selection of these functions easily leads to undesirable grid systems.

6.9   Incompressible Navier-Stokes Equations

The incompressible Navier-Stokes equations in primitive variable formulations and in the vorticity-stream function formulation are presented in Chapter One. In this section, the finite difference equations for the incompressible Navier-Stokes equations are summarized. Most equations are presented in two dimensions because the stream function formulation is limited to two dimensions and because the extension to three dimensions for primitive variable formulation is straightforward.

6.9.1 Vorticity-Stream Function Formulation

The vorticity transport equation may be expressed in either conservative or non-conservative form. The governing equation and the finite difference equations are provided in this section.

6.9.1.1 Vorticity transport equation in nonconservative, nondimensional form:

A. Governing equation

$$\frac{\partial \Omega}{\partial t} + u \frac{\partial \Omega}{\partial x} + v \frac{\partial \Omega}{\partial y} = \frac{1}{Re}\left(\frac{\partial^2 \Omega}{\partial x^2} + \frac{\partial^2 \Omega}{\partial y^2}\right) \tag{6.15}$$

B. Finite difference equation

1. FTCS explicit

$$\frac{\Omega_{i,j}^{n+1} - \Omega_{i,j}^{n}}{\Delta t} + u_{i,j}^{n}\frac{\Omega_{i+1,j}^{n} - \Omega_{i-1,j}^{n}}{2\Delta x} + v_{i,j}^{n}\frac{\Omega_{i,j+1}^{n} - \Omega_{i,j-1}^{n}}{2\Delta y}$$

$$= \frac{1}{Re}\left[\frac{\Omega_{i+1,j}^{n} - 2\Omega_{i,j}^{n} + \Omega_{i-1,j}^{n}}{(\Delta x)^2} + \frac{\Omega_{i,j+1}^{n} - 2\Omega_{i,j}^{n} + \Omega_{i,j-1}^{n}}{(\Delta y)^2}\right] \tag{6.16}$$

2. Upwind explicit

$$\frac{\Omega_{i,j}^{n+1} - \Omega_{i,j}^{n}}{\Delta t} + \frac{1}{2}(1-\epsilon_x)u_{i,j}^{n}\frac{\Omega_{i+1,j}^{n} - \Omega_{i,j}^{n}}{\Delta x} + \frac{1}{2}(1+\epsilon_x)\,u_{i,j}^{n}\frac{\Omega_{i,j}^{n} - \Omega_{i-1,j}^{n}}{\Delta x}$$

$$+ \frac{1}{2}(1-\epsilon_y)v_{i,j}^{n}\frac{\Omega_{i,j+1}^{n} - \Omega_{i,j}^{n}}{\Delta y} + \frac{1}{2}(1+\epsilon_y)\,v_{i,j}\frac{\Omega_{i,j}^{n} - \Omega_{i,j-1}^{n}}{\Delta y}$$

$$= \frac{1}{Re}\frac{\Omega_{i+1,j}^{n} - 2\Omega_{i,j}^{n} + \Omega_{i-1,j}^{n}}{(\Delta x)^2} + \frac{1}{Re}\frac{\Omega_{i,j+1}^{n} - 2\Omega_{i,j}^{n} + \Omega_{i,j-1}^{n}}{(\Delta y)^2} \tag{6.17}$$

Note the following features of the finite difference equation given by Equation (6.17) above: If u is positive, a backward approximation must be utilized. Thus, ϵ_x is set to 1. If u is negative, a forward approximation is used, and therefore ϵ_x is set equal to -1. The same analogy is applied to v and to the corresponding coefficient ϵ_y.

3. ADI

$$-\frac{1}{2}\left(\frac{1}{2}c_x + d_x\right)\Omega_{i-1,j}^{n+\frac{1}{2}} + (1+d_x)\Omega_{i,j}^{n+\frac{1}{2}} + \frac{1}{2}\left(\frac{1}{2}c_x - d_x\right)\Omega_{i+1,j}^{n+\frac{1}{2}} = D_x \tag{6.18}$$

and

$$-\frac{1}{2}\left(\frac{1}{2}c_y + d_y\right)\Omega_{i,j-1}^{n+1} + (1+d_y)\Omega_{i,j}^{n+1} + \frac{1}{2}\left(\frac{1}{2}c_y - d_y\right)\Omega_{i,j+1}^{n+1} = D_y \quad (6.19)$$

where the Courant number and the diffusion number are defined as

$$c_x = u\frac{\Delta t}{\Delta x} \quad , \quad c_y = v\frac{\Delta t}{\Delta y}$$

$$d_x = \frac{1}{Re}\frac{\Delta t}{(\Delta x)^2} \quad , \quad d_y = \frac{1}{Re}\frac{\Delta t}{(\Delta y)^2}$$

and

$$D_x = \frac{1}{2}\left(\frac{1}{2}c_y + d_y\right)\Omega_{i,j-1}^{n} + (1-d_y)\Omega_{i,j}^{n} + \frac{1}{2}\left(-\frac{1}{2}c_y + d_y\right)\Omega_{i,j+1}^{n}$$

$$D_y = \frac{1}{2}\left(\frac{1}{2}c_x + d_x\right)\Omega_{i-1,j}^{n+\frac{1}{2}} + (1-d_x)\Omega_{i,j}^{n+\frac{1}{2}} + \frac{1}{2}\left(-\frac{1}{2}c_x + d_x\right)\Omega_{i+1,j}^{n+\frac{1}{2}}$$

6.9.1.2 Vorticity transport equation in conservative, nondimensional form:

A. Governing equation

$$\frac{\partial\Omega}{\partial t} + \frac{\partial}{\partial x}(u\Omega) + \frac{\partial}{\partial y}(v\Omega) = \frac{1}{Re}\left(\frac{\partial^2\Omega}{\partial x^2} + \frac{\partial^2\Omega}{\partial y^2}\right) \quad (6.20)$$

B. Finite difference equations

1. FTCS explicit

$$\frac{\Omega_{i,j}^{n+1} - \Omega_{i,j}^{n}}{\Delta t} + \frac{u_{i+1,j}^{n}\Omega_{i+1,j}^{n} - u_{i-1,j}^{n}\Omega_{i-1,j}^{n}}{2\Delta x} + \frac{v_{i,j+1}^{n}\Omega_{i,j+1}^{n} - v_{i,j-1}^{n}\Omega_{i,j-1}^{n}}{2\Delta y}$$

$$= \frac{1}{Re}\left[\frac{\Omega_{i+1,j}^{n} - 2\Omega_{i,j}^{n} + \Omega_{i-1,j}^{n}}{(\Delta x)^2} + \frac{\Omega_{i,j+1}^{n} - 2\Omega_{i,j}^{n} + \Omega_{i,j-1}^{n}}{(\Delta y)^2}\right] \quad (6.21)$$

2. Upwind explicit

$$\frac{\Omega_{i,j}^{n+1} - \Omega_{i,j}^{n}}{\Delta t} + \frac{1}{2}(1-\epsilon_x)\frac{u_{i+1,j}\Omega_{i+1,j} - u_{i,j}\Omega_{i,j}}{\Delta x} + \frac{1}{2}(1+\epsilon_x)\frac{u_{i,j}\Omega_{i,j} - u_{i-1,j}\Omega_{i-1,j}}{\Delta x}$$

$$+\frac{1}{2}(1-\epsilon_y)\frac{v_{i,j+1}\Omega_{i,j+1} - v_{i,j}\Omega_{i,j}}{\Delta y} + \frac{1}{2}(1+\epsilon_y)\frac{v_{i,j}\Omega_{i,j} - v_{i,j-1}\Omega_{i,j-1}}{\Delta y}$$

$$= \frac{1}{Re}\frac{\Omega_{i+1,j} - 2\Omega_{i,j} + \Omega_{i-1,j}}{(\Delta x)^2} + \frac{1}{Re}\frac{\Omega_{i,j+1} - 2\Omega_{i,j} + \Omega_{i,j-1}}{(\Delta y)^2} \quad (6.22)$$

3. ADI

$$-\frac{1}{2}\left[\frac{1}{2}(1+\epsilon_x)c_{x(i-1,j)}+d_x\right]\Omega_{i-1,j}^{n+\frac{1}{2}}+\left(1+d_x+\frac{1}{2}\epsilon_x c_{x(i,j)}\right)\Omega_{i,j}^{n+\frac{1}{2}}$$

$$+\frac{1}{2}\left[\frac{1}{2}(1-\epsilon_x)c_{x(i+1,j)}-d_x\right]\Omega_{i+1,j}^{n+\frac{1}{2}}=D_x \tag{6.23}$$

and

$$-\frac{1}{2}\left[\frac{1}{2}(1+\epsilon_y)c_{y(i,j-1)}+d_y\right]\Omega_{i,j-1}^{n+1}+\left(1+d_y+\frac{1}{2}\epsilon_y c_{y(i,j)}\right)\Omega_{i,j}^{n+1}$$

$$+\frac{1}{2}\left[\frac{1}{2}(1-\epsilon_y)c_{y(i,j+1)}-d_y\right]\Omega_{i,j+1}^{n+1}=D_y \tag{6.24}$$

where

$$D_x=\frac{1}{2}\left[\frac{1}{2}(1+\epsilon_y)c_{y(i,j-1)}+d_y\right]\Omega_{i,j-1}^{n}+\left(1-d_y-\frac{1}{2}\epsilon_y c_{y(i,j)}\right)\Omega_{i,j}^{n}$$

$$+\frac{1}{2}\left[-\frac{1}{2}(1-\epsilon_y)c_{y(i,j+1)}-d_y\right]\Omega_{i,j+1}^{n}$$

and

$$D_y=\frac{1}{2}\left[\frac{1}{2}(1+\epsilon_x)c_{x(i-1,j)}+d_x\right]\Omega_{i-1,j}^{n+1}+(1-d_x-\frac{1}{2}\epsilon_x c_{x(i,j)})\Omega_{i,j}^{n+\frac{1}{2}}$$

$$+\frac{1}{2}\left[-\frac{1}{2}(1-\epsilon_x)c_{x(i+1,j)}+d_x\right]\Omega_{i+1,j}^{n+1}$$

6.9.2 Stream Function Equation

The stream function equation is

$$\frac{\partial^2\psi}{\partial x^2}+\frac{\partial^2\psi}{\partial y^2}=-\Omega \tag{6.25}$$

and is classified as an elliptic equation. Typically, iterative schemes are used for solution. Various solution schemes are provided in this section.

1. Point Gauss-Seidel

$$\psi_{i,j}^{k+1}=\frac{1}{2(1+\beta^2)}\left[(\Delta x)^2\Omega_{i,j}^k+\psi_{i+1,j}^k+\psi_{i-1,j}^{k+1}+\beta^2(\psi_{i,j+1}^k+\psi_{i,j-1}^{k+1})\right] \tag{6.26}$$

where $\beta=\dfrac{\Delta x}{\Delta y}$.

2. Line Gauss-Seidel

$$\psi_{i-1,j}^{k+1}-2(1+\beta^2)\psi_{i,j}^{k+1}+\psi_{i+1,j}^{k+1}=(\Delta x)^2\Omega_{i,j}^k-\beta^2\psi_{i,j+1}^k-\beta^2\psi_{i,j-1}^{k+1} \tag{6.27}$$

3. Point Successive Over-Relaxation

$$\psi_{i,j}^{k+1} = (1-\omega)\psi_{i,j}^{k} + \frac{\omega}{2(1+\beta^2)}\left[(\Delta x)^2\Omega_{i,j}^{k} + \psi_{i+1,j}^{k} + \psi_{i-1,j}^{k+1} + \beta^2(\psi_{i,j+1}^{k} + \psi_{i,j-1}^{k+1})\right]$$

$$(6.28)$$

6.9.3 Primitive Variable Formulation

6.9.3.1 Governing equations – physical space: The incompressible continuity equation is typically amended by a time-dependent pressure term. Thus the governing incompressible Navier-Stokes equations are expressed as

$$\frac{\partial p}{\partial t} + \beta\left(\frac{\partial u}{\partial x} + \frac{\partial v}{\partial y}\right) = 0 \tag{6.29}$$

$$\frac{\partial u}{\partial t} + \frac{\partial}{\partial x}(u^2 + p) + \frac{\partial}{\partial y}(uv) = \frac{1}{Re}\left(\frac{\partial^2 u}{\partial x^2} + \frac{\partial^2 u}{\partial y^2}\right) \tag{6.30}$$

$$\frac{\partial v}{\partial t} + \frac{\partial}{\partial x}(uv) + \frac{\partial}{\partial y}(v^2 + p) = \frac{1}{Re}\left(\frac{\partial^2 v}{\partial x^2} + \frac{\partial^2 v}{\partial y^2}\right) \tag{6.31}$$

Equations (6.29) through (6.31) can be written in a flux vector form as

$$\frac{\partial Q}{\partial t} + \frac{\partial E}{\partial x} + \frac{\partial F}{\partial y} = \frac{1}{Re}[N]\nabla^2 Q \tag{6.32}$$

where

$$Q = \begin{bmatrix} p \\ u \\ v \end{bmatrix} \quad , \quad E = \begin{bmatrix} \beta u \\ u^2 + p \\ uv \end{bmatrix} \quad , \quad F = \begin{bmatrix} \beta v \\ uv \\ v^2 + p \end{bmatrix}$$

and

$$N = \begin{bmatrix} 0 & 0 & 0 \\ 0 & 1 & 0 \\ 0 & 0 & 1 \end{bmatrix}$$

Equation (6.32) is now written in a delta form as

$$\frac{\Delta Q}{\Delta t} + \frac{\partial}{\partial x}(E^n + A\Delta Q) + \frac{\partial}{\partial y}(F^n + B\Delta Q) = \frac{1}{Re}[N]\nabla^2 Q$$

and rearranged as

$$\Delta Q + \Delta t\left[\frac{\partial}{\partial x}(A\Delta Q) + \frac{\partial}{\partial y}(B\Delta Q) - \frac{N}{Re}\left(\frac{\partial^2}{\partial x^2} + \frac{\partial^2}{\partial y^2}\right)\Delta Q\right]$$

$$= \Delta t\left[-\frac{\partial E^n}{\partial x} - \frac{\partial F^n}{\partial y} + \frac{N}{Re}\left(\frac{\partial^2 Q}{\partial x^2} + \frac{\partial^2 Q}{\partial y^2}\right)\right] \tag{6.33}$$

or

$$\left\{I + \Delta t\left[\frac{\partial A}{\partial x} + \frac{\partial B}{\partial y} - \frac{N}{Re}\left(\frac{\partial^2}{\partial x^2} + \frac{\partial^2}{\partial y^2}\right)\right]\right\}\Delta Q = \text{RHS} \tag{6.34}$$

6.9.3.2 *Finite difference equation: ADI formulation*

$$\left[-\left(\frac{\Delta t}{2\Delta x}\right)A_{i-1,j} - \left(\frac{N}{Re}\frac{\Delta t}{(\Delta x)^2}\right) + \epsilon_i\right]\Delta Q^*_{i-1,j} + \left[I + \left(\frac{2N}{Re}\frac{\Delta t}{(\Delta x)^2}\right) - 2\epsilon_i\right]\Delta Q^*_{i,j}$$

$$+ \left[\left(\frac{\Delta t}{2\Delta x}\right)A_{i+1,j} - \left(\frac{N}{Re}\frac{\Delta t}{(\Delta x)^2}\right) + \epsilon_i\right]\Delta Q^*_{i+1,j} = \text{RHS}_{i,j} \tag{6.35}$$

and

$$\left[-\left(\frac{\Delta t}{2\Delta y}\right)B_{i,j-1} - \left(\frac{N}{Re}\frac{\Delta t}{(\Delta y)^2}\right) + \epsilon_i\right]\Delta Q^*_{i,j-1} + \left[I + \left(\frac{2N}{Re}\frac{\Delta t}{(\Delta y)^2}\right) - 2\epsilon_i\right]\Delta Q^*_{i,j}$$

$$+ \left[\left(\frac{\Delta t}{2\Delta y}\right)B_{i,j+1} - \left(\frac{N}{Re}\frac{\Delta t}{(\Delta y)^2}\right) + \epsilon_i\right]\Delta Q^*_{i,j+1} = \Delta Q^*_{i,j} \tag{6.36}$$

where

$$\text{RHS}_{i,j} = \Delta t\left\{-\frac{E_{i+1,j} - E_{i-1,j}}{2\Delta x} - \frac{F_{i,j+1} - F_{i,j-1}}{2\Delta y} + \frac{N}{Re}\frac{Q_{i+1,j} - 2Q_{i,j} + Q_{i-1,j}}{(\Delta x)^2}\right.$$

$$+ \frac{N}{Re}\frac{Q_{i,j+1} - 2Q_{i,j} + Q_{i,j-1}}{(\Delta y)^2}\bigg\} - \epsilon_e\Big\{(Q_{i-2,j} - 4Q_{i-1,j} + 6Q_{i,j} - 4Q_{i+1,j}$$

$$\left. + Q_{i+2,j}) + (Q_{i,j-2} - 4Q_{i,j-1} + 6Q_{i,j} - 4Q_{i,j+1} + Q_{i,j+2})\right\}$$

6.9.3.3 *Governing equations – computational space:* The incompressible

Navier-Stokes equations in the computational space are given by Equation (5.279) as

$$\frac{\partial \bar{Q}}{\partial t} + \frac{\partial \bar{E}}{\partial \xi} + \frac{\partial \bar{F}}{\partial \eta} = \frac{\partial \bar{E}v}{\partial \xi} + \frac{\partial \bar{F}v}{\partial \eta} \tag{6.37}$$

Equation (6.37) can be written in a delta form as follows

$$\left\{I + \Delta t\left[\frac{\partial A}{\partial \xi} + \frac{\partial B}{\partial \eta} - \frac{\partial A_v}{\partial \xi} - \frac{\partial B_v}{\partial \eta}\right]\right\}\Delta \bar{Q} = -\Delta t\left[\frac{\partial \bar{E}}{\partial \xi} + \frac{\partial \bar{F}}{\partial \eta} - \frac{\partial \bar{E}_v}{\partial \xi} - \frac{\partial \bar{F}_v}{\partial \eta}\right]$$

$$\tag{6.38}$$

6.9.3.4 *Finite difference equations:*

a. ADI. Equation (6.38) is split and solved sequentially by the following equations

$$\left[-\left(\frac{\Delta t}{2\Delta\xi}\right)A_{i-1,j} + \left(\frac{\Delta t}{2\Delta\xi}\right)(A_v)_{i-1,j} + \varepsilon_i\right]\Delta\bar{Q}^*_{i-1,j} + [I - 2\varepsilon_i]\,\Delta\bar{Q}^*_{i,j}$$

$$+ \left[\left(\frac{\Delta t}{2\Delta\xi}\right)A_{i+1,j} - \left(\frac{\Delta t}{2\Delta\xi}\right)(A_v)_{i+1,j} + \varepsilon_i\right]\Delta\bar{Q}^*_{i+1,j} = \text{RHS}_{i,j} \quad (6.39)$$

and

$$\left[-\left(\frac{\Delta t}{2\Delta\eta}\right)B_{i,j-1} + \left(\frac{\Delta t}{2\Delta\eta}\right)(B_v)_{i,j-1} + \varepsilon_i\right]\Delta\bar{Q}_{i,j-1} + [I - 2\varepsilon_i]\,\Delta\bar{Q}_{i,j}$$

$$+ \left[\left(\frac{\Delta t}{2\Delta\eta}\right)B_{i,j+1} - \left(\frac{\Delta t}{2\Delta\eta}\right)(B_v)_{i,j+1} + \varepsilon_i\right]\Delta\bar{Q}_{i,j+1} = \Delta\bar{Q}^*_{i,j} \quad (6.40)$$

where

$$\text{RHS}_{i,j} = -\Delta t\left[\frac{\bar{E}^n_{i+1,j} - \bar{E}^n_{i-1,j}}{2\Delta\xi} + \frac{\bar{F}^n_{i,j+1} - \bar{F}^n_{i,j-1}}{2\Delta\eta} - \frac{(\bar{E}^n_v)_{i+1,j} - (\bar{E}^n_v)_{i-1,j}}{2\Delta\xi}\right.$$

$$\left. - \frac{(\bar{F}^n_v)_{i,j+1} - (\bar{F}^n_v)_{i,j-1}}{2\Delta\eta}\right] - \varepsilon_e D$$

D is a damping term, and ε_i and ε_e are the coefficients of implicit and explicit damping terms, respectively.

b. Flux Vector Splitting.

$$\left[\left(\frac{\Delta t}{\Delta\xi}\right)A^+_{i-1,j} + \left(\frac{\Delta t}{2\Delta\xi}\right)(A_v)_{i-1,j}\right]\Delta\bar{Q}^*_{i-1,j} + \left[I + \left(\frac{\Delta t}{\Delta\xi}\right)\left(A^+_{i,j} - A^-_{i,j}\right)\right]\Delta\bar{Q}^*_{i,j}$$

$$+ \left[\left(\frac{\Delta t}{\Delta\xi}\right)A^-_{i+1,j} - \left(\frac{\Delta t}{2\Delta\xi}\right)(A_v)_{i+1,j}\right]\Delta\bar{Q}^*_{i+1,j} = \text{RHS}_{i,j} \quad (6.41)$$

and

$$\left[-\left(\frac{\Delta t}{\Delta\eta}\right)B^+_{i,j-1} + \left(\frac{\Delta t}{2\Delta\eta}\right)(B_v)_{i,j-1}\right]\Delta\bar{Q}_{i,j-1} + \left[I + \left(\frac{\Delta t}{\Delta\eta}\right)\left(B^+_{i,j} - B^-_{i,j}\right)\right]\Delta\bar{Q}_{i,j}$$

$$+ \left[\left(\frac{\Delta t}{\Delta\eta}\right)B^-_{i,j+1} - \left(\frac{\Delta t}{2\Delta\eta}\right)(B_v)_{i,j+1}\right]\Delta\bar{Q}_{i,j+1} = \Delta\bar{Q}_{i,j+1} = \Delta\bar{Q}^*_{i,j} \quad (6.42)$$

where

$$\text{RHS}_{i,j} = -\Delta t\left[\frac{1}{\Delta\xi}\left(\bar{E}^+_{i,j} - \bar{E}^+_{i-1,j} + \bar{E}^-_{i+1,j} - \bar{E}^-{i,j}\right)\right.$$

$$\frac{1}{\Delta\eta}\left(\bar{F}^+_{i,j} - \bar{F}^+_{i,j-1} + \bar{F}^-_{i,j+1} - \bar{F}^-_{i,j}\right) - \frac{1}{2\Delta\xi}\left(Ev_{i+1,j} - \bar{E}v_{i-1,j}\right)$$

$$\left. - \frac{1}{2\Delta\eta}\left(\bar{F}v_{i,j+1} - \bar{F}v_{i,j-1}\right)\right]$$

6.10   Navier-Stokes Equations

The Navier-Stokes equations in computational space $(\xi,\ \eta,\ \zeta)$ are given by Equation (5.39) for a three-dimensional flow and by Equation (5.188) for a two-dimensional planar or axisymmetric flow. Selected finite difference formulations for the two-dimensional Navier-Stokes equations and the three-dimensional thin-layer Navier-Stokes equations are summarized below.

6.10.1   Two-Dimensional Planar or Axisymmetric Navier-Stokes Equations

6.10.1.1   *MacCormack explicit:*

Predictor

$$
\begin{aligned}
\bar{Q}_{i,j}^{*} = {} & \bar{Q}_{i,j}^{n} - \frac{\Delta\tau}{\Delta\xi}\left[\bar{E}_{i+1,j}^{n} - \bar{E}_{i,j}^{n}\right] - \frac{\Delta\tau}{\Delta\eta}\left[\bar{F}_{i,j+1}^{n} - \bar{F}_{i,j}^{n}\right] \\
& + \frac{\Delta\tau}{\Delta\xi}\left[(\bar{E}_v^n)_{i+1,j} - (\bar{E}_v^n)_{i,j}\right] + \frac{\Delta\tau}{\Delta\eta}\left[(\bar{F}_v^n)_{i,j+1} - (\bar{F}_v^n)_{i,j}\right] \\
& - \Delta\tau\alpha\left[\bar{H}_{i,j}^{n} - (\bar{H}_v^n)_{i,j}\right]
\end{aligned}
\tag{6.43}
$$

Corrector

$$
\begin{aligned}
\bar{Q}_{i,j}^{n+1} = {} & \frac{1}{2}\bigg\{\bar{Q}_{i,j}^{n} + \bar{Q}_{i,j}^{*} - \frac{\Delta\tau}{\Delta\xi}\left[\bar{E}_{i,j}^{*} - \bar{E}_{i-1,j}^{*}\right] - \frac{\Delta\tau}{\Delta\eta}\left[\bar{F}_{i,j}^{*} - \bar{F}_{i,j-1}^{*}\right] \\
& + \frac{\Delta\tau}{\Delta\xi}\left[(\bar{E}_v^*)_{i,j} - (\bar{E}_v^*)_{i-1,j}\right] + \frac{\Delta\tau}{\Delta\eta}\left[(\bar{F}_v^*)_{i,j} - (\bar{F}_v^*)_{i,j-1}\right] \\
& - \Delta\tau\alpha\left[\bar{H}_{i,j}^{*} - (\bar{H}_v^*)_{i,j}\right]\bigg\}
\end{aligned}
\tag{6.44}
$$

6.10.1.2   *Explicit flux vector splitting (first-order convective terms):*

$$
\begin{aligned}
\bar{Q}_{i,j}^{n+1} = {} & \bar{Q}_{i,j}^{n} - \Delta\tau\bigg[\frac{1}{\Delta\xi}\left(\bar{E}_{i,j}^{+} - \bar{E}_{i-1,j}^{+} + \bar{E}_{i+1,j}^{-} - \bar{E}_{i,j}^{-}\right) \\
& + \frac{1}{\Delta\eta}\left(\bar{F}_{i,j}^{+} - \bar{F}_{i,j-1}^{+} + \bar{F}_{i,j+1}^{-} - \bar{F}_{i,j}^{-}\right) + \alpha\bar{H}_{i,j}\bigg] \\
& + \frac{\Delta\tau}{2(\Delta\xi)^2}\left(\bar{r}_{i+1,j}R_{i+1,j} - \bar{r}_{i,j}R_{i,j} + \bar{r}_{i-1,j}R_{i-1,j}\right) \\
& + \frac{\Delta\tau}{2(\Delta\eta)^2}\left(\bar{s}_{i,j+1}S_{i,j+1} - \bar{s}_{i,j}S_{i,j} + \bar{s}_{i,j-1}S_{i,j-1}\right) + \alpha\Delta\tau\left(\bar{H}_v\right)_{i,j}
\end{aligned}
\tag{6.45}
$$

6.10.1.3 *Explicit flux vector splitting (second-order convective terms):*

$$\bar{Q}_{i,j}^{n+1} = \bar{Q}_{i,j}^n - \Delta\tau \Big[\frac{1}{2\Delta\xi} \left(\bar{E}_{i-2,j}^+ - 4\bar{E}_{i-1,j}^+ + 3\bar{E}_{i,j}^+ - 3\bar{E}_{i,j}^- + 4\bar{E}_{i+1,j}^- - \bar{E}_{i+2,j}^- \right)$$

$$+ \frac{1}{2\Delta\eta} \left(\bar{F}_{i,j-2}^+ - 4\bar{F}_{i,j-1}^+ + 3\bar{F}_{i,j}^+ - 3\bar{F}_{i,j}^- + 4\bar{F}_{i,j+1}^- - \bar{F}_{i,j+2}^- \right)$$

$$+ \alpha\bar{H}_{i,j} \Big] + \frac{\Delta\tau}{2(\Delta\xi)^2} \left(\bar{r}_{i+1,j} R_{i+1,j} - \bar{r}_{i,j} R_{i,j} + \bar{r}_{i-1,j} R_{i-1,j} \right)$$

$$+ \frac{\Delta\tau}{2(\Delta\eta)^2} \left(\bar{s}_{i,j+1} S_{i,j+1} - \bar{s}_{i,j} S_{i,j} + \bar{s}_{i,j-1} S_{i,j-1} \right) + \alpha\Delta\tau \left(\bar{H}_v \right)_{i,j} \qquad (6.46)$$

6.10.1.4 *Runge Kutta explicit:*

$$\bar{Q}_{i,j}^1 = \bar{Q}_{i,j}^n - \frac{\Delta\tau}{4} L\bar{Q}_{i,j}^n \qquad (6.47)$$

$$\bar{Q}_{i,j}^2 = \bar{Q}_{i,j}^n - \frac{\Delta\tau}{3} L\bar{Q}_{i,j}^1 \qquad (6.48)$$

$$\bar{Q}_{i,j}^3 = \bar{Q}_{i,j}^n - \frac{\Delta\tau}{2} L\bar{Q}_{i,j}^2 \qquad (6.49)$$

$$\bar{Q}_{i,j}^{n+1} = \bar{Q}_{i,j}^n - \Delta\tau L\bar{Q}_{i,j}^3 \qquad (6.50)$$

where $L\bar{Q}_{i,j} = \Delta_\xi \left(\bar{E}_v - \bar{E} \right) + \Delta\eta \left(\bar{F}_v - \bar{F} \right) + \alpha \left(\bar{H}_v - \bar{H} \right)$ is the difference operator. For example, if a second-order central difference operator is used, then

$$\Delta_\xi(\bar{E}_v - \bar{E}) = \frac{\left(\bar{E}_v \right)_{i+1,j} - \left(\bar{E}_v \right)_{i-1,j}}{2\Delta\xi} - \frac{\bar{E}_{i+1,j} - \bar{E}_{i-1,j}}{2\Delta\xi}$$

6.10.1.5 *Implicit flux vector splitting (first-order convective terms):*

$$\left\{ -\frac{\Delta\tau}{\Delta\xi} A_{i-1,j}^+ - \frac{\Delta\tau}{2(\Delta\xi)^2} \bar{r}_{i-1,j} R_{i-1,j} \right\} \Delta\bar{Q}_{i-1,j}^*$$

$$+ \left\{ I + \frac{\Delta\tau}{\Delta\xi} \left(A_{i,j}^+ - A_{i,j}^- \right) + \frac{\Delta\tau}{2(\Delta\xi)^2} \bar{r}_{i,j} R_{i,j} \right\} \Delta\bar{Q}_{i,j}^*$$

$$+ \left\{ \frac{\Delta\tau}{\Delta\xi} A_{i+1,j}^- - \frac{\Delta\tau}{2(\Delta\xi)^2} \bar{r}_{i+1,j} R_{i+1,j} \right\} \Delta\bar{Q}_{i+1,j}^* = \text{RHS} \qquad (6.51)$$

and

$$\left\{ -\frac{\Delta\tau}{\Delta\eta} B_{i,j-1}^+ - \frac{\Delta\tau}{2(\Delta\eta)^2} \bar{s}_{i,j-1} S_{i,j-1} \right\} \Delta\bar{Q}_{i,j-1} +$$

$$
+ \left\{ I + \frac{\Delta\tau}{\Delta\eta}\left(B_{i,j}^{+} - B_{i,j}^{-} \right) + \frac{\Delta\tau}{2(\Delta\eta)^2}\bar{s}_{i,j}S_{i,j} + \alpha\Delta\tau\left[C_{i,j} - (C_v)_{i,j} \right] \right\} \Delta\bar{Q}_{i,j}
$$

$$
+ \left\{ \frac{\Delta\tau}{\Delta\eta}B_{i,j+1}^{-} - \frac{\Delta\tau}{2(\Delta\eta)^2}\bar{s}_{i,j+1}S_{i,j+1} \right\} \Delta\bar{Q}_{i,j+1} = \Delta\bar{Q}_{i,j}^{*} \tag{6.52}
$$

where

$$
\text{RHS} = -\Delta\tau \left\{ \frac{1}{\Delta\xi}\left(\bar{E}_{i,j}^{+} - \bar{E}_{i-1,j}^{+} + \bar{E}_{i+1,j}^{-} - \bar{E}_{i,j}^{-} \right) + \frac{1}{\Delta\eta}\left(\bar{F}_{i,j}^{+} \right. \right.
$$

$$
\left. - \bar{F}_{i,j-1}^{+} + \bar{F}_{i,j+1}^{-} - \bar{F}_{i,j}^{-} \right) - \frac{1}{2(\Delta\xi)^2}\left(\hat{e}_{i+1,j}\hat{E}_{i+1,j} - \hat{e}_{i,j}\hat{E}_{i,j} + \hat{e}_{i-1,j}\hat{E}_{i-1,j} \right)
$$

$$
\left. - \frac{1}{2(\Delta\eta)^2}\left(\hat{f}_{i,j+1}\hat{F}_{i,j+1} - \hat{f}_{i,j}\hat{F}_{i,j} + \hat{f}_{i,j-1}\hat{F}_{i,j-1} \right) + \alpha\left(\bar{H}_{i,j} - (\bar{H}_v)_{i,j} \right) \right\}
$$

Note that in the formulation of the RHS given above, the mixed partial derivatives are excluded. In addition, the following notation is used

$$
\frac{\partial \bar{E}_v}{\partial \xi} = \frac{\partial}{\partial \xi}\left(\hat{e}\frac{\partial \hat{E}}{\partial \xi} \right) = \frac{1}{2(\Delta\xi)^2}\left[\hat{e}_{i+1,j}\hat{E}_{i+1,j} - \hat{e}_{i,j}\hat{E}_{i,j} + \hat{e}_{i-1,j}\hat{E}_{i-1,j} \right]
$$

6.10.1.6 *Implicit flux vector splitting (second-order convective terms):*

$$
\left\{ -2\frac{\Delta\tau}{\Delta\xi}A_{i-1,j}^{+} - \frac{\Delta\tau}{2(\Delta\xi)^2}\bar{r}_{i-1,j}R_{i-1,j} \right\}\Delta\bar{Q}_{i-1,j}^{*}
$$

$$
+ \left\{ I + \frac{3\Delta\tau}{2\Delta\xi}\left(A_{i,j}^{+} - A_{i,j}^{-} \right) + \frac{\Delta\tau}{2(\Delta\xi)^2}\bar{r}_{i,j}R_{i,j} \right\}\Delta\bar{Q}_{i,j}^{*}
$$

$$
+ \left\{ 2\frac{\Delta\tau}{\Delta\xi}A_{i+1,j}^{-} - \frac{\Delta\tau}{2(\Delta\xi)^2}\bar{r}_{i+1,j}R_{i+1,j} \right\}\Delta\bar{Q}_{i+1,j}^{*} =
$$

$$
\text{RHS}_{i,j} - \frac{\Delta\tau}{2\Delta\xi}\left[A_{i-2,j}^{+}\Delta\bar{Q}_{i-2,j}^{n-1} - A_{i+2,j}^{-}\Delta\bar{Q}_{i+2,j}^{n-1} \right] \tag{6.53}
$$

and

$$
\left\{ -2\frac{\Delta\tau}{\Delta\eta}B_{i,j-1}^{+} - \frac{\Delta\tau}{2(\Delta\eta)^2}\bar{s}_{i,j-1}S_{i,j-1} \right\}\Delta\bar{Q}_{i,j-1}
$$

$$
+ \left\{ I + \frac{3\Delta\tau}{2\Delta\eta}\left(B_{i,j}^{+} - B_{i,j}^{-} \right) + \frac{\Delta\tau}{2(\Delta\eta)^2}\bar{s}_{i,j}S_{i,j} + \alpha\Delta\tau\left(C_{i,j} - (C_v)_{i,j} \right) \right\}\Delta\bar{Q}_{i,j}
$$

$$
+ \left\{ 2\frac{\Delta\tau}{\Delta\eta}B_{i,j+1}^{-} - \frac{\Delta\tau}{2(\Delta\eta)^2}\bar{s}_{i,j+1}S_{i,j+1} \right\}\Delta\bar{Q}_{i,j+1}
$$

$$
= \Delta\bar{Q}_{i,j}^{*} - \frac{\Delta\tau}{2\Delta\eta}\left[B_{i,j-2}^{+}\Delta\bar{Q}_{i,j-2}^{*} - B_{i,j+2}^{-}\Delta\bar{Q}_{i,j+2}^{*} \right] \tag{6.54}
$$

where

$$\text{RHS}_{i,j} = -\Delta\tau\left\{\frac{1}{2\Delta\xi}\left(\bar{E}^+_{i-2,j} - 4\bar{E}^+_{i-1,j} + 3\bar{E}^+_{i,j} - 3\bar{E}^-_{i,j} + 4\bar{E}^-_{i+1,j}\right.\right.$$

$$\left. - \bar{E}^-_{i+2,j}\right) + \frac{1}{2\Delta\eta}\left(\bar{F}^+_{i,j-2} - 4\bar{F}^+_{i,j-1} + 3\bar{F}^+_{i,j} - 3\bar{F}^-_{i,j} + 4\bar{F}^-_{i,j+1}\right.$$

$$\left. - \bar{F}^-_{i,j+2}\right) - \frac{1}{2(\Delta\xi)^2}\left(\hat{e}_{i+1,j}\hat{E}_{i+1,j} - \hat{e}_{i,j}\hat{E}_{i,j} + \hat{e}_{i-1,j}\hat{E}_{i-1,j}\right)$$

$$\left. - \frac{1}{2(\Delta\eta)^2}\left(\hat{f}_{i,j+1}\hat{F}_{i,j+1} - \hat{f}_{i,j}\hat{F}_{i,j} + \hat{f}_{i,j-1}\hat{F}_{i,j-1}\right) + \alpha\left(\bar{H}_{i,j} - (\bar{H}_v)_{i,j}\right)\right\}$$

6.10.2 Three-Dimensional Thin-Layer Navier-Stokes Equations

6.10.2.1 *Explicit flux vector splitting (first-order convective terms):*

$$\bar{Q}^{n+1}_{i,j,k} = \bar{Q}^n_{i,j,k} - \Delta\tau\left[\frac{1}{\Delta\xi}\left(\bar{E}^+_{i,j,k} - \bar{E}^+_{i-1,j,k} + \bar{E}^-_{i+1,j,k} - \bar{E}^-_{i,j,k}\right)\right.$$

$$+ \frac{1}{\Delta\eta}\left(\bar{F}^+_{i,j,k} - \bar{F}^+_{i,j-1,k} + \bar{F}^-_{i,j+1,k} - \bar{F}^-_{i,j,k}\right) + \frac{1}{\Delta\zeta}(\bar{G}^+_{i,j,k} - \bar{G}^+_{i,j,k-1}$$

$$\left. + \bar{G}^-_{i,j,k+1} - \bar{G}^-_{i,j,k}) + \frac{\partial\bar{F}v_T}{\partial\eta}\right. \tag{6.55}$$

Second-order central difference formulation is used to evaluate $\frac{\partial\bar{F}v_T}{\partial\eta}$; a compact form of it can be written as

$$\left(\frac{\partial\bar{F}_{vT}}{\partial\eta}\right)_{i,j,k} = \frac{1}{2(\Delta\eta)^2}\begin{bmatrix} 0 \\[1ex] \displaystyle\sum_{m=1}^{3}\left(\bar{L}^m_{i,j+1,k}M^m_{i,j+1,k} - \bar{L}^m_{i,j,k}M^m_{i,j,k} + \bar{L}^m_{i,j-1,k}M^m_{i,j-1,k}\right) \\[2ex] \displaystyle\sum_{m=4}^{6}\left(\bar{L}^m_{i,j+1,k}M^m_{i,j+1,k} - \bar{L}^m_{i,j,k}M^m_{i,j,k} + \bar{L}^m_{i,j-1,k}M^m_{i,j-1,k}\right) \\[2ex] \displaystyle\sum_{m=7}^{9}\left(\bar{L}^m_{i,j+1,k}M^m_{i,j+1,k} - \bar{L}^m_{i,j,k}M^m_{i,j,k} + \bar{L}^m_{i,j-1,k}M^m_{i,j-1,k}\right) \\[2ex] \displaystyle\sum_{m=10}^{16}\left(\bar{L}^m_{i,j+1,k}M^m_{i,j+1,k} - \bar{L}^m_{i,j,k}M^m_{i,j,k} + \bar{L}^m_{i,j-1,k}M^m_{i,j-1,k}\right) \end{bmatrix}$$

$$\tag{6.56}$$

where

$$\bar{L}_{i,j+1,k} = L_{i,j,k} + L_{i,j+1,k}$$

$$\bar{L}_{i,j,k} = L_{i,j-1,k} + 2L_{i,j,k} + L_{i,j+1,k}$$

$$\bar{L}_{i,j-1,k} = L_{i,j,k} + L_{i,j-1,k}$$

and

m	1	2	3	4	5	6	7	8	9
$L^m / \left(\dfrac{\mu}{Re_\infty J} \right)$	b_1	b_5	b_7	b_5	b_2	b_6	b_7	b_6	b_3
M^m	u	v	w	u	v	w	u	v	w

m	10	11	12	13	14	15	16
$L^m / \left(\dfrac{\mu}{Re_\infty J} \right)$	$\dfrac{1}{2}b_1$	$\dfrac{1}{2}b_2$	$\dfrac{1}{2}b_3$	b_5	b_6	b_7	$\dfrac{\gamma}{Pr}b_4$
M^m	u^2	v^2	w^2	uv	vw	uw	e

6.10.2.2. *Beta formulation:*

$$\frac{\Delta \bar{Q}}{\Delta t} + \beta \left[\left(\frac{\partial \bar{E}}{\partial \xi} + \frac{\partial \bar{F}}{\partial \eta} + \frac{\partial \bar{G}}{\partial \zeta} - \frac{\partial \bar{F}_{vT}}{\partial \eta} \right)^{n+1} \right]$$

$$+ (1 - \beta) \left[\left(\frac{\partial \bar{E}}{\partial \xi} + \frac{\partial \bar{F}}{\partial \eta} + \frac{\partial \bar{G}}{\partial \zeta} - \frac{\partial \bar{F}_{vT}}{\partial \eta} \right)^{n} \right] = 0 \qquad (6.57)$$

where

$$\beta = 0 \quad ; \quad \text{the formulation is FTCS explicit}$$
$$\beta = \frac{1}{2} \quad ; \quad \text{the formulation is Crank-Nicolson implicit}$$

and

$$\beta = 1 \quad ; \quad \text{the formulation is Euler implicit}$$

For implicit formulations, Equation (6.57) is solved sequentially in three steps as follows

$$\left\{ I + \Delta t \beta \left[\frac{\partial}{\partial \xi}(A) + D_{\text{imp}(\xi)} \right] \right\} \Delta \bar{Q}^* = \text{RHS} + D_{\text{exp}} \qquad (6.58)$$

$$\left\{ I + \Delta t \beta \left[\frac{\partial}{\partial \eta}(B - B_v) + D_{\text{imp}(\eta)} \right] \right\} \Delta \bar{Q}^{**} = \Delta \bar{Q}^* \qquad (6.59)$$

$$\left\{ I + \Delta t \beta \left[\frac{\partial}{\partial \zeta}(C) + D_{\text{imp}(\zeta)} \right] \right\} \Delta \bar{Q} = \Delta \bar{Q}^{**} \qquad (6.60)$$

where second-order implicit and fourth-order explicit damping terms have been added. The fourth-order damping is generally given by

$$D_e = -\epsilon_e \Delta t \frac{1}{J} \left[(\nabla_\xi \Delta_\xi)^2 + (\nabla_\eta \Delta_\eta)^2 + (\nabla_\zeta \Delta_\zeta)^2 \right] J \bar{Q}^n$$

and the second-order damping is given by

$$D_{\mathrm{imp}(\xi)} = -\epsilon_i \Delta t \frac{1}{J} (\nabla_\xi \Delta_\xi J)$$

$$D_{\mathrm{imp}(\eta)} = -\epsilon_i \Delta t \frac{1}{J} (\nabla_\eta \Delta_\eta J)$$

$$D_{\mathrm{imp}(\zeta)} = -\epsilon_i \Delta t \frac{1}{J} (\nabla_\zeta \Delta_\zeta J)$$

The difference operators used in the equations above are associated with the central difference approximations of second- and fourth-order derivatives. For example,

$$(\nabla_\xi \Delta_\xi) Q = Q_{i+1} - 2Q_i + Q_{i-1}$$

and

$$(\nabla_\xi \Delta_\xi)^2 Q = Q_{i+2} - 4Q_{i+1} + 6Q_i - 4Q_{i-1} + Q_{i-2}$$

6.11   References

[6.1] Hoffmann, K. A., and Chiang, S. T., "Computational Fluid Dynamics for Engineers – Volume I," EES, 1993.

[6.2] Hirsch, C., "Numerical Computation of Internal and External Flows – Volume I," John Wiley & Sons, 1988.

Chapter 7

Auxiliary Relations and Fluid Properties

7.1   Introductory Remarks

The governing equations of fluid motion typically involve more unknowns than available conservation equations which include conservations of mass, momentum, and energy. Therefore, in order to close the system, additional relations such as the equation of state and relations for the transport properties such as viscosity and thermal conductivity must be introduced. This chapter will summarize the appropriate relations and will provide some limited values of fluid properties for selected gases and liquids.

7.2   Viscosity

Viscosity of a fluid is a property used to describe the response of a fluid to the imposed shearing forces. In a fluid flow, the rate of deformation is proportional to the shear stress. The constant of proportionality is called the *coefficient of dynamic viscosity*. Typically, it is referred to as the *coefficient of viscosity* or just as *viscosity*. If the relation between the shear stress and the rate of deformation is linear, then the fluid is called a *Newtonian fluid*. These relations are provided in Tables 1.5a and 1.5b.

In general, the coefficient of viscosity is a function of composition of the fluid, its temperature, and pressure. In most cases, the pressure dependency is negligible and the coefficient of viscosity is expressed as a function of temperature only. Pressure dependency, however, should be included at very high or at very low pressures. An interesting fact is that the coefficient of viscosity for a liquid (with the exception of water) decreases with increasing temperature, while the coefficient of viscosity for a gas always increases with temperature.

There are several expressions relating the coefficient of viscosity to the temperature. Perhaps the most commonly used relation for dilute gases is the Sutherland's law, which can be expressed as

$$\mu = c_1 \frac{T^{3/2}}{T + c_2} \tag{7.1}$$

where the constants c_1 and c_2 are given in Table 7.1 for selected gases.

Table 7.1: Sutherland's constants for various gases.

Gas		SI system		British system	
		$c_1 \times 10^6$ $\left(\dfrac{\text{kg}}{\text{sec m K}^{1/2}}\right)$	c_2 (K)	$c_1 \times 10^8$ $\left(\dfrac{\text{lbf sec}}{\text{ft}^2 \, {}^\circ\text{R}^{1/2}}\right)$	c_2 (°R)
Air		1.458	110.4	2.27	198.6
Carbon dioxide	CO_2	1.550	233.0	2.42	420.0
Carbon monoxide	CO	1.400	109.0	2.18	196.2
Freon-12	$C Cl_2 F_2$	1.480	317.0	2.31	570.0
Helium	He	1.520	97.8	2.36	176.0
Hydrogen	H_2	0.649	70.6	1.01	127.0
Methane	CH_4	0.983	155.0	1.53	279.0
Nitrogen	N_2	1.390	102.0	2.16	183.6
Oxygen	O_2	1.650	110.0	2.57	198.0

The temperature in Equation (7.1) must be in °R or K, providing the values of viscosity in the units of $\frac{\text{lbf sec}}{\text{ft}^2} = \frac{\text{slug}}{\text{ft sec}}$ or $\frac{\text{kg}}{\text{sec m}} = \frac{\text{N sec}}{\text{m}^2}$, respectively.

The Sutherland's law may also be expressed as

$$\frac{\mu}{\mu_o} = \left(\frac{T}{T_o}\right)^{3/2} \frac{T_o + S}{T + S} \tag{7.2}$$

or in a power law form as

$$\frac{\mu}{\mu_o} = \left(\frac{T}{T_o}\right)^n \tag{7.3}$$

where S is an effective temperature, called *Sutherland's constant*, which is characteristic of the gas. The values of S, n, and T_o are given in Table 7.2

Table 7.2: Constants used in Sutherland's law (7.2) and power law (7.3) for various gases.

Gas		T_o (°R)	n	μ_o (mP)	% error for temperature range (°R)	S (°R)	±2% error for temperature range (°R)
Air		491.6	0.666	0.1716	378–3420 ±4	199	300–3420
Argon	Ar	491.6	0.720	0.2125	360–2700 ±3	260	220–2700
Carbon dioxide	CO_2	491.6	0.710	0.1657	414–2700 ±2	245	234–2700
Carbon monoxide	CO	491.6	0.790	0.1370	377–3060 ±5	400	342–3060
Hydrogen	H_2	491.6	0.680	0.08411	144–1980 ±2	174	404–1980
Nitrogen	N_2	491.6	0.670	0.1663	400–2700 ±3	192	180–2700
Oxygen	O_2	491.6	0.690	0.1919	414–3600 ±2	250	335–3600
Steam		750.0	1.040	0.1703	504–2700 ±3	1550	650–2700

Viscosity of selected gases at atmospheric pressure is provided in Table 7.3.

Table 7.3: Viscosity of selected gases at atmospheric pressure $(\mu \times 10^4 \frac{\text{N·sec}}{\text{m}^2})$

Gas	T(K)									
	50	100	150	200	300	400	500	600	800	1000
Air		0.071	0.104	0.133	0.185	0.230	0.270	0.306	0.370	0.425
Argon		0.082	0.122	0.160	0.229	0.289	0.343			
Carbon dioxide					0.150	0.196	0.239	0.278	0.346	0.406
Carbon monoxide		0.067	0.100	0.129	0.178	0.221	0.258	0.291	0.350	0.402
Helium	0.064	0.098	0.125	0.151	0.199	0.243	0.284	0.322	0.394	0.462
Hydrogen	0.025	0.042	0.055	0.068	0.090	0.109	0.126	0.143	0.173	0.201
Methane		0.041	0.059	0.078	0.112	0.142	0.170	0.194	0.238	0.276
Nitrogen		0.068	0.100	0.129	0.180	0.223	0.261			
Oxygen		0.076	0.112	0.147	0.207					

7.2.1 Viscosity from the Point of View of Kinetic Theory

Shear stresses within a gas are the result of the momentum transport of the molecules due to their random motion. A simple analysis of this physical activity provides a relation for the viscosity as

$$\mu = \frac{1}{3}\rho a \lambda$$

A more accurate expression can be obtained if one considers the molecular quantities in which Maxwell's distribution function is used. That expression is

$$\mu = 0.499\,\rho a \lambda$$

where λ is the mean free path of molecules, and a is the average speed of a particle given by

$$a = \int_0^\infty v f(v)\,dv = \sqrt{\frac{8kT}{\pi m}}$$

where

$$m = \text{Mass of the molecule}$$

$$T = \text{Temperature}$$

$$k = \text{Boltzmann constant} = \begin{cases} 1.3803 \times 10^{-23}\ \frac{\text{J}}{\text{Molecule K}} \\[6pt] 1.3803 \times 10^{-16}\ \frac{\text{ergs}}{\text{Molecule K}} \\[6pt] 7.23 \times 10^{-27}\ \frac{\text{Btu}}{\text{Molecule }^\circ\text{R}} \end{cases}$$

$$f(v) = \text{Maxwell-Boltzmann velocity distribution function}$$

$$= \frac{4v^2}{\sqrt{\pi}}\left(\frac{m}{2kT}\right)^{3/2} \exp\left(-\frac{mv^2}{2kT}\right)$$

A generalized kinetic-theory formula for viscosity is given by

$$\mu = \frac{0.002\sqrt{MW\,T}}{\sigma^2 \Omega_v} \tag{7.4}$$

where

$$\sigma = \text{Collision diameter (Angstrom)}$$

$$MW = \text{Molecular weight of the gas}$$

$$\mu = \text{Viscosity (centipoise)}$$

$$T = \text{Absolute temperature (}^\circ\text{ R)}$$

and Ω_v is dimensionless and equal to unity for noninteracting molecules. Equation (7.4) may also be expressed as

$$\mu = \frac{2.6693 \times 10^{-5}\sqrt{MW\,T}}{\sigma^2 \Omega_v} \qquad (7.5)$$

where σ is in units of Angstrom, μ is in units of $\frac{g}{cm\;sec}$, and T is in K.

In general, Ω_v depends on the ratio of T/T_ϵ, where T_ϵ is an effective temperature characteristic of particular molecule and of the force potential selected. Some values of Ω_v are provided in Table 7.4, and values of both σ and T_ϵ are given in Table 7.5.

Table 7.4: The values of collision integrals for Stockmayer potential.

$T^* = T/T_\epsilon$	Ω_v
0.3	2.840
1.0	1.593
3.0	1.039
10.0	0.8244
30.0	0.7010
100.0	0.5887
400.0	0.4811

Table 7.5: Effective temperature and collision diameter for dilute gas.

Gas		T_ϵ (K)	σ (Å)
Air	Air	78.6	3.711
Argon	Ar	93.3	3.542
Methane	CH_4	148.6	3.758
Carbon monoxide	CO	91.7	3.690
Carbon dioxide	CO_2	195.2	3.941
Helium	He	10.22	2.551
Hydrogen	H_2	59.7	2.827
Nitrogen	N_2	71.4	3.798
Oxygen	O_2	106.7	3.467

Additional information on viscosity of gases is provided in Section 4.6.2.

7.2.2 Pressure and Temperature Dependency of Viscosity for Air

As described previously, no single functional relation $\mu = \mu(p, T)$ exists which can describe any large class of fluids. Typically, a set of tables and/or charts are used to determine the appropriate value of viscosity at a given pressure and temperature.

The tabulated values of viscosity for air are provided in Table 7.6, taken from Ref. [7.1].

Table 7.6: Coefficient of viscosity for air.

| T (K) | μ/μ_o | | | | | | | Reference viscosity $\mu_0 \times 10^{-6}$ | |
| | Pressure (atm) | | | | | | | | |
	100	10	1.0	0.1	0.01	0.001	0.0001	$\frac{\text{lb sec}}{\text{ft}^2}$	$\frac{\text{gm}}{\text{cm sec}}$
500	1.000	1.000	1.000	1.000	1.000	1.000	1.000	0.558	267
1000	1.000	1.000	1.000	1.000	1.000	1.000	1.000	0.868	416
1500	1.000	1.000	1.000	1.000	1.000	1.000	1.000	1.100	527
2000	1.000	1.000	1.000	1.000	1.000	1.000	1.000	1.293	619
2500	1.000	1.000	1.000	1.000	1.000	1.000	1.000	1.461	700
3000	1.000	1.000	1.000	1.000	1.000	1.000	1.000	1.612	772
3500	1.000	1.001	1.003	1.006	1.010	1.010	1.011	1.751	838
4000	1.003	1.008	1.016	1.020	1.022	1.024	1.032	1.879	899
4500	1.010	1.022	1.029	1.033	1.038	1.055	1.096	1.999	957
5000	1.022	1.036	1.043	1.051	1.074	1.128	1.181	2.11	1011
5500	1.036	1.052	1.060	1.086	1.146	1.209	1.227	2.22	1062
6000	1.050	1.067	1.090	1.148	1.228	1.257	1.256	2.32	1112
6500	1.072	1.090	1.139	1.229	1.276	1.286	1.271	2.42	1159
7000	1.089	1.124	1.208	1.294	1.317	1.303	1.264	2.52	1204
7500	1.112	1.175	1.283	1.332	1.337	1.307	1.210	2.61	1247
8000	1.143	1.238	1.342	1.371	1.347	1.280	1.072	2.69	1289
8500	1.185	1.307	1.386	1.386	1.343	1.207	0.826	2.78	1330
9000	1.238	1.368	1.425	1.396	1.314	1.068	0.517	2.86	1370
9500	1.298	1.418	1.438	1.393	1.251	0.853	0.261	2.94	1408
10000	1.361	1.468	1.445	1.375	1.143	0.595	0.118	3.02	1446
10500	1.418	1.496	1.448	1.335	0.983	0.361	0.055	3.10	1482
11000	1.467	1.501	1.442	1.267	0.782	0.200	0.029	3.17	1518
11500	1.509	1.511	1.424	1.168	0.571	0.108	0.018	3.24	1552
12000	1.549	1.520	1.394	1.040	0.387	0.063	0.012	3.31	1586
12500	1.577	1.516	1.342	0.881	0.249	0.036	0.009	3.38	1620
13000	1.581	1.508	1.274	0.711	0.158	0.024	0.008	3.45	1652
13500	1.594	1.492	0.187	0.347	0.100	0.018	0.007	3.52	1684
14000	1.599	1.468	1.082	0.408	0.067	0.015	0.007	3.58	1716
14500	1.601	1.415	0.940	0.268	0.042	0.013	0.008	3.65	1747
15000	1.604	1.387	0.828	0.212	0.016	0.012	0.008	3.71	1777

7.2.3 Viscosity of Liquids

Viscosity of liquids is also found to be primarily dependent on temperature and slightly dependent on pressure. Thus, in most applications, the pressure dependency is ignored. The viscosity of some liquids at various temperatures is provided in Table 7.7.

Table 7.7: Viscosity of selected liquids at atmospheric pressure
$(\mu \times 10^4 \ \frac{\text{N·sec}}{\text{m}^2})$

Liquid					T (K)					
	180	200	220	240	260	280	300	320	360	400
Ammonia		4.97	3.48	2.52	1.94	1.58	1.31	1.08	0.70	0.48
Benzene						8.10	5.88	4.53	2.97	2.03
Butane		4.83	3.68	2.91	2.36	1.94	1.60	1.32	0.89	0.59
Chlorine	8.94	7.00	5.71	4.84	4.19	3.70	3.35			
Ethanol	332.00	136.00	70.00	40.00	24.40	15.60	10.50	7.31	3.96	2.31
Methanol	102.00	43.90	24.10	15.10	10.10	7.17	5.34	4.10	2.54	1.66
Propane	3.86	2.98	2.38	1.92	1.55	1.24	0.98	0.77	0.46	
R 12	9.70	6.70	4.90	3.80	3.00	2.50	2.13	1.85	1.28	
Water						14.20	8.55	5.77	3.24	2.17

7.2.4 Kinematic Viscosity

Kinematic viscosity is defined as the ratio of dynamic viscosity to the density

$$\nu = \frac{\mu}{\rho}$$

7.2.5 Bulk Viscosity

Bulk viscosity represents the energy imbalance between the translational energy mode and other internal energy modes such as rotational and vibrational. It is also argued that bulk viscosity is a representative of rotational nonequilibrium. Therefore, the effect of bulk viscosity may be important for flowfields where thermal nonequilibrium is considerable. Bulk viscosity is given by

$$k = \frac{2}{3}\mu + \lambda$$

where λ is known as the second coefficient of viscosity. For most applications, the bulk viscosity is assumed to be zero. This assumption is known as the *Stokes hypothesis*.

7.3 Thermal Conductivity

The Fourier's heat conduction law states that the heat conduction per unit area is proportional to the normal gradient of temperature. Mathematically,

$$\frac{q}{A} \propto -\frac{\partial T}{\partial n} \tag{7.6}$$

Expression (7.6) can be expressed as an equation by introducing a constant of proportionality defined as *thermal conductivity k*. Therefore,

$$\frac{q}{A} = -k\frac{\partial T}{\partial n}$$

It should be observed that the equation above involving thermal conductivity has similar mathematical form as viscosity. In fact, thermal conductivity also depends on the composition of material, pressure, and temperature. Just as viscosity, the dependency of thermal conductivity at moderate pressures is secondary and, in those applications, it can be determined based on temperature alone. Just as for viscosity, Sutherland's law or the power law can be used to determine the thermal conductivity of dilute gases. The formulations are given by

$$\frac{k}{k_o} = \left(\frac{T}{T_o}\right)^{3/2} \frac{T_o + S}{T + S} \tag{7.7}$$

and, in a power law form, as

$$\frac{k}{k_o} = \left(\frac{T}{T_o}\right)^n \tag{7.8}$$

The values of S, T_o, k_o, and n for several gases are provided in Table 7.8.

<table>
<tr><td colspan="7">Table 7.8: Constants used in thermal conductivity relations (7.7) and (7.8).</td></tr>
<tr>
<td>Gas</td>
<td>T_o
(°R)</td>
<td>n</td>
<td>k_o
$\left(\frac{Btu}{hr\ ft\ °R}\right)$</td>
<td>% error
for
temperature
range (°R)</td>
<td>S
(°R)</td>
<td>±2%
error for
tempera-
ture range
(°R)</td>
</tr>
<tr><td>Air</td><td>491.6</td><td>0.81</td><td>0.01395</td><td>±3 375–1800</td><td>350</td><td>300–1800</td></tr>
<tr><td>Argon   Ar</td><td>491.6</td><td>0.73</td><td>0.009444</td><td>±4 385–2700</td><td>270</td><td>270–2700</td></tr>
<tr><td>Carbon
dioxide CO_2</td><td>491.6</td><td>0.85</td><td>0.01342</td><td>±2 370–1080</td><td>320</td><td>230–1080</td></tr>
<tr><td>Carbon
monoxide CO</td><td>491.6</td><td>1.38</td><td>0.008407</td><td>±2 324–1080</td><td>4000</td><td>324–1080</td></tr>
<tr><td>Hydrogen H_2</td><td>491.6</td><td>0.85</td><td>0.0940</td><td>±2 365–1260</td><td>300</td><td>325–1260</td></tr>
<tr><td>Nitrogen N_2</td><td>491.6</td><td>0.76</td><td>0.0140</td><td>±4 375–2160</td><td>300</td><td>260–2160</td></tr>
<tr><td>Oxygen O_2</td><td>491.6</td><td>0.86</td><td>0.01419</td><td>±2 390–1080</td><td>400</td><td>360–1080</td></tr>
<tr><td>Steam</td><td>491.6</td><td>1.20</td><td>0.01036</td><td>±2 540–1440</td><td>2300</td><td>540–1440</td></tr>
</table>

Thermal conductivity of several gases at atmospheric pressure is also provided in Table 7.9.

Gas	T (K)								
	100	150	200	300	400	500	600	800	1000
Air	0.0092	0.0138	0.0181	0.0261	0.0331	0.0395	0.0450	0.0569	0.0672
Argon	0.0065	0.0096	0.0124	0.0177	0.0223	0.0264	0.0301	0.0369	0.0427
Carbon dioxide			0.0095	0.0166	0.0244	0.0323	0.0403	0.0560	0.0680
Helium	0.0073	0.0095	0.1151	0.1500	0.1800	0.2110	0.2470	0.3070	0.3630
Hydrogen	0.0680	0.0990	0.1280	0.1820	0.2210	0.2560	0.2910	0.3600	0.4280
Methane	0.0106	0.0162	0.0218	0.0343	0.0484	0.0671	0.0948	0.1240	0.1690
Nitrogen	0.0094	0.0139	0.0183	0.0260	0.0325	0.0386	0.0441	0.0541	0.0631
Oxygen	0.0091	0.0138	0.0182	0.0267	0.0343	0.0412	0.0480	0.0603	0.0717

Table 7.9: Thermal conductivity of selected gases at atmospheric pressure, k (W/m K)

The thermal conductivity of air below 2000 K at atmospheric pressure may be calculated from the following relation as well.

$$k = 4.76 \times 10^{-6} \frac{T^{3/2}}{T + 112.0}$$

where T is in (K) and k is in $\left(\frac{cal}{cm\ sec\ K}\right)$.

A generalized kinetic-theory based formula for thermal conductivity is given by

$$k = \frac{0.000199}{\sigma^2 \Omega_k} \sqrt{\frac{T}{MW}}$$

In this formulation, T is in (K), providing units of $\left(\frac{cal}{cm\ sec\ K}\right)$ for thermal conductivity. The value of Ω_k is identical to that of Ω_v appearing in the expression for viscosity, i.e., (7.4). The values of σ for various gases is provided in Table 7.5, whereas the values of collision integrals are given in Table 7.4.

7.3.1 Pressure and Temperature Dependency of Thermal Conductivity for Air

A table similar to Table 7.6, which provided values of viscosity for air, is provided as Table 7.10 for the values of thermal conductivity.

<table>
<tr><td colspan="10" align="center">Table 7.10: Thermal conductivity of air.</td></tr>
<tr><td rowspan="3">T (K)</td><td colspan="7" align="center">k/k_o</td><td colspan="2" align="center">Reference thermal conductivity</td></tr>
<tr><td colspan="7" align="center">Pressure (atm)</td><td colspan="2" align="center">$k_0 \times 10^{-4}$</td></tr>
<tr><td>100</td><td>10</td><td>1.0</td><td>0.1</td><td>0.01</td><td>0.001</td><td>0.0001</td><td>$\frac{\text{Btu}}{\text{ft sec }^\circ\text{R}}$</td><td>$\frac{\text{watt}}{\text{cm K}}$</td></tr>
<tr><td>500</td><td>1.021</td><td>1.021</td><td>1.021</td><td>1.021</td><td>1.021</td><td>1.021</td><td>1.021</td><td>5.84</td><td>364</td></tr>
<tr><td>1000</td><td>1.100</td><td>1.100</td><td>1.100</td><td>1.100</td><td>1.100</td><td>1.100</td><td>1.100</td><td>9.10</td><td>567</td></tr>
<tr><td>1500</td><td>1.150</td><td>1.150</td><td>1.150</td><td>1.150</td><td>1.150</td><td>1.150</td><td>1.150</td><td>11.53</td><td>719</td></tr>
<tr><td>2000</td><td>1.177</td><td>1.177</td><td>1.177</td><td>1.251</td><td>1.460</td><td>2.090</td><td>3.990</td><td>13.55</td><td>844</td></tr>
<tr><td>2500</td><td>1.256</td><td>1.317</td><td>1.619</td><td>2.500</td><td>4.630</td><td>7.670</td><td>5.500</td><td>15.31</td><td>954</td></tr>
<tr><td>3000</td><td>1.421</td><td>1.928</td><td>3.200</td><td>5.480</td><td>5.020</td><td>2.190</td><td>1.465</td><td>16.90</td><td>1053</td></tr>
<tr><td>3500</td><td>1.941</td><td>3.150</td><td>4.720</td><td>3.960</td><td>1.719</td><td>2.110</td><td>3.710</td><td>18.35</td><td>1143</td></tr>
<tr><td>4000</td><td>2.690</td><td>3.940</td><td>2.990</td><td>1.600</td><td>2.910</td><td>6.040</td><td>15.030</td><td>19.69</td><td>1227</td></tr>
<tr><td>4500</td><td>3.220</td><td>3.060</td><td>1.714</td><td>3.320</td><td>7.340</td><td>17.650</td><td>30.500</td><td>21.00</td><td>1305</td></tr>
<tr><td>5000</td><td>3.070</td><td>1.997</td><td>3.290</td><td>7.180</td><td>16.630</td><td>25.800</td><td>11.840</td><td>22.10</td><td>1379</td></tr>
<tr><td>5500</td><td>2.460</td><td>2.910</td><td>5.990</td><td>13.710</td><td>22.200</td><td>11.400</td><td>3.540</td><td>23.30</td><td>1449</td></tr>
<tr><td>6000</td><td>1.930</td><td>4.530</td><td>10.190</td><td>18.740</td><td>13.090</td><td>3.960</td><td>6.140</td><td>24.30</td><td>1516</td></tr>
<tr><td>6500</td><td>3.350</td><td>6.980</td><td>14.500</td><td>15.390</td><td>5.490</td><td>5.920</td><td>12.990</td><td>25.40</td><td>1580</td></tr>
<tr><td>7000</td><td>4.690</td><td>9.970</td><td>15.690</td><td>8.320</td><td>3.280</td><td>10.950</td><td>26.900</td><td>26.40</td><td>1642</td></tr>
<tr><td>7500</td><td>6.310</td><td>12.480</td><td>12.240</td><td>5.920</td><td>8.620</td><td>19.970</td><td>51.900</td><td>27.30</td><td>1701</td></tr>
<tr><td>8000</td><td>8.210</td><td>13.190</td><td>7.800</td><td>3.420</td><td>13.990</td><td>34.700</td><td>84.800</td><td>28.20</td><td>1759</td></tr>
<tr><td>8500</td><td>9.860</td><td>11.550</td><td>5.100</td><td>9.720</td><td>22.100</td><td>56.300</td><td>107.600</td><td>29.10</td><td>1814</td></tr>
<tr><td>9000</td><td>10.900</td><td>8.790</td><td>3.260</td><td>14.030</td><td>34.100</td><td>78.800</td><td>93.800</td><td>30.00</td><td>1868</td></tr>
<tr><td>9500</td><td>10.880</td><td>6.380</td><td>6.660</td><td>20.300</td><td>49.100</td><td>92.600</td><td>54.900</td><td>30.80</td><td>1921</td></tr>
<tr><td>10000</td><td>9.870</td><td>4.070</td><td>10.950</td><td>28.200</td><td>64.800</td><td>84.800</td><td>25.600</td><td>31.60</td><td>1972</td></tr>
<tr><td>10500</td><td>8.330</td><td>6.500</td><td>14.600</td><td>37.600</td><td>76.600</td><td>59.900</td><td>11.450</td><td>32.40</td><td>2020</td></tr>
<tr><td>11000</td><td>6.840</td><td>8.470</td><td>19.170</td><td>48.300</td><td>78.600</td><td>35.200</td><td>5.810</td><td>33.20</td><td>2070</td></tr>
<tr><td>11500</td><td>5.590</td><td>10.480</td><td>24.500</td><td>58.200</td><td>68.600</td><td>19.040</td><td>3.400</td><td>34.00</td><td>2120</td></tr>
<tr><td>12000</td><td>4.790</td><td>12.960</td><td>30.900</td><td>65.400</td><td>51.600</td><td>10.750</td><td>2.290</td><td>34.70</td><td>2160</td></tr>
<tr><td>12500</td><td>3.340</td><td>15.550</td><td>37.000</td><td>67.300</td><td>34.800</td><td>6.090</td><td>1.807</td><td>35.50</td><td>2210</td></tr>
<tr><td>13000</td><td>5.780</td><td>18.770</td><td>43.300</td><td>63.400</td><td>22.500</td><td>4.080</td><td>1.622</td><td>36.20</td><td>2250</td></tr>
<tr><td>13500</td><td>9.950</td><td>22.300</td><td>48.700</td><td>54.500</td><td>14.280</td><td>3.180</td><td>1.461</td><td>36.90</td><td>2250</td></tr>
<tr><td>14000</td><td>11.440</td><td>25.900</td><td>52.300</td><td>43.500</td><td>9.580</td><td>2.760</td><td>1.586</td><td>37.60</td><td>2340</td></tr>
<tr><td>14500</td><td>13.730</td><td>31.400</td><td>54.200</td><td>30.400</td><td>6.250</td><td>2.430</td><td>1.670</td><td>38.20</td><td>2380</td></tr>
<tr><td>15000</td><td>14.950</td><td>33.800</td><td>52.400</td><td>24.300</td><td>3.040</td><td>2.330</td><td>1.754</td><td>38.90</td><td>2420</td></tr>
</table>

7.3.2 Thermal Conductivity of Liquids

Just as viscosity, the thermal conductivity of liquids are primarily temperature dependent and are only slightly influenced by pressure. Therefore, in most applica-

tions the pressure effect is ignored. The thermal conductivity of selected liquids at
atmospheric pressure is provided in Table 7.11.

<table>
<tr><td colspan="11" align="center">Table 7.11: Thermal conductivity of selected liquids at atmo-
spheric pressure, k (W/m · K)</td></tr>
<tr><td rowspan="2">Liquid</td><td colspan="10" align="center">T(K)</td></tr>
<tr><td>180</td><td>200</td><td>220</td><td>240</td><td>260</td><td>280</td><td>300</td><td>320</td><td>360</td><td>400</td></tr>
<tr><td>Ammonia</td><td></td><td>0.657</td><td>0.620</td><td>0.581</td><td>0.542</td><td>0.504</td><td>0.467</td><td>0.425</td><td>0.337</td><td>0.230</td></tr>
<tr><td>Benzene</td><td></td><td></td><td></td><td></td><td></td><td>0.153</td><td>0.143</td><td>0.134</td><td>0.125</td><td>0.114</td></tr>
<tr><td>Butane</td><td>0.161</td><td>0.151</td><td>0.141</td><td>0.131</td><td>0.121</td><td></td><td></td><td></td><td></td><td></td></tr>
<tr><td>Chlorine</td><td>0.189</td><td>0.181</td><td>0.172</td><td>0.163</td><td>0.154</td><td>0.144</td><td>0.134</td><td>0.124</td><td>0.101</td><td>0.072</td></tr>
<tr><td>Methanol</td><td>0.242</td><td>0.234</td><td>0.227</td><td>0.220</td><td>0.213</td><td>0.207</td><td>0.200</td><td>0.193</td><td>0.183</td><td>0.175</td></tr>
<tr><td>Propane</td><td>0.165</td><td>0.152</td><td>0.139</td><td>0.123</td><td>0.115</td><td>0.106</td><td>0.096</td><td>0.087</td><td>0.068</td><td></td></tr>
<tr><td>R 12</td><td>0.112</td><td>0.105</td><td>0.097</td><td>0.091</td><td>0.083</td><td>0.076</td><td>0.068</td><td>0.061</td><td>0.047</td><td></td></tr>
<tr><td>Water</td><td></td><td></td><td></td><td></td><td></td><td>0.582</td><td>0.613</td><td>0.640</td><td>0.674</td><td>0.688</td></tr>
</table>

7.4 Specific Heats

Specific heat represents the amount of heat δQ required to produce a variation
dT in the temperature of a substance. Specific heat may be defined under a constant
pressure process or a constant volume process. Specific heat at constant pressure is
defined as

$$c_p = \left(\frac{\partial h}{\partial T}\right)_p \tag{7.9}$$

and specific heat at constant volume is defined as

$$c_v = \left(\frac{\partial e}{\partial T}\right)_v \tag{7.10}$$

The ratio of specific heats is defined as

$$\gamma = \frac{c_p}{c_v} \tag{7.11}$$

In general, the enthalpy h and the internal energy e are functions of two thermodynamic properties such as pressure and temperature. If one assumes that h and e
are only functions of temperature, then the gas is called *thermally perfect gas*. For a
thermally perfect gas, relations (7.9) and (7.10) can be expressed as

$$c_p = \frac{dh}{dT} \qquad \text{or} \qquad dh = c_p\, dT \tag{7.12}$$

and

$$c_v = \frac{de}{dT} \qquad \text{or} \qquad de = c_v \, dT \tag{7.13}$$

In general, specific heats are a function of both pressure and temperature. Typically, the influence of pressure is less significant, and, therefore, it may be ignored. That is particularly the case for liquids. The effect of pressure on specific heats of liquids becomes important only at extremely high pressures. Even the temperature dependency of specific heats is slight for relatively moderate changes in temperature. However, the temperature effect on the specific heats of gases is more appreciable, and some pressure dependency at extreme values of pressure is present. Typically, the influence of pressure is less at higher temperatures.

For a calorically perfect gas, the specific heats c_p and c_v are constants. Thus, relations (7.12) and (7.13) can be written as

$$h = c_p \, T \tag{7.14}$$

and

$$e = c_v \, T \tag{7.15}$$

For a thermally perfect gas, the following relations hold

$$c_p - c_v = R \tag{7.16}$$

$$c_p = \frac{\gamma R}{\gamma - 1} \tag{7.17}$$

$$c_v = \frac{R}{\gamma - 1} \tag{7.18}$$

where R is the gas constant, defined in Sec. 7.5. According to Kinetic Theory, the ratio of specific heats is

$$\gamma = \frac{5}{3} = 1.67 \qquad \text{for monatomic gases}$$

$$\gamma = \frac{7}{5} = 1.40 \qquad \text{for diatomic gases}$$

$$\gamma = \frac{8}{6} = 1.33 \qquad \text{for polyatomic gases}$$

A simple temperature-dependent expression for c_p can be established by the curve fit of experimental data. For example, a polynomial form of the following may be used to determine c_p.

$$c_p = (a + bT + cT^2 + dT^3 + eT^4)R \tag{7.19}$$

The units of c_p are the same as R, i.e., $(Nm)/(kg\ K)$, and T is in K. The values of coefficients are presented in Table 7.12, as provided in Ref. [7.2].

Table 7.12: The coefficients used in expression (7.19).
(Source: Ref. [7.2])

Gas	Temperature Range, K	a	$b \times 10^3$	$c \times 10^6$	$d \times 10^9$	$e \times 10^{12}$
Argon	300–1000	2.50				
	1000–5000	2.50				
Carbon dioxide	300–1000	2.40078	8.73510	−6.60709	2.00219	0.000632740
	1000–5000	4.46080	3.09817	−1.23926	0.227413	−0.0155260
Carbon monoxide	300-1000	3.71009	−1.61910	3.69236	−2.03197	0.239533
	1000–5000	2.98407	.1.48914	−0.578997	0.103646	−0.00693536
Hydrogen	300–1000	3.05745	2.67652	−5.80992	5.52104	−1.81227
	1000–5000	3.10019	0.511195	0.0526442	−0.0349100	0.00369453
Methane	300–1000	3.82619	−3.97946	24.5583	−22.7329	6.96270
	1000–5000	1.50271	10.4168	−3.91815	0.677779	−0.0442837
Nitrogen	300–1000	3.67483	−1.20815	2.32401	−0.632176	−0.225773
	1000–5000	2.89632	1.51549	−0.572353	0.0998074	−0.00652236
Oxygen	300–1000	3.62560	−1.87822	7.05545	−6.76351	2.15560
	1000–5000	3.62195	0.736183	−0.196522	0.0362016	−0.00289456
Steam	300–1000	4.07013	−1.10845	4.15212	−2.96374	0.807021
	1000–5-000	2.71676	2.94514	−0.802243	0.102267	−0.00484721

The values of specific heat at constant pressure for selected gases and liquids are provided in Tables 7.13 and 7.14. Note that the values in Table 7.13 are in terms of c_p/R, where R is the gas constant; its value is provided in Table 7.15 in both SI and British units.

Table 7.13: c_p/R of selected gases at atmospheric pressure.

Gas	T(K)				
	100	200	300	400	500
Air	3.5824	3.5062	3.5059	3.5333	3.5882
Argon	2.5000	2.5000	2.5000	2.5000	2.5000
Carbon dioxide	3.5010	3.5010	3.5050	3.5290	3.5830
Carbon monoxide	3.5120	3.8810	4.4600	4.9520	5.4360
Helium	2.5000	2.5000	2.5000	2.5000	2.5000
Hydrogen					3.5200
Methane	4.0000	4.0260	4.2950	4.8710	5.5740
Nitrogen	3.5000	3.5010	3.5030	3.5180	3.5580
Oxygen	3.5010	3.5030	3.5340	3.6210	3.7390

Gas	T(K)				
	600	800	1000	2000	3000
Air	3.6626	3.8280	3.9790	4.6620	9.9600
Argon	2.5000	2.5000	2.5000	2.5000	2.5000
Carbon dioxide	3.6610	3.8370	3.9910	4.3580	4.4710
Carbon monoxide	5.6690	6.1630	6.5090	7.2420	7.4600
Helium	2.5000	2.5000	2.5000	2.5000	2.5000
Hydrogen	3.5270	3.5620	3.6320	4.1120	4.4210
Methane	6.2820	7.5690	8.6350	11.3540	12.1940
Nitrogen	3.6210	3.7810	3.9320	4.3250	4.4500
Oxygen	3.8600	4.0570	4.1940	4.5390	4.7940

Table 7.14: Specific heat at constant pressure of selected liquids, c_p (kJ/kg K).

Liquid	T(K)									
	180	200	220	240	260	280	300	320	360	400
Ammonia		4.610	4.350	4.43	4.540	4.660	4.810	5.040	6.040	24.10
Benzene						1.690	1.730	1.780	1.910	2.09
Butane	2.02	2.050	2.100	2.16	2.230	2.320	2.440	2.580	2.950	
Chlorine	0.95	0.940	0.930	0.93						
Methanol		2.190	2.230	2.28	2.340	2.420	2.530	2.670	3.010	3.47
Propane	2.06	2.120	2.190	2.27	2.370	2.500	2.670	2.970	4.170	
R 12		0.855	0.873	0.89	0.913	0.942	0.979	1.041	1.390	

7.5   Equation of State

A relation between the density, pressure, and temperature of a fluid is known as the *equation of state*. This relation may be presented in the form of tables, charts, or an equation. For a thermally perfect gas, the equation of state is

$$p = \rho R T$$

where R is the gas constant defined by the universal gas constant $\mathcal{R}$ divided by the molecular weight, i.e.,

$$R = \frac{\mathcal{R}}{MW}$$

The universal gas constant is

$$\mathcal{R} = \begin{cases} 8314.34 \; \frac{\text{N·m}}{\text{kg mole K}} & = 1.987 \; \frac{\text{cal}}{\text{gm mole K}} \\[2ex] 1545.33 \; \frac{\text{ft lbf}}{\text{lbm mole °R}} & = 4.9723 \times 10^4 \frac{\text{ft lbf}}{\text{Slug mole °R}} \end{cases}$$

The gas constant for air is

$$R = \begin{cases} 287.05 \; \frac{\text{N·m}}{\text{kg K}} \\[2ex] 53.34 \; \frac{\text{ft lbf}}{\text{lbm °R}} & = 1716.16 \; \frac{\text{ft·lbf}}{\text{Slug °R}} \end{cases}$$

The gas constants for various gases are provided in Table 7.15.

Table 7.15: Molecular weight and gas constants of selected gases.

Gas		MW	$R\left(\frac{\text{N m}}{\text{kg K}}\right)$	$R\left(\frac{\text{ft lbf}}{\text{lbm }^\circ\text{R}}\right)$
Air		28.966	287.04	53.34
Argon	Ar	39.948	208.13	38.73
Carbon dioxide	CO_2	44.010	188.92	35.11
Carbon monoxide	CO	28.010	296.83	55.19
Freon 12	$C\,Cl_2\,F_2$	120.91	68.76	12.78
Helium	He	4.003	2077.00	386.00
Hydrogen	H_2	2.016	4124.18	767.00
Methane	CH_4	16.043	518.25	96.30
Nitrogen	N_2	28.013	296.80	55.15
Oxygen	O_2	31.999	260.00	48.29

An equation of state for real gases is the Van der Waals equation provided in Sec. 4.2.2 by Equation (4.1).

7.6 Speed of Sound

The speed of sound is the speed at which an infinitessimal pressure pulse propagates through a fluid. Since the changes in the fluid properties are infinitessimal, the process is considered to be isentropic. The speed of sound is related to the changes in pressure and density according to

$$a^2 = \left(\frac{\partial p}{\partial \rho}\right)_s \tag{7.20}$$

which, by the use of isentropic relations, may be written as

$$a^2 = \gamma \frac{p}{\rho} \tag{7.21}$$

If one assumes a perfect gas, then Equation (7.21) becomes

$$a^2 = \gamma RT \tag{7.22}$$

The speed of sound for air where $\gamma = 1.4$ can be determined from the following relations

$$a = 49.02\sqrt{T} \tag{7.23}$$

where T is in $^\circ$R and a is in fps, or

$$a = 20.047\sqrt{T} \qquad (7.24)$$

where T is in K and a is in m/s.

The speed of sound for an equilibrium chemically reacting flow or for a real gas may be determined from the following relation

$$a = \left[\frac{1 + \dfrac{1}{p}\left(\dfrac{\partial e}{\partial v}\right)_T}{1 - \rho\left(\dfrac{\partial h}{\partial p}\right)_T}\right](\gamma RT) \qquad (7.25)$$

The speed of sound for liquids is expressed in terms of bulk modulus and is given by

$$a = \sqrt{\frac{E}{\rho}} \qquad (7.26)$$

The density and bulk modulus of water at different temperatures at atmospheric pressure are given in Table 7.16.

Table 7.16: Density and bulk modulus of water.

English units			SI units		
T ($^\circ$F)	ρ (slugs/ft^3)	$E \times 10^{-3}$ (lbf/in^2)	T ($^\circ$C)	ρ (kg/m^3)	$E \times 10^{-6}$ (kPa)
32	1.940	287	0	999.8	1.98
40	1.940	296	5	1000.0	2.05
60	1.938	313	10	999.7	2.10
80	1.934	324	20	998.2	2.17
100	1.927	331	30	995.7	2.25
120	1.918	332	50	988.0	2.29
140	1.908	330	70	977.8	2.25
160	1.896	326	80	971.8	2.20
180	1.883	318	90	965.3	2.14
200	1.868	308	100	958.4	2.07

7.7 Prandtl Number

Prandtl number is defined as *the ratio of kinematic viscosity to thermal diffusivity,*

$$Pr = \frac{\nu}{\alpha} = \frac{\mu c_p}{k} \tag{7.27}$$

The Prandtl number represents the ratio of diffusion of momentum by viscosity to the diffusion of heat by conduction. The Prandtl number for gases at moderate temperatures can be estimated from the following relation,

$$Pr = \frac{4\gamma}{7.08\gamma - 1.8} \tag{7.28}$$

It should be noted that the Prandtl number is also pressure- and temperature-dependent. In fact, that is easily recognized by considering relation (7.27), because μ, k, and c_p are all pressure- and temperature-dependent. Since expressions or tables are available to estimate the values of μ, k, and c_p, they can be used to determine the Prandtl number. Nonetheless, the values of the Prandtl number for air can be determined from Table 7.17, which includes both pressure- and temperature-dependency. The values of the Prandtl number for selected gases and liquids are provided in Tables 7.18 and 7.19, respectively.

Table 7.17: Prandtl number for air

T(K)	Pressure (atm)						
	100	10	1.0	0.1	0.01	0.001	0.0001
500	0.738	0.738	0.738	0.738	0.738	0.738	0.738
1000	0.756	0.756	0.756	0.756	0.756	0.756	0.756
1500	0.767	0.767	0.767	0.767	0.767	0.767	0.767
2000	0.773	0.773	0.773	0.766	0.724	0.668	0.614
2500	0.762	0.751	0.696	0.645	0.611	0.654	0.771
3000	0.740	0.680	0.627	0.636	0.740	0.745	0.714
3500	0.678	0.631	0.660	0.744	0.737	0.658	0.606
4000	0.640	0.662	0.762	0.759	0.619	0.580	0.587
4500	0.654	0.743	0.752	0.610	0.578	0.611	0.764
5000	0.702	0.767	0.611	0.581	0.624	0.799	0.993
5500	0.748	0.620	0.583	0.617	0.785	0.989	0.871
6000	0.763	0.592	0.602	0.736	0.969	0.891	0.455
6500	0.610	0.592	0.673	0.906	0.955	0.464	0.392
7000	0.593	0.620	0.796	0.986	0.830	0.404	0.361
7500	0.595	0.688	0.927	0.969	0.424	0.371	0.342
8000	0.620	0.788	0.983	0.648	0.387	0.351	0.322
8500	0.666	0.891	0.943	0.411	0.363	0.335	0.279
9000	0.730	0.961	0.807	0.382	0.348	0.316	0.200
9500	0.806	0.966	0.497	0.364	0.336	0.279	0.114
10000	0.886	0.872	0.429	0.348	0.319	0.216	0.0576
10500	0.937	0.532	0.404	0.339	0.295	0.145	0.0314
11000	0.955	0.463	0.382	0.327	0.254	0.0877	0.0213
11500	0.947	0.434	0.369	0.312	0.201	0.0524	0.0167
12000	0.908	0.412	0.355	0.292	0.146	0.0346	0.0143
12500	0.728	0.396	0.343	0.263	0.101	0.0238	0.0129
13000	0.525	0.383	0.333	0.227	0.0688	0.0190	0.0121
13500	0.438	0.369	0.319	0.185	0.470	0.0162	0.0110
14000	0.421	0.360	0.302	0.144	0.0345	0.0149	0.0108
14500	0.401	0.349	0.277	0.0986	0.0245	0.0130	0.0109
15000	0.394	0.341	0.253	0.0819	0.0129	0.0120	0.0110

Table 7.18: Prandtl number of selected gases

| Gas | T(K) | | | | | | |
---	100	200	300	400	500	600	800
Air			0.712	0.702	0.698	0.698	0.704
Argon			0.67				
Carbon dioxide			0.78	0.75	0.73	0.71	0.71
Carbon monoxide			0.78				
Helium			0.68	0.67	0.67	0.66	0.65
Hydrogen	0.68		0.73	0.748	0.73	0.74	
Methane			0.72	0.73	0.74	0.74	
Nitrogen		0.74	0.72	0.71	0.71	0.71	0.73
Oxygen		0.71	0.71	0.713	0.72	0.726	0.734

Table 7.19: Prandtl number of selected liquids

| Liquid | T(K) | | | | | | | | | |
---	180	200	220	240	260	280	300	320	360	400
Ammonia		3.5	2.4	2.0	1.68	1.48	1.38	1.29	1.27	
Benzene						8.9	7.1	5.9	4.5	
Butane	8.4	6.2	5.2	4.4	3.8	3.4	3.2	3.1		
Ethanol					29.0	21.0	15.0	12.0	8.0	5.9
Methanol		41.0	24.0	16.0	11.0	8.3	6.8	5.7	4.1	3.1
Propane	5.3	4.5	3.9	3.4	3.2	3.0	3.1	3.4		
R 12		5.3	4.3	3.7	3.3	3.1	3.0	3.2	4.0	
Water							5.81			0.980

7.7.1 Turbulent Prandtl Number

In order to express the turbulent shear stress and the turbulent heat flux in a similar fashion as their laminar counterpart, quantities such as eddy viscosity ν_t and eddy diffusivity α_t have been introduced, e.g.,

$$\tau_\ell + \tau_t = \rho\nu\frac{\partial \bar{u}}{\partial y} - \rho\overline{u'v'} = \rho(\nu + \nu_t)\frac{\partial \bar{u}}{\partial y} = (\mu + \mu_t)\frac{\partial \bar{u}}{\partial y}$$

and

$$q_\ell + q_t = -\rho c_p \alpha \frac{\partial \bar{T}}{\partial y} - \rho c_p \overline{v'T'} = -\rho c_p (\alpha + \alpha_t) \frac{\partial \bar{T}}{\partial y} = -(k + k_t) \frac{\partial \bar{T}}{\partial y}$$

Similar to the nondimensional parameter defined as the Prandtl number, a nondimensional qunatity based on eddy viscosity and eddy conductivity can be defined. The nondimensional parameter is known as the *turbulent* Prandtl number and is given by

$$Pr_t = \frac{c_p \mu_t}{k_t}$$

Unlike the dynamic viscosity and thermal conductivity, which are fluid properties, eddy viscosity and thermal conductivity, and, therefore, the Prandtl number are not fluid properties.

For most applications, the turbulent Prandtl number is assumed to be a constant. However, it should be noted that the turbulent Prandtl number could very well vary within the domain, depending on the local conditions. For applications involving air, a turbulent Prandtl number of 0.86–0.9 is typically used.

7.8 U.S. Standard Atmosphere

A steady state, idealized representation of the earth's atmosphere is presented in the U.S. Standard Atmosphere, 1976 [7.3]. Selected data from Ref. [7.3] is presented in Table 7.20.

Table 7.20: U.S. Standard Atmosphere, 1962: Metric Units

Altitude [m]	Pressure [N/m²]		Temperature [K]	Density [kg/m³]	Speed of Sound [m/s]	Viscosity [(N·s)/m²]	Thermal Conductivity [W/(m·K)]
0	1.01325	E+00	288.150	1.2250	340.29	1.7894 E−05	2.5362 E−05
500	9.5460	E+04	284.900	1.1673	338.37	1.7737 E−05	2.5106 E−05
1000	8.9874	E+04	281.650	1.1116	336.43	1.7578 E−05	2.4849 E−05
1500	8.4555	E+04	278.400	1.0581	334.49	1.7419 E−05	2.4591 E−05
2000	7.9495	E+04	275.150	1.0065	332.53	1.7260 E−05	2.4332 E−05
2500	7.4682	E+04	271.900	0.95686	330.56	1.7099 E−05	2.4073 E−05
3000	7.0108	E+04	268.650	0.90912	328.58	1.6937 E−05	2.3812 E−05
3500	6.5764	E+04	265.400	0.86323	326.58	1.6775 E−05	2.3551 E−05
4000	6.1640	E+04	262.150	0.81913	324.58	1.6611 E−05	2.3289 E−05
4500	5.7728	E+04	258.900	0.7767	322.56	1.6447 E−05	2.3026 E−05
5000	5.4019	E+04	255.650	0.73612	320.53	1.6281 E−05	2.2763 E−05
5500	5.0506	E+04	252.400	0.69711	318.49	1.6115 E−05	2.2498 E−05
6000	4.7181	E+04	249.150	0.65970	316.43	1.5947 E−05	2.2233 E−05
6500	4.4034	E+04	245.900	0.62384	314.36	1.5779 E−05	2.1966 E−05
7000	4.1060	E+04	242.650	0.58950	312.27	1.5610 E−05	2.1699 E−05
7500	3.8251	E+04	239.400	0.55662	310.18	1.5439 E−05	2.1431 E−05
8000	3.5599	E+04	236.150	0.52517	308.06	1.5268 E−05	2.1163 E−05
8500	3.3099	E+04	232.900	0.49509	305.94	1.5095 E−05	2.0893 E−05
9000	3.0742	E+04	229.650	0.46635	303.79	1.4922 E−05	2.0623 E−05
9500	2.8523	E+04	226.400	0.43890	301.64	1.4747 E−05	2.0351 E−05
10000	2.6436	E+04	223.150	0.41271	299.46	1.4571 E−05	2.0079 E−05
10500	2.4474	E+04	219.900	0.38773	297.27	1.4394 E−05	1.9806 E−05
11000	2.2632	E+04	216.650	0.36392	295.07	1.4216 E−05	1.9533 E−05
11500	2.0916	E+04	216.650	0.33633	295.07	1.4216 E−05	1.9533 E−05
12000	1.9330	E+04	216.650	0.31083	295.07	1.4216 E−05	1.9533 E−05
12500	1.7864	E+04	216.650	0.28726	295.07	1.4216 E−05	1.9533 E−05
13000	1.6510	E+04	216.650	0.26548	295.07	1.4216 E−05	1.9533 E−05
13500	1.5258	E+04	216.650	0.24536	295.07	1.4216 E−05	1.9533 E−05
14000	1.4101	E+04	216.650	0.22675	295.07	1.4216 E−05	1.9533 E−05
14500	1.3032	E+04	216.650	0.20956	295.07	1.4216 E−05	1.9533 E−05
15000	1.2044	E+04	216.650	0.19367	295.07	1.4216 E−05	1.9533 E−05
15500	1.1131	E+04	216.650	0.17899	295.07	1.4216 E−05	1.9533 E−05
16000	1.0287	E+04	216.650	0.16542	295.07	1.4216 E−05	1.9533 E−05
16500	9.5074	E+03	216.650	0.15288	295.07	1.4216 E−05	1.9533 E−05
17000	8.7866	E+03	216.650	0.14129	295.07	1.4216 E−05	1.9533 E−05

(continued on next page)

Table 7.20: Continued

Altitude [m]	Pressure [N/m^2]		Temperature [K]	Density [kg/m^3]	Speed of Sound [m/s]	Viscosity [(N·S)/m^2]	Thermal Conductivity [W/(m·K)]
17500	8.1205	E+03	216.650	0.13058	295.07	1.4216 E−05	1.9533 E−05
18000	7.5048	E+03	216.650	0.12068	295.07	1.4216 E−05	1.9533 E−05
18500	6.9358	E+03	216.650	0.11153	295.07	1.4216 E−05	1.9533 E−05
19000	6.4100	E+03	216.650	0.10307	295.07	1.4216 E−05	1.9533 E−05
19500	5.9240	E+03	216.650	0.09526	295.07	1.4216 E−05	1.9533 E−05
20000	5.4748	E+03	216.650	0.08804	295.07	1.4216 E−05	1.9533 E−05
20500	5.0602	E+03	217.150	0.08118	295.41	1.4244 E−05	1.9575 E−05
21000	4.6778	E+03	217.650	0.07487	295.75	1.4271 E−05	1.9617 E−05
21500	4.3251	E+03	218.150	0.06907	296.09	1.4298 E−05	1.9659 E−05
22000	3.9997	E+03	218.650	0.06373	296.43	1.4326 E−05	1.9701 E−05
22500	3.6995	E+03	219.150	0.05881	296.77	1.4353 E−05	1.9743 E−05
23000	3.4224	E+03	219.650	0.05428	297.11	1.4381 E−05	1.9785 E−05
23500	3.1666	E+03	220.150	0.05011	297.44	1.4408 E−05	1.9827 E−05
24000	2.9304	E+03	220.650	0.04627	297.78	1.4435 E−05	1.9869 E−05
24500	2.7124	E+03	221.150	0.04273	298.12	1.4662 E−05	1.9911 E−05
25000	2.5110	E+03	221.650	0.03947	298.45	1.4490 E−05	1.9953 E−05
25500	2.3249	E+03	222.150	0.03646	298.79	1.4517 E−05	1.9995 E−05
26000	2.1530	E+03	222.650	0.03369	299.13	1.4544 E−05	2.0037 E−05
26500	1.9942	E+03	223.150	0.03113	299.46	1.4571 E−05	2.0079 E−05
27000	1.8474	E+03	223.650	0.02878	299.80	1.4598 E−05	2.0121 E−05
27500	1.7117	E+03	224.150	0.02660	300.13	1.4625 E−05	2.0163 E−05
28000	1.5862	E+03	224.650	0.02460	300.47	1.4652 E−05	2.0205 E−05
29000	1.3629	E+03	225.650	0.02104	301.14	1.4706 E−05	2.0289 E−05
30000	1.1718	E+03	226.650	0.01801	301.80	1.4760 E−05	2.0372 E−05
31000	1.0082	E+03	227.650	0.01543	302.47	1.4814 E−05	2.0456 E−05
32000	8.6801	E+02	228.650	0.01323	303.13	1.4868 E−05	2.0539 E−05
33000	7.4822	E+02	231.450	0.01126	304.98	1.5018 E−05	2.0773 E−05
34000	6.4612	E+02	234.250	0.00961	306.82	1.5167 E−05	2.1005 E−05
35000	5.5892	E+02	237.050	0.00821	308.65	1.5315 E−05	2.1237 E−05
36000	4.8431	E+02	239.850	0.00703	310.47	1.5463 E−05	2.1469 E−05
37000	4.2036	E+02	242.650	0.00604	312.27	1.5610 E−05	2.1699 E−05
38000	3.6545	E+02	245.450	0.00519	314.07	1.5756 E−05	2.1929 E−05
39000	3.1822	E+02	248.250	0.00447	315.86	1.5901 E−05	2.2159 E−05
40000	2.7752	E+02	251.050	0.00385	317.63	1.6045 E−05	2.2388 E−05
45000	1.4313	E+02	265.050	0.00188	326.37	1.6757 E−05	2.3523 E−05
50000	7.5944	E+01	270.650	0.00098	329.80	1.7037 E−05	2.3973 E−05
55000	3.9969	E+01	259.450	0.00054	322.90	1.6475 E−05	2.3071 E−05
60000	2.0314	E+01	245.450	0.00029	314.07	1.5756 E−05	2.1929 E−05
65000	9.9220	E+00	231.450	0.00015	304.98	1.5018 E−05	2.0773 E−05
70000	4.6342	E+00	217.450	0.00007	295.61	1.4260 E−05	1.9600 E−05
75000	2.0679	E+00	206.650	0.00003	288.18	1.3661 E−05	1.8685 E−05
80000	8.8627	E−01	196.650	0.00002	281.12	1.3095 E−05	1.7830 E−05

7.9 References

[7.1] Hansen, C. F., "Approximations for the Thermodynamic and Transport Properties of High Temperature Air," NASA TR R-50, 1959

[7.2] Zucrow, M. J., and Hoffman, J. D., "Gas Dynamics—Volume I," Wiley, New York, 1976.

[7.3] U. S. Standard Atmosphere, 1976, U.S. Government Printing Office, Washington, D. C., Oct. 1976.

Chapter 8
Conversion Factors

Table 8.1. Conversion factors for quantities having dimensions of [L]: length

Multiply		by	to obtain	
centimeter	cm	3.2808 E−02	feet	ft
		3.9370 E−01	inches	in
		1.0000 E−05	kilometers	km
		1.0000 E−02	meters	m
		3.9370 E+02	mils	
		6.2140 E−06	miles	mi
		1.0000 E+01	millimeters	mm
		1.0936 E−02	yards	yd
foot	ft	3.0480 E+01	centimeters	cm
		1.2000 E+01	inches	in
		3.0480 E−04	kilometers	km
		1.6450 E−04	nautical miles	
		3.0480 E−01	meters	m
		1.2000 E+04	mils	
		1.8939 E−04	miles	mi
		3.0480 E+02	millimeters	mm
		3.3333 E−01	yards	yd

(Continued on next page)

Table 8.1. (continued)

Multiply		by	to obtain	
inch	in	2.5400	centimeters	cm
		8.3333 E−02	feet	ft
		2.5400 E−05	kilometers	km
		2.5400 E−02	meters	m
		1.0000 E+03	mils	
		1.5780 E−05	miles	mi
		2.5400 E+01	millimeters	mm
		2.7778 E−02	yards	yd
kilometer	km	1.0000 E+05	centimeters	cm
		3.2808 E+03	feet	ft
		3.9370 E+04	inches	in
		5.3960 E−01	nautical miles	
		1.0000 E+03	meters	m
		3.9370 E+07	mils	
		6.2137 E−01	miles	mi
		1.0000 E+06	millimeters	mm
		1.0936 E+03	yards	yd
nautical mile		1.8530 E+05	centimeters	cm
		6.0803 E+03	feet	ft
		7.2960 E+04	inches	in
		1.8530	kilometers	km
		1.8530 E+03	meters	m
		1.1516	miles	mi
		2.0270 E+03	yards	yd
meter	m	1.0000 E+02	centimeters	cm
		3.2808	feet	ft
		3.9370 E+01	inches	in
		1.0000 E−03	kilometers	km
		5.3960 E−04	nautical miles	
		3.9370 E+04	mils	
		6.2137 E−04	miles	mi
		1.0000 E+03	millimeters	mm
		1.0936	yards	yd

(Continued on next page)

Table 8.1. (continued)			
Multiply	by	to obtain	
mil	2.5400 E−03	centimeters	cm
	8.333 E−05	feet	ft
	1.0000 E−03	inches	in
	2.5400 E−08	kilometers	km
	2.5400 E−02	millimeters	mm
	2.7780 E−05	yards	yd
mile mi	1.6093 E+05	centimeters	cm
	5.2800 E+03	feet	ft
	6.3360 E+04	inches	in
	1.6093 E+00	kilometers	km
	8.6840 E−01	nautical miles	
	1.6093 E+03	meters	m
	1.7600 E+03	yards	yd
millimeter mm	1.0000 E−01	centimeters	cm
	3.2808 E−03	feet	ft
	3.9370 E−02	inches	in
	1.0000 E−06	kilometers	km
	1.0000 E−03	meters	m
	3.9370 E+01	mils	
	6.2137 E−07	miles	mi
	1.0936 E−03	yards	yd
yard yd	9.1440 E+01	centimeters	cm
	3.0000	feet	ft
	3.6000 E+01	inches	in
	9.1440 E−04	kilometers	km
	4.9340 E−04	nautical miles	
	9.1440 E−01	meters	m
	3.6000 E+04	mils	
	5.6818 E−04	miles	mi
	9.1440 E+02	millimeters	mm

Table 8.2. Conversion factors for quantities having dimensions of $[L^2]$: area

Multiply		by	to obtain	
acre	acre	4.3560 E+04	square feet	ft^2
		6.2726 E+06	square inches	in^2
		4.0469 E−03	square kilometers	km^2
		4.0469 E+03	square meters	m^2
		1.5625 E−03	square miles	mi^2
		4.8400 E+03	square yards	yd^2
		4.0469 E−01	hectars	
hectar		2.4711	acres	
		1.0000 E+04	square meters	m^2
		3.8610 E−03	square miles	mi^2
square centimeter	cm^2	1.0764 E−03	square feet	ft^2
		1.5500 E−01	square inches	in^2
		1.0000 E−10	square kilometers	km^2
		1.0000 E−04	square meters	m^2
		3.8610 E−11	square miles	mi^2
		1.0000 E+02	square millimeters	mm^2
		1.1960 E−04	square yards	yd^2
square foot	ft^2	2.2960 E−05	acres	acre
		9.2903 E+02	square centimeters	cm^2
		1.4400 E+02	square inches	in^2
		9.2903 E−08	square kilometers	km^2
		9.2903 E−02	square meters	m^2
		3.5870 E−08	square miles	mi^2
		9.2903 E+04	square millimeters	mm^2
		1.1110 E−01	square yards	yd^2
square inch	in^2	6.4516	square centimeters	cm^2
		6.9444 E−03	square feet	ft^2
		6.4516 E−10	square kilometers	km^2
		6.4516 E−04	square meters	m^2
		6.4516 E+02	square millimeters	mm^2
		7.7160 E−04	square yards	yd^2

(Continued on next page)

Table 8.2. (continued)

Multiply		by	to obtain	
square kilometer	km^2	2.4711 E+02	acres	acre
		1.0000 E+10	square centimeters	cm^2
		1.0764 E+07	square feet	ft^2
		1.5500 E+09	square inches	in^2
		1.0000 E+06	square meters	m^2
		3.8610 E−01	square miles	mi^2
		1.0000 E+12	square millimeters	mm^2
		1.1960 E+06	square yards	yd^2
square meter	m^2	2.4711 E−04	acres	acre
		1.0000 E+04	square centimeters	cm^2
		1.0764 E+01	square feet	ft^2
		1.5500 E+03	square inches	in^2
		1.0000 E−06	square kilometers	km^2
		3.8610 E−07	square miles	mi^2
		1.0000 E+06	square millimeters	mm^2
		1.1960	square yards	yd^2
square mile	mi^2	6.4000 E+02	acres	acre
		2.5900 E+10	square centimeters	cm^2
		2.7878 E+07	square feet	ft^2
		4.0145 E+09	square inches	in^2
		2.5900	square kilometers	km^2
		2.5900 E+06	square meters	m^2
		3.0976 E+06	square yards	yd^2
square millimeter	mm^2	1.0000 E−02	square centimeters	cm^2
		1.0764 E−05	square feet	ft^2
		1.5500 E−03	square inches	in^2
		1.0000 E−12	square kilometers	km^2
		1.0000 E−06	square meters	m^2
		3.8610 E−13	square miles	mi^2
		1.1960 E−06	square yards	yd^2

(Continued on next page)

Table 8.2. (continued)

Multiply		by	to obtain	
square yard	yd^2	2.0661 E−04	acres	acre
		8.3613 E+03	square centimeters	cm^2
		9.0000	square feet	ft^2
		1.2960 E+03	square inches	in^2
		8.3613 E−07	square kilometers	km^2
		8.3613 E−01	square meters	m^2
		3.2283 E−07	square miles	mi^2
		8.3613 E+05	square millimeters	mm^2

Table 8.3. Conversion factors for quantities having dimensions of $[L^3]$: volume

Multiply		by	to obtain	
cubic centimeters	cm^3	3.5315 E−05	cubic feet	ft^3
		6.1024 E−02	cubic inches	in^3
		1.0000 E−06	cubic meters	m^3
		1.3080 E−06	cubic yards	yd^3
		2.6420 E−04	gallons	gal
		1.0000 E−03	liters	l
		2.1130 E−03	pints	pt
		1.0570 E−03	quarts	qt
cubic feet	ft^3	2.8317 E+04	cubic centimeters	cm^3
		1.7280 E+03	cubic inches	in^3
		2.8317 E−02	cubic meters	m^3
		3.7037 E−02	cubic yards	yd^3
		7.4810	gallons	gal
		28.3200	liters	l
		5.9840 E+01	pints	pt
		2.9920 E+01	quarts	qt
cubic inch	in^3	1.6387 E+01	cubic centimeters	cm^3
		5.7870 E−04	cubic feet	ft^3
		1.6387 E−05	cubic meters	m^3
		2.1433 E−05	cubic yards	yd^3
		4.3290 E−03	gallons	gal
		1.6387 E−02	liters	l
		3.4630 E−02	pints	pt
		1.7320 E--02	quarts	qt
cubic meters	m^3	1.0000 E+06	cubic centimeters	cm^3
		3.5315 E+01	cubic feet	ft^3
		6.1024 E+04	cubic inches	in^3
		1.3080	cubic yards	yd^3
		2.6420 E+02	gallons	gal
		1.0000 E+03	liters	l
		2.1130 E+03	pints	pt
		1.0570 E+03	quarts	qt

(Continued on next page)

Table 8.3. (continued)

Multiply		by	to obtain	
cubic yard	yd^3	7.6455 E+05	cubic centimeters	cm^3
		2.7000 E+01	cubic feet	ft^3
		4.6656 E+04	cubic inches	in^3
		7.6455 E−01	cubic meters	m^3
		2.0200 E+02	gallons	gal
		7.6460 E+02	liters	l
		1.6160 E+03	pints	pt
		8.0790 E+02	quarts	qt
gallon	gal	1.3368 E−01	cubic feet	ft^3
		3.7854	liters	l
		3.7854 E−03	cubic meters	m^3
		8.0000	pints	pt
		4.0000	quarts	qt
liter	l	1.0000 E+03	cubic centimeters	cm^3
		3.5315 E−02	cubic feet	ft^3
		6.1024 E+01	cubic inches	in^3
		1.0000 E−03	cubic meters	m^3
		1.3080 E−03	cubic yards	yd^3
		2.6417 E−01	gallons	gal
		2.11345	pints	pt
		1.0567	quarts	qt
pint	pt	4.7317 E+02	cubic centimeters	cm^3
		1.6710 E−02	cubic feet	ft^3
		2.8870 E+01	cubic inches	in^3
		4.7317 E−04	cubic meters	m^3
		6.1890 E−04	cubic yards	yd^3
		1.2500 E−01	gallons	gal
		4.7317 E−01	liters	l
		5.0000 E−01	quarts	qt

(Continued on next page)

Table 8.3. (continued)		
Multiply	by	to obtain
quart qt	9.4635 E+02	cubic centimeters	cm^3
	3.3421 E−02	cubic feet	ft^3
	5.7750 E+01	cubic inches	in^3
	9.4635 E−04	cubic meters	m^3
	1.2380 E−03	cubic yards	yd^3
	2.5000 E−01	gallons	gal
	9.4635 E−01	liters	l
	2.0000	pints	pt

Table 8.4. Conversion factors for quantities having dimensions of [L/t]: linear velocity

Multiply		by	to obtain	
foot/second	ft/sec	3.0480 E+01	centimeters/second	cm/sec
		1.0973	kilometers/hour	km/hr
		5.9248 E−01	knots	kt
		3.0480 E−01	meters/second	m/sec
		6.8182 E−01	miles/hour	mi/hr
		1.1360 E−02	miles/minute	mi/min
kilometer/ hour	km/hr	9.1130 E−01	feet/second	ft/sec
		5.3960 E−01	knots	kt
		2.7778 E−01	meters/second	m/sec
		6.2137 E−01	miles/hour	mi/hr
		1.0360 E−02	miles/minute	mi/min
knot	kt	1.6890	feet/second	ft/sec
		1.8532	kilometers/hour	km/hr
		5.1480 E−01	meters/second	m/sec
		1.1516	miles/hour	mi/hr
		1.9190 E−02	miles/minute	mi/min
meter/second	m/sec	3.2808	feet/second	ft/sec
		3.6000	kilometers/hour	km/hr
		1.9438	knots	kt
		2.2369	miles/hour	mi/hr
		3.7280 E−02	miles/minute	mi/min
mile/hour	mi/hr	1.4667	feet/second	ft/sec
		1.6093	kilometers/hour	km/hr
		8.6898 E−01	knots	kt
		4.4704 E−01	meters/second	m/sec
		1.6667 E−02	miles/minute	mi/min
mile/minute	mi/min	8.8000 E+01	feet/second	ft/sec
		9.6540 E+01	kilometers/hour	km/hr
		5.2100 E+01	knots	kt
		2.6820 E+01	meters/second	m/sec
		6.0000 E+01	miles/hour	mi/hr

Table 8.5. Conversion factors for quantities having dimensions of [1/t]: angular velocity

Multiply		by	to obtain	
degree/sec	deg/sec	1.7453 E–02	radians/sec	rad/sec
		1.6667 E–01	revolutions/min	rev/min
		2.7778 E–03	revolutions/sec	rev/sec
radian/sec	rad/sec	5.7296 E+01	degrees/sec	deg/sec
		9.5493	revolutions/min	rev/min
		1.5915 E–01	revolutions/sec	rev/sec
revolution/min	rev/min	6.0	degrees/sec	deg/sec
		1.0472 E–01	radians/sec	rad/sec
		1.6667 E–02	revolutions/sec	rev/sec
revolution/sec	rev/sec	3.6000 E+02	degrees/sec	deg/sec
		6.2832	radians/sec	rad/sec
		6.000 E+01	revolutions/min	rev/min

Table 8.6. Conversion factors for quantities having dimensions of [M]: mass

Multiply		by	to obtain	
gram	g	1.0000 E−03	kilograms	kg
		1.0000 E+03	milligrams	
		3.5274 E−02	ounces	oz
		2.2046 E−03	pounds	lb_m
		9.8420 E−07	tons (long)	
		1.0000 E−06	tons (metric)	
		1.1020 E−06	tons (short)	
kilogram	kg	1.0000 E+03	grams	g
		1.0000 E+06	milligrams	
		3.5274 E+01	ounces	oz
		2.2046	pounds	lb_m
		9.8420 E−04	tons (long)	
		1.0000 E−03	tons (metric)	
		1.1020 E−03	tons (short)	
milligram		1.0000 E−03	grams	g
		1.0000 E−06	kilograms	kg
		3.5274 E−05	ounces	oz
		2.2046 E−06	pounds	lb_m
		9.8420 E−10	tons (long)	
		1.0000 E−09	tons (metric)	
		1.1020 E−09	tons (short)	
ounce	oz	2.8349 E+01	grams	g
		2.8349 E−02	kilograms	kg
		2.8349 E+04	milligrams	
		6.2500 E−02	pounds	lb_m
		1.9428 E−03	slugs	
		2.7900 E−05	tons (long)	
		2.8350 E−05	tons (metric)	
		3.1250 E−05	tons (short)	
pound	lb_m	4.5359 E+02	grams	g
		4.5359 E−01	kilograms	kg
		4.5359 E+05	milligrams	
		1.6000 E+01	ounces	oz

(Continued on next page)

<table>
<tr><td colspan="4" align="center">Table 8.6. (continued)</td></tr>
<tr><td align="center">Multiply</td><td align="center">by</td><td colspan="2" align="center">to obtain</td></tr>
<tr><td>pound lb_m</td><td>4.4640 E−04</td><td>tons (long)</td><td></td></tr>
<tr><td></td><td>4.5360 E−04</td><td>tons (metric)</td><td></td></tr>
<tr><td></td><td>5.0000 E−04</td><td>tons (short)</td><td></td></tr>
<tr><td>ton (long)</td><td>1.0160 E+06</td><td>grams</td><td>g</td></tr>
<tr><td></td><td>1.0160 E+03</td><td>kilograms</td><td>kg</td></tr>
<tr><td></td><td>1.0160 E+09</td><td>milligrams</td><td></td></tr>
<tr><td></td><td>3.5840 E+04</td><td>ounces</td><td>oz</td></tr>
<tr><td></td><td>2.2400 E+03</td><td>pounds</td><td>lb_m</td></tr>
<tr><td></td><td>1.0160</td><td>tons (metric)</td><td></td></tr>
<tr><td></td><td>1.1200</td><td>tons (short)</td><td></td></tr>
<tr><td>ton (metric)</td><td>1.0000 E+06</td><td>grams</td><td>g</td></tr>
<tr><td></td><td>1.0000 E+03</td><td>kilograms</td><td>kg</td></tr>
<tr><td></td><td>1.0000 E+09</td><td>milligrams</td><td></td></tr>
<tr><td></td><td>3.5270 E+04</td><td>ounces</td><td>oz</td></tr>
<tr><td></td><td>2.2050 E+03</td><td>pounds</td><td>lb_m</td></tr>
<tr><td></td><td>9.8420 E−01</td><td>tons (long)</td><td></td></tr>
<tr><td></td><td>1.1020</td><td>tons (short)</td><td></td></tr>
<tr><td>ton (short)</td><td>9.0720 E+05</td><td>grams</td><td>g</td></tr>
<tr><td></td><td>9.0720 E+02</td><td>kilograms</td><td>kg</td></tr>
<tr><td></td><td>9.0720 E+08</td><td>milligrams</td><td></td></tr>
<tr><td></td><td>3.2000 E+04</td><td>ounces</td><td>oz</td></tr>
<tr><td></td><td>2.0000 E+03</td><td>pounds</td><td>lb_m</td></tr>
<tr><td></td><td>8.9290 E−01</td><td>tons (long)</td><td></td></tr>
<tr><td></td><td>9.0720 E−01</td><td>tons (metric)</td><td></td></tr>
</table>

Table 8.7. Conversion factors for quantities having dimensions of [F] or $[ML/t^2]$: force

Multiply		by	to obtain	
dyne	dyne	1.0000 E−05	Newtons	N
		2.2481 E−06	pounds	lb_f
		7.2330 E−05	poundals	poundal
pound	lb_f	4.4482	Newtons	N
		4.4482 E+05	dynes	dyne
		3.2174 E+01	poundals	poundal
poundal	poundal	1.3826 E+04	dynes	dyne
		1.3826 E−01	Newtons	N
		3.1081 E−02	pounds	lb_f
Newton	N	1.0000 E+05	dynes	dyne
		2.2481 E−01	pounds	lb_f
		7.2330	poundals	poundal
kip (1000 lb)		4.4482 E+03	Newtons	N
ton (2000 lb)		8.8964 E+03	Newtons	N

Table 8.8. Conversion factors for quantities having dimensions of $[M/L^3]$: density

Multiply		by	to obtain	
gram per cubic centimeter	g/cm^3	1.0000 E+03	kilograms per cubic meter	kg/m^3
		6.2430 E+01	pounds per cubic foot	lb_m/ft^3
		3.6130 E$-$02	pounds per cubic inch	lb_m/in^3
kilogram per cubic meter	kg/m^3	1.0000 E$-$03	grams per cubic centimeter	g/cm^3
		6.2430 E$-$02	pounds per cubic foot	lb_m/ft^3
		3.6130 E$-$05	pounds per cubic inch	lb_m/in^3
pound per cubic foot	lb_m/ft^3	1.6020 E$-$02	grams per cubic centimeter	g/cm^3
		1.6020 E+0 1	kilograms per cubic meter	kg/m^3
		5.7870 E$-$04	pounds per cubic inch	lb_m/in^3
pound per cubic inch	lb_m/in^3	2.7680 E+01	grams per cubic centimeter	g/cm^3
		2.7680 E+04	kilograms per cubic meter	kg/m^3
		1.7280 E+03	pounds per cubic foot	lb_m/ft^3

Table 8.9. Conversion factors for quantities having dimensions of $[F/L^2]$ or $[M/Lt^2]$: pressure, momentum flux

Multiply		by	to obtain	
dyne/square centimeter	dyne/cm^2	2.0886 E−03	pounds/square feet	lb$_f$/ft^2
		1.4504 E−05	pounds/square inches	lb$_f$/in^2
		9.8692 E−07	atmospheres	atm
		7.5006 E−04	millimeters of mercury (0°C)	mm Hg
		2.9530 E−05	inches of mercury (0°C)	in Hg
Newton/square meter	N/m^2	2.0886 E−02	pounds/square feet	lb$_f$/ft^2
		1.4504 E−04	pounds/square inches	lb$_f$/in^2
		9.8692 E−06	atmospheres	atm
		7.5006 E−03	millimeters of mercury (0°C)	mm Hg
		2.9530 E−04	inches of mercury (0°C)	in Hg
pound/square foot	lb$_f$/ft^2	4.7880 E+02	dynes/square centimeters	dyne/cm^2
		4.7880 E+01	Newtons/square meters	N/m^2
		6.9444 E−03	pounds/square inches	lb$_f$/in^2
		4.7254 E−04	atmospheres	atm
		3.5913 E−01	millimeters of mercury (0°C)	mm Hg
		1.4139 E−02	inches of mercury (0°C)	in Hg
pound/square inch	lb$_f$/in^2	6.8947 E+04	dynes/square centimeters	dyne/cm^2
		6.8947 E+03	Newtons/square meters	N/m^2
		4.6333 E+03	poundals/square feet	poundals/ ft^2

(Continued on next page)

<table>
<tr><td colspan="3" align="center">Table 8.9. (continued)</td></tr>
<tr><td align="center">Multiply</td><td align="center">by</td><td align="center">to obtain</td></tr>
<tr><td>pound/square inch lb_f/in^2</td><td>6.8046 E−02</td><td>atmospheres   atm</td></tr>
<tr><td></td><td>5.1715 E+01</td><td>millimeters of mercury (0°C)   mm Hg</td></tr>
<tr><td></td><td>2.0360</td><td>inches of mercury (0°C)   in Hg</td></tr>
<tr><td>poundal/ square foot   poundal/ft²</td><td>1.4882 E+01</td><td>dynes/square centimeters   dyne/cm²</td></tr>
<tr><td></td><td>1.4882</td><td>Newtons/square meters   N/m²</td></tr>
<tr><td></td><td>3.1081 E−02</td><td>pounds/ square feet lb_f/ft^2</td></tr>
<tr><td></td><td>2.1584 E−04</td><td>pounds/ square inches lb_f/in^2</td></tr>
<tr><td></td><td>1.4687 E−05</td><td>atmospheres   atm</td></tr>
<tr><td></td><td>1.1162 E−02</td><td>millimeters of mercury (0°C)   mm Hg</td></tr>
<tr><td></td><td>4.3945 E−04</td><td>inches of mercury (0°C)   in Hg</td></tr>
<tr><td>atmosphere   atm</td><td>1.0133 E+06</td><td>dynes/square centimeters   dyne/cm²</td></tr>
<tr><td></td><td>1.0133 E+05</td><td>Newtons/square meters   N/m²</td></tr>
<tr><td></td><td>6.8087 E+04</td><td>poundals/ square feet   poundals/ ft²</td></tr>
<tr><td></td><td>2.1162 E+03</td><td>pounds/ square feet lb_f/ft^2</td></tr>
<tr><td></td><td>1.4696 E+01</td><td>pounds/ square inches lb_f/in^2</td></tr>
<tr><td></td><td>7.6000 E+02</td><td>millimeters of mercury (0°C)   mm Hg</td></tr>
<tr><td></td><td>2.9921 E+01</td><td>inches of mercury (0°C)   in Hg</td></tr>
</table>

(Continued on next page)

<table>
<tr><td colspan="4" align="center">Table 8.9. (continued)</td></tr>
<tr><td align="center">Multiply</td><td align="center">by</td><td colspan="2" align="center">to obtain</td></tr>
<tr><td rowspan="7">millimeter of
mercury (0°C)   mm Hg</td><td>1.3332 E+03</td><td>dynes/square
  centimeters</td><td>dyne/cm^2</td></tr>
<tr><td>1.3332 E+02</td><td>Newtons/square
  meters</td><td>N/m^2</td></tr>
<tr><td>8.9588 E+01</td><td>poundals/
  square feet</td><td>poundals/
  ft^2</td></tr>
<tr><td>2.7845</td><td>pounds/square feet</td><td>lb$_f$/ft^2</td></tr>
<tr><td>1.9337 E−02</td><td>pounds/square inches</td><td>lb$_f$/in^2</td></tr>
<tr><td>1.3158 E−03</td><td>atmospheres</td><td>atm</td></tr>
<tr><td>3.9370 E−02</td><td>inches of
  mercury (0°C)</td><td>in Hg</td></tr>
<tr><td rowspan="7">inch of mercury
  (0°C)   in Hg</td><td>3.3864 E+04</td><td>dynes/square
  centimeters</td><td>dyne/cm^2</td></tr>
<tr><td>3.3864 E+03</td><td>Newtons/square
  meters</td><td>N/m^2</td></tr>
<tr><td>2.2756 E+03</td><td>poundals/
  square feet</td><td>poundals/
  ft^2</td></tr>
<tr><td>7.0727 E+01</td><td>pounds/square feet</td><td>lb$_f$/ft^2</td></tr>
<tr><td>4.9116 E−01</td><td>pounds/square inches</td><td>lb$_f$/in^2</td></tr>
<tr><td>3.3421 E−02</td><td>atmospheres</td><td>atm</td></tr>
<tr><td>2.5400 E+01</td><td>millimeters of
  mercury (0°C)</td><td>mm Hg</td></tr>
<tr><td rowspan="8">inch of water
  at 4°C   in H$_2$O</td><td>2.4584 E−03</td><td>atmospheres</td><td>atm</td></tr>
<tr><td>2.4910 E−03</td><td>dynes/square
  centimeters</td><td>dyne/cm^2</td></tr>
<tr><td>1.8683 E−01</td><td>centimeters of
  mercury (0°C)</td><td>cm Hg</td></tr>
<tr><td>7.3556 E−02</td><td>inches of
  mercury (0°C)</td><td>in Hg</td></tr>
<tr><td>1.8683</td><td>millimeters of
  mercury (0°C)</td><td>mm Hg</td></tr>
<tr><td>5.2040</td><td>pounds/square feet</td><td>lb$_f$/ft^2</td></tr>
<tr><td>3.6128 E−02</td><td>pounds/square inches</td><td>lb$_f$/in^2</td></tr>
<tr><td>2.4910 E+02</td><td>Newtons/square
  meters</td><td>N/m^2</td></tr>
</table>

Table 8.10. Conversion factors for quantities having dimensions of [FL] or [ML^2/t^2]: energy, work, or moment

Multiply		by	to obtain	
erg	erg	1.0000 E−07	absolute Joules	J
		2.3730 E−06	foot-poundals	ft-poundal
		7.3756 E−08	foot-pounds	ft-lb$_f$
		2.3901 E−08	calories	cal
		9.4783 E−11	Btu	Btu
		3.7251 E−14	horsepower-hours	hp-hr
		2.7778 E−14	kilowatt-hours	kwh
Absolute Joule	J	1.0000 E+07	ergs	erg
		2.3730 E+01	foot-poundals	ft-poundal
		7.3756 E−01	foot-pounds	ft-lb$_f$
		2.3901 E−01	calories	cal
		9.4783 E−04	Btu	Btu
		3.7251 E−07	horsepower-hours	hp-hr
		2.7778 E−07	kilowatt-hours	kwh
foot-poundal	ft-poundal	4.2140 E+05	ergs	erg
		4.2140 E−02	absolute Joules	J
		3.1081 E−02	foot-pounds	ft-lb$_f$
		1.0072 E−02	calories	cal
		3.9942 E−05	Btu	Btu
		1.5698 E−08	horsepower-hours	hp-hr
		1.1706 E−08	kilowatt-hours	kwh
foot-pound	ft-lb$_f$	1.3558 E+07	ergs	erg
		1.3558	absolute Joules	J
		3.2174 E+01	foot-poundals	ft-poundal
		3.2405 E−01	calories	cal
		1.2851 E−03	Btu	Btu
		5.0505 E−07	horsepower-hours	hp-hr
		3.7662 E−07	kilowatt-hours	kwh
calorie	cal	4.1840 E+07	ergs	erg
		4.1840	absolute Joule	J
		9.9287 E+01	foot-poundals	ft-poundal
		3.0860	foot-pounds	ft-lb$_f$

(Continued on next page)

Table 8.10. (continued)

Multiply		by	to obtain	
calorie	cal	3.9657 E−03	Btu	Btu
		1.5586 E−06	horsepower-hours	hp-hr
		1.1622 E−06	kilowatt-hours	kw-hr
Btu	Btu	1.0550 E+10	ergs	erg
		1.0550 E+03	absolute Joules	J
		2.5036 E+04	foot-poundals	ft-poundal
		7.7816 E+02	foot-pounds	ft-lb$_f$
		2.5216 E+02	calories	cal
		3.9301 E−04	horsepower-hours	hp-hr
		2.9307 E−04	kilowatt-hours	kwh
horsepower-hour	hp-hr	2.6845 E+13	ergs	erg
		2.6845 E+06	absolute Joules	J
		6.3705 E+07	foot-poundals	ft-poundal
		1.9800 E+06	foot-pounds	ft-lb$_f$
		6.4162 E+05	calories	cal
		2.5445 E+03	Btu	Btu
		7.4570 E−01	kilowatt-hours	kwh
kilowatt-hour	kwh	3.6000 E+13	ergs	erg
		3.6000 E+06	absolute Joules	J
		8.5429 E+07	foot-poundals	ft-poundal
		2.6552 E+06	foot-pounds	ft-lb$_f$
		8.6045 E+05	calories	cal
		3.4122 E+03	Btu	Btu
		1.3410	horsepower-hours	hp-hr

Table 8.11. Conversion factors for quantities having dimensions of [FL/t] or [ML²/t³]: power or rate of work

Multiply		by	to obtain	
Btu/minute	Btu/min	1.7548 E+08	ergs/second	ergs/sec
		7.7816 E+02	foot-pounds/minute	ft-lb$_f$/min
		1.2970 E+01	foot-pounds/second	ft-lb$_f$/sec
		2.3581 E−02	horsepower	hp
		1.7548 E−02	kilowatts	kw
		1.7548 E+01	watts	w
erg/second	erg/sec	5.6869 E−09	Btu/minute	Btu/min
		2.3885 E−08	calories/second	cal/sec
		4.4260 E−06	foot-pounds/minute	ft-lb$_f$/min
		7.3756 E−08	foot-pounds/second	ft-lb$_f$/sec
		1.3410 E−10	horsepower	hp
		1.0000 E−10	kilowatts	kw
		1.0000 E−07	watts	w
foot-pound/ minute	ft-lb$_f$/min	1.2850 E−03	Btu/minute	Btu/min
		2.2590 E+05	ergs/second	erg/sec
		1.6670 E−02	foot-pounds/second	ft-lb$_f$/sec
		3.0300 E−05	horsepower	hp
		2.2600 E−05	kilowatts	kw
		2.2600 E−02	watts	w
foot-pound/ second	ft-lb$_f$/sec	7.7104 E−02	Btu/minute	Btu/min
		3.2383 E−01	calories/second	cal/sec
		1.3558 E+07	ergs/second	erg/sec
		6.0000 E+01	foot-pounds/minute	ft-lb$_f$/min
		1.8182 E−03	horsepower	hp
		1.3558 E−03	kilowatts	kw
		1.3558	watts	w
horsepower	hp	4.2436 E+01	Btu/minute	Btu/min
		7.4570 E+09	ergs/sec	erg/sec
		3.3000 E+04	foot-pounds/minute	ft-lb$_f$/min

(Continued on next page)

Table 8.11. (continued)

Multiply		by	to obtain	
horsepower	hp	5.5000 E+02	foot-pounds/second	ft-lb$_f$/sec
		7.4570 E$-$01	kilowatts	kw
		7.4570 E+02	watts	w
kilowatt	kw	5.6890 E+01	Btu/minute	Btu/min
		1.0000 E+10	ergs/sec	erg/sec
		4.4260 E+04	foot-pounds/minute	ft-lb$_f$/min
		7.3756 E+06	foot-pounds/second	ft-lb$_f$/sec
		1.3410	horsepower	hp
		1.0000 E+03	watts	w
watt	w	5.6890 E$-$02	Btu/minute	Btu/min
		2.3900 E$-$01	calories/second	cal/sec
		1.0000 E+07	ergs/second	erg/sec
		4.4260 E+01	foot-pounds/minute	ft-lb$_f$/min
		7.3756 E$-$01	foot-pounds/second	ft-lb$_f$/sec
		1.3410 E$-$03	horsepower	hp
		1.0000 E$-$03	kilowatts	kw

Table 8.12. Conversion factors for quantities having dimensions of $[Ft/L^2]$ or $[M/Lt]$: viscosity

Multiply		by	to obtain	
poise		1.0000 E−01	kilograms/meter second	kg/(m sec)
		6.7197 E−02	pounds/foot second	lb_m/(ft sec)
		2.0886 E−03	pounds second/ square foot	lb_f sec/(ft^2)
		1.0000 E+02	centipoises	
		2.4191 E+02	pounds/foot hour	lb_m/(ft hr)
kilogram/meter second	kg/m sec	1.0000 E+01	poises	
		6.7197 E−01	pounds/foot seconds	lb_m/(ft sec)
		2.0886 E−02	pounds second/ square foot	lb_f sec/(ft^2)
		1.0000 E+03	centipoises	
		2.4191 E+03	pounds/foot hour	lb_m/(ft hr)
pound/foot second	lb_m/ft sec	1.4882 E+01	poises	
		1.4882	kilograms/meter second	kg/(m sec)
		3.1081 E−02	pounds second/ square foot	lb_f sec/(ft^2)
		1.4882 E+03	centipoises	
		3.6000 E+03	pounds/foot hour	lb_m/(ft hr)
pound second/ square foot	lb_f sec/ft^2	4.7780 E+02	poises	
		4.7780 E+01	kilograms/meter second	kg/(m sec)
		3.2174 E+01	pounds/foot second	lb_m/(ft sec)
		4.7880 E+04	centipoises	
		1.1583 E+05	pounds/foot hour	lb_m/(ft hr)

(Continued on next page)

Table 8.12. (continued)

Multiply	by	to obtain	
centipoise $lb_f\ sec/ft^2$	1.0000 E−02	poises	
	1.0000 E−03	kilograms/meter second	kg/(m sec)
	6.7197 E−04	pounds/foot second	lb_m/(ft sec)
	2.0886 E−05	pounds second/ square foot	$lb_f\ sec/ft^2$
	2.4191	pounds/foot hour	lb_m/(ft hr)
pound/foot hour lb_m/ft hr	4.1338 E−03	poises	
	4.1338 E−04	kilograms/meter second	kg/(m sec)
	2.7778 E−04	pounds/foot second	lb_m/(ft sec)
	8.6336 E−06	pounds second/ square foot	$lb_f\ sec/ft^2$
	4.1338 E−01	centipoises	

Table 8.13. Conversion factors for quantities having dimensions of [F/Tt] or [ML/Tt³]: thermal conductivity

Multiply		by	to obtain	
erg/second centimeter Kelvin	erg/sec cm K	1.0000 E−05	watts/meters Kelvin	W/(m K)
		4.0183 E−05	foot-pounds/ second³ degrees Fahrenheit	ft lb$_m$/(sec³ °F)
		1.2489 E−06	pounds/ second degrees Fahrenheit	lb$_f$/(sec °F)
		2.3901 E−08	calories/ second centimeters Kelvin	cal/(sec cm K)
		5.7780 E−06	Btu/hour foot degrees Fahrenheit	Btu/(hr ft °F)
watt/meter Kelvin	W/m K	1.0000 E+05	ergs/second centimeters Kelvin	ergs/(sec cm K)
		4.0183	foot-pounds/ second³ degrees Fahrenheit	ft lb$_m$/(sec³ °F)
		1.2489 E−01	pounds/ second degrees Fahrenheit	lb$_f$/(sec °F)
		2.3901 E−03	calories/ second centimeters Kelvin	cal/(sec cm K)
		5.7780 E−01	Btu/hour foot degrees Fahrenheit	Btu/(hr ft °F)

(Continued on next page)

<table>
<tr><td colspan="3" align="center">Table 8.13. (continued)</td></tr>
<tr><td align="center">Multiply</td><td align="center">by</td><td align="center">to obtain</td></tr>
<tr>
<td>foot-pound/
  centimeter
second³ degree
Fahrenheit   ft lb$_m$/(sec³ °F)</td>
<td>2.4886 E+04</td>
<td>ergs/second
  centimeters
  Kelvin   erg/(sec cm K)</td>
</tr>
<tr>
<td></td>
<td>2.4886 E−01</td>
<td>watts/meters
  Kelvin   W/(m K)</td>
</tr>
<tr>
<td></td>
<td>3.1081 E−02</td>
<td>pounds/
  second
  degrees
  Fahrenheit   lb$_f$/(sec °F)</td>
</tr>
<tr>
<td></td>
<td>5.9479 E−04</td>
<td>calories/
  second
  centimeters
  Kelvin   cal/(sec cm K)</td>
</tr>
<tr>
<td></td>
<td>1.4379</td>
<td>Btu/hour
  foot degrees
  Fahrenheit   Btu/(hr ft °F)</td>
</tr>
<tr>
<td>pound/
  second degree
Fahrenheit   lb$_f$/(sec °F)</td>
<td>8.0068 E+05</td>
<td>ergs/second
  centimeters
  Kelvin   erg/(sec cm K)</td>
</tr>
<tr>
<td></td>
<td>8.0068</td>
<td>watts/meters
  Kelvin   W/(m K)</td>
</tr>
<tr>
<td></td>
<td>3.2174 E+01</td>
<td>foot-pounds/
  second³
  degree   ft-lb$_m$/
  Fahrenheit   (sec³ °F)</td>
</tr>
<tr>
<td></td>
<td>1.9137 E−02</td>
<td>calories/
  second
  centimeters
  Kelvin   cal/(sec cm K)</td>
</tr>
<tr>
<td></td>
<td>4.6263</td>
<td>Btu/hours
  foot degrees
  Fahrenheit   Btu/(hr ft °F)</td>
</tr>
</table>

(Continued on next page)

<table>
<tr><td colspan="3" align="center">Table 8.13. (continued)</td></tr>
<tr><td align="center">Multiply</td><td align="center">by</td><td align="center">to obtain</td></tr>
<tr>
<td>calorie/second
  centimeter
  Kelvin   cal/(sec cm K)</td>
<td>4.1840 E+07</td>
<td>ergs/second
  centimeters
  Kelvin   erg/(sec cm K)</td>
</tr>
<tr>
<td></td>
<td>4.1840 E+02</td>
<td>watts/meters
  Kelvin   W/(m K)</td>
</tr>
<tr>
<td></td>
<td>1.6813 E+03</td>
<td>foot-pounds/
  second^3
  degrees
  Fahrenheit   ft-lb$_m$/(sec^3 °F)</td>
</tr>
<tr>
<td></td>
<td>5.2256 E+01</td>
<td>pounds/
  second
  degrees
  Fahrenheit   lb$_f$/(sec °F)</td>
</tr>
<tr>
<td></td>
<td>2.4175 E+02</td>
<td>Btu/hour
  foot degrees
  Fahrenheit   Btu/(hr ft °F)</td>
</tr>
<tr>
<td>Btu/hour
  foot degree
  Fahrenheit   Btu/(hr ft °F)</td>
<td>1.7307 E+05</td>
<td>ergs/second
  centimeters
  Kelvin   erg/(sec cm K)</td>
</tr>
<tr>
<td></td>
<td>1.7307</td>
<td>watts/meters
  Kelvin   W/(m K)</td>
</tr>
<tr>
<td></td>
<td>6.9546</td>
<td>foot-pounds/
  second^3
  degrees
  Fahrenheit   ft-lb$_m$/(sec^3 °F)</td>
</tr>
<tr>
<td></td>
<td>2.1616 E−01</td>
<td>pounds/
  second
  degrees
  Fahrenheit   lb$_f$/(sec °F)</td>
</tr>
<tr>
<td></td>
<td>4.1365 E−03</td>
<td>calories/
  second
  centimeters
  Kelvin   cal/(sec cm K)</td>
</tr>
</table>

Table 8.14. Conversion factors for quantities having dimensions of $[L^2/t]$: thermal diffusivity, momentum diffusivity

Multiply		by	to obtain	

square centimeter/
  second
	cm^2/sec	1.0000 E−04	square meters/
			  second
				m^2/sec

		3.8750	square feet/ hour	ft^2/hr
		1.0000 E+02	centistokes	

square meter/
  second
	m^2/sec	1.0000 E+04	square centimeters/
			  second
				cm^2/sec

		3.8750 E+04	square feet/ hour	ft^2/hr
		1.0000 E+06	centistokes	

square foot/
  hour
	ft^2/hr	2.5807 E−01	square centimeters/
			  second
				cm^2/sec

		2.5807 E−05	square meters/ second	m^2/sec
		2.5807 E+01	centistokes	
centistokes		1.0000 E−02	square centimeters/ second	cm^2/sec
		1.0000 E−06	square meters/ second	m^2/sec
		3.8750 E−02	square feet/ hour	ft^2/hr

Table 8.15. Conversion factors for quantities having dimensions of $[M/t^3T]$ or $[F/LtT]$: heat transfer coefficient

Multiply	by	to obtain	
gram/second³ Kelvin g/(sec³K)	1.0000 E−03	watts/square meters Kelvin	watts/(m²K)
	1.2248 E−03	pounds/second³ degree Fahrenheit	lb$_m$/(sec³ °F)
	3.8068 E−05	pounds/feet second degree Fahrenheit	lb$_f$/(ft sec °F)
	2.3901 E−08	calories/square centimeters seconds Kelvin	cal/ (cm²sec K)
	1.0000 E−07	watts/square centimeters Kelvin	watts/ (cm² K)
	1.7611 E−04	Btu/square feet hours degree Fahrenheit	Btu/ (ft²hr °F)
kilogram/ second³ Kelvin kg/(sec³K)	1.0000 E+03	grams/ second³ Kelvin	g/(sec³K)
	1.2248	pounds/second³ degree Fahrenheit	lb$_m$/(sec³ °F)
	3.8068 E−02	pounds/feet second degree Fahrenheit	lb$_f$/(ft sec °F)
	2.3901 E−05	calories/square centimeters seconds Kelvin	cal/ (cm²sec K)
	1.0000 E−04	watts/square centimeters Kelvin	watts/ (cm² K)
	1.7611 E−01	Btu/square feet hours degree Fahrenheit	Btu/ (ft²hr °F)

(Continued on next page)

	Table 8.15. (continued)	
Multiply	by	to obtain
calorie/square centimeter second Kelvin cal/(cm² sec K)	4.1840 E+07	grams/second³ Kelvin g/(sec³ K)
	4.1840 E+04	watt/ square meter Kelvin watt/(m² K)
	5.1245 E+04	pound/second³ degree Fahrenheit lb_m/(sec³ °F)
	1.5928 E+03	pounds/feet seconds degree Fahrenheit lb_f/(ft sec °F)
	4.1840	watts/square centimeters watts/ Kelvin (cm² K)
	7.3686 E+03	Btu/square feet hours degree Fahrenheit Btu/(ft²hr °F)
watt/square centimeter Kelvin watts/(cm² K)	1.0000 E+07	grams/second³ Kelvin g/(sec³ K)
	1.0000 E+04	watts/square meters Kelvin watts/(m² K)
	1.2248 E+04	pounds/second³ degree Fahrenheit lb_m/(sec³ °F)
	3.8068 E+02	pounds/feet second degree Fahrenheit lb_f/(ft sec °F)
	2.3901 E−01	calories/square centimeters seconds Kelvin cal/(cm²sec K)
	1.7611 E+03	Btu/square feet hours degree Fahrenheit Btu/(ft²hr °F)

(Continued on next page)

<table>
<tr><th colspan="4" align="center">Table 8.15. (continued)</th></tr>
<tr><th colspan="2" align="center">Multiply</th><th align="center">by</th><th align="center">to obtain</th></tr>
<tr>
<td>pound/
  second^3
  degree
  Fahrenheit</td>
<td>$\mathrm{lb}_m/(\sec^3\,{}^\circ\mathrm{F})$</td>
<td>8.1647 E+02</td>
<td>grams/second^3
  Kelvin $\mathrm{g}/(\sec^3\,\mathrm{K})$</td>
</tr>
<tr><td></td><td></td><td>8.1647 E−01</td><td>watts/square
  meters
  Kelvin $\mathrm{watts}/(\mathrm{m}^2\,\mathrm{K})$</td></tr>
<tr><td></td><td></td><td>3.1081 E−02</td><td>pounds/feet
  second degree
  Fahrenheit $\mathrm{lb}_f/(\mathrm{ft}\,\sec\,{}^\circ\mathrm{F})$</td></tr>
<tr><td></td><td></td><td>1.9514 E−05</td><td>calories/square
  centimeters   cal/
  second Kelvin $(\mathrm{cm}^2\sec\,\mathrm{K})$</td></tr>
<tr><td></td><td></td><td>8.1647 E−05</td><td>watts/square
  centimeters   watts/
  Kelvin $(\mathrm{cm}^2\,\mathrm{K})$</td></tr>
<tr><td></td><td></td><td>1.4379 E−01</td><td>Btu/square feet
  hours degree
  Fahrenheit $\mathrm{Btu}/(\mathrm{ft}^2\mathrm{hr}\,{}^\circ\mathrm{F})$</td></tr>
<tr>
<td>pound/foot
  second degree
  Fahrenheit</td>
<td>$\mathrm{lb}_f/(\mathrm{ft}\,\sec\,{}^\circ\mathrm{F})$</td>
<td>2.6269 E+04</td>
<td>grams/second^3
  Kelvin $\mathrm{g}/(\sec^3\,\mathrm{K})$</td>
</tr>
<tr><td></td><td></td><td>2.6269 E+01</td><td>watts/square
  meters
  Kelvin $\mathrm{watts}/(\mathrm{m}^2\,\mathrm{K})$</td></tr>
<tr><td></td><td></td><td>3.2174 E+01</td><td>pounds/second^3
  degree
  Fahrenheit $\mathrm{lb}_m/(\sec^3\,{}^\circ\mathrm{F})$</td></tr>
<tr><td></td><td></td><td>6.2784 E−04</td><td>calories/square
  centimeters
  seconds   cal/
  Kelvin $(\mathrm{cm}^2\sec\,\mathrm{K})$</td></tr>
<tr><td></td><td></td><td>2.6269 E−03</td><td>watts/square
  centimeters   watts/
  Kelvin $(\mathrm{cm}^2\,\mathrm{K})$</td></tr>
</table>

(Continued on next page)

<table>
<tr><td colspan="3" align="center">Table 8.15. (continued)</td></tr>
<tr><td align="center">Multiply</td><td align="center">by</td><td align="center">to obtain</td></tr>
<tr>
<td>pound/
  foot second
  degree
  Fahrenheit $\mathrm{lb}_f/(\mathrm{ft\,sec\,°F})$</td>
<td>4.6263</td>
<td>Btu/square feet
  hours degree
  Fahrenheit $\mathrm{Btu}/(\mathrm{ft}^2\mathrm{hr\,°F})$</td>
</tr>
<tr>
<td rowspan="6">Btu/square
  foot hour
  degree
  Fahrenheit $\mathrm{Btu}/(\mathrm{ft}^2\mathrm{hr\,°F})$</td>
<td>5.6782 E+03</td>
<td>grams/second^3
  Kelvin $\mathrm{g}/(\mathrm{sec}^3\,\mathrm{K})$</td>
</tr>
<tr>
<td>5.6782</td>
<td>watts/
  square meters
  Kelvin $\mathrm{watts}/(\mathrm{m}^2\,\mathrm{K})$</td>
</tr>
<tr>
<td>6.9546</td>
<td>pounds/second^3
  degree
  Fahrenheit $\mathrm{lb}_m/(\mathrm{sec}^3\,\mathrm{°F})$</td>
</tr>
<tr>
<td>2.1616 E−01</td>
<td>pounds/feet
  seconds degree
  Fahrenheit $\mathrm{lb}_f/(\mathrm{ft\,sec\,°F})$</td>
</tr>
<tr>
<td>1.3571 E−04</td>
<td>calories/square
  centimeters cal/
  second Kelvin $(\mathrm{cm}^2\mathrm{sec\,K})$</td>
</tr>
<tr>
<td>5.6782 E−04</td>
<td>watts/square
  centimeters watts/
  Kelvin $(\mathrm{cm}^2\,\mathrm{K})$</td>
</tr>
</table>

Table 8.16. Conversion factors for quantities having dimensions of $[M/L^2t]$ or $[Ft/L^3]$: mass transfer coefficient

Multiply		by	to obtain	
gram/square centimeter second	$g/(cm^2sec)$	1.0000 E+01	kilograms/square meters seconds	$kg/(m^2sec)$
		2.0482	pounds/square feet seconds	$lb_m/(ft^2sec)$
		6.3659 E−02	pounds seconds/ cubic feet	$lb_f sec/ft^3$
		7.3724 E+03	pounds/square feet hours	$lb_m/(ft^2hr)$
kilogram/ square meter second	$kg/(m^2sec)$	1.0000 E−01	grams/square centimeters seconds	$g/(cm^2sec)$
		2.0482 E−01	pounds/square feet seconds	$lb_m/(ft^2sec)$
		6.3659 E−03	pounds seconds/ cubic feet	$lb_f sec/ft^3$
		7.3734 E+02	pounds/square feet hours	$lb_m/(ft^2hr)$
pound/square foot second	$lb_m/(ft^2sec)$	4.8824 E−01	grams/square centimeters seconds	$g/(cm^2sec)$
		4.8824	kilograms/square meter seconds	$kg/(m^2sec)$
		3.1081 E−02	pounds seconds/ cubic feet	$lb_f sec/ft^3$
		3.6000 E+03	pounds/square feet hours	$lb_m/(ft^2hr)$
pound second/ cubic foot	$lb_f sec/(ft^3)$	1.5709 E+01	grams/square centimeters seconds	$g/(cm^2sec)$
		1.5709 E+02	kilograms/square meter seconds	$kg/(m^2sec)$
		3.2174 E+01	pounds/square feet seconds	$lb_m/(ft^2sec)$
		1.1583 E+05	pounds/square feet hours	$lb_m/(ft^2hr)$

(Continued on next page)

Table 8.16. (continued)

Multiply	by		to obtain
pound/ square foot hour $\mathrm{lb}_m/(\mathrm{ft}^2\mathrm{hr})$	1.3562 E−04	grams/square centimeters seconds	$\mathrm{g}/(\mathrm{cm}^2\mathrm{sec})$
	1.3562 E−03	kilograms/square meter seconds	$\mathrm{kg}/(\mathrm{m}^2\mathrm{sec})$
	2.7778 E−04	pounds/square feet seconds	$\mathrm{lb}_m/(\mathrm{ft}^2\mathrm{sec})$
	8.6336 E−06	pounds seconds/ cubic feet	$\mathrm{lb}_f\mathrm{sec}/(\mathrm{ft}^3)$

Table 8.17. Temperature Conversions

$$°F = \left(°C \times \frac{9}{5}\right) + 32 = (°C + 40) \times \frac{9}{5} - 40$$

$$°C = (F - 32) \times \frac{5}{9} = (°F + 40) \times \frac{5}{9} - 40$$

$$°R = °F + 459.69$$

$$K = °C + 273.16$$

Chapter 9

Mathematical Relations

9.1 Algebra

9.1.1 Binomial Expansions

$$(a \pm b)^2 = a^2 \pm 2ab + b^2$$

$$(a \pm b)^3 = a^3 \pm 3a^2b + 3ab^2 \pm b^3$$

$$(a \pm b)^4 = a^4 \pm 4a^3b + 6a^2b^2 \pm 4ab^3 + b^4$$

$$(a \pm b)^n = a^n \pm na^{n-1}b + \frac{n(n-1)}{(1)(2)}a^{n-2}b^2 \pm \frac{n(n-1)(n-2)}{(1)(2)(3)}a^{n-3}b^3$$

$$+ (\pm 1)^m \frac{n(n-1)\cdots((n-m+1)}{m!}a^{n-m}b^m + \ldots$$

When n is a positive integer, the series is finite with $(n+1)$ terms. When n is negative or fractional, the series is infinite.

9.1.2 Algebraic Equations

9.1.2.1 Quadratic Equations

$$ax^2 + bx + c = 0$$

The roots of this equation are

$$x = \frac{-b \pm \sqrt{b^2 - 4ac}}{2a}$$

9.1.2.2   Binomial Equations

$$x^n + a = 0$$

The n roots of this equation are:

$$x = \sqrt[n]{-a}\left[\cos\frac{(2m+1)\pi}{n} + \sqrt{-1}\,\sin\frac{(2m+1)\pi}{n}\right]$$

$$x^n - a = 0$$

The n roots of this equation are:

$$x = \sqrt[n]{a}\left[\cos\frac{2m\pi}{n} + \sqrt{-1}\,\sin\frac{2m\pi}{n}\right]$$

where $m = 0,\ 1,\ 2,\ 3,\ \ldots,\ n-1$.

9.1.2.3   Cubic Equations

$$ax^3 + bx^2 + cx + d = 0$$

Any cubic equation of the form above can be reduced to the form

$$y^3 + Ay + B = 0$$

with the substitution of $x = y - (b/3a)$, and the following definitions

$$A = \frac{c}{a} - \frac{b^2}{3a^2} \qquad\qquad B = \frac{2}{27}\frac{b^3}{a^3} - \frac{bc}{3a^2} + \frac{d}{a}$$

9.1.2.3.1   Algebraic Solution

$$y^3 + Ay + B = 0$$

The three roots of this equation are:

$$y_1 = P + Q$$

$$y_2 = -\frac{1}{2}(P + Q) + \frac{\sqrt{-3}}{2}(P - Q)$$

$$y_3 = -\frac{1}{2}(P + Q) - \frac{\sqrt{3}}{2}(P - Q)$$

where

$$P = \sqrt[3]{-\frac{1}{2}A + \left(\frac{A^3}{27} + \frac{B^2}{4}\right)^{\frac{1}{2}}}$$

and

$$Q = \sqrt[3]{-\frac{1}{2}A - \left(\frac{A^3}{27} + \frac{B^2}{4}\right)^{\frac{1}{2}}}$$

9.1.2.3.2 Trigonometric Solution

$$y^3 + Ay + B = 0$$

If $\dfrac{A^3}{27} + \dfrac{B^2}{4} < 0$, then a trigonometric solution is useful. Determine the value of ϕ according to

$$\phi = \cos^{-1}\left(\frac{-\dfrac{B}{2}}{\sqrt{|A|^3/27}}\right)$$

Then, the roots of the equation are:

$$y_1 = 2\sqrt{\frac{|A|}{3}}\cos\frac{\phi}{3}$$

$$y_2 = -2\sqrt{\frac{|A|}{3}}\cos\frac{\phi+\pi}{3}$$

$$y_3 = -2\sqrt{\frac{|A|}{3}}\cos\frac{\phi-\pi}{3}$$

9.1.3 Series Expansions

9.1.3.1 Common Functions

$$e = 1 + \frac{1}{1!} + \frac{1}{2!} + \frac{1}{3!} + \dots$$

$$e^x = 1 + x + \frac{x^2}{2!} + \frac{x^3}{3!} + \dots$$

$$e^{-x^2} = 1 - x^2 + \frac{x^4}{2!} - \frac{x^6}{3!} + \frac{x^8}{4!} - \dots$$

$$a^x = 1 + x\ln a + \frac{(x\ln a)^2}{2!} + \frac{(x\ln a)^3}{3!} + \dots$$

$$\ln x = (x-1) - \frac{1}{2}(x-1)^2 + \frac{1}{3}(x-1)^3 - \dots \qquad 0 < x \le 2$$

$$\ln x = \frac{x-1}{x} + \frac{1}{2}\left(\frac{x-1}{x}\right)^2 + \frac{1}{3}\left(\frac{x-1}{x}\right)^3 + \dots \qquad x > \frac{1}{2}$$

$$\ln x = 2\left[\frac{x-1}{x+1} + \frac{1}{3}\left(\frac{x-1}{x+1}\right)^3 + \frac{1}{5}\left(\frac{x-1}{x+1}\right)^5 + \dots\right] \qquad x > 0$$

$$\ln(1+x) = x - \frac{x^2}{2} + \frac{x^3}{3} - \frac{x^4}{4} + \dots$$

$$\ln\left(\frac{1+x}{1-x}\right) = 2\left(x + \frac{x^3}{3} + \frac{x^5}{5} + \frac{x^7}{7} + \dots\right) \qquad x^2 < 1$$

$$\ln\left(\frac{1+x}{1-x}\right) = 2\left[\frac{1}{x} + \frac{1}{3x^3} + \frac{1}{5x^5} + \frac{1}{7x^7} + \dots\right] \qquad x^2 > 1$$

$$\ln\left(\frac{1+x}{x}\right) = 2\left[\frac{1}{2x+1} + \frac{1}{3(2x+1)^3} + \frac{1}{5(2x+1)^5} + \dots\right] \qquad x > 0$$

$$\ln(a+x) = \ln a + 2\left[\frac{x}{2a+x} + \frac{1}{3}\left(\frac{x}{2a+x}\right)^3 + \frac{1}{5}\left(\frac{x}{2a+x}\right)^5 + \dots\right]$$

$$a > 0 \quad \text{and} \quad -a < x < -\infty$$

$$\sin x = x - \frac{x^3}{3!} + \frac{x^5}{5!} - \frac{x^7}{7!} + \dots$$

$$\cos x = 1 - \frac{x^2}{2!} + \frac{x^4}{4!} - \frac{x^6}{6!} + \dots$$

$$\tan x = x + \frac{x^3}{3} + \frac{2x^5}{15} + \frac{17x^7}{315} + \frac{62x^9}{2835} \qquad x^2 < \frac{\pi^2}{4}$$

$$\sin^{-1} x = x + \frac{1}{2}\frac{x^3}{3} + \frac{1}{2}\frac{3}{4}\frac{x^5}{5} + \frac{1}{2}\frac{3}{4}\frac{5}{6}\frac{x^7}{7} + \dots \qquad x^2 < 1$$

$$\tan^{-1} x = x - \frac{x^3}{3} + \frac{x^5}{5} - \frac{x^7}{7} + \dots \qquad x^2 < 1$$

$$\tan^{-1} x = \frac{\pi}{2} - \frac{1}{x} + \frac{1}{3x^3} - \frac{1}{5x^5} + \dots \qquad x^2 > 1$$

$$\sinh x = x + \frac{x^3}{3!} + \frac{x^5}{5!} + \frac{x^7}{7!} + \dots$$

$$\cosh x = 1 + \frac{x^2}{2!} + \frac{x^4}{4!} + \frac{x^6}{6!} + \dots$$

$$\tanh x = x - \frac{x^3}{3} + \frac{2x^5}{15} - \frac{17x^7}{315} + \dots$$

$$\sinh^{-1} x = x - \frac{1}{2}\frac{x^3}{3} + \frac{1}{2}\frac{3}{4}\frac{x^5}{5} - \frac{1}{2}\frac{3}{4}\frac{5}{6}\frac{1}{6x^6} - \dots \qquad x^2 < 1$$

$$\sinh^{-1} x = \ln 2x - \frac{1}{2}\frac{1}{2x^2} - \frac{1}{2}\frac{3}{4}\frac{1}{4x^4} + \frac{1}{2}\frac{3}{4}\frac{5}{6}\frac{1}{6x^2} - \dots \qquad x > 1$$

$$\cosh^{-1}x = \ln 2x - \frac{1}{2}\frac{1}{2x^2} - \frac{1}{2}\frac{3}{4}\frac{1}{4x^4} - \frac{1}{2}\frac{3}{4}\frac{5}{6}\frac{1}{6x^6} - \cdots$$

$$\tan^{-1}x = x - \frac{x^3}{3} + \frac{x^5}{5} - \frac{x^7}{7} + \cdots \qquad x^2 \le 1$$

9.1.3.2 Binomial Series

$$(a+x)^n = a^n + na^{n-1}x + \frac{n(n-1)}{2!}a^{n-2}x^2 + \frac{n(n-1)(n-2)}{3!}a^{n-3}x^3 + \cdots$$

$$x^2 < a^2$$

For n positive integer, the series has $(n+1)$ terms; when n is negative, the series has infinite terms.

9.1.3.3 Taylor Series

A function $f(x)$ can be expanded about a point a, if (1) the function $f(x)$ is continuous, and (2) all derivatives exist and have finite values at a, then

$$f(x) = f(a) + \frac{(x-a)}{1!}f'(a) + \frac{(x-a)^2}{2!}f''(a) + \frac{(x-a)^3}{3!}f'''(a) + \cdots$$

$$+ \frac{(x-a)^n}{n!}f^n(a) + R_n$$

As $n \to \infty$, $R_n \to 0$.

9.2 Trigonometric Relations

9.2.1 Basic Relations

$$\sec\alpha = \frac{1}{\cos\alpha}$$

$$\csc\alpha = \frac{1}{\sin\alpha}$$

$$\cot\alpha = \frac{1}{\tan\alpha}$$

$$\sin^2\alpha + \cos^2\alpha = 1$$

$$1 + \tan^2\alpha = \sec^2\alpha$$

$$1 + \cot^2\alpha = \csc^2\alpha$$

$$\sin(\pm\alpha) = \pm\sin\alpha \qquad\qquad \cos(\pm\alpha) = \cos\alpha$$

$$\tan(\pm\alpha) = \pm\tan\alpha \qquad\qquad \cot(\pm\alpha) = \pm\cot\alpha$$

$$\sin\!\left(\frac{\pi}{2} \pm \alpha\right) = \cos\alpha \qquad\qquad \cos\!\left(\frac{\pi}{2} \pm \alpha\right) = \mp\sin\alpha$$

$$\tan\!\left(\frac{\pi}{2} \pm \alpha\right) = \mp\cot\alpha \qquad\qquad \cot\!\left(\frac{\pi}{2} \pm \alpha\right) = \mp\tan\alpha$$

$$\sin(\pi \pm \alpha) = \mp\sin\alpha \qquad\qquad \cos(\pi \pm \alpha) = -\cos\alpha$$

$$\tan(\pi \pm \alpha) = \pm\tan\alpha \qquad\qquad \cot(\pi \pm \alpha) = \pm\cot\alpha$$

$$\sin(2n\pi \pm \alpha) = \pm\sin\alpha \qquad\qquad \cos(2n\pi \pm \alpha) = \cos\alpha$$

$$\tan(2n\pi \pm \alpha) = \pm\tan\alpha \qquad\qquad \cot(2n\pi \pm \alpha) = \pm\cot\alpha$$

n an integer

9.2.2 Functions of Half Angles

$$\sin\frac{\alpha}{2} = \sqrt{\frac{1 - \cos\alpha}{2}} \qquad\qquad \cos\frac{\alpha}{2} = \sqrt{\frac{1 + \cos\alpha}{2}}$$

$$\tan\frac{\alpha}{2} = \frac{1 - \cos\alpha}{\sin\alpha} = \frac{\sin\alpha}{1 + \cos\alpha} = \sqrt{\frac{1 - \cos\alpha}{1 + \cos\alpha}}$$

$$\cot\frac{\alpha}{2} = \frac{1 + \cos\alpha}{\sin\alpha} = \frac{\sin\alpha}{1 - \cos\alpha} = \sqrt{\frac{1 + \cos\alpha}{1 - \sin\alpha}}$$

9.2.3 Functions of Multiple Angles

$$\sin 2\alpha = 2\sin\alpha\cos\alpha$$

$$\cos 2\alpha = \cos^2\alpha - \sin^2\alpha = 1 - 2\sin^2\alpha = 2\cos^2\alpha - 1$$

$$\tan 2\alpha = \frac{2\tan\alpha}{1 - \tan^2\alpha} \qquad\qquad \cot 2\alpha = \frac{\cot^2\alpha - 1}{2\cot\alpha}$$

$$\sin 3\alpha = 3\sin\alpha - 4\sin^3\alpha \qquad\qquad \cos 3\alpha = 4\cos^3\alpha - 3\cos\alpha$$

$$\sin 4\alpha = 8\cos^3\alpha\sin\alpha - 4\cos\alpha\sin\alpha \qquad\qquad \cos 4\alpha = 8\cos^4\alpha - 8\cos^2\alpha + 1$$

$$\sin n\alpha = 2\sin(n-1)\alpha\cos\alpha - \sin(n-2)\alpha \qquad \cos n\alpha = 2\cos(n-1)\alpha\cos\alpha - \cos(n-2)\alpha$$

9.2.4 Power of Functions

$$\sin^2 \alpha = \frac{1}{2}(1 - \cos 2\alpha) \qquad\qquad \cos^2 \alpha = \frac{1}{2}(1 + \cos 2\alpha)$$

$$\sin^3 \alpha = \frac{1}{4}(3 \sin \alpha - \sin 3\alpha) \qquad\qquad \cos^3 \alpha = \frac{1}{4}(\cos 3\alpha + 3 \cos \alpha)$$

$$\sin^4 \alpha = \frac{1}{8}(\cos 4\alpha - 4 \cos 2\alpha + 3) \qquad\qquad \cos^4 \alpha = \frac{1}{8}(\cos 4\alpha + 4 \cos 2\alpha + 3)$$

$$\sin^5 \alpha = \frac{1}{16}(10 \sin \alpha - 5 \sin 3\alpha + \sin 5\alpha) \qquad \cos^5 \alpha = \frac{1}{16}(10 \cos \alpha + 5 \cos 3\alpha + \cos 5\alpha)$$

9.2.5 Functions of Sum or Difference of Two Angles

$$\sin(\alpha \pm \beta) = \sin \alpha \cos \beta \pm \cos \alpha \sin \beta$$

$$\cos(\alpha \pm \beta) = \cos \alpha \cos \beta \mp \sin \alpha \sin \beta$$

$$\tan(\alpha \pm \beta) = \frac{\tan \alpha \pm \tan \beta}{1 \mp \tan \alpha \tan \beta}$$

$$\cot(\alpha \pm \beta) = \frac{\cot \alpha \cot \beta \mp 1}{\cot \beta \pm \cot \alpha}$$

9.2.6 Products, Sums, and Differences of Two Functions

$$\sin \alpha \sin \beta = \frac{1}{2} \cos(\alpha - \beta) - \frac{1}{2} \cos(\alpha + \beta)$$

$$\cos \alpha \cos \beta = \frac{1}{2} \cos(\alpha - \beta) + \frac{1}{2} \cos(\alpha + \beta)$$

$$\sin \alpha \cos \beta = \frac{1}{2} \sin(\alpha - \beta) + \frac{1}{2} \sin(\alpha + \beta)$$

$$\sin \alpha + \sin \beta = 2 \sin \frac{1}{2}(\alpha + \beta) \cos \frac{1}{2}(\alpha - \beta)$$

$$\sin \alpha - \sin \beta = 2 \cos \frac{1}{2}(\alpha + \beta) \sin \frac{1}{2}(\alpha - \beta)$$

$$\cos \alpha + \cos \beta = 2 \cos \frac{1}{2}(\alpha + \beta) \cos \frac{1}{2}(\alpha - \beta)$$

$$\cos \alpha - \cos \beta = -2 \sin \frac{1}{2}(\alpha + \beta) \sin \frac{1}{2}(\alpha - \beta)$$

$$\tan\alpha \pm \tan\beta = \frac{\sin(\alpha \pm \beta)}{\cos\alpha\cos\beta}$$

$$\cot\alpha \pm \cot\beta = \pm\frac{\sin(\alpha \pm \beta)}{\sin\alpha\sin\beta}$$

$$\sin^2\alpha - \sin^2\beta = \sin(\alpha + \beta)\sin(\alpha - \beta)$$

$$\cos^2\alpha - \cos^2\beta = -\sin(\alpha + \beta)\sin(\alpha - \beta)$$

$$\cos^2\alpha - \sin^2\beta = \cos(\alpha + \beta)\cos(\alpha - \beta)$$

9.2.7 Properties of Plane Triangles

Law of sines: $\quad \dfrac{a}{\sin\alpha} = \dfrac{b}{\sin\beta} = \dfrac{c}{\sin\gamma}$

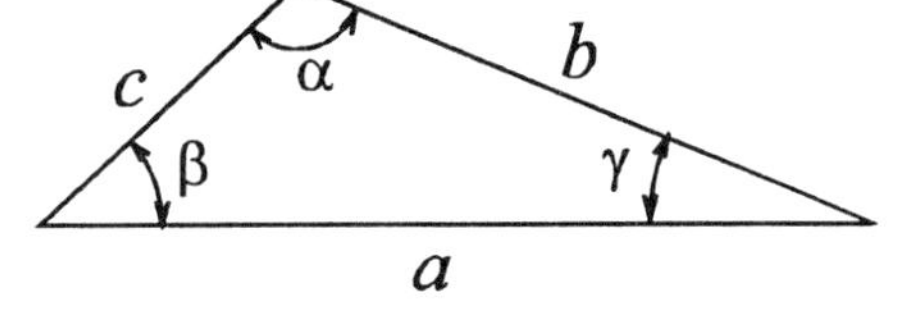

Law of cosines: $\quad a^2 = b^2 + c^2 - 2bc\cos\alpha$

Law of tangents: $\quad \dfrac{a - b}{a + b} = \dfrac{\tan\frac{1}{2}(\alpha - \beta)}{\tan\frac{1}{2}(\alpha + \beta)}$

Area: $\quad A = \dfrac{1}{2}ab\sin\gamma = \dfrac{a^2\sin\beta\sin\gamma}{2\sin\alpha} = \sqrt{s(s - a)(s - b)(s - c)}$

where $\quad s = \dfrac{1}{2}(a + b + c)$

$$\sin\alpha = \frac{2}{bc}\sqrt{s(s - a)(s - b)(s - c)} \qquad \sin\beta = \frac{b}{a}\sin\alpha \qquad \sin\gamma = \frac{c}{a}\sin\alpha$$

$$\cos\alpha = \frac{1}{2bc}(b^2 + c^2 - a^2) \qquad\qquad \cos\beta = \frac{1}{2ac}(a^2 + c^2 - b^2)$$

$$\cos\gamma = \frac{1}{2ab}(a^2 + b^2 - c^2)$$

$$\sin\frac{\alpha}{2} = \sqrt{\frac{(s - b)(s - c)}{bc}} \qquad \sin\frac{\beta}{2} = \sqrt{\frac{(s - c)(s - a)}{ac}} \qquad \sin\frac{\gamma}{2} = \sqrt{\frac{(s - a)(s - b)}{ab}}$$

9.3   Hyperbolic Functions

9.3.1 Definitions

Hyperbolic functions are combinations of exponential functions e^x and e^{-x}, defined as follow.

$$\sinh x = \frac{1}{2}(e^x - e^{-x})$$

$$\cosh x = \frac{1}{2}(e^x + e^{-x})$$

$$\tanh x = \frac{e^x - e^{-x}}{e^x + e^{-x}}$$

$$\coth x = \frac{1}{\tanh x} = \frac{e^x + e^{-x}}{e^x - e^{-x}}$$

$$\operatorname{csch} x = \frac{1}{\sinh x} = \frac{2}{e^x - e^{-x}}$$

$$\operatorname{sech} x = \frac{1}{\cosh x} = \frac{2}{e^x + e^{-x}}$$

$$\sinh x + \cosh x = e^x$$

$$\sinh x - \cosh x = -e^{-x}$$

9.3.2 Basic Relations

$$\sinh(-x) = -\sinh x \qquad \cosh(-x) = \cosh x$$

$$\tanh(-x) = -\tanh x \qquad \coth(-x) = -\coth x$$

$$\operatorname{sech}(-x) = \operatorname{sech} x \qquad \operatorname{csch}(-x) = -\operatorname{csch} x$$

$$\operatorname{sech}^2 x + \tanh^2 x = 1$$

$$\cosh^2 x - \sinh^2 x = 1$$

$$\coth^2 x - \operatorname{csch}^2 x = 1$$

$$\sinh^2 x + \cosh^2 x = \cosh 2x$$

$$(\sinh x \pm \cosh x)^2 = \cosh 2x \pm \sinh 2x$$

$$\sinh (x \pm y) = \sinh x \cosh y \pm \cosh x \sinh y$$

$$\cosh (x \pm y) = \cosh x \cosh y \pm \sinh x \sinh y$$

$$\tanh (x \pm y) = \frac{\tanh x \pm \tanh y}{1 \pm \tanh x \tanh y}$$

$$\coth (x \pm y) = \frac{\coth x \coth y \pm 1}{\coth y \pm \coth x}$$

$$\sinh 2x = 2\sinh x \cosh x \qquad \cosh 2x = \sinh^2 x + \cosh^2 x$$

$$\tanh 2x = \frac{2\tanh x}{1 + \tanh^2 x} \qquad \coth 2x = \frac{1 + \coth^2 x}{2\coth x}$$

$$\sinh^2 x = \frac{1}{2}(\cosh 2x - 1) \qquad \cosh^2 x = \frac{1}{2}(\cosh 2x + 1)$$

$$\tanh^2 x = \frac{\cosh 2x - 1}{\cosh 2x + 1} \qquad \coth^2 x = \frac{\cosh 2x + 1}{\cosh 2x - 1}$$

$$\sinh x + \sinh y = 2\sinh \frac{x + y}{2} \cosh \frac{x - y}{2}$$

$$\sinh x - \sinh y = 2\cosh \frac{x + y}{2} \sinh \frac{x - y}{2}$$

$$\cosh x + \cosh y = 2\cosh \frac{x + y}{2} \cosh \frac{x - y}{2}$$

$$\cosh x - \cosh y = 2\sinh \frac{x + y}{2} \sinh \frac{x - y}{2}$$

$$\tanh x + \tanh y = \frac{\sinh (x + y)}{\cosh x \cosh y}$$

$$\tanh x - \tanh y = \frac{\sinh (x - y)}{\cosh x \cosh y}$$

$$\coth x + \coth y = \frac{\sinh (x + y)}{\sinh x \sinh y}$$

$$\coth x - \coth y = \frac{\sinh (x - y)}{\sinh x \sinh y}$$

$$\sinh (x + y) + \sinh (x - y) = 2\sinh x \cosh y$$

$$\sinh (x + y) - \sinh (x - y) = 2\cosh x \sinh y$$

$$\cosh (x + y) + \cosh (x - y) = 2\cosh x \cosh y$$

$$\cosh (x + y) - \cosh (x - y) = 2\sinh x \sinh y$$

$$\tanh (x + y) - \tanh (x - y) = \frac{\sinh 2x}{\cosh (x + y)\cosh (x - y)}$$

$$\tanh (x + y) - \tanh (x - y) = \frac{\sinh 2y}{\cosh (x + y)\cosh (x - y)}$$

$$\coth (x + y) + \coth (x - y) = \frac{\sinh 2x}{\sinh (x + y)\sinh (x - y)}$$

$$\coth (x + y) - \coth (x - y) = \frac{-\sinh 2x}{\sinh (x + y)\sinh (x - y)}$$

9.4   Functions of Complex Variables

9.4.1 Basic Relations

$$e^{ix} = \cos x + i \sin x$$

$$e^{-ix} = \cos x - i \sin x$$

$$\sin x = \frac{e^{ix} - e^{-ix}}{2i}$$

$$\cos x = \frac{e^{ix} + e^{-ix}}{2}$$

$$(\cos x \pm i \sin x)^n = \cos nx \pm i \sin x$$

$$\sinh ix = \frac{e^{ix} - e^{-ix}}{2}$$

$$\cosh ix = \frac{e^{ix} + e^{-ix}}{2}$$

$$e^{2n\pi i} = 1 \qquad\qquad e^{(2n+1)\pi i} = -1$$

$$e^{\theta + 2n\pi i} = e^{\theta} \qquad\qquad e^{\theta + (2n+1)\pi i} = -e^{\theta}$$

where $n = 0, \pm 1, \pm 2, \ldots$.

9.4.2 Trigonometric/Hyperbolic Functions

$$\sin ix = i \sinh x \qquad\qquad \sin x = -i \sinh ix$$

$$\cos ix = \cosh x \qquad\qquad \cos x = \cosh ix$$

$$\tan ix = i \tanh x \qquad\qquad \tan x = -i \tanh ix$$

$$\cot ix = -i \coth x \qquad\qquad \cot x = i \coth ix$$

$$\sec ix = \operatorname{sech} x \qquad\qquad \sec x = \operatorname{sech} ix$$

$$\csc ix = -i \operatorname{csch} x \qquad\qquad \csc x = i \operatorname{csch} ix$$

$$\sinh ix = i \sin x \qquad\qquad \sinh x = -i \sin ix$$

$$\cosh ix = \cos x \qquad\qquad \cosh x = \cos ix$$

$$\tanh ix = i \tan x \qquad\qquad \tanh x = -i \tan ix$$

$$\coth ix = -i \cot x \qquad\qquad \coth x = i \cot ix$$

$$\operatorname{sech} ix = \sec x \qquad\qquad \operatorname{sech} x = \sec ix$$

$$\operatorname{csch} ix = -i \csc x \qquad\qquad \operatorname{csch} x = i \csc ix$$

$$\sin(x \pm iy) = \sin x \cosh y \pm i \cos x \sinh y$$

$$\cos(x \pm iy) = \cos x \cosh y \mp i \sin x \sinh y$$

$$\tan(x \pm iy) = \frac{\sin 2x \pm i \sinh 2y}{\cos 2x + \cosh 2y}$$

$$\sinh(x \pm iy) = \sinh x \cos y \pm i \cosh x \sin y$$

$$\cosh(x \pm iy) = \cosh x \cos y \pm i \sinh x \sin y$$

$$\tanh(x \pm iy) = \frac{\sinh 2x \pm i \sin 2y}{\cosh 2x + \cos 2y}$$

9.5 Vector Algebra

9.5.1 Cartesian Coordinate System

The derivative operators

$$\nabla = \vec{i}\frac{\partial}{\partial x} + \vec{j}\frac{\partial}{\partial y} + \vec{k}\frac{\partial}{\partial z}$$

$$\nabla^2 = \frac{\partial^2}{\partial x^2} + \frac{\partial^2}{\partial y^2} + \frac{\partial^2}{\partial z^2}$$

$$\nabla^4 = \frac{\partial^4}{\partial x^4} + \frac{\partial^4}{\partial y^4} + \frac{\partial^4}{\partial z^4} + 2\frac{\partial^4}{\partial x^2 \partial y^2} + 2\frac{\partial^4}{\partial x^2 \partial z^2} + 2\frac{\partial^4}{\partial y^2 \partial z^2}$$

Total derivative

$$\frac{d}{dt} = \frac{\partial}{\partial t} + u\frac{\partial}{\partial x} + v\frac{\partial}{\partial y} + w\frac{\partial}{\partial z}$$

Stress

$$\tau = \left\{ \begin{array}{ccc} \tau_{xx} & \tau_{xy} & \tau_{xz} \\ \tau_{yx} & \tau_{yy} & \tau_{yz} \\ \tau_{zx} & \tau_{zy} & \tau_{zz} \end{array} \right\}$$

Operations

$$\nabla \cdot \vec{V} = \frac{\partial u}{\partial x} + \frac{\partial v}{\partial y} + \frac{\partial w}{\partial z}$$

$$\nabla u = \frac{\partial u}{\partial x}\vec{i} + \frac{\partial u}{\partial y}\vec{j} + \frac{\partial u}{\partial z}\vec{k}$$

$$\nabla^2 u = \frac{\partial^2 u}{\partial x^2} + \frac{\partial^2 u}{\partial y^2} + \frac{\partial^2 u}{\partial z^2}$$

$$\nabla^2 \vec{V} = \left(\frac{\partial^2 u}{\partial x^2} + \frac{\partial^2 u}{\partial y^2} + \frac{\partial^2 u}{\partial z^2} \right) \vec{i} + \left(\frac{\partial^2 v}{\partial x^2} + \frac{\partial^2 v}{\partial y^2} + \frac{\partial^2 v}{\partial z^2} \right) \vec{j}$$

$$+ \left(\frac{\partial^2 w}{\partial x^2} + \frac{\partial^2 w}{\partial y^2} + \frac{\partial^2 w}{\partial z^2} \right) \vec{k}$$

$$\vec{V} \cdot \nabla \vec{V} = \left(u \frac{\partial u}{\partial x} + v \frac{\partial u}{\partial y} + w \frac{\partial u}{\partial z} \right) \vec{i} + \left(u \frac{\partial v}{\partial x} + v \frac{\partial v}{\partial y} + w \frac{\partial w}{\partial z} \right) \vec{j}$$

$$+ \left(u \frac{\partial w}{\partial x} + v \frac{\partial w}{\partial y} + w \frac{\partial w}{\partial z} \right) \vec{k}$$

$$\nabla \times \vec{V} = \begin{vmatrix} \vec{i} & \vec{j} & \vec{k} \\ \frac{\partial}{\partial x} & \frac{\partial}{\partial y} & \frac{\partial}{\partial z} \\ u & v & w \end{vmatrix} = \left(\frac{\partial w}{\partial y} - \frac{\partial v}{\partial z} \right) \vec{i} + \left(\frac{\partial u}{\partial z} - \frac{\partial w}{\partial x} \right) \vec{j} + \left(\frac{\partial v}{\partial x} - \frac{\partial u}{\partial y} \right) \vec{k}$$

$$\nabla \cdot \tau = \left(\frac{\partial \tau_{xx}}{\partial x} + \frac{\partial \tau_{xy}}{\partial y} + \frac{\partial \tau_{xz}}{\partial z} \right) \vec{i} + \left(\frac{\partial \tau_{yx}}{\partial x} + \frac{\partial \tau_{yy}}{\partial y} + \frac{\partial \tau_{yz}}{\partial z} \right) \vec{j}$$

$$+ \left(\frac{\partial \tau_{zx}}{\partial x} + \frac{\partial \tau_{zy}}{\partial y} + \frac{\partial \tau_{zz}}{\partial z} \right) \vec{k}$$

$$\tau \cdot \nabla \vec{V} = \tau_{xx} \frac{\partial u}{\partial x} + \tau_{yy} \frac{\partial v}{\partial y} + \tau_{zz} \frac{\partial w}{\partial z}$$

$$+ \tau_{xy} \left(\frac{\partial u}{\partial y} + \frac{\partial v}{\partial x} \right) + \left(\frac{\partial v}{\partial z} + \frac{\partial w}{\partial y} \right) + \tau_{zx} \left(\frac{\partial w}{\partial x} + \frac{\partial u}{\partial z} \right)$$

9.5.2 Cylindrical Coordinate System

The derivative operators

$$\nabla = \vec{e}_r \frac{\partial}{\partial r} + \vec{e}_\theta \frac{1}{r} \frac{\partial}{\partial \theta} + \vec{e}_z \frac{\partial}{\partial z}$$

$$\nabla^2 = \frac{1}{r} \frac{\partial}{\partial r} \left(r \frac{\partial}{\partial r} \right) + \frac{1}{r^2} \frac{\partial^2}{\partial \theta^2} + \frac{\partial^2}{\partial z^2}$$

Total derivative

$$\frac{d}{dt} = \frac{\partial}{\partial t} + u_r \frac{\partial}{\partial r} + \frac{u_\theta}{r} \frac{\partial}{\partial \theta} + u_z \frac{\partial}{\partial z}$$

Stress

$$\tau = \left\{ \begin{array}{ccc} \tau_{rr} & \tau_{r\theta} & \tau_{rz} \\ \tau_{\theta r} & \tau_{\theta\theta} & \tau_{\theta z} \\ \tau_{zr} & \tau_{z\theta} & \tau_{zz} \end{array} \right\}$$

Operations

$$\nabla \cdot \vec{V} = \frac{1}{r}\frac{\partial}{\partial r}(ru_r) + \frac{1}{r}\frac{\partial u_\theta}{\partial \theta} + \frac{\partial u_z}{\partial z}$$

$$\nabla u_r = \frac{\partial u_r}{\partial r}\vec{e}_r + \frac{1}{r}\frac{\partial u_r}{\partial \theta}\vec{e}_\theta + \frac{\partial u_r}{\partial z}\vec{e}_z$$

$$\nabla^2 u_r = \frac{1}{r}\frac{\partial}{\partial r}\left(r\frac{\partial u_r}{\partial r}\right) + \frac{1}{r^2}\frac{\partial^2 u_r}{\partial \theta^2} + \frac{\partial^2 u_r}{\partial z^2}$$

$$\nabla^2 \vec{V} = \left\{\frac{\partial}{\partial r}\left[\frac{1}{r}\frac{\partial}{\partial r}(ru_r)\right] + \frac{1}{r^2}\frac{\partial^2 u_r}{\partial \theta^2} + \frac{\partial^2 u_r}{\partial z^2} - \frac{2}{r^2}\frac{\partial u_\theta}{\partial \theta}\right\}\vec{e}_r$$

$$+ \left\{\frac{\partial}{\partial r}\left[\frac{1}{r}\frac{\partial}{\partial r}(ru_\theta)\right] + \frac{1}{r^2}\frac{\partial^2 u_\theta}{\partial \theta^2} + \frac{\partial^2 u_\theta}{\partial z^2} + \frac{2}{r^2}\frac{\partial u_r}{\partial \theta}\right\}\vec{e}_\theta$$

$$+ \left\{\frac{1}{r}\frac{\partial}{\partial r}\left(r\frac{\partial u_z}{\partial r}\right) + \frac{1}{r^2}\frac{\partial^2 u_z}{\partial \theta^2} + \frac{\partial^2 u_z}{\partial z^2}\right\}\vec{e}_z$$

$$\vec{V}\cdot\nabla\vec{V} = \left(u_r\frac{\partial u_r}{\partial r} + \frac{u_\theta}{r}\frac{\partial u_r}{\partial \theta} + u_z\frac{\partial u_r}{\partial z} - \frac{u_\theta^2}{r}\right)\vec{e}_r$$

$$+ \left(u_r\frac{\partial u_\theta}{\partial r} + \frac{u_\theta}{r}\frac{\partial u_\theta}{\partial \theta} + u_z\frac{\partial u_\theta}{\partial z} + \frac{u_r u_\theta}{r}\right)\vec{e}_\theta$$

$$+ \left(u_r\frac{\partial u_z}{\partial r} + \frac{u_\theta}{r}\frac{\partial u_z}{\partial \theta} + u_z\frac{\partial u_z}{\partial z}\right)\vec{e}_z$$

$$\nabla \times \vec{V} = \frac{1}{r}\begin{vmatrix} \vec{e}_r & r\vec{e}_\theta & \vec{e}_z \\ \frac{\partial}{\partial r} & \frac{\partial}{\partial \theta} & \frac{\partial}{\partial z} \\ u_r & ru_\theta & u_z \end{vmatrix}$$

$$\nabla \cdot \tau = \left[\frac{1}{r}\frac{\partial}{\partial r}(r\tau_{rr}) + \frac{1}{r}\frac{\partial}{\partial \theta}(\tau_{r\theta}) + \frac{\partial}{\partial z}(\tau_{rz}) - \frac{\tau_{\theta\theta}}{r}\right]\vec{e}_r$$

$$+ \left[\frac{1}{r}\frac{\partial}{\partial \theta}(\tau_{\theta\theta}) + \frac{\partial}{\partial r}(\tau_{\theta r}) + \frac{\partial}{\partial z}(\tau_{\theta z}) + \frac{2}{r}\tau_{\theta r}\right]\vec{e}_\theta$$

$$+ \left[\frac{1}{r}\frac{\partial}{\partial r}(r\tau_{zr}) + \frac{1}{r}\frac{\partial \tau_{z\theta}}{\partial \theta} + \frac{\partial \tau_{zz}}{\partial z}\right]\vec{e}_z$$

$$\tau\cdot\nabla\vec{V} = \tau_{rr}\frac{\partial u_r}{\partial r} + \tau_{\theta\theta}\left(\frac{1}{r}\frac{\partial u_\theta}{\partial \theta} + \frac{u_r}{r}\right) + \tau_{zz}\left(\frac{\partial u_z}{\partial z}\right)$$

$$+ \tau_{r\theta}\left[r\frac{\partial}{\partial r}\left(\frac{u_\theta}{r}\right) + \frac{1}{r}\frac{\partial u_r}{\partial \theta}\right] + \tau_{\theta z}\left(\frac{1}{r}\frac{\partial u_\theta}{\partial \theta} + \frac{\partial u_\theta}{\partial z}\right) + \tau_{zz}\left(\frac{\partial u_z}{\partial r} + \frac{\partial u_r}{\partial z}\right)$$

9.5.3 Spherical Coordinate System

The derivative operators

$$\nabla = \frac{\partial}{\partial r}\vec{e}_r + \frac{1}{r}\frac{\partial}{\partial \theta}\vec{e}_\theta + \frac{1}{r\sin\theta}\frac{\partial}{\partial \phi}\vec{e}_\phi$$

$$\nabla^2 = \frac{1}{r^2}\frac{\partial}{\partial r}\left(r^2\frac{\partial}{\partial r}\right) + \frac{1}{r^2\sin\theta}\frac{\partial}{\partial \theta}\left(\sin\theta\frac{\partial}{\partial \theta}\right) + \frac{1}{r^2\sin^2\theta}\frac{\partial^2}{\partial \phi^2}$$

Total derivative

$$\frac{d}{dt} = \frac{\partial}{\partial t} + u_r\frac{\partial}{\partial r} + \frac{u_\theta}{r}\frac{\partial}{\partial \theta} + \frac{u_\phi}{r\sin\theta}\frac{\partial}{\partial \phi}$$

Operations

$$\nabla \cdot \vec{V} = \frac{1}{r^2}\frac{\partial}{\partial r}(r^2 u_r) + \frac{1}{r\sin\theta}\frac{\partial}{\partial \theta}(\sin\theta u_\theta) + \frac{1}{r\sin\theta}\frac{\partial u_\phi}{\partial \phi}$$

$$\nabla u_r = \frac{\partial u_r}{\partial r}\vec{e}_r + \frac{1}{r}\frac{\partial u_r}{\partial \theta}\vec{e}_\theta + \frac{1}{r\sin\theta}\frac{\partial u_r}{\partial \phi}\vec{e}_\phi$$

$$\nabla^2 u_r = \frac{1}{r^2}\frac{\partial}{\partial r}\left(r^2\frac{\partial u_r}{\partial r}\right) + \frac{1}{r^2\sin\theta}\frac{\partial}{\partial \theta}\left(\sin\theta\frac{\partial u_r}{\partial \theta}\right) + \frac{1}{r^2\sin^2\theta}\frac{\partial^2 u_r}{\partial \phi^2}$$

$$\nabla^2\vec{V} = \left(\nabla^2 u_r - \frac{2u_r}{r^2} - \frac{2}{r^2}\frac{\partial u_\theta}{\partial \theta} - \frac{2u_\theta\cot\theta}{r^2} - \frac{2}{r^2\sin\theta}\frac{\partial u_\phi}{\partial \phi}\right)\vec{e}_r$$

$$+ \left(\nabla^2 u_\theta + \frac{2}{r^2}\frac{\partial u_r}{\partial \theta} - \frac{u_\theta}{r^2\sin^2\theta} - \frac{2\cos\theta}{r^2\sin^2\theta}\frac{\partial u_\phi}{\partial \phi}\right)\vec{e}_\theta$$

$$+ \left(\nabla^2 u_\phi - \frac{u_\phi}{r^2\sin^2\theta} + \frac{2}{r^2\sin\theta}\frac{\partial u_r}{\partial \phi} + \frac{2\cos\theta}{r^2\sin^2\theta}\frac{\partial u_\theta}{\partial \phi}\right)\vec{e}_\phi$$

$$\vec{V}\cdot\nabla\vec{V} = \left(u_r\frac{\partial u_r}{\partial r} + \frac{u_\theta}{r}\frac{\partial u_r}{\partial \theta} + \frac{u_\phi}{r\sin\theta}\frac{\partial u_r}{\partial \phi} - \frac{u_\theta^2 + u_\phi^2}{r}\right)\vec{e}_r$$

$$+ \left(u_r\frac{\partial u_\theta}{\partial r} + \frac{u_\theta}{r}\frac{\partial u_\theta}{\partial \theta} + \frac{u_\phi}{r\sin\theta}\frac{\partial u_\theta}{\partial \phi} + \frac{u_r u_\theta}{r} - \frac{u_\phi^2\cot\theta}{r}\right)\vec{e}_\theta$$

$$+ \left(u_r\frac{\partial u_\phi}{\partial r} + \frac{u_\theta}{r}\frac{\partial u_\phi}{\partial \theta} + \frac{u_\phi}{r\sin\theta}\frac{\partial u_\phi}{\partial \phi} + \frac{u_\phi u_r}{r} + \frac{u_\theta u_\phi\cot\theta}{r}\right)\vec{e}_\phi$$

$$\nabla \times \vec{V} = \frac{1}{r^2\sin\theta}\begin{vmatrix} \vec{e}_r & r\vec{e}_\theta & r\sin\theta\vec{e}_\phi \\ \frac{\partial}{\partial r} & \frac{\partial}{\partial \theta} & \frac{\partial}{\partial \phi} \\ u_r & ru_\theta & r\sin\theta u_\phi \end{vmatrix}$$

$$= \left[\frac{1}{r\sin\theta}\frac{\partial}{\partial \theta}(\sin\theta u_\phi) - \frac{1}{r\sin\theta}\frac{\partial u_\theta}{\partial \phi}\right]\vec{e}_r + \left[\frac{1}{r\sin\theta}\frac{\partial u_r}{\partial \phi} - \frac{1}{r}\frac{\partial}{\partial r}(ru_\phi)\right]\vec{e}_\theta$$

$$+ \left[\frac{1}{r} \frac{\partial}{\partial r}(ru_\theta) - \frac{1}{r} \frac{\partial u_r}{\partial \theta} \right] \vec{e}_\phi$$

$$\nabla \cdot \tau = \left[\frac{1}{r^2} \frac{\partial}{\partial r}(r^2 \tau_{rr}) + \frac{1}{r \sin \theta} \frac{\partial}{\partial \theta}(\sin \theta \tau_{r\theta}) + \frac{1}{r \sin \theta} \frac{\partial \tau_{r\phi}}{\partial \phi} - \frac{\tau_{\theta\theta} + \tau_{\phi\phi}}{r} \right] \vec{e}_r$$

$$+ \left[\frac{1}{r^2} \frac{\partial}{\partial r}(r^2 \tau_{r\theta}) + \frac{1}{r \sin \theta} \frac{\partial}{\partial \theta}(\sin \theta \tau_{\theta\theta}) + \frac{1}{r \sin \theta} \frac{\partial \tau_{\theta\phi}}{\partial \phi} + \frac{\tau_{r\theta}}{r} - \frac{\cot \theta}{r} \tau_{\phi\phi} \right] \vec{e}_\theta$$

$$+ \left[\frac{1}{r^2} \frac{\partial}{\partial r}(r^2 \tau_{r\phi}) + \frac{1}{r} \frac{\partial \tau_{r\phi}}{\partial \theta} + \frac{1}{r \sin \theta} \frac{\partial \tau_{\phi\phi}}{\partial \phi} + \frac{\tau_{r\phi}}{r} + \frac{2 \cot \theta}{r} \tau_{\theta\theta} \right] \vec{e}_\phi$$

$$\tau \cdot \nabla \vec{V} = \tau_{rr} \frac{\partial u_r}{\partial r} + \tau_{\theta\theta} \left(\frac{1}{r} \frac{\partial u_\theta}{\partial \theta} + \frac{u_r}{r} \right) + \tau_{\phi\phi} \left(\frac{1}{r \sin \theta} \frac{\partial u_\phi}{\partial \phi} + \frac{u_r}{r} + \frac{u_\theta \cot \theta}{r} \right)$$

$$+ \tau_{r\theta} \left(\frac{\partial u_\theta}{\partial r} + \frac{1}{r} \frac{\partial u_r}{\partial \theta} - \frac{u_\theta}{r} \right) + \tau_{r\phi} \left(\frac{\partial u_\phi}{\partial r} + \frac{1}{r \sin \theta} \frac{\partial u_r}{\partial \phi} - \frac{u_\phi}{r} \right)$$

$$+ \tau_{\theta\phi} \left(\frac{1}{r} \frac{\partial u_\phi}{\partial \theta} + \frac{1}{r \sin \theta} \frac{\partial u_\theta}{\partial \phi} - \frac{\cot \theta}{r} u_\phi \right)$$

9.5.4 General Orthogonal Curvilinear Coordinate System

The derivative operators

$$\nabla = \vec{e}_1 \frac{1}{h_1} \frac{\partial}{\partial x_1} + \vec{e}_2 \frac{1}{h_2} \frac{\partial}{\partial x_2} + \vec{e}_3 \frac{1}{h_3} \frac{\partial}{\partial x_3}$$

$$\nabla^2 = \frac{1}{h_1 h_2 h_3} \left[\frac{\partial}{\partial x_1} \left(\frac{h_2 h_3}{h_1} \frac{\partial}{\partial x_1} \right) \frac{\partial}{\partial x_2} \left(\frac{h_1 h_3}{h_2} \frac{\partial}{\partial x_2} \right) + \frac{\partial}{\partial x_3} \left(\frac{h_1 h_2}{h_3} \frac{\partial}{\partial x_3} \right) \right]$$

Total derivative

$$\frac{D}{Dt} = \frac{\partial}{\partial t} + \frac{u_1}{h_1} \frac{\partial}{\partial x_1} + \frac{u_2}{h_2} \frac{\partial}{\partial x_2} + \frac{u_3}{h_3} \frac{\partial}{\partial x_3}$$

Operations

$$\nabla \cdot \vec{V} = \frac{1}{h_1 h_2 h_3} \left[\frac{\partial}{\partial x_1}(h_2 h_3 u_1) + \frac{\partial}{\partial x_2}(h_1 h_3 u_2) + \frac{\partial}{\partial x_3}(h_1 h_2 u_3) \right]$$

$$\nabla u_1 = \frac{1}{h_1} \frac{\partial u_1}{\partial x_1} \vec{e}_1 + \frac{1}{h_2} \frac{\partial u_2}{\partial x_2} \vec{e}_2 + \frac{1}{h_3} \frac{\partial u_3}{\partial x_3} \vec{e}_3$$

$$\nabla^2 u_1 = \frac{1}{h_1 h_2 h_3} \left[\frac{\partial}{\partial x_1} \left(\frac{h_2 h_3}{h_1} \frac{\partial u_1}{\partial x_1} \right) + \frac{\partial}{\partial x_2} \left(\frac{h_1 h_3}{h_2} \frac{\partial u_1}{\partial x_2} \right) + \frac{\partial}{\partial x_3} \left(\frac{h_1 h_2}{h_3} \frac{\partial u_1}{\partial x_3} \right) \right]$$

$$\nabla \times \vec{V} = \frac{1}{h_1 h_2 h_3} \begin{vmatrix} h_1 \vec{e}_1 & h_2 \vec{e}_2 & h_3 \vec{e}_3 \\ \dfrac{\partial}{\partial x_1} & \dfrac{\partial}{\partial x_2} & \dfrac{\partial}{\partial x_3} \\ h_1 u_1 & h_2 u_2 & h_3 u_3 \end{vmatrix}$$

$$
\begin{aligned}
\nabla \cdot \tau = & \left\{ \frac{1}{h_1 h_2 h_3} \left[\frac{\partial}{\partial x_1}(h_2 h_3 \tau_{11}) + \frac{\partial}{\partial x_2}(h_3 h_1 \tau_{12}) + \frac{\partial}{\partial x_3}(h_1 h_2 \tau_{13}) \right] \right. \\
& \left. + \tau_{12} \frac{1}{h_1 h_2} \frac{\partial h_1}{\partial x_2} + \tau_{31} \frac{1}{h_1 h_3} \frac{\partial h_1}{\partial x_3} - \tau_{22} \frac{1}{h_1 h_2} \frac{\partial h_2}{\partial x_1} - \tau_{33} \frac{1}{h_1 h_3} \frac{\partial h_3}{\partial x_3} \right\} \vec{e}_1 \\[2mm]
& + \left\{ \frac{1}{h_1 h_2 h_3} \left[\frac{\partial}{\partial x_1}(h_2 h_3 \tau_{12}) + \frac{\partial}{\partial x_2}(h_3 h_1 \tau_{22}) + \frac{\partial}{\partial x_3}(h_1 h_2 \tau_{23}) \right] \right. \\
& \left. + \tau_{23} \frac{1}{h_2 h_3} \frac{\partial h_2}{\partial x_3} + \tau_{12} \frac{1}{h_2 h_1} \frac{\partial h_2}{\partial x_1} - \tau_{33} \frac{1}{h_2 h_3} \frac{\partial h_3}{\partial x_2} - \tau_{11} \frac{1}{h_1 h_2} \frac{\partial h_1}{\partial x_2} \right\} \vec{e}_2 \\[2mm]
& + \left\{ \frac{1}{h_1 h_2 h_3} \left[\frac{\partial}{\partial x_1}(h_2 h_3 \tau_{13}) + \frac{\partial}{\partial x_2}(h_3 h_1 \tau_{23}) + \frac{\partial}{\partial x_3}(h_1 h_2 \tau_{33}) \right] \right. \\
& \left. + \tau_{31} \frac{1}{h_3 h_1} \frac{\partial h_3}{\partial x_1} + \tau_{23} \frac{1}{h_2 h_3} \frac{\partial h_3}{\partial x_2} - \tau_{11} \frac{1}{h_1 h_3} \frac{\partial h_1}{\partial x_3} - \tau_{22} \frac{1}{h_2 h_3} \frac{\partial h_2}{\partial x_3} \right\} \vec{e}_3
\end{aligned}
$$